Privileged Scaffolds in Medicinal Chemistry
Design, Synthesis, Evaluation

RSC Drug Discovery Series

Titles in the Series:

1: Metabolism, Pharmacokinetics and Toxicity of Functional Groups
2: Emerging Drugs and Targets for Alzheimer's Disease; Volume 1
3: Emerging Drugs and Targets for Alzheimer's Disease; Volume 2
4: Accounts in Drug Discovery
5: New Frontiers in Chemical Biology
6: Animal Models for Neurodegenerative Disease
7: Neurodegeneration
8: G Protein-Coupled Receptors
9: Pharmaceutical Process Development
10: Extracellular and Intracellular Signaling
11: New Synthetic Technologies in Medicinal Chemistry
12: New Horizons in Predictive Toxicology
13: Drug Design Strategies: Quantitative Approaches
14: Neglected Diseases and Drug Discovery
15: Biomedical Imaging
16: Pharmaceutical Salts and Cocrystals
17: Polyamine Drug Discovery
18: Proteinases as Drug Targets
19: Kinase Drug Discovery
20: Drug Design Strategies: Computational Techniques and Applications
21: Designing Multi-Target Drugs
22: Nanostructured Biomaterials for Overcoming Biological Barriers
23: Physico-Chemical and Computational Approaches to Drug Discovery
24: Biomarkers for Traumatic Brain Injury
25: Drug Discovery from Natural Products
26: Anti-Inflammatory Drug Discovery
27: New Therapeutic Strategies for Type 2 Diabetes: Small Molecules
28: Drug Discovery for Psychiatric Disorders
29: Organic Chemistry of Drug Degradation
30: Computational Approaches to Nuclear Receptors

31: Traditional Chinese Medicine
32: Successful Strategies for the Discovery of Antiviral Drugs
33: Comprehensive Biomarker Discovery and Validation for Clinical Application
34: Emerging Drugs and Targets for Parkinson's Disease
35: Pain Therapeutics; Current and Future Treatment Paradigms
36: Biotherapeutics: Recent Developments using Chemical and Molecular Biology
37: Inhibitors of Molecular Chaperones as Therapeutic Agents
38: Orphan Drugs and Rare Diseases
39: Ion Channel Drug Discovery
40: Macrocycles in Drug Discovery
41: Human-based Systems for Translational Research
42: Venoms to Drugs: Venom as a Source for the Development of Human Therapeutics
43: Carbohydrates in Drug Design and Discovery
44: Drug Discovery for Schizophrenia
45: Cardiovascular and Metabolic Disease: Scientific Discoveries and New Therapies
46: Green Chemistry Strategies for Drug Discovery
47: Fragment-Based Drug Discovery
48: Epigenetics for Drug Discovery
49: New Horizons in Predictive Drug Metabolism and Pharmacokinetics
50: Privileged Scaffolds in Medicinal Chemistry: Design, Synthesis, Evaluation

How to obtain future titles on publication:
A standing order plan is available for this series. A standing order will bring delivery of each new volume immediately on publication.

For further information please contact:
Book Sales Department, Royal Society of Chemistry, Thomas Graham House, Science Park, Milton Road, Cambridge, CB4 0WF, UK
Telephone: +44 (0)1223 420066, Fax: +44 (0)1223 420247,
Email: booksales@rsc.org
Visit our website at www.rsc.org/books

Privileged Scaffolds in Medicinal Chemistry
Design, Synthesis, Evaluation

Edited by

Stefan Bräse
Karlsruhe Institute of Technology, Karlsruhe, Germany
Email: stefan.braese@kit.edu

ROYAL SOCIETY OF CHEMISTRY

THE QUEEN'S AWARDS FOR ENTERPRISE: INTERNATIONAL TRADE 2013

RSC Drug Discovery Series No. 50

Print ISBN: 978-1-78262-030-3
PDF eISBN: 978-1-78262-224-6
ISSN: 2041-3203

A catalogue record for this book is available from the British Library

Published by The Royal Society of Chemistry,
Thomas Graham House, Science Park, Milton Road,
Cambridge CB4 0WF, UK

Registered Charity Number 207890

For further information see our web site at www.rsc.org

Printed in the United Kingdom by CPI Group (UK) Ltd, Croydon, CR0 4YY, UK

Preface

Privileged scaffolds – when I was approached two years ago by the RSC editorial board to edit a book on this topic, I was immediately thrilled as it was very timely.

I would like to quote from an interview given by Brent Stockwell, one of the leading figures in drug discovery:

> "In 1957, during a cleanup of his lab at the pharmaceutical company Roche, Sternbach discovered an old flask containing a chemical he had synthesized previously, but discarded for lack of interest. On a lark, he decided to have it tested for its anti-anxiety potential, a therapeutic area he had become interested in. The chemical, of an unknown identity, had striking anti-anxiety activity that was superior to the existing marketed drugs of that era. Within three years, Sternbach was able to figure out the identity of the chemical and had it approved for use in patients – a remarkable success, considering that drug development nowadays takes 10 to 15 years. This chemical was the first of the benzodiazepines, a class of drugs with a specific shape and structure that is powerful for treating anxiety. Some years later, Ben Evans and his colleagues at Merck discovered that benzodiazepine derivatives were also effective in treating other diseases and in interacting with other types of proteins. He suggested that this class of molecules is **privileged**, in the sense that it is especially effective at interacting with proteins and altering the course of disease. Other privileged molecular structures have since been discovered, and these molecules might be the key to addressing the undruggable proteins. Since these **privileged compounds** are so effective at interacting with numerous classes of proteins, they may be an effective starting point to look for new drugs against the supposedly undruggable protein targets."

RSC Drug Discovery Series No. 50
Privileged Scaffolds in Medicinal Chemistry: Design, Synthesis, Evaluation
Edited by Stefan Bräse

Published by the Royal Society of Chemistry, www.rsc.org

With this notion in mind, I asked a number of colleagues both in industry and academia to give their own opinion for a given compound class. In all cases, a short overview into syntheses is given.

The book is organized by compound classes with a general lead-in and a chapter addressing computational approaches. It should be noted that during the preparation, the Baell/Holloway paper in *Science* (2014) addressing "Pan Assay Interference Compounds - PAINS" was published, which is also a very valuable source for addressing privileged structures from a toxophoric standpoint.

Stefan Bräse
Karlsruhe Institute of Technology
Germany

Contents

RSC Drug Discovery Series No. 50
Privileged Scaffolds in Medicinal Chemistry: Design, Synthesis, Evaluation
Edited by Stefan Bräse

Published by the Royal Society of Chemistry, www.rsc.org

CHAPTER 1

Privileged Scaffolds in Medicinal Chemistry: An Introduction

ELIEZER J. BARREIRO

Laboratório de Avaliação e Síntese de Substâncias Bioativas, Universidade Federal do Rio de Janeiro, CCS, Cidade Universitária, PO Box 68.006, ZIP 21941-910, Rio de Janeiro, RJ, Brazil
Email: ejbarreiro@ccsdecania.ufrj.br

1.1 Introduction

The 20th century has seen significant technological advances, as demonstrated by comparing technology's impact on everyday life at the beginning and end of the century. Many agree that this evolution can hardly have been predicted, nor the drastic changes to several scientific concepts. In many sectors, technological and scientific advancements made throughout the century were spectacular, in particular, in the ways in which we communicate, which is probably due to the evolution of computer science, among others.

The drug discovery process has also undergone huge changes and when we compare, even superficially, the stage that was achieved by the end of the century with that of earlier years, it is clear that there are significant differences. For example, at the end of the 19th century and beginning of the 20th century, when acetylsalicylic acid (ASA **1**; Figure 1.1), which may be considered the first drug to be industrially produced, was discovered, there was a completely different scientific environment to that of 1997, when

RSC Drug Discovery Series No. 50
Privileged Scaffolds in Medicinal Chemistry: Design, Synthesis, Evaluation
Edited by Stefan Bräse

Published by the Royal Society of Chemistry, www.rsc.org

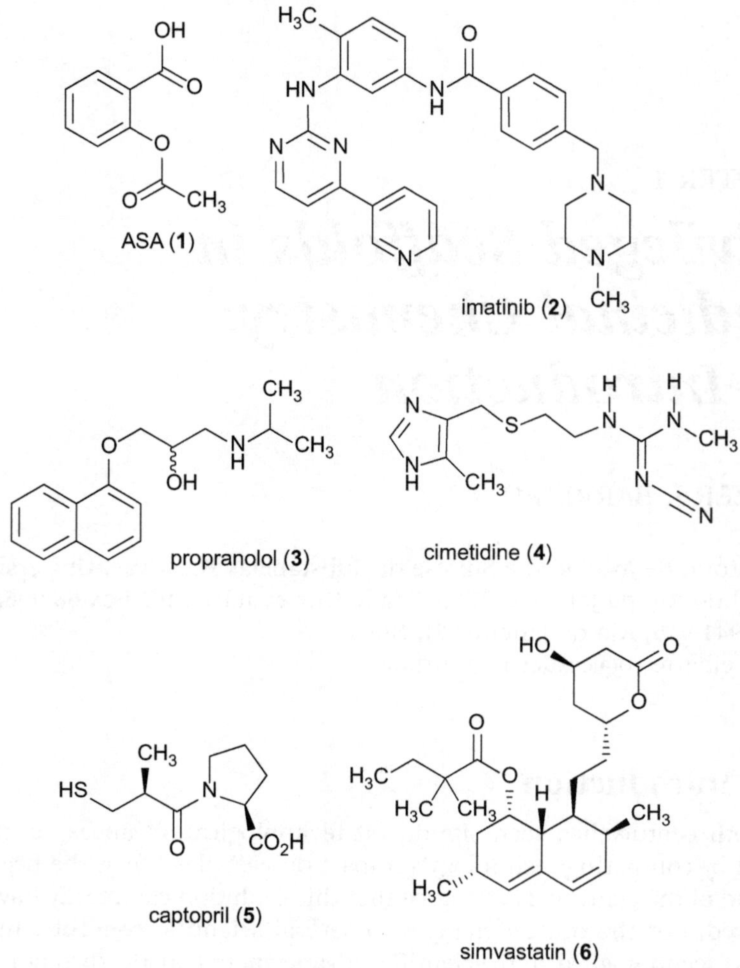

Figure 1.1 Structures of ASA, imatinib, propranolol, cimetidine, captopril, simvastatin.

imatinib (**2**, Figure 1.1), a powerful tyrosine-kinase (TK) inhibitor was created in Basel, Switzerland, in the Ciba-Geigy laboratories (currently Novartis)[1,2] and was launched in 2001 for the treatment of chronic myeloid leukemia. In the interval between both discoveries, we can see scientific and technological achievements that altered the paradigms of the drug discovery process. Obviously, most drugs that are now part of the contemporary therapeutic arsenal were created in the past century. Significant innovative examples include:

- propranolol (**3**, Figure 1.1), created by Black and co-workers[3] in the ICI laboratories in England in 1964;
- cimetidine (**4**, Figure 1.1),[4] created in 1975 at Smith, Kline & French (SK&F);

- captopril (**5**, Figure 1.1), created by Ondetti and Cushman[5,6] at Squibb laboratories; and,
- simvastatin (**6**, Figure 1.1), created by Patchett and collaborators[7] at Merck in 1998.[8]

All of these examples are the result of research efforts conducted in industrial laboratories and represent first-in-class drugs that are significant therapeutic innovations.

In addition to these discoveries, imatinib was a fantastic therapeutic innovation at the turn of the century (**2**).[1,2,9] It is used now in cancer chemotherapy, and was also created in an industrial research laboratory, involving modern medicinal chemistry strategies supported with HTS techniques. We understand that its discovery in the laboratories of Ciba-Geigy unraveled a new paradigm in which it was realized that multifactor diseases, generally chronic ones, need multitarget drugs. This new way of thinking among medicinal chemists, the discoverers of new drugs, has influenced the adoption of new approaches and the development of new terminology, in the latter half of the last century.

In 1988, Evans[10] published an article which mentioned the term 'privileged structures',[11] describing them as simple structural subunits present in the molecules of several drugs, with distinctive therapeutic uses, or affinities to several different receptors. This terminology has widened in its use, maybe in an excessively liberal way, and terms like 'molecular framework', 'chemotype', 'molecular fragment', and 'molecular scaffold', all of them synonymous, were created. In summary, some of these terms acquired different meanings, and due to current challenges in medicinal chemistry, they may be applied concurrently with other drug discovery techniques, such as molecular docking of fragments elected for the virtual screening in the search of new ligands of determinate targets, or in the construction of intelligent chemical libraries for use in HTS approaches, or to identify ligands, now called hits.[12] The identification of a new hit has widened the notion of molecular optimization through the use of classic medicinal chemistry techniques, to increase the affinity for the target in question, whether in potency or in selectivity. This establishes a certain hierarchy of the initial hit for the ligand, still without proof of concept for the prototype, now with pharmadynamic and pharmakinetic properties identified in functional pharmacological models.

Often the use of the terms 'privileged structure', 'fragment', or 'molecular scaffold' is mixed with the unique identity of each term being determined by molecular weight (in the case of fragments) or by the higher level of molecular simplification of a specific structural subunit for the use of molecular scaffold, here referring to cyclic structural subunits, aromatic or not. Both terms, however, refer to privileged structures. The bio IT experts use each term in a more precise way, which is mainly due to the function of the form or the elected molecular topology for each study.[13]

The evolution observed in the area of drug design and discovery throughout the last century may enable us to consider medicines as one of the biggest inventions of that century, because practically the entire contemporary therapeutic arsenal was invented or discovered then, with few examples of drugs being created and introduced in the 21st century.[14] The drug discovery process has seen changes throughout the last century, going beyond research laboratories of large pharmaceutical companies and reaching partnerships or multimember consortiums, involving university laboratories or high technology companies, or company–company joint ventures.[15]

Throughout the 20th century, or at least until its last decade, several drug discovery strategies were based on the paradigm inspired by the pioneering and masterful work of Hermann Emil Fischer and Paul Ehrlich, German Nobel Prize winners who established the basis for thought in this field throughout the 20th century. In 1902, Fischer was the first organic chemist to receive the Nobel Prize in Chemistry, mainly for the excellence of his work with carbohydrates, which inspired the key-lock model. The model explains, empirically, the differences observed in organoleptic properties among some sugars, with them being substances of similar chemical structures. This concept, together with Ehrlich's magic bullet,[16,17] for which he was awarded the Nobel Prize in Medicine in 1908, has inspired the thought of generations of scientists who were part of the discovery/invention process of new drugs throughout the 20th century.[18]

The Fischer–Ehrlich paradigm foresaw a few fundamental concepts for the design of new drugs, like that of complementary and molecular recognition between the bioreceptor and the drug, as well as the selectivity by a receptor as an attribute of efficacy and safety in the use of drugs. It was taken that, as corollary to safety in the use of drugs, its selectiveness for the therapeutic target and the possible future adverse side effects of a drug being related to lower selectivity or affinity for several receptors, or possible promiscuity. These ideas governed the thought of researchers in the area throughout most of the 20th century.[19]

1.2 The Privileged Scaffolds in Drug Discovery

Medicinal chemistry has as its main mission the understanding of molecular reasons for the activity of a drug or drug candidate. In this understanding, a few structural subunits of a certain bioactive molecule may be more relevant to a specific pharmacologic activity, governing the main interactions with a receptor. Those are the pharmacophoric contributions or pharmacophoric molecular groupings. Not unusually, the structures of drugs or their precursors have several functional groups, as well as the pharmacophoric ones, and all of them are called auxophoric subunits. Evidently, they all contribute to the total free energy of the drug-receptor complex, distinctively influencing the activity. Therefore, we may understand that some molecular scaffolds may have pharmacophoric characteristics for a certain type of receptor and not for others.[20] Some scaffolds may have privileged characteristics, being recognized molecularly by distinctive receptors without being important pharmacophores.[21]

An example of a pharmacophoric scaffold[22] can be identified in the class of first generation β-lactamic antibiotics (Chapter 2), where we will find penicillins and cephalosporins, represented by the 7-oxo-4-thia-1-aza-bicyclo[3.2.0]heptane (7) ring present in penicillin-G (**8**) (Figure 1.2) (Chapters 3 & 7).

Another classical molecular pharmacophoric scaffold is the system cyclopenteneperhydrophenantrene (**9**), which is present in several natural hormones such as testosterone (**10**) and synthetic drugs like prednisolone (**11**), a synthetic glucocorticoid, as shown in Figure 1.3.

Additionally, other important natural privileged scaffolds are represented by the systems of chalcone (**12**),[23] 1,4-benzopyrone (**13**),[24] isoflavone (**14**), coumarin (**15**)[25] (Chapter 11) among those oxygenated and structural sub-units that characterize several groups of alkaloids with distinctive pharmacological properties, like the quinoline ring (**16**),[26] isoquinoline (**17**) (Chapter 7), indole (**18**),[27] pyrrolidine (**19**) and other different possible combinations (Figure 1.4).[28–30]

The indole nucleus (**18**),[27] present in several natural and synthetic compounds (Chapters 13 & 14), is recognized as a central active scaffold, in several ergot alkaloids (*e.g.* ergotamine **20**) (Chapter 14) or in synthetic 3-carboxamide derivatives as ORG-28312 (**21**),[31] which presents agonistic affinity for CB1 receptors (Figure 1.4). The 3-carboxamide indole isosteres 4- and 6-azaindole ring appears in the structure of distinct active synthetic derivatives as **22** and **23**, described as potent renin inhibitors (Figure 1.5).[32]

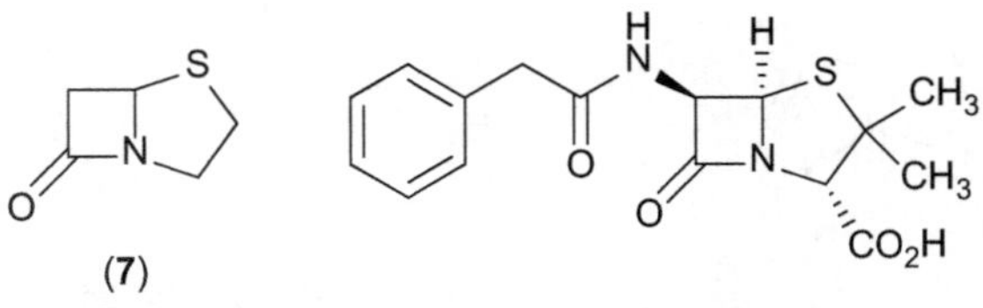

(**7**)

penicillin-G (**8**)

Figure 1.2 Structures of penicillin and its bicyclo system.

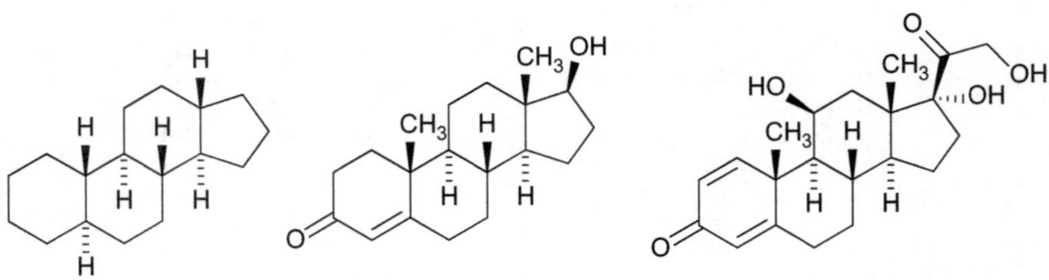

(**9**) testosterone (**10**) prednisolone (**11**)

Figure 1.3 Structures of testosterone, prednisolone and its tetracyclo system.

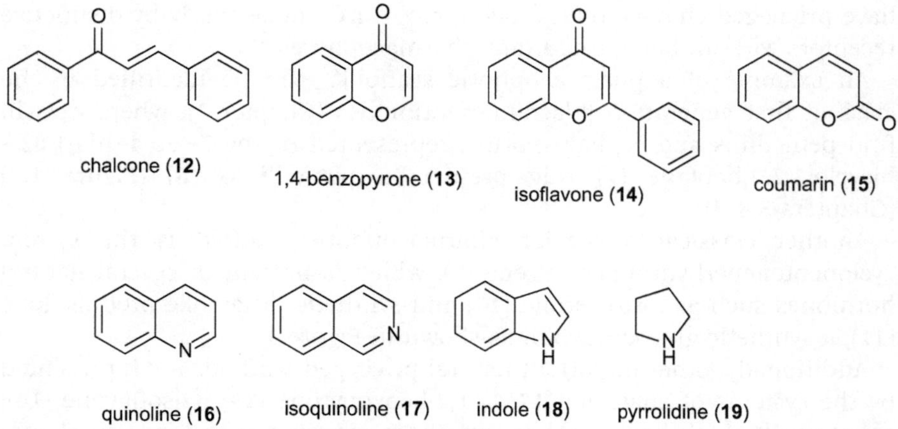

Figure 1.4 Representative natural privileged scaffolds.

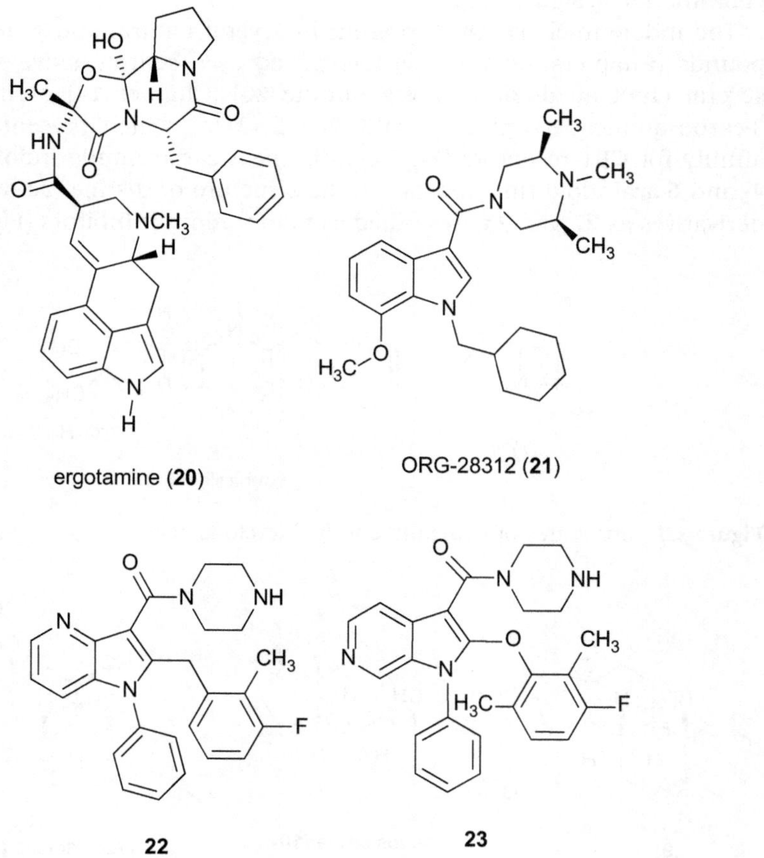

Figure 1.5 Structures of ergotamine, ORG-28312 and synthetic renin inhibitors.

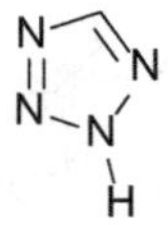

tetrazole (**24**)

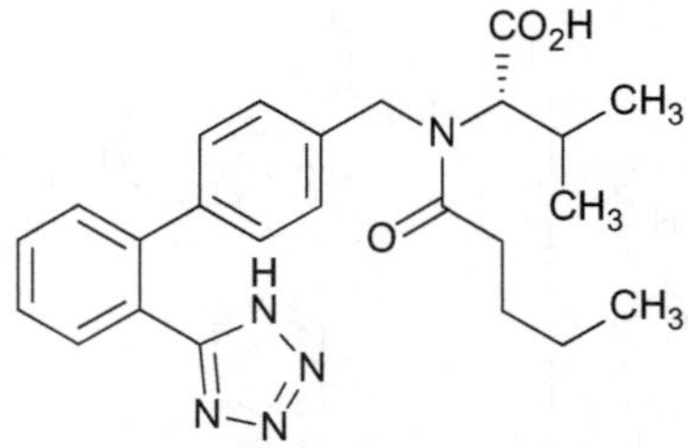

valsartan (**25**)

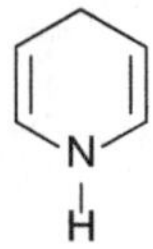

1,4-dihydropyridine (**26**)

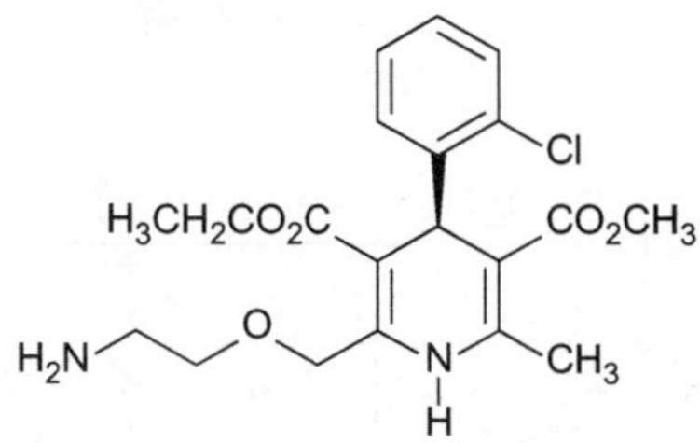

amlodipine (**27**)

Figure 1.6 Structuers of valsartan and amlodipine with its heterocycles scaffolds.

Among synthetic drugs, the presence of tetrazole scaffold **24** in several synthetic drugs with selective antagonist properties of AT1 receptors, characterizes the sartan group of antihypertensive drugs as valsartan (**25**)[33] (Figure 1.6), while the 1,4-dihydropyridine scaffold **26** present in several Ca^{++} channel blockers such as amlodipine (**27**), an important blockbuster drug belonging to a secondary class of antihypertensive drugs.[34]

The *N*-phenylpyrazole scaffold **28** (Chapter 5) is present in a great number of drugs or drug candidates[35] as the recent disclosed direct factor Xa inhibitor apixaban (**29**; BMS-562247-01, Eliquis™),[36] an anticoagulant agent indicated for the treatment of venous thromboembolic disease, where this structural subunit is included in the dihydropyrazolo[5,4-*c*]pyridine-3-carboxamide moiety **30**. The celecoxib (**31**),[37] a selective COX-2 inhibitor also has this privileged scaffold **28** in its structure, included in the terphenyl like motif **32**. This system *per se* also represents an important privileged scaffold, present in the bestselling statin compound atorvastatin **33**. In addition, the terphenyl-like scaffold represented by the pyridinylimidazole system (Chapter 4), is present also in MAPKp38 inhibitor SB-203580 (**34**) (Scheme 6). This compound, presents in the central ring of the terphenyl-like system an imidazole ring representing a bioisosteric[38] system (**35**) of the 1,5-diarylpyrazole motif **32** (Figure 1.7) (Chapter 5).

The 1*H*-pyrazolo[3,4-*d*]pyridine scaffold **36** (Chapter 5) is present in numerous bioactive derivatives, as demonstrated by the derivative BAY418543 (**37**),[39,40] which is described as soluble guanylate cyclase stimulators (sGC), and is useful to control pulmonary hypertension disease (Figure 1.8) and in

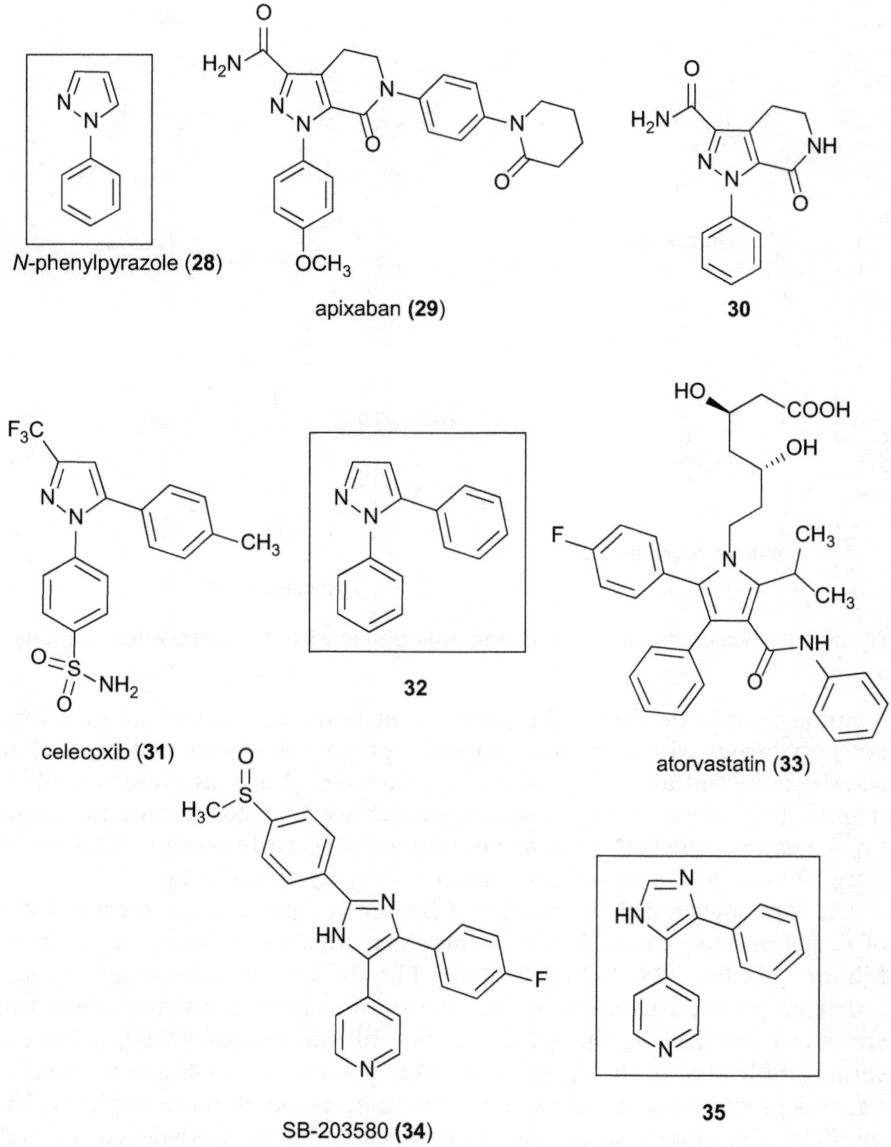

Figure 1.7 Structure apixaban, celecoxib, atrovastatin, SB-203580 and important scaffolds *N*-phenylpyrazole and terphenyl.

the B-Raf^{V600E} inhibitor **38**, recently described by Wenglowsky and co-workers[41] as a potent agent in preclinical evaluation to treat primary and metastatic melanomas (Figure 1.8).[42] This 7-azaindazole compound has an isostere system as present in PLX4032 (**39**),[43] a difluorophenylsulfonamide substructure with the pyrrolo-pyridine scaffold **40**, a 7-azaindole ring,[44] described as being useful to control metastatic melanoma.

Several isosteres of azabicyclic aromatic[45] represent important privileged scaffolds **41–47**, present in numerous drugs such as the one in the bioactive compounds in Figure 1.9.

The scaffold **41** is present in the compound **48** a pyrimido[4,5-*b*]indole derivative possessing a thiophenyl moiety,[46] described as a dual agent acting as a kinase inhibitor on EGFR and PDGFR-β, with $IC_{50} = 10{,}41$ and 40,3 μM, respectively, promoting antiangiogenic effect. This scaffold in a tautomeric form **42** (Figure 1.9) is present in ruxolitinib (**49**),[47] described as being an antimyelofibrosis (MF) agent acting also as a dual inhibitor of Janus kinase JAK-1 and JAK-2.

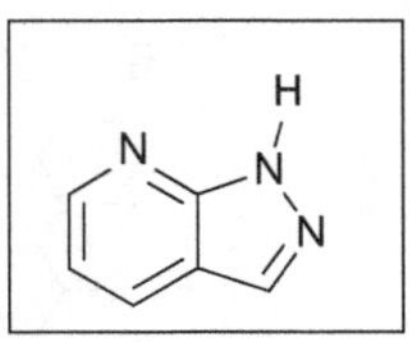

1*H*-pyrazolo[3,4-*d*]pyridine (**36**)

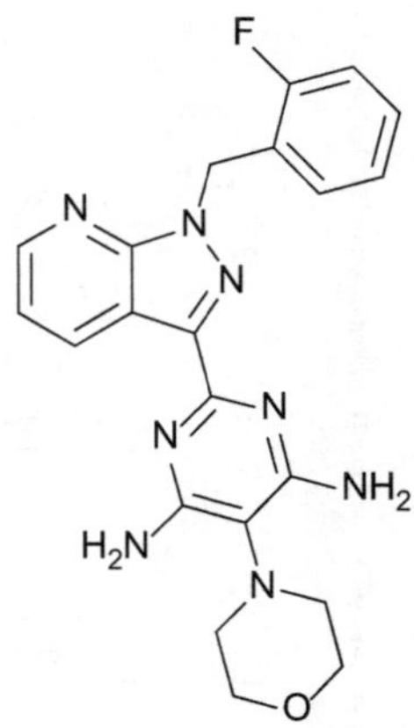

BAY 418543 (**37**)

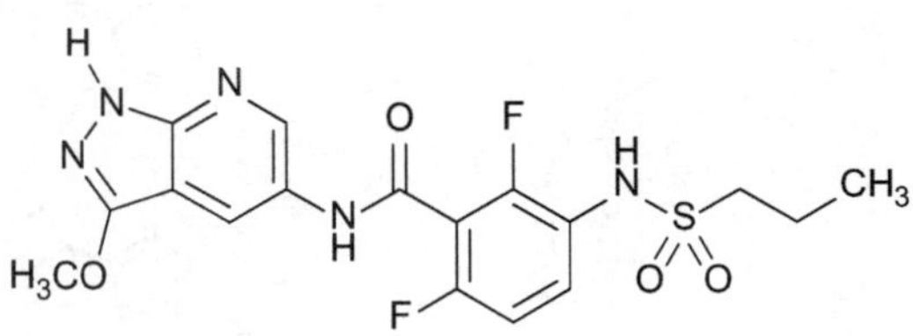

38

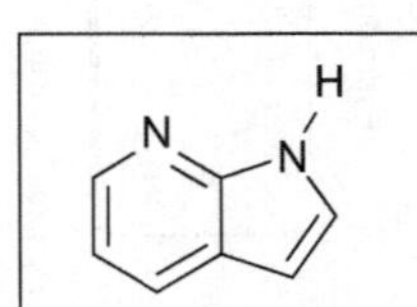

pyrrolo-pyridine (**40**)

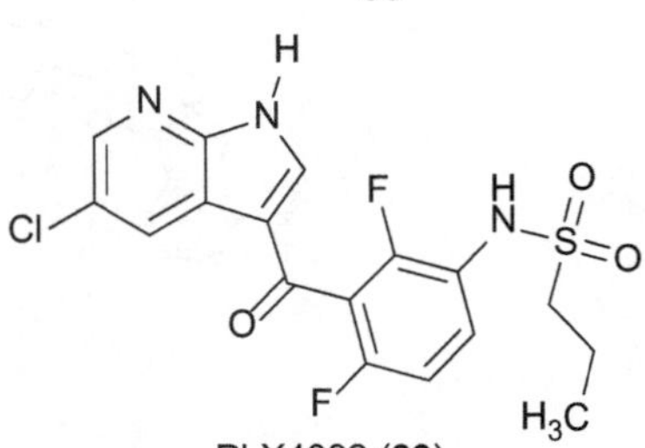

PLX4032 (**39**)

Figure 1.8 Structure of important heterocyclic scaffold (**36** and **40**) and compounds BAY 418543, PLX4032 and **38**.

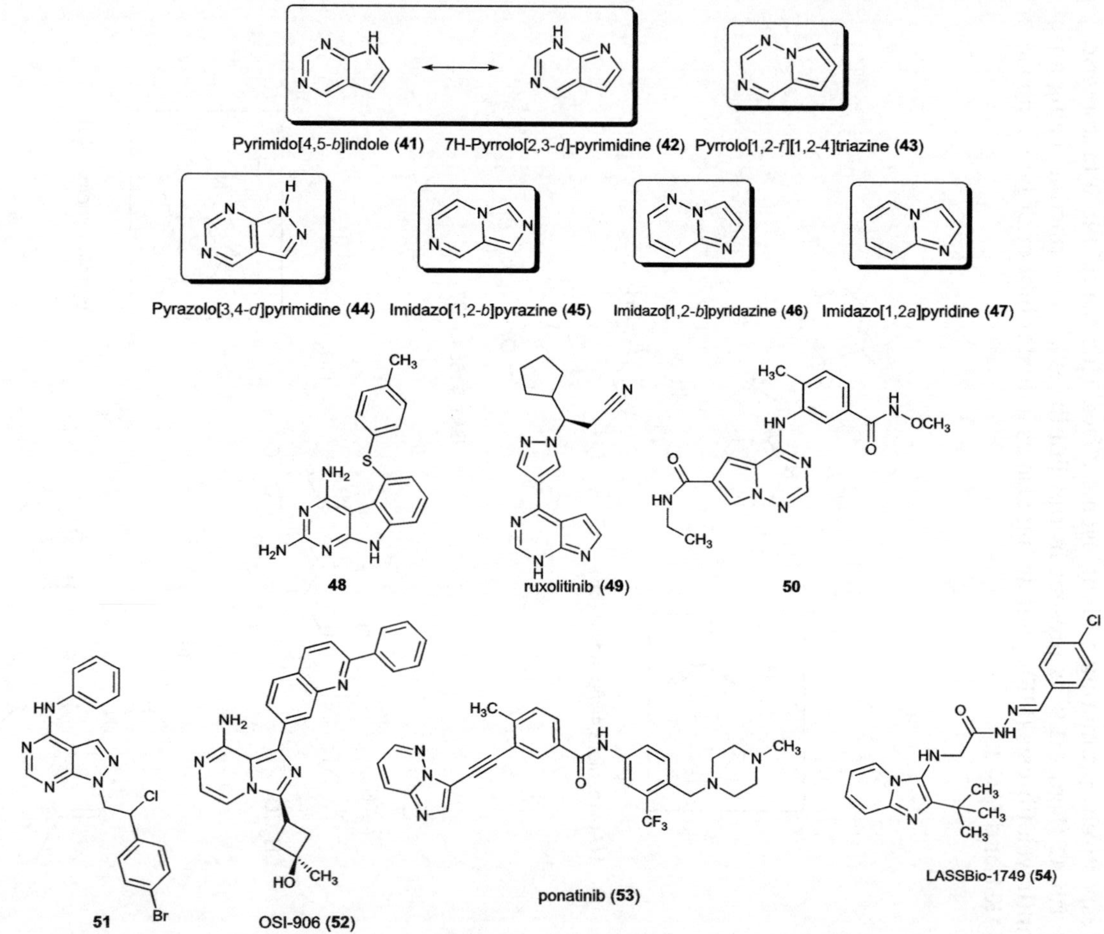

Figure 1.9 Structure of important azaheterocyclic scaffolds with examples.

The scaffold **43** (Figure 1.9), belonging to pyrrolo[1,2-*f*][1,2,4]triazine system,[48] appears in derivative **50** (BMS-582949),[49] with an *N*-methoxybenzamide moiety and was described as a potent multiple p38 MAP kinase inhibitor. Compound **51** (Figure 1.8) was discovered applying a cross-docking approach on a library of the pyrazolo[3,4-*d*]pyrimidine privileged scaffold **44** (Chapter 5).[50] This derivative having a bromine atom in the *para* position of the side chain phenyl ring was active at submicromolar potency against T315I Bcr-Abl expressing cells.[51]

The derivative OSI-906 (**52**),[52] shown in Figure 1.9, is a six-membered compound, possessing the imidazo[1,2*b*]pyrazine scaffold **45** (Chapter 4). This derivative is in Phase III clinical trials as a selective dual antagonist of insulin and IGF-I receptor with IC_{50} 0.024 μM in LISN cells.[53] The imidazo[1,2*b*]pyridazine scaffold **46** (Figure 1.9) appears in ponatinib (**53**),[54] an oral drug approved by the US Food Drug Administration in 2012, for patients with resistant or intolerant chronic myeloid leukemia (CML). It is a multi-targeted tyrosine-kinase inhibitor derivative with an acetylenic benzamide chemotype. The only bisazo isosteric aromatic scaffold shown in Figure 1.8, the imidazo[1,2*a*]pyridine **47**, is present in the *N*-acylhydrazone compound LASSBio-1749 (**54**), described by Lacerda and co-workers as being a potent anti-TNFα agent.[55]

1.3 Conclusion

This introductory chapter provides a brief perspective about the privileged scaffold concept use in medicinal chemistry. This approach can be used alone or as a combined strategy, mixing other molecular design techniques such as bioisosterism.

The reader can find several more important details with a major number of examples of this useful strategy of drug design and discovery, in the following chapters of this book.

Acknowledgements

The author acknowledges Professor Dr Stefan Bräse (Karlsruher Institut für Technologie, KIT), editor of this book, for the kind invitation to contribute with this introductory chapter.

References

1. A. S. Kesselheim and J. Avorn, *Nat. Rev. Drug Discovery*, 2013, **12**, 425.
2. R. Capdeville, E. Buchdunger, J. Zimmermann and A. Matter, *Nat. Rev. Drug Discovery*, 2002, **1**, 493.
3. J. W. Black, A. F. Crowther, R. G. Shanks, L. H. Smith and A. C. Dornhorst, *Lancet*, 1964, **283**, 1080.
4. R. W. Brimblecombe, W. A. M. Duncan, G. J. Durant, C. R. Ganellin, M. E. Parsons and J. W. Black, *Br. J. Pharmacol.*, 1975, **53**, 435.

5. D. W. Cushman, H. S. Cheung, E. F. Sabo and M. A. Ondetti, *Am. J. Cardiol.*, 1982, **49**, 1390.
6. M. A. Ondetti, B. Rubin and D. W. Cushman, *Science*, 1977, **196**, 441.
7. C. H. Kuo, A. A. Patchett and N. L. Wendler, *J. Org. Chem.*, 1983, **48**, 1991.
8. M. J. Mol, D. W. Erkelens, J. A. Gevers Leuven, J. A. Schouten and A. F. Stalenhoef, *Atherosclerosis*, 1988, **69**, 131.
9. P. W. Manley and J. Zimmermann, in *Polypharmacology in Drug Discovery*, J.-U. Peters, John Wiley & Sons, New Jersey, 2012, ch. 20, pp. 409–419.
10. B. E. Evans, K. E. Rittle, M. G. Bock, R. M. DiPardo, R. M. Fredinger, W. L. Whitter, G. F. Lundell, D. F. Veber, P. S. Anderson, R. S. L. Chang, V. J. Lotti, D. J. Cerino, T. B. Chen, P. J. Kling, K. A. Kunkel, J. P. Springer and J. Hirshfield, *J. Med. Chem.*, 1988, **31**, 2235.
11. C. D. Duarte, E. J. Barreiro and C. A. M. Fraga, *Mini-Rev. Med. Chem.*, 2007, 7, 1108.
12. J. Polanski, A. Kurczyk, A. Bak and R. Musiol, *Curr. Med. Chem.*, 2012, **19**, 1921; L. Costantino and D. Barlocco, *Curr. Med. Chem.*, 2006, **13**, 65; G. W. Bemis and M. A. Murcko, *J. Med. Chem.*, 1996, **39**, 2887.
13. R. W. DeSimone, K. S. Currie, S. A. Mitchell, J. W. Darrow and D. A. Pippin, *Comb. Chem. High Throughput Screening*, 2004, 7, 473.
14. B. Munos, *Nat. Rev. Drug Discovery*, 2009, **8**, 959; D. Swinney and J. Anthony, *Nat. Rev. Drug Discovery*, 2011, **10**, 507.
15. D. M. Hyurin, *ACS Med. Chem. Lett.*, 2013, **4**, 313.
16. K. Strebhardt and A. Ullrich, *Nat. Rev. Cancer*, 2008, **8**, 473.
17. R. S. Schwartz, *New Engl. J. Med.*, 2004, **350**, 1079.
18. D. Jürgen, *Nat. Rev. Drug Discovery*, 2006, **5**, 635.
19. H. P. Hang, *Br. J. Pharmacol.*, 2006, **147**, S9.
20. H.-J. Böhm, A. Flohr and M. Stahl, *Drug Discovery Today: Technol.*, 2004, **1**, 217.
21. C. M. Marson, *Chem. Soc. Rev.*, 2011, **40**, 5514; Y. Hu, D. Stumpfe and J. Bajorath, *J. Chem. Inf. Mod.*, 2011, **51**, 1742; P. Ert, S. Jelfs, J. Mühlbacher, A. Schuffenhauer and P. Selzer, *J. Med. Chem.*, 2006, **49**, 4568; K. Bondensgaard, M. Ankersen, H. Thøgersen, B. S. Hansen, B. S. Wulff and R. P. Bywater, *J. Med. Chem.*, 2004, **47**, 888.
22. N. Brown, *Scaffold Hopping in Medicinal Chemistry*, ed. N. Brown Wiley-VCH Verlag GmbH & Co. KGaA, 1st edn, 2014, pp. 3–13.
23. N. K. Sahu, S. S. Balbhadra and D. V. Kohli, *Curr. Med. Chem.*, 2012, **19**, 209.
24. A. Gaspar, M. J. Matos, J. Garrido, E. Uriarte and F. Borges, *Chem. Rev.*, 2014, **114**, 4960.
25. M. J. Matos, D. Viña, S. Vasquez-Rodriguez, E. Uriarte and L. Santana, *Curr. Top. Med. Chem.*, 2012, **12**, 2210.
26. I. Balderas-Renteria, P. Gonzalez-Barranco, A. Garcia, B. K. Banik and G. Rivera, *Curr. Med. Chem.*, 2012, **19**, 4377; V. R. Solomon and H. Lee, *Curr. Med. Chem.*, 2011, **18**, 1488.

27. F. R. de Sá Alves, E. J. Barreiro and C. A. Fraga, *Mini Rev. Med. Chem.*, 2009, **9**, 782.
28. M. E. Welsch, S. A. Snyder and B. R. Stockwell, *Curr. Opin. Chem. Biol.*, 2010, **14**, 1.
29. M. Singh, M. Kaur and O. Silakari, *Eur. J. Med. Chem.*, 2014, **84**, 206.
30. R. S. Keri, S. Budagumpi, R. K. Pai and R. G. Balakrishna, *Eur. J. Med. Chem.*, 2014, **78**, 340.
31. J. M. Adam, J. Cairns, W. Caulfield, P. Cowley, I. Cumming, M. Easson, D. Edwards, M. Ferguson, R. Goodwin, F. Jeremiah, T. Kiyoi, A. Mistry, E. Moir, R. Morphy, J. Tierney, M. York, J. Baker, J. E. Cottney, A. K. Houghton, P. J. Westwood and G. Walker, *Med. Chem. Commun.*, 2010, **1**, 54.
32. B. Scheiper, H. Matter, H. Steinhagen, Z. Böcskei, V. Fleury and G. McCort, *Bioorg. Med. Chem. Lett.*, 2011, **21**, 5480.
33. E. Carosati, P. Ioan, M. Micucci, F. Broccatelli, G. Cruciani, B. S. Zhorov, A. Chiarini and R. Budriesi, *Curr. Med. Chem.*, 2012, **19**, 4306; P. Ioan, E. Carosati, M. Micucci, G. Cruciani, F. Broccatelli, B. S. Zhorov, A. Chiarini and R. Budriesi, *Curr. Med. Chem.*, 2011, **18**, 4901.
34. P. Naik, P. Murumkar, R. Giridhar and M. R. Yadav, *Bioorg. Med. Chem.*, 2010, **18**, 8418.
35. H. Kumar, D. Saini, S. Jain and N. Jain, *Eur. J. Med. Chem.*, 2013, **70**, 248; A. A. Bekhit, A. Hymete, A. El-Din, A. Bekhit, A. Damtew and H. Y. Aboul-Enein, *Mini-Rev. Med. Chem.*, 2010, **10**, 1014.
36. D. J. P. Pinto, M. J. Orwat, S. Koch, K. A. Rossi, R. S. Alexander, A. Smallwood, P. C. Wong, A. R. Rendina, J. M. Luettgen, R. M. Knabb, K. He, B. Xin, R. R. Wexler and P. Y. S. Lam, *J. Med. Chem.*, 2007, **50**, 5339.
37. T. D. Penning, J. J. Talley, S. R. Bertenshaw, J. S. Carter, P. W. Collins, S. Docter, M. J. Graneto, L. F. Lee, J. W. Malecha, J. M. Miyashiro, R. S. Rogers, D. J. Rogier, S. S. Yu, G. D. Anderson, E. G. Burton, J. N. Cogburn, S. A. Gregory, C. M. Koboldt, W. E. Perkins, K. Seibert, A. W. Veenhuizen, Y. Y. Zhang and P. C. Isakson, *J. Med. Chem.*, 1997, **40**, 1347.
38. *Bioisosteres in Medicinal Chemistry*, ed. N. Brown, Wiley-VCH Verlag GmbH, Weinheim, 2012; Y. Hamada and Y. Kiso, *Expert Opin. Drug Discov.*, 2012, **7**, 903; N. A. Meanwell, *J. Med. Chem.*, 2011, **54**, 2529; S. R. Langdon, P. Ertl and N. Brown, *Mol. Inf.*, 2010, **29**, 366; L. M. Lima and E. J. Barreiro, *Curr. Med. Chem.*, 2005, **12**, 23; P. H. Olsen, *Curr. Opin. Drug Discov. Dev.*, 2001, **4**, 471; G. A. Patani and E. J. LaVoie, *Chem. Rev.*, 1996, **96**, 3147; A. Burger, *Prog. Drug Res.*, 1991, **37**, 287; C. A. Lipinski, *Annu. Rep. Med. Chem.*, 1986, **21**, 283; W. Thornber, *Chem. Soc. Rev.*, 1979, **8**, 563.
39. J. A. Watts, M. A. Gellar, M. B. Fulkerson and J. A. Kline, *Pulm. Pharmacol. Ther.*, 2013, **26**, 205.
40. N. Griebenow, H. Schirok, J. Mittendorf, A. Straub, M. Follmann, J.-P. Stasch, A. Knorr, K.-H. Schlemmer and G. Redlich, *Bioorg. Med. Chem. Lett.*, 2013, **23**, 1197.

41. S. Wenglowsky, L. Ren, K. A. Ahrendt, E. R. Laird, I. Aliagas, B. Alicke, A. J. Buckmelter, E. F. Choo, V. Dinkel, B. Feng, S. L. Gloor, S. E. Gould, S. Gross, J. Gunzner-Toste, J. D. Hansen, G. Hatzivassiliou, B. Liu, K. Malesky, S. Mathieu, B. Newhouse, N. J. Raddatz, Y. Ran, S. Rana, N. Randolph, T. Risom, J. Rudolph, S. Savage, L. T. Selby, M. Shrag, K. Song, H. L. Sturgis, W. C. Voegtli, Z. Wen, B. S. Willis, R. D. Woessner, W.-I. Wu, W. B. Young and J. Grina, *ACS Med. Chem. Lett.*, 2011, **2**, 342.
42. C. Zhang and G. Bollag, *Curr. Op. Genet. Dev.*, 2010, **20**, 79.
43. K. T. Flaherty, I. Puzanov, K. B. Kim, A. Ribas, G. A. McArthur, J. A. Sosman, P. J. O'Dwyer, R. J. Lee, J. F. Grippo, K. Nolop and P. B. Chapman, *New Engl. J. Med.*, 2010, **363**, 809.
44. G. Bollag, J. Tsai, J. Zhang, C. Zhang, P. Ibrahim, K. Nolop and P. Hirth, *Nature Rev. Drug Discovery*, 2012, **11**, 873; H. Matter, B. Scheiper, H. Steinhagen, Z. Böcskei, V. Fleury and G. McCort, *Bioorg. Med. Chem. Lett.*, 2011, **18**, 5487.
45. For examples of triazolo isosteres compounds, see: J-F. Brazeau and G. Rosse, *ACS Med. Chem. Lett.*, 2014, **5**, 1069
46. A. Gangjee, N. Zaware, S. Raghavan, M. Ihnat, S. Shenoy and R. L. Kisliuk, *J. Med. Chem.*, 2010, **53**, 1563.
47. R. A. Mesa, U. Yasothan and P. Kirkpatrick, *Nat. Rev. Drug Discovery*, 2009, **11**, 103.
48. For examples of 1,2,3-triazine scaffold, see:R. Kumar, A. D. Singh, J. Singh, H. Singh, R. K. Roy and A. Chaudhary, *Mini-Rev. Med. Chem.*, 2014, **14**, 72.
49. C. Liu, J. Lin, S. T. Wrobleski, S. Lin, J. Hynes Jr., H. Wu, A. J. Dyckman, T. Li, J. Wityak, K. M. Gillooly, S. Pitt, D. R. Shen, R. F. Zhang, K. W. McIntyre, L. Salter-Cid, D. J. Shuster, H. Zhang, P. H. Marathe, A. M. Doweyko, J. S. Sack, S. E. Kiefer, K. F. Kish, J. A. Newitt, M. McKinnon, J. H. Dodd, J. C. Barrish, G. L. Schieven and K. Leftheris, *J. Med. Chem.*, 2010, **53**, 6629.
50. M. Radi, C. Tintori, F. Musumeci, C. Brullo, C. Zamperini, E. Dreassi, A. L. Fallacara, G. Vignaroli, E. Crespan, S. Zanoli, I. Laurenzana, I. Filippi, G. Maga, S. Schenone, A. Angelucci and M. Botta, *J. Med. Chem.*, 2013, **56**, 5382.
51. G. Vignaroli, M. Mencarelli, D. Sementa, E. Crespan, M. Kissova, G. Maga, S. Schenone, M. Radi and M. Botta, *ACS Comb. Sci.*, 2014, **16**, 168.
52. M. J. Mulvihill, A. Cooke, M. Rosenfeld-Franklin, E. Buck, K. Foreman, D. Landfair, M. O'Connor, C. Pirritt, Y. Sun, Y. Yao, L. D. Arnold, N. W. Gibson and Q.-S. Ji, *Future Med. Chem.*, 2009, **1**, 1153.
53. M. J. Mulvihill and E. Buck, in *Accounts in Drug Discovery Case Studies in Medicinal Chemistry*, ed. J. C. Barrish, P. H. Carter, P. T. W. Cheng and R. Zahler, RSC Publishing, London, 2011, pp. 71–102.
54. W. S. Huang, C. A. Metcalf, R. Sundaramoorthi, Y. Wang, D. Zou, R. M. Thomas, X. Zhu, L. Cai, D. Wen, S. Liu, J. Romero, J. Qi, I. Chen,

G. Banda, S. P. Lentini, S. Das, Q. Xu, J. Keats, F. Wang, S. Wardwell, Y. Ning, J. T. Snodgrass, M. I. Broudy, K. Russian, T. Zhou, L. Commodore, N. I. Narasimhan, Q. K. Mohemmad, J. Iuliucci, V. M. Rivera, D. C. Dalgarno, T. K. Sawyer, T. Clackson and W. C. Shakespeare, *J. Med. Chem.*, 2010, **53**, 4701.

55. R. B. Lacerda, N. M. Sales, L. L. Silva, R. Tesch, A. L. P. Miranda, E. J. Barreiro, P. D. Fernandes and C. A. M. Fraga, *PLoS One*, 2014, **9**, e91660.

CHAPTER 2

Privileged Scaffolds in Medicinal Chemistry – A Computational Approach

PRIYA ANAND,[a,†] SHALINI JOHN,[b,†] IRENE MELICIANI,[a,‡] ALEXANDER SCHUG[b] AND WOLFGANG WENZEL*[a]

[a] Institute of Nanotechnology, Karlsruhe Institute of Technology, Karlsruhe, Germany; [b] Steinbuch Centre for Computing, Karlsruhe Institute of Technology, Karlsruhe, Germany
*Email: Wolfgang.wenzel@kit.edu

2.1 Introduction

In the era of medicinal chemistry, computational methodologies play a crucial role in "lead generation", which is a key early step in the drug discovery process. Computer-aided drug design is an effective preliminary step used in pharmaceutical research during the development and screening of new drug molecules.[1,2] Because of advancements in experiment and protein structure prediction, many three-dimensional (3D) structures are available. Understanding the 3D conformation of a protein is highly important in medicine and biotechnology, as it aids in clarifying the properties, activities and biological characteristics of the protein, including protein–ligand interactions, protein–protein interactions, drug function and drug design.[3–5]

[†]Equally contributed as first authors.
[‡]Current address: Intelligent Pharma, C/Baldiri Reixac 4, 08028 Barcelona, Spain.

RSC Drug Discovery Series No. 50
Privileged Scaffolds in Medicinal Chemistry: Design, Synthesis, Evaluation
Edited by Stefan Bräse

Published by the Royal Society of Chemistry, www.rsc.org

In this chapter, we discuss the modeling methods used for the prediction of protein structure and receptor- or protein–ligand interactions in detail. Following decades of research in medicinal chemistry, it has been widely accepted that the therapeutic activity of a drug molecule is related to the affinity of the ligand to the binding pocket of the target macromolecule or receptor.[6,7] Therefore, understanding the structures of and interactions between the receptor and the ligand is of great interest.

2.2 Scope of the Study

Computational chemistry and molecular modeling have become crucial components of the modern drug discovery process. Based on the increased number of protein crystal structures available in the PDB (www.rcsb.org), structure-based virtual screening techniques, as well as ligand-based techniques, have been widely used for "lead" discovery.[8,9] Predicting the binding sites of proteins and their interactions with small molecules is crucial for structure-based drug design. For many proteins involved in disease, the biophysical mechanisms that activate potential binding partners remain poorly understood. Molecular docking is an effective tool used to identify the ligands that optimally fit into the binding sites of proteins in energetically and geometrically stable conformations. When the 3D protein structure is available, the prediction of energetically favorable binding orientations and the affinity of small molecules for the targeted binding site can be achieved using different docking methodologies.

Over the last two decades, many docking algorithms using similar methodologies but distinct searching and scoring algorithms and varying computational speed have been developed. Recent improvements in docking algorithms and performance have made it possible to dock thousands of ligands within a timescale that is applicable for pharmaceutical development processes. Several interesting reviews[8,10–12] have provided insight into the application of docking to the drug design process. In recent years, many new programs have been developed, and older programs are being updated using new technology. Because of this constant development of new programs and algorithms, the identification of an appropriate program for different datasets and target molecules is necessary. This review provides a brief introduction to the available 3D protein structure prediction methods, docking algorithms, scoring functions and applications of the docking approach for fragment-based drug design and structure-based virtual screening.

2.3 Privileged Scaffolds

In 1988, Evans *et al.*[13] recognized the efficiency of certain structural motifs that frequently occur in drug molecules and can be used as templates to search for novel receptor agonists and antagonists. Thus, these frequently occurring structural motifs were defined as "privileged scaffolds". Benzodiazepine motifs were originally referred to as privileged scaffolds due to their ability to mimic the structure of beta turns.[14] However, the original concept has been further developed over several decades based on research in both academic and industrial settings. Currently, a privileged scaffold is defined as a single molecular framework that can serve as a ligand for more than one type of enzyme or receptor target based on prudent structural modification.[15] In general, privileged scaffolds are drug-like molecules used to generate drug-like databases that can be used as effective tools for the discovery of novel lead compounds. Importantly, the presence of a privileged scaffold in a compound library does not mean that all the compounds are biologically active and non-toxic.

2.4 Molecular Docking

Binding of a small molecule to the active or allosteric site of a receptor often triggers a conformational change, including refolding or dimerization, which may be associated with many other biological responses involving intra- and intermolecular interactions. Consideration of such interactions is required to predict the preferred orientation and binding affinity between a small molecule and its target. The most important inter-atomic interactions mediating the binding of small molecules to their receptors are electrostatic and van-der-Waals interactions. Other important factors, such as entropy, changes in the solvent molecules and intramolecular contributions resulting from the flexibility of the receptor, contribute to the protein–ligand affinity.[16,17] Molecular docking is a computational methodology used to study molecular interactions and predict the binding orientation between a protein/receptor and a set of potentially interesting compounds.[18] The success rate of a molecular docking method depends on the accuracy of the structural model of the complex and the scoring algorithm of the docking program. Here, we discuss the receptor structure prediction methods and various molecular docking algorithms available at present.

2.5 Protein Structure Prediction

Proteins are composed of long chains of amino acid residues (see Figure 2.1) encoded by the DNA/RNA sequence. Genes encode the specific amino acid sequence which, when translated, folds into a unique 3D conformation. Proteins are the molecular workhorses in cells and represent the largest molecular constituent (besides water) of cells. Proteins play a crucial role in all biological functions; they act as structural elements, signal receptors, transporters and antibodies, as well as enzymes, which catalyze biochemical reactions. All these functions are directly dependent on the appropriate 3D protein structure, which is determined by the sequence of the amino acids in the protein polymer. Therefore, understanding protein function, regulation and interaction is one of the most important goals of biochemistry. Computer simulation is a valuable tool for investigating the structure, folding and interactions of a protein with other biomolecules, because this approach

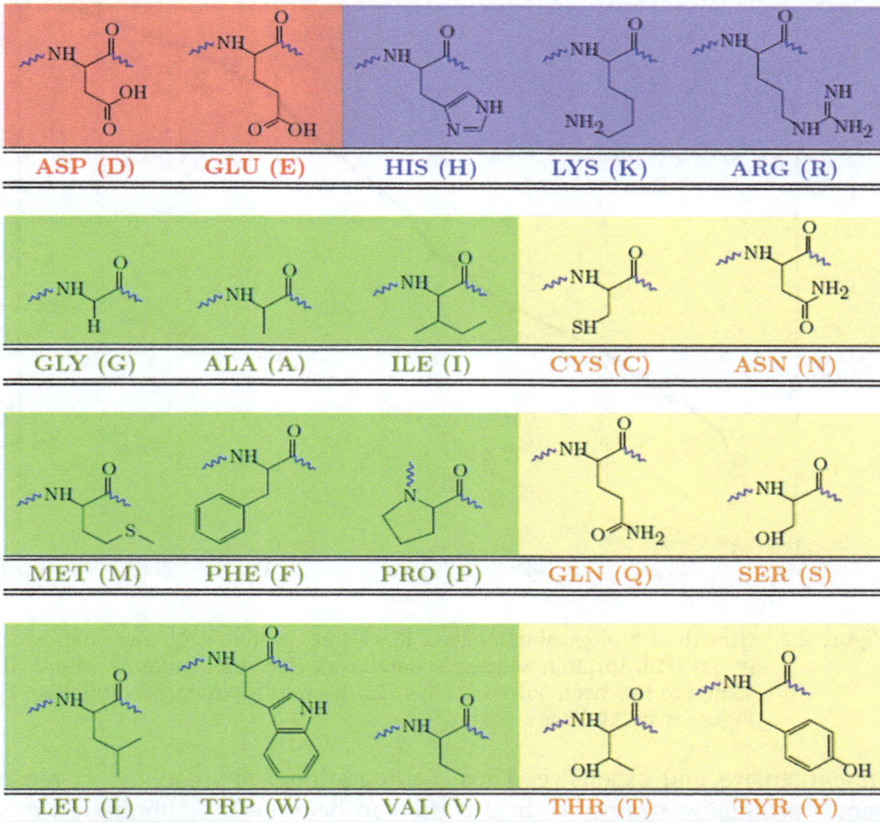

Figure 2.1 Twenty naturally occurring amino acids found in eukaryotes, grouped according to their side-chains.
Figure from Anand (2013).[330]

can explore regimes beyond the experimental capability. At present, 3D protein structure prediction, *i.e.*, the prediction of the final folded conformation (the tertiary structure of the protein) based on the amino acid sequence (the primary structure of the protein) is one of the most important unsolved problems in biophysics.

Experimental methods such as X-ray crystallography and NMR spectroscopy have led to the deposition of more than 53 000 experimentally solved structures in the Protein Data Bank (PDB; http://www.rcsb.org), but there are far more protein/gene sequences reported in Swiss-Prot (>400 000), DDBJ/EMBL/GenBank (more than 100 million nucleotide sequences),[19–21] UniProtKB (5 million nucleotide sequences translated into amino acid sequences)[22] and TrEMBL (>7 500 000)[23] (Figure 2.2). The number of known protein/gene sequences is increasing exponentially compared to the number of 3D structures submitted in the PDB database. The number of protein/gene sequences far outstrips the capacity of experimentalists to determine the 3D structures of the proteins, as experimental methods are

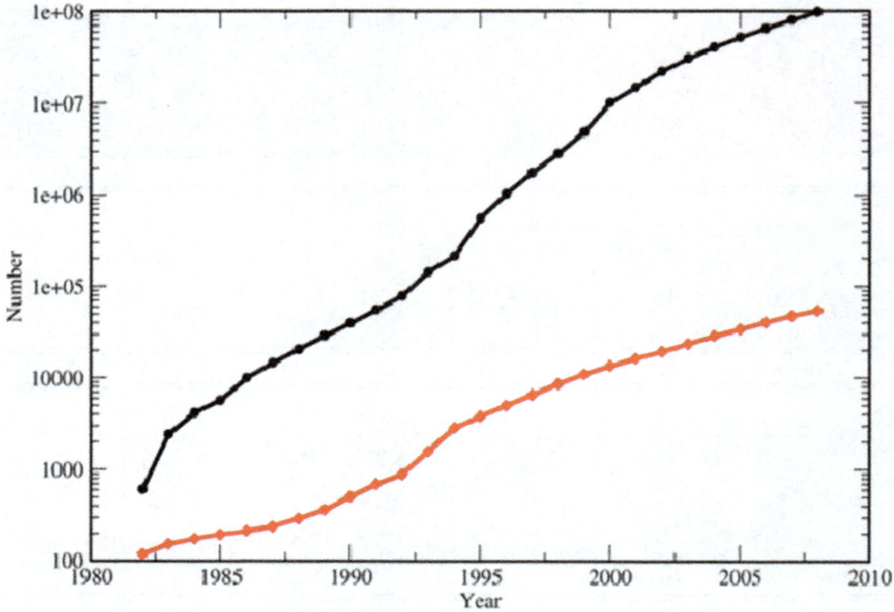

Figure 2.2 Growth of biological databases. Black line: protein sequences deposited in TrEMBL (protein sequence database). Red line: proteins whose 3D structure has been solved; 75 594 (PDB) protein structures (30/08/2011). Figure from Meliciani (2011).[16]

time-intensive and expensive. Thus, computational approaches[24,25] represent a promising strategy to bridge the gap between the submission of a protein sequence and the prediction of its 3D structure.

Computational 3D protein structure prediction methods fall into two broad categories: comparative modeling (also referred to as template-based modeling (TBM)), which includes homology modeling (HM) and the threading method, and *ab initio* modeling (also referred to as free modeling (FM)).[26,27] Despite the progress in *ab initio* protein structure prediction methods, comparative modeling is a more reliable method for predicting the 3D structure of a protein at an accuracy that is comparable to a low-resolution experimentally determined structure. Similarly, genomic information can be statistically analyzed for co-evolving amino acids using methods such as DCA[28,29] or PSICOV.[30] As co-evolution infers spatial adjacency, this information can be exploited to provide spatial constraints for structure prediction. The successful application of this approach has been demonstrated for the analysis of globular proteins,[31] protein complexes,[32] trans-membrane proteins,[33] and different conformations of the same protein.[34]

2.5.1 Comparative Modeling

Many different methods for protein structure prediction have been proposed,[35] but the most widely used methodology is homology modeling.

The 3D model of the query protein is constructed from the amino acid sequence or experimental 3D structure of a related homologous protein defined as the "template". Homology modeling benefits from the fact that there is a limited number of protein folding conformations in nature,[36] resulting in a more evolutionarily conserved 3D protein structure.[16]

The primary goal of homology modeling is to identify one or more known structures/sequences that are similar to the structure of the query sequence, followed by the alignment of the residues of the query sequence with the residues of the template sequence.[37–39] The search for template protein structures or sequences is performed using data from the PDB (http://www.rcsb.org/pdb/),[40] SCOP (http://scop2.mrc-lmb.cam.ac.uk/),[41] DALI[42] and CATH (http://www.cathdb.info/)[43] databases. The simplest search algorithms are based on: (1) pair-wise sequence comparison of the query and the template using BLAST[44] (http://ncbi.nlm.nih.gov/BLAST/); (2) profile-sequence alignment using PSI-BLAST[45] (http:// ncbi.nlm.nih.gov/blast/psiblast.cgi), or (3) profile-profile alignment using SALIGN, or hidden Markov models (HMMs), such as SAM-T02.[39,44–49]

After identifying a template, it is possible to construct an all-atom model of the query protein based on the sequence alignment between the template and query protein sequences. Standard sequence alignment programs use methods such as the Needleman-Wunsch[50] and Smith-Waterman[51] methods. These algorithms calculate an alignment score using substitution scoring matrices, such as BLOSUM[52] or PAM.[53,54]

The higher the score of the alignment (*i.e.*, > 25%) between the query sequence and the template, the more significant is the predicted homology model. However, gaps in the sequence alignment or the template structure decrease the quality of the generated model. The high-quality homology models can be used for a variety of applications, including functional annotation, ligand binding site prediction, virtual screening, docking of small molecules, molecular replacement, drug design, protein–protein interaction, and protein–protein docking prediction.[55–57] "Low-resolution" homology models can also be useful for these purposes, as they contain sufficient information about the spatial arrangement of important residues[57] in the protein to direct the design of new experiments. For example, the design of site-directed mutagenesis experiments could be considerably improved if such "low-resolution" model structures were used. The inaccuracies in a low-resolution model are most frequently located in the protein loop regions, which are more variable even between closely related proteins. Alternatively, the active sites in a protein tend to be highly conserved and thus are more accurately modeled using homology modeling.

RaptorX,[58] 3D-JIGSAW,[59] Biskit,[60] FoldX,[61] HHpred,[62] MODELLER,[63,64] Robetta,[65] SWISS-MODEL,[66] Bhageerath,[67] and YASARA[68] are some of the programs that are commonly used for homology modeling. Rosetta@Home[69] and POEM@HOME[70] are widely used distributed computing projects for protein structure prediction. Rosetta@Home predicts protein–protein docking and designs new proteins with the assistance of 60 000 volunteered computers processing at 83 teraFLOPS. POEM@HOME is hosted by the Karlsruhe Institute of Technology and runs on the Berkeley Open Infrastructure for Network Computing (BONIC) software platform, which includes 6650 active users and 70 599 computers (http://boinc.fzk.de/poem/). Table 2.1 summarizes the software that is commonly used for protein structure prediction, including homology modeling, protein threading, and *ab initio* prediction methods.

2.5.2 Threading (Fold Recognition)

Threading, also known as fold recognition, is another common method used to generate 3D models of proteins. This method maps query sequences directly onto 3D structures of template proteins to detect query-template pairs that may not display evolutionary relationships. Fold recognition is a widely used and effective method because there are a limited number of different protein folds in nature due to constraints imposed by the fundamental physical and chemical properties of polypeptide chains and due to evolution. Therefore, there is a high probability (70–80%) that a fold shared by both the template and query proteins has already been studied using experimental techniques such as X-ray crystallography or nuclear magnetic resonance (NMR) spectroscopy and is already represented in the protein databases. At present, there are nearly 2300 different known protein folds (CATH database), and new folds are being discovered every year as a significant component of ongoing structural genomics projects.

Threading techniques use sequence profile–profile alignments,[71–74] structural profile alignments,[75] HMMs[46,62] and machine learning methods. RaptorX,[58] HHpred,[62] 3D-PSSM,[76] I-TASSER,[77] GenTHREADER,[78] Phyre2,[79] MUSTER,[80] SPARKS,[81] and BioShell[82] are some popular threading servers. In CASP7, HHsearch,[62] a HMM–HMM alignment method was rated as the best single threading server.

2.5.3 Model Building and Refinement

Model building involves the prediction of the atomic coordinates of the query structure using residue equivalences defined in the sequence/structure alignment. Sali *et al.* (1990)[64] proposed 3D modeling based on the satisfaction of spatial restraints obtained from the alignment of the query sequence with a homologous protein of known structure. Havel & Snow (1991)[83] and Srinivasan (1993)[84] described an elegant distance geometry approach for constructing an all-atom model according to distance constraints. Bohr *et al.* (1990)[85] used neural networks and optimization in Cartesian space to construct a model from the Cα distance plot of a homologous protein. Snow (1993) performed comparative modeling *via* the optimization of a potential function constructed from the alignment of the sequence with related structures. Several other methods for constructing full backbone coordinates from the positions of the Cα atoms alone were

Table 2.1 List of broadly used homology modeling, loop modeling and side chain modeling programs.[a]

No.	Program name	Methods	Availability
1	3D-JIGSAW	Rigid-body assembly	bmm.cancerresearchuk.org/~3djigsaw/
2	Biskit	Combines external programs into an automated workflow	http://biskit.pasteur.fr/
3	CABS	Reduced modeling tool	biocomp.chem.uw.edu.pl/tools/cabsflex/
4	CPHModel	Fragment assembly	www.cbs.dtu.dk/services/CPHmodels/
5	EasyModeller	GUI to MODELLER	easymodeller.blogspot.de
6	ESyPred3D	Template detection, alignment, 3D modeling	www.unamur.be/sciences/biologie/urbm/bioinfo/esypred/
7	FoldX	Energy calculations and protein design	http://foldx.crg.es/
8	Genesilico	Consensus template search/fragment assembly	https://genesilico.pl/meta2/
9	HHpred	Template detection, alignment, 3D modeling	http://toolkit.tuebingen.mpg.de/hhpred
10	MODELLER	Spatial restraints	https://salilab.org/modeller
11	LOMETS	Local meta threading server	http://zhanglab.ccmb.med.umich.edu/LOMETS/
12	Phyre/Phyre2	Remote template detection, alignment, 3D modeling, multi-templates, *ab initio*	www.sbg.bio.ic.ac.uk/phyre2/
13	Protinfo CM	Comparative modeling of protein structure using minimum perturbation and loop building	http://protinfo.compbio.washington.edu/abcm/
14	ROBETTA	Rosetta homology modeling and *ab initio* fragment assembly with Ginzu domain prediction	http://robetta.bakerlab.org/
15	BHAGEERATH-H	Combination of *ab initio* folding and homology methods	www.scfbio-iitd.res.in/bhageerath/
16	RaptorX	Remote homology detection, protein 3D modeling, binding site prediction	http://raptorx.uchicago.edu/
17	SWISS-MODEL	Rigid-body assembly	www.swissmodel.expasy.org
18	Yasara	Detection of templates, alignment, modeling, including ligands and oligomers, hybridization of model fragments	http://www.yasara.org/
19	TIP-STRUCTFAST	Automated comparative modeling	http://amazon-tip64.eidogen-sertanty.com/
20	WHATIF	Rigid-body assembly	http://swift.cmbi.ru.nl/whatif/
21	SCHRODINGER	Rigid-body assembly	http://www.schrodinger.com
22	NEST	Artificial evolution	http://trantor.bioc.columbia.edu/programs/jackal/
23	COMPOSER	Rigid-body assembly	www-cryst.bioc.cam.ac.uk/software
24	CONGEN	Rigid-body assembly	www.congenomics.com/congen/
25	SEGMOD	Segment matching	Module in Look, sold to Celera in 1999
26	DRAGON	Spatial restraints	Contact Robin Munro at ku.ca.crm.rmin@ornumr
27	ICM	Rigid-body assembly	www.molsoft.com/
28	Builder	Self-consistent mean	Contact Koehl P. at ude.drofnats.bsc@lheok

Table 2.1 (*Continued*)

No.	Program name	Methods	Availability
29	PrISM	Rigid-body assembly	https://bhapp.c2b2.columbia.edu/software/cgi-bin/software.pl?input = PrISM:Scap_Rotamer_Library
30	I-TASSER	Combination of *ab initio* folding and threading methods	http://zhanglab.ccmb.med.umich.edu/I-TASSER/
31	LOOPP@BioHPC	Multiple methods	http://cbsuapps.tc.cornell.edu/loopp.aspx
32	mGenTHREADER/GenTHREADER	Sequence profile and predicted secondary structure	http://bioinf.cs.ucl.ac.uk/psipred/ or www.chemogenomix.com/chemogenomix/GenThreader.html
33	MUSTER	Profile-profile alignment	http://zhanglab.ccmb.med.umich.edu/MUSTER/
34	SUPERFAMILY	HMM	http://supfam.org/SUPERFAMILY/
35	QUARK	MC fragment assembly	http://zhanglab.ccmb.med.umich.edu/QUARK/
36	LOOPY (Loop Modeling)	Colony energy with *ab initio* conformation sampling and torsional space minimization	https://bhapp.c2b2.columbia.edu/software/cgi-bin/software.pl?input = Loopy
37	PLOP (Loop Modeling)	Extensive conformation sampling, OPLS energy, sufficient energy minimization	www.jacobsonlab.org/plop_manual/plop_overview.htm
38	COILS (Loop Modeling)	Scans the database of known loops from the PDB	www.ch.embnet.org/software/COILS_form.html
39	CODA (Loop Modeling)	Combines the database and *ab initio* approaches for loop modeling	www-cryst.bioc.cam.ac.uk/coda/
40	SCAP (Side Chain Modeling)	Colony energy method using simple energy and the large Cartesian coordinate rotamer library	Wiki.c2b2.columbia.edu/honiglab_public/index.php/Software:Scap_Rotamer_Library
41	SCWRL (Side Chain Modeling)	Simple energy using the backbone dependent rotamer library	http://dunbrack.fccc.edu/SCWRL3.php
42	SMOL (Side Chain Modeling)	Optimized scoring function using the extended backbone-dependent rotamer library and the MC search method	Contact Grishin N.V. at Nikolai.Grichine@UTSouthwestern.edu
43	SCCOMP (Side Chain Modeling)	Optimized scoring function and Gibbs sampling-like algorithm	ignmtest.ccbb.pitt.edu/cgi-bin/sccomp/sccomp1.cgi
44	RAMP (Side Chain Modeling)	Knowledge-based potentials and small rotamer library	Software.compbio.washington.edu/ramp/
45	Confmat (Side Chain Modeling)	Self-consistent mean force field and small rotamer library	Contact Koehl P at koehl@csb.stanford.edu
46	Maxsprout (Side Chain Modeling)	Rough energy function and small rotamer library	www.ebi.ac.uk/services/teams/maxsprout

[a]Source: Wikipedia and *Xiang et al.* (2006).[25]

described by Bassolino-Klimas & Bruccoleri (1992),[86] Correa (1990),[87] Holm & Sander (1991),[88] Levitt (1992),[89] Luo (1992),[90] and Payne (1993).[91]

Given the sequence alignment, there are four principal methods used for model construction: (1) the spatial restraint method (SSR),[92] (2) the segment matching method (SMM),[89] (3) the multiple template method (MTM),[93,94] and (4) artificial evolution (AE).[95] In SSR, a 3D model is obtained by optimally satisfying the spatial restraints derived from the alignment, such as the Cα–Cα distances, the primary chain N–O distances, and the primary chain and side chain dihedral angles. SMM compares a sliding nine-residue fragment of the query sequence to all such fragments in the database and retains up to 150 candidate fragments displaying the highest profile–profile scores. Then, a dot plot of the positions of the fragments in the query sequence against the positions of the matched fragments in the candidate sequence is generated to model the query sequence. One of the most frequently used homology modeling programs, MODELLER,[63] uses this method for model building.

In MTM, several solved 3D protein structures are used to build the query protein model. The query sequence is optimally aligned with multiple templates to build the model. The advantage of using a MTM is that it increases the alignment coverage and also incorporates the best single template, which is dependent on the alignment accuracy[96] and the template complementarity.[97,98] The MTM has been implemented in several packages, such as SWISS-MODEL,[66] MOE,[99] and 3D-JIGSAW.[59]

In AE, the alignment of the sequences of the query and the template is achieved using evolutionary characteristics, such as mutations, insertions, and deletions. The model of the query protein is built by editing the template structure based on the alignment. The aligned residues are mutated, which either positively or negatively alters the energy burden. Model building in AE begins with the operation with the least energy cost and so on. Each operation is followed by a slight energy minimization to remove atom clashes. When there is no significant energy penalty, the operation is considered successful.[95] This method is implemented in the NEST program, which is a module of the JACKAL package.[25]

2.5.4 Loop Modeling

The most difficult task in model building is the prediction of loop regions and side-chain conformations. Loop modeling is a challenge for protein structure prediction and is used to predict the conformations of highly variable loop regions in proteins. Loop modeling methods can be classified into two major approaches: (1) knowledge-based methods (or database methods) and (2) energy-based methods (or *ab initio* methods). Knowledge-based methods for loops longer than 5 residues often display difficulties with identifying near-native conformations in the template library, as there is a limited number of relevant loop structures from known protein structures, thus limiting their utility for these cases.[100] ArchDB, a database of structural motifs consisting of one loop and its bracing secondary structures, was developed[101] and evaluated using two different sequence profiles. ArchDB, which currently contains 12 665 super-secondary elements classified into 1496 motif subclasses, is located at http://sbi.imim.es/archdb.

Energy-based methods use a *de novo* energy function to evaluate loop conformation.[102,103] *De novo* loop modeling is an effective method for loops that are not longer than 12 residues.[104] Rapp and Friesner[105] used the generalized Born solvation model and the AMBER94 force field. Fiser *et al.*[106] utilized a scoring function that included the CHARMM22 force field and statistical preferences taken from the protein databases. Zhang *et al.*[107] used the DFIRE-based statistical potential for loop discrimination to perform an extensive *ab initio* study of a dataset of loops. Their results suggested that a single-term DFIRE statistical energy function provides an accurate loop prediction that is equivalent to more rigorous physical–chemical energy functions.[108] A loop conformational search can be performed using tools such as a local move Monte Carlo (LMMC),[109] a torsion angle conformational search,[110] LoopBuilder,[111] replica exchange,[112] or a dihedral angle-based buildup procedure in hierarchical loop prediction (HLP).[104] LOOPER[113] facilitates a systematic and efficient sampling strategy to search for loop conformers displaying optimal interactions between the loop backbone and the rest of the atoms in the protein. For the final ranking of the candidate loop structures, the CHARMM energy scoring function using a Generalized Born solvation term is used.[113] Additional loop modeling software that can be easily obtained from the web include LOOPY, PLOP, COILS, CODA, and MODELLER (Table 2.1).

2.5.5 Side Chain Modeling

The majority of the methods used for side chain prediction utilize rotamer libraries, which are constructed based on statistical knowledge of protein 3D structures, such as the Grow-to-Fit molecular dynamics method (G2FMD),[114] statistical machine learning methods,[115] and Iterative REduction of Conformational Space (IRECS).[116] Ponder and Richards[117] developed the first library of side-chain rotamers for the 17 types of residues containing dihedral angle degrees of freedom in their side chains based on 10 high-resolution protein structures determined *via* X-ray crystallography. Subsequently, several additional libraries have been developed. Baker *et al.*[118] developed a "solvated rotamer" approach for side chain packing at the protein–protein interface. This approach uses a rotamer library that includes solvated rotamers in which one or more water molecules are fixed to polar functional groups in probable hydrogen bond orientations, together with a simple energetic description of water-mediated hydrogen bonds. SCAP, SCWRL, SMOL, SCCOMP, RAMP, SMD, Confmat, and Maxsprout (Table 2.1) are several of the publically available side chain prediction programs. The program SCWRL,[119] the best program available, uses backbone-dependent rotamer libraries *via* a search method based on graph theory.

2.5.6 Model Building

Automated model building web servers include I-TASSER,[77] ROBETTA,[120] Pmodeller-Pcons,[121] and HHPred3.[62] I-TASSER[122,123] searches the entire PDB library to detect appropriate protein fragments from which the global structure is assembled by combining the aligned fragments. For portions for which no alignment matches are found, the 3D structure is built using *de novo* simulations. The final refinement of the model is performed by searching for the lowest energy conformation. Model building in Pcons is performed using Pfrag,[96] a modified SegMod homology modeling program, and the final refinement is achieved using the ENCAD force field.[121] Model prediction using ROBETTA involves extensive and computationally expensive conformational sampling and all-atom energy refinement.[124] Other web servers are Multiple Mapping Method with Multiple Templates (M4T)[125] and PROTEUS2.[126]

2.5.7 Model Quality Assessment

The final step in comparative modeling is the evaluation of the predicted structure. Quality assessment (QA) programs rate the quality of the model based on statistical evidence. There are statistical and physico-chemical scoring functions used for the evaluation of protein models, which are based on their alignment with a single template or multiple templates or with meta-server results. The QA method generates a local score as a function of the residue or the residue window[127–131] or a global score[132–135] that may be based on one or multiple assessment criteria. Computational methods such as PROCHECK,[136] Sub-AQUA,[137] SFCHECK,[138] Squid, WHATCHECK, PROSAII, Verify3D,[139] ERRAT,[140] ANOLEA,[141] Probe, PROVE, GRASP2, and MolProbity,[142] evaluate the stereochemistry of the predicted structure, including the bond lengths and angles, peptide bond and side-chain ring planarity, chirality, primary chain and side chain torsion angles, and clashes between non-bonded pairs of atoms. These methods also validate the fit of a sequence to a structure. Another option for QA is the ModFOLD[130] server, which combines scores from ModSSEA,[143] MODCHECK,[144] and ProQ.[145] The local quality of a structure can also be quantified using ProQres,[127] which relies on 3D properties, such as the secondary structure, solvent accessibility, and atom–atom and residue-residue contacts, to measure the local quality. APOLLO/ModelEvaluator quantifies the absolute quality of a protein model using support vector regression (SVR).[146,147]

2.5.8 *Ab Initio* Prediction

If protein templates are not available, it is necessary to build the 3D model from scratch. This procedure has been referred to by several names, *e.g.*, *ab initio* modeling,[148–150] *de novo* modeling, and physics-based modeling. *Ab initio* modeling conducts a conformational search under the guidance of a designed energy function and generates several possible conformations (structure decoys), from which it selects the final model. The success of *ab initio* modeling depends on three factors: (1) an accurate energy function in which the native structure of a protein corresponds to the most thermodynamically stable state compared to all possible decoy structures; (2) an efficient search method that can rapidly identify the low-energy states *via* a conformational search; and (3) the selection of native-like models from a pool of decoy structures.[151,152]

Energy functions are categorized into two groups: (a) physics-based energy functions and (b) knowledge-based energy functions. Examples of physics-based force fields include AMBER,[153–155] CHARMM,[156] GROMOS96,[157] and OPLS.[158] Knowledge-based energy refers to the empirical energy terms that are derived from the statistics of the solved protein structures deposited in the PDB. These energy terms are classified into two types:[159] (1) generic and sequence-independent terms, such as the hydrogen bonding and the local backbone stiffness of a polypeptide chain,[160] and (2) amino acid- or protein sequence-dependent terms, *e.g.*, pairwise residue contact potential,[161] distance-dependent atomic contact potential,[74,162–164] and secondary structure propensities.[160,165] ROSETTA[166] and TASSER[167] are the two FM programs that construct 3D models based on a purely knowledge-based approach.

Methods for *ab initio* prediction include molecular dynamics (MD) simulations of proteins and protein-substrate complexes to provide a detailed and dynamic picture of the nature of the inter-atomic interactions with regards to protein structure and function; Monte Carlo (MC) simulations that do not use forces but rather compare energies using Boltzmann probabilities; genetic algorithms (GAs) that attempt to improve the sampling and convergence of MC approaches; and exhaustive and semi-exhaustive lattice-based approaches based on using a crude/approximate fold representation (such as two residues per lattice point) and then exploring all or much of the conformational space given the crude representation. Global optimization techniques can efficiently search high-dimensional spaces for the global minimum, indicating the native fold of the protein.[168–178] GROMACS,[168] CHARMM,[170] NAMD,[176] DESMOND,[177] LAMMPS,[175] and AMBER[169] are a few of the available MD simulation and modeling programs. Rosetta,[173] Profasi,[171] Simple Molecular Mechanics for Proteins (SMMP),[172] and SIMONA[174,179] are MC-based simulation packages.

2.5.9 Critical Assessment of Techniques for Protein Structure Prediction (CASP)

Since 1994, advancements in protein secondary structure prediction have been assessed *via* the CASP (www.predictioncenter.org) experiment. CASP is conducted every two years by the protein structure prediction community in the categories of TBM and FM. Groups working on computational methods in protein structure prediction submit their predictions for a set of targets/queries released during the competition, whose experimental structures have been determined but are not yet available. CASP has become a good indicator of the progress of protein-based structure prediction methodologies over the years. Based on CASP6-10, improvements have been made, primarily for medium- or high-difficulty targets.[180,181] The performance of fully automated servers has exhibited great improvement based on the CASP experiments. The best predictions have been those that use human expertise;[182] of the best six predicted structures in CASP7 and CASP8, 29% were predicted by automated servers, and the rest were predicted by humans.

2.6 Molecular Docking Methodology

The following section focuses on the different types of algorithms and scoring functions commonly utilized in molecular docking programs. The standards of docking programs are based on their (1) representation of the receptor, (2) search algorithm, and (3) scoring function.

2.6.1 Receptor Representation

Although there are number of docking programs (Table 2.2) available, there is a continuous need to critically examine the estimations and strategies implemented in the existing algorithms. The representation of a protein or receptor molecule provided by a docking program is a very important estimation.[8] Most docking methods use rigid protein docking. A great challenge in medicinal chemistry is the integration of the conformational flexibility of receptor molecules during the docking process. To meet this challenge, different types of representations, such as grid, atomic, and surface, of the receptor molecules are implemented in the current docking software. DOCK[183] and FLOG[184] are the two docking programs that perform rigid ligand and rigid receptor docking. Flexscreen[55,69,70,185,186] is another docking program that employs a force field-based scoring function (similar to Autodock[187]) and an MC-based search algorithm based on the stochastic tunneling method.[188] This program implements a novel backbone reconstruction algorithm that can modify the conformation of preselected extended backbone regions with high efficiency during the simulations.[186]

DOCK was the first automated docking program developed by the Kuntz group[183] for small molecule docking and it is continuously under development. Based on the lock and key method[189,190] of binding, it assumes that proteins are rigid bodies and only the ligand molecules are flexible, displaying large conformational degrees of freedom to fit well into the binding site of the protein. This assumption was challenged by Koshland in 1958[191] *via* the induced fit model, which proposes that during the binding process, the conformational flexibility of the receptors and the ligand molecules is needed to form low-energy stable interactions.[191] Therefore, according to the induced-fit perspective, it is necessary to consider the conformational flexibility of both the receptors and the ligand molecules to achieve the

Table 2.2 List of broadly used docking programs.

No.	Program name	Docking algorithm	Scoring function	License terms for academic use	Website
1	Dock	Shape fitting, IC	Force field-based, ChemScore, GB/SA solvation scoring	Free	http://dock.compbio.ucsf.edu
2	Autodock	GA, MC	Force field-based methods	Free	http://autodock.scripps.edu
3	FlexX	Incremental construction	PLP, ScreenScore, DrugScore, FlexXScore	Commercial	http://www.biosolveit.de/flexx
4	GOLD	GA	Force field-based, ChemScore, GoldScore	Commercial	http://www.ccdc.cam.ac.uk/products/life_sciences/gold
5	Glide	MC	GlideScore	Commercial	http://www.schrodinger.com
6	Surflex	Incremental construction	Empirical methods	Commercial	http://www.tripos.com/index.php
7	ICM	MC	ICM Score	Commercial	http://www.molsoft.com/docking.html
8	FRED	Shape fitting	PLP, Gaussian shape score, ScreenScore	Free	http://www.eyesopen.com/products/applications/fred.html
9	LigandFit	MC	PLP, PMF, LigScore	Commercial	http://accelrys.com/products/discovery-studio
10	eHiTS	Incremental construction	eHiTS Score	Commercial	http://www.simbiosys.ca/ehits/index.html

minimum-energy interaction. However, consideration of the flexibility of the receptor during the binding process is computationally expensive. To overcome this limitation in computational capacity, different approximations have been developed and adopted by various docking programs. These approximations include flexible ligand and rigid, partially flexible and fully flexible receptor docking. AutoDock,[192] FlexScreen,[193–195] and FlexX[196] adopt the flexible ligand and rigid receptor docking methodology. AutoDock algorithms have been continuously under development, and AutoDock 4.0 is capable of allowing side chain flexibility of the receptor molecule during the docking process.[197]

Currently, different methodologies are available to incorporate the flexibility of the receptor. Soft-docking, the first estimation that incorporates the partial flexibility of the protein, was introduced by Jiang and Kim.[198] The aim of the soft-docking methodology is to decrease the Van der Waals repulsion energy value between atoms in the ligand and in the binding site. Therefore, this method has the advantage of computational efficiency based on adjustment of the Van der Waals parameter. The smooth potential of AutoDock 3.0 and the LJ 8-4 potential of the GOLD docking programs are based on this methodology. Another method to incorporate the flexibility of the receptor is to generate ensembles of rigid protein conformations merged together depending on the selected method to represent the conformational feasibility of the protein.[199–202] These ensembles of protein conformations are further used to dock libraries of small molecules using the rigid-protein method implemented in DOCK. This method has been further optimized and adopted by many different docking programs.[203]

2.6.2 Docking Algorithms

The two key components involved in small molecule docking are search algorithms and scoring functions.[204] Different types of search algorithms are utilized in small molecule docking to allow the ligand to explore its proper binding site and its optimal conformational degrees of freedom at the binding site. The search algorithm in some docking programs uses full flexibility, whereas some programs use partial flexibility. In the first docking algorithm, DOCK, the ligand and the protein molecules were considered to be rigid bodies, in which the ligands were only allowed to explore six translational and rotational degrees of freedom in the receptor binding site without allowing any other degrees of freedom, *i.e.,* dihedrals and angles.[183] Based on the flexibility of the ligand and the protein, the search algorithms are classified into three major categories: systematic search methods, simulation methods, and stochastic methods.

2.6.2.1 Systematic Search Methods

Systematic search is an efficient algorithm that attempts to explore all of the degrees of freedom of the ligand, either at the binding site or on the surface of the whole receptor molecule, by matching the descriptors. Thereby, the number of possible orientations of a ligand at the active site increases exponentially. This search can be performed using various methods, including a *de novo* ligand design strategy or docking with the rigid backbone and a partially flexible side chain of the ligand.[205] The DOCK[203] program utilizes this algorithm by incorporating the structural complementarity of the rigid backbone, increasing the side chain flexibility and systematically exploring the possible space in the active site of the receptor molecule. Subsequently, the pruning algorithm carefully examines and removes the unfavorable conformations, thereby ensuring the efficiency of the docking program. FlexX,[206] DOCK,[207] and PhDOCK[208] use a similar descriptor search algorithm in which the interaction between the protein and the ligand are based on physico-chemical properties, such as a hydrogen bond acceptors and donors and hydrophobic interactions.

2.6.2.2 Molecular Dynamics Simulation Methods

MD simulation, the most well known simulation approach, uses Newton's laws of motion to generate configurations of the system in a time-dependent manner. There are many programs that perform MD simulations, such as GROMACS,[209,210] NAMD,[176] AMBER,[154] and CHARMM.[156] The standard MD simulations are often trapped in local minima and are unable to cross the high-energy conformational barriers within the simulation time. Moreover, the simulations should be long enough to converge the binding site, thereby limiting the computational pertinence. Therefore, observing the global energy minimum of the docked conformation is tedious by standard MD simulation, and the quality of the MD results highly depends on the initial conformation of the molecule.[153,211] To overcome the limitations of standard MD simulation techniques, alternative methods have been developed to enhance the docking process and reduce the complexity.

One such approach is to simulate different components of the docked system at various temperatures; another approach is to initiate different simulations using distinct ligand docking conformations. Mangoni *et al.*[212] introduced the concept of flexible ligand docking to a flexible receptor, complementing the original study by Di Nola *et al.*,[213] which allows the local freezing of various degrees of freedom. Wang and Pak[214] have further developed the flexible docking method using a well-jumping technique. However, most of these techniques used a single structure for docking simulation. McCammon and colleagues utilized MD snapshots of a protein instead of a single flexible protein structure, and this approach was more effective for virtual screening.[215]

2.6.2.3 Stochastic Search Methods

The stochastic search methods are often referred to as stochastic techniques. The Monte Carlo[216,217] and the Genetic Algorithm (GA)[192,218] are the two known algorithms that use random search methods. The metropolis Monte Carlo method applies random Cartesian moves to either single or multiple ligands and obtains ligands that are either accepted or rejected based on the Boltzmann probability distribution function. Monte Carlo techniques display significant advantages over MD simulation methods, which are primarily efficient for local optimization.

AutoDock[216] and FlexScreen[219–223] uses a Monte Carlo technique in conjunction with biomolecular force fields in which the flexible ligand molecules are docked to the binding site of a rigid protein followed by a simulated annealing procedure using grid-based approximation of the energy. Prodock[224] performs a Monte Carlo technique based on AMBER and ECEPP/3 force fields. This methodology, which is slightly modified from the standard Monte Carlo, is based on the execution of a local gradient-based minimization after every random movement. Then, the results are selected based on Boltzmann acceptance criteria. The Internal Coordinates Mechanics (ICM) program uses a Monte Carlo minimization method for internal coordinates.[225] This algorithm performs random movements based on a biased probability method according to the side chain flexibility of the protein. MCDOCK[225] implements a Monte Carlo minimization method and uses a stepwise strategy to dock a flexible ligand to a rigid protein. The DockVision[226] program utilizes a Monte Carlo-based algorithm to perform rigid ligand and rigid protein docking. QXP,[227] a component of the FLO96 package, uses a grid-based representation of a protein and the metropolis Monte Carlo method to perform flexible ligand and rigid protein docking. Glide and Affinity[228] are the commercial programs that use the Monte Carlo method for docking.

A GA is an evolutionary algorithm that is a popular optimizing tool that generates a population of possible structures from the initial structure based on genetic operators, such as crossover, mutations, and migrations, to obtain the optimal solution. GOLD[218] is a docking program that uses a GA for docking flexible ligands to the binding site of a protein to explore the full range of ligand conformational flexibility, including the side chain flexibility of the protein. Autodock 3.0,[192] DIVALI,[229] and DARWIN[230] implement a GA as the search method, with slight modifications.

2.6.3 Scoring Functions

The efficiency of a docking program greatly depends on its scoring function, which evaluates and ranks the predicted conformations of each ligand. The scoring function plays a crucial role in finalizing the docked conformation, and its design is very important. If the scoring function is not able to differentiate between the biological and docked conformations, then obtaining the biological binding conformation is impossible.[11] Rigorous calculations of the binding free energy for protein–ligand complexes are computationally expensive. Therefore, the relative binding energies can be obtained using free energy perturbation (FEP)[231] and thermodynamics integration (TI)[232] methodologies. However, these techniques remain expensive, less accurate and time consuming because obtaining adequate sampling of intermediate states for each and every protein–ligand complex is not practical for many systems.

Currently, the four types of scoring functions utilized to calculate binding free energies are the molecular mechanics force field methods, semi-empirical methods, empirical scoring functions and knowledge-based functions. These scoring functions are often simplified to avoid calculating all the physical phenomena, such as the entropic effect, to determine the binding free energy.

2.6.3.1 Force Field-based Methods

The force field-based scoring functions are developed based on physical properties, such as electrostatic and van der Waals interactions and bond stretching, bending and torsional forces.[233] AMBER,[155] OPLS,[158] CHARMM,[156] and TIRPOSE force fields[234] are various types of force field methods. In general, most of the functions and parameters of these force field methods are derived from experimental data and *ab initio* quantum mechanical calculations. Therefore, different tools use distinct parameter sets, *e.g.*, DOCK and AutoDock use the Amber force field, G-Score uses the TIRPOSE force field, *etc.*

Water molecules at the binding site of a receptor or protein play a major role in and are often the driving force of protein–ligand complex formation. The major limitation in the force field-based scoring function is the consideration of the water molecule or the desolvation effect. The standard molecular mechanics force field scoring function was originally designed to include an enthalpic gas-phase contribution and to ignore the solvation and entropic effects of structure and energetics. As a consequence of ignoring the desolvation effect, the Coulombic electrostatic interactions are biased to overestimate the charge–charge interactions that would tend to choose highly charged ligands.[235] Several algorithms have been developed to incorporate the effect of the solvent into the MD model. In the GOLD docking program, both the Chemscore and Goldscore scoring functions incorporate the bridging potential of water molecules or the displacement of water molecules by the ligand in the system. FlexX uses a particle approach to account for the contribution of water molecules to docking. This algorithm uses the phantom particles that were generated with the protein structure, and the interaction between the protein and the ligand molecule is calculated by turning the phantom particles on or off.

The FEP and TI techniques use explicit solvent models to account for the solvation effect and to calculate the binding free energies of a protein–ligand complex. However, due to the computational expense, particularly when using large databases for virtual screening, and the prediction accuracy of the force field, alternative methods using less sampling and greater accuracy were desired. Recently, new strategies using an accelerated force field-containing continuum solvent model have been developed to include the effect of the solvent when scoring and post-processing the binding conformation of ligands during molecular docking. The enhanced scoring functions included in implicit solvent models, such as the Poisson-Boltzmann/surface area (PB/SA) and the Generalized-Born/surface area (GB/SA) models, display binding energies that approximate the experimental data.[236,237]

Molecular mechanics force fields generally omit intermolecular forces between the atoms in the protein molecule rather than calculate the interaction energy between the protein and the ligand. Because the molecular mechanics force field computes only the enthalpic forces and considers only the conformation of a single protein, the estimated energy is similar to the biological binding free energy, which simplifies the scoring and reduces the simulation time. However, there are some major limitations to force field-based methods. The standard force fields were originally designed for enthalpic terms and energetics between a protein and a ligand in the gas phase, but not for the entropic terms. Therefore, interactions with the solvents cannot be calculated. The standard force fields also require a threshold distance value that arbitrarily selects and affects the non-bonded long-range interactions involved in ligand binding. A modified version of the force field-based scoring function was developed to include a torsional entropic term and explicit protein–ligand hydrogen bonding terms. This modified force field is implemented in the G-Score, GOLD and AutoDock scoring functions.

2.6.3.2 Semi-empirical Methods

The linear interaction energy (LIE) method[238] is a semi-empirical method used for lead optimization. This method was used to calculate the binding free energy by simulating the initial and final structure of the receptor using MD or Monte Carlo simulations, rather than using extensive sampling of intermediate structures, as in FEP and TI methodologies. However, this method remains computationally expensive and cannot be used for lead identification.[239,240]

2.6.3.3 Empirical Methods

An empirical scoring function calculates the binding affinity of a protein–ligand complex based on weighted energy terms. These methods are capable of reproducing the experimental data based on different parameterized functions. Empirical scoring functions are often simpler than force field scoring functions due to their simple energy terms. The coefficients of individual terms, such as electrostatics, van der Waals energy, hydrogen bonds, hydrophobicity, entropy, *etc.*, that contribute to the binding affinity of a protein–ligand complex are determined *via* regression analysis of a training set of complexes with 3D structure information.[241] Although its terms are easy to evaluate, there are several disadvantages of using this method. The major limitation of this method is the estimation of the binding affinity, which largely depends on the training set used for the fitting and regression analyses. Different docking programs, such as Surflex[242] and FlexX,[196] use distinct energy terms. The non-bonded energy terms of this scoring function are calculated in a different manner. The ChemScore function, implemented in the GOLD program, calculates the hydrophobic interactions and does not distinguish the types of hydrogen bonds, whereas the LUDI function distinguishes the hydrogen bond types into ionic and neutral and calculates the hydrophobic interactions based on the molecular surface area. Most of these methods calculate the hydrogen bonding interactions based on distances and angles. Because this scoring function largely depends on the dataset used for parameterization, the conformational space is limited for ligands at the active site due to poor sampling results achieved using generalized scores, rather than specific scores.

2.6.3.4 Knowledge-based Methods

Knowledge-based scoring functions, also known as statistical potential-based scoring functions, have been developed to score the protein–ligand interaction based on the potential energy derived from experimentally determined structures.[243] The basic principle of this method is based on the statistical assumption that the pairwise potentials that are more frequently observed in the atom pairs are important for stable complex formation. This statistical assumption is based on the Boltzmann principle and statistical mechanics. The number of atom pair interactions is based on the molecular environment.

This method is rapid and computationally efficient for screening large databases. Therefore, many different knowledge-based scoring functions have been developed and implemented in various docking programs. DrugScore,[244] potential mean forces (PMF)[245,246] and SMOG[247] are knowledge-based scoring functions that are used to calculate protein–ligand interactions based on pair-wise potentials. Many recent studies have extended this scoring function to calculate the binding affinity and have implemented this function in quantitative structure activity relationship (QSAR) and machine learning techniques.[248] The major advantages of this technique compared to force field-based and empirical methods are the computational simplicity, the rapid scoring process and the high accuracy due to the extraction of the potentials from the structure rather than reproducing the affinity *via* a fitting process.

2.6.3.5 Consensus Scoring

Although different scoring functions have been developed, each of them has their own merits and demerits. Because of their deficits in general applicability and accuracy, the consensus score has been introduced. The consensus scoring function combines the scores from different scoring functions to balance their limitations and utilize the advantages of individual scores to improve the probability of selecting a true ligand. The consensus scoring function is implemented in SCS,[249] GFscore,[250] SeleX-CS,[251] MultiScore,[252] and X-score.[241]

2.6.3.6 Evaluation of the Scoring Function

The effectiveness of the scoring function is evaluated based on its ability to identify the appropriate binding mode of a known protein–ligand complex, retrieve true ligands *via* virtual database screening, choose the biologically relevant binding mode from other possibilities in a decoy set and predict the binding affinity. The optimal scoring function should distinguish the known ligands displaying the highest calculated binding scores from the decoy ligand set.

The measure used in docking applications to determine the effectiveness of the prediction is the root mean square deviation (RMSD) between the experimental conformation and the top predicted conformation of the ligand molecule.[235] An RMSD value of ≤2 Å is typically considered a successful prediction. However, this criterion is limited because small or closely symmetric ligands placed randomly at the active site result in a lower RMSD, whereas larger ligands display a high RMSD due to the solvent-exposed region. Furthermore, non-crucial interactions hinder the accuracy of the binding mode prediction. To overcome these limitations, conformation evaluation methods have been developed to distinguish the docked conformations of a true ligand from those of a false ligand. Different datasets have been prepared to validate and improve docking studies. A well-organized dataset is primarily focused on crystallographic reliability, which was collected for the validation of GOLD.[218] Various programs, such as FlexX, GOLD, and DOCK, have clearly displayed higher scores for ligands with an X-ray-based conformation compared to those with a modeled conformation.[253]

The next important criterion to determine the effectiveness of a scoring function is to predict the binding affinities of ligands at the active sites of proteins. The Pearson correlation between the experimental and calculated scores and the Spearman correlation coefficient, which calculates the difference between two sets of rankings, are effective coefficients to predict the binding affinity. However, the assessment of the binding affinity is more promising compared to that of the binding prediction due to the uncertainty of the experimental data measured by different groups under different assay conditions.[235]

Another method to evaluate the effectiveness of a scoring function is *in silico* structure-based virtual screening. The purpose of virtual screening is to retrieve potential hit compounds from massive chemical databases that contain ligands that bind well to the target protein. This application is used for computer-aided drug design to retrieve likely drug-like molecules from the chemical databases.[254–257] Maybridge, Chembridge, Asinex, NCI, and

Zinc are the major chemical databases available for virtual screening.[258] These databases include chemical compounds from either commercial vendors or pharmaceutical companies or virtual compounds that have yet to be synthesized. Virtual screening can also be used to retrieve experimentally active compounds from the inactive compounds in a chemical database. The enrichment factor (EF) is a measure that is used to calculate the percentage of active compounds discriminated from inactive compounds by the docking program.[259] Therefore, the EF is used as an assessment tool to validate the performance of a virtual screening process. Decoys are compounds that are assumed to be inactive against the target of interest. The physical properties, such as the calculated Log P values and molecular weights, of the decoy compounds should be similar to those of active compounds even though their chemical and structural properties are distinct.

The Directory of Useful Decoys (DUD)[260] is the largest database containing known active compounds and decoys for 40 different target proteins and is the standard database currently in use for benchmarking virtual screening and molecular docking methods. However, the DUD set has several drawbacks, such as a restricted level of physicochemical similarity between the active and decoy compounds, a lower synthetic feasibility of small molecules in the set, and the risk of an increased number of decoys; furthermore, some decoys may interact with the protein target. Several developments have helped to overcome these drawbacks and to improve the databases for *in silico* virtual screening. Due to the limitations of decoys, the calculated EF does not always reflect a true assumption, and it also varies for different targets and scoring algorithms, rendering the correct EF difficult to identify. Another method to evaluate virtual screening is based on the measurement of receiver operating characteristic (ROC) curves.[261] The advantage of this method is that it is appropriate when the number of active and inactive compounds is comparable but is independent of the ratio of active to decoy compounds. The disadvantage of this method is that it is not efficient in early recognition. Specifically, for a database containing 500 000 compounds, this approach screened very few thousand compounds, *i.e.*, only the top 0.1% of the active compounds in the database. Therefore, an effective scoring function must be able to fulfill all the aforementioned criteria using multiple datasets.

2.7 Fragment-based Drug Design

Fragment-based drug design (FBDD) is primarily focused on identifying lead molecules that exhibit biological activity for a particular target, which can further be used during the drug discovery process. Enhancements in high-throughput screening and combinatorial chemistry have considerably increased the size of the fragment libraries over the years.[262,263] The fundamental aim of FBDD is the examination of fragments of small molecules obtained *via* library screening. Those that display weak interactions with the target protein can be used for further lead optimization. The weakly interacting fragments are optimized by either combining them or elongating them to identify a strongly interacting lead molecule. Once the exact binding modes of the fragments are identified, the structure-based lead optimization methodology is applied to build the fragments into a drug-like molecule.

The molecular docking approach is a significant tool used in FBDD[264] to screen libraries of fragment molecules. The retrieved fragments are developed into lead molecules *via* two different approaches. First, in fragment evolution, the identified fragments are used to interact with neighboring sites on a protein surface. Second, in fragment linking, the identified fragments are linked together by an optimal linker molecule. The resulting molecules can be either directly entered into the drug-design process or replaced with similar molecules to optimize their properties aside from affinity, such as solubility and specificity. This process is defined as fragment-based optimization.

Though the docking approach of FBDD is very important in drug discovery, some challenges exist, such as the number of false positives, the problem of incorrect binding modes and difficulty in scoring. Therefore, many recent efforts have focused on the evaluation of the methods available for fragment-based docking to improve fragment-specific approaches.[265,266]

A validation study of the Glide and GOLD programs indicated that Glide is suitable for fragment docking, although it has several shortcomings. For example, the predicted binding free energy did not strongly correlate to the experimental values.[267] GOLD is suitable for both fragment and drug-like molecule docking. Several techniques have been developed for fragment-specific docking, and many different studies have demonstrated that fragment docking depends on not only the docking algorithm and the sampling and scoring functions but also the nature of the receptor molecules.[268–270] Most computational *de novo* design tools use fragments that are complementary to the binding site of a protein to design lead molecules from the scratch. The implementation of implicit solvent models (GBSA) in multiple-copy simultaneous search (MCSS) or GOLD docking has resulted in an increased success rate of fragment docking. Thus, several approaches are available to develop lead compounds from fragment molecules. Therefore, fragment docking is an efficient method to investigate receptor-fragment interactions.[271,272]

2.8 Structure-based Virtual Screening

Docking programs play a crucial role in structure-based drug design methodologies. The identification of the position and interactions of a potential lead molecule at the binding site of a receptor is an important step in the early stage of the drug discovery process. The completion of the Human Genome Project has provided a wealth of information about various attractive therapeutic targets.[273] The advances in X-ray crystallography and NMR spectroscopy techniques have provided the structural details of target proteins and their interactions with their ligand molecules.

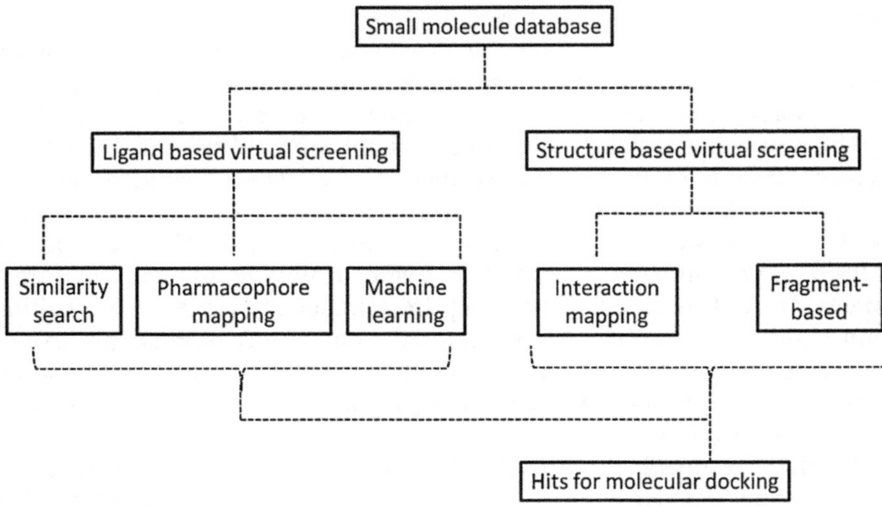

Figure 2.3 General overview of virtual screening methodologies. Different ligand-based and structure-based virtual screening methodologies.

High-throughput screening (HTS) is a conventional experimental technique that is widely used in the pharmaceutical industry. Although conventional HTS is a well-established technique, it requires a large amount of experimental resources; therefore, HTS is highly expensive, time consuming and suffers severely from a low hit rate. These drawbacks have led to the development of a computational alternative known as *in silico* virtual screening. Virtual screening is a more rapid, cost-effective and successful screening approach that significantly increases the efficiency of the lead discovery process.[274] There are two types of virtual screening methods: ligand-based and structure-based screening (Figure 2.3). Ligand-based methods, such as QSAR[275] and pharmacophore modeling, are used to retrieve similar molecules from the database when the biological activity of a ligand molecule is known.[256,276,277] The retrieved molecules are refined using conventional drug-related filters,[278] such as Lipinski's rule of five and the adsorption, distribution, metabolism, excretion, and toxicity (ADMET) properties to obtain additional drug-like compounds.[279]

The structure-based methodology is used when the atomic resolution of the receptor and ligand is available to evaluate the binding affinity. Structure-based virtual screening is performed by docking the small molecules from a chemical database to the binding site of a receptor molecule. The 3D structure of the receptor is used for the construction of pharmacophore models that are developed based on the structural complementarity of essential amino acids at the binding site by considering the flexibility of the protein.[1,201,254,255,280] Thereby, the geometry and chemical properties of the binding site residues match the geometry of the ligand molecules.[281] The compounds retrieved using this method are more accurate due to the implementation of 3D filters, along with 1D and 2D filter functions that reduce the number of scoring and docking processes. The success rate of structure-based virtual screening is reportedly high, as it performs better than the ligand-based approach and is one of the most widely used methodologies to identify novel lead compounds. Many different docking programs can be used for virtual screening; Dock,[203] Autodock,[192] GOLD,[218] Glide,[282] LigandFit,[283] Surflex,[242] eHITS,[284] and ICM[285] are the most widely used programs.

One of the purposes of virtual screening is to select the compounds displaying drug-like properties to filter out the reactive, undesirable functional groups and toxic compounds. In general, compounds retrieved from the database *via* ligand- and structure-based virtual screening are examined further to evaluate their drug-like properties, focusing on those displaying the desired activity. These refined drug-like compounds are the starting structures for lead optimization.[7,286] The example discussed later on in this book clearly explains the purpose and the importance of lead optimization.

2.9 Applications of Modeling to Privileged Scaffolds

A scaffold is regarded as privileged if it appears as a ligand for more than one array of receptors. While studying the biological activity of benzodiazepines, such as cholecystokinin antagonists, Evans *et al.* considered all secondary metabolites to be privileged structures.[13,15] However, the privilege associated with a scaffold depends on not only the commonality of its function, but also the functional or side chain groups that surround it. Differences in the side chain decoration significantly influence the biological activity of a privileged scaffold. Privileged scaffolds are distinct from "frequent hitters", which interact randomly with a plethora of targets and interfere with biological assays.[287] Privileged scaffolds are a significant source of lead compounds, and a variety of privileged scaffolds have been reported so far.[288] Screening of privileged scaffold-based combinatorial libraries might provide active lead compounds for a variety of receptors. Privileged scaffold library construction was initiated in the early 1990s upon the collection of 1,4-benzodiazapenes containing 192 molecules displaying diversity in their amine, acid, phenol, indole, and amide functionalities.[289] Later, Kim and colleagues utilized a 1,4-pyrazolodiazepin-8-one scaffold (Chapter 5) to construct a library with 3 points of diversity.[290] Recently, many different scaffolds, such as benzopyran, purine, glycoside, and benzimidazole scaffolds, have been utilized by different groups for library construction.[15]

2.9.1 Benzimidazole

Heterocyclic benzimidazole is a well-known structural motif used for the development of biological lead molecules. To date, thousands of benzimidazole derivatives have been synthesized and are recognized as privileged scaffolds due to their therapeutic activity.[291] Benzimidazole exists as a backbone for several drug molecules that possess a wide range of biological activities, such as antiviral, antifungal, anti-inflammatory, antiulcer, anti-parasitic, antihypertensive, anti-cancer, antihistamine, anthelmintic, cardiovascular, and antihistamine activities.[292] In this book, the chapter on (benz)imidazole by Pfau (Chapter 4) also discusses some examples of market drugs that utilize imidazole and benzimidazole as scaffolds.

Many articles in the literature have explained the usefulness of docking for understanding the binding mechanism of benzimidazole with different targets.[293] For example, Robinson *et al.*[294] examined the binding mode of benzimidazole at the active site of β-tubulin. Benzimidazole is an effective anthelmintic drug that binds to a broad spectrum of parasites, including nematodes, trematodes and cestodes. Benzimidazole exerts its function by binding to helminth β-tubulin and disrupts the microtubule-based processes of the nematode.[295,296] Although the therapeutic effect of benzimidazole is experimentally known, the details of the exact site and mode of binding to the active site of β-tubulin were not well understood. Therefore, Robinson *et al.* performed a molecular docking simulation to identify the binding mode of benzimidazole at the active site of β-tubulin using information available from previous experimental studies.[297–299] As a result, a detailed explanation of the mechanism by which benzimidazole interacts with the active site residues of β-tubulin was determined. Moreover, these docking results provided a structural explanation for the benzimidazole-resistant nature of parasitic nematodes, which developed as a result of a natural mutation in β-tubulin, and the species specificity of benzimidazoles.

Sharma *et al.*[300] modeled the 3D structure of *Brugia malayi* β-tubulin using homology modeling and performed docking studies. A set of ten anti-filarial drugs, including benzimidazole, were screened for their drug-like properties. Additional drug-like compounds were assessed for molecular docking using AutoDock4.0 followed by MD simulation studies. The docking results revealed that Gly10, Cys12, and Ser138 are the crucial residues involved in the hydrogen bond interaction with the ligands. Furthermore, Ala9, Gln11, Gly140, Gly142, Gly144, and Thr143 are the important residues that form extensive van der Waals and hydrophobic interactions with the ligands at the active site of *B. malayi* β-tubulin.

Sessions *et al.*[301] synthesized derivatives scaffolds of benzimidazole and benzoxazole and assayed them as novel selective inhibitors of Rho kinase II. They performed MD simulations to determine the binding modes and interactions of these inhibitors with the essential amino acids of Rho kinase II and the role of water molecules at the active site of Rho kinase II. These interactions are responsible for the improved microsomal stability, potency, and selectivity of these compounds against PKA, as well as their reduced cytochrome P-450 inhibitory activity.

Sundarapandian *et al.*[255] developed docking-enabled pharmacophore models for the identification of potent HDAC8 inhibitors. Overexpression of histone deacetylases (HDACs) is responsible for the suppression of the expression of a variety of genes, including cancer suppressor genes. Therefore, the inhibition of these enzymes may be a valid approach for the treatment of cancer.[302,303] Derivatives of benzimidazole,[304] together with different known inhibitors of HDAC8, were used as training set compounds for docking at the active site of HDAC8 using GOLD 4.1. Compounds displaying higher GOLD fitness scores and more favorable binding orientations relative to the experimentally determined crystal structure of the bound ligands were selected and used for the generation of ligand-based pharmacophore models. The best model was utilized as a 3D query for the virtual screening of different chemical databases. Subsequently, the retrieved hits were filtered based on their drug-like properties, followed by molecular docking studies. Finally, three compounds displaying higher scores, more favorable binding orientations and stronger interactions with key active site residues were selected and screened for lead optimization. The outcome of this study provided a set of novel virtual leads that can be utilized for designing novel HDAC inhibitors. Moreover, this study exemplifies the application of molecular docking, pharmacophore modeling, virtual screening, and lead optimization during the drug design process.

2.9.2 Coumarins

Coumarins, also known as benzopyrones, are naturally occurring synthetically feasible oxygen-containing heterocyclic compounds found abundantly in plants[305] (Chapter 11). The molecular framework of coumarin corresponds to 2*H*-1-benzopiran-2-one. The derivatives of coumarins are classified into (1) simple coumarins, (2) pyranocoumarins, (3) furanocoumarins, (4) coumarinolignans and (5) bis- and tris-coumarins. The structural diversity of natural and synthetic coumarins enables them to interact with a wide range of enzymes and receptors and to exert pharmacological effects.[305,306] Simple coumarins derivatives belong to the major class of coumarins and are involved in many biological functions. Coumarin derivatives are well known for their anti-oxidant, anti-inflammatory, anti-coagulant, anti-thrombotic, anti-HIV, anti-cancer, and lipid-lowering effects.[307–309]

The structural and biological diversity of coumarins make them a promising scaffold and potent drug family in medicinal chemistry (see chapter by Gläser *et al.*, Chapter 11). Many reports have described the binding of coumarins to different targets.[307–312] For example, John *et al.*[7] utilized coumarin derivatives for the discovery of potent human cholesterol esterase (CEase) inhibitors. Pancreatic CEase, also known as bile salt-activated lipase, is involved in the hydrolysis of various substrates, including dietary cholesterol esters, fat-soluble vitamins, triglycerides, and phospholipids. CEase is present in several mammals, including humans.[313,314] A lack of CEase activity can lead to the incomplete digestion of milk fat and the accumulation of enterocytes in the ileum of newborn mice.[315] Apart from its role in fat digestion, CEase is directly involved in lipoprotein metabolism by catalyzing the conversion of larger and less atherogenic low-density lipoprotein into

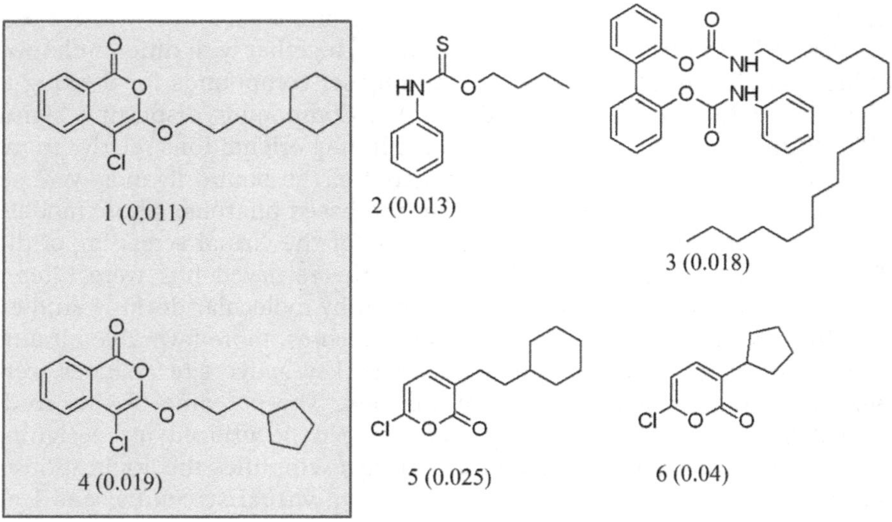

Figure 2.4 Chemical structures of the training set compounds together with their experimental K_i values in μM used for pharmacophore generation. The coumarin derivatives are highlighted by a green box.
Figure from John *et al.* (2011).[7]

smaller and more atherogenic low-density lipoprotein subspecies, and it may also regulate the serum cholesterol level.[316,317] Due to its important roles, CEase has emerged as a potential target, particularly for the development of hypocholesterolemic agents.

Coumarin derivatives are among the known inhibitors of human CEase (hCEase).[7,318] Thus, in one study,[7] coumarin derivatives were used to extract structural and pharmacological information that could be used for the identification of novel ligands. Pharmacophore modeling is an effective technique that describes the molecular properties of a ligand that are necessary for its interaction with a target macromolecule. John *et al.* utilized derivatives of coumarins (Figure 2.4), including 4-chloro-3-(3-cyclopentylpropoxy)-1*H*-isochromen-1-one and 4-chloro-3-(4-cyclohexylbutoxy)-1*H*-isochromen-1-one, as a training set to generate pharmacophore models based on common characteristics.[7] In general, these models are developed by comparing a set of conformational models and a variety of 3D configurations of the chemical characteristics that are shared between the training set compounds. Therefore, the common characteristics of the pharmacophores are based on the common chemical characteristics available among the known inhibitors or activators of the corresponding targets. The reliability of the generated pharmacophore models was validated using different methods. The refined best model "Hypo1" (Figure 2.5) was utilized for virtual screening of databases such as Maybridge, Chembridge, and NCI2000, which contain 59 652, 50 000, and 238 819 compounds, respectively, to identify new scaffolds to be utilized for the design of novel

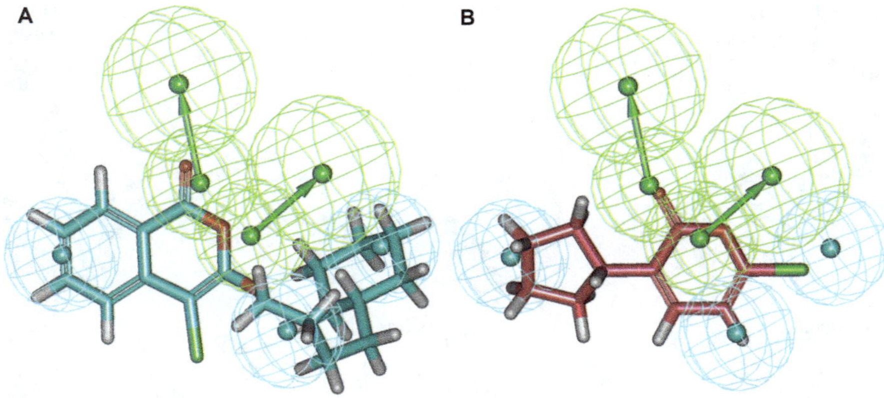

Figure 2.5 Compound 1 (A) and compound 6 (B) in the training set mapped to "Hypo 1". Green color represents hydrogen bond acceptor (HBA), and cyan color represents hydrophobic (HY) groups.
Figure from John *et al.* (2011).[7]

hCEase inhibitors. The retrieved compounds were filtered according to Lipinski's rule of five and the ADMET properties to remove non-drug-like compounds.

A molecular docking study was performed using GOLD 4.1 and Autodock 4.2 to reduce the number of false positives and to further refine the hit compounds. The docked compounds were filtered by selecting the compounds displaying a GOLD fitness score greater than that of any training set compound, as well as by selecting the compounds that interacted with active site amino acids and exhibited structural diversity. As a result, 57 compounds were selected out of 353 hit compounds from the database. Finally, the top four representative compounds, SEW00846, NCI0040784, GK03167, and CD10645, were used for lead optimization (Figure 2.6). Although these four compounds displayed more favorable characteristics than the training set compounds, enhancement of their side chains would further improve their binding affinities to the catalytically active amino acids. The lead optimization technique was adopted in this study by adding different substitutions to the side chains of the hit compounds; *e.g.,* SEW00846 fits well in the active site of hCEase, but extending its interaction towards Asp320 and other HY amino acids increased its binding affinity. Therefore, various substitutions were performed on the side chains of the four hit compounds, depending on the size of the active site, leading to a total of 104 optimized compounds. These compounds were docked to the active site of hCEase using GOLD according to the same parameters used to dock the direct database hits. The results were analyzed based on the highest GOLD fitness score compared to their respective precursor, their interaction with the essential amino acids and the binding modes of the compounds at the active site of hCEase. The GOLD predictions of the optimized compounds were further evaluated using AutoDock.

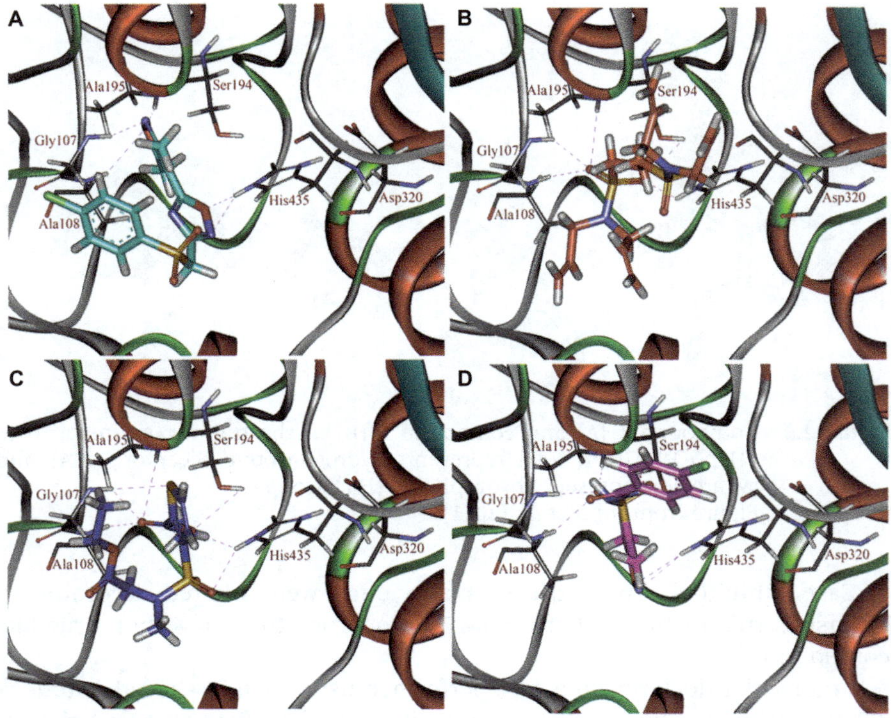

Figure 2.6 Binding orientations of the database hit compounds: (A) SEW00846, (B) NCI0040784, (C) GK03167 and (D) CD10645 are shown in cyan, red, blue and magenta, respectively. Hydrogen bonds are shown in dotted lines. Figure from John *et al.* (2011).[7]

The resulting compounds were validated for their synthetic accessibility using SYLVIA 1.0 from the Molecular Networks group. SYLVIA v1.0[319,320] was employed to calculate the synthetic accessibility of these optimized compounds. The estimation of the synthetic accessibility using SYLVIA provides a number between 1 and 10, which represents the range from easy to difficult to synthesize. Finally, the top 10 optimized compounds based on their GOLD fitness scores, AutoDock binding energies and the SYLVIA scores were selected as possible virtual leads for the design of hCEase inhibitors.

2.9.2.1 Interaction between Cannabinoid Receptors and Coumarin Derivatives

Meliciani *et al.*[56] investigated the interaction between the cannabinoid receptors 1/2 (CB1/2), which are members of the membrane-bound G protein-coupled receptor superfamily and are involved in neuroinflammatory and neurodegenerative disorders, such as Huntington's and Alzheimer's diseases.[321,322] These authors used a novel set of coumarin derivatives to

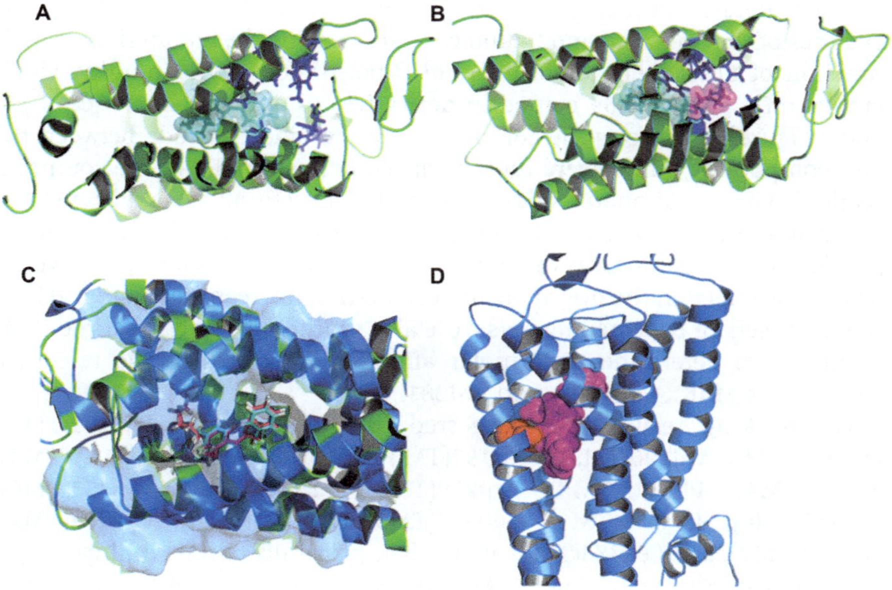

Figure 2.7 (A) Homology model of the CB1 receptor illustrating the hydrophobic pocket formed by Ala198, Cys264, Trp279, Trp356, Leu359, Met363, Phe379 and Cys386 is interacting with the reference ligand (A) AM281 or (B) CP55940. (C) Binding of the new compound (ligand27:(3-[(2-phenylchlorophenyl)methyl]-5-methoxy-7-methyl-2*H*-chromen-2-one), in magenta) and a structurally related compound MAK15 (in blue) shows that substitution of the aliphatic group at position R7 exerts little effect on the ligand orientation. (D) The local effect of mutating Gly197, an interacting amino acid in the CB1 receptor, to Ala.
Figure from Meliciani *et al.*[56]

determine the binding mode of the ligands and to design a novel set of compounds to inhibit these receptors.

The homology models for CB1 (residues 80 to 439) and CB2 (residues 1 to 349) receptors for ligand-receptor docking were constructed based on the crystal structure of bovine rhodopsin (PDB code 1U19)[323] as a template structure (Figure 2.7 (A&B)). Templates were selected using the PHYRE server,[79] and sequence alignment between the receptors and the template was conducted using ClustalW.[324] Based on the sequence alignment, models were built using the MOE program. The CB1/2 homology model revealed a structure typical of the G-protein coupled receptor family, which was characterized by an extracellular N-terminus, followed by seven transmembrane (7-TM) α-helices (from TM-1 to TM-7) connected by three intracellular and three extracellular loops, and finally an intracellular C-terminus. The seven transmembrane helices form a cavity within the plasma membrane that serves as a ligand-binding domain.

For the docking simulations, the FlexScreen[193,222] receptor-ligand docking software with a SASA-based implicit solvation model[325] was used. The binding energies for the ligand were computed as the difference between the energies of the unbound and bound complexes using the biophysical scoring function of Flexscreen. Meliciani *et al.*[56] performed ligand binding simulations on a family of 39 coumarin derivatives to design a novel set of ligands. Figure 2.7C shows the comparison of the binding mode between the new compounds, which were functionalized at position R7 of the coumarin scaffold with an aliphatic side group and a structurally related reference compound MAK15. The results suggested that R7 substitution with an aliphatic side group improves the binding energy by 60 Kcal mol^{-1}, however, it has a little effect on the overall ligand orientation (as seen in Figure 2.7C. All the novel sets of designed ligands were also synthesized and experimentally examined to determine their molar affinities compared to the reference compounds MAKK15, NV88, and AM281.

The docking results also suggested that amino acids Phe191(TM3), Lys192(TM3), Val196(TM3), Thr197(TM3), Phe200(TM3), Trp241 (TM4), Ala244(TM4), Phe278(TM5), Trp279(TM5), Arg340(TM6), Cys355(TM6), Trp356(TM6), Leu359(TM6), Leu360(TM6), Met363(TM6), Cys386(TM7), Leu387(TM7), and Leu388(TM7) are in close proximity to the binding site (as illustrated in Figure 2.7 (A&B)). Mutations at these positions significantly reduced the binding energies and significant loss of affinity for CP55940 and win55,[326–328] which was in agreement with the experimental studies reporting that Ala substitution of Lys192, Phe191, Gly197 (Figure 2.7D), Trp279 and Trp3502. Docking studies by Behrenswerth *et al.*[328] using AM81, CP55490 and win55 also suggested that the ligand is bind CB1/2 within the transmembrane region (Figure 2.7 (A&B)).

2.10 Outlook and Challenges

Considerable progress is being made in the field of protein/receptor structure prediction, particularly regarding knowledge-based approaches, such as comparative modeling and threading. Many different programs and web servers are now available for these tasks, which differ with respect to the force fields used, alignment algorithms selected, *etc.* Continuous improvement in the algorithms for template searching, sequence alignment and model evaluation has contributed to the success of these methods for protein 3D structure prediction. However, the accuracy of homology modeling is highly dependent on the sequence identity between the targets and the templates. The model is typically of high quality when templates displaying >40% similarity are used for model generation. Further advancement in *ab initio* modeling will not only aid in protein 3D structure prediction but also provide a better understanding of the underlying principles of protein folding in nature. The development of composite methods using knowledge-based and physics-based energy terms may represent a promising approach for 3D structure prediction. However, some problems remain to be addressed. For protein monomers, structure-based homology modeling is by far the most important prediction method, but homology modeling often fails to predict the relative orientation of multi-domain proteins.[329] Furthermore, there is a need to develop high-quality repositories for experimental and computational data to improve virtual screening and drug design.

The molecular docking technique has been widely used in the pharmaceutical industry for more than a decade to identify new active compounds for a particular target protein. Receptor-ligand molecular docking is not a standalone technique; instead, its application is always associated with *in silico* structure-based virtual screening and experimental investigation. Many docking algorithms and scoring functions have been discussed in this chapter. Despite a plethora of existing docking algorithms and scoring functions, the predominant challenge is the appropriate sampling of proteins and ligands. Receptor flexibility is very important in the docking process; some algorithms consider the flexibility of the receptor, but these algorithms are far from accurate. Due to the absence of protein flexibility, knowledge of the actual physical characteristics of binding is lacking, which may limit the identification of potent drug molecules. Therefore, appropriately accounting for receptor flexibility, including the backbone and the side chain of the protein, is required. In addition, the ligand sampling issues must be resolved.

In many scoring functions, the reliable scoring and ranking of a test set of compounds remains the predominant challenge in lead optimization and structure-based virtual screening. In spite of the available algorithms, an improved algorithm that can extract the biological conformation of a ligand is critically needed. The primary limitation of the existing scoring functions is the lack of balance between electrostatic and entropic effects, which is due to the assumption that the implemented electrostatic, entropic, or solvation terms are relevant and transferable to different protein systems. Nevertheless, significant results have been produced *via* structure-based screening, although additional emphasis on target-specific approaches and functions should be considered. Currently, due to the significant increase in computational resources, the development of a new and more efficient algorithm is feasible. However, it is often not possible to achieve predictive accuracy using available scoring functions, in part due to neglect of entropy terms. Therefore, an accurate and reliable solution that reduces the shortcomings in the docking programs remains a challenge. Regardless, the application of docking to virtual screening techniques and lead optimization has produced impressive results and has extended the conventional approach to structure-based design.

References

1. T. Cheng, Q. Li, Z. Zhou, Y. Wang and S. Bryant, *AAPS J.*, 2012, **14**, 133–141.
2. I. Muegge and M. Rarey, Small Molecule Docking and Scoring, in *Reviews in Computational Chemistry*, ed. K. B. Lipkowitz and D. B. Boyd, John Wiley & Sons, Inc., New York, USA, 2001, vol. 17.

3. W. Ilona, *MATCH Communications in Mathematical and in Computer Chemistry*, 2006, **55**, 271–278.
4. H. M. Berman, T. N. Bhat, P. E. Bourne, Z. Feng, G. Gilliland, H. Weissig and J. Westbrook, *Nat. Struct. Mol. Biol.*, 2001, 957–959.
5. J. Westbrook, Z. Feng, L. Chen, H. Yang and H. M. Berman, *Nucleic Acids Res.*, 2003, **31**, 489–491.
6. N.-N. Wei and A. Hamza, *J. Chem. Inf. Model.*, 2013, **54**, 338–346.
7. S. John, S. Thangapandian, S. Sakkiah and K. W. Lee, *J. Enzyme Inhib. Med. Chem.*, 2011, **26**, 535–545.
8. D. B. Kitchen, H. Decornez, J. R. Furr and J. Bajorath, *Nat. Rev. Drug Discovery*, 2004, **3**, 935–949.
9. M. Smieško, *J. Chem. Inf. Model.*, 2013, **53**, 1415–1423.
10. R. D. Taylor, P. J. Jewsbury and J. W. Essex, *J Comput.-Aided Mol. Des.*, 2002, **16**, 151–166.
11. X. Y. Meng, H. X. Zhang, M. Mezei and M. Cui, *Curr. Comput.-Aided Drug Des.*, 2011, **7**, 146–157.
12. E. Yuriev and P. A. Ramsland, *J. Mol. Recognit.*, 2013, **26**, 215–239.
13. B. E. Evans, K. E. Rittle, M. G. Bock, R. M. DiPardo, R. M. Freidinger, W. L. Whitter, G. F. Lundell, D. F. Veber and P. S. Anderson, *J. Med. Chem.*, 1988, **31**, 2235–2246.
14. W. C. Ripka, G. V. De Lucca, A. C. Bach, Ii, R. S. Pottorf and J. M. Blaney, *Tetrahedron*, 1993, **49**, 3593–3608.
15. R. W. DeSimone, K. S. Currie, S. A. Mitchell, J. W. Darrow and D. A. Pippin, *Comb. Chem. High Throughput Screening*, 2004, 7, 473–494.
16. I. Meliciani, PhD, Karlsruher Institut für Technologie (KIT), 2011.
17. H. Gohlke and G. Klebe, *Angew. Chem., Int. Ed.*, 2002, **41**, 2644–2676.
18. W. P. Walters, M. T. Stahl and M. A. Murcko, *Drug Discovery Today*, 1998, **3**, 160–178.
19. D. A. Benson, I. Karsch-Mizrachi, D. J. Lipman, J. Ostell and D. L. Wheeler, *Nucleic Acids Res.*, 2003, **31**, 23–27.
20. S. Miyazaki, H. Sugawara, T. Gojobori and Y. Tateno, *Nucleic Acids Res.*, 2003, **31**, 13–16.
21. T. Kulikova, P. Aldebert, N. Althorpe, W. Baker, K. Bates, P. Browne, A. van den Broek, G. Cochrane, K. Duggan, R. Eberhardt, N. Faruque, M. Garcia-Pastor, N. Harte, C. Kanz, R. Leinonen, Q. Lin, V. Lombard, R. Lopez, R. Mancuso, M. McHale, F. Nardone, V. Silventoinen, P. Stoehr, G. Stoesser, M. A. Tuli, K. Tzouvara, R. Vaughan, D. Wu, W. Zhu and R. Apweiler, *Nucleic Acids Res.*, 2004, **32**, D27–D30.
22. A. Bairoch, R. Apweiler, C. H. Wu, W. C. Barker, B. Boeckmann, S. Ferro, E. Gasteiger, H. Huang, R. Lopez, M. Magrane, M. J. Martin, D. A. Natale, C. O'Donovan, N. Redaschi and L. S. L. Yeh, *Nucleic Acids Res.*, 2005, **33**, D154–D159.
23. E. Jain, A. Bairoch, S. Duvaud, I. Phan, N. Redaschi, B. Suzek, M. Martin, P. McGarvey and E. Gasteiger, *BMC Bioinf.*, 2009, **10**, 136.
24. K. Ginalski, *Curr. Opin. Struct. Biol.*, 2006, **16**, 172–177.
25. Z. Xiang, *Curr. Protein Peptide Sci.*, 2006, 7, 217–227.

26. Y. Zhang, *Curr. Opin. Struct. Biol.*, 2008, **18**, 342–348.
27. R. Bonneau and D. Baker, *Ann. Rev. Biophys. Biomol. Struct.*, 2001, **30**, 173–189.
28. M. Weigt, R. A. White, H. Szurmant, J. A. Hoch and T. Hwa, *Proc. Natl. Acad. Sci.*, 2008, **106**, 67–72.
29. F. Morcos, A. Pagnani, B. Lunt, A. Bertolino, D. S. Marks, C. Sander, R. Zecchina, J. N. Onuchic, T. Hwa and M. Weigt, *Proc. Natl. Acad. Sci.*, 2011, **108**, E1293–E1301.
30. D. T. Jones, D. W. A. Buchan, D. Cozzetto and M. Pontil, *Bioinformatics*, 2012, **28**, 184–190.
31. J. I. Sułkowska, F. Morcos, M. Weigt, T. Hwa and J. N. Onuchic, *Proc. Natl. Acad. Sci.*, 2012, **109**, 10340–10345.
32. A. Schug, M. Weigt, J.N Onuchic, T. Hwa and H. Szurmant, *Proc. Natl. Acad. Sci. U. S. A.*, 2009, **106**, 22124–22129.
33. T. H. Walther, C. Gottselig, S. L. Grage, M. Wolf, A. V. Vargiu, M. J. Klein, S. Vollmer, S. Prock, M. Hartmann, S. Afonin, E. Stockwald, H. Heinzmann, O. V. Nolandt, W. Wenzel, P. Ruggerone and A. S. Ulrich, *Cell*, 2013, **152**, 316–326.
34. A. E. Dago, A. Schug, A. Procaccini, J. A. Hoch, M. Weigt and H. Szurmant, *Proc. Natl. Acad. Sci.*, 2012, **109**, E1733–E1742.
35. M. J. Zaki and C. Bystroff, *Methods in Molecular Biology*, Humana Press, 2008, vol. 413.
36. C. Chothia, *Nature*, 1992, **357**, 543–544.
37. C. L. Tang, L. Xie, I. Y. Y. Koh, S. Posy, E. Alexov and B. Honig, *J. Mol. Biol.*, 2003, **334**, 1043–1062.
38. T. Ohlson, B. Wallner and A. Elofsson, *Proteins*, 2004, **57**, 188–197.
39. M. A. Marti-Renom, M. S. Madhusudhan and A. Sali, *Protein Sci.*, 2004, **13**, 1071–1087.
40. F. C. Bernstein, T. F. Koetzle, G. J. Williams, E. F. Meyer, Jr., M. D. Brice, J. R. Rodgers, O. Kennard, T. Shimanouchi and M. Tasumi, *Eur. J. Biochem.*, 1977, **80**, 319–324.
41. L. Lo Conte, B. Ailey, T. J. P. Hubbard, S. E. Brenner, A. G. Murzin and C. Chothia, *Nucleic Acids Res.*, 2000, **28**, 257–259.
42. L. Holm and P. Rosenström, *Nucleic Acids Res.*, 2010, **38**, W545–W549.
43. C. A. Orengo, A. D. Michie, S. Jones, D. T. Jones, M. B. Swindells and J. M. Thornton, *Structure*, 1997, **5**, 1093–1108.
44. S. F. Altschul, W. Gish, W. Miller, E. W. Myers and D. J. Lipman, *J. Mol. Biol.*, 1990, **215**, 403–410.
45. S. F. Altschul, T. L. Madden, A. A. Schäffer, J. Zhang, Z. Zhang, W. Miller and D. J. Lipman, *Nucleic Acids Res.*, 1997, **25**, 3389–3402.
46. K. Karplus, C. Barrett and R. Hughey, *Bioinformatics*, 1998, **14**, 846–856.
47. K. Karplus, R. Karchin, J. Draper, J. Casper, Y. Mandel-Gutfreund, M. Diekhans and R. Hughey, *Proteins: Struct., Funct., Bioinf.*, 2003, **53**, 491–496.
48. W. R. Pearson, *Methods Enzymol.*, 1990, **183**, 63–98.

49. R. Karchin, M. Cline, Y. Mandel-Gutfreund and K. Karplus, *Proteins: Struct., Funct., Bioinf.*, 2003, **51**, 504–514.
50. S. B. Needleman and C. D. Wunsch, *J. Mol. Biol.*, 1970, **48**, 443–453.
51. T. F. Smith and M. S. Waterman, *J. Mol. Biol.*, 1981, **147**, 195–197.
52. D. W. Mount, *Cold Spring Harbor Protocols*, 2008, vol. 2008, p. pdb.top39.
53. M. O. Dayhoff, R. M. Schwartz and B. C. Orcutt, *Atlas of Protein Sequence and Structure*, 1978, **5**, 345–351.
54. S. Henikoff and J. G. Henikoff, *Proc. Natl. Acad. Sci. U. S. A.*, 1992, **89**, 10915–10919.
55. H. Merlitz and W. Wenzel, in *Computational Science and Its Applications - ICCSA 2004,* ed. A. Laganá, M. Gavrilova, V. Kumar, Y. Mun, C. J. K. Tan and O. Gervasi, Springer, Berlin Heidelberg, 2004, ch. 48, vol. 3045, pp. 465–472.
56. I. Meliciani, N. Volz, V. Rempel, S. Hinz, T. Karcz, C. E. Müller, S. Bräse and W. Wenzel, *Biophys. J.*, 2011, **100**, 548a–549a.
57. I. Meliciani, PhD, University of Karlsruhe, 2011.
58. M. Källberg, H. Wang, S. Wang, J. Peng, Z. Wang, H. Lu and J. Xu, *Nat. Protoc.*, 2012, 7, 1511–1522.
59. P. A. Bates, L. A. Kelley, R. M. MacCallum and M. J. E. Sternberg, *Proteins: Struct., Funct., Bioinf.*, 2001, **45**, 39–46.
60. R. Grünberg, M. Nilges and J. Leckner, *Bioinformatics*, 2007, **23**, 769–770.
61. J. Schymkowitz, J. Borg, F. Stricher, R. Nys, F. Rousseau and L. Serrano, *Nucleic Acids Res.*, 2005, **33**, W382–388.
62. J. Soding, *Bioinformatics*, 2005, **21**, 951–960.
63. N. Eswar, B. Webb, M. A. Marti-Renom, M. S. Madhusudhan, D. Eramian, M. Y. Shen, U. Pieper and A. Sali, *Current Protocols in Bioinformatics*, 2006, ch. 5, Unit 5 6.
64. A. Sali and T. L. Blundell, *J. Mol. Biol.*, 1993, **234**, 779–815.
65. A. Leaver-Fay, M. Tyka, S. M. Lewis, O. F. Lange, J. Thompson, R. Jacak, K. Kaufman, P. D. Renfrew, C. A. Smith, W. Sheffler, I. W. Davis, S. Cooper, A. Treuille, D. J. Mandell, F. Richter, Y. E. Ban, S. J. Fleishman, J. E. Corn, D. E. Kim, S. Lyskov, M. Berrondo, S. Mentzer, Z. Popovic, J. J. Havranek, J. Karanicolas, R. Das, J. Meiler, T. Kortemme, J. J. Gray, B. Kuhlman, D. Baker and P. Bradley, *Methods Enzymol.*, 2011, **487**, 545–574.
66. T. Schwede, J. Kopp, N. Guex and M. C. Peitsch, *Nucleic Acids Res.*, 2003, **31**, 3381–3385.
67. B. Jayaram, P. Dhingra, B. Lakhani and S. Shekhar, *J. Chem. Sci.*, 2012, **124**, 83–91.
68. E. Krieger and G. Vriend, *Bioinformatics*, 2002, **18**, 315–318.
69. D. B. Kokh and W. Wenzel, *J. Med. Chem.*, 2008, **51**, 5919–5931.
70. H. Merlitz and W. Wenzel, *Chem. Phys. Lett.*, 2002, **362**, 271–277.
71. K. Ginalski, J. Pas, L. S. Wyrwicz, M. v. Grotthuss, J. M. Bujnicki and L. Rychlewski, *Nucleic Acids Res.*, 2003, **31**, 3804–3807.

72. J. Skolnick, D. Kihara and Y. Zhang, *Proteins*, 2004, **56**, 502–518.
73. L. Jaroszewski, L. Rychlewski, Z. Li, W. Li and A. Godzik, *Nucleic Acids Res.*, 2005, **33**, W284–288.
74. H. Zhou and Y. Zhou, *Proteins*, 2005, **58**, 321–328.
75. J. Shi, T. L. Blundell and K. Mizuguchi, *J. Mol. Biol.*, 2001, **310**, 243–257.
76. L. A. Kelley, R. M. MacCallum and M. J. E. Sternberg, *J. Mol. Biol.*, 2000, **299**, 501–522.
77. H. Zhou, S. B. Pandit and J. Skolnick, *Proteins*, 2009, **77**(Suppl 9), 123–127.
78. D. T. Jones, *J. Mol. Biol.*, 1999, **287**, 797–815.
79. L. A. Kelley and M. J. E. Sternberg, *Nat. Protoc.*, 2009, **4**, 363–371.
80. S. Wu and Y. Zhang, *Proteins: Struct., Funct., Bioinf.*, 2008, **72**, 547–556.
81. Y. Yang, E. Faraggi, H. Zhao and Y. Zhou, *Bioinformatics*, 2011, **27**, 2076–2082.
82. D. Gront and A. Kolinski, *Bioinformatics*, 2006, **22**, 621–622.
83. T. F. Havel and M. E. Snow, *J. Mol. Biol.*, 1991, **217**, 1–7.
84. S. Srinivasan, C. J. March and S. Sudarsanam, *Protein Sci.*, 1993, **2**, 277–289.
85. H. Bohr, J. Bohr, S. Brunak, R. M. J. Cotterill, H. Fredholm, B. Lautrup and S. B. Petersen, *FEBS Lett.*, 1990, **261**, 43–46.
86. D. Bassolino-Klimas and R. E. Bruccoleri, *Proteins: Struct., Funct., Bioinf.*, 1992, **14**, 465–474.
87. P. E. Correa, *Proteins: Struct., Funct., Bioinf.*, 1990, **7**, 366–377.
88. L. Holm and C. Sander, *J. Mol. Biol.*, 1991, **218**, 183–194.
89. M. Levitt, *J. Mol. Biol.*, 1992, **226**, 507–533.
90. Y. Luo, X. Jiang, L. Lai, C. Qu, X. Xu and Y. Tang, *Protein Eng.*, 1992, **5**, 147–150.
91. P. W. Payne, *Protein Sci.*, 1993, **2**, 315–324.
92. A. Sali, *Mol. Med. Today*, 1995, **1**, 270–277.
93. C. Chothia, A. M. Lesk, M. Levitt, A. G. Amit, R. A. Mariuzza, S. E. Phillips and R. J. Poljak, *Science*, 1986, **233**, 755–758.
94. T. L. Blundell, B. L. Sibanda, M. J. Sternberg and J. M. Thornton, *Nature*, 1987, **326**, 347–352.
95. D. Petrey, Z. Xiang, C. L. Tang, L. Xie, M. Gimpelev, T. Mitros, C. S. Soto, S. Goldsmith-Fischman, A. Kernytsky, A. Schlessinger, I. Y. Y. Koh, E. Alexov and B. Honig, *Proteins: Struct., Funct., Bioinf.*, 2003, **53**, 430–435.
96. P. Larsson, B. Wallner, E. Lindahl and A. Elofsson, *Protein Sci.*, 2008, **17**, 990–1002.
97. S. Chakravarty, S. Godbole, B. Zhang, S. Berger and R. Sanchez, *BMC Struct. Biol.*, 2008, **8**, 31–44.
98. J. Peng and J. Xu, *Proteins: Struct., Funct., Bioinf.*, 2011, **79**, 1930–1939.
99. S. Boyd, *Chem. World*, 2005, **2**, 66.
100. H.-P. Peng and A.-S. Yang, *Bioinformatics*, 2007, **23**, 2836–2842.
101. J. Espadaler, N. Fernandez Fuentes, A. Hermoso, E. Querol, F. X. Aviles, M. J. E. Sternberg and B. Oliva, *Nucleic Acids Res.*, 2004, **32**, D185–D188.

102. C. S. Soto, M. Fasnacht, J. Zhu, L. Forrest and B. Honig, *Proteins: Struct., Funct., Bioinf.*, 2008, **70**, 834–843.
103. K. Zhu, D. L. Pincus, S. Zhao and R. A. Friesner, *Proteins: Struct., Funct., Bioinf.*, 2006, **65**, 438–452.
104. M. P. Jacobson, D. L. Pincus, C. S. Rapp, T. J. F. Day, B. Honig, D. E. Shaw and R. A. Friesner, *Proteins: Struct., Funct., Bioinf.*, 2004, **55**, 351–367.
105. C. S. Rapp, *Proteins*, 1999, **35**, 173–183.
106. A. Fiser, R. K. G. Do and A. Sali, *Protein Sci.*, 2000, **9**, 1753–1773.
107. C. Zhang, S. Liu and Y. Zhou, *Protein Sci.*, 2004, **13**, 391–399.
108. J. Lee, D. Lee, H. Park, E. A. Coutsias and C. Seok, *Proteins: Struct., Funct., Bioinf.*, 2010, **78**, 3428–3436.
109. M. Cui, M. Mezei and R. Osman, *Prot. Eng. Des. Sel.*, 2008, **21**, 729–735.
110. A. K. Felts, E. Gallicchio, D. Chekmarev, K. A. Paris, R. A. Friesner and R. M. Levy, *J. Chem. Theory Comput.*, 2008, **4**, 855–868.
111. Z. Xiang, C. S. Soto and B. Honig, *Proc. Natl. Acad. Sci.*, 2002, **99**, 7432–7437.
112. M. A. Olson, M. Feig and C. L. Brooks, *J. Comput. Chem.*, 2008, **29**, 820–831.
113. V. Z. Spassov, P. K. Flook and L. Yan, *Protein Eng., Des. Sel.*, 2008, **21**, 91–100.
114. W. Zhang and Y. Duan, *Protein Eng., Des. Sel.*, 2006, **19**, 55–65.
115. A. Yan, A. Kloczkowski, H. Hofmann and R. L. Jernigan, *J. Biomol. Struct. Dyn.*, 2007, **25**, 275–288.
116. C. Hartmann, I. Antes and T. Lengauer, *Protein Sci.*, 2007, **16**, 1294–1307.
117. J. W. Ponder and F. M. Richards, *J. Mol. Biol.*, 1987, **193**, 775–791.
118. L. Jiang, B. Kuhlman, T. Kortemme and D. Baker, *Proteins: Struct., Funct., Bioinf.*, 2005, **58**, 893–904.
119. G. G. Krivov, M. V. Shapovalov and R. L. Dunbrack, Jr., *Proteins*, 2009, 77, 778–795.
120. S. Raman, R. Vernon, J. Thompson, M. Tyka, R. Sadreyev, J. Pei, D. Kim, E. Kellogg, F. DiMaio, O. Lange, L. Kinch, W. Sheffler, B.-H. Kim, R. Das, N. V. Grishin and D. Baker, *Proteins: Struct., Funct., Bioinf.*, 2009, 77, 89–99.
121. B. Wallner, P. Larsson and A. Elofsson, *Nucleic Acids Res.*, 2007, **35**, W369–374.
122. Y. Zhang, *BMC Bioinf.*, 2008, **9**, 40.
123. Y. Zhang, *Proteins*, 2007, **69**(Suppl 8), 108–117.
124. D. Chivian, D. E. Kim, L. Malmström, P. Bradley, T. Robertson, P. Murphy, C. E. M. Strauss, R. Bonneau, C. A. Rohl and D. Baker, *Proteins: Struct., Funct., Bioinf.*, 2003, **53**, 524–533.
125. N. Fernandez-Fuentes, C. J. Madrid-Aliste, B. K. Rai, J. E. Fajardo and A. Fiser, *Nucleic Acids Res.*, 2007, **35**, W363–W368.
126. S. Montgomerie, J. A. Cruz, S. Shrivastava, D. Arndt, M. Berjanskii and D. S. Wishart, *Nucleic Acids Res.*, 2008, **36**, W202–W209.

127. B. Wallner and A. Elofsson, *Protein Sci.*, 2006, **15**, 900–913.
128. X. Gao, D. Bu, S. C. Li, J. Xu and M. Li, *Genome Inform*, 2007, **19**, 27–39.
129. M. Wiederstein and M. J. Sippl, *Nucleic Acids Res.*, 2007, **35**, W407–410.
130. L. J. McGuffin, *Bioinformatics*, 2008, **24**, 586–587.
131. P. Mereghetti, M. Ganadu, E. Papaleo, P. Fantucci and L. De Gioia, *BMC Bioinf.*, 2008, **9**, 66.
132. D. Eramian, M. Y. Shen, D. Devos, F. Melo, A. Sali and M. A. Marti-Renom, *Protein Sci.*, 2006, **15**, 1653–1666.
133. P. Benkert, S. C. Tosatto and D. Schomburg, *Proteins*, 2008, **71**, 261–277.
134. J. Qiu, W. Sheffler, D. Baker and W. S. Noble, *Proteins*, 2008, **71**, 1175–1182.
135. A. Randall and P. Baldi, *BMC Struct. Biol.*, 2008, **8**, 52.
136. A. L. Morris, M. W. MacArthur, E. G. Hutchinson and J. M. Thornton, *Proteins: Struct., Funct., Bioinf.*, 1992, **12**, 345–364.
137. R. A. Laskowski, J. A. Rullmannn, M. W. MacArthur, R. Kaptein and J. M. Thornton, *J. Biomol. NMR*, 1996, **8**, 477–486.
138. A. A. Vaguine, J. Richelle and S. J. Wodak, *Acta Cryst. Sect. D*, 1999, **55**, 191–205.
139. J. Bowie, R. Luthy and D. Eisenberg, *Science*, 1991, **253**, 164–170.
140. C. Colovos and T. O. Yeates, *Protein Sci.*, 1993, **2**, 1511–1519.
141. F. Melo, D. Devos, E. Depiereux and E. Feytmans, *Proc. Int. Conf. Intell. Syst. Mol. Biol., 5th*, 1997, **5**, 187–190.
142. I. W. Davis, A. Leaver-Fay, V. B. Chen, J. N. Block, G. J. Kapral, X. Wang, L. W. Murray, W. B. Arendall 3rd, J. Snoeyink, J. S. Richardson and D. C. Richardson, *Nucleic Acids Res.*, 2007, **35**, W375–383.
143. L. McGuffin, *BMC Bioinf.*, 2007, **8**, 345.
144. C. S. Pettitt, L. J. McGuffin and D. T. Jones, *Bioinformatics*, 2005, **21**, 3509–3515.
145. B. Wallner, H. Fang and A. Elofsson, *Proteins: Struct., Funct., Bioinf.*, 2003, **53**, 534–541.
146. Z. Wang, J. Eickholt and J. Cheng, *Bioinformatics*, 2011, **27**, 1715–1716.
147. Z. Wang, A. N. Tegge and J. Cheng, *Proteins: Struct., Funct., Bioinf.*, 2009, **75**, 638–647.
148. J. L. Klepeis, Y. Wei, M. H. Hecht and C. A. Floudas, *Proteins: Struct., Funct., Bioinf.*, 2005, **58**, 560–570.
149. A. Liwo, M. Khalili and H. A. Scheraga, *Proc. Natl. Acad. Sci. U. S. A.*, 2005, **102**, 2362–2367.
150. S. Wu, J. Skolnick and Y. Zhang, *BMC Biol.*, 2007, **5**, 17.
151. M. Jacobson and A. Sali, *Annual Reports in Medicinal Chemistry,* Academic Press, 2004, vol. 39, pp. 259–276.
152. J. Lee, S. Wu and Y. Zhang, in *From Protein Structure to Function with Bioinformatics*, ed. D. Rigden, Springer, Netherlands, 2009, ch. 1, pp. 3–25.
153. W. D. Cornell, P. Cieplak, C. I. Bayly, I. R. Gould, K. M. Merz, D. M. Ferguson, D. C. Spellmeyer, T. Fox, J. W. Caldwell and P. A. Kollman, *J. Am. Chem. Soc.*, 1995, **117**, 5179–5197.

154. D. A. Pearlman, D. A. Case, J. W. Caldwell, W. S. Ross, T. E. Cheatham Iii, S. DeBolt, D. Ferguson, G. Seibel and P. Kollman, *Comput. Phys. Commun.*, 1995, **91**, 1–41.
155. P. K. Weiner and P. A. Kollman, *J. Comput. Chem.*, 1981, **2**, 287–303.
156. B. R. Brooks, R. E. Bruccoleri, B. D. Olafson, D. J. States, S. Swaminathan and M. Karplus, *J. Comput. Chem.*, 1983, **4**, 187–217.
157. W. F. van Gunsteren, S. R. Billeter, A. A. Eising, P. H. Hünenberger, P. Krüger, A. E. Mark, W. R. P. Scott and I. G. Tironi, *Biomolecular Simulation: The {GROMOS96} Manual and Userguide*, Hochschuleverlag AG an der ETH Zürich1996.
158. W. L. Jorgensen and J. Tirado-Rives, *J. Am. Chem. Soc.*, 1988, **110**, 1657–1666.
159. J. Skolnick, *Curr. Opin. Struct. Biol.*, 2006, **16**, 166–171.
160. Y. Zhang, A. Kolinski and J. Skolnick, *Biophys. J.*, 2003, **85**, 1145–1164.
161. J. Skolnick, A. Godzik, L. Jaroszewski and A. Kolinski, *Protein Sci.*, 1997, **6**, 676–688.
162. R. Samudrala and J. Moult, *J. Mol. Biol.*, 1998, **275**, 895–916.
163. H. Lu and J. Skolnick, *Proteins: Struct., Funct., Bioinf.*, 2001, **44**, 223–232.
164. M.-Y. Shen and A. Sali, *Protein Sci.*, 2006, **15**, 2507–2524.
165. Y. Zhang and J. Skolnick, *Proc. Natl. Acad. Sci. U. S. A.*, 2005, **102**, 1029–1034.
166. K. T. Simons, C. Kooperberg, E. Huang and D. Baker, *J. Mol. Biol.*, 1997, **268**, 209–225.
167. Y. Zhang and J. Skolnick, *Proc. Natl. Acad. Sci. U. S. A.*, 2004, **101**, 7594–7599.
168. B. Hess, C. Kutzner, D. van der Spoel and E. Lindahl, *J. Chem. Theory Comput.*, 2008, **4**, 435–447.
169. D. A. Case, T. E. Cheatham, T. Darden, H. Gohlke, R. Luo, K. M. Merz, A. Onufriev, C. Simmerling, B. Wang and R. J. Woods, *J. Comput. Chem.*, 2005, **26**, 1668–1688.
170. B. R. Brooks, C. L. Brooks, A. D. Mackerell, L. Nilsson, R. J. Petrella, B. Roux, Y. Won, G. Archontis, C. Bartels, S. Boresch, A. Caflisch, L. Caves, Q. Cui, A. R. Dinner, M. Feig, S. Fischer, J. Gao, M. Hodoscek, W. Im, K. Kuczera, T. Lazaridis, J. Ma, V. Ovchinnikov, E. Paci, R. W. Pastor, C. B. Post, J. Z. Pu, M. Schaefer, B. Tidor, R. M. Venable, H. L. Woodcock, X. Wu, W. Yang, D. M. York and M. Karplus, *J. Comput. Chem.*, 2009, 30, 1545-1614.
171. A. Irbäck and S. Mohanty, *J. Comput. Chem.*, 2006, **27**, 1548–1555.
172. F. Eisenmenger, U. H. E. Hansmann, S. Hayryan and C.-K. Hu, *Comput. Phys. Commun.*, 2001, **138**, 192–212.
173. C. A. Rohl, C. E. M. Strauss, K. M. S. Misura and D. Baker, in *Methods in Enzymology*, ed. B. Ludwig and L. J. Michael, Academic Press, 2004, vol. 383, pp. 66–93.
174. T. Herges and W. Wenzel, *Biophys. J.*, 2004, **87**, 3100–3109.
175. S. Plimpton, *J. Comput. Phys.*, 1995, **117**, 1–19.

176. J. C. Phillips, R. Braun, W. Wang, J. Gumbart, E. Tajkhorshid, E. Villa, C. Chipot, R. D. Skeel, L. Kalé and K. Schulten, *J. Comput. Chem.*, 2005, **26**, 1781–1802.
177. K. J. Bowers, E. Chow, H. Xu, R. O. Dror, M. P. Eastwood, B. A. Gregersen, J. L. Klepeis, I. Kolossvary, M. A. Moraes, F. D. Sacerdoti, J. K. Salmon, Y. Shan and D. E. Shaw, Scalable Algorithms for Molecular Dynamics Simulations on Commodity Clusters, Tampa, Florida, 2006.
178. H. E. P. Sanchez, I. Kondov, J. M. García, K. V. Klenin and W. Wenzel, *3rd International Workshop on Science Gateways for Life Sciences (IWSG 2011)*, 2011.
179. T. Strunk, M. Wolf, M. Brieg, K. Klenin, A. Biewer, F. Tristram, M. Ernst, P. J. Kleine, N. Heilmann, I. Kondov and W. Wenzel, *J. Comput. Chem.*, 2012, **33**, 2602–2613.
180. A. Kryshtafovych, K. Fidelis and J. Moult, *Proteins*, 2007, **69**(Suppl 8), 194–207.
181. A. Kryshtafovych, K. Fidelis and J. Moult, *Proteins*, 2009, **77**(Suppl 9), 217–228.
182. J. N. Battey, J. Kopp, L. Bordoli, R. J. Read, N. D. Clarke and T. Schwede, *Proteins*, 2007, **69**(Suppl 8), 68–82.
183. I. D. Kuntz, J. M. Blaney, S. J. Oatley, R. Langridge and T. E. Ferrin, *J. Mol. Biol.*, 1982, **161**, 269–288.
184. M. Miller, S. Kearsley, D. Underwood and R. Sheridan, *J. Comput.-Aided Mol. Des.*, 1994, **8**, 153–174.
185. B. Fischer, K. Fukuzawa and W. Wenzel, *Proteins: Struct., Funct., Bioinf.*, 2008, **70**, 1264–1273.
186. J. Flick, F. Tristram and W. Wenzel, *J. Comput. Chem.*, 2012, **33**, 2504–2515.
187. G. Morris, D. Goodsell, R. Huey and A. Olson, *J. Comput.-Aided Mol. Des.*, 1996, **10**, 293–304.
188. W. Wenzel and K. Hamacher, *Phys. Rev. Lett.*, 1999, **82**, 3003–3007.
189. C. A. Sotriffer, *Curr. Top. Med. Chem.*, 2011, **11**, 179–191.
190. E. Fischer, *Ber Dtsch. Chem. Ges.*, 1894, **27**, 2985–2993, DOI: 10.1002/cber.18940270364.
191. D. E. Koshland, *Proc. Natl. Acad. Sci.*, 1958, **44**, 98–104.
192. G. M. Morris, D. S. Goodsell, R. S. Halliday, R. Huey, W. E. Hart, R. K. Belew and A. J. Olson, *J. Comput. Chem.*, 1998, **19**, 1639–1662.
193. B. Fischer, S. Basili, H. Merlitz and W. Wenzel, *Proteins: Struct., Funct., Bioinf.*, 2007, **68**, 195–204.
194. H. Merlitz, B. Burghardt and W. Wenzel, *Chem. Phys. Lett.*, 2004, **390**, 500–505.
195. H. Perez-Sanchez, B. Fischer, D. Kokh, H. Merlitz and W. Wenzel, *J. Cheminf.*, 2010, **2**, P23.
196. M. Rarey, B. Kramer, T. Lengauer and G. Klebe, *J. Mol. Biol.*, 1996, **261**, 470–489.

197. G. M. Morris, R. Huey, W. Lindstrom, M. F. Sanner, R. K. Belew, D. S. Goodsell and A. J. Olson, *J. Comput. Chem.*, 2009, **30**, 2785–2791.
198. F. Jiang and S.-H. Kim, *J. Mol. Biol.*, 1991, **219**, 79–102.
199. R. M. A. Knegtel, I. D. Kuntz and C. M. Oshiro, *J. Mol. Biol.*, 1997, **266**, 424–440.
200. H. A. Carlson, K. M. Masukawa, K. Rubins, F. D. Bushman, W. L. Jorgensen, R. D. Lins, J. M. Briggs and J. A. McCammon, *J. Med. Chem.*, 2000, **43**, 2100–2114.
201. S. John, S. Thangapandian and K. W. Lee, *J. Biomol. Struct. Dyn.*, 2012, **29**, 921–936.
202. S. Thangapandian, S. John, Y. Lee, S. Kim and K. W. Lee, *Int. J. Mol. Sci.*, 2011, **12**, 9440–9462.
203. T. A. Ewing, S. Makino, A. G. Skillman and I. Kuntz, *J. Comput.-Aided Mol. Des.*, 2001, **15**, 411–428.
204. N. Brooijmans, in *Structural Bioinformatics*, ed. Jenny Gu and P. E. Bourne, John Wiley & Sons, Inc, 2nd edn, 2009.
205. A. R. Leach, Molecular Modelling: Principles and Applications, Longman, 1996.
206. M. Rarey, S. Wefing and T. Lengauer, *J. Comput.-Aided Mol. Des.*, 1996, **10**, 41–54.
207. D. Moustakas, P. T. Lang, S. Pegg, E. Pettersen, I. Kuntz, N. Brooijmans and R. Rizzo, *J. Comput.-Aided Mol. Des.*, 2006, **20**, 601–619.
208. D. Joseph-McCarthy, B. E. Thomas, M. Belmarsh, D. Moustakas and J. C. Alvarez, *Proteins: Struct., Funct., Bioinf.*, 2003, **51**, 172–188.
209. H. J. C. Berendsen, D. van der Spoel and R. van Drunen, *Comput. Phys. Commun.*, 1995, **91**, 43–56.
210. E. Lindahl, B. Hess and D. van der Spoel, *J. Mol. Model.*, 2001, 7, 306–317.
211. N. Brooijmans and I. D. Kuntz, *Annu. Rev. Biophys. Biomol. Struct.*, 2003, **32**, 335–373.
212. M. Mangoni, D. Roccatano and A. Di Nola, *Proteins: Struct., Funct., Bioinf.*, 1999, **35**, 153–162.
213. A. Di Nola, D. Roccatano and H. J. C. Berendsen, *Proteins: Struct., Funct., Bioinf.*, 1994, **19**, 174–182.
214. Y. Pak and S. Wang, *J. Phys. Chem. B*, 1999, **104**, 354–359.
215. S. E. Nichols, R. Baron, A. Ivetac and J. A. McCammon, *J. Chem. Inf. Model.*, 2011, **51**, 1439–1446.
216. D. S. Goodsell, H. Lauble, C. D. Stout and A. J. Olson, *Proteins: Struct., Funct., Bioinf.*, 1993, **17**, 1–10.
217. N. Metropolis, A. W. Rosenbluth, M. N. Rosenbluth, A. H. Teller and E. Teller, *J. Chem. Phys.*, 1953, **21**, 1087–1092.
218. G. Jones, P. Willett, R. C. Glen, A. R. Leach and R. Taylor, *J. Mol. Biol.*, 1997, **267**, 727–748.
219. B. Fischer, H. Merlitz and W. Wenzel, *Biophys. J.*, 2007, 367a–367a.
220. B. Fischer, S. Basili, H. Merlitz and W. Wenzel, *Proteins: Struct., Funct., Bioinf.*, 2007, 195–204.

221. H. Merlitz and W. Wenzel, Computational Science and Its Applications – Iccsa 2004, Pt 3, 2004, **3045**, 465–472.
222. H. Merlitz, B. Burghardt and W. Wenzel, *Chem. Phys. Lett.*, 2003, **370**, 68–73.
223. H. Merlitz and W. Wenzel, *Chem. Phys. Lett.*, 2002, **362**, 271–277.
224. J.-Y. Trosset and H. A. Scheraga, *J. Comput. Chem.*, 1999, **20**, 412–427.
225. M. Liu and S. Wang, *J. Comput.-Aided Mol. Des.*, 1999, **13**, 435–451.
226. T. N. Hart and R. J. Read, *Proteins: Struct., Funct., Bioinf.*, 1992, **13**, 206–222.
227. C. McMartin and R. Bohacek, *J. Comput.-Aided Mol. Des.*, 1997, **11**, 333–344.
228. B. A. Luty, Z. R. Wasserman, P. F. W. Stouten, C. N. Hodge, M. Zacharias and J. A. McCammon, *J. Comput. Chem.*, 1995, **16**, 454–464.
229. K. P. Clark and Ajay, *J. Comput. Chem.*, 1995, **16**, 1210–1226.
230. J. S. Taylor and R. M. Burnett, *Proteins: Struct., Funct., Bioinf.*, 2000, **41**, 173–191.
231. P. Kollman, *Chem. Rev.*, 1993, **93**, 2395–2417.
232. S. A. Adcock and J. A. McCammon, *Chem. Rev.*, 2006, **106**, 1589–1615.
233. N. Huang, C. Kalyanaraman, J. J. Irwin and M. P. Jacobson, *J. Chem. Inf. Model.*, 2005, **46**, 243–253.
234. B. Kramer, M. Rarey and T. Lengauer, *Proteins: Struct., Funct., Bioinf.*, 1999, **37**, 228–241.
235. S.-Y. Huang, S. Z. Grinter and X. Zou, *Phys. Chem. Chem. Phys.*, 2010, **12**, 12899–12908.
236. Y. Li, Z. Liu and R. Wang, *J. Chem. Inf. Model.*, 2010, **50**, 1682–1692.
237. X. Zou, Yaxiong and I. D. Kuntz, *J. Am. Chem. Soc.*, 1999, **121**, 8033–8043.
238. J. Åqvist, C. Medina and J.-E. Samuelsson, *Protein Eng.*, 1994, **7**, 385–391.
239. K. B. Ljungberg, J. Marelius, D. Musil, P. Svensson, B. Norden and J. Åqvist, *Eur. J. Pharm. Sci.*, 2001, **12**, 441–446.
240. R. C. Rizzo, J. Tirado-Rives and W. L. Jorgensen, *J. Med. Chem.*, 2000, **44**, 145–154.
241. R. Wang, L. Lai and S. Wang, *J. Comput.-Aided Mol. Des.*, 2002, **16**, 11–26.
242. A. N. Jain, *J. Med. Chem.*, 2003, **46**, 499–511.
243. W. T. M. Mooij and M. L. Verdonk, *Proteins: Struct., Funct., Bioinf.*, 2005, **61**, 272–287.
244. H. F. G. Velec, H. Gohlke and G. Klebe, *J. Med. Chem.*, 2005, **48**, 6296–6303.
245. I. Muegge, in *Virtual Screening: An Alternative or Complement to High Throughput Screening?*, ed. G. Klebe, Springer, Netherlands, 2002, ch. 7, vol. 20, pp. 99–114.
246. I. Muegge, *J. Med. Chem.*, 2005, **49**, 5895–5902.

247. R. S. DeWitte, A. V. Ishchenko and E. I. Shakhnovich, *J. Am. Chem. Soc.*, 1997, **119**, 4608–4617.
248. P. J. Ballester and J. B. O. Mitchell, *Bioinformatics*, 2010, **26**, 1169–1175.
249. R. Teramoto and H. Fukunishi, *J. Chem. Inf. Model.*, 2007, **47**, 526–534.
250. S. Betzi, K. Suhre, B. Chétrit, F. Guerlesquin and X. Morelli, *J. Chem. Inf. Model.*, 2006, **46**, 1704–1712.
251. S. Bar-Haim, A. Aharon, T. Ben-Moshe, Y. Marantz and H. Senderowitz, *J. Chem. Inf. Model.*, 2009, **49**, 623–633.
252. G. E. Terp, B. N. Johansen, I. T. Christensen and F. S. Jørgensen, *J. Med. Chem.*, 2001, **44**, 2333–2343.
253. M. Kontoyianni, L. M. McClellan and G. S. Sokol, *J. Med. Chem.*, 2003, **47**, 558–565.
254. S. Thangapandian, S. John, S. Sakkiah and K. W. Lee, *Eur. J. Med. Chem.*, 2010, **45**, 4409–4417.
255. T. Sundarapandian, S. John, S. Sugunadevi and L. K. Woo, *J. Mol. Graphics Modell.*, 2010, **29**, 382–395.
256. S. Thangapandian, S. John and K. W. Lee, *Mol. Simul.*, 2012, **38**, 1143–1156.
257. S. Thangapandian, S. John, S. Sakkiah and K. W. Lee, *Eur. J. Med. Chem.*, 2011, **46**, 1593–1603.
258. S. Thangapandian, S. John, S. Sakkiah and K. W. Lee, *Chemical Biology & Drug Design*, 2011, **78**, 289–300.
259. A. Good and T. Oprea, *J. Comput.-Aided Mol. Des.*, 2008, **22**, 169–178.
260. N. Huang, B. K. Shoichet and J. J. Irwin, *J. Med. Chem.*, 2006, **49**, 6789–6801.
261. A. Jain, *J. Comput.-Aided Mol. Des.*, 2000, **14**, 199–213.
262. N. Hird, *Drug Discovery Today*, 2000, **5**, 307–308.
263. H.-J. Böhm, *Perspect. Drug Discovery Des.*, 1995, **3**, 21–33.
264. Z. D. Konteatis, *Expert Opin. Drug Discovery*, 2010, **5**, 1047–1065.
265. M. Sandor, R. Kiss and G. M. Keseru, *J. Chem. Inf. Model.*, 2010, **50**, 1165–1172.
266. A. D. Favia, G. Bottegoni, I. Nobeli, P. Bisignano and A. Cavalli, *J. Chem. Inf. Model.*, 2011, **51**, 2882–2896.
267. M. L. Verdonk, I. Giangreco, R. J. Hall, O. Korb, P. N. Mortenson and C. W. Murray, *J. Med. Chem.*, 2011, **54**, 5422–5431.
268. H. Niu and P. J. Matthew, *PLoS One*, 2010, **5**, e10109.
269. R. Ward, *J. Mol. Model.*, 2010, **16**, 1833–1843.
270. F.-Y. Lin and Y. J. Tseng, *J. Chem. Inf. Model.*, 2011, **51**, 1703–1715.
271. M. K. Haider, H.-O. Bertrand and R. E. Hubbard, *J. Chem. Inf. Model.*, 2011, **51**, 1092–1105.
272. D. Huang and A. Caflisch, *J. Mol. Recognit.*, 2010, **23**, 183–193.
273. A. L. Hopkins and C. R. Groom, *Nat. Rev. Drug Discovery*, 2002, **1**, 727–730.
274. P. Ripphausen, B. Nisius, L. Peltason and J. Bajorath, *J. Med. Chem.*, 2010, **53**, 8461–8467.

275. S. Thangapandian, S. John, M. Son, V. Arulalapperumal and K. W. Lee, *Future Med. Chem.*, 2012, **5**, 27–40.
276. S. John, S. Thangapandian, S. Sakkiah and K. W. Lee, *Eur. J. Med. Chem.*, 2010, **45**, 4004–4012.
277. S. Thangapandian, S. John, S. Sakkiah and K. W. Lee, *J. Chem. Inf. Model.*, 2010, **51**, 33–44.
278. S. John, S. Thangapandian, M. Arooj, J. Hong, K. Kim and K. Lee, *BMC Bioinf.*, 2011, **12**, S4.
279. S. John, S. Thangapandian, S. Sakkiah and K. Lee, *BMC Bioinf.*, 2011, **12**, 1–11.
280. S. Thangapandian, S. John, M. Arooj and K. W. Lee, *PLoS One*, 2012, **7**, e34593.
281. H. van de Waterbeemd and E. Gifford, *Nat. Rev. Drug Discovery*, 2003, **2**, 192–204.
282. R. A. Friesner, J. L. Banks, R. B. Murphy, T. A. Halgren, J. J. Klicic, D. T. Mainz, M. P. Repasky, E. H. Knoll, M. Shelley, J. K. Perry, D. E. Shaw, P. Francis and P. S. Shenkin, *J. Med. Chem.*, 2004, **47**, 1739–1749.
283. C. M. Venkatachalam, X. Jiang, T. Oldfield and M. Waldman, *J. Mol. Graphics Modell.*, 2003, **21**, 289–307.
284. Z. Zsoldos, I. Szabo, Z. Szabo and A. Peter Johnson, *J. Mol. Struct.: THEOCHEM*, 2003, **666–667**, 659–665.
285. R. Abagyan, M. Totrov and D. Kuznetsov, *J. Comput. Chem.*, 1994, **15**, 488–506.
286. J. Bajorath, *Nat. Rev. Drug Discovery*, 2002, **1**, 882–894.
287. O. Roche, P. Schneider, J. Zuegge, W. Guba, M. Kansy, A. Alanine, K. Bleicher, F. Danel, E.-M. Gutknecht, M. Rogers-Evans, W. Neidhart, H. Stalder, M. Dillon, E. Sjögren, N. Fotouhi, P. Gillespie, R. Goodnow, W. Harris, P. Jones, M. Taniguchi, S. Tsujii, W. von der Saal, G. Zimmermann and G. Schneider, *J. Med. Chem.*, 2001, **45**, 137–142.
288. J. Poupaert, P. Carato and E. Colacino, *Curr. Med. Chem.*, 2005, **12**, 877–885.
289. B. A. Bunin, M. J. Plunkett and J. A. Ellman, *Proc. Natl. Acad. Sci.*, 1994, **91**, 4708–4712.
290. J.-Y. Lee and Y.-C. Kim, *ChemMedChem*, 2009, **4**, 733–737.
291. M. Gaba, S. Singh and C. Mohan, *Eur. J. Med. Chem.*, 2014, **76**, 494–505.
292. Datani, H. Radhika, S. G. Kini and M. Mubeen, *J. Comput. Methods Mol. Des.*, 2012, **2**, 149.
293. F. Arjmand, S. Parveen, M. Afzal and M. Shahid, *J. Photochem. Photobiol., B*, 2012, **114**, 15–26.
294. M. W. Robinson, N. McFerran, A. Trudgett, L. Hoey and I. Fairweather, *J. Mol. Graphics Modell.*, 2004, **23**, 275–284.
295. E. Lacey, *Int. J. Parasitol.*, 1988, **18**, 885–936.
296. G. W. Lubega and R. K. Prichard, *Mol. Biochem. Parasitol.*, 1990, **38**, 221–232.
297. R. Prichard, *Trends Parasitol.*, 2001, **17**, 445–453.

298. M. Robinson, A. Trudgett, I. Fairweather and N. McFerran, *Trends Parasitol.*, 2002, **18**, 153–154.
299. M. W. Robinson, E. M. Hoey, I. Fairweather, J. P. Dalton, S. McGonigle and A. Trudgett, *Int. J. Parasitol.*, 2001, **31**, 1264–1268.
300. O. Sharma, A. Pan, S. L. Hoti, A. Jadhav, M. Kannan and P. Mathur, *Med. Chem. Res.*, 2012, **21**, 2415–2427.
301. E. H. Sessions, Y. Yin, T. D. Bannister, A. Weiser, E. Griffin, J. Pocas, M. D. Cameron, C. Ruiz, L. Lin, S. C. Schürer, T. Schröter, P. LoGrasso and Y. Feng, *Bioorg. Med. Chem. Lett.*, 2008, **18**, 6390–6393.
302. C. B. Yoo and P. A. Jones, *Nat. Rev. Drug Discovery*, 2006, **5**, 37–50.
303. S. Minucci and P. G. Pelicci, *Nat. Rev. Cancer*, 2006, **6**, 38–51.
304. S. P. Olesen, L. H. Jensen and P. Moldt, Patent Application *EP* 617023, 1994.
305. J. R. S. Hoult and M. Paya, *General Pharmacology: The Vascular System*, 1996, **27**, 713–722.
306. F. Borges, F. Roleira and N. Milhazes, in *Frontiers in Medicinal Chemistry*, ed. E. U. L. Santana, Bentham Science, 2009, vol. 4, pp. 23–85.
307. S. Han, V. Zhou, S. Pan, Y. Liu, M. Hornsby, D. McMullan, H. E. Klock, J. Haugen, S. A. Lesley, N. Gray, J. Caldwell and X.-j. Gu, *Bioorg. Med. Chem. Lett.*, 2005, **15**, 5467–5473.
308. H.-C. Lin, S.-H. Tsai, C.-S. Chen, Y.-C. Chang, C.-M. Lee, Z.-Y. Lai and C.-M. Lin, *Biochem. Pharmacol.*, 2008, **75**, 1416–1425.
309. X.-H. Liu, H.-F. Liu, J. Chen, Y. Yang, B.-A. Song, L.-S. Bai, J.-X. Liu, H.-L. Zhu and X.-B. Qi, *Bioorg. Med. Chem. Lett.*, 2010, **20**, 5705–5708.
310. J. Shobini, A. K. Mishra, K. Sandhya and N. Chandra, *Spectrochim. Acta, Part A*, 2001, **57**, 1133–1147.
311. L. Piazzi, A. Cavalli, F. Colizzi, F. Belluti, M. Bartolini, F. Mancini, M. Recanatini, V. Andrisano and A. Rampa, *Bioorg. Med. Chem. Lett.*, 2008, **18**, 423–426.
312. Z. Wang, B. Ling, R. Zhang and Y. Liu, *J. Phys. Chem. B*, 2008, **112**, 10033–10040.
313. C.-S. Wang and J. A. Hartsuck, *Biochim. Biophys. Acta, Lipids Lipid Metab.*, 1993, **1166**, 1–19.
314. D. Lombardo, *Biochim. Biophys. Acta,- Mol. Cell Biol. Lipids*, 2001, **1533**, 1–28.
315. C. S. Wang, M. E. Martindale, M. M. King and J. Tang, *Am. J. Clin. Nutr.*, 1989, **49**, 457–463.
316. J. Brodt-Eppley, P. White, S. Jenkins and D. Y. Hui, *Biochim. Biophys. Acta, Mol. Basis Dis.*, 1995, **1272**, 69–72.
317. H. Falt, O. Hernell and L. Blackberg, *Pediatr. Res.*, 2002, **52**, 509–515.
318. J. J. Heynekamp, L. A. Hunsaker, T. A. Vander Jagt, R. E. Royer, L. M. Deck and D. L. Vander Jagt, *Bioorg. Med. Chem.*, 2008, **16**, 5285–5294.
319. K. Boda, T. Seidel and J. Gasteiger, *J. Comput.-Aided Mol. Des.*, 2007, **21**, 311–325.

320. A. Zaliani, K. Boda, T. Seidel, A. Herwig, C. Schwab, J. Gasteiger, H. Claußen, C. Lemmen, J. Degen, J. Pärn and M. Rarey, *J. Comput.-Aided Mol. Des.*, 2009, **23**, 593–602.
321. T. Bisogno and V. D. Marzo, *CNS Neurol. Disord.: Drug Targets*, 2010, **9**, 10.
322. K. Van Laere, C. Casteels, I. Dhollander, K. Goffin, I. Grachev, G. Bormans and W. Vandenberghe, *J. Nucl. Med.*, 2010, **51**, 1413–1417.
323. T. Okada, M. Sugihara, A.-N. Bondar, M. Elstner, P. Entel and V. Buss, *J. Mol. Biol.*, 2004, **342**, 571–583.
324. J. D. Thompson, D. G. Higgins and T. J. Gibson, *Nucleic Acids Res.*, 1994, **22**, 4673–4680.
325. B. Lee and F. M. Richards, *J. Mol. Biol.*, 1971, **55**, 379–IN374.
326. T. Tuccinardi, P. L. Ferrarini, C. Manera, G. Ortore, G. Saccomanni and A. Martinelli, *J. Med. Chem.*, 2006, **49**, 984–994.
327. J.-Y. Shim, W. J. Welsh and A. C. Howlett, *Pept. Sci.*, 2003, **71**, 169–189.
328. A. Behrenswerth, N. Volz, J. Toräng, S. Hinz, S. Bräse and C. E. Müller, *Bioorg. Med. Chem.*, 2009, **17**, 2842–2851.
329. G. Launay and T. Simonson, *BMC Bioinf.*, 2008, **9**, 427.
330. P. Anand, PhD, University of Heidelberg, 2013.

CHAPTER 3

The β-Lactam (Azetidin-2-one) as a Privileged Ring in Medicinal Chemistry

JED F. FISHER AND SHAHRIAR MOBASHERY*

Department of Chemistry and Biochemistry, 251 Nieuwland Science Hall, University of Notre Dame, Notre Dame, IN 46556, USA
*Email: mobashery@nd.edu

3.1 Introduction

The amide bond is primordial. It is the unifying functional group of the proteins, and a key functional group of the nucleic acids, the coenzymes and the saccharides. In the vernacular of the organic chemist, a cyclic amide—the functional group formed by the condensation union of an amine "tail" and a carboxylic acid "head" into a ring—is called a lactam. Within the universe of the lactams, one particular lactam class is recognized by the medicinal chemist as both "distinctive"[1] and "enchanted".[2] These terms describe the four-membered lactam, or the azetidin-2-one (systematic nomenclature) ring. This ring is far more commonly known as the β-lactam, where the β-prefix denotes that the ring is formed by the cyclization of an amine found on the β-carbon of a carboxylic acid, to the carboxylic acid itself. The basis for the admiration of this ring by medicinal chemists is simple. Fleming discovered in 1928 that certain fungi biosynthesized natural products with potent cytotoxic activity against bacteria. Some seventeen years later, after heroic efforts were made to bring these antibiotics into the practice of medicine and after countless lives had been saved, the

RSC Drug Discovery Series No. 50
Privileged Scaffolds in Medicinal Chemistry: Design, Synthesis, Evaluation
Edited by Stefan Bräse

Published by the Royal Society of Chemistry, www.rsc.org

cornerstone functionality of these antibiotics was confirmed as the β-lactam ring. The subsequent history of antibacterial chemotherapy is dominated by the discovery and incorporation into clinical use of successive generations of these β-lactam antibiotics. Although three sub-families of β-lactam antibiotics—the penicillins, the cephalosporins, and the carbapenems—are bulwarks of modern antibacterial chemotherapy, persistent investigation by medicinal chemists over the past seventy years has shown that antibacterial activity is preserved over a great breadth of β-lactam structure. Indeed, as a consequence of the relentless development of antibiotic resistance to all classes of antibacterials, especially including the β-lactams, the current frontier for β-lactam antibacterial discovery is the understanding of the breadth of this structural diversity.

Notwithstanding the central place of the β-lactam as an antibacterial, the intensive study of the antibacterial β-lactams over the decades, combined with a modern interest in the use of small ring heterocycles to expand medicinal chemistry structural space, have led to the recognition that the β-lactam ring engages numerous biological targets, both prokaryotic and eukaryotic, and thus targets beyond its antibacterial protein targets. Indeed, the appearance of a β-lactam core in a non-antibacterial, and extensively (over the past decade) clinically used drug, ezetimibe, alone establishes the β-lactam as a durable (*vide infra*) small-ring heterocycle with value as a target-specific motif. This perspective—the present and future roles of the β-lactam as a privileged ring in drug discovery—is the theme of this chapter. Recent complementary perspectives on this same topic are acknowledged.[3–7]

3.2 Stability and Reactivity of the β-Lactam

The extraordinary circumstances that comprise the early scientific history of the β-lactam-containing penicillins—the deductive inference of their presence as powerful fungal-derived antibiotics by Fleming; the intense reduction to practice by Florey, Chain, Heatley and a host of other scientific talent in the midst of world war; and the contentious but ultimate proof of its β-lactam structure by *inter alia* Woodward and Hodgkin—have been retold on numerous occasions (see[2,8,9] and references cited therein). A key characteristic of the penicillin discerned in these early studies was its relative chemical instability, particularly in aqueous acid. The presumption that this hydrolytic instability directly reflected an intrinsic reactivity of its β-lactam core persisted not only through the studies of Sheehan and Henery-Logan on the synthesis of penicillins by the carbodiimide-dependent ring closure of the β-amino acid,[10] but to this day. This presumption is not without validity—there are indeed very reactive β-lactams—but it is misleading from the broader perspective of the β-lactam as a privileged medicinal motif. Indeed, the successful formulation, for both parenteral and oral administration, of legions of penicillin, cephalosporin, carbapenem and other β-lactam structures proves that all of these β-lactams have sufficient chemical stability for development and application as drugs. Within this realm of β-lactam antibacterials, physical organic studies confirm a wide breadth of intrinsic reactivity, controlled by the precise presentation and substitution of the β-lactam core within the monocyclic and bicyclic motifs of these structures.[11,12] In sharp contrast to the twisted amide,[13] the intrinsic reactivity of the spare β-lactam core is hardly greater than that of the acyclic amide of the polypeptide (approx. $t_{1/2}$ of 4 y at ambient temperature).[14] The strain of the four-membered β-lactam ring is not well expressed in the barrier accompanying nucleophile addition to its carbonyl, and the energetic quality of its amide resonance (in the simple β-lactam) is hardly compromised by its ring constraint.[15] Efforts to correlate of the antibacterial activity of diverse β-lactam structures to chemical reactivity, to distinctive spectroscopic (such as IR carbonyl frequency) or to structural characteristics (such as the out-of-plane displacement of the nitrogen) have been unsuccessful.

A useful contrast between β-lactam structure and solvolytic reactivity is provided by the contrasts between ezetimibe (a clinically-approved inhibitor of cholesterol absorption) and an exploratory class of β-lactamase inhibitors exemplified by the bicyclic 3-benzylidene oxapenem (Figure 3.1). Ezetimibe **1** decomposes in alkaline solution primarily through an unusual intramolecular displacement on the β-lactam by its alkoxide conjugate base.[16] The pH-dependence of this reaction gives a value for the pseudo-first order rate constant for the specific base-dependent hydrolysis of the β-lactam ring at pH 7 (39 °C) of 6×10^{-9} s^{-1} (approx. $t_{1/2}$ of 3 y). This value is typical for simple monocyclic β-lactams. A counterpoint to this excellent aqueous stability is the instability of a recent exploratory class of β-lactamase inhibitors. The β-lactamase enzymes are resistance enzymes found increasingly expressed by notorious bacterial pathogens, and as resistance enzymes they contribute significantly to the diminishing clinical efficacy of past-generation β-lactam antibacterials. The appearance of new generations of β-lactamases, termed carbapenemases, with high activity against the carbapenem antibacterials is an important current clinical concern. The *E*-3-benzylidene oxapenem structure **2** is a powerful and broad-spectrum β-lactamase inhibitor, which could not be clinically developed as a result of its high photochemical (approx. $t_{1/2}$ of 0.3 h) and hydrolytic reactivity (pH 7.4 at 37 °C, approx. $t_{1/2}$ of 3 h).[17] The *Z*-stereoisomer (a poorer β-lactamase inhibitor compared to the *E*-stereoisomer) had comparable photochemical instability but greater hydrolytic reactivity (approx. $t_{1/2}$ of 0.5 h). The β-lactamase inhibitor BLI-489 **3** and the well-known (and clinically important) β-lactamase inhibitor clavulanate **4** both show stability of >50 h in this same hydrolytic assay. These examples illustrate the breadth of β-lactam reactivity, and emphasize a complex relationship between the complete β-lactam structure and overall β-lactam stability.

Nonetheless, any perspective of the β-lactam as a medicinal chemistry motif must consider the β-lactam as both a structural motif, and as an acylating motif when it encounters a biological macromolecule capable of catalysis of this event. It is instructive to recall the mechanistic requirements for acyl transfer of a β-lactam to a nucleophile (Scheme 3.1). As the stability

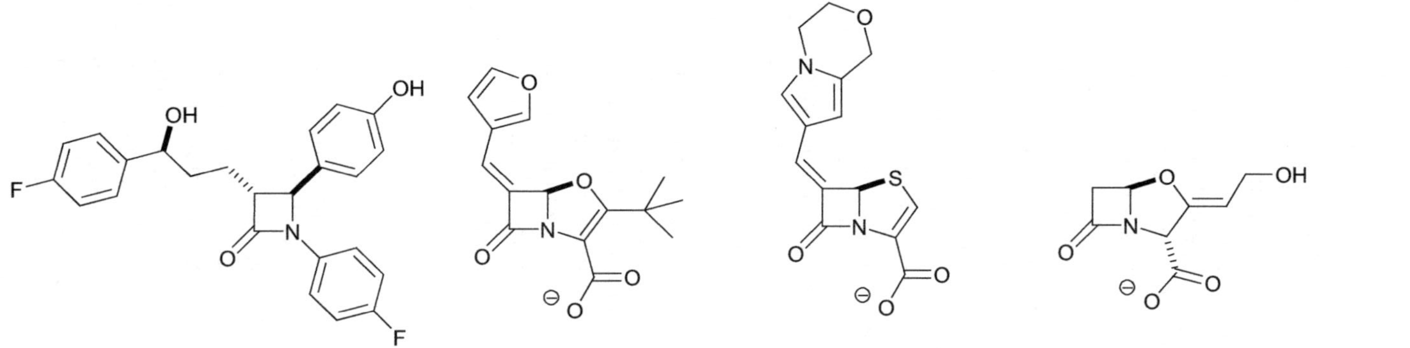

1 [163222-33-1]

2 [1616475-11-6], Na$^+$ Salt [1616528-05-2], Free acid

3 [623564-40-9], Na$^+$ salt [635322-76-8], Free acid

4 [57943-81-4], Na$^+$ Salt [58001-44-8], Free acid

Figure 3.1 The structures of four β-lactams with very different hydrolytic stabilities: ezetimibe **1**, with excellent hydrolytic stability; an exploratory β-lactamase inhibitor **2**, with unacceptably poor (with respect to clinical development) hydrolytic reactivity; and β-lactamase inhibitors **3** and **4**, with acceptable hydrolytic reactivities. Compound **4**, clavulanate, in combination with an antibacterial β-lactam has been used in the clinic for three decades for the treatment of bacterial infections. The CAS Registry Number for each of these structures, and for selected structures in the following schemes, is also given.

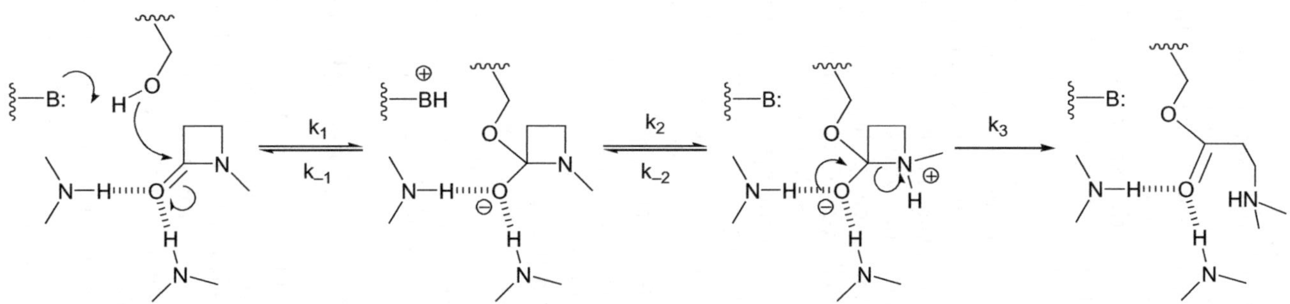

Scheme 3.1 Generalized mechanism for the general base-catalyzed addition of an alcohol to a β-lactam (acyl transfer to the alcohol). This mechanism is used by the antibacterial β-lactams during their inactivation of their penicillin-binding protein targets of bacteria, and in the destruction of the β-lactam as a resistance mechanism used by the β-lactamase enzymes.

of the simple β-lactam is comparable to that of the simple acylic amide (peptide), the catalytic requirements for acyl transfer from each are identical. The carbonyl of the lactam (amide) engages a specific set of hydrogen bonds (the "oxyanion hole")[18] that stabilize the accumulation of negative charge on the carbonyl oxygen. This stabilization allows general-base catalysis for the addition of the nucleophile (here, a primary alcohol, such as is provided by the amino-acid serine) to the carbonyl of the lactam. The rate constant for this addition is k_1. As the collapse of the resulting tetrahedral species to return to the lactam (k_{-1}) is favorable from every vantage (stereoelectronics, enthalpy, entropy), progression toward acyl-transfer requires a proton relay that transfers the proton acquired by the general base, to the newly basic nitrogen of the initial tetrahedral intermediate. This proton relay must occur on a time scale (set by the k_2 rate constant) that is competitive with the k_{-1} rate constant. Neither the design of the oxyanion hole, nor the design of the proton relay, is trivial. The energetically favorable collapse of the zwitterionic tetrahedral, with concomitant opening of the lactam, completes acyl transfer to the nucleophile. Irreversible acylation of the active-site serine of the penicillin-binding proteins by the antibacterial β-lactams is the key mechanistic event in the inactivation of these enzymes; transient acylation of the serine of the active site of the (serine) β-lactamases gives the characteristic "acyl-enzyme" intermediate of these mechanistically related enzymes. As both enzymes are bacterial, one might conclude that incidental β-lactam acylation of eukaryotic proteins is improbable. This conclusion is not justified. The universe of enzymes engaged in amide-bond transfer and hydrolysis is substantial. For this reason, exploratory β-lactams must be regarded as possessing stable structure, yet also be capable of acyl transfer to select biological macromolecules. While the expectation of such dual character (binding to a biological target, metabolism to a "reactive" metabolite) is customary with all drug candidates, activity-based profiling of β-lactams (Section 3.5) emphasizes a particular ability of the β-lactams to acylate a diversity of biological targets.

3.3 Synthesis of the β-Lactams

An enormous variety of synthetic transformations creates the β-lactam ring. Its privileged value is the result not just of its structural stability and its cryptic ability as an acylating agent, but no less importantly as a result of its ease of synthesis. The classical methods (now undergoing continuous refinement with respect to stereochemical optimization) include intramolecular acyl transfer to the nitrogen of a β-aminocarbonyl, and the intermolecular reaction of a ketene with an imine (the Staudinger reaction[19,20]). More recent methods include the intermolecular reaction of a nitrone with a copper acetylide: (the Kinugasa-Hashimoto reaction[21,22]), C–H amidation,[23] and the intramolecular insertion of a β-carbenoid of an N-alkylamide into the C–H bond of the alkyl carbon bonded to the nitrogen.[21] The staggering breadth of these (and other) synthetic methods that yield the β-lactam is beyond the scope of this review. Comprehensive and recent reviews on this topic are available, however.[24–26] The practical application of these methods is exemplified by the key step of the classic synthesis of a penicillin; of the aspartate-derived β-lactam; of the taxoid side chain; of the bicyclic β-lactam (1*S*,5*R*)-6-azabicyclo[3.2.0]hept-3-en-7-one; of recent syntheses of ezetimibe; and of the carbapenems.

3.3.1 The Sheehan and Henery–Logan Synthesis of Protected 6-Aminopenicillanic Acid

In his book, "The Enchanted Ring", Sheehan gives a personal account of the efforts in his group toward the complete synthesis of penicillins (and the ensuing legal entanglements). The realization of this objective was achieved in the late 1950s and described in detail in a full paper published in 1962.[10] The key step in these syntheses was the use of diisopropyl carbodiimide for the intramolecular closure of the suitably protected β-amino acid to give the intact β-lactam of the penicillin in good yield (Scheme 3.2). The selection of *N*-trityl protection for amine protection (in order to suppress the "azlactonization" reaction, and to favor β-lactam formation) was critical to this success. Although β-amino acid closure to the β-lactam remains a classic means for its synthesis, many more recent syntheses use acyl transfer from a β-amino ester to the nitrogen, as exemplified by several of the following examples.

3.3.2 Synthesis of 4-Oxo-2-azetidinecarboxylic Acid from Aspartate

The β-lactam derived from the common β-amino acid aspartate has been used extensively in synthesis (notably of the carbapenems, *vide infra*). A recent synthesis of this β-lactam (nomenclature: 4-oxo-2-azetidinecarboxylic acid) involves bis-benzyl esterification of aspartate, *N*-silyl protection of the primary amine, and base-catalyzed β-lactam formation by intramolecular acyl transfer in 68% yield. Benzyl deprotection by hydrogenolysis gives the *N*-silyl lactam, the enolate of which can be stereoselectively alkylated prior to *N*-desilylation (Scheme 3.3).[27] Reduction of

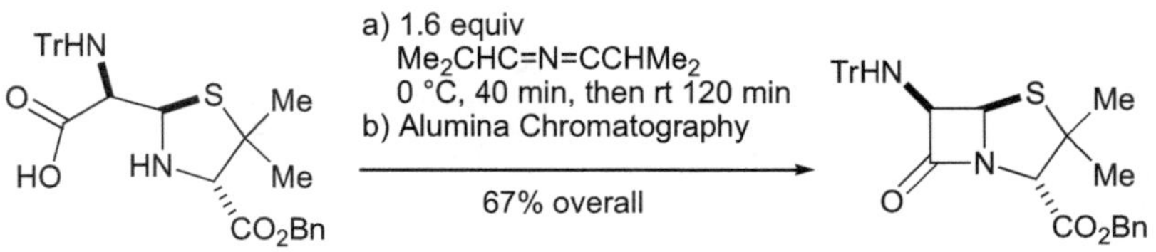

Scheme 3.2 The carbodiimide-mediated closure of a β-amino acid to a β-lactam, as exemplified by the key step of the Sheehan and Henery-Logan synthesis of an intermediate in their synthesis of penicillin.

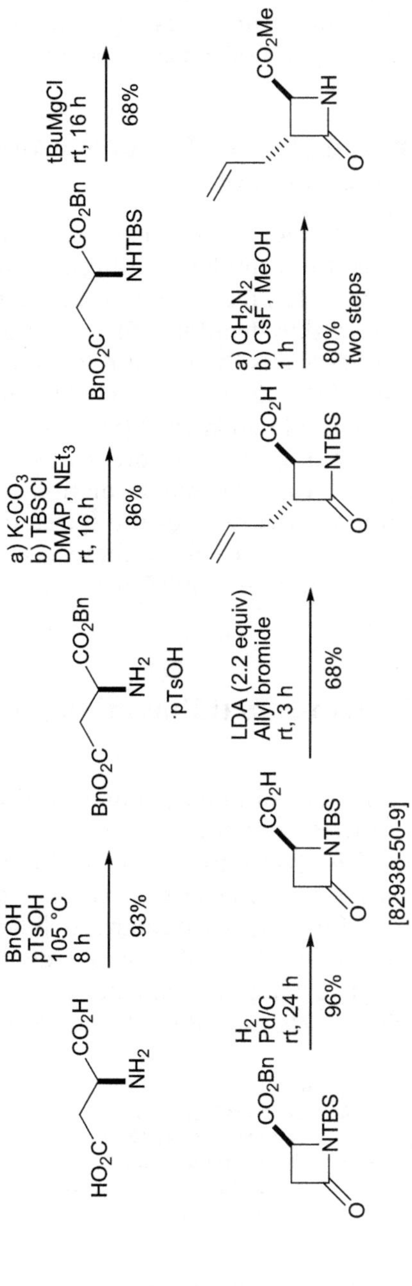

Scheme 3.3 Transformation of *S*-aspartic acid into a useful β-lactam intermediate.

azetidin-2-ones to the azetidine is usually a reliable transformation. In this example, however, exhaustive effort identified $LiAlH_4$ as the best reductant, although only it only achieved moderate yield on a small scale (57%) and yet lower yields on a multi-mmol scale.[27]

3.3.3 Synthesis of the Protected Taxoid Sidechains: (3*R*,4*S*)-3-Hydroxy-2-oxo-1-Azetidinecarboxylic Acid Esters

The discovery and development of Taxol™ for use in cancer chemotherapy is one of the premier medicinal chemistry achievements of the late 20th century. Its success has stimulated extensive medicinal chemistry efforts to identify analogs with improved physicochemical properties and greater efficacy (as exemplified[28,29]). This objective led to process chemistry efforts to reliably control the structure and stereochemistry of the ester segment required (for biological activity) at C-13 of the baccatin core. A robust solution is the "β-lactam synthetic method" developed by Ojima *et al.*[30,31] wherein judicious selection of auxiliaries enables stereospecific ketene-imine (Staudinger) access to appropriately protected (3*R*,4*S*)-3-hydroxy-4-substituted (alkyl, aryl) β-lactams. Imide activation of these β-lactams enables high-yielding acylation by the β-lactam of an alkali metal alkoxide to achieve synthesis of the requisite ester in moderate (50%) to good (>80%) yields, depending on the steric demands of *R* (Scheme 3.4).[5]

3.3.4 Synthetic Application of the 6-Azabicyclo[3.2.0]hept-3-en-7-one Enantiomers

The β-lactam ring has been used for the stereocontrolled synthesis of legions of biologically active β-amino acid derivatives.[32] An instructive example is 6-azabicyclo[3.2.0]hept-3-en-7-one, obtained from the regioselective reaction of cyclopentadiene with chlorosulfonyl isocyanate. Lipase-catalyzed hydrolytic kinetic resolution of this racemic β-lactam, on a multi-gram scale, gives exceptional yield and optical purity for both products (Scheme 3.5). Synthetic manipulations (starting from either product) provide access to both *cis*- and *trans*-cyclic, as well as *syn*- and *anti*-acyclic, β-amino acid derivatives.[33,34]

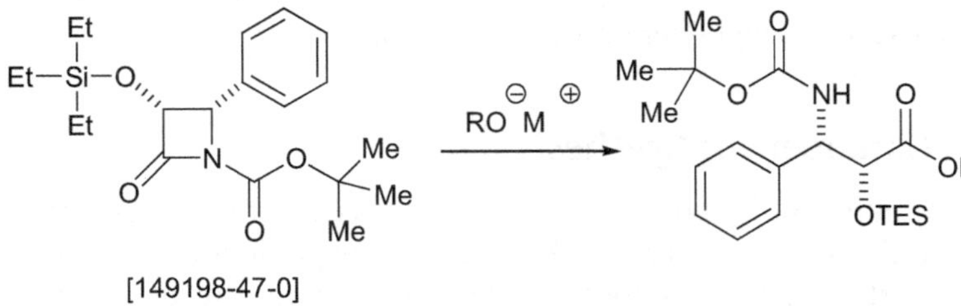

Scheme 3.4 The use of a β-lactam intermediate to control the stereochemistry of the α-hydroxy-β-amino acid segment used to functionalize baccatin to yield biologically active taxoids.

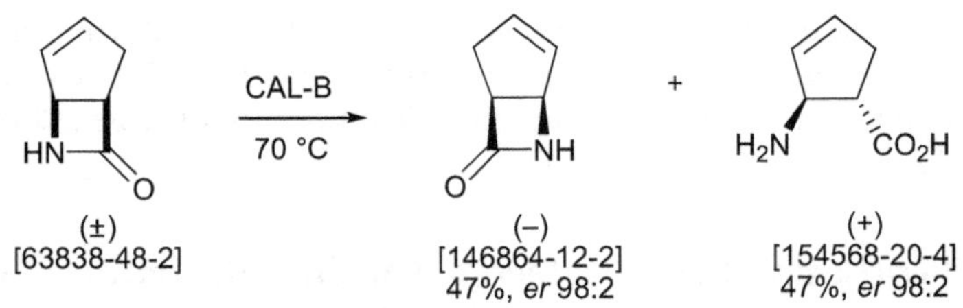

Scheme 3.5 Lipase-catalyzed kinetic resolution of a racemic β-lactam intermediate used especially for the synthesis of nucleoside analogs.

3.3.5 Recent Syntheses of Ezetimibe

Ezetimibe is a β-lactam-containing structure used both as a monotherapy (ZETIA™) and as a combination (with simvastatin) therapy (VYTORIN™) to improve serum lipid profiles. Notwithstanding continuing debate as to its clinical efficacy (*vide infra*), its structure has been the focus of numerous synthetic efforts to set the three stereocenters. Three of these efforts are highlighted.

An imaginative laboratory-scale synthesis of ezetimibe was disclosed by Śniezek *et al.*,[35] wherein the functionalities for the key late-stage assembly of the β-lactam, by intramolecular acyl transfer, were set by an $Sc(OTf)_3$-catalyzed dipolar cycloaddition between a nitrone and a dihydropyranone (Scheme 3.6). A process-scale synthesis addresses the challenge of the stereochemistry of the two stereogenic carbons of the β-lactam of ezetimibe by auxiliary control of a highly diastereoselective Mannich-like synthesis of a β-aminoimide.[36] Subsequent acyl transfer, with release of the chiral auxiliary, gives the β-lactam ring (Scheme 3.7). The final stereocenter (of the benzyl alcohol) is set by asymmetric reduction of the ketone.

Extensive structure-activity studies toward ezetimibe analogs with improved biological activity further exemplify the synthesis of substituted β-lactam rings. An optimized (by slow addition), auxiliary-controlled stereospecific Staudinger ketene-imine cycloaddition realizes an intermediate reminiscent of the key intermediate in the preceding ezetimibe synthesis. Upon warming in the presence of TBAF, intramolecular acyl-transfer gives the β-lactam. A process-scale synthesis using an analogous strategy prepared an ezetimibe analog. Even after optimization, the disclosed yield for the closure to the β-lactam of 52% is disappointing.[37]

3.3.6 Carbapenem Synthesis

The dominant (but by no means exclusive) classes of the β-lactams used in the chemotherapy of bacterial infections are the penicillins, cephalosporins, and the carbapenems. Of these three classes, only the penicillin class is robustly available by means of fermentation. Both other classes derive, by extensive and extraordinary skill and efficiency, from the synthetic chemical manipulation of the penicillins. The key reaction used to transform the

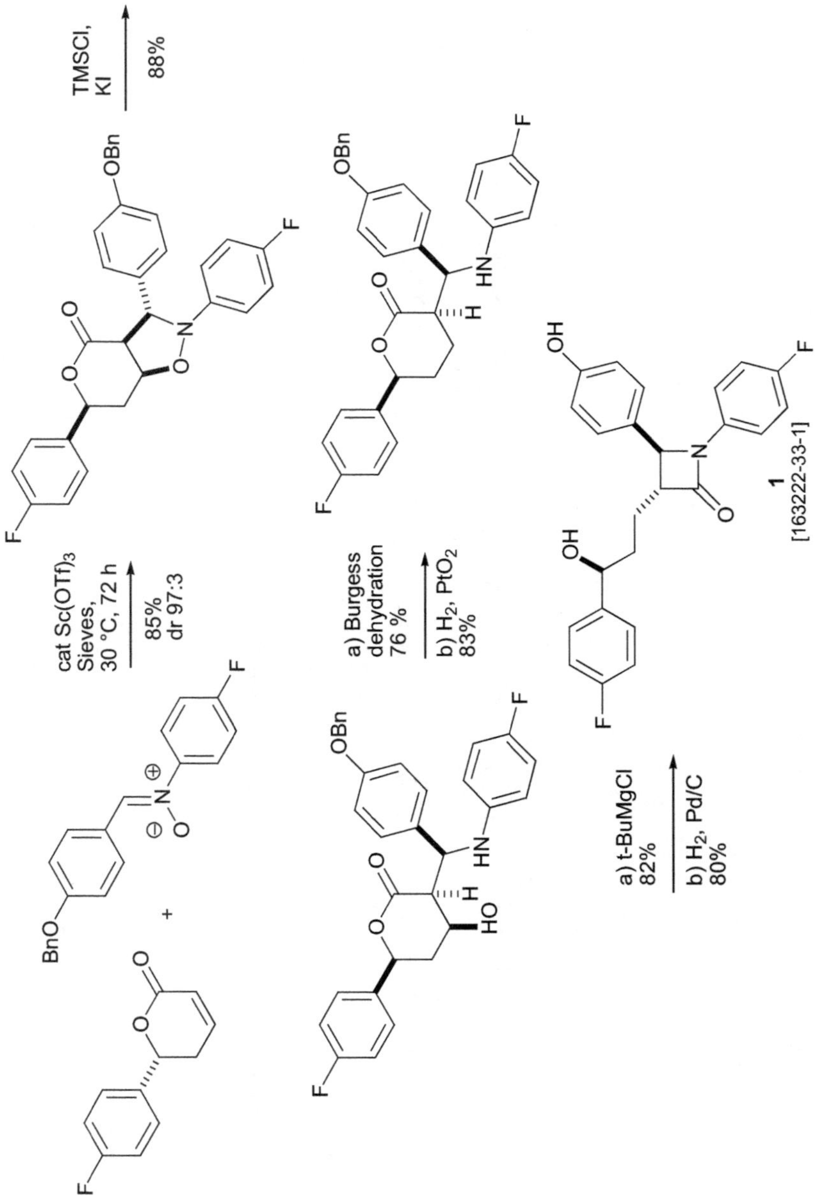

Scheme 3.6 A laboratory-scale synthesis of ezetimibe.[35]

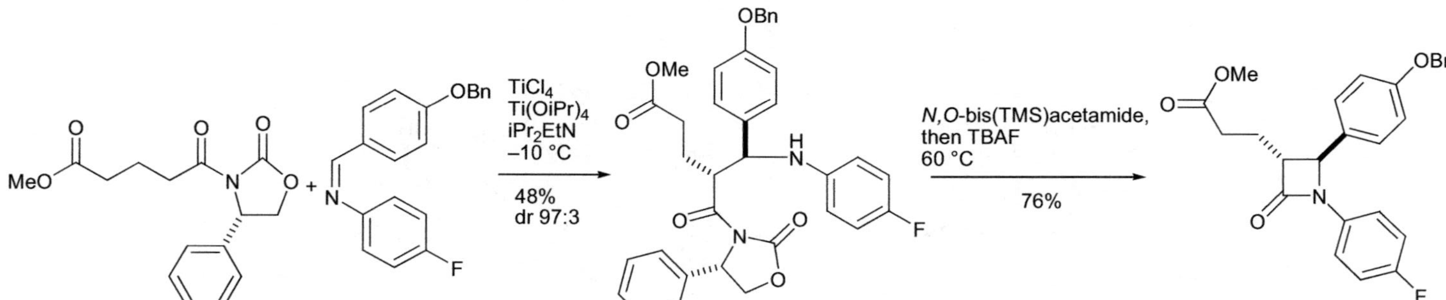

[204589-80-0]

Scheme 3.7 A process-scale synthesis of ezetimibe—key step.[36]

penicillin class to the cephalosporin class is the so-called Morin rearrangement, disclosed in 1963 by the Lilly β-lactam team[38] and studied intensively over the next decade.[39] The Morin rearrangement is the acid-catalyzed, thermal transformation of a penicillin sulfoxide to the cephalosporin (in the original report, in 15% yield). The mechanism[40] of this "abnormal Pummerer" reaction—a mechanism with important parallels to the biosynthesis of the cephalosporins from the penicillins—is shown in Scheme 3.8.

The synthetic chemistry involved in the production of the carbapenems is significantly more complex, as exemplified by the key steps of the 1980 synthesis of thienamycin disclosed by the Merck β-lactam team.[41] Key steps in this synthesis are Grignard-induced closure of *N*-silyl asparate diester followed by homologation to a β-lactam having a β-ketoester sidechain (Scheme 3.9). Diazo transfer to the α-carbon of the β-ketoester followed by catalytic $Rh(OAc)_2$-catalyzed carbenoid insertion into the N–H bond to secure ring-closure. Ketone transformation to the vinylphosphate, followed by amine conjugate addition followed by phosphate elimination gave the protected thienamycin derivative. The strategy exemplified by these final steps has proven durable across decades of structure-activity studies. Exemplifying a refined variation on this strategy is a 2009 synthesis of thienamycin.[42] Following the use of the Lectka asymmetric Staudinger reaction for formation of the β-lactam starting material, a series of moderate-yielding steps (note especially the desired inversion of configuration at C-6 accompanying the $Pb(OAc)_4$ oxidation) gave the key β-lactam intermediate (Scheme 3.10, boxed structure), which was subsequently transformed to thienamycin. This β-lactam is a commercial material and is used extensively in both patent and open literature syntheses of carbapenem and penem structures.[43,44] Numerous different routes converge on this key β-lactam starting material.[45,46]

3.4 Structure of the β-Lactams

In contrast to the well-known pucker of the cyclobutane, the orbital overlap constraint of the amide resonance confers a planar ring structure to the β-lactam, as exemplified by the crystal structure of ezetimibe (Figure 3.2).[35,47] As discussed previously, the solvolytic reactivity of the β-lactam depends greatly on its substitution. In contrast to the planar ring of simple (monocyclic) β-lactams such as ezetimibe, the β-lactam rings of the bicyclic β-lactam antibiotics—exemplified in Figure 3.1 by a penicillin (methicillin methyl ester),[48] a cephalosporin (cefaclor dihydrate),[49] and a carbapenem (biapenem)[50] are not planar. This non-planarity is typically expressed as the distance *h* of the offset of the nitrogen of the β-lactam from the plane defined by the three carbon atoms.[51] The value for this offset in carbapenems, the most non-planar and most reactive of the three, is approximately 0.46 Å.[52] Notwithstanding this distance, the crystal structures of the β-lactam antibiotics of Figure 3.1 demonstrate the subtlety of this deformation. As noted previously, the β-lactam *h* value, while having value for assessing relative β-lactam reactivity[51,53] is not useful guidance for antibacterial design.

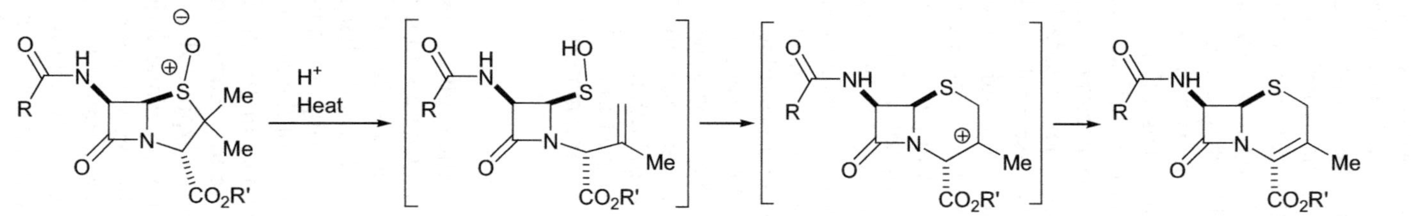

Scheme 3.8 The Morin rearrangement of a penicillin sulfoxide to give a cephem (cephalosporin) product is arguably one of the most important chemical reactions with respect to the preservation of human health.

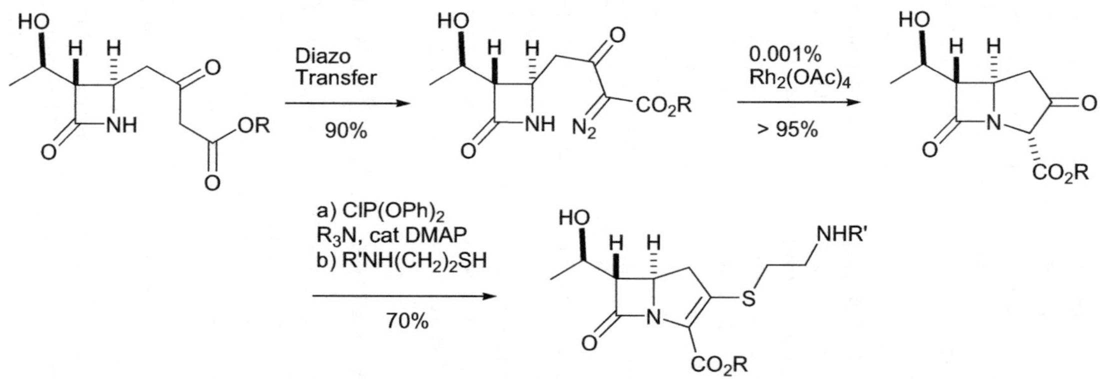

Scheme 3.9 The key ring-closing diazo insertion reaction used in the synthesis of thienamycin (and numerous other carbapenem antibiotics).[41]

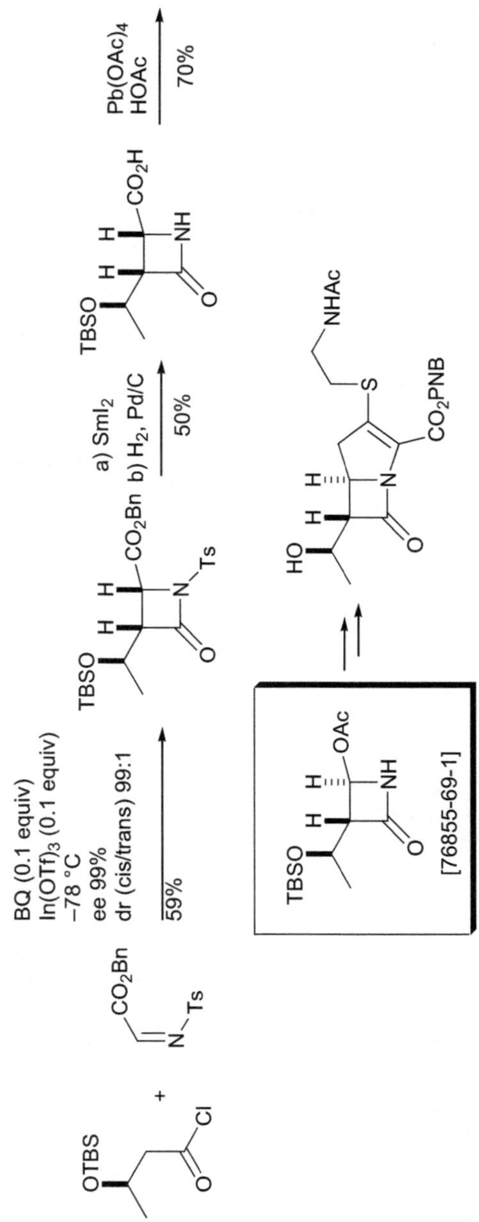

Scheme 3.10 Exemplification of the key β-lactam-forming and transforming steps used to access carbapenem analogs.[42]

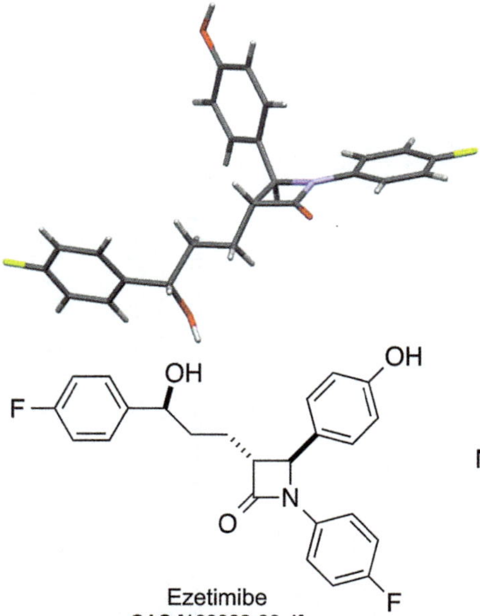

Ezetimibe
CAS [163222-33-1]
(3*R*,4*S*)-1-(4-fluorophenyl)-3-[(3*S*)-3-(4-fluorophenyl)-3-hydroxypropyl]-4-(4-hydroxyphenyl)-, 2-azetidinone

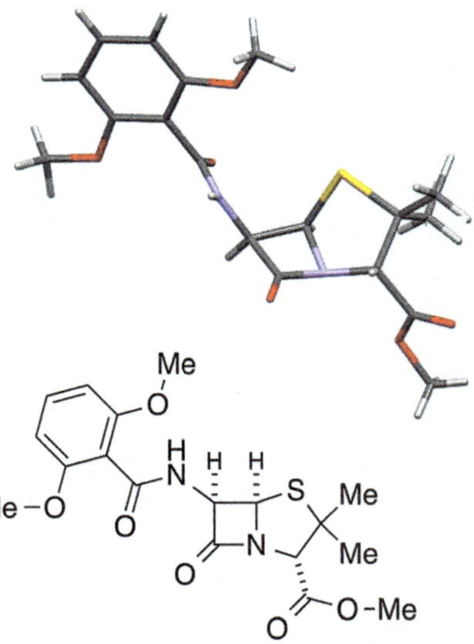

Methicillin, methyl ester
CAS [22668-52-6]
(2*S*,5*R*,6*R*)-6-[(2,6-Dimethoxybenzoyl)amino]-3,3-dimethyl-7-oxo-4-thia-1-azabicyclo[3.2.0]heptane-2-carboxylic acid, methyl ester

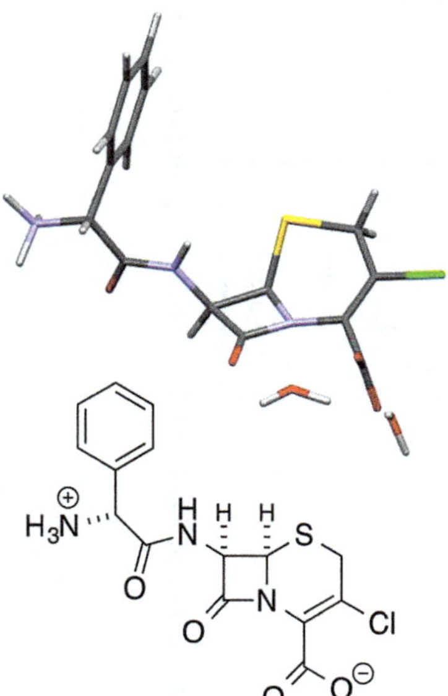

Cefaclor
CAS [53994-73-3]
(6*R*,7*R*)-7-[[(2*R*)-2-amino-2-phenylacetyl]amino]-3-chloro-8-oxo-5-thia-1-azabicyclo[4.2.0]oct-2-ene-2-carboxylic acid, inner salt

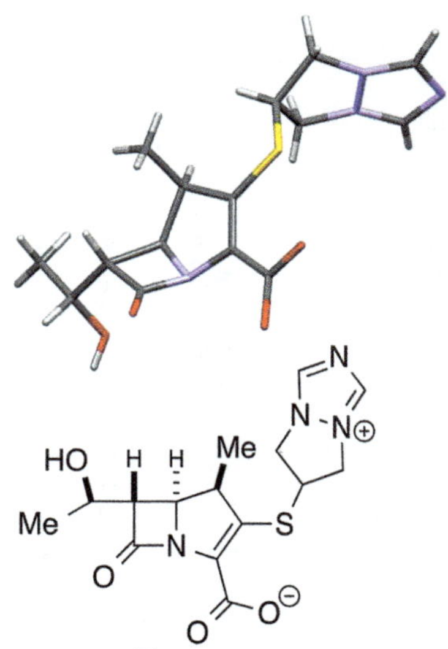

Biapenem
CAS [120410-24-4]
5H-Pyrazolo[1,2-*a*][1,2,4]triazol-4-ium, 6-[[(4*R*,5*S*,6*S*)-2-carboxy-6-[(1*R*)-1-hydroxyethyl]-4-methyl-7-oxo-1-azabicyclo [3.2.0]hept-2-en-3-yl]thio]-6,7-dihydro-, inner salt

3.5 Biological Target Profiling of the β-Lactam

The β-lactam offers an opportunity for target protein recognition either as a stable structural motif or as a latent electrophile. This latter characteristic of the β-lactam (and of the closely related β-lactone and β-sultam ring systems)[54,55] has enabled their use in the activity-based profiling of proteomes.[56,57] In this method, the selected functional group, either tagged with a reporter group (chromophore, fluorophore) or an affinity tag (such as an azide, an alkyne, or biotin), is exposed to the proteome and the proteins labeled are retrieved and identified. Although the concept of activity-based profiling is modern, the methodology is conceptually identical to classic methodologies, as are exemplified by the identification of a family of penicillin-binding proteins (using radioactivity as the reporting method)[58] as the molecular targets of the penicillins.[54]

The driving force for the method is the rapidity (in principle) that target identification can be integrated into the genomic (operon) and proteomic (pathway) character of the targeted proteome. Understandably, studies using β-lactams have focused on bacterial proteomes.[59,60] While the preeminent targets of these studies are the penicillin-binding proteins and the evolutionarily related β-lactamase resistance enzymes, using β-lactam structures complementary to classic antibacterial β-lactam structure identified additional proteins (having nucleophilic serines or cysteines) subject to β-lactam acylation. Specificity among these proteins is assessed as a function of probe concentration and the rapidity of the reaction. Among the notable findings is the β-lactam (and especially β-lactone) reactivity of the ATP-dependent virulence protease ClpP.[61–63] Extension of this methodology to other (non-bacterial) proteomes confirms the similarities between the β-lactam and β-lactone profiles, and identifies a substantial list of proteins undergoing acylation on a serine, cysteine or threonine nucleophile.[64] As many of these proteins exhibit selectivity for the β-lactam (β-lactone) structure and in some cases correlate to a disease state, there is opportunity to exploit the β-lactam as a privileged motif either for diagnosis or therapy.[54,64] Lastly, evidence has been obtained supporting the assertion that the copper-catalyzed Kinugasa synthesis of β-lactams from nitrone and alkyne reaction partners (CuANCR) has the same potential for cell-based target identification as the much more well-known copper-catalyzed azide–alkyne cycloaddition.[21]

Figure 3.2 Crystal structures of four biologically active β-lactams. Ezetimibe (top left), an inhibitor of cholesterol transport (see Section 3.7), exhibits a planar β-lactam ring. The remaining three structures are representative antibacterial β-lactams: a penicillin, methicillin (top right, crystal structure is of its methyl ester); a cephalosporin, cefaclor (bottom left, crystal structure is of a dihydrate); and a carbapenem, biapenem (bottom right). The β-lactam ring of each of these three is non-planar (relative to the plane of the three carbon atoms, the nitrogen of the β-lactam is depressed below the plane). The three carbon atoms bonded to this same nitrogen are more closely planar in the cephalosporin compared to either the penicillin or the carbapenem.

3.6 The Antibacterial β-Lactam

The targets of the antibacterial β-lactam are a family of bacterial enzymes, known as the penicillin-binding proteins (PBPs), which catalyze the biosynthesis and maintenance of the bacterial cell wall. All eubacteria have several PBPs (some, such as the Gram-positive soil bacterium *Bacillus subtilis*, have more than twenty). As the preservation of the integrity of their cell wall is essential, so is the preservation of the catalytic activity of many (but usually not all) of the PBP enzymes. The mechanism of the antibacterial β-lactam is irreversible acylation, with concomitant opening of the β-lactam ring, of the active-site serine of the PBP enzyme. As these enzymes are unique to bacteria, the antibacterial β-lactams have a basis for a wide safety margin for their clinical use. Two structural requirements confer antibacterial activity to the β-lactam: the β-lactam ring itself, and the presence of a negative charge proximal to the nitrogen of the β-lactam (Scheme 3.11, structure **5** where *n* is the number of atoms to the atom with the negative charge). The β-lactam is required for the mechanism-based inactivation of their target, and the negative charge for recognition by the target. In practice, a legion of additional criteria ultimately defines the structure of clinically efficacious β-lactam. These criteria include intrinsic chemical stability, potency, broad-spectrum or narrow-spectrum antibacterial activity, oral or parenteral delivery, selectivity toward the essential PBP enzymes of the bacterium, and stability to an ensemble of resistance enzymes and pathways. Particular emphasis rests upon this latter criterion. A half-century of

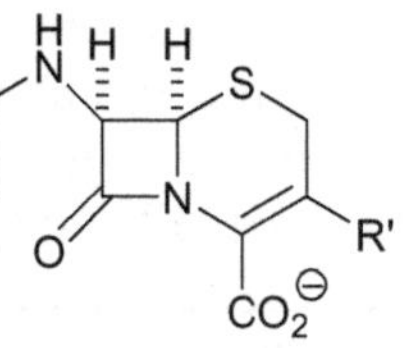

5

The antibacterial β-lactam

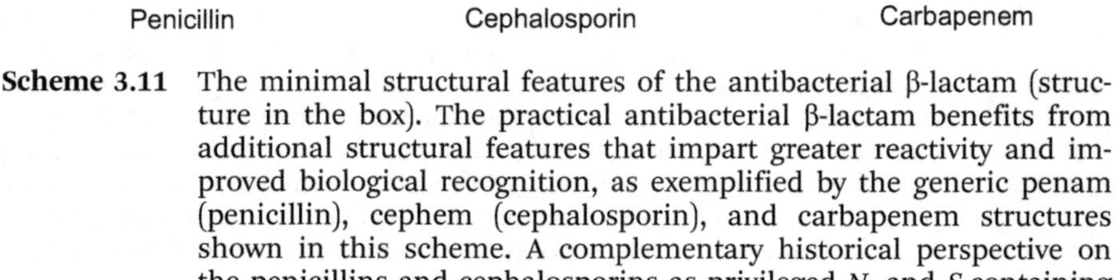

Scheme 3.11 The minimal structural features of the antibacterial β-lactam (structure in the box). The practical antibacterial β-lactam benefits from additional structural features that impart greater reactivity and improved biological recognition, as exemplified by the generic penam (penicillin), cephem (cephalosporin), and carbapenem structures shown in this scheme. A complementary historical perspective on the penicillins and cephalosporins as privileged *N*- and *S*-containing heterocycles is presented in Section 11.2 of Chapter 11.

extensive clinical use of the β-lactams has resulted in the selection, acquisition, and refinement of a host of resistance mechanisms. The primary resistance mechanism in Gram-positive (single membrane) bacteria is target modification (PBPs with reduced susceptibility to the irreversible β-lactam acylation). The primary resistance mechanism in the Gram-negative (dual membrane) bacteria is acquisition of an enzyme that catalyzes the hydrolytic deactivation of the β-lactam. Each of these mechanisms may be further abetted by reduced target access (a thickened cell wall, deletion of the protein "porins" used to define small molecule access to the bacterium), and especially in the Gram-negative bacteria facilitated depletion by active transport of the antibacterial that has penetrated the organism out of the bacterium. Indeed, the same half-century that has seen herculean efforts to define and refine generations of β-lactam structures, has also seen the emergence of extensively antibacterial-resistant (not just to β-lactams) bacteria. A concern today is the possibility that the β-lactams may have no further new structures to present as efficacious against infections caused by these bacteria.[65–71] The concern is real, but the conclusion is premature. There are certainly new generations of β-lactams awaiting discovery, and existing generations of β-lactams capable of preservation, as clinically effective antibiotics. The barrier to both objectives is not chemical intellect but rather an entrenched set of outdated policies juxtaposed against powerful economic disincentives for antibiotic discovery and development.[72] Promising new β-lactams (and β-lactam combinations) are well represented among the ensemble of antibacterial structures in clinical development.[73–75] These structures (and the medicinal chemistry strategies that they exemplify) are the focus of this concise perspective on the privileged antibacterial β-lactam structure.

3.6.1 The Bicyclic β-Lactam Antibacterials

The preeminent (but certainly not exclusive in their importance) sub-families of the antibacterial β-lactams are the penicillins (a penam), the cephalosporin (a cephem), and the carbapenems. These three structures are exemplified in Figure 3.2, and their generic structures are shown in Scheme 3.11. The carbapenem is distinguished from the two earlier (in terms of the timeline for drug development) penicillin and cephalosporin sub-families by its *trans* stereochemistry for the β-lactam substituent relative to the ring juncture, and the non-amide character of this substituent. In all three classes, ring annulation to the β-lactam advantageously enhances the intrinsic reactivity of the β-lactam. Penicillins offer one (primary) point for diversification (represented by the "R"), cephalosporins two points, and carbapenems one point. As would be anticipated, the optimal R groups differ for each sub-family. While the core structure and stereochemistry of each is dictated by the dual criteria of biosynthesis[76–80] and target recognition, the sparser generic antibacterial β-lactam structure of Scheme 3.11 implies that substantially greater variation in these structures should retain antibacterial activity. This implication is verified both by Nature (other natural β-lactam antibiotic structures) and by synthetic exploration. For example, the medicinal chemist easily would conceptualize (among many other possible isostere variations) replacing the sulfur of a cephalosporin with a carbon (carbacephems) atom or an oxygen atom ("oxacephems"), and the carbon atom of a carbapenem with a sulfur atom ("penems"). All possess varying degrees of antibacterial activity. Indeed, structures reflecting continued optimization of the primary "R" sub-structures within especially the cephalosporin and carbapenem sub-families, as well as β-lactam structures outside of these three sub-families, are represented among the β-lactams in current clinical development. An analysis of *N*-heterocycles in medicinal chemistry affirms the central place of the β-lactams as a medicinal chemistry motif, and further summarizes the key "R" variations used across generations of β-lactam structures.[81]

Here, our perspective on the β-lactam as a privileged antibacterial motif is exemplified by six β-lactam structures, presented as three pairs of point and counterpoint structures. The first pair is cefoxitin and temocillin. Neither structure is new. Cefoxitin is a synthetic (prepared by synthetic manipulation of a penicillin starting material) 7α-methoxy-substituted cephalosporin, inspired by the 7α-methoxy substitution of the cephamycin class of antibiotics.[82,83] The 7α-methoxy substituent protects the β-lactam against β-lactamase hydrolysis, while preserving the ability of its β-lactam to inactivate PBPs. Nonetheless, cefoxitin has had limited success as an antibacterial, largely as a consequence of its ability in Gram-negative bacteria (as a result of the selection of the PBPs it inhibits) to induce the expression of β-lactamase resistance enzymes, and other resistance and virulence pathways. The advantageous properties conferred by the methoxy substituent of cefoxitin (β-lactamase stability with antibacterial activity) when this substituent is transferred to the penam (penicillin) scaffold.

Temocillin, a 6α-methoxy penicillin, has had limited success as an antibacterial as a result of its limited spectrum of activity against Gram-negative bacteria (notably, a lack of activity against *Pseudomonas aeruginosa*). This lack of broad-spectrum activity, previously seen as a disadvantage, is now receiving reassessment, in the anticipation that improved diagnostics may allow narrow spectrum antibacterials to be used to counter infection with better sparing of the microbiome. As a result of its β-lactamase stability, and efficacy against the particularly troubling carbapenemase-expressing Gram-negative bacteria of the *Enterobacteriaceae* family, temocillin is being reassessed both as a diagnostic as well as an option for the therapy of infections caused by these bacteria.[84–86]

The second pair in Scheme 3.12 is cefuroxime (a second-generation cephalosporin) and ceftaroline (a current-generation cephalosporin, with broad-spectrum activity, notably including the β-lactam-resistant Gram-positive pathogens *Staphylococcus aureus* and *Streptococcus pneumoniae*).[65,87] Cefuroxime exemplifies a notable functional group pairing with its β-lactam structure: its eponymous oxime functional group, placed immediately

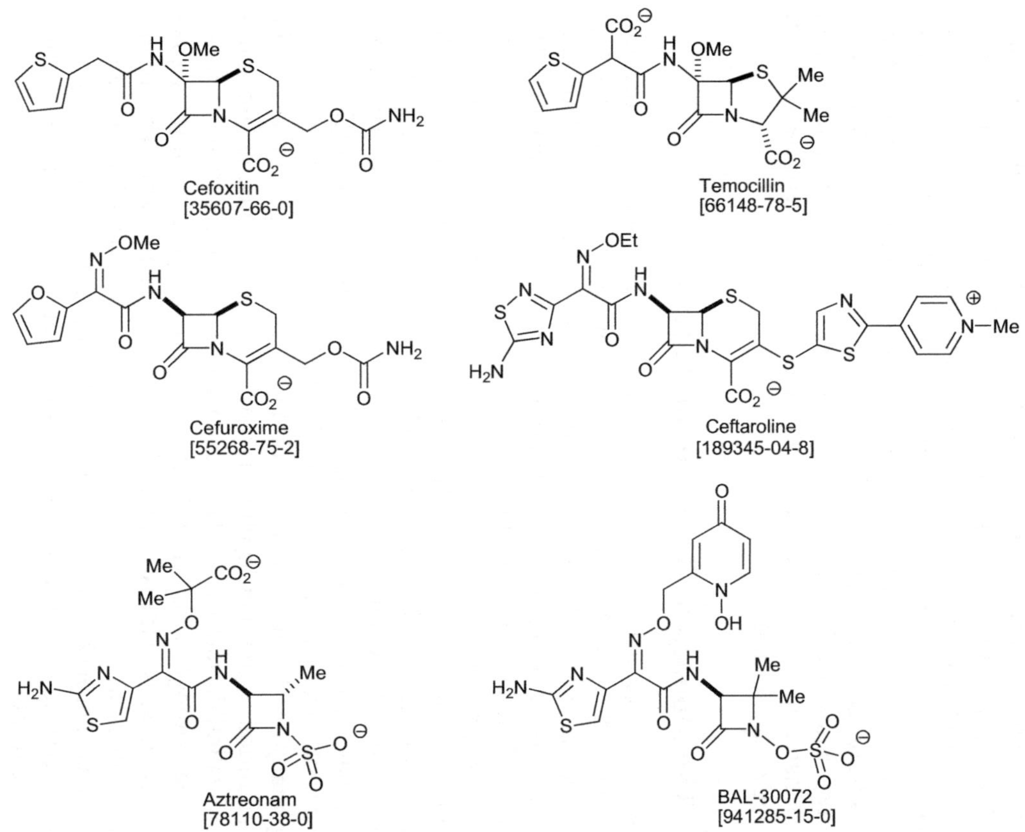

Scheme 3.12 Three pairs of β-lactam structures exemplifying key structure-activity features: the methoxy group of cefoxitin and temocillin; the oxime group of cefuroxime and ceftaroline; and the monocyclic β-lactams aztreonam and BAL-30072. BAL-30072 exemplifies siderophore mimicry that contributes to its advantageous antibacterial activity.

adjacent to the carbonyl of the 7β-amide side chain. This oxime appears, with varying *O*-substitution (as further exemplified in the third pair of structures of Scheme 3.12), across multiple generations of β-lactam structures as a result of its remarkable ability (similar to that of the methoxy group of cefoxitin and temocillin) to confer β-lactamase stability, while preserving the efficacy of PBP inactivation. Ceftaroline is distinguished from the earlier generation cephalosporins by the pyridinium-thiazole biheteroaryl of its right (relative to the perspective shown in Scheme 3.12) side chain. The incorporation of a positively charged heterocycle into this side chain is a universal characteristic of all new generation cephalosporins and carbapenems: in some unknown capacity this positive charge (or alternatively, an overall zwitterionic character to the β-lactam) is a structure-activity advantage. A key mechanistic advantage possessed by ceftaroline is its ability to allosterically predispose its principal penicillin-binding protein target to inactivation (by the customary mechanism of irreversible acylation of the active-site serine).[88–90]

3.6.2 The Monocyclic β-Lactam Antibacterials

The third pair of Scheme 3.12 is the monocyclic β-lactam structures aztreonam (a monobactam, discovered 35 years ago) and BAL-30072 (a monosulfactam, an antibacterial currently in early clinical development). A marked structural departure from the bicyclic β-lactams is evident. The "activating" group is a heteroatom bond to the nitrogen of the β-lactam, and the proximal negative charge is not a carboxylate. Aztreonam is a highly structurally optimized synthetic β-lactam having impressive Gram-negative activity, whose structural conception derives from a class of modestly antibacterial natural β-lactams. Nonetheless, the clinical impact of aztreonam over the past decades was modest. This assessment is changing. The emergence of metallo-β-lactamases as a key β-lactam resistance mechanism among many of the most potent Gram-negative pathogens has sharply drawn attention to the monocyclic β-lactams as markedly poorer substrates of the metallo-β-lactamases.[91] Aztreonam alone is not efficacious against these same pathogens, as these bacteria invariably possess other serine-based β-lactamases with strong activity toward aztreonam. The recent emergence of new serine β-lactamase inhibitor structures (Section 3.6.3), however, has sharply brought attention to the therapeutic promise of combining monocyclic β-lactam antibacterials with one or more of these serine β-lactamase inhibitors.[92–96]

An emerging concept in the design of antibacterial β-lactams is the incorporation into the β-lactam structure of segments that mimic motifs used by the Gram-negative bacterium for nutrient acquisition. The best studied of this "Trojan horse" strategy appropriates the iron-binding segments of the bacterial siderophores.[97–102] BAL-30072, wherein the siderophore-derived segment is incorporated into the oxime, is a structural exemplification.[103] This (mono)sulfactam possesses an advantageous spectrum for target inactivation among the members of the Gram-negative PBP family and synergizes with other β-lactams and in doing so suppresses resistance development.[104,105] A key contributor to this outcome is facilitated internalization of this β-lactam as a result of its siderophore sub-structure. While the longer-term efficacy of the siderophore Trojan horse strategy is uncertain—Gram-negative bacteria can use several siderophore pathways, and altering this balance might be a successful resistance strategy[106]—the initial experience is sufficiently positive as to justify its clinical development.

3.6.3 β-Lactamase Inhibitors

The combination of a β-lactam antibacterial with a serine β-lactamase inhibitor (either clavulanate, or one of the two —sulbactam or tazobactam— penam sulfones) has been a resounding clinical success for the past 30 years. Each of these is a β-lactam (Scheme 3.13). Although each possesses the structural requirements for antibacterial activity, their antibacterial activities are weak. The mechanism used by each is efficient acylation of the serine of the β-lactamase to give an acyl-enzyme that subsequently transforms to a new, and hydrolytically stable, acyl-enzyme species. As a consequence the β-lactamase is inactivated.[107,108] Despite the clinical value of these β-lactam-β-lactamase inhibitor combinations, the continuing emergence of new serine β-lactamases as well as new metallo-β-lactamases, each with a diverse substrate spectrum, has refocused attention on the possible value of alternate β-lactamase inhibitor structures. A criterion in this search is non-β-lactam structure so as to mitigate inactivation by a β-lactamase. Two structural classes have emerged from this extensive effort, both of which are now undergoing evaluation as β-lactam/serine β-lactamase inhibitor combinations for Gram-negative infection.[75] The first class exploits boronic acid complexation of the serine of the β-lactamase as its mechanism. The second class[109,110] is exemplified by the avibactam (formerly NXL104) and MK-7615 structures shown in Scheme 3.13. These two structures have a common bicyclic *N*-sulfonylurea motif that acylates, similar to a β-lactam inactivator, the serine nucleophile of the serine β-lactamases. The acyl-enzyme is sufficiently long-lived as to effectively inactivate these enzymes. Their bicyclic motif represents remarkable β-lactam mimicry. These inhibitors do not inactivate PBPs. Nonetheless, given the common evolutionary heritage of the PBPs and the serine β-lactamases,[111] the intriguing possibility exists that structure-based transformation of these inhibitors into antibacterial structures may be possible.

3.6.4 Non-PBP Targeting by Antibacterial β-Lactam Structures

The substrate used by the PBPs is the D-Ala-D-Ala terminus of the peptide stem attached to the carbohydrate of the bacterial cell wall. Penicillin was conjectured famously as a structural mimetic of this terminus.[112] As the

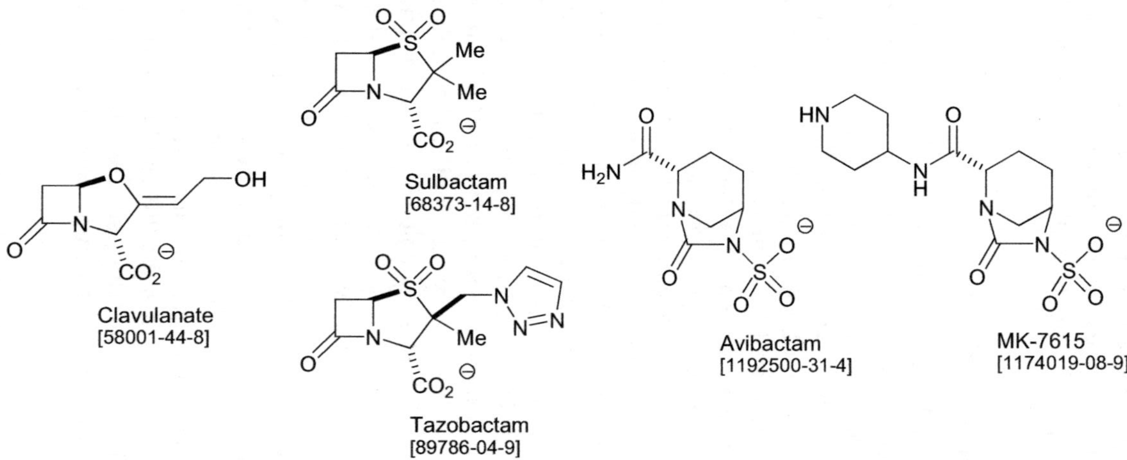

Scheme 3.13 Structures (depicted as the conjugate base) of β-lactam (clavulanate, sulbactam, tazobactam) and non-β-lactam (avibactam and MK-7615) β-lactamase inhibitors.

D-Ala-D-Ala motif is uniquely bacterial, and is recognized by bacterial enzymes other than PBPs, one might anticipate that within the universe of β-lactam structures might be found structure(s) that inhibit other bacterial enzymes. This outcome has happened. Several Gram-positive pathogens (notably, *Enterococcus faecium* and *Mycobacterium tuberculosis*) use a cysteine-dependent L,D-transpeptidase to complete the synthesis of their cell walls. This transpeptidase is inhibited, as a result of acylation of this cysteine, by carbapenems.[113–116] This outcome has led to the suggestion that appropriate β-lactam combination therapy might achieve clinical value against these pathogens.

An unexpected activity discovered for the penem β-lactams is inhibition of the signal peptidase involved in protein translocation from the cytoplasm to the periplasm of Gram-negative bacteria.[117,118] Here as well, the suggestion has been made that combination with PBP-targeting β-lactams could have advantageous antibacterial activity.[119] Turos and colleagues have examined extensively *N*-thiolated β-lactam structures with narrow spectrum Gram-positive antibacterial activity (and, in some cases, anticancer activity) that appear to interfere with the use by the bacterium of Coenzyme A.[120]

3.7 The Non-antibacterial β-Lactam in Medicinal Chemistry

Notwithstanding the depth with which the β-lactam has been explored as a foundational ring for antibacterial activity, and the breadth of β-lactam structures with other biological activities, it is only within the past decade—with the discovery and development of ezetimibe—that the β-lactam has received broader recognition as a scaffold suitable for drug discovery and development. Two explanations contribute to this delayed recognition. The first explanation, dating from the discovery of the penicillins, is the misperception of the β-lactam as a reactive functional group. The allergic reaction that follows (in some individuals) from the covalent modification (haptenation) of serum proteins by penicillins surely contributed to this belief.[121–123] Yet (as we have discussed), the antibacterial β-lactams are a more than usually reactive subset of the β-lactam-containing molecules. The second possible explanation for this perception is the belief that the β-lactam is unusually metabolically reactive. The ability of the antibacterial β-lactams to sustain an appreciable serum concentration through multi-day regimens alone argues against this perception. Every functional group has the potential for metabolic reactivity, and even though some functional groups are surely more predisposed than others, it is always the complete guise of the molecule that determines its metabolism and pharmacokinetics. An example of this phenomenon is provided by the first carbapenem, thienamycin. This carbapenem showed extensive hydrolytic metabolism catalyzed by the dehydropeptidase I enzyme. This liability was addressed by the inclusion with thienamycin of a specific inhibitor of this enzyme (PRIMAXIN™).[124] Subsequent generations of the carbapenems address this liability by introduction of a 1β-methyl (the traditional numbering for penicillins and carbapenems; using systematic numbering the methyl is 4*R*: refer to the structure of biapenem, Figure 3.1) substituent. This methyl substituent suppresses this hydrolytic metabolism, and additionally improves the selection of targets within the PBP family. For these reasons, it is used in all new carbapenem structures. The potential for metabolic reactivity of the β-lactam is real (as we discuss in Section 3.5) but can be improved, or controlled, by classical structure-activity refinement.

Ezetimibe (synthesis discussed in Section 3.3.5) exemplifies the realized potential of the β-lactam as a drug scaffold.[125] The history of this compound is a textbook example[126] of successful medicinal chemistry development of a screening lead through (primarily) the introduction of conformational constraint followed by metabolism-guided structure-activity refinement.[127] A screening effort at Schering-Plough to identify inhibitors of the enzyme acyl-CoA cholesterol acyltransferase (ACAT), as a means of beneficially interfering with cholesterol trafficking, gave a small series of *N*-acyl 1,4-diphenylethan-1-amine leads. The conception of the conformational constrain in terms of aryl-substituted azetidine and azetidinone (β-lactam) structures realized improved pharmacological activity but diminished (or loss of) ACAT inhibition.[128,129]

Subsequent pharmacological study identified ezetimibe as an inhibitor of the transmembrane protein receptor used for both dietary and biliary absorption of cholesterol. A mutational defect in this protein was implicated previously (before its function was known) as causative in Niemann-Pick disease.[130] Ezetimibe was approved ultimately for the treatment of human hyperlipidemia, both as a monotherapy and as a combination with a statin, on the basis of its pharmacological efficacy in serum cholesterol reduction and serum LDL reduction.[125] Until quite recently the human health benefit of ezetimibe therapy was contentious. However, new pharmacological and clinical data affirm this benefit.[131–133]

Several other observations suggest a robust future for ezetimibe-related β-lactam drug discovery. The Nieman-Pick cholesterol receptor is directly contributory to viral (HCV,[134] filovirus[135,136]) invasion of human cells. β-Lactam structures closely related to ezetimibe (exemplified by the β-lactam structure of Scheme 3.14) have nanomolar affinity for the human estrogen receptor, and show antiproliferative activity against a breast cancer cell culture.[137] Moreover, this β-lactam, and related β-lactams conceptualized as conformationally constrained combretastatin analogs,[138] have potent activity as a tubulin depolymerizer. These observations suggest the stilbene (combretastatin) and aryl-β-lactam motifs share the ability to mimic key features used for recognition by steroid-binding receptors.

Numerous other examples of exploratory β-lactam structures addressing a diversity of targets, including a serine (human) leukocyte elastase, thrombin, tryptase β-herpesvirus protease and cysteine (cathepsin) proteases, as well as antiviral and anticancer activities, are described.[139,140] Perhaps the most interesting area for the future non-antibacterial medicinal chemistry of the

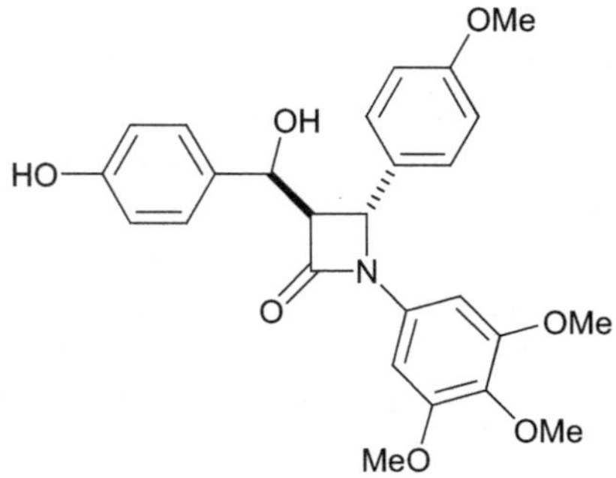

Scheme 3.14 An exploratory β-lactam structure with high affinity for the human estrogen receptor. The similarity of this structure to ezetimibe **1** suggests the diaryl-substituted β-lactam represents a general structural mimetic for biological recognition of steroids.

β-lactams is application to diseases of the central nervous system (CNS). As a result of their extensive clinical use, several antibacterial β-lactams have been observed to have CNS activity. The perhaps most closely identified of these phenomena is an infrequent (1 in 500) incidence of seizure associated with carbapenem chemotherapy.[141] The basis for this observation may relate to a close association between the effect of certain other β-lactams—notably, the cephalosporin ceftriaxone and the β-lactamase inhibitors clavulanate, sulbactam, and tazobactam (Scheme 3.13) - have on the uptake of the neurotransmitter, glutamate.[142–145] These latter β-lactams show a neuroprotective effect as a result of increased expression of the glutamate transporter.[146,147] As the β-lactamase inhibitors have greater blood-brain permeability compared to ceftriaxone, they have emerged as the focus for the study of their possible neuroprotective ability across a breadth of CNS diseases.[148]

3.8 Resurgence of the β-Lactam

The history of the antibacterial β-lactams until this century was one of successive generations of structure, with each generation sharply demarcated from the previous by a structural accomplishment that addressed activity, spectrum, safety, and/or evasion of resistance. This sense of generational transitions is no longer evident. The last recent "glorious" period of β-lactam innovation—the discovery, development, process-scale syntheses, and clinical introduction of clavulanate and sulbactam, thienamycin (carbapenem), cefoxitin (cephamycin), and the oxime-containing cephalosporins—was forty years ago. Subsequent structural innovations in the antibacterial β-lactams now occur more slowly, and are more nuanced. It is not that these nuances have been unimportant. Rather, the circumstances (and requirements) for successful antibacterial drug discovery and development progressively have changed, and virtually all of these changes have made meeting these objectives more difficult.[149,150] Yet the clinical value of the antibacterial to modern medicine is undiminished, and the dissemination and adaptation by bacteria of resistance mechanisms is constant.[72,151,152] A β-lactam resurgence is surely needed. The transition of the search for new antibiotics from nature by major pharma to academia has not diminished the excitement of the search.[153–156] Likewise, many of the potentially new generation-establishing discoveries (such as siderophore-mediated β-lactam delivery and DBO combination chemotherapy) are no longer happening in major pharma, but in academia and biotechnology. Here, too, exciting discoveries are being made. It is not at all evident, however, that we may expect that the future translation of these discoveries to major pharma, for clinical development, will be routine. Both the value of the exclusivity of the patent,[157] and the expertise required for successful clinical design and implementation, are diminishing.[158] The antibacterial future—for β-lactams and for all other antibacterial structures—is not secure.

In contrast, the resurgent value of the β-lactam (as well as other small rings) to the broader realm of medicinal chemistry is sharply evident. For example, one can easily profile a proteome for β-lactone (or for β-lactam or β-sultam) reactivity.[54,55,159] A β-lactone will always be a challenging substructure for drug development (as exemplified by salinosporamide, whose β-lactone is central to the anticancer activity of this proteasome-inactivating natural product). One may now credibly inquire, however, whether replacement of the β-lactone of an intriguing molecule (such as salinosporamide) by the more stable β-lactam will retain the biological activity.[160,161] We have reached an era where the appearance of intriguing biological activity in a small ring heterocycle—as occurred once for the penicillins and more recently for the 1,2-diazetidin-3-ones ("aza-β-lactam" inhibition of protein phosphatase methylesterase)[162,163] will no longer be a matter of astonishment. This deeper understanding of the β-lactam is the basis for a broad-based resurgence of interest in its medicinal chemistry.

References

1. A. K. Bose, M. S. Manhas, B. K. Banik and V. Srirajan, β-Lactams: Cyclic Amides of Distinction, in *The Amide Linkage*, ed. A. Greenberg, C. M. Breneman and J. F. Liebman, John Wiley & Sons, New York, 2000, pp. 157–214.
2. J. C. Sheehan, *The Enchanted Ring*, MIT Press, Cambridge, 1982.
3. B. K. Banik, *Topics in Heterocyclic Chemistry (Heterocyclic Scaffolds I: β-Lactams)*, Springer-Verlag, Berlin Heidelberg, 2010.
4. P. D. Mehta, N. P. S. Sengar and A. K. Pathak, *Eur. J. Med. Chem.*, 2010, **45**, 5541.
5. A. Kamath and I. Ojima, *Tetrahedron*, 2012, **68**, 10640.
6. N. Arya, A. Y. Jagdale, T. A. Patil, S. S. Yeramwar, S. S. Holikatti, J. Dwivedi, C. J. Shishoo and K. S. Jain, *Eur. J. Med. Chem.*, 2014, **74**, 619.
7. M. I. Konaklieva, *Antibiotics*, 2014, **3**, 128.
8. R. Bentley, *J. Chem. Educ.*, 2004, **81**, 1462.

9. S. A. Testero, J. F. Fisher and S. Mobashery, *Burger's Medicinal Chemistry, Drug Discovery and Development*, 7th edn, 2010, vol. 7, p. 259.
10. J. C. Sheehan and K. R. Henery-Logan, *J. Am. Chem. Soc.*, 1962, **84**, 2983.
11. M. I. Page, *Curr. Pharm. Des.*, 1999, **5**, 895.
12. M. I. Page, *Amino Acids, Pept. Proteins Org. Chem.*, 2011, **3**, 169.
13. M. Szostak and J. Aubé, *Chem. Rev.*, 2013, **113**, 5701.
14. D. Kahne and W. C. Still, *J. Am. Chem. Soc.*, 1988, **110**, 7529.
15. S. A. Glover and A. A. Rosser, *J. Org. Chem.*, 2012, **77**, 5492.
16. J. Batova, A. Imramovsky, J. HajIcek, L. Hejtmankova and J. Hanusek, *J. Pharm. Sci.*, 2014, **103**, 2240.
17. M. D. Miller, M. Kale, K. Reddy, S. Tentarelli, M. Zambrowski, M. Zhang, T. Palmer, J. Breen, S. Lahiri, P. S. Shirude and J. C. Verheijen, *ACS Med. Chem. Lett.*, 2014, **5**, 915.
18. L. Simón and J. M. Goodman, *Org. Biomol. Chem.*, 2012, **10**, 1905.
19. D. H. Paull, A. Weatherwax and T. Lectka, *Tetrahedron*, 2009, **65**, 6771.
20. R. Tuba, *Org. Biomol. Chem.*, 2013, **11**, 5976.
21. B. Alcaide, P. Almendros and A. Luna, *RSC Adv.*, 2014, **4**, 1689.
22. S. Stecko, B. Furman and M. Chmielewski, *Tetrahedron*, 2014, **70**, 7817.
23. Q. Zhang, K. Chen and B.-F. Shi, *Synlett*, 2014, **25**, 1941.
24. J. Xu, *Tetrahedron*, 2012, **68**, 10696.
25. C. R. Pitts and T. Lectka, *Chem. Rev.*, 2014, **114**, 7930.
26. P. A. Magriotis, *Eur. J. Org. Chem.*, 2014, **2014**, 2647.
27. A. Soicke, C. Reuter, M. Winter, J.-M. Neudörfl, N. Schlörer, R. Kühne and H.-G. Schmalz, *Eur. J. Org. Chem.*, 2014, **2014**, 6467.
28. K. C. Nicolaou and R. A. Valiulin, *Org. Biomol. Chem.*, 2013, **11**, 4154.
29. I. Ojima, K. Kumar, D. Awasthi and J. G. Vineberg, *Bioorg. Med. Chem.*, 2014, **22**(18), 5060.
30. I. Ojima, C. M. Sun, Y. H. Zucco Park, O. Duclos and S. Kuduk, *Tetrahedron Lett.*, 1993, **34**, 4149.
31. I. Ojima, *Acc. Chem. Res.*, 1995, **28**, 383.
32. L. Kiss and F. Fülöp, *Chem. Rev.*, 2014, **114**, 1116.
33. M. Cherepanova, L. Kiss, E. Forró and F. Fülöp, *Eur. J. Org. Chem.*, 2014, **2014**, 403–409.
34. L. Kiss, M. Nonn, E. Forró, R. Sillanpää, S. Fustero and F. Fülöp, *Eur. J. Org. Chem.*, 2014, **2014**, 4070.
35. M. Śniezek, S. Stecko, I. Panfil, B. Furman and M. Chmielewski, *J. Org. Chem.*, 2013, **78**, 7048.
36. C. H. V. Sasikala, P. Reddy Padi, V. Sunkara, P. Ramayya, P. K. Dubey, V. B. R. Bhaskar and C. Praveen, *Org. Process Res. Dev.*, 2009, **13**, 907.
37. S. Karlsson and J. H. Sörensen, *Org. Process Res. Dev.*, 2012, **16**, 586.
38. R. B. Morin, B. G. Jackson, R. A. Mueller, E. R. Lavagnino, W. B. Scanlon and S. L. Andrews, *J. Am. Chem. Soc.*, 1963, **85**, 1896.
39. R. D. G. Cooper and D. O. Spry, Rearrangements of Cephalosporins and Penicillins, in *Cephalosporins and Penicillins: Chemistry and Biology*, ed. E. H. Flynn, Academic Press, New York, 1972, pp. 183–254.

40. J. D. Freed, D. J. Hart and N. A. Magomedov, *J. Org. Chem.*, 2001, **66**, 839.
41. T. N. Salzmann, R. W. Ratcliffe, B. G. Christensen and F. A. Bouffard, *J. Am. Chem. Soc.*, 1980, **102**, 6161.
42. M. J. Bodner, R. M. Phelan and C. A. Townsend, *Org. Lett.*, 2009, **11**, 3606.
43. S. J. Brenek, S. Caron, E. Chisowa, R. Colon-Cruz, M. P. Delude, M. T. Drexler, R. E. Handfield, B. P. Jones, D. V. Nadkarni, J. D. Nelson, M. Olivier, R. M. Weekly, G. C. A. Bellinger, Z. Brkic, N. Choi, J. Desneves, M. A.-P. Lee, W. Pearce and J. K. Watson, *Org. Process Res. Dev.*, 2012, **16**, 1338.
44. S. J. Brenek, S. Caron, E. Chisowa, M. P. Delude, M. T. Drexler, M. D. Ewing, R. E. Handfield, N. D. Ide, D. V. Nadkarni, J. D. Nelson, M. Olivier, H. H. Perfect, J. E. Phillips, J. J. Teixeira, R. M. Weekly and J. P. Zelina, *Org. Process Res. Dev.*, 2012, **16**, 1348.
45. B. Long and J. Xiang, *Synth. Commun.*, 2009, **39**, 4019.
46. B. Grzeszczyk, S. Stecko, L. Mucha, O. Staszewska-Krajewska, M. Chmielewski and B. Furman, *J. Antibiot.*, 2013, **66**, 161.
47. T. Dražić, K. Molčanov, V. Sachdev, M. Malnar, S. Hećimović, J. V. Patankar, S. Obrowsky, S. Levak-Frank, I. Habuš and D. Kratky, *Eur. J. Med. Chem.*, 2014, **87**, 722.
48. P. Blanpain, M. Melebeck and F. Durant, *Acta Crystallogr., Sect. B: Struct. Crystallogr. Cryst. Chem.*, 1977, **33**, 580.
49. J. van de Streek, J. Rantanen and A. D. Bond, *Acta Crystallogr., Sect. C: Cryst. Struct. Commun.*, 2013, **69**, 1229.
50. T. Kumagai, S. Tamai, T. Abe, H. Matsunaga, K. Hayashi, I. Kishi, M. Shiro and Y. Nagao, *J. Org. Chem.*, 1998, **63**, 8145.
51. R. B. Woodward, *Philos. Trans. R. Soc., B*, 1980, **289**, 239.
52. K. Yanagi, Y. Takeuchi and M. Sunagawa, *Acta Crystallogr., Sect. C: Cryst. Struct. Commun.*, 1992, **48**, 1737.
53. A. Nangia and G. R. Desiraju, *J. Mol. Struct.*, 1999, **474**, 65.
54. T. Bottcher and S. A. Sieber, *Med. Chem. Commun.*, 2012, **3**, 408.
55. R. Kolb, N. C. Bach and S. A. Sieber, *Chem. Commun.*, 2013, **50**, 427.
56. T. Bottcher, M. Pitscheider and S. A. Sieber, *Angew. Chem., Int. Ed.*, 2010, **49**, 2680.
57. M. J. Niphakis and B. F. Cravatt, *Annu. Rev. Biochem.*, 2014, **83**, 341.
58. N. Kresge, R. D. Simoni and R. L. Hill, *J. Biol. Chem.*, 2007, **282**, e25.
59. I. Staub and S. A. Sieber, *J. Am. Chem. Soc.*, 2008, **130**, 13400.
60. A. W. Puri and M. Bogyo, *Biochemistry*, 2013, **52**, 5985.
61. I. Staub and S. A. Sieber, *J. Am. Chem. Soc.*, 2009, **131**, 6271.
62. E. Zeiler, V. S. Korotkov, K. Lorenz-Baath, T. Bottcher and S. A. Sieber, *Bioorg. Med. Chem.*, 2012, **20**, 583.
63. C. L. Compton, K. R. Schmitz, R. T. Sauer and J. K. Sello, *ACS Chem. Biol.*, 2013, **8**, 2669.
64. N. Nasheri, C. S. McKay, K. Fulton, S. Twine, M. H. Powdrill, A. R. Sherratt and J. P. Pezacki, *ChemBioChem*, 2014, **15**, 2195.

65. K. Bush, *Curr. Opin. Pharmacol.*, 2012, **12**, 527.
66. M. G. P. Page and J. Heim, *IDrugs*, 2009, **12**, 561.
67. L. I. Llarrull, S. A. Testero, J. F. Fisher and S. Mobashery, *Curr. Opin. Microbiol.*, 2010, **13**, 551.
68. G. H. Talbot, *Ann. N. Y. Acad. Sci.*, 2013, **1277**, 76–83.
69. R. J. Worthington and C. Melander, *J. Org. Chem.*, 2013, **78**, 4207.
70. E. Leemans, J. F. Fisher and S. Mobashery, in *Antimicrobials: New and Old Molecules in the Fight Against Multi-resistant Bacteria*, ed. F. Marinelli and O. Genilloud, Springer-Verlag, Berlin Heidelberg, 2014, pp. 59–84.
71. W. Qin, M. Panunzio and S. Biondi, *Antibiotics*, 2014, **3**, 193.
72. M. Metz and D. M. Shlaes, *Antimicrob. Agents Chemother.*, 2014, **58**, 4253.
73. M. S. Butler, M. A. Blaskovich and M. A. Cooper, *J. Antibiot.*, 2013, **66**, 571.
74. Y. He, J. Lawrence, C. Liu and N. Yin, *Annu. Rep. Med. Chem.*, 2014, **49**, 209.
75. M. G. P. Page and K. Bush, *Curr. Opin. Pharmacol.*, 2014, **18**, 91.
76. K. Tahlan and S. E. Jensen, *J. Antibiot.*, 2013, **66**, 401.
77. R. B. Hamed, J. R. Gomez-Castellanos, L. Henry, C. Ducho, M. A. McDonough and C. J. Schofield, *Nat. Prod. Rep.*, 2013, **30**, 21.
78. N. M. Gaudelli and C. A. Townsend, *Nat. Chem. Biol.*, 2014, **10**, 251.
79. R. Cox, *Nat. Prod. Rep.*, 2014, **31**, 1405.
80. W. C. Chang, Y. Guo, C. Wang, S. E. Butch, A. C. Rosenzweig, A. K. Boal, C. Krebs and J. M. Bollinger Jr, *Science*, 2014, **343**, 1140.
81. E. Vitaku, D. T. Smith and J. T. Njardarson, *J. Med. Chem.*, 2014, **57**, 10257.
82. R. Nagarajan, L. D. Boeck, M. Gorman, R. L. Hamill, C. E. Higgens, M. M. Hoehn, W. M. Stark and J. G. Whitney, *J. Am. Chem. Soc.*, 1971, **93**, 2308.
83. S. Karady, S. H. Pines, L. M. Weinstock, F. E. Roberts, G. S. Brenner, A. M. Hoinowski, T. Y. Cheng and M. Sletzinger, *J. Am. Chem. Soc.*, 1972, **94**, 1410.
84. H. Giamarellou, *Clin. Microbiol. Infect.*, 2008, **14**(Suppl. 1), 194.
85. D. van Duin, K. S. Kaye, E. A. Neuner and R. A. Bonomo, *Diagn. Microbiol. Infect. Dis.*, 2013, **75**, 115.
86. N. Woodford, R. Pike, D. Meunier, R. Loy, R. Hill and K. L. Hopkins, *J. Antimicrob. Chemother.*, 2014, **69**, 564.
87. T. E. Long and J. T. Williams, *Expert Opin. Invest. Drugs*, 2014, **23**, 1375.
88. L. H. Otero, A. Rojas-Altuve, L. I. Llarrull, C. Carrasco-López, M. Kumarasiri, E. Lastochkin, J. Fishovitz, M. Dawley, D. Hesek, M. Lee, J. W. Johnson, J. F. Fisher, M. Chang, S. Mobashery and J. A. Hermoso, *Proc. Natl. Acad. Sci. U. S. A.*, 2013, **110**, 16808.
89. J. Fishovitz, A. Rojas-Altuve, L. H. Otero, M. Dawley, C. Carrasco-Lopez, M. Chang, J. A. Hermoso and S. Mobashery, *J. Am. Chem. Soc.*, 2014, **136**, 9814.

90. J. Fishovitz, J. A. Hermoso, M. Chang and S. Mobashery, *IUBMB Life*, 2014, **66**, 572.
91. P. Nordmann, L. Poirel, M. A. Toleman and T. R. Walsh, *J. Antimicrob. Chemother.*, 2011, **66**, 689.
92. M. E. Falagas, D. E. Karageorgopoulos and P. Nordmann, *Future Microbiol.*, 2011, **6**, 653.
93. D. M. Livermore, S. Mushtaq, M. Warner, J. Zhang, S. Maharjan, M. Doumith and N. Woodford, *Antimicrob. Agents Chemother.*, 2011, **55**, 390.
94. S. Shakil, E. I. Azhar, S. Tabrez, M. A. Kamal, N. R. Jabir, A. M. Abuzenadah, G. A. Damanhouri and Q. Alam, *J. Chemother.*, 2011, **23**, 263.
95. B. Shen, Y. Yu, H. Chen, X. Cao, X. Lao, Y. Fang, Y. Shi, J. Chen and H. Zheng, *PLoS One*, 2013, **8**, e62955.
96. X. Wang, F. Zhang, C. Zhao, Z. Wang, W. W. Nichols, R. Testa, H. Li, H. Chen, W. He, Q. Wang and H. Wang, *Antimicrob. Agents Chemother.*, 2014, **58**, 1774.
97. A. Górska, A. Sloderbach and M. P. Marszall, *Trends Pharmacol. Sci.*, 2014, **35**(9), 442.
98. S. Han, R. P. Zaniewski, E. S. Marr, B. M. Lacey, A. P. Tomaras, A. Evdokimov, J. R. Miller and V. Shanmugasundaram, *Proc. Natl. Acad. Sci. U. S. A.*, 2010, **107**, 22002.
99. G. L. A. Mislin and I. J. Schalk, *Metallomics*, 2014, **6**, 408.
100. C. Ji, P. A. Miller and M. J. Miller, *J. Am. Chem. Soc.*, 2012, **134**, 9898.
101. M. G. P. Page, *Ann. N. Y. Acad. Sci.*, 2013, **1277**, 115–126.
102. T. A. Wencewicz, T. E. Long, U. Mollmann and M. J. Miller, *Bioconjugate. Chem.*, 2013, **24**, 473.
103. M. G. P. Page, C. Dantier and E. Desarbre, *Antimicrob. Agents Chemother.*, 2010, **54**, 2291.
104. B. Hofer, C. Dantier, K. Gebhardt, E. Desarbre, A. Schmitt-Hoffmann and M. G. P. Page, *J. Antimicrob. Chemother.*, 2013, **68**, 1120.
105. M. Hornsey, L. Phee, W. Stubbings and D. W. Wareham, *Int. J. Antimicrob. Agents*, 2013, **42**, 343.
106. C. van Delden, M. G. P. Page and T. Köhler, *Antimicrob. Agents Chemother.*, 2013, **57**, 2095.
107. S. M. Drawz, K. M. Papp-Wallace and R. A. Bonomo, *Antimicrob. Agents Chemother.*, 2014, **58**, 1835.
108. R. R. Watkins, K. M. Papp-Wallace, S. M. Drawz and R. A. Bonomo, *Front. Microbiol.*, 2013, **4**, 392.
109. T. A. Blizzard, H. Chen, S. Kim, J. Wu, R. Bodner, C. Gude, J. Imbriglio, K. Young, Y. W. Park, A. Ogawa, S. Raghoobar, N. Hairston, R. E. Painter, D. Wisniewski, G. Scapin, P. Fitzgerald, N. Sharma, J. Lu, S. Ha, J. Hermes and M. L. Hammond, *Bioorg. Med. Chem. Lett.*, 2014, **24**, 780.
110. S. D. Lahiri, M. R. Johnstone, P. L. Ross, R. E. McLaughlin, N. B. Olivier and R. A. Alm, *Antimicrob. Agents Chemother.*, 2014, **58**, 5704.

111. I. Massova and S. Mobashery, *Antimicrob. Agents Chemother.*, 1998, **42**, 1.
112. D. J. Tipper and J. L. Strominger, *Proc. Natl. Acad. Sci. U. S. A.*, 1965, **54**, 1133.
113. J. L. Mainardi, J. E. Hugonnet, F. Rusconi, M. Fourgeaud, L. Dubost, A. N. Moumi, V. Delfosse, C. Mayer, L. Gutmann, L. B. Rice and M. Arthur, *J. Biol. Chem.*, 2007, **282**, 30414.
114. V. Dubée, S. Triboulet, J. L. Mainardi, M. Etheve-Quelquejeu, L. Gutmann, A. Marie, L. Dubost, J. E. Hugonnet and M. Arthur, *Antimicrob. Agents Chemother.*, 2012, **56**, 4189.
115. H. S. Kim, J. Kim, H. N. Im, J. Y. Yoon, D. R. An, H. J. Yoon, J. Y. Kim, H. K. Min, S. J. Kim, J. Y. Lee, B. W. Han and S. W. Suh, *Acta Crystallogr., Sect. D: Biol. Crystallogr.*, 2013, **69**, 420.
116. L. Lecoq, V. Dubée, S. Triboulet, C. Bougault, J. E. Hugonnet, M. Arthur and J. P. Simorre, *ACS Chem. Biol.*, 2013, **8**, 1140.
117. M. Paetzel, R. E. Dalbey and N. C. Strynadka, *Nature*, 1998, **396**, 186.
118. D. A. Harris, M. E. Powers and F. E. Romesberg, *Bioorg. Med. Chem. Lett.*, 2009, **19**, 3787.
119. A. G. Therien, J. L. Huber, K. E. Wilson, P. Beaulieu, A. Caron, D. Claveau, K. Deschamps, R. G. Donald, A. M. Galgoci, M. Gallant, X. Gu, N. J. Kevin, J. Lafleur, P. S. Leavitt, C. Lebeau-Jacob, S. S. Lee, M. M. Lin, A. A. Michels, A. M. Ogawa, R. E. Painter, C. A. Parish, Y. W. Park, L. L. Benton-Perdomo, M. Petcu, J. W. Phillips, M. A. Powles, K. I. Skorey, J. Tam, C. M. Tan, K. Young, S. Wong, S. T. Waddell and L. Miesel, *Antimicrob. Agents Chemother.*, 2012, **56**, 4662.
120. B. Bhattacharya and E. Turos, *Tetrahedron*, 2012, **68**, 10665.
121. A. Ariza, D. Garzon, D. R. Abanades, V. de los Rios, G. Vistoli, M. J. Torres, M. Carini, G. Aldini and D. Pérez-Sala, *J. Proteomics*, 2012, **77**, 504.
122. D. Garzon, A. Ariza, L. Regazzoni, R. Clerici, A. Altomare, F. R. Sirtori, M. Carini, M. J. Torres, D. Pérez-Sala and G. Aldini, *Chem. Res. Toxicol.*, 2014, **27**, 1566.
123. R. E. Jenkins, F. S. Yaseen, M. M. Monshi, P. Whitaker, X. Meng, J. Farrell, J. Hamlett, J. P. Sanderson, S. El-Ghaiesh, D. Peckham, M. Pirmohamed, B. K. Park and D. J. Naisbitt, *Chem. Res. Toxicol.*, 2013, **26**, 963.
124. E. H. Cordes, *Hallelujah Moments: Tales of Drug Discovery*, Oxford University Press, New York, 2014.
125. J. W. Clader, *J. Med. Chem.*, 2004, **47**, 1.
126. R. Ganellin, S. Roberts and R. Jefferis, *Introduction to Biological and Small Molecule Drug Research and Development: Theory and Case Studies*, Academic Press, 2013, ch. 15.
127. A. F. Stepan, V. Mascitti, K. Beaumont and A. S. Kalgutkar, *Med. Chem. Commun.*, 2013, **4**, 631.
128. D. A. Burnett, M. A. Caplen, H. R. Davis Jr, R. E. Burrier and J. W. Clader, *J. Med. Chem.*, 1994, **37**, 1733.

129. S. B. Rosenblum, T. Huynh, A. Afonso, H. R. Davis Jr, N. Yumibe, J. W. Clader and D. A. Burnett, *J. Med. Chem.*, 1998, **41**, 973.
130. L. J. Wang and B. L. Song, *Biochim. Biophys. Acta*, 2012, **1821**, 964.
131. N. O. Stitziel, H. H. Won, A. C. Morrison, G. M. Peloso, R. Do, L. A. Lange, P. Fontanillas, N. Gupta, S. Duga, A. Goel, M. Farrall, D. Saleheen, P. Ferrario, I. Konig, R. Asselta, P. A. Merlini, N. Marziliano, M. F. Notarangelo, U. Schick, P. Auer, T. L. Assimes, M. Reilly, R. Wilensky, D. J. Rader, G. K. Hovingh, T. Meitinger, T. Kessler, A. Kastrati, K. L. Laugwitz, D. Siscovick, J. I. Rotter, S. L. Hazen, R. Tracy, S. Cresci, J. Spertus, R. Jackson, S. M. Schwartz, P. Natarajan, J. Crosby, D. Muzny, C. Ballantyne, S. S. Rich, C. J. O'Donnell, G. Abecasis, S. Sunyaev, D. A. Nickerson, J. E. Buring, P. M. Ridker, D. I. Chasman, E. Austin, Z. Ye, I. J. Kullo, P. E. Weeke, C. M. Shaffer, L. A. Bastarache, J. C. Denny, D. M. Roden, C. Palmer, P. Deloukas, D. Y. Lin, Z. Z. Tang, J. Erdmann, H. Schunkert, J. Danesh, J. Marrugat, R. Elosua, D. Ardissino, R. McPherson, H. Watkins, A. P. Reiner, J. G. Wilson, D. Altshuler, R. A. Gibbs, E. S. Lander, E. Boerwinkle, S. Gabriel and S. Kathiresan, *N. Engl. J. Med.*, 2014, **371**, 2072.
132. M. A. Blazing, R. P. Giugliano, C. P. Cannon, T. A. Musliner, A. M. Tershakovec, J. A. White, C. Reist, A. McCagg, E. Braunwald and R. M. Califf, *Am. Heart J.*, 2014, **168**, 205.
133. A. Bhardwaj, *Sci. Transl. Med.*, 2014, **6**, 262ec193.
134. B. Sainz Jr, N. Barretto, D. N. Martin, N. Hiraga, M. Imamura, S. Hussain, K. A. Marsh, X. Yu, K. Chayama, W. A. Alrefai and S. L. Uprichard, *Nat. Med.*, 2012, **18**, 281.
135. J. E. Carette, M. Raaben, A. C. Wong, A. S. Herbert, G. Obernosterer, N. Mulherkar, A. I. Kuehne, P. J. Kranzusch, A. M. Griffin, G. Ruthel, P. Dal Cin, J. M. Dye, S. P. Whelan, K. Chandran and T. R. Brummelkamp, *Nature*, 2011, **477**, 340.
136. M. Cote, J. Misasi, T. Ren, A. Bruchez, K. Lee, C. M. Filone, L. Hensley, Q. Li, D. Ory, K. Chandran and J. Cunningham, *Nature*, 2011, **477**, 344.
137. N. M. O'Boyle, J. K. Pollock, M. Carr, A. J. S. Knox, S. M. Nathwani, S. Wang, L. Caboni, D. M. Zisterer and M. J. Meegan, *J. Med. Chem.*, 2014, **57**, 9370.
138. N. M. O'Boyle, M. Carr, L. M. Greene, N. O. Keely, A. J. Knox, T. McCabe, D. G. Lloyd, D. M. Zisterer and M. J. Meegan, *Eur. J. Med. Chem.*, 2011, **46**, 4595.
139. L. Troisi, C. Granito and E. Pindinelli, *Top. Heterocycl. Chem.*, 2010, **22**, 101.
140. B. K. Banik, I. Banik and F. F. Becker, *Top. Heterocycl. Chem.*, 2010, **22**, 349.
141. J. P. Cannon, T. A. Lee, N. M. Clark, P. Setlak and S. A. Grim, *J. Antimicrob. Chemother.*, 2014, **69**, 2043.
142. J. D. Rothstein, S. Patel, M. R. Regan, C. Haenggeli, Y. H. Huang, D. E. Bergles, L. Jin, M. Dykes Hoberg, S. Vidensky, D. S. Chung,

S. V. Toan, L. I. Bruijn, Z. Z. Su, P. Gupta and P. B. Fisher, *Nature*, 2005, **433**, 73.
143. S. G. Lee, Z. Z. Su, L. Emdad, P. Gupta, D. Sarkar, A. Borjabad, D. J. Volsky and P. B. Fisher, *J. Biol. Chem.*, 2008, **283**, 13116.
144. S. M. Rawls, F. Karaca, I. Madhani, V. Bhojani, R. L. Martinez, M. Abou-Gharbia and R. B. Raffa, *Neurosci.*, 2010, **169**, 1800.
145. X. Cui, L. Li, Y. Y. Hu, S. Ren, M. Zhang and W. B. Li, *Mol. Neurobiol.*, 2015, **51**, 1322.
146. M. I. Konaklieva, B. J. Plotkin and T. Herbert, *Anti-Infect. Agents Med. Chem.*, 2009, **8**, 28.
147. B. A. Rasmussen, D. A. Baron, J. K. Kim, E. M. Unterwald and S. M. Rawls, *Amino Acids*, 2011, **40**, 761.
148. J. A. Schroeder, N. G. Tolman, F. F. McKenna, K. L. Watkins, S. M. Passeri, A. H. Hsu, B. R. Shinn and S. M. Rawls, *Drug Alcohol Depend.*, 2014, **142**, 41.
149. M. A. Cooper and D. Shlaes, *Nature*, 2011, **472**, 32.
150. D. M. Shlaes and B. Spellberg, *Curr. Opin. Pharmacol.*, 2012, **12**, 522.
151. J. L. Martínez, T. M. Coque and F. Baquero, *Nat. Rev. Microbiol.*, 2015, **13**, 116.
152. J. F. Fisher, J. W. Johnson and S. Mobashery, in *Handbook of Antimicrobial Resistance*, ed M. Götte, A. Berghuis, G. Matlashewski, D. Sheppard and M. Wainberg, Springer, 2015.
153. R. M. Phelan, B. J. DiPardo and C. A. Townsend, *ACS Chem. Biol.*, 2012, 7, 835.
154. R. Li, E. P. Lloyd, K. A. Moshos and C. A. Townsend, *ChemBioChem*, 2014, **15**, 320.
155. C. T. Walsh and T. A. Wencewicz, *J. Antibiot.*, 2014, **67**, 7.
156. G. D. Wright, *Can. J. Microbiol.*, 2014, **60**, 147.
157. J. Brougher and K. M. Linnik, *Nat. Biotechnol.*, 2014, **32**, 877.
158. J. G. Bartlett, D. N. Gilbert and B. Spellberg, *Clin. Infect. Dis.*, 2013, **56**, 1445.
159. M. Gersch, J. Kreuzer and S. A. Sieber, *Nat. Prod. Rep.*, 2012, **29**, 659.
160. P. C. Hogan and E. J. Corey, *J. Am. Chem. Soc.*, 2005, **127**, 15386.
161. W. M. Hewitt, M. Egger, N. B. Zuckerman and J. P. Konopelski, *Tetrahedron*, 2014, **70**, 5283.
162. D. A. Bachovchin, J. T. Mohr, A. E. Speers, C. Wang, J. M. Berlin, T. P. Spicer, V. Fernandez-Vega, P. Chase, P. S. Hodder, S. C. Schurer, D. K. Nomura, H. Rosen, G. C. Fu and B. F. Cravatt, *Proc. Natl. Acad. Sci. U. S. A.*, 2011, **108**, 6811.
163. A. M. Zuhl, J. T. Mohr, D. A. Bachovchin, S. Niessen, K. L. Hsu, J. M. Berlin, M. Dochnahl, M. P. Lopez-Alberca, G. C. Fu and B. F. Cravatt, *J. Am. Chem. Soc.*, 2012, **134**, 5068.

CHAPTER 4

(Benz)imidazoles

ROLAND PFAU

Boehringer Ingelheim Pharma GmbH & Co. KG, 88397 Biberach an der Riss, Germany
Email: Roland.Pfau@Boehringer-Ingelheim.com

4.1 General Considerations About (Benz)imidazoles

4.1.1 Physico-chemical Properties of (Benz)imidazoles

Imidazole is an aromatic five-membered ring system containing two nitrogen atoms in positions 1 and 3. One of the nitrogens is a weak base (pK_a1: 6.92), and in case the other nitrogen is not substituted, imidazole is also a very weak acid (pK_a2: 14.2), with the proton being able to tautomerize.

Benzimidazole is an aromatic bicycle consisting of an imidazole ring fused to a benzene ring *via* its two adjacent carbon atoms. While the basicity is decreased, the acidity is increased compared to imidazole (pK_a1: 5.53; pK_a2: 12.8).

Tuning acidity and basicity by suitable substituents can have remarkable effects for potential interactions (see Section 4.2.2). Proper substitution also influences which tautomer is preferred.

Due to their basicity, both scaffolds are capable of salt formation with strong acids, which is beneficial for the identification of proper galenic formulations. Additionally, the solubility of compounds with (benz)imidazole moieties in acidic media (like gastric juice) is increased. For benzimidazoles with free NH moieties, also salts of strong bases can be an option for galenic formulations.

RSC Drug Discovery Series No. 50
Privileged Scaffolds in Medicinal Chemistry: Design, Synthesis, Evaluation
Edited by Stefan Bräse

Published by the Royal Society of Chemistry, www.rsc.org

While imidazole itself is already fairly polar (log P: −0.08), it offers the option to attach lipophilic substituents for achieving an overall acceptable lipophilicity. Benzimidazole is more lipophilic (log P: 1.50), so the choice of substituents has to be more balanced, providing the chance to include some peripheric polar groups. To decrease the lipophilicity of the benzimidazole scaffold, exchanging one of the –CH= through –N= in the benzene part might be an option to fine tune lipophilicity.

4.1.2 (Benz)imidazoles As Scaffolds: Geometry and Options For Interaction

The imidazole scaffold offers four adjacent positions for substitution on its five-membered ring system. The remaining basic nitrogen can also be substituted, leading to permanently charged compounds, which might be beneficial for some niche indications to prevent systemic exposure. All positions for substitution are within the aromatic ring plane, at an angle of around 72° to each other. By interaction with the aromatic system, the attached substituents might be pre-organized for a preferred conformer and thus the spatial sector for an attached residue. If neither nitrogen atom is substituted, tautomerism is possible and H-bond formation with the target can be obtained *via* induced fit.

Imidazole itself already offers a variety of potential options for important interaction:

- H-bond donor (if N is not substituted; flexible through tautomerism),
- H-bond acceptor,
- coordination of metal ions (*e.g.* in cytochromes),
- lipophilic interaction with unsubstituted CH-groups and
- CH–π-interaction of the aromatic system with alkyl groups.

H-bonds can be formed between the compound and its target or intramolecularly with an appropriate substituent, thereby preforming a certain conformation of the drug substance (see Section 4.2.4).

Imidazole itself is a bioisostere of a carboxamide unit. Thus, it might be interpreted as peptide backbone unit isostere. Depending on the substituents and their substitution pattern, small oligo-peptide-mimetics with regular *trans*- as well as *cis*-locked[1] conformations can be formed (Figures 4.1 and 4.2).

Benzimidazole can be viewed as indole bioisostere as well as extended imidazole scaffold, sharing the potential options for target interactions. As bicyclic ring system, it has additional positions for substituents within the same ring plane as the imidazole, adding options for the spatial positioning of such substituents relative to each other. Via the additional aromatic ring, benzimidazoles can interact by π-stacking (Figure 4.3).

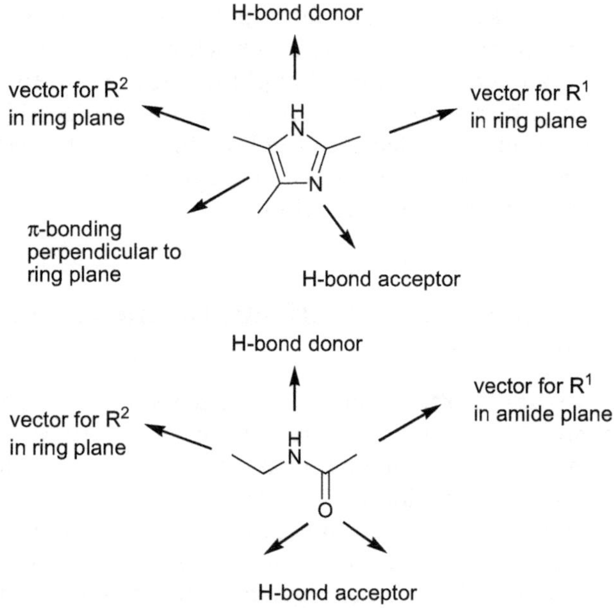

Figure 4.1 Imidazole as *trans*-amide isostere.

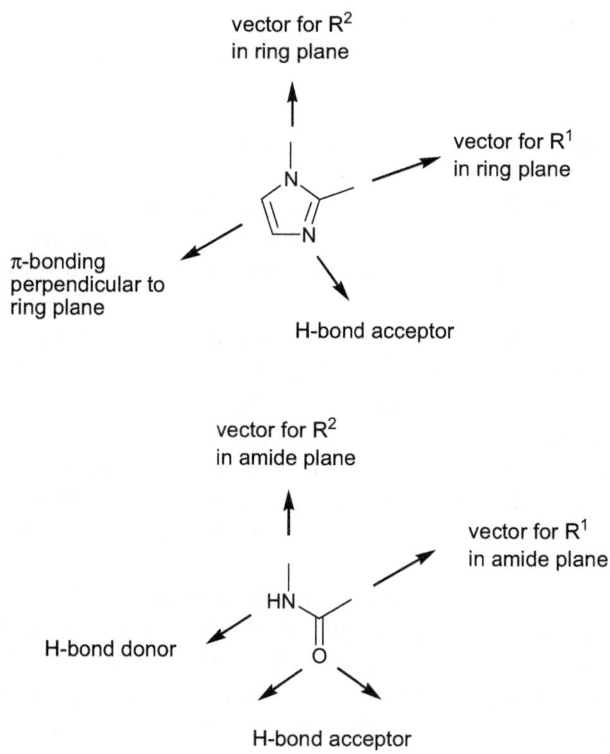

Figure 4.2 Imidazole as *cis*-amide isostere.

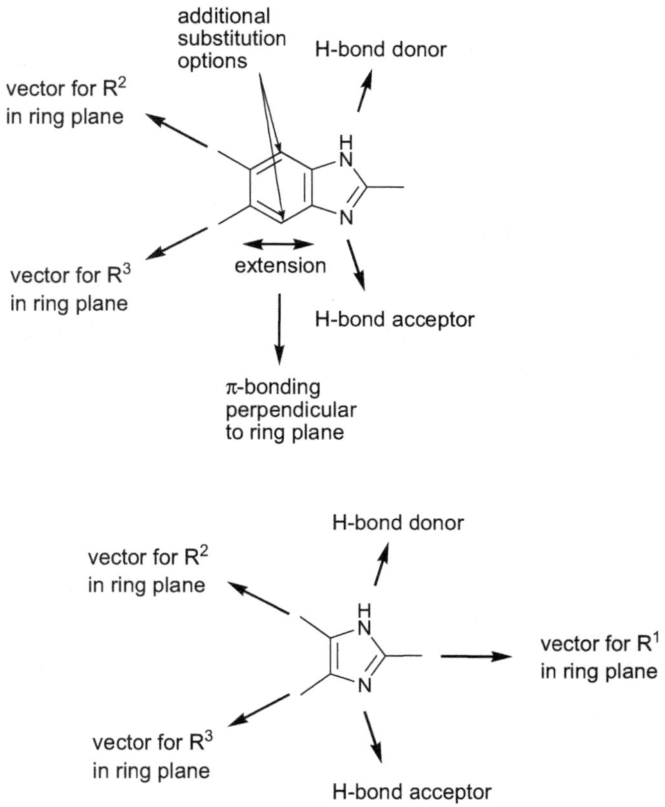

Figure 4.3 Benzimidazoles *vs.* imidazoles.

4.1.3 Synthesis of (Benz)imidazoles

Various methods for synthesis of imidazoles are known (Scheme 4.1).[2] However, the suiting starting materials often are not readily available, and certain limitations in scope regarding the kind of substituents on the imidazole core prevent full coverage of the theoretical structural space. If identified as hits, imidazoles are at risk to be cumbersome regarding further optimization, which might lead to their down-prioritization if other alternatives with easier accessibility and modification options were identified.

For benzimidazoles, very reliable and robust methods for synthesis exist (Scheme 4.2),[3] leading to good yields in most cases, which are prerequisites for the efficient synthesis of large libraries. If not readily available, the starting materials are often very easy to prepare. As a consequence, lead optimization of such hits seems rather easy compared to imidazoles.

4.1.4 Natural Products Containing (Benz)imidazoles

Imidazoles as well as benzimidazoles are structural motifs in numerous natural products. Although they play important roles as part of catalytic

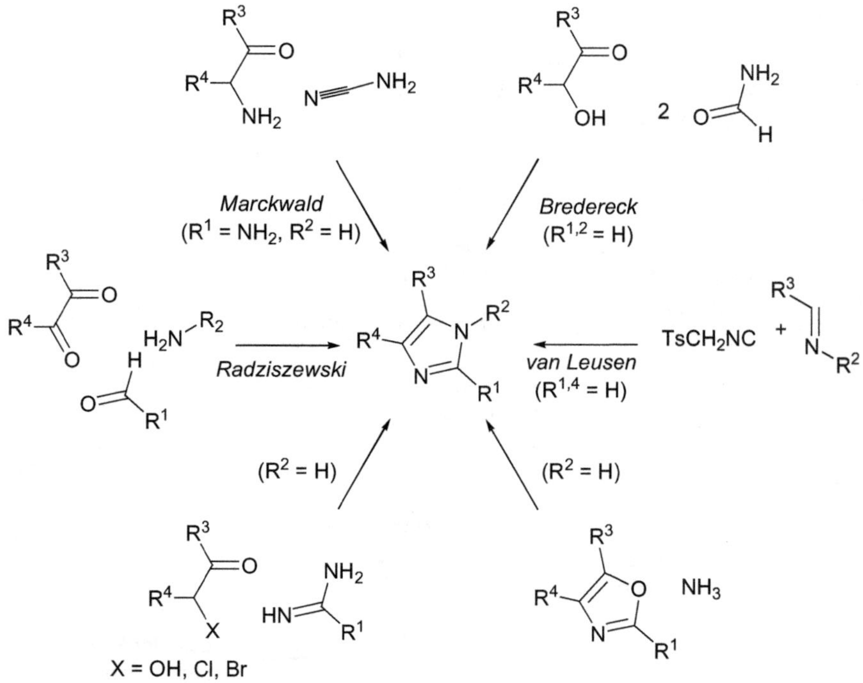

Scheme 4.1 Various syntheses of imidazoles.

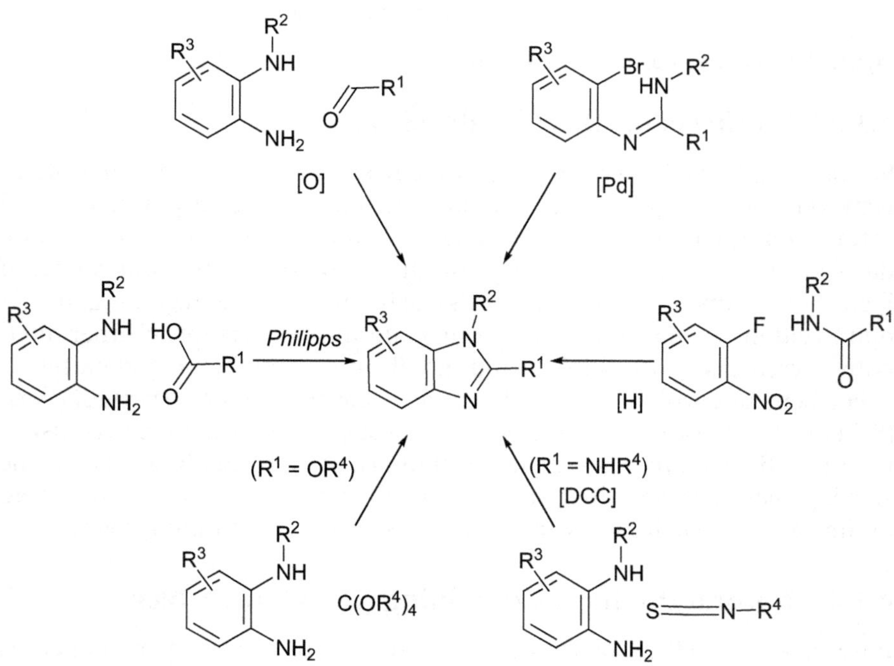

Scheme 4.2 Various syntheses of benzimidazoles.

centers within enzymes or for metal ion coordination due to their options for interaction, they are mostly attached as residues.

Imidazole is part of the amino acid histidine. In serine proteases, it is part of the catalytic triad, together with serine and aspartic acid. In cysteine proteases, it can either form a diad, together with cysteine, or a triad with additional aspartic or glutamic acid, asparagine or glutamine (Figure 4.4). Imidazole is utilized for the proton transfer, leading to amide bond cleavage of the targeted protein.

Decarboxylation of histidine leads to histamine, which is a neurotransmitter involved in sleep-wake-regulation and also playing regulatory functions during inflammation (see Section 4.2.3) and for gastric acid release. It is metabolized *via* *N*-methylation by an *N*-methyl-transferase followed by degradation by mono-aminooxidase-B and aldehyde dehydrogenase 2 (Scheme 4.3).

Imidazole is contained in various alkaloids (*e.g.* oroidin, hymenidin, most nagelamides, sceptrin, stylissazoles, dihydrosventrin, (bromo)ageliferin, 1,9-dideoxy-preaxinellamine).[4] In some alkaloids, imidazole is even a core group in a scaffold-like fashion (*e.g.* (nor)topsentines, (iso)naamines) (Figure 4.5).

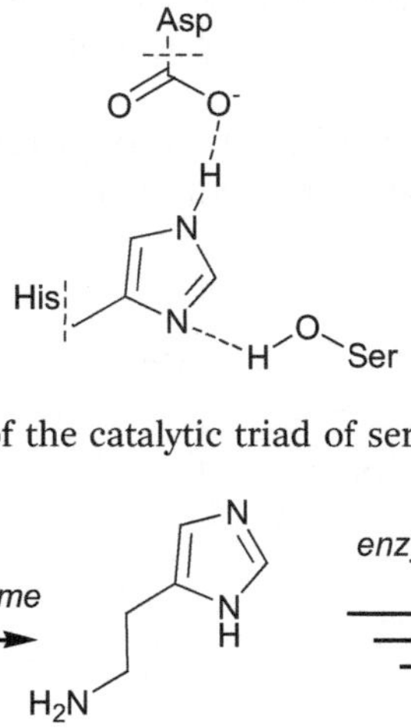

Figure 4.4 Schematic view of the catalytic triad of serine proteases.

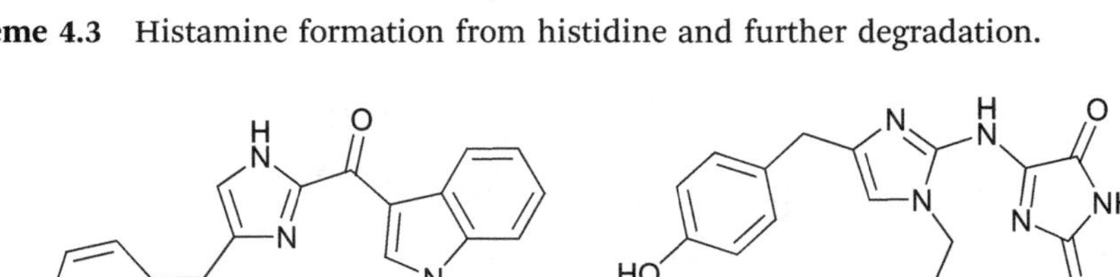

Scheme 4.3 Histamine formation from histidine and further degradation.

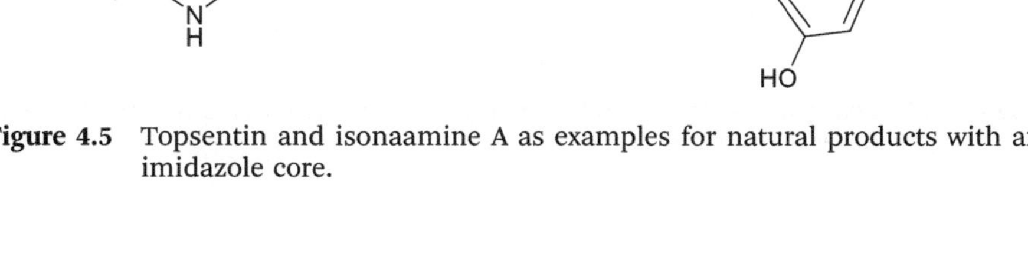

Figure 4.5 Topsentin and isonaamine A as examples for natural products with an imidazole core.

Fewer examples of natural products containing benzimidazole are known, like the cobalamines with vitamin B_{12} or coenzyme B_{12} as two important members. In the latter cases, the benzimidazole moiety coordinates the central cobalt ion within the corrine ring to which it is attached *via* a linker.

4.2 Case Studies of Marketed Drugs

4.2.1 Angiotensin II Receptor Antagonists

To block the angiotensin AT1 receptor from binding its natural substrate, angiotensin II, and stabilizing its inactive form is a concept for lowering high blood pressure. It offers better tolerability than other concepts for treatment of hypertension.

With S-8307, Takeda identified a starting point for the lead optimization towards potent, AT1-selective and non-peptide Angiotensin II receptor antagonists with an imidazole scaffold (Figure 4.6). This compound was part of a 1-benzyl-imidazole-5-acetic acid series, inspiring DuPont as well as SmithKline Beecham to use the imidazole scaffold for further pharmacophore-based optimization, ultimately leading to Losartan and Eprosartan (Figures 4.7 and 4.8).[5]

Later, Daiichi Sankyo developed olmesartan based on the imidazole scaffold, offering a longer human half-life which is beneficial for

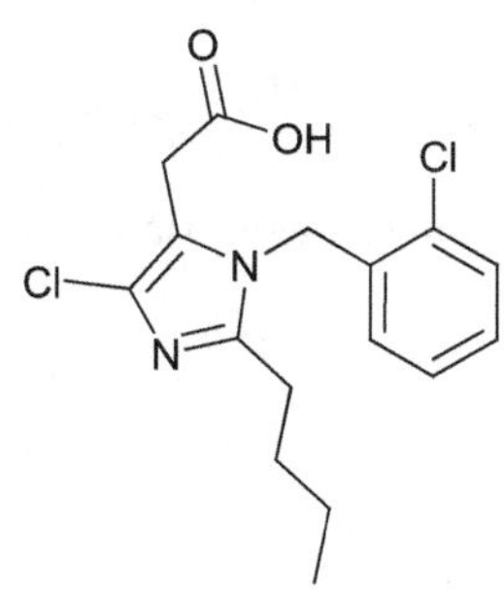

Figure 4.6 S-8307: Takeda 1982 screening hit.

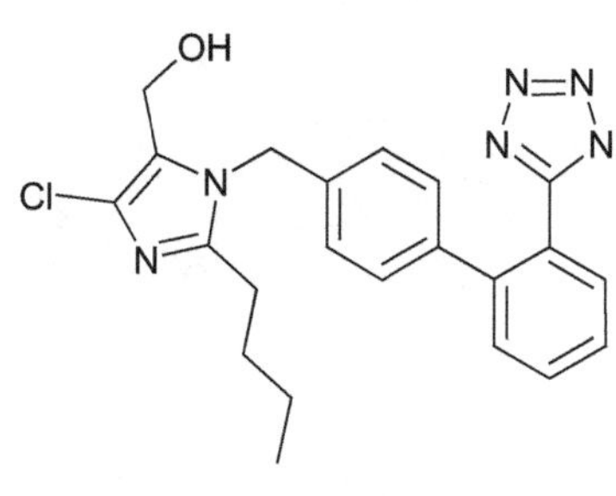

Figure 4.7 Losartan: DuPont/Merck, first registration 1994, $t_{1/2}$: 2 h, BA: 33%, daily dose: 50–100 mg.

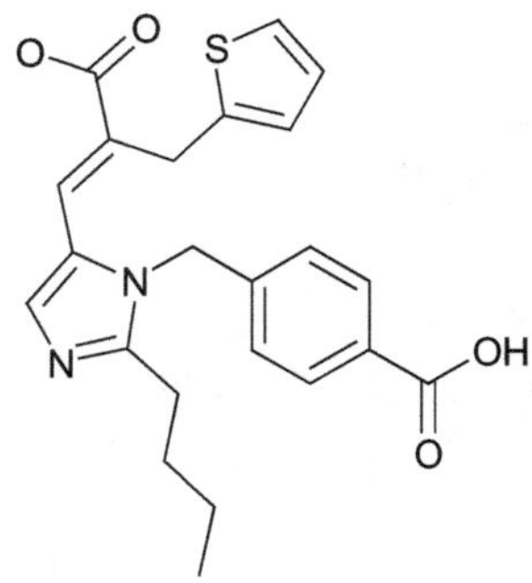

Figure 4.8 Eprosartan: SmithKline Beecham, first registration 1997, $t_{1/2}$: 5 h, BA: 13%, daily dose: 400–800 mg.

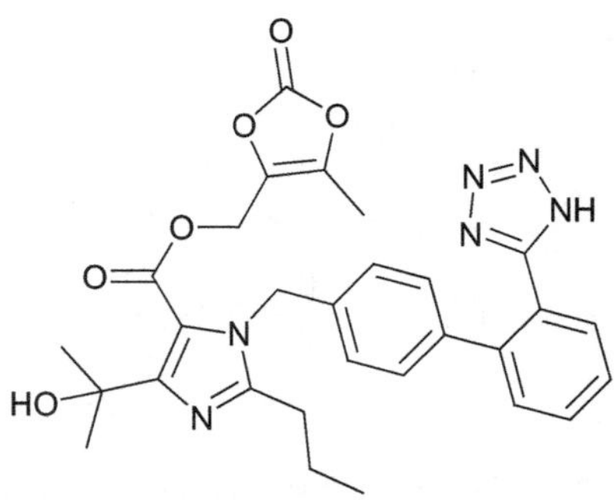

Figure 4.9 Olmesartan (medoxomil): Daiichi Sankyo, first registration 2002, $t_{1/2}$: 14–16 h, BA: 29%, daily dose: 10–40 mg.

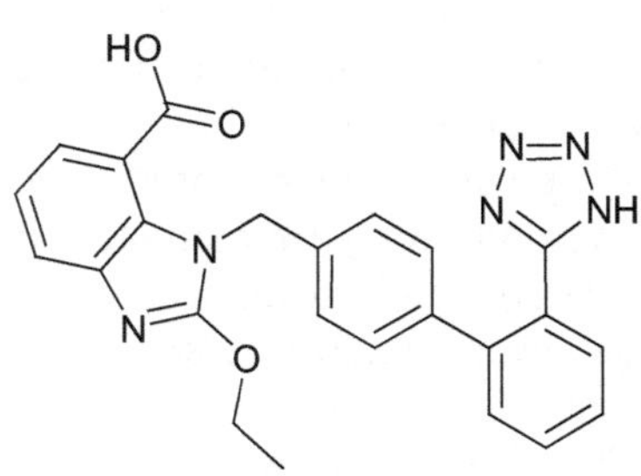

Figure 4.10 Candesartan (cilexetil): Takeda, first registered 1997, $t_{1/2}$: 9–12 h, BA: 15%, daily dose: 8–32 mg.

convenience, efficacy and safety. It is given as a prodrug (medoxomil), which is readily hydrolyzed during intestinal absorption (Figure 4.9).

Switching from imidazole to benzimidazole, Takeda's candesartan as well as Boehringer Ingelheim's telmisartan entered the market at about the same time. Candesartan is also administered as ester prodrug (Figure 4.10). Telmisartan shows the longest half-life of all approved AT1 blockers currently available (Figure 4.11).

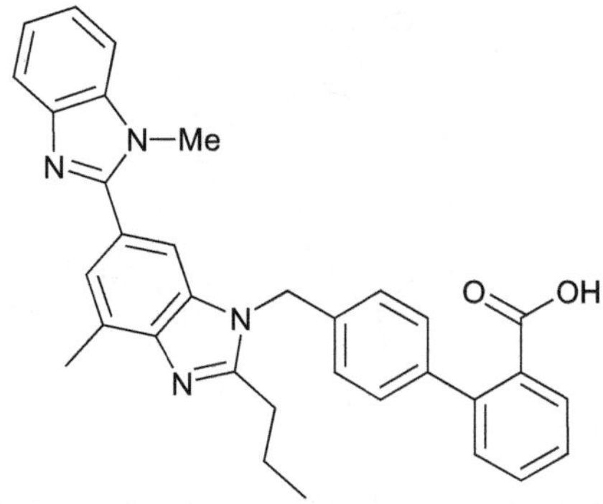

Figure 4.11 Telmisartan: Boehringer Ingelheim, first registration 1998, $t_{1/2}$: 24 h, BA: 42–100%, daily dose: 40–80 mg.

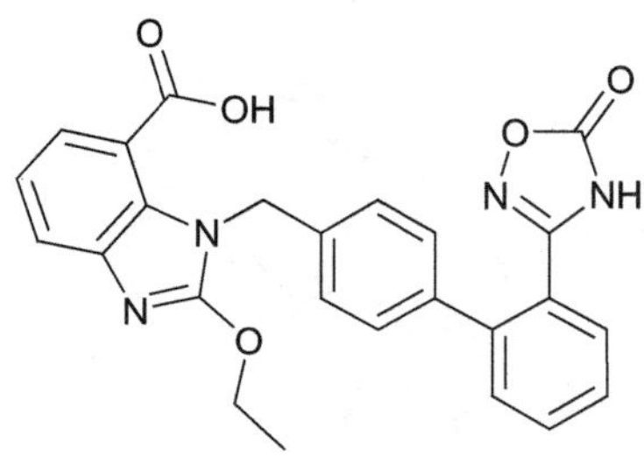

Figure 4.12 Azilsartan: Takeda, first registered 2011, daily dose: 20–40 mg.

Recently, Takeda registered Azilsartan, which also relies on the benzimidazole scaffold and is masked as prodrug (Figure 4.12).

Angiotensin II receptor antagonists serve as a very instructive example of how related and even interchangeable imidazole and benzimidazole can be as scaffolds for a suitable target. Given the option that with imidazole as scaffold, both the 4- and 5-position are tolerated to be substituted, fusing the phenyl ring and exploring potentially new substitution options of the newly formed benzimidazole might lead to an interesting new structural class offering new options and even benefits *versus* the original imidazole lead. Of course, the reverse path is also an option, deriving an imidazole-based structural class from a benzimidazole lead.

As similar as the shown structures might seem, a recent reassessment of the binding mode of the shown Angiotensin II receptors concludes from mutagenesis as well as modelling studies that the binding mode of each of them is unique (Figure 4.13).[6]

4.2.2 H^+, K^+-ATPase Inhibitors

For medical conditions related to gastric acid secretion, blocking the proton pump helps to improve the symptoms. The prazoles are a group of covalent H^+, K^+-ATPase inhibitors with widespread use as antacids. The role of the

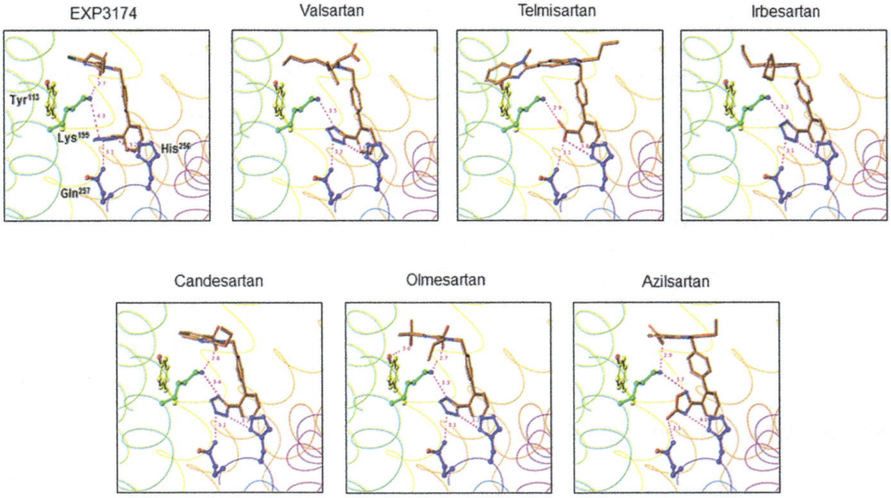

Figure 4.13 Molecular modeling of a close-up view of the interactions between seven angiotensin II receptor blockers and the AT_1 receptor (taken from ref. 6 under Creative Commons Attribution (CC BY) license).

benzimidazole contained in all prazoles is unique, and it is worthwhile to look into its role in activating the prazoles for binding. The basicity of the benzimidazole as well as the nucleophilicity and electron-donating substitution pattern of the pyridine are of crucial importance to make sure that activation of the prazole takes place at the actual target, in an area of elevated acidity (parietal cell's secretory canaliculus). The prazoles selectively bind to an H^+, K^+-ATPases thiol group of surface-exposed cysteine, inactivating it thereby. The prazole's pK_a seem to correlate with the rapidity of the onset of action as well as the achievable gastric pH shift and duration of action.[7]

Both the pyridine and the sulfoxide moieties are capable of forming stabilizing internal H-bonds, preforming or stabilizing an intermediate, thereby lowering the activation energy for the following process (Scheme 4.4).

AstraZeneca's racemic omeprazole was the first of this class to hit the market and became a great success (Figure 4.14). Later, AstraZeneca introduced esomeprazole, the magnesium salt of the S-enantiomer of omeprazole, due to its superior pharmacokinetics (Figure 4.14).[8]

Lansoprazole and pantoprazole were introduced four years after omeprazole (Figure 4.15). Compared to omeprazole, lansoprazole offers higher bioavailability and faster onset of action, while pantaprazole offers longer duration of action at comparable dosages (Figure 4.16). The latter effect might be based on the differences of chemical stability in acidic media, with pantaprazole showing slightly reduced basicity. As result of higher stability, differences of the pattern of bound cysteins are discussed, possibly being the reason for differences in duration of action: omeprazole is discussed to bind

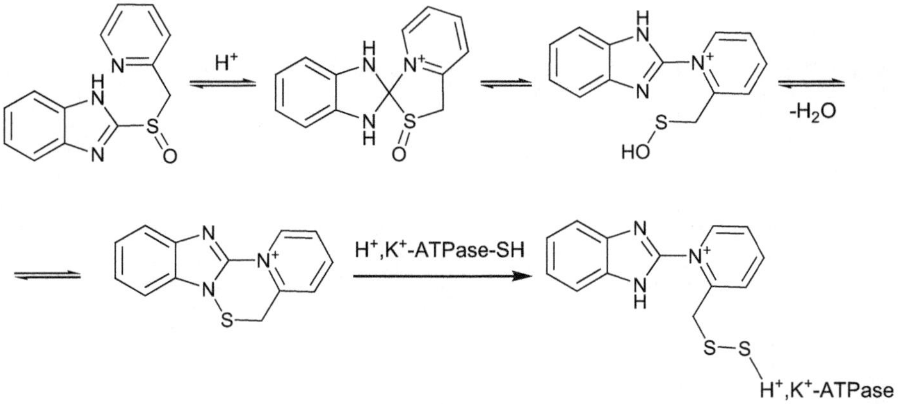

Scheme 4.4 Prazole-mechanism.

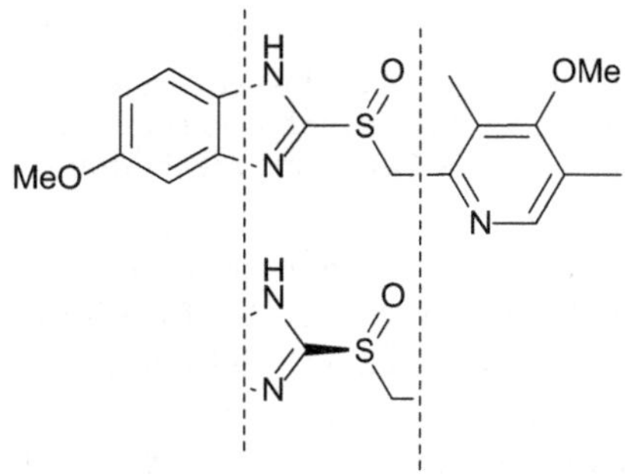

Figure 4.14 Omeprazole (AstraZeneca, first registered 1988) and esomeprazole (AstraZeneca, first registered 2000).

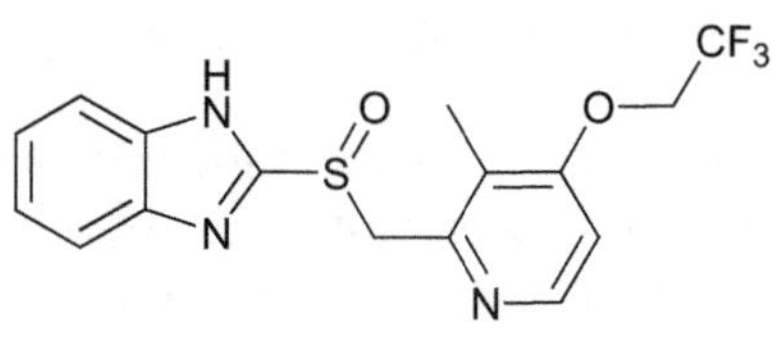

Figure 4.15 Lansoprazole (Takeda, first registered 1993).

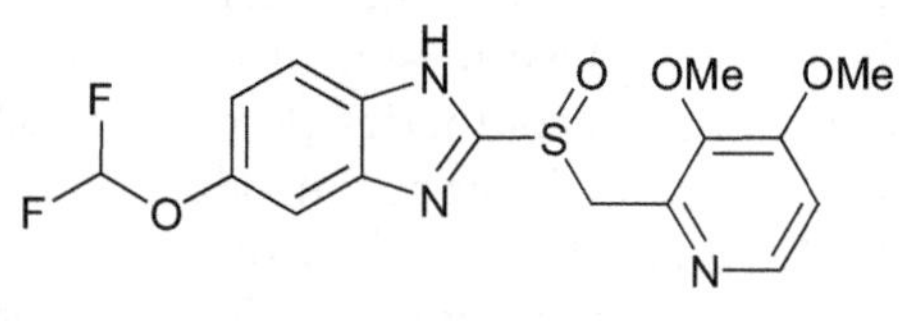

Figure 4.16 Pantoprazole (Takeda/Nycomed/Byk Gulden, first registered 1994).

at Cys-892 and Cys-813, with only the latter blocking the proton pump, but in reach of detoxicating glutathione. Pantoprazole binds to Cys-813 and Cys-822, with the latter more hidden from glutathione access.[9]

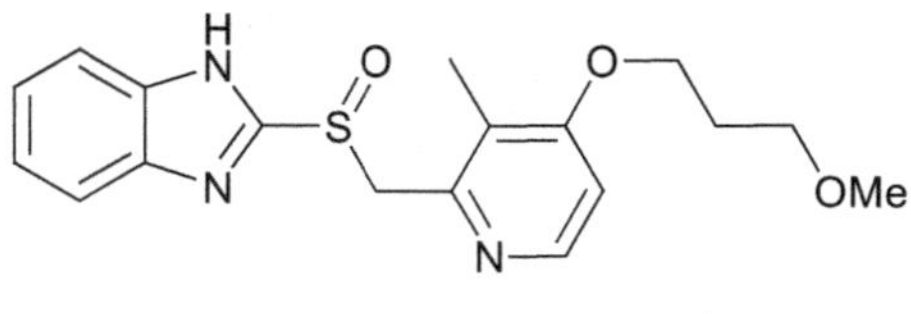

Figure 4.17 Rabeprazole (Eisei, first registered 1998).

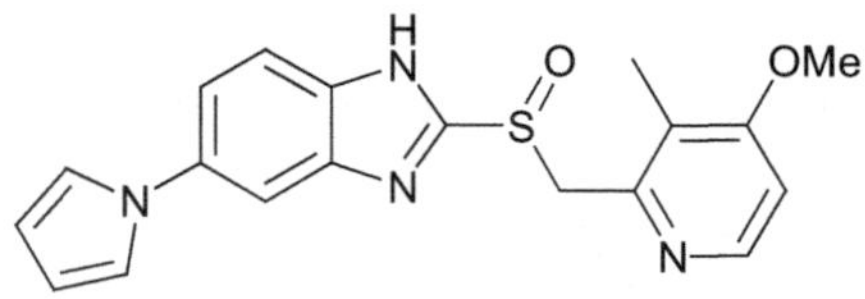

Figure 4.18 Ilaprazole (Il-Yang, first registered 2008).

Another four years later, Eisei introduced rabeprazole, which is claimed to be activated at the highest pH levels and to react faster at a given pH level compared to all previously marketed PPIs (PPI: proton pump inhibitor), due to its higher pK_a (Figure 4.17). Moreover, its metabolism is not so dependent from CYP2C19 unlike all former PPIs, and thus more predictable.[7]

Ilaprazole is the most recent PPI to reach the market. It is claimed that it has a longer half-life than other PPIs and is independent of the CYP2C19 metabolism (Figure 4.18).[10]

4.2.3 H1-antihistamines

Histamine, a small messenger molecule, plays a key role in the regulation of various physiological functions, such as inflammatory response. For enhancing inflammatory response, it has to activate the H1 receptor. In case of medical conditions like allergic rhinitis, allergic conjunctivitis or urticarial, H1-antihistamines act as inverse agonists to suppress an overshooting immunological response by stabilization of the H1-receptor's inactive state.[11]

Cetrizine (UCB/various, first registered 1987) and loratadine (Schering-Plough, first registered 1988) are the current most widespread active ingredients in oral antihistaminic drugs, both members of the "second generation" H1-antihistamines with improved side-effect profile.

Two more recent oral antihistamines utilizing benzimidazole as scaffold are mizolastine and bilastine. Both of them show high affinity and selectivity for the H1-receptor, and bilastine shows high metabolic stability (Figures 4.19 and 4.20).[12]

Direct ophthalmic administration generally leads to a quicker onset of action in patients suffering from allergic conjunctivitis. Among the active ingredients of drugs formulated for ophthalmic use, two more recent examples utilize benzimidazole and imidazole: emedastine and alcaftadine. Both of them show some affinity to other histamine-subtype-receptors in addition to high affinity to the H1-receptor: emedastine towards H2- and H3-receptor,[13] alcaftadine towards H2- and H4-receptor (Figures 4.21 and 4.22).[14]

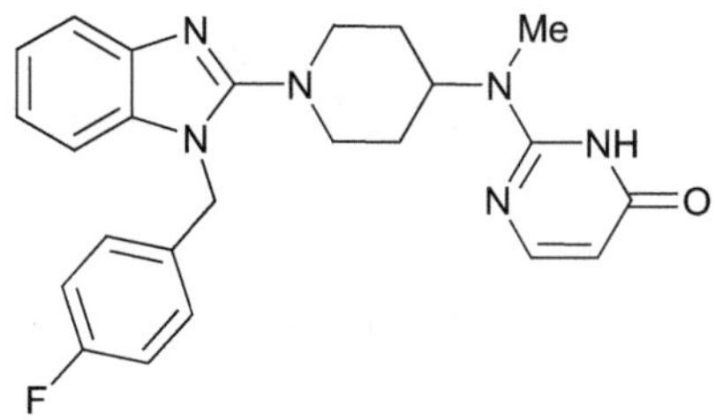

Figure 4.19 Mizolastine (Sanofi-Synthelabo, first registered 1997).

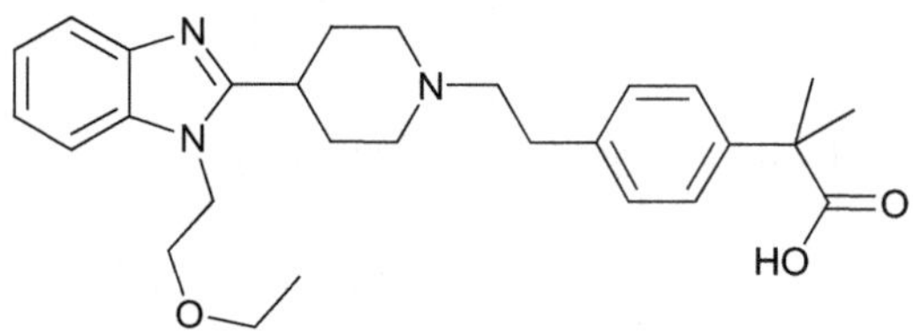

Figure 4.20 Bilastine (FAES Farma SA, first registered 2009).

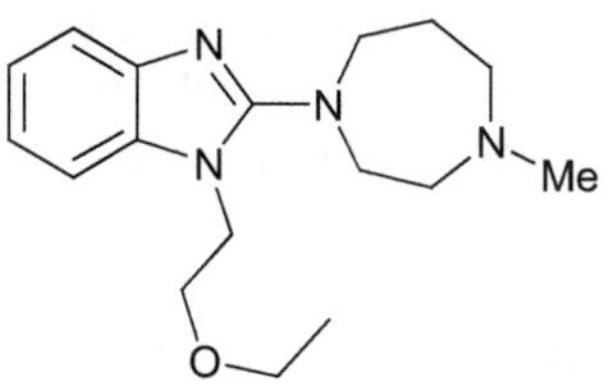

Figure 4.21 Emedastine (Kanebo, first registered 1996).

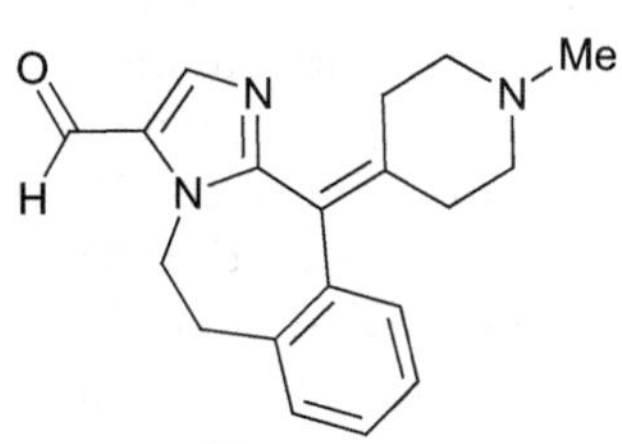

Figure 4.22 Alcaftadine (Vistakon/Janssen Cilag, first registered 2010).

4.2.4 Anthelmintics

Targeting not a human's, but a vital parasite's physiological pathway is the basis for efficient anti-parasitic drugs. Bendazoles are β-tubulin binders, taking advantage of the differences between human and helminthic β-tubulin. They utilize benzimidazole as scaffold. The mechanism of action is the prevention of glucose absorption in parasites by interfering with mitosis of intestinal cells through binding to β-tubulin, as well as direct glucose uptake inhibition.[15]

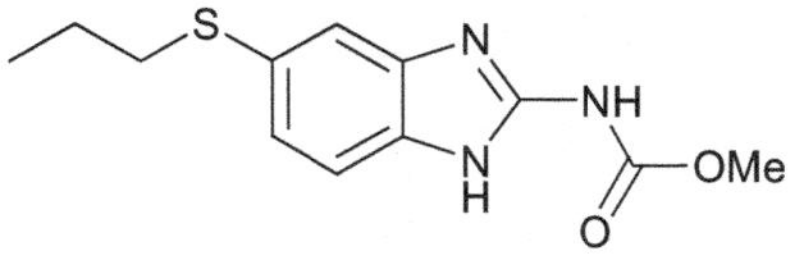

Figure 4.23 Albendazole (SmithKline Beecham, first registered 1987, β-tubulin inhibitor, parasitic infection).

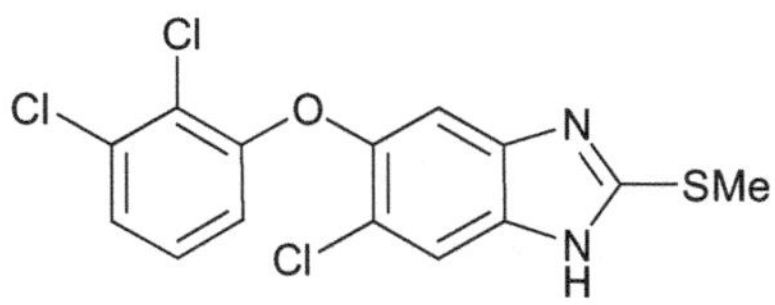

Figure 4.24 Triclabendazole (Novartis, first registered 1989, β-tubulin inhibitor, parasitic infection).

Albendazole is a broad-spectrum anti-parasitic drug. The amino-benzimidazole portion of the compound interacts with Glu198 of helminthic β-tubulin.[16] The thioether is rapidly metabolized to the sulfoxide and sulfone as active metabolites (Figure 4.23).

Triclabendazole is a more specific anti-parasitic drug for the treatment of fascioliasis and paragonimiasis. Rapid metabolization of the thioether leads to sulfoxide and sulfone, which are also active and carry the pharmacological effect (Figure 4.24). The methyl group of the thio-substituent leaves the ring plane of the benzimidazole after being metabolized due to an internal H-bond of the sulfoxide (or the sulfone) with the NH of the benzimidazole. In contrast, in albendazole the internal H-bond between the carbamate's carbonyl and the NH of the benzimidazole keeps the group in the imidazole's ring plane. This difference in shape is discussed as the main reason for the differences in selectivity.[17]

4.2.5 Miscellaneous

In the following, some unique examples of marketed drugs from various indications are described, to demonstrate the widespread utility of (benz)imidazoles as scaffolds.

Zolpidem is a hypnotic agent acting on the ω_1 GABA A receptor subtype in the brain.[18] It is the most widespread treatment against insomnia, showing less serious dependency and rebound issues than the classical benzodiazepines. The latter is probably due to differences in the binding pattern to the subunits of the pentameric GABA A receptor (Figure 4.25).

The N1-position of the imidazole is integrated into the imidazo-pyridine ring, thereby modifying on the vector of the overall molecular dipole. In the meantime, however, it has been shown that the benzimidazole-analogue of Zolpidem shows similar *in vivo* efficacy.[19]

Dabigatran etexilate is a double prodrug of a direct reversible thrombin inhibitor, dabigatran, for the prevention of deep vein thrombosis and risk

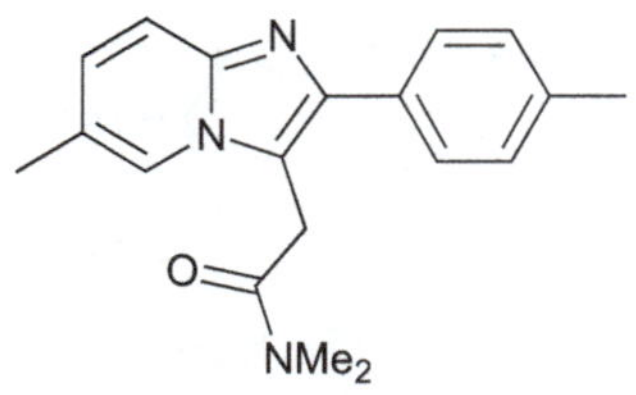

Figure 4.25 Zolpidem (Synthelabo, first registered 1988, GABA A receptor agonist, insomnia).

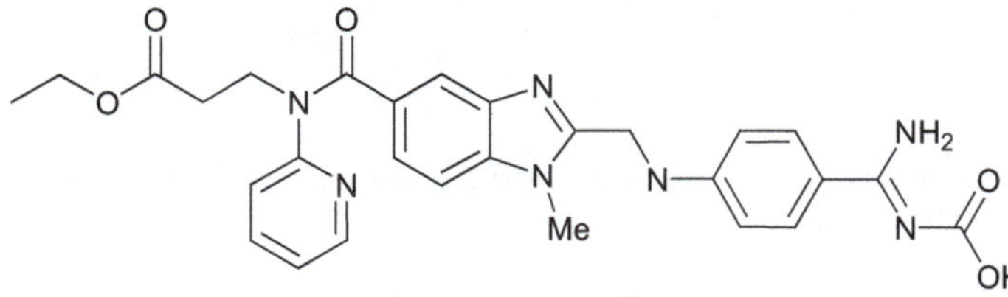

Figure 4.26 Dabigatran etexilate (Boehringer Ingelheim, first registered 2008, Factor IIa antagonist).

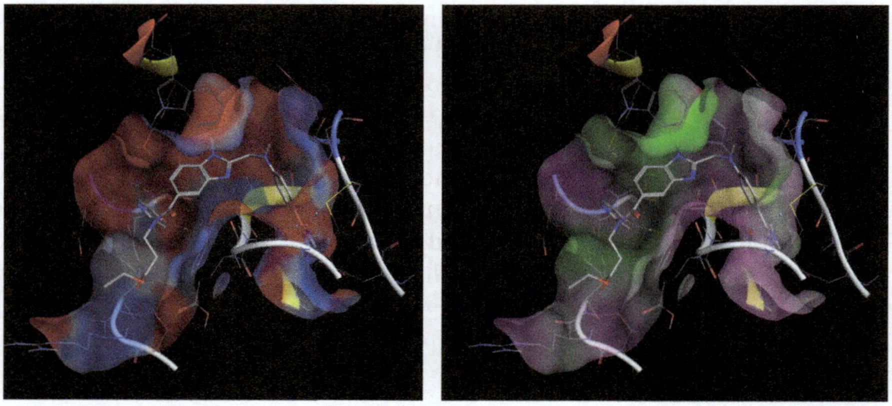

Figure 4.27 Binding site of dabigatran-ethyl ester in thrombin, including electrostatic and lipophilic potential surface (based on PDB ID: 1KTS).

reduction of stroke in humans suffering from atrial fibrillation, utilizing benzimidazole as core scaffold (Figure 4.26). In case of dabigatran, the benzimidazole serves to place its three substituents correctly for most efficient interaction with three different binding pockets. The core itself contributes to the binding energy by lipophilic interaction with thrombin. The *N*-methyl group on the benzimidazole interacts with the imidazole ring system of His57 (part of the catalytic triade) *via* lipophilic and CH-π-interaction (Figure 4.27).[20]

Pimobendan is a PDE 3 inhibitor and putative Ca^{2+} sensitizer for treatment of heart failure by increasing on contractility (Figure 4.28). From a docking study, the benzimidazole is discussed to contribute to the binding

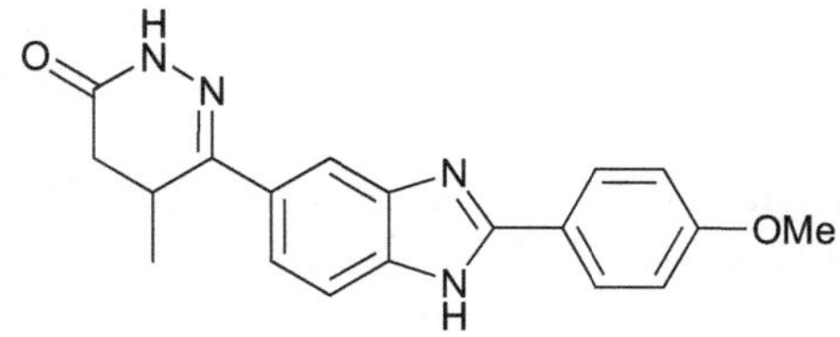

Figure 4.28 Pimobendan (Boehringer Ingelheim, first registered 1994, PDE 3 inhibitor).

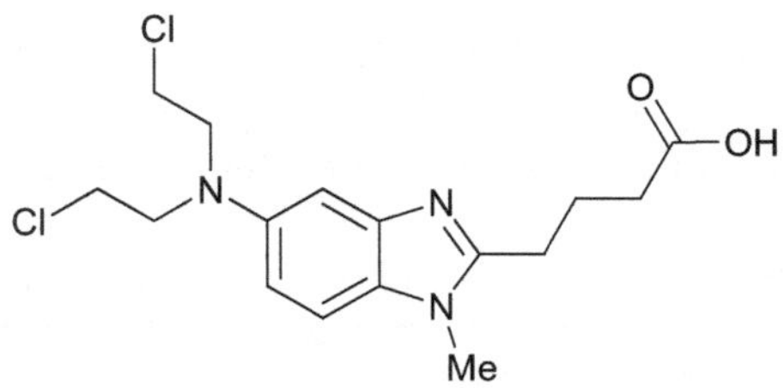

Figure 4.29 Bendamustine (Jenapharm, first registered 1971, DNA alkylating agent).

by hydrophobic interaction. In addition, the substituents are arranged to interact efficiently with the lateral binding pockets.[21]

Bendamustine is an alkylating agent from the class of nitrogen mustards. It works by intra-strand as well as inter-strand DNA crosslinking, which is hard to repair by cellular mechanisms (Figure 4.29). The N-assisted cleavage and electrophilic attack to DNA bases might be supported by the special electronic situation of the benzimidazole compared to phenyl-analogue chlorambucil. The benzimidazole is also discussed to be the reason for the beneficial pharmacokinetic properties of this agent, as it is acting amphoteric due to the basicity of the benzimidazole. It is used for treatment of chronic lymphocytic leukemia and indolent non-Hodgkin's lymphomas.[22]

References

1. S. Petit, C. Fruit and L. Bischoff, *Org. Lett.*, 2010, **12**, 4928.
2. M. R. Grimmet, *Science of Synthesis*, ed. R. Neier, Thieme Chemistry, Stuttgart, 2002, vol. 12, ch. 3, pp. 325–528.
3. M. R. Grimmet, *Science of Synthesis*, ed. R. Neier, Thieme Chemistry, Stuttgart, 2002, vol. 12, ch. 4, pp. 529–612.
4. Z. Jin, *Nat. Prod. Rep.*, 2011, **28**, 1143.
5. R. R. Wexler, W. J. Greenlee, J. D. Irvin, M. R. Goldberg, K. Prendergast, R. D. Smith and P. B. M. W. M. Timmermanns, *J. Med. Chem.*, 1996, **39**, 625.
6. S.-I. Miura, N. Nakao, H. Hanazawa, Y. Matsuo, K. Saku and S. S. Karnik, *PLoS One*, 2013, **8**, 1.
7. F. Pace, S. Pallotta, S. Casalini and G. B. Porro, *Ther. Clin. Risk Manage.*, 2007, **3**, 363.

8. P. Lindberg, *Comprehensive Medicinal Chemistry II*, ed. J. B. Taylor, Elsevier Science & Technology, Amsterdam, 2006, vol. 8, ch. 17, pp. 213–225.
9. G. Sachs and J. M. Shin, *Suchasna Gastroenterol.*, 2008, **5**, 66.
10. L. Zhou, J. Li and Z. Zhang, *Yixue Zongshu*, 2012, **18**, 1550.
11. F. E. R. Simons and K. J. Simons, *J. Allergy Clin. Immunol.*, 2011, **128**, 1139.
12. O. D. Wolthers, *BioMed Res. Int.*, 2013, 1.
13. N. A. Sharif, S. X. Su and J. M. Yanni, *J. Ocul. Pharmacol. Ther.*, 1994, **10**, 653.
14. R. Namdar and C. Valdez, *Drugs Today*, 2011, **47**, 883.
15. K. Leder, *Manual of Clinical Microbiology*, ed. P. R. Murray, American Society of Microbiology, Washington, 2007, vol. 2, ch. 150, pp. 2221–2239.
16. R. Aguayo-Ortiz, O. Méndez-Lucio, A. Romo-Mancillas, R. Castillo, L. Yépez-Mulia, J. L. Medina-Franco and A. Hernández-Campos, *J. Mol. Graphics Modell.*, 2013, **45**, 26.
17. K. B. Lipkowitz and R. O. McCracken, *J. Parasitol.*, 1991, **77**, 998.
18. H. D. Langtry, *Drugs*, 1990, **40**, 291.
19. J. L. Falcó, M. Piqué, M. González, I. Buira, E. Méndez, J. Terencio, C. Pérez, M. Príncep, A. Palomer and A. Guglietta, *Eur. J. Med. Chem.*, 2006, **41**, 985.
20. N. Hauel, H. Nar, H. Priepke, U. Ries, J. M. Stassen and W. Wienen, *J. Med. Chem.*, 2002, **45**, 1757.
21. P. Fossa, F. Giordanetto, G. Menozzi and L. Mosti, *Quant. Struct.-Act. Relat.*, 2002, **21**, 267.
22. M. Eichbaum, E. Bischofs, K. Nehls, A. Schneeweiss and C. Sohn, *Drugs Today*, 2009, **45**, 431.

CHAPTER 5

Pyrazoles

CARSTEN S. KRAMER

EMBL - The European Molecular Biology Laboratory, 69117 Heidelberg, Germany
Email: carsten.kramer@embl.de

5.1 General Remarks about Pyrazoles

5.1.1 Physicochemical Properties of Pyrazoles

Pyrazoles are five-membered heterocycles containing two adjacent nitrogens in positions 1 and 2 (Figure 5.1). The term 'pyrazole' was given by Ludwig Knorr in 1883.[1,2] Knorr was able to synthesize the first pyrazole derivative from 3-oxobutanoate and phenylhydrazone,[2] later in 1887 the structure of this compound was determined to be 3-methyl-1-phenyl-1*H*-pyrazol-5-ol).[3]

In general, pyrazoles can be drawn in three tautomeric forms: whereby 1*H*-pyrazole has aromatic character, 3*H*- and 4*H*-pyrazoles are non-aromatic heterocycles that can be obtained as 3,3- and 4,4-disubstituted derivatives (Scheme 5.1).

The cyclic 1,2-diazoles contain a pyrrole-like and a pyridine-like N-atom.[4] Therefore, the pyrazole moiety can act as a hydrogen-donor and a hydrogen-acceptor. The dipole moment for the pyrazole molecule was calculated to be 1,92 D in benzene solution, thereby the dipole moment is directed from the ring center to the bond between position 2 and 3.[4] Like its structural isomer imidazole, pyrazole follows Hückel's rule and exhibits aromaticity. By annulation of the pyrazole core, an indazole or isoindazole nucleus can be formed (Scheme 5.1).

RSC Drug Discovery Series No. 50
Privileged Scaffolds in Medicinal Chemistry: Design, Synthesis, Evaluation
Edited by Stefan Bräse

Published by the Royal Society of Chemistry, www.rsc.org

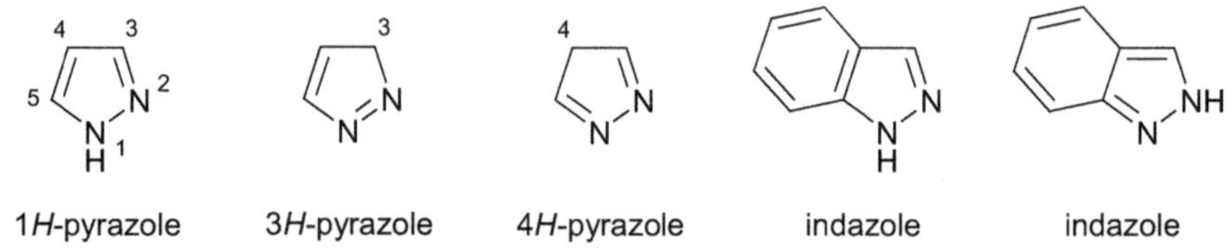

Figure 5.1 Tautomeric forms and derivatives of pyrazole.

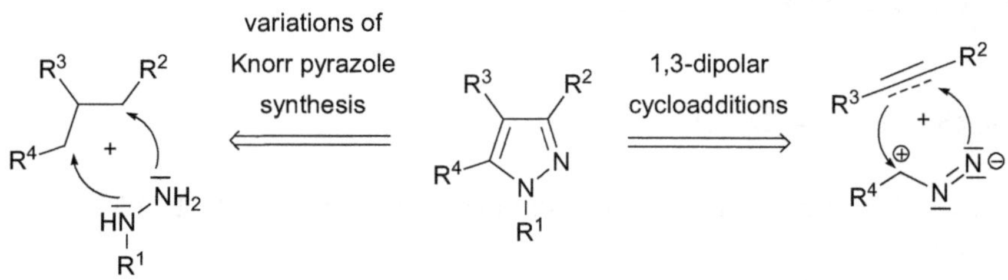

Scheme 5.1 Pyrazoles are accessible *via* Knorr-type reactions and 1,3-dipolar cycloadditions.

In comparison to imidazole, pyrazole is the weaker base (pyrazole: $pK_a = 2.57$, imidazole: $pK_a = 7.00$), since the charge in the positive pyrazolium ion is less delocalized than in the imidazolium ion.[4] 1*H*-Pyrazoles can also act as acids, hereby the N-H proton is abstracted by bases or metallic sodium to yield the corresponding salt. The N-H acidity is in the same order of magnitude as 1*H*-imidazoles (pyrazole: $pK_a = 14.21$, imidazole: $pK_a = 14.52$).[4] Pyrazoles and imidazoles share, not only commonalities in their physical properties and chemical reactivities, but they can also be transformed into imidazoles by photochemical rearrangement.[4]

5.1.2 Synthesis of Pyrazoles

Until now, countless approaches have been developed to synthesize pyrazoles from various starting materials and these have already been reviewed in depth.[4–9] Many strategies have been established to master the main challenges in the synthesis of this heterocyclic system.

1. Since the pyrazole moiety can bear up to five different substituents, convenient methods for the direct creation of a highly substituted core needed to be developed.
2. The synthesis of asymmetric substituted pyrazoles *via* intermolecular reactions should proceed in high regioselectivity.
3. Less-toxic substitutes for hydrazine should be used, for safety reasons. Also, diazo-compounds should be replaced due to their explosive nature.
4. General demands: all starting materials should be easy accessible and one-pot multicomponent reactions are preferred since large libraries

can be synthesized quickly. For the same reason, procedures should be transferrable on solid support synthesis. Finally, direct syntheses of metallated (like Sn-, B-, Si-)pyrazoles give access to attractive building blocks for following cross coupling reactions.

The most common approach for the synthesis of pyrazoles is a cyclization reaction in a [3 + 2]-manner. In general, two main subgroups of these reactions can be described (Scheme 5.1):

1. Variations of the Knorr pyrazole synthesis include the cyclocondensation between 1,3-dielectrophilic compounds (like 1,3-dicarbonyl compounds, α,β-unsaturated carbonyl compounds (with a leaving group) with hydrazines (Scheme 5.2).
2. 1,3-Dipolar cycloadditions between 1,3-dipoles (like diazoalkanes, nitrilimines or azomethine imines) and unsaturated building blocks (alkynes, olefins) give pyrazoles or corresponding pyrazole precursors (Scheme 5.3). In principle, these type of reactions can also be performed as a multicomponent reaction (MCR): in a multicomponent reaction more than two fragments are combined, leading to the desired heterocyclic product in a single step. By variation of the reacting partners, parallel conducted multicomponent reactions lead to a highly diverse library of pyrazoles. An example for a one pot, four-component reaction in pyrazole synthesis is presented in Scheme 5.4.[10]

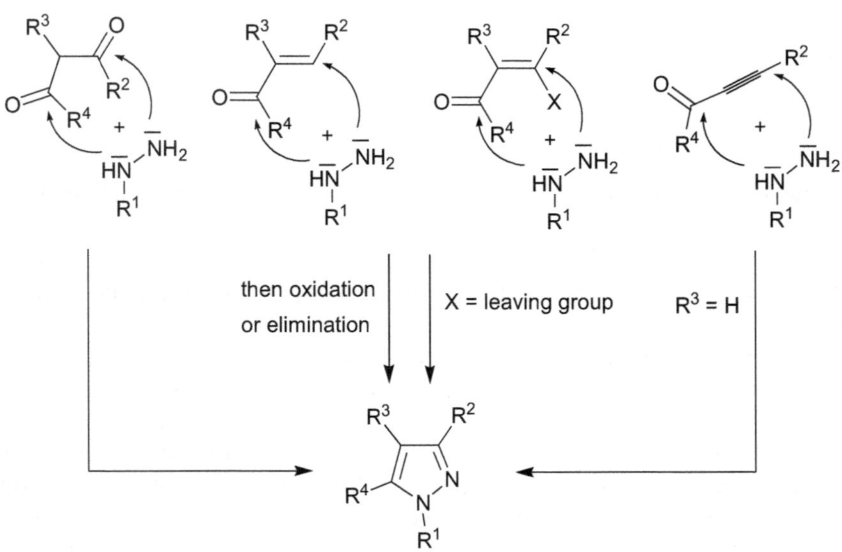

Scheme 5.2 Synthesis of pyrazoles by cyclocondensation between 1,3-dielectrophilic compounds and hydrazines.

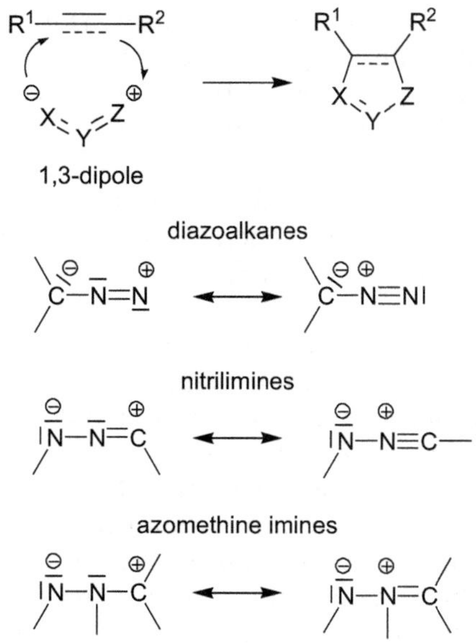

Scheme 5.3 Synthesis of pyrazoles by cyclocondensation between 1,3-dielectrophilic compounds and hydrazines.

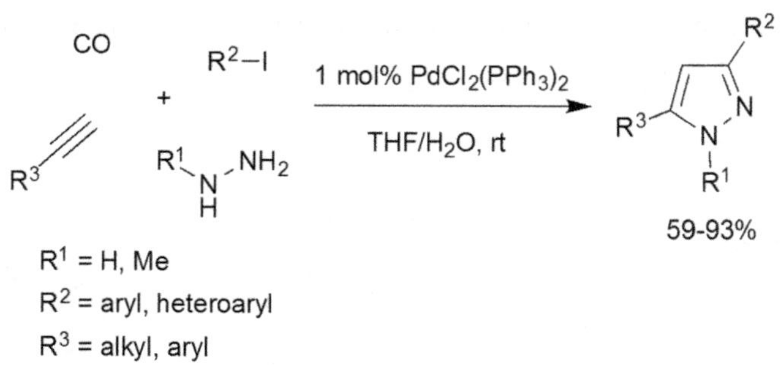

Scheme 5.4 Different substituted pyrazoles can be created in short time by MCR.[10]

Also, several examples were reported applying a [4 + 1]-cyclization approach for the formation of the pyrazole core. An example is shown in Scheme 5.5.[11]

Another strategy involves the cyclization of an acyclic substrate. The main advantage of this is that there are no issues regarding regioselectivity due the intramolecular nature of the reaction. During cyclization, a C–C-, a N–N- or a C–N-bond has to be created. The latter approach was applied in recent developed methodologies: Highly substituted pyrazoles could be synthesized by 5-*exo*/*endo-dig* cyclization of metal-activated alkynes by nucleophilic addition of nitrogen derivatives. One example of a 5-*exo*-dig hydroamination of an *in situ* formed alkyne is shown in Scheme 5.6.[12]

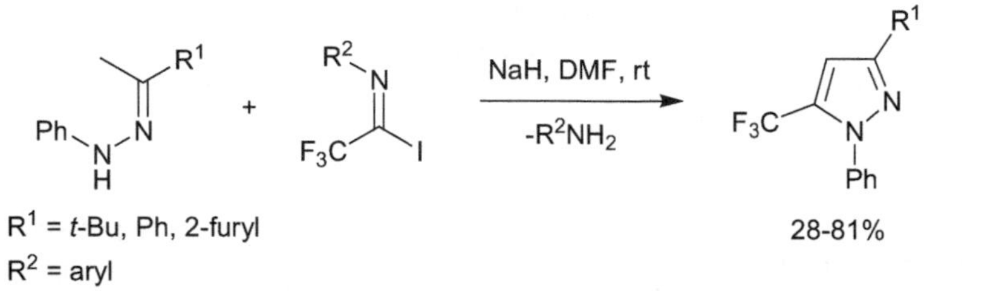

Scheme 5.5 Synthesis of different substituted pyrazoles in a [4 + 1]-cyclization.[11]

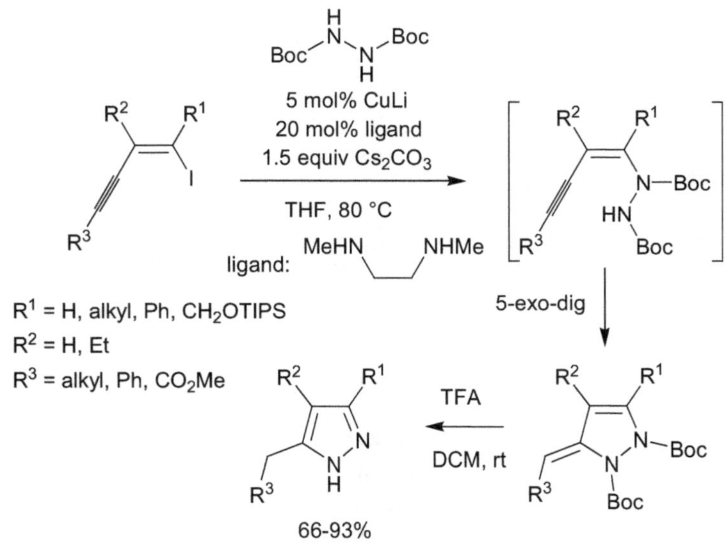

Scheme 5.6 Pyrazole synthesis *via* 5-*exo*-dig cyclization of an *in situ* formed acyclic substrate.[12]

Other strategies for the formation of the pyrazole moiety include: ring contraction (of six-[13] or seven-membered cycles),[14] ring enlargement (of three-[15–17] or four-membered cyclic systems), conversion of five-membered rings into pyrazoles (with size-retention),[18] and aromatization of dihydropyrazoles (by *e.g.* oxidation)[19] or aromatization of 3,3-disubstituted 3*H*-pyrazoles (by rearrangement).[20] Most of these strategies have already been systematically reviewed before and some examples are provided in Scheme 5.7.[5]

The so-constructed pyrazole moiety can be further substituted if the positions are accessible. Three positions are prone for a direct one-step conversion (Scheme 5.8).[6,9]

1. Position N1 can easily be alkylated or acylated by the use of the corresponding alkyl or acyl halide and base. Arylation on N1 can be performed by copper(I)-catalyzed reactions with various (hetero)aryl-halides.
2. Position C4 can be attacked by electrophiles to form, for example, the corresponding halo-, nitro-, or acyl-derivatives.

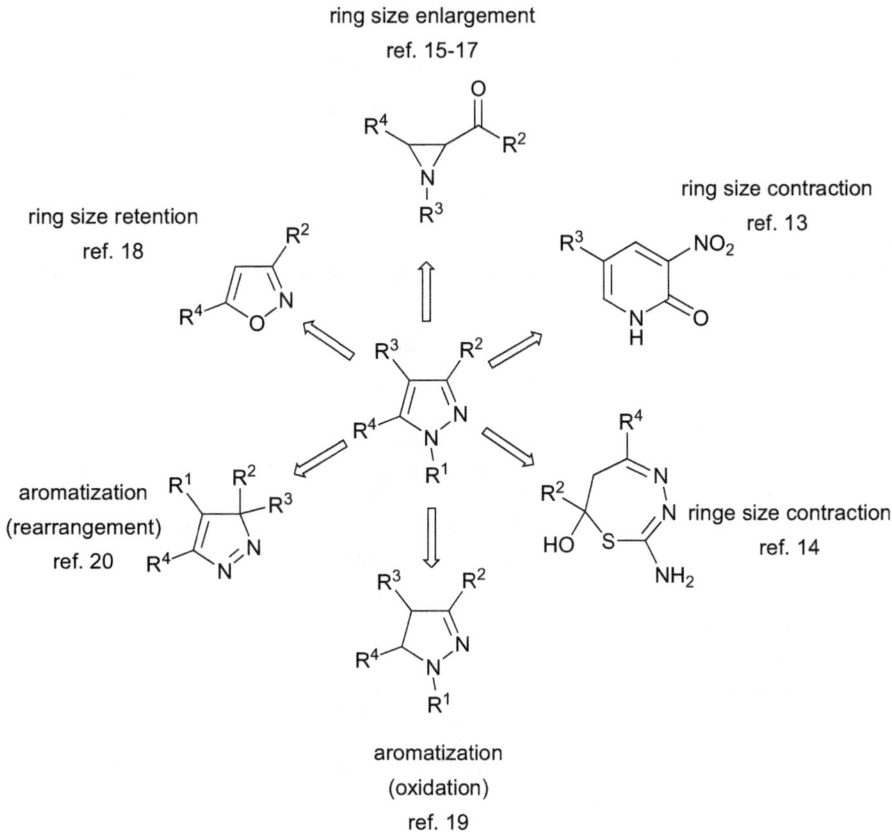

Scheme 5.7 Pyrazoles can be obtained from different classes of heterocycles.

3. Position C5 is accessible after metallation with *n*-BuLi (therefore the N1-positon has to be protected). Afterwards, the lithium organyls can be converted *via* metal–metal exchange to useful cross-coupling building blocks like boronates.

As shown, positions 1, 4 and 5 of the pyrazole core are directly selectable. Thus it might not be necessary to perform a *de novo* synthesis of a desired 1,3,4,5-substituted pyrazole core by assembling appropriate substituted building blocks. Instead, it might be possible to obtain the desired product through successive transformation of the pyrazole core. For instance, Knochel and co-workers could obtain different fully substituted pyrazoles synthesized by successive metallation of *N*-protected pyrazoles.[21]

5.1.3 Natural Products Containing Pyrazoles

Since the pyrazole moiety is in contrast to imidazole not readily available for organisms, natural products containing pyrazoles are rare. It seems that just

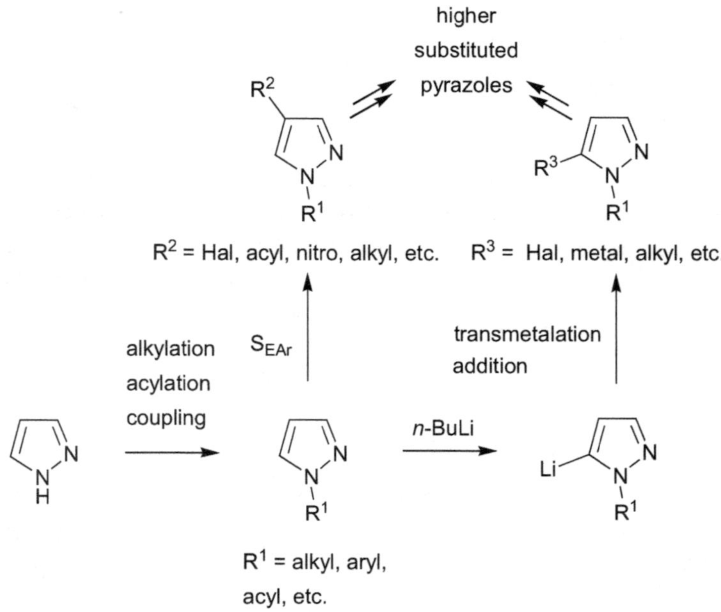

Scheme 5.8 Certain positions within the pyrazole core can be activated for further derivatization.

a few enzymes are able to form the N–N bond during the *de novo* synthesis of a pyrazole core.[4] This finding is quite interesting: whereby chemists could establish a huge number of different routes to the pyrazole moiety and could in this way synthesize many pyrazole-based lead structures, nature is remarkably limited in methods.

A recent review from Kumar and co-workers summarizes all known pyrazole-containing natural products along with their synthesis.[22] By now, the following pyrazole-containing natural products were isolated: L-α-Amino-β-(pyrazolyl-*N*)-propanoic acid, Withasomnine, 4-Hydroxywithasomnine, 4-Methoxywithasomnine, Pyrazofurin, Pyrazofurin B, Formycin (Formycin A), Formycin B, Oxoformycin B, Nostocine A, Fluviol A-E, Pyrazole-3(5)-carboxylic acid, 4-Methyl pyrazole-3(5)-carboxylic acid and 3-*n*-Nonylpyrazole. Some representatives are shown in Figure 5.2.

5.2 (Former) Marketed Drugs

5.2.1 Anti-inflammatory Drugs

5.2.1.1 Non-Steroidal Anti-Inflammatory Drugs (NSAID)

In addition to anti-inflammatory actions of Lonazolac (Figure 5.3), it has analgesic, antipyretic, and platelet-inhibitory actions. Just as other acetic acid derivatives like Indomethacin or Diclofenac, the drug blocks the synthesis of prostaglandins by inhibiting the cyclooxygenase.[23]

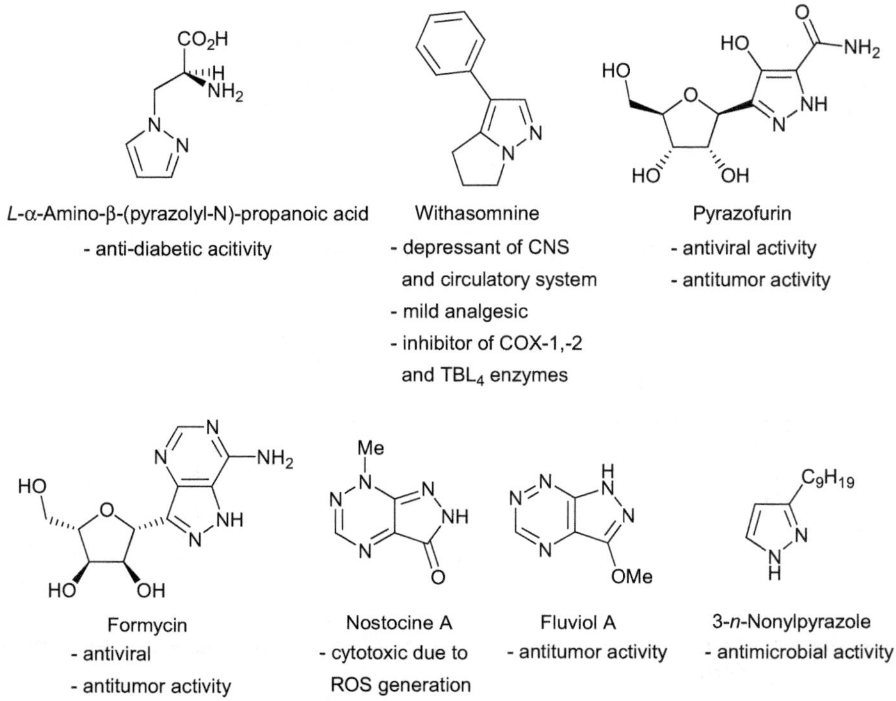

Figure 5.2 Pyrazole-containing natural products and their biological activities.[22]

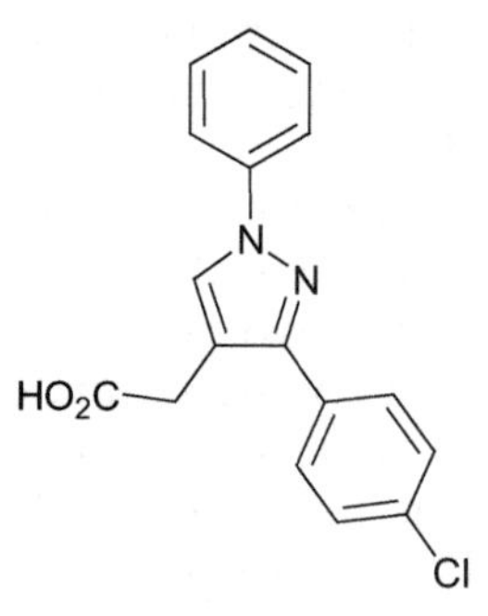

Lonazolac

Figure 5.3 Structure of Lonazolac.

Celecoxib (Figure 5.4) is a NSAID used in the treatment of osteoarthritis, rheumatoid arthritis, Morbus Bechterew, painful menstruation, acute pain, and menstrual symptoms, and to reduce numbers of colon and rectum polyps in patients with familial adenomatous polyposis (FAP).[24] It is marketed by Pfizer under the brand names Celebrex and Celebra. Celecoxib was branded in the European Union under the name Onsenal for the adjuvant therapy of FAP, but it was withdrawn from Pfizer in 2011. The analgesic

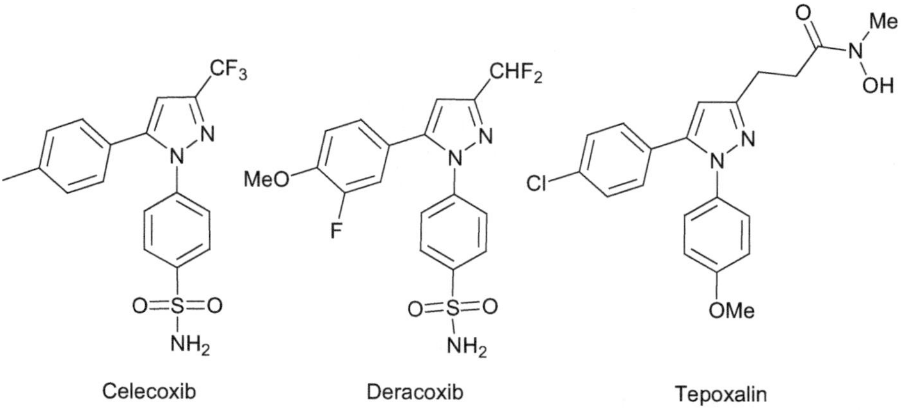

Figure 5.4 Structures of Celecoxib, Deracoxib and Tepoxalin.

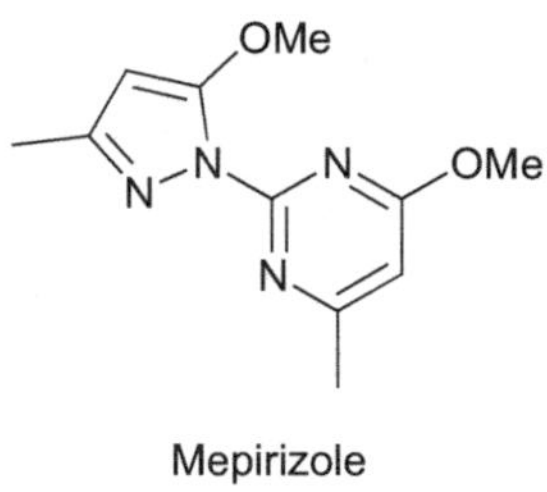

Figure 5.5 Structure of Mepirizole.

effect of Celecoxib is caused by the selective inhibition of the COX-2 enzyme and therefore decreased production of prostaglandins. It was intended to reduce adverse effects by selective inhibition of COX-2 and omitting the function of COX-1, but in general this strategy did not pay off. The structural prerequisites necessary for the COX-2-selectivity are well described in the literature.[25]

The molecule Deracoxib, shown in Figure 5.4 shares the same pharmacodynamic effects and is used for pain treatment in dogs. The structurally similar drug Tepoxaline is also used in veterinary medicine as a pain reliever and is believed to also be an inhibitor of the 5-LOX pathway.[26]

The pyrimidyl-pyrazole Mepirizole (or Epirizole) (Figure 5.5) exhibits antipyretic, analgesic and anti-inflammatory activity and was introduced from Daiichi Seiyaku in the middle 1970s.[27]

5.2.1.2 Glucocorticoids

Cortivazol is a high affinity ligand for the glucocorticoid receptor and is used in anti-inflammatory therapy.[28] The drug shows a high structural similarity to the natural receptor agonist cortisol (Figure 5.6).

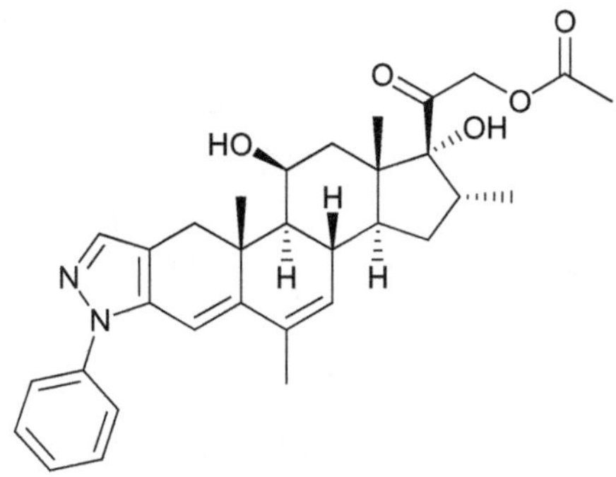

Figure 5.6 Structure of Cortivazol.

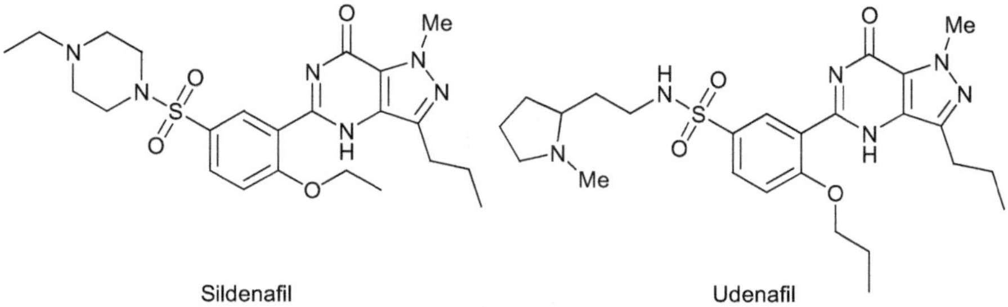

Figure 5.7 Structures of Sildenafil and Udenafil.

5.2.2 Vasodilators

Unlike any other drug, Sildenafil – marketed in 1998 by Pfizer under the brand name Viagra – is one of the most popular drugs of our time (Figure 5.7). By inhibition of the cGMP-degrading enzyme phosphodiesterase type 5 (PDE5), intracellular levels of the second messenger cGMP are increased, which leads to relaxation of smooth muscles in the wall of certain blood vessels.[24] In the treatment of erectile dysfunction, the resulting vasodilatation improves perfusion of the cavernous bodies from the penis. Patients with pulmonary arterial hypertension (PAH) can also benefit from the muscle relaxing effect within the pulmonary circulation system. Therefore, Sildenafil (branded as Revatio) received orphan drug designation as a medicine for PAH.[29]

Udenafil has been approved only in South Korea and is marketed under the brand name Zydena.[24]

Regadenoson (Figure 5.8) is basically a pyrazole-adenosine conjugate and is, like the natural substrate, an A2A adenosine receptor agonist.[24] Binding on the receptor causes hyperemia of the cardiac muscle due to coronary vasodilation, therefore Regadenoson can be used for myocardial perfusion imagining (MPI).[30]

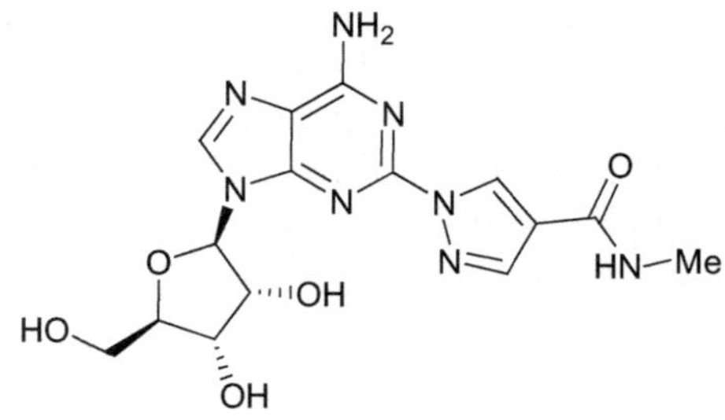

Figure 5.8 Structure of Regadenoson.

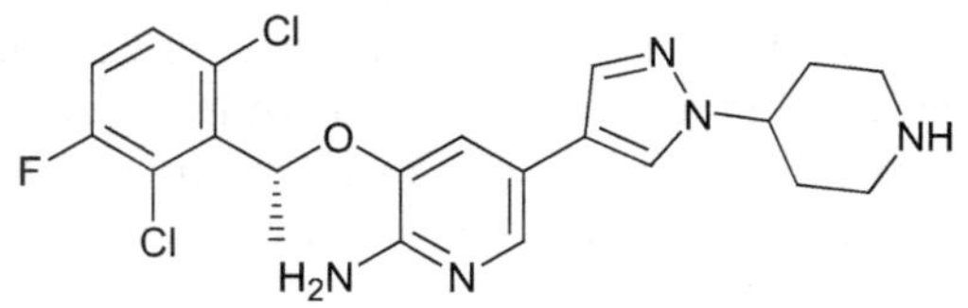

Figure 5.9 Structure of Crizotinib.

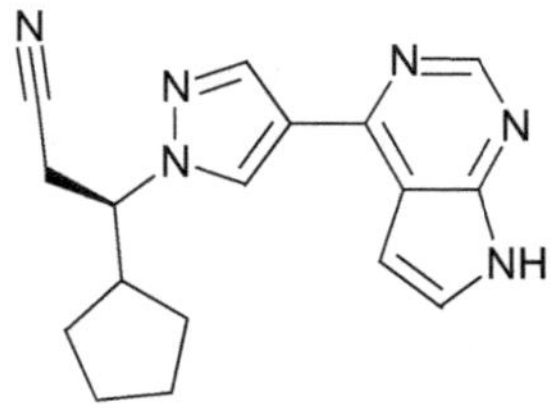

Figure 5.10 Structure of Ruxolitinib.

5.2.3 Tyrosine-kinase-inhibitors

Crizotinib (Figure 5.9) is used to treat patients with a type of lung cancer called non-small-cell lung cancer (NSCLC). It is only used if the NSCLC is 'ALK-positive', which means that the cancer cells contain certain defects affecting the gene responsible for the ALK-protein (ALK = anaplastic lymphoma kinase).[29] ALK belongs to the receptor tyrosine kinase (RTK)-family, which is involved in tumor growth, metastasis and the development of new blood vessels that supply the tumor.[29]

Crizotinib acts as an inhibitor of several tyrosine kinases, namely ALK, c-MET and ROS1 (c-ros oncogene 1).[31,32] The drug was branded as Xalkori by Pfizer and authorized by the FDA in 2011. Currently, just conditional approval is granted by EMA.

Ruxolitinib (Figure 5.10) is a drug for the treatment of intermediate or high-risk myelofibrosis, including primary myelofibrosis, post-polycythemia vera (post-PV) myelofibrosis and post-essential thrombocythemia (post-ET) myelofibrosis.[24] The drug acts as a kinase inhibitor that selectively diminishes the activity of Janus Associated Kinases (JAK) 1 and 2.[33] In the case of myelofibrosis, patients have abnormal JAK1 and JAK2 activity, which can be corrected by treatment with Ruxolitinib. Ruxolitinib (brand name Jakafi) was authorized by the FDA in 2011 as an orphan drug for the treatment of the aforementioned conditions.

5.2.4 Cannabinoid-receptor-antagonists

Rimonabant (Figure 5.11) acts as a selective inverse agonist on the CB1-receptor (CB = cannabinoid) and exhibits a fully substituted pyrazole core.[24] The drug was marketed in 2006 by Sanofi-Aventis in Europe as an anorectic drug for body mass reduction, but it was withdrawn later due to serious psychiatric problems.[29,34] The highly similar bromo-derivative Surinabant (SR147778) is currently being examined in clinical trials as an aid to smoking cessation and as an anti-obesity drug.[35]

5.2.5 Antibacterial Agents

Cefoselis (Figure 5.12) is a fourth-generation cephalosporin and has a broad spectrum of antibacterial activity against Gram-positive and Gram-negative bacteria, including *Pseudomonas aeruginosa* and methicillin-resistant *Staphylococcus aureus*.[36–38]

Sulfaphenazole (Figure 5.13) belongs to the group of antibacterial phenylpyrazoles. It is known to be a specific and strong inhibitor of CYP2C9.[39]

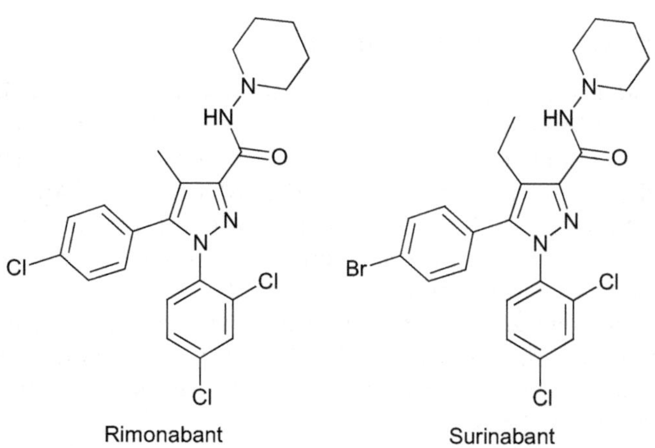

Figure 5.11 Structures of Rimonabant and Surinabant.

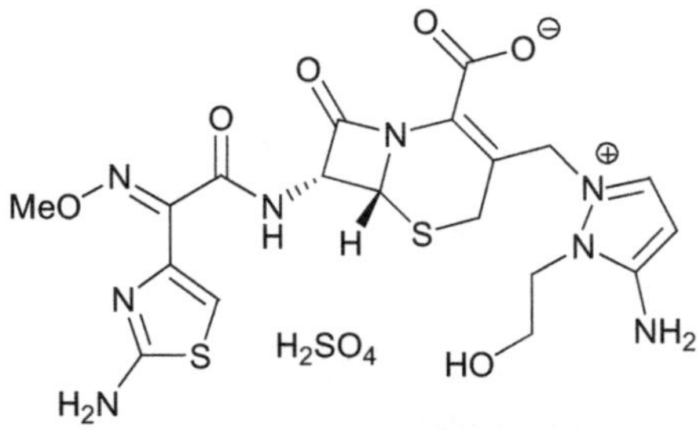

Figure 5.12 Structure of Cefoselis.

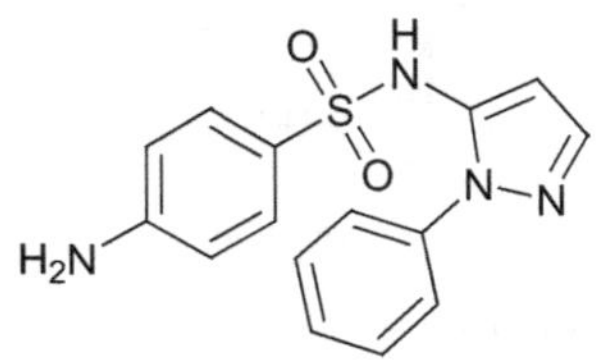

Figure 5.13 Structure of Sulfaphenazole.

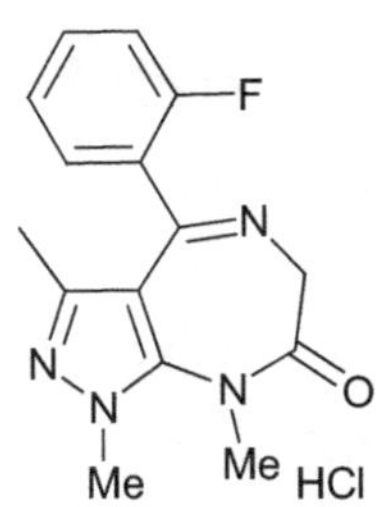

Figure 5.14 Structure of Zolazepam.

5.2.6 Miscellaneous

The benzodiazepine-related drug Zolazepam (Figure 5.14) is used in a 1 : 1 mixture with Tiletamine as an anesthetic in veterinary medicine (branded as Telazol).[23,40]

Quinpirole (Figure 5.15) is a D2/3 receptor agonist and is used in neurological research.[23]

Betazole (or Histalog, Figure 5.16) is a selective H2 agonist and it is used clinically to examine the gastric secretory function in the so-called (maximal) Histalog test.[23,41]

Stanozolol (Figure 5.17) is an anabolic dihydrotestosterone-analogue and was marketed as Winstrol in the early 1960s.[27] Because of severe adverse effects, Stanozolol is nearly obsolete apart from its positive effects as

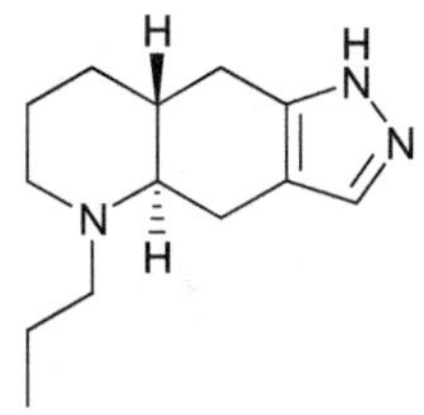

Figure 5.15 Structure of Quinpirole.

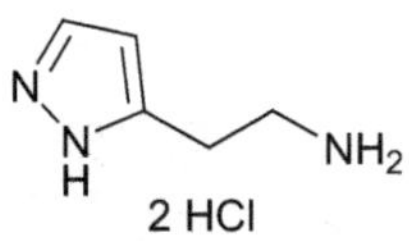

Figure 5.16 Structure of Betazole.

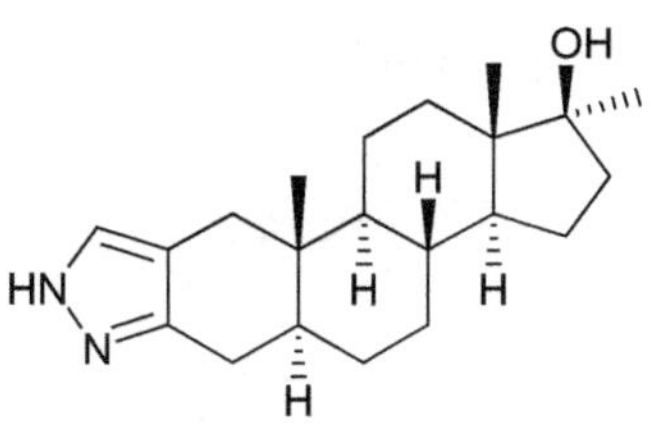

Figure 5.17 Structure of Stanozolol.

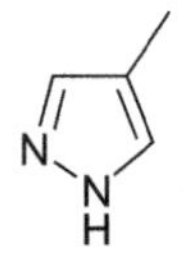

Figure 5.18 Structure of Fomepizole.

long-term therapy for hereditary angioedema.[42] The main applications of this drug remain in veterinary medicine. Also, like other anabolic steroids, Stanozolol is abused in the body building scene.

Fomepizole (4-Methylpyrazole, Figure 5.18) is used as an antidote for methanol or ethylene glycol intoxication by acting as a competitive inhibitor of alcohol dehydrogenase (ADH).[43] ADH catalyzes the generation of toxic metabolites by oxidation of either methanol or ethylene glycol.

5.2.7 Pyrazole and Pyrazolyl-ligands in Biological Active Metal Complexes

Metal complexes like Cisplatin or Carboplatin have been extensively studied and show great potential in cancer therapy. In such bioactive metal complexes, pyrazole and pyrazolyl-ligands play an increasingly important role. In a recent review, Keter and Darkwa summarized the scientific efforts on such pyrazole-containing metal complexes for the use as anticancer, antiviral, antibacterial, and antiparasitic agents.[44] To the best of the authors' knowledge, none of these complexes are marketed yet.

References

1. A. Chauhan, P. K. Sharma and N. Kaushik, *Int. J. ChemTech Res.*, 2011, **3**, 11–17.
2. L. Knorr, *Ber. Dtsch. Chem. Ges.*, 1883, **16**, 2597–2599.
3. L. Knorr, *Justus Liebigs Ann. Chem.*, 1887, **238**, 137–219.
4. T. Eicher, A. Speicher, S. Hauptmann and ebrary Inc, Wiley-VCH, Weinheim, 3., completely rev. and enl. edn., 2012.
5. B. Stanovnik and J. Svete, *Sci. Synth.*, 2002, **12**, 15–225.
6. S. Fustero, M. Sanchez-Rosello, P. Barrio and A. Simon-Fuentes, *Chem. Rev.*, 2011, **111**, 6984–7034.
7. S. Fustero, A. Simon-Fuentes and J. F. Sanz-Cervera, *Org. Prep. Proced. Int.*, 2009, **41**, 253–290.
8. A. R. Katritzky, *Comprehensive Heterocyclic Chemistry III*, Elsevier, Amsterdam, New York, 2008.
9. J. J. Li, *Heterocyclic Chemistry in Drug Discovery*, John Wiley & Sons, Hoboken, N.J., 2013.
10. M. S. Mohamed Ahmed, K. Kobayashi and A. Mori, *Org. Lett.*, 2005, **7**, 4487–4489.
11. H. B. Yu and W. Y. Huang, *Synlett*, 1997, 679–680.
12. R. Martin, M. Rodriguez Rivero and S. L. Buchwald, *Angew. Chem., Int. Ed.*, 2006, **45**, 7079–7082.
13. Y. M. Yutilov and N. N. Smolyar, *Khim. Geterotsikl.*, 1986, 1136.
14. T. E. Glotova, A. S. Nakhmanovich, M. V. Sigalov, T. N. Komarova, E. I. Kositsina, V. Y. Vitkovskii and I. D. Kalikhman, *Bull. Acad. Sci. USSR. Chem.*, 1987, **36**, 194–196.
15. N. H. Cromwell and M. A. Graff, *J. Org. Chem.*, 1952, **17**, 414–425.
16. N. H. Cromwell and H. Hoeksema, *J. Am. Chem. Soc.*, 1949, **71**, 716–719.
17. N. H. Cromwell and R. J. Mohrbacher, *J. Am. Chem. Soc.*, 1953, **75**, 6252–6256.
18. A. Alberola, C. Andres, A. G. Ortega and R. Pedrosa, *J. Heterocycl. Chem.*, 1984, **21**, 1575–1576.
19. W. A. F. Gladston and R. O. C. Norman, *J. Chem. Soc. C*, 1966, 1536–1540.
20. M. W. Majchrzak, E. Jefferson and J. Warkentin, *J. Am. Chem. Soc.*, 1990, **112**, 2449–2451.

21. C. Despotopoulou, L. Klier and P. Knochel, *Org. Lett.*, 2009, **11**, 3326–3329.
22. V. Kumar, K. Kaur, G. K. Gupta and A. K. Sharma, *Eur. J. Med. Chem.*, 2013, **69**, 735–753.
23. PubChem-Database, https://pubchem.ncbi.nlm.nih.gov/, Accessed 05/20/2014.
24. DrugBank-Database, www.drugbank.ca/, Accessed 05/20/2014.
25. T. D. Penning, J. J. Talley, S. R. Bertenshaw, J. S. Carter, P. W. Collins, S. Docter, M. J. Graneto, L. F. Lee, J. W. Malecha, J. M. Miyashiro, R. S. Rogers, D. J. Rogier, S. S. Yu, E. G. AndersonGd Burton, J. N. Cogburn, S. A. Gregory, C. M. Koboldt, W. E. Perkins, K. Seibert, A. W. Veenhuizen, Y. Y. Zhang and P. C. Isakson, *J. Med. Chem.*, 1997, **40**, 1347–1365.
26. D. C. Argentieri, D. M. Ritchie, M. P. Ferro, T. Kirchner, M. P. Wachter, D. W. Anderson, M. E. Rosenthale and R. J. Capetola, *J. Pharmacol. Exp. Ther.*, 1994, **271**, 1399–1408.
27. FDA, www.accessdata.fda.gov/Scripts/cder/drugsatfda/index.cfm, Accessed 05/20/2014.
28. J. A. Schlechte, S. S. Simons Jr, D. A. Lewis and E. B. Thompson, *Endocrinology*, 1985, **117**, 1355–1362.
29. EMA, www.ema.europa.eu, Accessed 05/20/2014.
30. M. D. Cerqueira, *Am. J. Cardiol.*, 2004, **94**, 33D–40D, discussion 40D-42D.
31. M. M. Awad, R. Katayama, M. McTigue, W. Liu, Y. L. Deng, A. Brooun, L. Friboulet, D. Huang, M. D. Falk, S. Timofeevski, K. D. Wilner, E. L. Lockerman, T. M. Khan, S. Mahmood, J. F. Gainor, S. R. Digumarthy, J. R. Stone, M. Mino-Kenudson, J. G. Christensen, A. J. Iafrate, J. A. Engelman and A. T. Shaw, *N. Engl. J. Med.*, 2013, **368**, 2395–2401.
32. J. J. Cui, M. Tran-Dube, H. Shen, M. Nambu, P. P. Kung, M. Pairish, L. Jia, J. Meng, L. Funk, I. Botrous, M. McTigue, N. Grodsky, K. Ryan, E. Padrique, G. Alton, S. Timofeevski, S. Yamazaki, Q. Li, H. Zou, J. Christensen, B. Mroczkowski, S. Bender, R. S. Kania and M. P. Edwards, *J. Med. Chem.*, 2011, **54**, 6342–6363.
33. R. A. Mesa, *IDrugs*, 2010, **13**, 394–403.
34. T. M. Fong and S. B. Heymsfield, *Int. J. Obes.*, 2009, **33**, 947–955.
35. ChlinicalTrialsRegister.eu, Accessed 05/21/2014.
36. T. Kuriyama, T. Karasawa, K. Nakagawa, S. Nakamura and E. Yamamoto, *Oral Microbiol. Immunol.*, 2002, **17**, 285–289.
37. M. W. Climo, S. M. Markowitz, D. S. Williams, C. G. Hale-Cooper and G. L. Archer, *J. Antimicrob. Chemother.*, 1997, **40**, 59–66.
38. P. Zalewski and J. Cielecka-Piontek, *Ann. Acad. Med. Silesiensis*, 2011, **65**, 77–81.
39. A. Mancy, S. Dijols, S. Poli, P. Guengerich and D. Mansuy, *Biochemistry*, 1996, **35**, 16205–16212.
40. H. A. DeWald, S. Lobbestael and D. E. Butler, *J. Med. Chem.*, 1977, **20**, 1562–1569.

41. L. D. Wruble, A. J. Cummins, J. Goldenberg and H. Schapiro, *Am. J. Dig. Dis.*, 1967, **12**, 1087–1090.
42. D. E. Sloane, C. W. Lee and A. L. Sheffer, *J. Allergy Clin. Immunol.*, 2007, **120**, 654–658.
43. M. J. Casavant, *Pediatrics*, 2001, **107**, 170.
44. F. K. Keter and J. Darkwa, *Biometals*, 2012, **25**, 9–21.

CHAPTER 6

Quinolines: Privileged Scaffolds in Medicinal Chemistry

ARANTXA ENCINAS LÓPEZ

Institute of Organic Chemistry, Karlsruhe Institute of Technology (KIT), Fritz-Haber-Weg 6, 76131 Karlsruhe, Germany
Email: arantxa.lopez@partner.kit.edu

6.1 Introduction

Nitrogen heterocycles play a central role both in medicinal chemistry and drug discovery. Among them, quinoline is a privileged nucleus,[1] since it is one of the key building elements for many naturally occurring compounds displaying a broad range of biological activities.[2] As an example, Quinine is a well-known antimalarial agent, Campthotecin and Luotonin A are antitumoral derivatives, and Graveolinine has antitubercular properties (Figure 6.1).

The most simple of these natural products, Quinoline, was isolated from coal tar by Friedlieb Ferdinand Runge in the 19th century. Shortly after, Charles Frédéric Gerhart obtained it from the decomposition of the alkaloid Cinchonidine (a Cinchona alkaloid), from where it got its current name. Quinoline occurs widely in coal tar, oil shale, and petroleum, and can also be produced by combustion of a number of substances, including tobacco. Nevertheless, coal tar currently remains the principal source of commercial quinoline.

From the chemical point of view, Quinoline is a colorless liquid at room temperature with a molecular formula of C_9H_7N and a molecular weight of 129.16. It has a rigid planar structure and displays similar reactions as pyridine or benzene (Figure 6.2).

RSC Drug Discovery Series No. 50
Privileged Scaffolds in Medicinal Chemistry: Design, Synthesis, Evaluation
Edited by Stefan Bräse

Published by the Royal Society of Chemistry, www.rsc.org

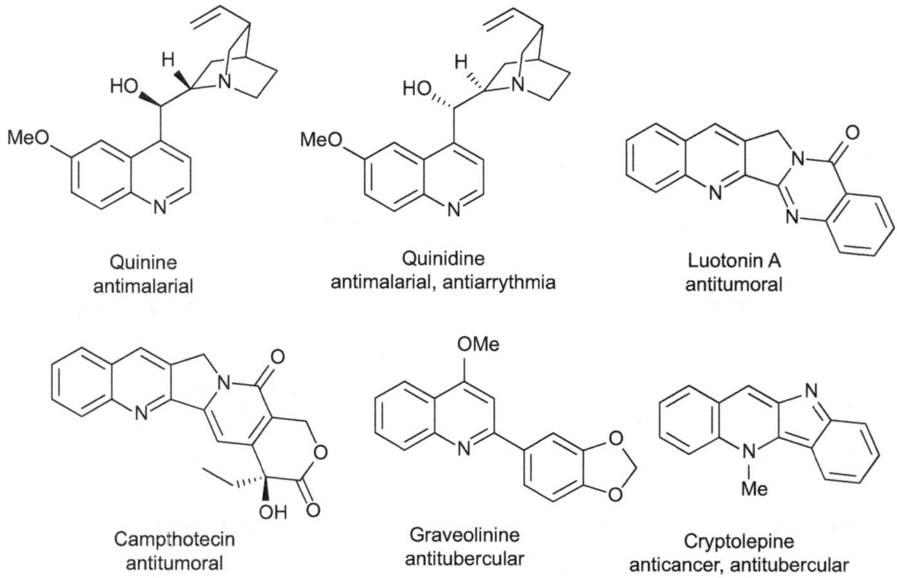

Figure 6.1 Some natural-occurring quinolines.

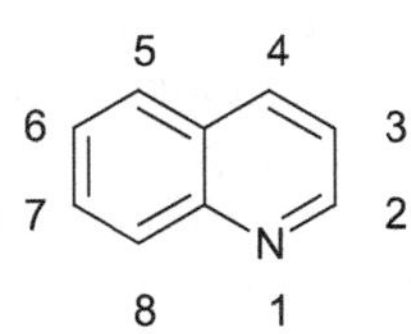

Figure 6.2 Quinoline chemical structure.

This versatile heterocycle has attracted the attention of both medicinal and synthetic chemists. Therefore, over the years, a large number of researchers have focused on the design of new bioactive molecules. As a consequence, several synthetic methods have been developed in order to obtain adequate diversity by substitution on the quinoline ring system, with the idea that different derivatives can show different biological effects.

6.2 Synthesis of Quinolines

The diverse chemical, biological, and pharmacological properties associated with quinoline compounds have encouraged the development of several synthetic methods for accessing this privileged scaffold.[3,4] The classic synthetic methods for the synthesis of the quinoline nucleus were developed at the end of the 19th century. These (nowadays well-known named reactions) include Skraup,[5,6] Doebner-von Miller,[7–9] Friedländer,[10,11] Pfitzinger,[12,13] and Combes[9,14] syntheses. All of them basically consist of a cyclo-condensation

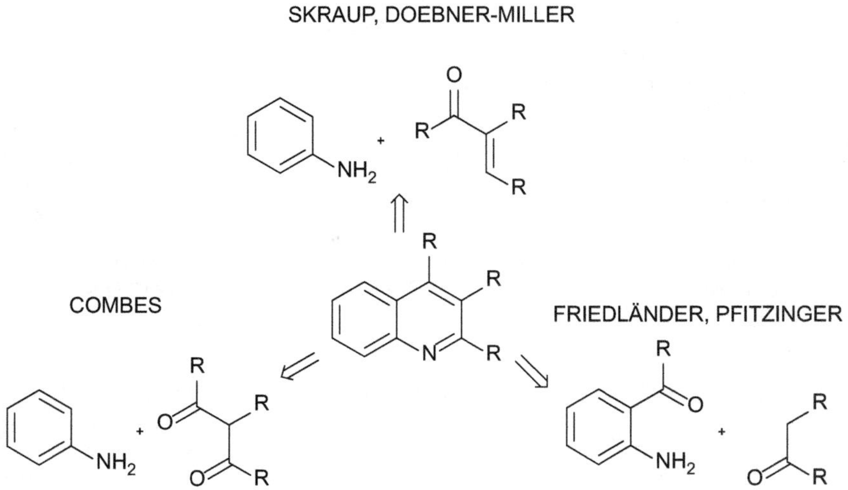

Scheme 6.1 Classic synthetic methods towards quinoline framework.

of aniline (or an aniline derivative) with a carbonyl compound followed by an aromatization with dehydration/oxidation reactions. The difference between these methods resides in the starting materials and the reaction conditions employed (acid or basic). In Combes, Doebner–Miller and Skraup[15] synthesis, the condensation takes place in acid conditions between an unsubstituted aniline and: (a) a β-diketone (in Combes) or (b) an α,β-unsaturated carbonyl in Doebner–Miller and Skraup synthesis, whereas in both Friedländer and Pfitzinger reactions, the condensation takes place between 2-aminoaryl ketones acting as the aniline starting material (2-aminobenzaldehyde or isatic acid,[16] respectively) and α methylene ketones (Scheme 6.1).

Although many of these conventional methods are very effective and still often employed, they suffer several important drawbacks in terms of environmental impact. Most of them need harsh conditions: strong bases or acids, high temperatures, or very long reaction times. They also produce large amounts of waste and, in some cases, expensive or toxic metal catalysts are needed. Therefore, in recent years, the development of alternative and efficient eco-friendly methodologies to new quinoline-based structures has not only been an important goal, but also a challenge for both synthetic and medicinal chemists.

As a result of this effort, different green synthetic approaches to the quinoline nucleus have been designed.[17] Some of them focus on the use of novel efficient sources of energy, like microwave synthesis,[18] photo-catalytic synthesis,[19] or ultrasound synthesis.[20] Others have opted for the elimination of solvents[21] in the chemical reaction or the replacement of hazardous solvents with environmentally benign solvents like water,[22] polyethylene glycol[23] or ionic liquids.[24,25] The replacement of toxic catalysts by so called green catalysts[26] have also been successfully employed.

All these new approaches offer significant advantages over the conventional strategies, by reducing waste production and energy consumption. They are also less hazardous, avoid tedious workup procedures and can be performed using renewable materials. All in all, the new synthetic methods towards quinoline scaffold are both economically and environmentally favorable and therefore the development of new synthetic methods fulfilling these premises remains a challenging active research area.

6.3 Biological Activity

Heterocyclic systems with a quinoline nucleus represent privileged moieties in medicinal chemistry, since this scaffold occurs in various natural products (especially in alkaloids), and in pharmacologically active substances. Quinoline based compounds exhibit a wide spectrum of biological activities, including antimalarial,[27,28] anticancer,[29] antitubercular,[30] antiviral,[31] anti-inflammatory,[32] antileshmanial,[33–35] antifungal,[36–38] antidepressant,[39] and antibiotic,[40] among others. Therefore, the quinoline skeleton has been chosen for the design of new bioactive molecules by several researchers, leading to the generation of a large number of synthetic and semisynthetic quinoline-based derivatives with important pharmacological applications (Figure 6.3).

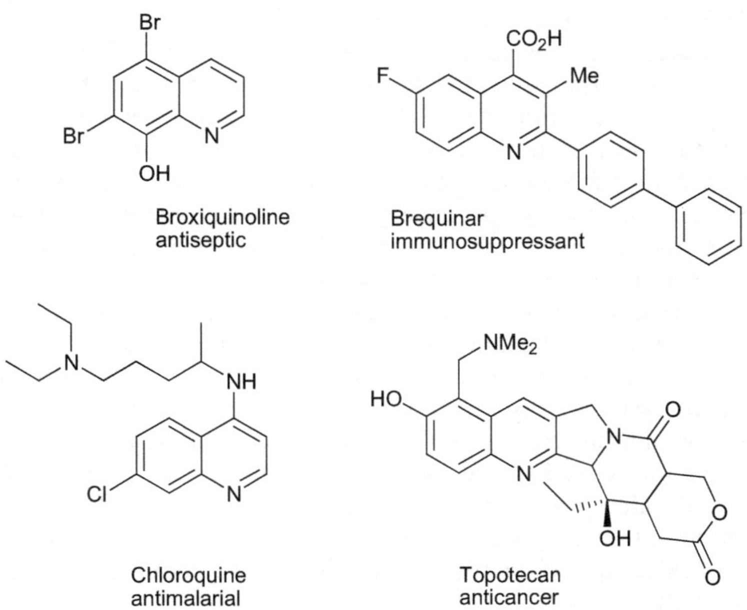

Figure 6.3 Synthetic and semisynthetic biologically active quinoline compounds.

6.3.1 Antimalarial

Malaria is the most lethal human parasitic infection. It is caused by five species of parasite that belong to the genus Plasmodium: *P. falciparum, P. vivax*, *P. ovale*, *P. malariae* and *P. knowlesi*. Of all of them, *P. falciparum* and *P. vivax* are considered the most important, *P. Falciparum* being the most virulent malaria parasite.

Quinoline-related compounds are historically among the most important antimalarial drugs ever used.[41] Furthermore, they are still one of the four major drug classes currently used to treat malaria (together with antifolates, artemisinin derivatives, and antimicrobials). Within the quinoline based antimalarials, three main substitution patterns can be found: the 4-aminoquinolines (for example, Chloroquine or Amodiaquine), 4-methanolquinolines (like Quinine, Quinidine, Mefloquine) and 8-aminoquinolines (Pamaquine, Primaquine). From all of them, only Quinine and Quinidine are naturally occurring drugs, whereas Chloroquine, Amodiaquine, Mefloquine, and Primaquine are synthetic compounds (Figure 6.4.).

The first, and probably the most important, quinoline-based antimalarial drug was Quinine, a naturally occurring quinoline related alkaloid, used broadly for its antimicrobial potency.[42] This alkaloid is present in the bark of cinchona trees, and it is believed that the Quechua people of Peru were already employing its extracts as a remedy for fever. The isolation, purification, and naming of Quinine was performed by Pelletier and Caventou in 1820.[43] Since that moment, Quinine has been used successfully for the treatment of malaria and it remains the antimalarial drug of choice. In the 1930s, and due to the decrease in the availability of natural sources and the poor synthetic availability of Quinine,[44] an intensive search for synthetic alternatives to Quinine started. As a result, in 1925 the 8-aminoquinoline[45] Pamaquine was synthesized. Regrettably, due to its toxicity, this therapy had to be abandoned.

Shortly after, Chloroquine[46] (a 4-aminoquinoline) was chemically synthesized for the first time. It is worth mentioning that Chloroquine has become the most famous drug for the treatment of malaria so far, due to its excellent clinical efficacy, low toxicity, ease of use and simple, cost-effective synthesis. Sadly, however, this therapeutic success has led to abuse in the use of this derivative. As a consequence, parasite resistance to Chloroquine emerged, resulting in the apparent uselessness of Chloroquine in several parts of the world.[47] Thus, through the 20th century, investigations focused on the development of alternative molecules to overcome Chloroquine resistance. By systematic synthetic modifications around the quinoline ring, several new antimalarials were discovered: the 4-aminoquinoline (Amodiaquine) was introduced in the 1940s and has been used for years. Another 8-aminoquinoline (Primaquine) has also been used since the 1950s for the eradication of liver stages in course of *P. vivax* infections; and the quinoline methanol (Mefloquine) or the Bisquinoline[48] (Piperaquine) were both developed in the 1980s.

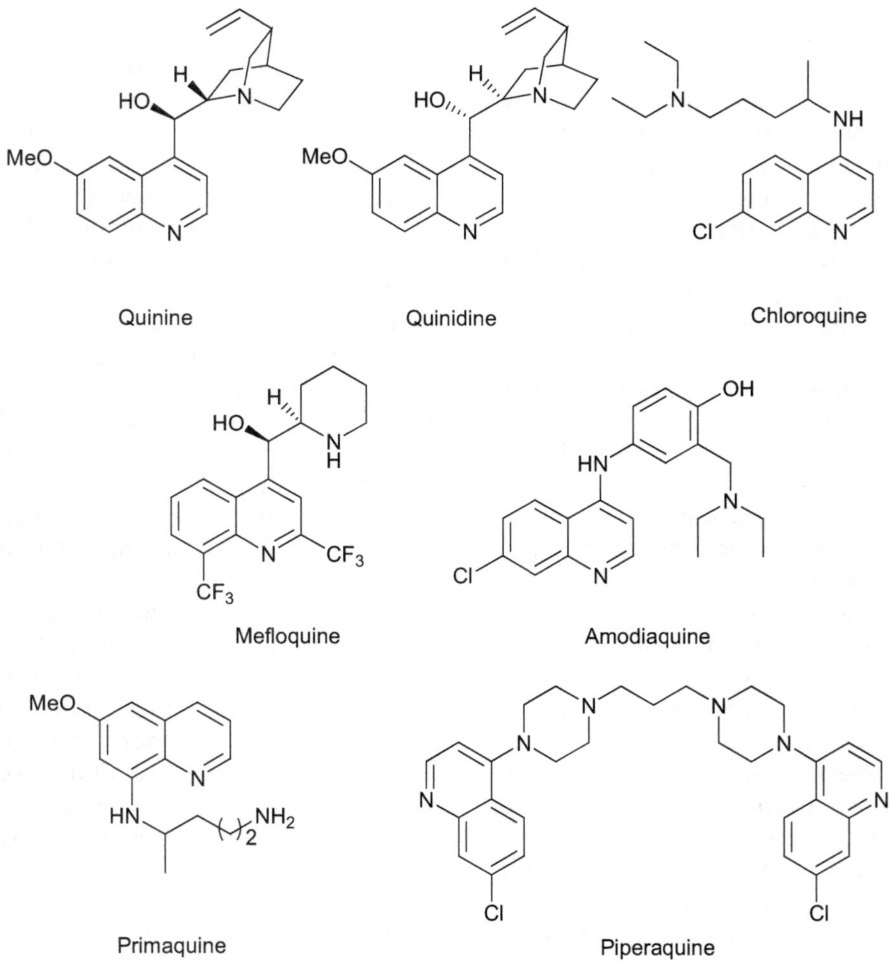

Figure 6.4 Quinoline antimalarial drugs.

Although many effective antimalarial compounds have been introduced during the 20th and 21st centuries, Quinine, the original quinoline-based antimalarial, is still used to treat the disease in certain critical circumstances, such as severe malaria in poor regions and in areas with Chloroquine resistance.[49]

Likewise Chloroquine, despite its drawbacks, is nowadays still in use as treatment for malaria caused by *P. ovale*, *P. malariae*, and, in most regions, *P. vivax*.[50]

In any case, there is still an urgent need for new antimalarial agents that are active against drug-resistant malaria strains. New quinoline derivatives for the treatment of malaria are therefore still being searched.[51]

6.3.2 Antitumoral

Cancer is a leading cause of death worldwide. There are several naturally occurring, as well as synthetic and semisynthetic, quinoline-based molecules that have been reported to have antiproliferative and antitumor activity. For example, the natural alkaloid Camptothecin[52] and its semisynthetic analogues Topotecan (Hycamptin®) and Irinotecan (Camptosar®). Luotonin A,[53] an alkaloid that is structurally similar to Camptothecine, indoloquinoline[54–56] Cryptolepine and Dofequidar (MS-209) are some examples of cytotoxic quinolines with established antitumor activity (Figure 6.5).

The development of anticancer quinoline derivatives is of great interest, since these molecules follow different mechanisms of action, such as inhibition of (a) the DNA enzyme topoisomerase I or II,[57] (b) tubulin,[58] (c) vascular endothelial growth factor (VEGFR),[59] (d) carbonic anhydrase,[60] (e) cMet kinase.[61] They also target tumor hypoxia,[62] regulate free-radicals, and increase the activity of superoxide dismutase,[63] among others.

Because the quinoline framework has long been considered for the design of novel anticancer agents, research for new synthetic quinoline based molecules is an ever-active field.[64–67]

Regrettably, as in other therapeutic areas, there are serious limitations in the treatment of cancer. The emergence of tumor cells exhibiting multidrug resistance (MDR) and the low tumor selectivity represents a serious medical problem.[68] Thus, the design of novel compounds with high efficacy but also specificity has been of great interest for the treatment of tumors. One approach to solve these therapeutic issues has been the "repositioning" of existing drugs. This allows the process of drug development to be accelerated, since many existing drugs have already been approved by regulatory agencies. Therefore, it is possible to directly focus on the improvement of their efficacy and specificity. Following this principle, it has been reported that Chloroquine (known for its antimalarial activity) has anticancer properties in combination with radiation or Akt inhibitors. In this sense, some Chloroquine derivatives with modified substitution pattern have been synthesized and also tested as anticancer molecules.[69]

Similarly, the antimalarial quinidine has been employed for the treatment of cancer[70]

6.3.3 Antitubercular

Tuberculosis is a lung infection caused mainly by *Mycobacterium tuberculosis*. It is the leading bacterial infectious agent and the second leading infectious cause of mortality behind only HIV/AIDS. The development of resistance to the existing drugs together with the association of tuberculosis and HIV infection has prompted the spread of this disease. These facts and the adverse effects showed by the first- and second-line antituberculosis drugs have led to tuberculosis becoming a major health threat to humankind. Thus, in the last few years there has been a strong interest in

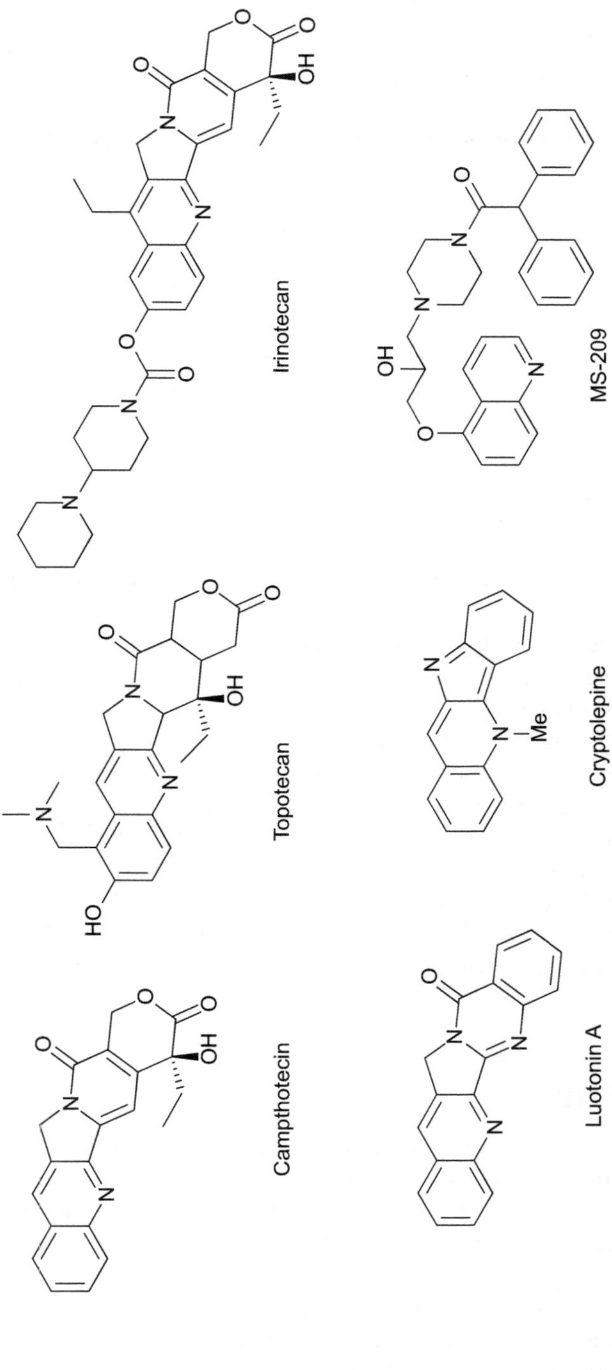

Figure 6.5 Quinoline-based anticancer derivatives.

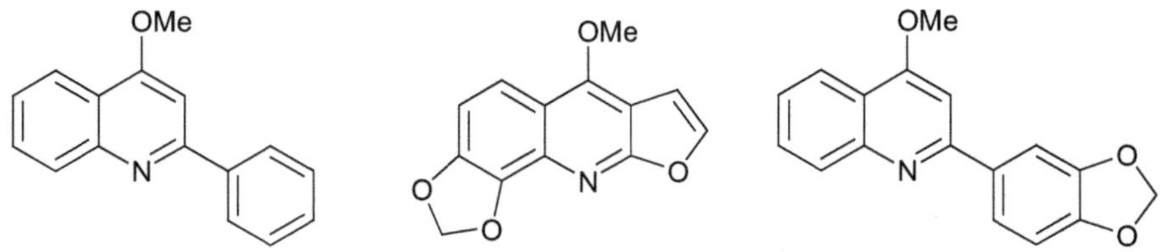

Figure 6.6 *Lunasia Amara Blanco* alkaloids with antitubercular activity.

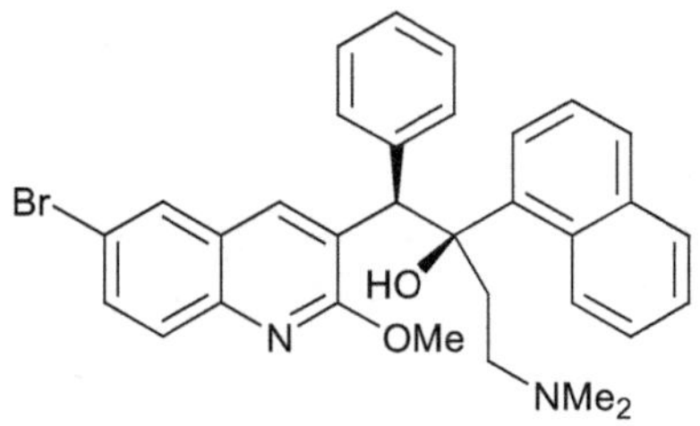

Figure 6.7 Bedaquiline (Sirturo™).

developing new antitubercular drugs agents with novel mechanisms of action.[71,72]

Quinoline-based compounds are known to show antitubercular activity, such as the quinoline alkaloids isolated from the leaves of *Lunasia amara Blanco* (family *Rutaceae*)[73] (Figure 6.6.)

Moreover, the majority of quinoline-based antimalarial drugs have shown antitubercular properties.[74] The use of quinoline derivatives as antitubercular agents continues to attract interest among scientific groups, not only in academia but also in the pharmaceutical industry.[75,76] This effort has recently led to great success. At the end of 2012, the quinoline-based compound Bedaquiline (Sirtruro™) was approved by the FDA as antitubercular drug for pulmonary multidrug-resistant tuberculosis, making it the first agent in a new class of anti-TB drugs to be introduced in 40 years[77] (Figure 6.7).

6.3.4 Anti-HIV

Acquired immunodeficiency syndrome (AIDS), caused by infection with the human immunodeficiency virus (HIV) continues to be a worldwide epidemic. The emergence of drug-resistant virus strains, and the undesired side effects of current drugs, has limited the utility of many of the conventional antiretroviral drugs. Therefore, the identification of novel targets and new lead molecules has become urgent. Due to their broad biological activity, quinoline compounds have been considered as good starting materials for the search of novel anti-HIV agents.[78,79]

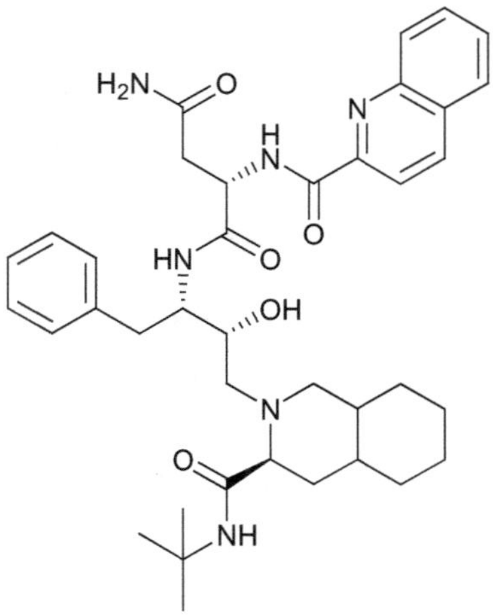

Figure 6.8 Saquinavir.

In the 1990s, a quinoline-containing drug (Saquinavir), was the first protease inhibitor approved by the FDA. It is a highly specific inhibitor of HIV-1 and HIV-2 proteases (Figure 6.8).

6.3.5 Miscellaneous

In addition to the antimalarial, antitumoral, anti-HIV, anti-inflammatory, antileshmanial, antifungal, antidepressant, and antibacterial activities described above, quinoline molecules have also recently been reported to act as antioxidants,[80] selective CB2 receptor agonists,[81] inhibitors of several enzymes including COXs,[82] phosphodiesterases (PDEs)[83,84] LRRK2,[85] farnesyl pyrophosphate synthase (FPPS),[86] or diacylglycerol acyltransferase (DGAT),[87] to name a few.

In summary, and taking in account the broad activities shown by quinoline based molecules, it is clear that the quinoline skeleton is a privileged scaffold in medicinal chemistry. Important therapeutic areas like cancer, HIV, malaria and neurodegenerative diseases (Alzheimer's, Parkinson's *etc.*) can be addressed with molecules based on this privileged structure.

6.4 Prominent Commercialized Drugs with Quinoline Scaffold

Apart from the already discussed antimalarials (Quinine, Quinidine, Chloroquine, Mefloquine, Amodiaquinine, Primaquine: see Figure 6.4), antitumorals (Camptothecin, Irinotecan, Topotecan. See Figure 6.5), antitubercular drugs (Bedaquiline, see Figure 6.7) and anti-HIV drugs (Saquinavir, see Figure 6.8), there are other prominent FDA approved quinoline-containing drugs like the anaesthetic Dibucain, the antitumoral Lenvatinib, the antiviral Clioquinol, and the antiasthmathic Montelukast (Figure 6.9).

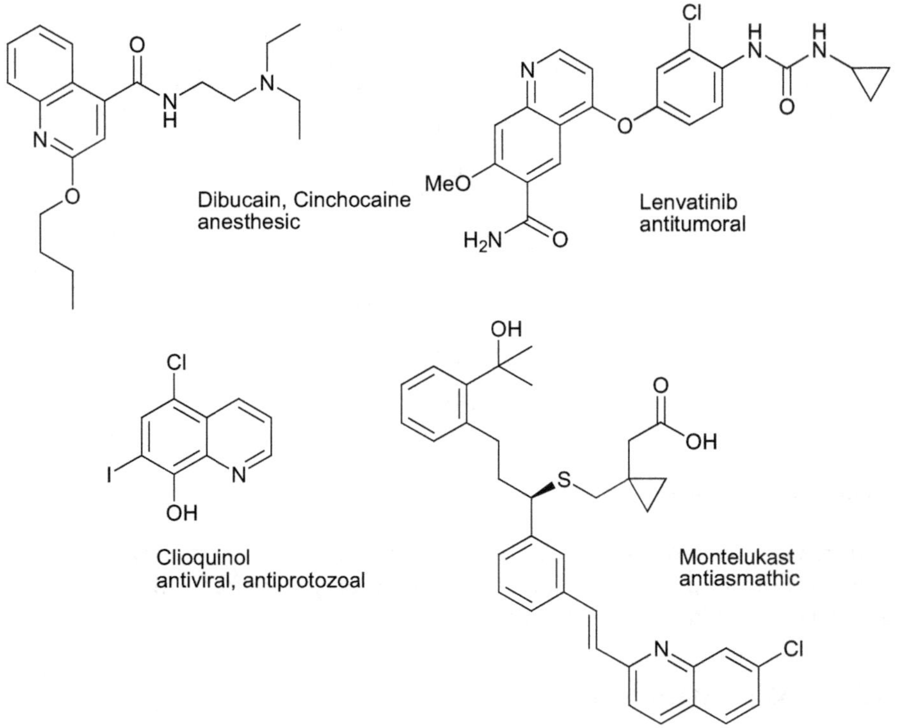

Figure 6.9 Prominent FDA-approved quinoline-containing drugs.

References

1. M. E. Welsch, S. A. Snyder and B. R. Stockwell, *Curr. Opin. Chem. Biol.*, 2010, **14**, 1.
2. S. Kumar, S. Bawa and H. Gupta, *Mini-Rev. Med. Chem.*, 2009, **9**, 1648.
3. R. H. Manske, *Chem. Rev.*, 1942, **30**, 113.
4. G. Jones, *Chemistry of Heterocyclic Compounds: Quinolines, Part I*, John Wiley & Sons, Inc., 2008, vol. 32, pp. 93–318.
5. Z. H. Skraup, *Ber. Dtsch. Chem. Ges.*, 1880, **13**, 2086.
6. R. H. F. Manske and M. Kulka, *Org. React.*, 1953, 7, 59.
7. O. Doebner and W. von Miller, *Ber. Dtsch. Chem. Ges.*, 1881, **14**, 2812.
8. O. Doebner and W. von Miller, *Ber.*, 1883, **16**, 2464.
9. F. W. Bergstrom, *Chem. Rev.*, 1944, **35**, 77.
10. P. Friedländer, *Ber. Dtsch. Chem. Ges.*, 1882, **15**, 2572.
11. P. Friedländer and C. F. Gohring, *Ber. Dtsch. Chem. Ges.*, 1883, **16**, 1833.
12. W. Pfitzinger, *J. Prakt. Chem.*, 1886, **33**, 100.
13. G. Jones, *The Chemistry of Heterocyclic Compounds*, ed. A. Weissberger and E.C. Taylor, John Wiley and Sons, Chichester, 1977, vol. 32, pp. 93–318.
14. A. Combes, *Bull. Soc. Chim. Fr.*, 1883, **49**, 89.

15. In the Skraup reaction the carbonyl derivative (acrolein) is generated in situ by dehydration of glycerol.
16. In the Pfitzinger reaction the isatic acid is generated from isatin.
17. S. M. Prajapati, K. D. Patel, R. H. Vekariya, S. N. Panchal and H. D. Patel, *RSC Adv.*, 2014, **4**, 24463.
18. H. Saggadi, D. Luart, N. Thiebault, I. Polaert, L. Estel and C. Len, *Catal. Commun.*, 2014, **44**, 15.
19. M. Austin, O. J. Egan, R. Tully and A. C. Pratt, *Org. Biomol. Chem.*, 2007, **5**, 3778.
20. J. V. Prasad, J. S. Reddy, N. R. Kumar, K. A. Solomon and G. Gopikrishna, *J. Chem. Sci.*, 2011, **123**, 673.
21. M. K. Chaudhuri and S. Hussain, *J. Chem. Sci.*, 2006, **118**, 199.
22. Q. Shen, L. Wang, J. Yu, M. Liu, J. Qiu, L. Fang, F. Guo and J. Tang, *Synthesis*, 2012, **44**, 389.
23. K. Karnakar, S. Narayana Murthy, K. Ramesh, G. Satish, J. Babu Nanubolu and Y. V. D. Nageswar, *Tetrahedron Lett.*, 2012, **53**, 2897.
24. A. S. Amarasekara and M. A. Hasan, *Tetrahedron Lett.*, 2014, **55**, 3319.
25. E. T. Kermani, H. Khabazzadeh and T. Jazinizadeh, *J. Heterocycl. Chem.*, 2011, **48**, 1192.
26. R. P. Pandit and Y. R. Lee, *RSC Adv.*, 2013, **3**, 22039.
27. K. Kaur, M. Jain, R. P. Reddy and R. Jain, *Eur. J. Med. Chem.*, 2010, **45**, 3245.
28. W.-J. Lu, K. J. Wicht, L. Wang, K. Imai, Z.-W. Mei, M. Kaise, I. E. T. E. Sayeda, T. J. Eganb and T. Inokuchia, *Eur. J. Med. Chem.*, 2013, **64**, 498.
29. W. A. Denny, W. R. Wilson, D. C. Ware, G. J. Atwell, J. B. Milbank and R. J. Stevenson, *U.S. Pat.* 7064117, 2006.
30. A. Lilienkampf, J. Mao, B. Wan, Y. Wang, S. G. Franzblau and A. P. Kozikowski, *J. Med.Chem.*, 2009, **52**, 2109.
31. R.-H. Guo, Q. Zhang, Y.-B. Ma, X.-Y. Huang, J. Luo, L.-J. Wang, C.-A. Geng, X.-M. Zhang, J. Zhou, Z.-Y. Jiang and J.-J. Chen, *Bioorg. Med. Chem.*, 2011, **19**, 1400.
32. S. Mukherjee and M. Pal, *Curr. Med. Chem.*, 2013, **20**, 4386.
33. V. S. Gopinath, J. Pinjari, R. T. Dere, A. Verma, P. Vishwakarma, R. Shivahare, M. Moger, P. S. K. Goud, V. Ramanathan, P. Bose, M. V. S. Rao, S. Gupta, S. K. Puri, D. Launay and D. Martin, *Eur. J. Med. Chem.*, 2013, **69**, 527.
34. D. Bompart, J. Núñez-Durán, D. Rodríguez, V. V. Kouznetsov, C. M. Meléndez Gómez, F. Sojo, F. Arvelo, G. Visbal, A. Alvarez, X. Serrano-Martín and Y. García-Marchán, *Bioorg. Med. Chem.*, 2013, **21**, 4426.
35. N. Campos Vieira, C. Herrenknecht, J. Vacus, A. Fournet, C. Bories, B. Figadère, L. S. Espindola and P. M. Loiseau, *Biomed. Pharmacother.*, 2008, **62**, 684.
36. R. Musiol, M. Serda, S. Hensel-Bielowka and J. Polanski, *Curr. Med. Chem.*, 2010, **17**, 1960.

37. V. V. Kouznetsov, C. M. Meléndez Gómez, M. G. Derita, L. Svetaz, E. del Olmo and S. A. Zacchino, *Bioorg. Med. Chem.*, 2012, **20**, 6506.
38. S Vandekerckhove, H. G. Tran, T. Desmet and M. D'hooghe, *Bioorg. Med. Chem. Lett.*, 2013, **23**, 4641.
39. M. Tran and G. P. Stack, Antidepressant Azaheterocyclymethyl derivatives of 2, 3- dihydro-1,4- dioxino(2,3-F)quinoline, *US Pat.* 0165245 A1, 2002.
40. A. Mahamoud, J. Chevalier, A. Davin-Regli, J. Barbe and J.-M. Pages, *Curr. Drug Targets*, 2006, 7, 843.
41. K. Kaur, M. Jain, R. P. Reddy and R. Jain, *Eur. J. Med. Chem.*, 2010, **45**, 3245.
42. F. Rocco, *The Miraculous Fever-Tree. The Cure that Changed the World*, Harper Collins, San Francisco, 2004.
43. M. Delepine, *J. Chem. Educ.*, 1951, **28**, 454.
44. T. S. Kaufman and E. A. Rúveda, *Angew. Chem., Int. Ed.*, 2005, **44**, 854.
45. N. Vale, R. Moreira and P. Gomes, *Eur. J. Med. Chem.*, 2009, **44**, 937.
46. R. F Loeb, W. M. Clarke, G. R. Coateney, L. T. Coggeshall, F. R. Dieuaide, A. R. Dochez, E. G. Hakansson, E. K. Marshall, S. C. Marvel, O. R. McCoy, J. J. Sapero, W. H. Serbell, J. A. Shannon and G. A Carden, *J. Am. Med. Assoc.*, 1946, **130**, 1069.
47. M. Foley and L. Tilley, *Int. J. Parasitol.*, 1997, **27**, 231.
48. S. Vangapandu, M. Jain, K. Kaur, S. Patel, P. Patil and R. Jain, *Med. Res. Rev.*, 2007, **27**, 65.
49. J. Achan, A. O. Talisuna, A. Erhart, A. Yeka, J. K. Tibenderana, F. N. Baliraine, P. J. Rosenthal and U. D'Alessandro, *Malar. J.*, 2011, **10**, 144.
50. C. C. Hand and S. R. Meshnick, *J. Infect. Dis.*, 2011, **203**, 11.
51. S. Vandekerckhove, C. Müller, D. Vogt, C. Lategan, P. J. Smith, K. Chibale, N. De Kimpe and M. D'hooghe, *Bioorg. Med. Chem. Lett.*, 2013, **23**, 318.
52. M. E. Wall, M. C. Wani, C. E. Cook, K. H. Palmer, A. T. McPhail and G. A. Sim, *J. Am. Chem. Soc.*, 1966, **88**, 3888.
53. S. Dallavalle, L. Merlini, G. L. Beretta, S. Tinelli and F. Zunino, *Bioorg. Med. Chem. Lett.*, 2004, **14**, 5757.
54. J. Lavrado, R. Moreira and A. Paulo, *Curr. Med. Chem.*, 2010, **17**, 2348.
55. K. Suresh, V. K. Eyunni, J. R. Etukala and S. Y. Ablordeppey, *Mini-Rev. Med. Chem.*, 2008, **8**, 538.
56. C.-M. Lu, Y.-L. Chen, H.-L. Chen, C.-A. Chen, P.-J. Lu, C.-N. Yang and C.-C. Tzeng, *Bioorg. Med. Chem.*, 2010, **18**, 1948.
57. W. Luniewski, J. Wietrzyk, J. Godlewska, M. Switalska, M. Piskozub, W. Peczynska-Czoch and L. Kaczmarek, *Bioorg. Med. Chem. Lett.*, 2012, **22**, 6103.
58. H.-Y Lee, J.-Y. Chang, C.-Y. Nien, C-C. Kuo, K.-H. Shih, C.-H. Wu, C.-Y. Chang, W.-Y. Lai and J.-P. Liou, *J. Med. Chem.*, 2011, **54**, 8517.
59. Y. Yang, L. Shi, Y. Zhou, H.-Q. Li, Z.-W. Zhu and H.-L. Zhu, *Bioorg. Med. Chem. Lett.*, 2010, **20**, 6653.

60. M. M. Ghorab, F. A. Ragab and M. M. Hamed, *Arzneimittelforsch.*, 2010, **60**, 141.
61. S. Li, Y. Zhao, K. Wang, Y. Gao, J. Han, B. Cui and P. Gong, *Bioorg. Med. Chem.*, 2013, **21**, 2843.
62. R. G. Boyle and S. Travers, *Anti-cancer Agents Med. Chem.*, 2006, **6**, 281.
63. A. E. Rashad, W. A. El-Sayed, A. M. Mohamed and M. M. Ali, *Arch. Pharm. Chem. Life Sci.*, 2010, **8**, 440.
64. V. R. Solomon and H. Lee, *Curr. Med. Chem.*, 2011, **18**, 1488.
65. R. K. Arafa, G. H. Hegazy, G. A. Piazza and A. H. Abadi, *Eur. J. Med. Chem.*, 2013, **63**, 826.
66. E. J. Koh, M. I. El-Gamal, C.-H. Oh, S. H. Lee, T. Sim, G. Kim, H. Choi, J. H. Hong, S.-g. Lee and K. H. Yoo, *Eur. J. Med. Chem.*, 2013, **70**, 10.
67. A. R. Ellanki, A. Islam, V. S. Rama, R. P. Pulipati, D. Rambabu, G. R. Krishna, C. M. Reddy, K. Mukkanti, G. R. Vanaja, A. M. Kalle, K. S. Kumar and M. Pal, *Bioorg. Med. Chem. Lett.*, 2012, **22**, 3455.
68. A. Ganguly, K. Banerjee, P. Chakraborty, S. Das, A. Sarkar, A. Hazra, M. Banerjee, A. Maity, M. Chatterjee, N. B. Mondal and S. K. Choudhuri, *Biomed. Pharmacother.*, 2011, **65**, 387.
69. V. R. Solomon, Ch. Hu and H. Lee, *Bioorg. Med. Chem.*, 2010, **18**, 1563.
70. Q. Zhou, Z. K. Melkoumian, A. Lucktong, M. Moniwa, J. R. Davie and J. S. Strobl, *J. Biol Chem.*, 2000, **275**, 35256.
71. A. Nayyar, S. R. Patel and R. Jain, *Front. Med. Chem.*, 2009, **4**, 506.
72. A. Koul, E. Arnoult, N. Lounis, J. Guillemont and K. Andries, *Nature*, 2011, **469**, 483.
73. A. M. Aguinaldo, L. T. Byrne, F. Abe and S. G. Franzblau, *Int. J. Antimicrob. Agents*, 2007, **29**, 744.
74. S. Jayaprakash, Y. Iso, B. Wan, S. G. Franzblau and A. P. Kozikowski, *Chem. Med. Chem.*, 2006, **1**, 593.
75. K. D. Thomas, A. V. Adhikari, S. Telkar, I. H. Chowdhury, R. Mahmood, N. K. Pal, G. Row and E. Sumesh, *Eur. J. Med. Chem.*, 2011, **46**, 5283.
76. M. Tukulula, S. Little, J. Gut, P. J. Rosenthal, B. Wan, S. G. Franzblau and K. Chibalea, *Eur. J. Med. Chem.*, 2011, **46**, 5283.
77. A. Mullard, *Nat. Rev. Drug Discovery*, 2013, **12**, 87.
78. L. M. Bedoya, M. J. Abad, E. Calongea, L. A. Saavedrac, C. M. Gutierrez, V. V. Kouznetsovd, J. Alcamia and P. Bermejo, *Antiviral Res.*, 2010, **87**, 338.
79. N. Ahmed, K. G. Brahmbhatt, S. Sabde, D. Mitra, I. P. Singh and K. K Bhutani, *Bioorg. Med. Chem.*, 2010, **18**, 2872.
80. W. G. Salgueiro, M. C. D. F. Xavier, L. F. B. Duarte, D. F. Câmara, D. A. Fagundez, A. T. G. Soares, G. Perin, D. Alves and D. S. Avila, *Eur. J. Med. Chem.*, 2014, **75**, 448.
81. R. Saari, J.-C. Törmä and T. Nevalainen, *Bioorg. Med. Chem.*, 2011, **19**, 939.
82. A. Zarghi, R. Ghodsi, E. Azizi, B. Daraie, M. Hedayati and O. G. Dadrass, *Bioorg. Med. Chem.*, 2009, **17**, 5312.

83. K. S. Kumar, S. K. Kumar, B. Y. Sreenivas, D. R. Gorja, R. Kapavarapu, D. Rambabu, G. R. Krishna, C. M. Reddy, M. V. B. Rao, K. V. L. Parsa and M. Pal, *Bioorg. Med. Chem.*, 2012, **20**, 2199.
84. J. Fiorito, F. Saeed, H. Zhang, A. Staniszewski, Y. Feng, Y. I. Francis, S. Rao, D. M. Thakkar, S.-X. Deng, D. W. Landry and O. Arancio, *Eur. J. Med. Chem.*, 2013, **60**, 285.
85. A. W. Garofalo, M. Adler, D. L. Aubele, E. F. Brigham, D. Chian, M. Franzini, E. Goldbach, G. T. Kwong, R. Motter, G. D. Probst, K. P. Quinn, L. Ruslim, H. L. Sham, D. Tam, P. Tanaka, A. P. Truong, X. M. Ye and Z. Ren, *Bioorg. Med. Chem. Lett.*, 2013, **23**, 1974.
86. J. Liu, W. Liu, H. Ge, J. Gao, Q. He, L. Su, J. Xu, L.-q. Gu, Z.-s. Huang and D. Li, *Biochim. Biophys. Acta*, 2014, **1840**, 1051.
87. G. Zhou, P. C. Ting, G. Wishart, N. Zorn, R. G. Aslanian, M. Lin, M. Smith, S. S. Walker, J. Cook, M. Van Heek and J. Lachowicz, *Bioorg. Med. Chem. Lett.*, 2014, **24**, 1790.

CHAPTER 7

Isoquinolines

ESTHER S. ROESCH

Pforzheim University of Applied Sciences, Institute for Materials and Material Technologies (IMMT), Tiefenbronner Strasse 65, 75175 Pforzheim, Germany
Email: esther.roesch@hs-pforzheim.de

7.1 Introduction

The use of natural herbs containing alkaloids for disease treatment dates back to ancient medicine. The oldest known cuneiform script records on pharmaceutical products mentioning opium, among others, were found near the Sumerian city of Nippur and relate back to the 4th millennium BC.[1–5] Other cultures, such as the Greeks and Egyptians also had a good knowledge of the application of herbal medicine without understanding its biochemical principle. Over the last 6000 years, knowledge of natural compounds has been refined and is ongoing to date.

In recent decades, medicinal chemistry has evolved as a scientific discipline for drug development. As far as the development of medicinal chemistry is concerned, we are only beginning to understand the complex correlations between disease, targets and molecular structure. Alkaloids have been intensively investigated in this context. Compared to the long history of herbal drug use, scientists only recently (about 200 years ago) started to elucidate the molecular structures of the pharmaceutically active compounds and manipulate naturally given scaffolds. The work of Gadamer (1867–1928) shows impressively how he and his group were able to isolate several new plant alkaloids. However, lacking sophisticated analytical methods, it took almost 50 years from discovery to structure elucidation of

RSC Drug Discovery Series No. 50
Privileged Scaffolds in Medicinal Chemistry: Design, Synthesis, Evaluation
Edited by Stefan Bräse

Published by the Royal Society of Chemistry, www.rsc.org

some compounds.[6] Nowadays, researchers benefit from the wealth of modern technologies and can categorize the different alkaloids according to their chemical constitution. High-throughput screening identified natural product scaffolds as potent hits for certain targets, attracting the attention of medicinal chemists to those compounds as drug leads. Furthermore, modern techniques such as structure–activity relationship (SAR) or X-ray structure based modelling of active sites of enzymes allow engineering of therapeutically active molecules from scratch.

Literature research reveals that, currently, to the best of the author's knowledge, only few drugs or drug candidates have isoquinoline moieties (Table 7.1), "although isoquinoline alkaloids are one of the most widely distributed alkaloids with proven therapeutic potential".[7] The information is scattered over many databases and fragmented into many documents with partly limited access and search functionalities. Because pharmaceutical research and development is a high-profit business,[8] information is well protected, proprietary and tightly regulated by the authorities. Many drug candidates in the pharmaceutical pipeline are no longer in-house developments, but are licensed from other companies by mergers or acquisitions. This results in renamed molecules and it can sometimes make it hard to track the chemical structure.

It became evident that isoquinoline scaffolds are commonly used for SAR analyses because building blocks are commercially readily available for fast side chain modifications. However, the modification of natural product scaffolds was a great challenge since the complex molecular backbone and the occurrence of multiple stereogenic centres presented a major synthetic effort.[9] Thus, in the majority of cases, the natural products underwent only subtle side chain derivatizations and were discarded once undesired side effects were observed. Modern semisynthetic and total synthetic approaches allow better, but still time-consuming access to natural product derivatives. Hence, in this chapter, many of the molecules discussed are synthetically accessible small molecules derived from mono- and polysubstituted isoquinoline scaffolds.

This chapter reveals that, compared to the large number of isolated and characterized natural isoquinoline products, isoquinoline scaffolds are scarce in the later stages of clinical development. There are, for instance, about 2500 known structures of benzylisoquinoline alkaloids[10] compared to very few approved isoquinoline drugs such as dimethisoquin/quinisocain (1), fasudil (2), papaverine (3), and berberine (4) (Figure 7.1).

Isoquinoline moieties could therefore have an expandable potential for drug products. This chapter presents both the natural scaffolds with their chemical derivatives and the synthetically engineered small molecules, containing isoquinoline moieties, for therapeutic applications.

7.2 Synthesis of Isoquinolines – An Overview

The synthesis of the isoquinoline core has been of great importance to the pharmaceutical industry. Commercial drugs such as papaverine (**3**),

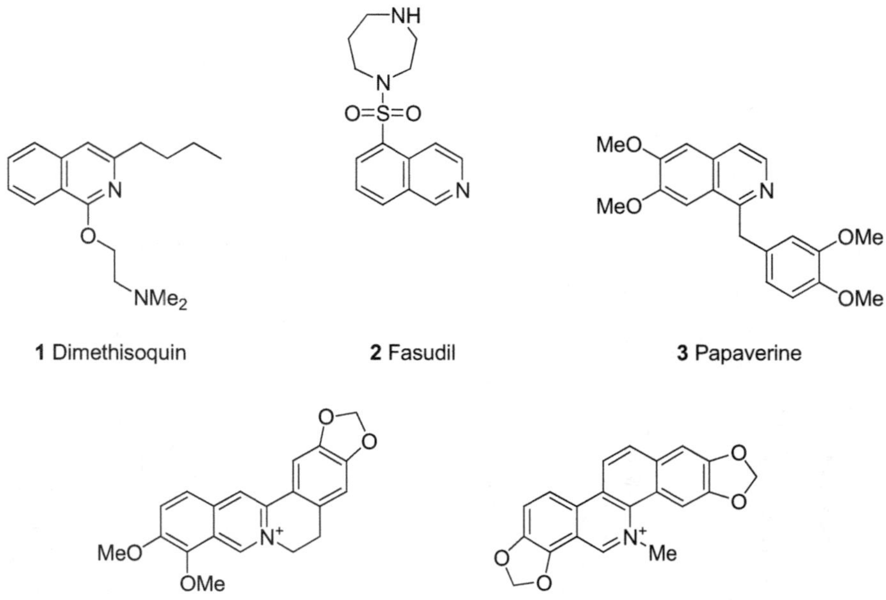

1 Dimethisoquin **2** Fasudil **3** Papaverine

4 Berberine **5** Sanguinarine

Figure 7.1 Examples of therapeutically active isoquinoline drugs.

a natural product, or dimethisoquin (**1**), a natural product derivative, made a synthetic access necessary.[11] Traditional approaches such as Pomeranz-Fritsch,[12–15] Pictet-Gams,[16] Pictet-Spengler,[15,17,18] Bischler-Napieralski,[15,19,20] and Schlittler-Müller,[21,22] allow the preparation of derivatized isoquinoline scaffolds (Scheme 7.1 a–e). It is, however, common to these well established procedures to be limited to electron rich carbocycles.[15] Transition metal catalysts have been studied to facilitate the activation and functionalization of C–H bonds[11] using *e.g.* Cu(I),[23] Rh(III),[24] Ru(II),[11] or Pd(II)[15] (Scheme 7.1 f–j). Konno *et al.* describe a highly regioselective annulation reaction of fluoroalkylated alkynes.[25] Donohoe *et al.* reported the successful palladium-catalyzed α-arylation of ketones with polysubstituted isoquinolines as the product. The advantage of this synthetic route is the rapid access to polysubstituted isoquinoline compounds.[15] Other approaches include the cyclization of *ortho*-ethynylbenzaldehyde derivatives using either an ammonia source, to obtain 3-substituted isoquinolines,[26] or β-cyanocarbenes, which involves the formation of an isobenzofuran followed by an intramolecular Diels–Alder reaction with the nitrile.[27] Alonso *et al.* reported a light-induced formation of isoquinolines by radical intramolecular cyclization using acyloximes as the starting material (Scheme 7.1 i–k).[28] Thus, the modern synthetic routes allow the regioselective mono- and polyderivatization of positions 3, 4 and 1 without being limited to electron-rich moieties.[15,29]

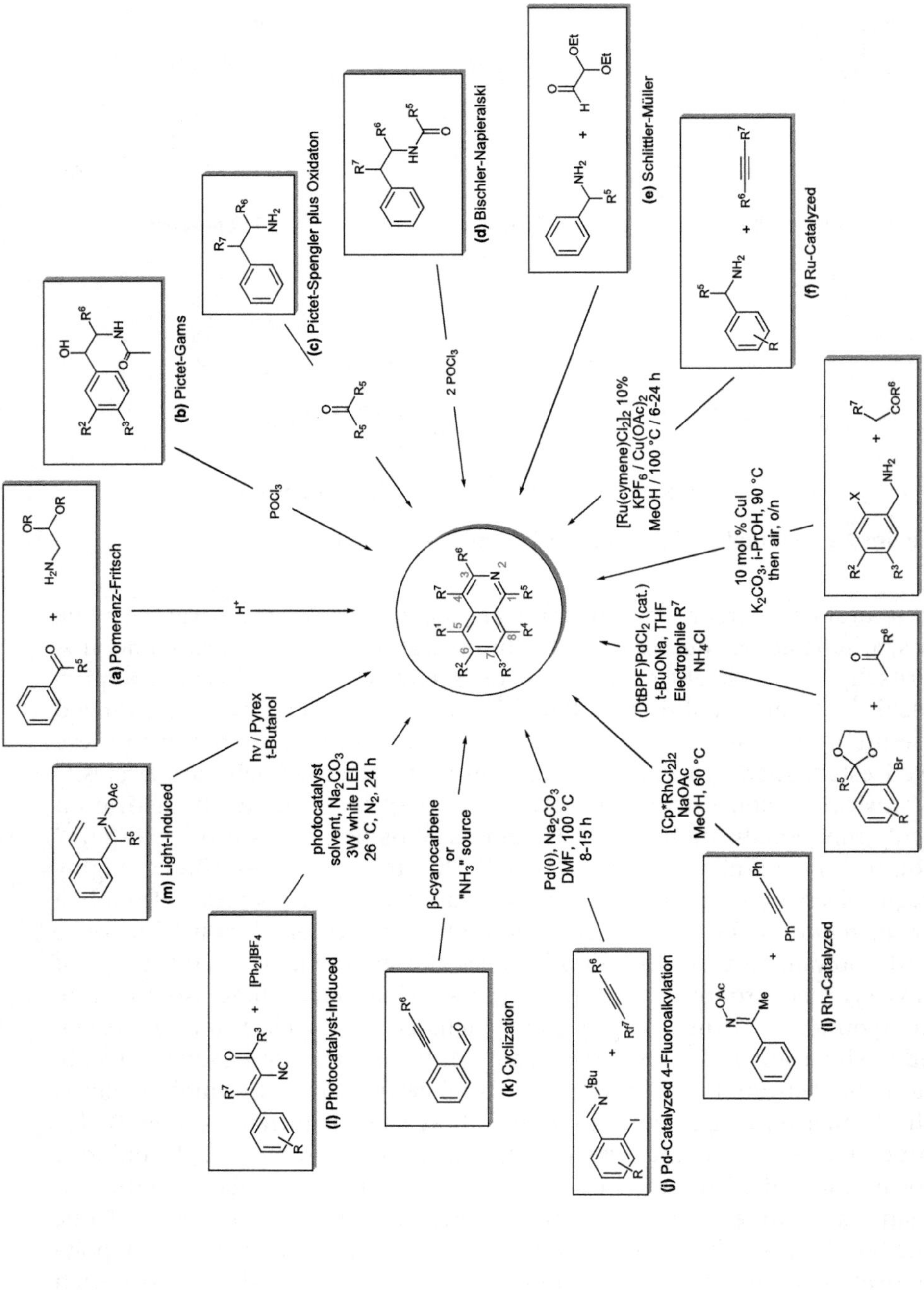

Scheme 7.1 Synthesis of isoquinolines.[11–28,30]

7.3 Drug Candidates and Drugs

Many drug candidates or commercially available drugs are derived from natural products. Often, products cannot be used in their native form, but serve as lead structures and undergo modifications to optimize safety and efficacy. The pool of bioactive natural compounds offers great inspiration for drug development. Table 7.1 provides an overview of isoquinoline containing drug scaffolds in clinical development. The lack of bioavailability or druggability might be an explanation for the observation that only few examples of isoquinoline containing compounds succeeded in entering the clinical pipeline.[31,32]

7.4 Natural Isoquinoline Derivatives

Abundance and ease of access facilitated the traditional use and research of plant alkaloids. However, in recent years other sources of novel pharmacologically active compounds such as terrestrial animals, insects[90] and marine organisms[91] have attracted increasing attention. Examples of therapeutically active isoquinoline derivatives are sanguinarine (5) and berberine (4), two potent antimicrobials, and papaverine (3), a muscle relaxant.[10]

Although alkaloids are structurally very diverse, the isoquinoline motif is abundant among these natural products. Even though the term isoquinoline (**6**) is often incorrectly used to describe dihydro- (**7**), tetrahydro- (**8**), and perhydroisoquinoline (**9**) moieties (Figure 7.2), the emphasis of this chapter will be solely on isoquinoline (**6**) derivatives in medicinal chemistry.

The prefix *epi* is used for interchanged 2,3 and 9,10 substitution patterns (*e.g.* berberine (**4**), epiberberine (**10**).[92] The prefix *pseudo* describes a 2,3,10,11 substitution pattern (*e.g.* pseudopalmatine (**11**)) as a regioisomer of the 2,3,9,10 derivative.[92] The prefix *nor* indicates that the compound is the tertiary amine analogue of the quarternary amine alkaloid due to the lacking *N*-methyl group (*e.g.* nitidine (**12**), nornitidine (**13**)) (Figure 7.3).

7.4.1 Protoberberine

Isoquinoline alkaloids are one of the most widespread structural motifs in the plant kingdom.[7] Protoberberines (Figure 7.4) are naturally occurring isoquinoline alkaloids, of which berberine (**4**) is probably the most widely distributed.[92]

Quarternary protoberberine alkaloids represent approximately 25% of all naturally occurring alkaloids with a protoberberine skeleton.[92] Due to the vast amount of natural compounds, they are classified according to common structural motifs. Protoberberine alkaloids are derivatives of 5,6-dihydrodibenzo[*a*,*g*]quinolizium salts, which include quarternary alkaloids such as berberine (**4**), pseudoberberine (**14**), jatrorrhizine (**15**), coptisine (**18**), and palmatine (**17**) (Figure 7.5).

Table 7.1 Overview of drug scaffolds containing isoquinolines in clinical development.[a]

Indication	Compound
Experimental phase	
T[c]	Ariprazole isoquinoline derivatives[33]
F[c]	H-1152P (**165**)[37,38]
B[d]	TMC-120B (**72**)[43,44]
U[b,d]	Fagaronine (**19**)[50]
U[b,d]	Nitidine (**12**)[50]
A[d]	RO0509347 (**89**)[54]
U[d]	Hystatin 1 (**146**)[55,56]
Preclinical phase/phase 0	
X[c]	A-425619 (**181**)[58,59]
Ö[b,c]	Asunaprevir (**194**) (BMS-605339)[60,61]
U[d]	S 30972-1 (**65**)[62,63]
U[d]	Pazelliptine (PZE) (**54**)[34,64]
U, Ä[d]	Sanguinarine (**5**)[3–5,65]
U[d]	Chelerythrine (**16**)[66]
U[b,d]	Ellipticine (**52**)[34,67,68] (NSC-71795)
X[b,c]	GRC 6211 (**182**)[58]
T[b,d]	Papaverine (**3**)[45]
Phase 1	
U[d]	T-215 (**64**)[34] (clinical phase not disclosed)
U[b,d]	Retelliptine (**58**)[34,39,71] (SR 95325 B, NSC D-626717-W)
U[d]	Elliprabin (**57**)[34] (Sun-4599)
G[c]	Asunaprevir (**194**)[45] (BMS-650032)
U[d]	NK314 (**51**)[50,77–79]
U[b,d]	NK109 (**50**)[82]
S[d]	Ethaverine (**73**)[83,84]
U[b,d]	Ditercalinium (**59**)[34,87] (NSC 335153)
U[d]	S 16020-2 (**60**)[89] (9-hydroxy-olivacine)
A[d]	Berberine (**4**)[45]
N[c]	PK-11195 (**192**)[45]
H[c]	FG-4592 (**195**)[45] (Roxadustat)
Phase 2	
U[b,d]	Datelliptium (DHE) (**56**)[34–36] (SR 95156, NSC 626718)
U[b,d]	Elliptinium acetate (EA) (**62**)[39–42] (Celiptium)
F[c]	AR-12286 (**176**)[45–49]
V[c]	Pralnacasan (**162**)[51] (HMR3480, VX-740)
W[d]	MHE (**53**)[52,53] (2*N*-methyl-9-hydroxy-ellipticine)
G[c]	Asunaprevir (**194**) (BMS-650032)[45]
H[c]	FG-4592 (**195**)[45] (Roxadustat)
O, L, R[c]	Fasudil (**2**)[45,57] (HA-1077)
N[c]	PK-11195 (**192**)[45]
K, A, B[d]	Berberine (**4**)[45]
U[d]	Papaverine (**3**)[45]
F[d]	Moxaverine (**85**)[45]
Phase 3	
F[d]	Moxaverine (**85**)[45]
Q, P[c]	Fasudil (**2**)[45,57] (HA-1077)
G[c]	Asunaprevir (**194**) (BMS-650032)[45,60]
H[c]	FG-4592 (**195**)[45] (Roxadustat)
F[c]	K-115 (**193**)[37,47,69,70] (Ripasudil)
A, B[d]	Berberine (**4**)[45]
Phase 4/approved	
U[d]	Elliptinium acetate (EA) (**62**)[72,73] (Celiptium)
Y[c]	Dimethisoquin Hydrochloride (**1**)[74–76]
Ü[c]	Fasudil (**2**)[80,81] (HA-1077)
B, M, C, L[d]	Berberine (**4**)[45]
Z[d]	Papaverine Hydrochloride (**3**)[85,86]
J[d]	Sanguinarine (**5**)[7,88]

[a]**A:** diabetes mellitus, **B:** hyperlipidemia, dyslipidemia, **C:** diarrhoea, **D:** bronchial asthma, **E:** metabolic syndrome, **F:** ocular blood flow/glaucoma, **G:** CHC (chronic hepatitis C), **H:** CDK (chronic kidney disease), **I:** polycystic ovary syndrome (PCOS), **J:** antiplaque, **K:** non-alcoholic fatty liver disease, **L:** hypercholesterolemia, **M:** cardiovascular risk/coronary artery disease, **N:** PET Ligand, **O:** amyotrophic lateral sclerosis, **P:** Raynaud's Phenomenon, **Q:** diabetic macular oedema, **R:** atherosclerosis, **S:** peripheral vasodilator, **T:** neuroscience, **U:** cancer, **V** rheumatoid arthritis (RA), **W:** HIV, **X:** pain, **Y:** anaesthetic, **Z:** smooth muscle relaxant, **Ä:** supragingival plaque control, **Ö:** hepatitis C virus (HCV), **Ü:** cerebral vasospasm.

[b]Discontinued.

[c]Synthetic compound.

[d]Natural product derivative.

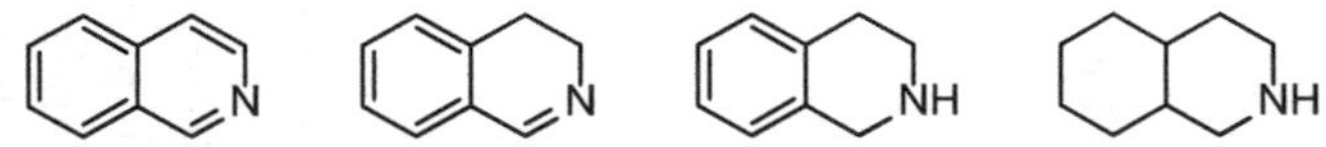

6 7 8 9

Figure 7.2 Isoquinoline scaffolds: isoquinoline (**6**), dihydroisoquinoline (7), tetrahydroisoquinoline (**8**), perhydroisoquinoline (**9**).

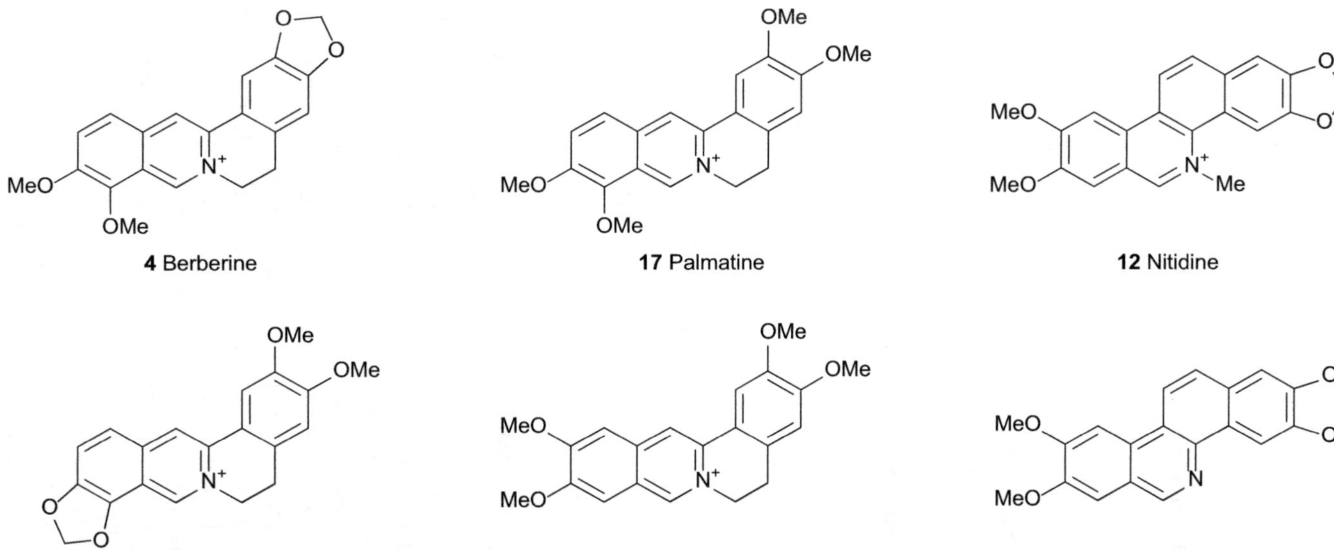

4 Berberine **17** Palmatine **12** Nitidine

10 Epiberberine **11** Pseudopalmatine **13** Nornitidine

Figure 7.3 Nomenclature of natural products.

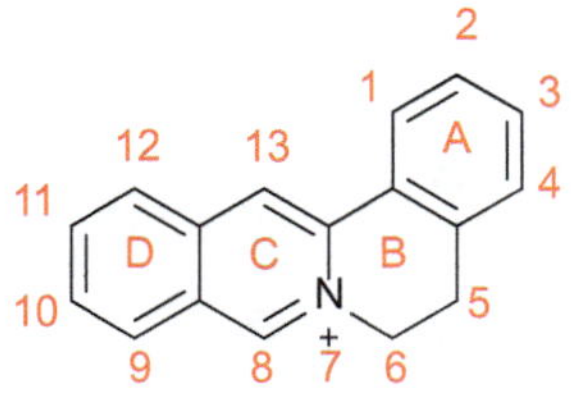

Figure 7.4 Protoberberine scaffold.[7]

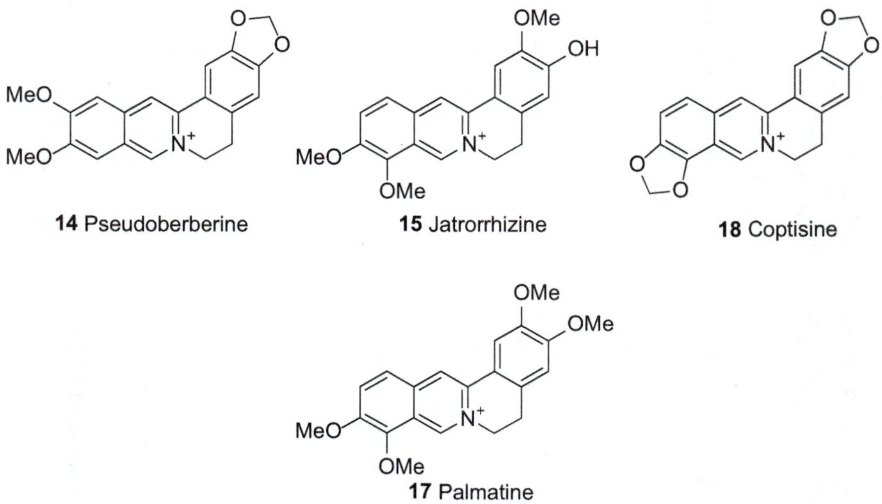

Figure 7.5 Protoberberine alkaloids.

Usually, the Protoberberines subsume substituents in positions 2,3,9,10 (*e.g.* jatrorrhizine (**15**), coptisine (**18**), palmatine (**17**)).

Like berberine (**4**), other protoberberine compounds also show biological and pharmacological activity. Some reported representatives of the protoberberine alkaloids are berberine (**4**), jatrorrhizine (**15**), coptisine (**18**), pseudoberberine (**14**), palmatine (**17**), coralyne (**20**), dehydrocorydamine (**21**), groenlandicine (**22**), epiberberine (**10**), dehydrocorydaline (**23**), columbamine (**24**), thalifendine (**25**), stepharanine (**26**), dehydrodiscretamine (**27**), dehydrocheilanthifoline (**28**), demethyleneberberine (**29**), pseudopalmatine (**11**), dehydrocavidine (**30**), dehydroapocavidine (**31**), dehydroisoapocavidine (**32**), dehydreothalidastine (**33**), thalidastine (**35**), 5-hydroxy-coptisine (**37**), and chemical derivative analogues thereof (*e.g.* **34**) (Figure 7.6). The following will provide a brief description of specific characteristics of the main constituents.

Of the compounds in the protoberberine class, berberine (**4**) is the most thoroughly studied alkaloid,[93,94] with researchers' enthusiasm hardly on the wane (Scifinder indicates 12 044 references from inception to March 2014 containing "berberine" as entered; see Figure 7.7). A reason might be its

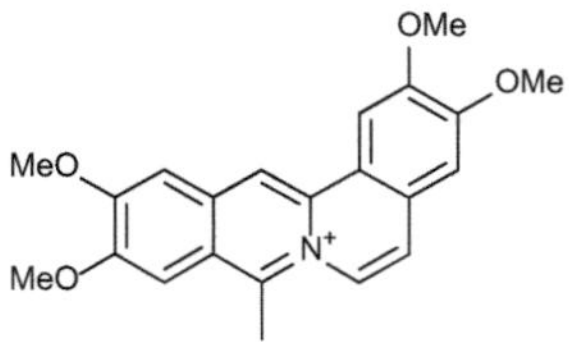

20 Coralyne

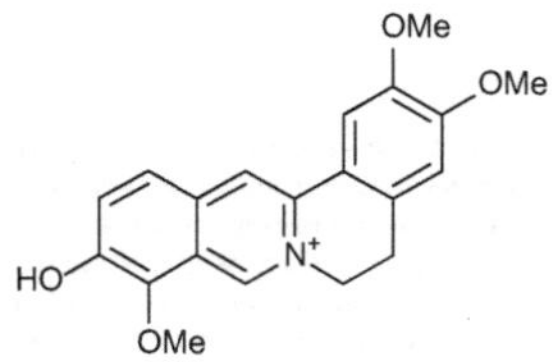

21 Dehydrocorydalmine

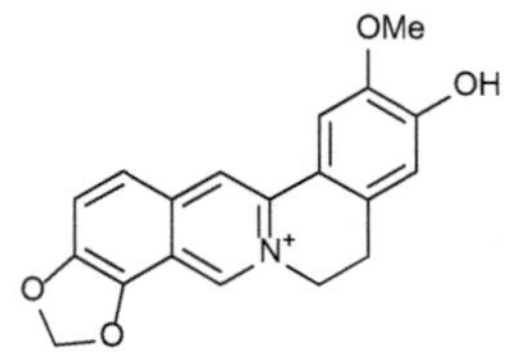

22 Groenlandicine

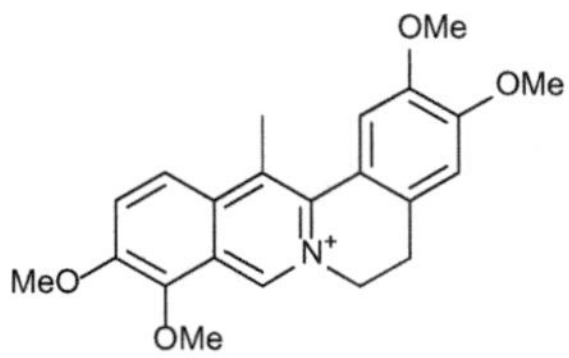

23 Dehydrocorydaline

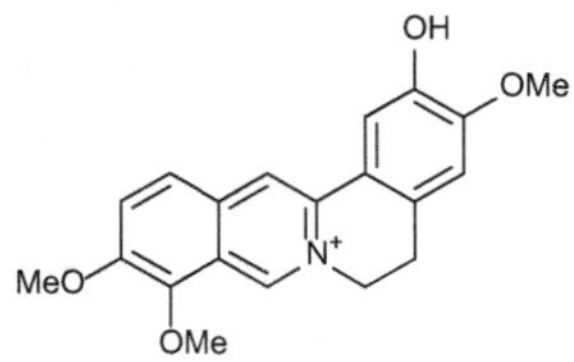

24 Columbamine

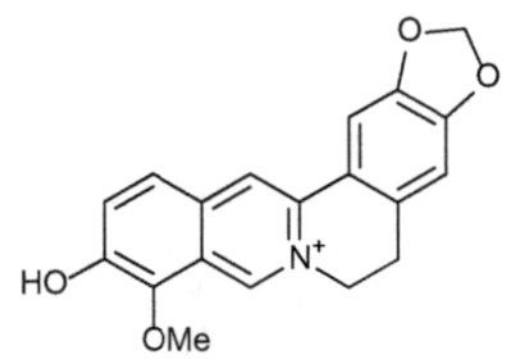

25 Thalifendine

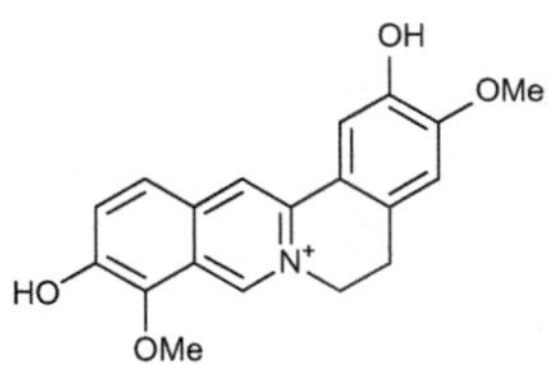

26 Stepharanine

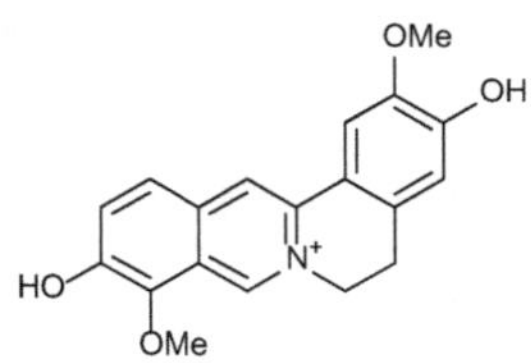

27 Dehydrodiscretamine

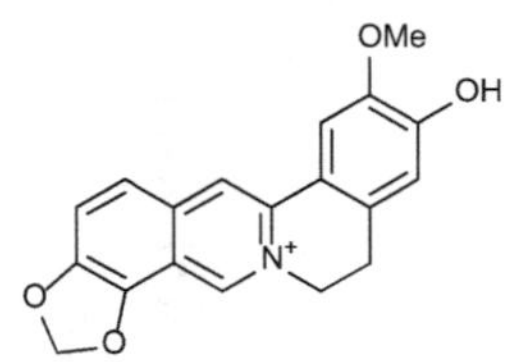

28 Dehydrocheilanthifoline

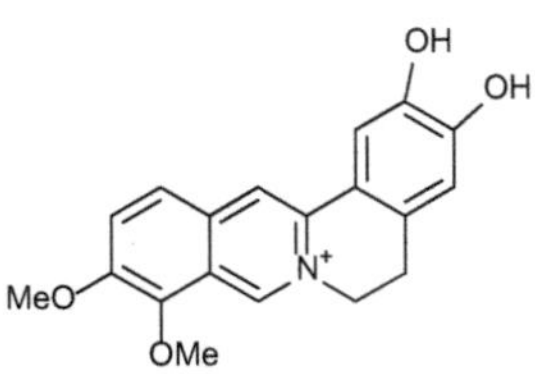

29 Demethyleneberberine

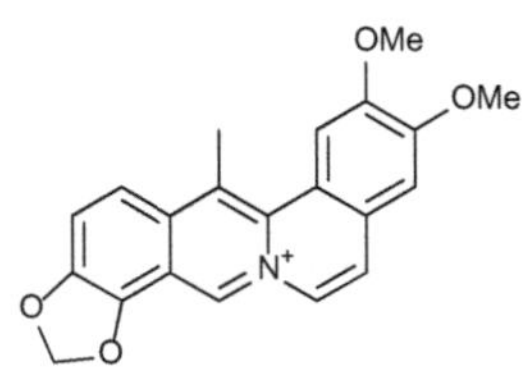

30 Dehydrocavidine

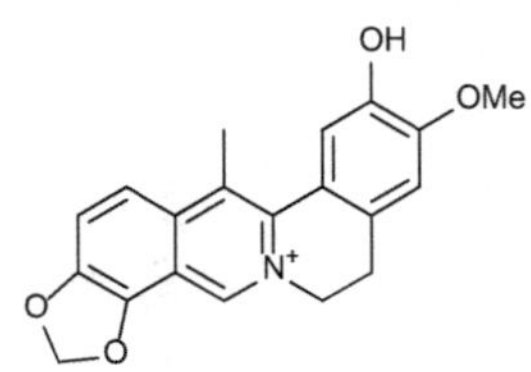

31 Dehydroapocavidine

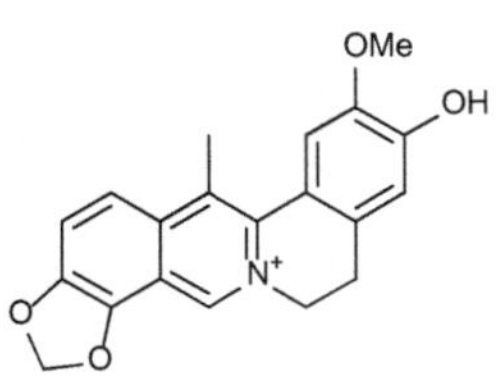

32 Dehydroisoapocavidine

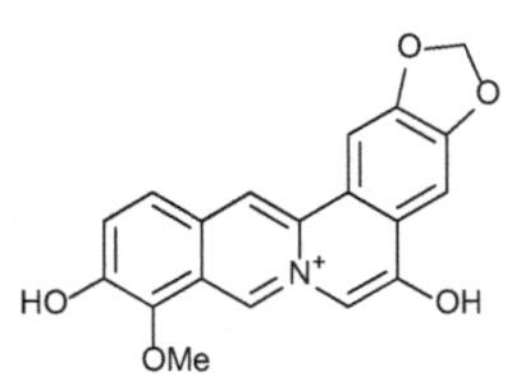

33 Dehydrothalidastine

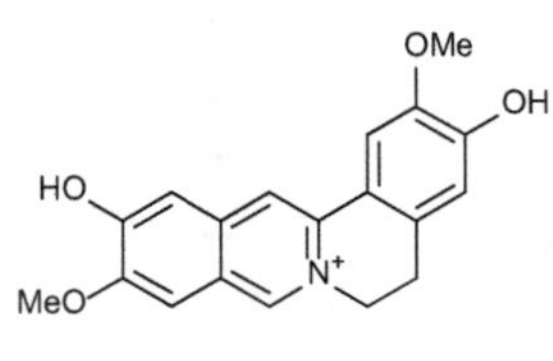

34 3-O-demethyldehydrocorytenchirine

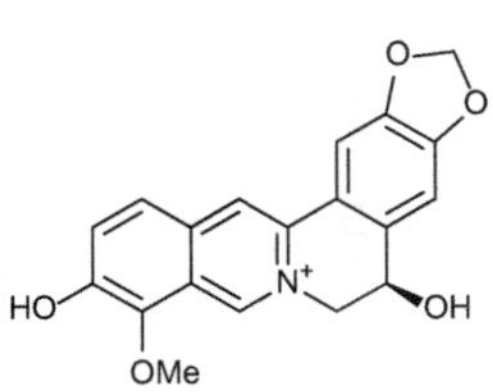

35 Thalidastine

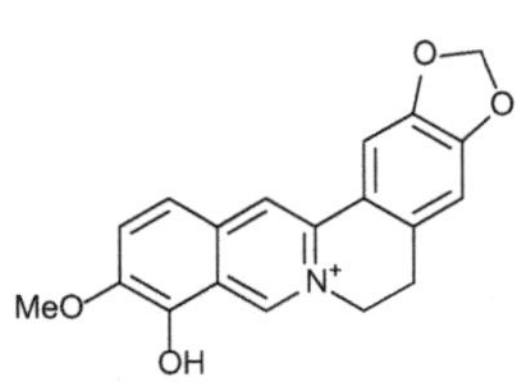

36 Berberrubine

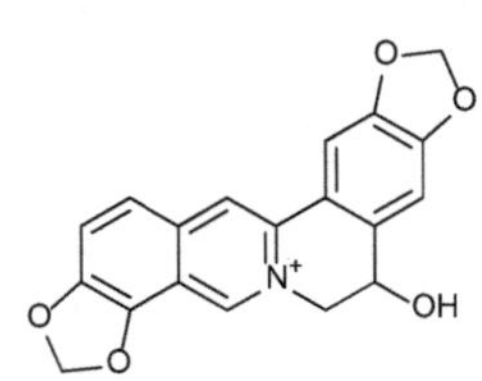

37 5-Hydroxy-coptisine

Figure 7.6 Chemical structures of protoberberine derivatives.

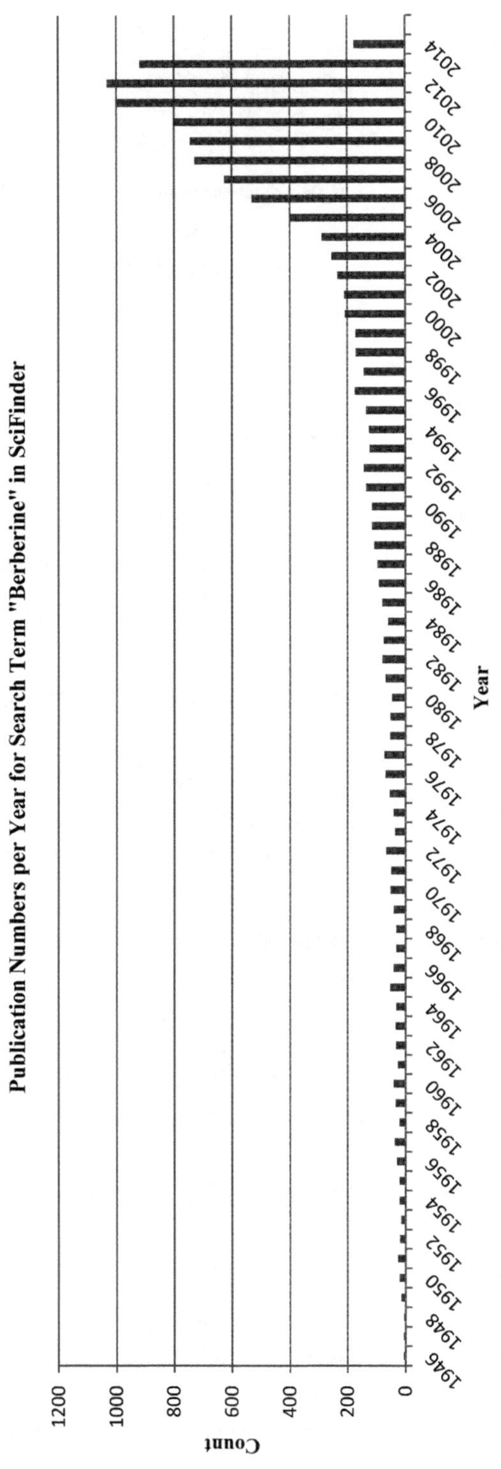

Figure 7.7 Publication numbers per year containing "berberine" as entered in SciFinder.

widespread occurrence in plants such as *Hydrastis canadensis* (goldenseal), *Arcangelisia flava* (Menispermaceae), *Berberis aquifolium* (Oregon grape), *Berberis aristata* (tree turmeric),[93,95] *Coptis chinensis* (Ranunculaceae),[93,96] *Berberis vulgaris* (barberry), *Berberis thunbergii* (Red Barberry),[97] *Phellodendron amurense*, *Tinospora cordifolia* and to a smaller extent in *Argemone mexicana* and *Eschscholzia californica*, mainly found in their roots, rhizomes, stems, and bark.[98] Some 12-substituted berberine derivatives have been found to be effective in the treatment of type 2 diabetes mellitus.[99] 9-substituted derivatives have been studied with respect to Alzheimer disease[100–107] and cancer, which is related to DNA binding,[108,109] 2, 3, 7, 9, 10, 11, and 13-substituted derivatives as cholesterol-lowering agents,[110,111] 8-substituted analogues as antitubercular agents,[112] and 13-substituted analogues as antifungal agents.[113] Early reports of clinical investigations of compounds in the protoberberine class date back to 1930[114] and are still being studied in ongoing clinical trials.[45] The main focus of interventions is dyslipidemia,[98,115] metabolic syndrome,[116] diabetes mellitus,[59,60] and related diseases.[45]

Thalifendine (**25**) is a naturally occurring alkaloid extracted from *Berberis darwinii*[117] or *Thalictrum foliolosum.*[118] Thalifendine (**25**) and its conjugates were also found to be metabolites both in humans and in rats with orally administered berberine.[97,119–126] Dehydrothalidastine (**33**) was reported as an artefact formed by dehydration of thalidastine (**35**).[92]

The synthesis of thalifendine (**25**) and related protoberberine alkaloids was reported by Yang *et al.* using homopiperonylamine and 2-methoxy-3-hydroxybenzaldehyde as starting materials.[127,128] Thalifendine (**25**) showed good up-regulatory activities for both low density lipoprotein receptor (LDLR) and insulin receptor (InsR) mRNA expression.[127,129]

Berberrubine (**36**) is a yellow[130] to red[131] amorphous powder, recently isolated from *Dicranostigma leptopodum* (Maxim) Fedde by Zhong *et al.*, who published detailed spectroscopic data of the compound.[130] It was found to have positive effects on ocular diseases by inhibiting *in vitro* expression of interleukin-8 (IL-8) and monocyte chemotactic protein-1 (MCP-1) levels and mRNA expression in cultured human retinal pigment epithelial cells (ARPE-19) stimulated with interleukin-1β (IL-1β) or tumour necrosis factor α (TNF-α).[132,133] 12-(substituted aminomethyl)berberrubine derivatives showed stimulating effects on glucose uptake of the insulin-resistant 3T3-L1 adipocytes and L6 muscle cell myotubes. These compounds could therefore be considered for type 2 diabetes mellitus, obesity, and fatty liver disease.[99] Berberrubine (**36**) exhibited higher Dock Score than controls as a three-in-one agonist for PPAR enzymes (peroxisome proliferator-activated receptors). These receptors have three forms (alpha, gamma, and delta) and are known as key regulators of adipogenesis, lipid, and carbohydrate metabolism, which makes them a potential drug target for metabolic syndrome.[134] Das *et al.* describe a facile preparation route with inexpensive berberine (**4**) as a starting material. Microwave irradiation followed by selective demethylation of the C-9 methoxy group resulted in the formation of berberrubine (**36**) (Scheme 7.2).[126,131]

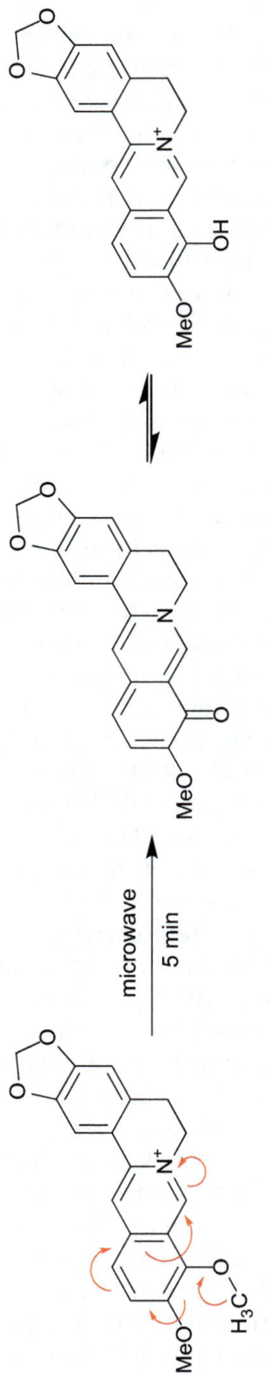

4 Berberine

36 Berberrubine

Scheme 7.2 Facile microwave-assisted synthesis of berberrubine (**36**).[126,131]

Li *et al.* reported a synthetic route to 12-(substituted aminomethyl)berberrubine, which displays antidiabetic properties.[99] Additionally, they isolated eight protoberberine-type alkaloids, of which three, namely dehydrocavidine (**30**), dehydroapocavidine (**31**), and dehydroisoapocavidine (**32**), exhibited inhibitory activity against the Hepatitis B virus (HBV), using Hepatitis B surface antigen (HBsAg) and Hepatitis B excretory antigen (HBeAg) as established marker proteins to indicate viral replication.[135] Dehydrocheilanthifoline (**28**), isolated in 1972,[136] also showed an inhibitory effect on HBsAg and HBeAg secretion on two human hepatocarcinoma cell lines, Hep G and Hep 2,2,15.[31,137]

Coralyne (**20**) is a synthetic protoberberine analogue.[7] Its partially saturated ring makes it a planar molecule. Protoberberines are prone to interact with DNA due to their structural motif. Kan *et al.* developed a simple, cost-effective colorimetric assay for the sensitive and selective detection of coralyne (**20**), based on target-induced split G-quadruplex formation.[138] Coralyne (**20**) is known to form strong complexes with polyadenine with a length of at least eight adenine stretches.[139] It exhibits interesting fluorescence properties that can be exploited for selective detections. Halide ions such as bromide or iodide quench the fluorescence of unbound coralyne (**20**). Once coralyne (**20**) is bound to cyclic diadenosine monophosphate (c-di-AMP) it becomes highly fluorescent, as it protects coralyne from the halide quenchers.[139] Coralyne (**20**) binding to phenylalanine-specific transfer RNA ($tRNA^{phe}$) has been studied and job plot analysis revealed a stoichiometry of around 5 nucleotides per coralyne (**20**) molecule.[140] It was found that the binding of planar coralyne (**20**) (together with planar sanguinarine (**5**)) to DNA is thermodynamically more favoured than the binding of the buckled structure of *e.g.* berberine (**4**) to DNA.[141] The affinity of coralyne (**20**) to specific DNA sequences can be exploited for logic gates on a molecular level. Lin *et al.* reported a method using coralyne (**20**) to construct a molecular set of two-input (AND, OR, INHIBIT, NAND, NOR, REVERSE IMPLICATION) and three-output (OR, NOR, AND) logic gates.[142] The manifold metabolic distribution of protoberberines such as berberine (**4**), palmatine (**17**), or coralyne (**20**) can be supported by the findings of Hazra *et al.* who studied the binding of three isoquinoline alkaloids to haemoglobin. Despite the structural similarities, the findings indicate the following: due to significant differences between the behaviour of berberine (**4**), palmatine (**17**), and coralyne (**20**), coralyne (**20**) is bound at a site different from berberine (**4**) and palmatine (**17**).[143] Zee-Cheng *et al.* described the preparation of coralyne chloride.[144] Coralyne (**20**) crystals have a yellow appearance and are soluble in ethanol. The melting point of the chloride salt is 250–252 °C.[145] The compound is effective against L1210 lymphoid and P388 lymphocytic leukemia. It inhibits enzymes, for instance, tRNA methyltransferase and the reverse transcriptase activity of RNA tumour viruses.[7] It also exhibits selective cytotoxicity against culture cell lines such as SF-268 glioblastoma cells. Coralyne (**20**) was shown to intercalate in DNA, to act as potent poison for the Topo-I and II enzymes and to cause efficient photoinduced DNA damage in pBR322.[7]

Although groenlandicine (**22**) was isolated and characterized long ago,[146] little systematic research has been conducted to evaluate its pharmacological effects. Recent studies have begun to include groenlandicine (**22**) due to noteworthy observations. Jung *et al.* reported positive inhibitory effects of six Coptidis Rhizoma protoberberine alkaloids on several Alzheimer disease (AD) related enzymes such as acetylcholinesterase (AChE), butyrylcholinesterase (BChE) and β-site amyloid precursor protein cleaving enzyme 1 (BACE1).[147] With IC_{50} (AChE) = 0.54 ± 0.01 μM, IC_{50} (BChE) = 3.32 ± 0.01 μM and IC_{50} (BACE1) = 19.68 ± 1.42 μM inhibition concentrations, groenlandicine (**22**) was suggested to be a promising Alzheimer disease treatment.[147] Ma *et al.* investigated the pharmacochemistry and integrated pharmacokinetics of six protoberberine alkaloids including groenlandicine (**22**) after oral administration in rats, reporting major differences in plasma levels despite the chemically akin structures.[148] Groenlandicine (**22**) was found to exhibit significant peroxynitrite ($ONOO^-$) scavenging effects in a dose-dependent manner with IC_{50} values of 0.84 and 0.78 μM. Computational pharmaceutical analysis determined a higher Dock Score for groenlandicine (**22**) than Donepenzil, one of the four currently prescribed AD drugs, in binding to AChE. This is in accordance with experimental findings.[149] Groenlandicine (**22**) exhibits hypoglycemic effects and inhibits mammalian DNA topoisomerase I.[148]

Jatrorrhizine (**15**) is often investigated together with other protoberberine compounds, especially in the context of herbal extracts with therapeutic effects. In these cases, sophisticated analytical methods must be established for the simultaneous determination of different protoberberines in blood plasma. Deng *et al.* reported a sensitive liquid chromatography mass spectrometry (LC-MS) tandem method for pharmacokinetic studies in rat plasma.[150] Recent studies reported a highly sensitive electrochemical method for the analytical determination of jatrorrhizine (**15**) using an electrochemically pretreated glassy carbon electrode (EPGCE).[151] A hindering factor for analytics is the metabolization resulting in manifold metabolites similar to the native compounds. Jatrorrhizine (**15**), for instance, is not only found in herbs, but is also a phase I metabolite of berberine found in rat bile, urine, and faeces after oral administration.[124] In contrast, oral administration of berberine chloride (**4**) in healthy volunteers showed one metabolite, jatrorrhizine-3-sulfate, in urine, but no phase I metabolites.[119] Qiu *et al.* identified jatrorrhizine-3-*O*-β-D-glucuronide as another berberine metabolite in humans and rats.[121]

Jatrorrhizine (**15**) is a water-soluble, orange, crystalline solid with a melting point of 208–210 °C. The EC_{50} value in mice was found to be ≥98 g mL^{-1}.[7] Jatrorrhizine (**15**) showed inhibition of α 1- and α 2-adenoreceptors, which could be linked to its hypotensive and antiarrhythmic actions.[7] The compound has antiplasmodial activity and was found to be effective against almost all fungal species. Furthermore, it was reported to have hypoglycemic and antimicrobial activity.[7] Wu *et al.* concluded that Jatrorrhizine could potentially become a safe and potent antihypercholesterolemic agent by up-regulating the mRNA and protein expression of low density lipoprotein receptor (LDLR) and cholesterol 7α-hydroxylase (CYP7A1).[152] Jatrorrhizine (**15**) was found to non-competitively inhibit both monoamine oxidase A (MAO-A) and -B (MAO-B) from rat brain mitochondria with IC_{50} values of 4 and 62 μM, respectively, indicating that enzyme inhibition is dependent on the substitution pattern of ring A.[96,153,154] Jatrorrhizine (**15**) elicited a concentration-dependent suppression of the acridine orange (AO) induced mutagenicity notable although almost three orders of magnitude lower than its close analogue berberine.[155] Furthermore, anti-protozoal,[156] anti-phototoxicity against UVB light,[157] anti-radical and antioxidant activities[158] were reported. The formation of the chemically derivatized analogue bisjatrorrhizine was synthesized by an *ortho*-oxidative coupling of the phenolic group of jatrorrhizine (**15**).[92,159] The synthesis of jatrorrhizine homodimers and berberine-jatrorrhizine heterodimers was reported with the aim of modulating binding affinities to DNA with varying spacer length and attaching positions.[160,161]

In the literature, there is conflicting data regarding the melting point of coptisine (**18**). Preininger *et al.* reported that coptisine (**18**) has a melting point of 320–333 °C,[162] whereas Awe *et al.* describe the formation of colourless needles with a melting point of 212–217 °C.[163] Colombo *et al.* studied both synthetically obtained and naturally extracted coptisine (**18**). Natural and synthetic coptisine (**18**) showed similar cytotoxicity, being very toxic on colon carcinoma HT 29 cells and antiproliferative on the human tumour colon cell line LoVo. Coptisine (**18**) was also toxic on the murine leukaemia cell line L1210, but nevertheless, berberine (**4**) showed a significantly lower IC_{50} value. The study suggests that natural coptisine (**18**) tends to be more active than the synthetic compound. One possible explanation is that both compounds differed in their salt forms (natural coptisine (**18**) was recrystallized as chloride whereas the synthetic salt was obtained as iodide salt).[164] Hirano *et al.* conducted structure-activity-relationship analysis by synthesizing derivatives with a partial coptisine (**18**) structure. The results indicate that the partial charge of the catechol skeleton has an effect on the investigated gastric-mucous membrane protection activity.[165] Similar to groenlandicine (**22**), computational pharmaceutical analysis resulted in a higher Dock Score to AChE of coptisine (**18**) than Donepenzil to AChE.[149] Experimental studies showed that coptisine (**18**) has inhibitory effects on AChE, BChE, and BACE1, but other protoberberines such as berberine (**4**), palmatine (**17**), jatrorrhizine (**15**), and groenlandicine (**22**) exhibited a higher potency.[147] Coptisine (**18**) and berberine (**4**) displayed strong antihepatoma activity.[166] Further studies provided evidence that coptisine (**18**) selectively prevents vascular smooth muscle cell proliferation (VSMC) in contrast to berberine and palmatine (**17**).[167] A safety evaluation of coptisine (**18**) showed slight toxicity with a LD_{50} value of 852.12 mg kg^{-1} when orally administered to mice.[168] Due to the increasing interest in protoberberine compounds, a highly sensitive and selective LC-MS detection method was developed by Zhou *et al.*[169] Colombo *et al.* described a four-step procedure for the synthesis of coptisine (**18**).[164]

Recently, a new quaternary protoberberine alakloid was isolated from *Dicranostigma leptopodum* (Maxim) Fedde, named 5-hydroxy-coptisine (**37**).[130] 5-hydroxy-coptisine (**37**) is a yellow amorphous powder with a melting point of 258–259 °C.[130] Zhong *et al.* published detailed spectroscopic data of the newly characterized compound.[130]

Palmatine (**17**) is found in berberine containing plant families, but to a much lesser extent. The colour of palmatine chloride (**17**) is canary yellow, with a melting point of 221 °C.[141] The compound is soluble in ethanol. The IC_{50} value in mice is 65 mg kg^{-1}.[141] Palmatine (**17**) is known as a treatment of jaundice, dysentery, hypertension, inflammation, and liver-related diseases.[7] Hepasor, a medication containing 65% palmatine (**17**), 20% jatrorrhizine (**15**), and 15% columbamine (**24**),[170] prevented liver damage from chemically induced traumatization and promoted the healing process of the hepatic injury.[171] Palmatine (**17**) inhibits I_{CRAC} (Ca^{2+} release-activated Ca^{2+} current) and therefore protects hepatocytes from calcium overload.[7] Palmatine (**17**) was also shown to easily bind to DNA. It is furthermore useful as a sedative due to its inhibition of the dopamine biosynthesis.[7,153] Palmatine (**17**) showed significant antitumour activity against HL-60 leukemic cells[7] and anti-protozoal activity.[156] A safety evaluation of palmatine (**17**) categorized it as slightly toxic, with an LD_{50} value of 1533.68 mg kg^{-1} after oral administration to mice.[168] It was also investigated in the context of Alzheimer disease (AD) along with berberine (**4**) and chelerythrine (**16**), the latter being the most promising candidate.[172]

Epiberberine (**10**) is a naturally occurring alkaloid isolated from *Nandina domestica*[173] or *Coptidis rhizoma.*[147] Pharmacokinetic studies of herb extract indicated a maximum plasma concentration of epiberberine (**10**) of 8.849 μg mL^{-1} and a time to reach maximum plasma concentration of 210 min. The half-life of absorption was 300.91 min., evidencing the therapeutic effect of the compound.[148] A safety evaluation of epiberberine (**10**) showed slight toxicity with a LD_{50} value of 1360 mg kg^{-1} after oral administration to mice.[168] Similar to other known protoberberine alkaloids such as berberine (**4**), palmatine (**17**), groenlandicine (**22**), jatrorrhizine (**15**) or coptisine (**18**), epiberberine (**10**) also showed potent AChE inhibitory effects and good BACE1 inhibition.[147] Docking studies support the experimentally observed AChE inhibitory effects of epiberberine (**10**).[149]

7.4.2 Benzo[*c*]phenanthridines

Benzophenanthridines comprise the second largest group of alkaloids among isoquinolines.[7] In 1984, Krane *et al.* reported on benzophenanthridine alkaloids, stating that "presently, nearly 80 naturally occurring compounds of this type are known".[32]

The subgroup of quaternary benzo[*c*]phenanthridine alkaloids (Figure 7.8) constitutes a relatively small class of isoquinoline alkaloids.[175,176] The most important benzophenanthridine alkaloids are sanguinarine (**5**), chelerythrine (**16**), fagaronine (**19**), and nitidine (**12**).[140] It is noteworthy that most

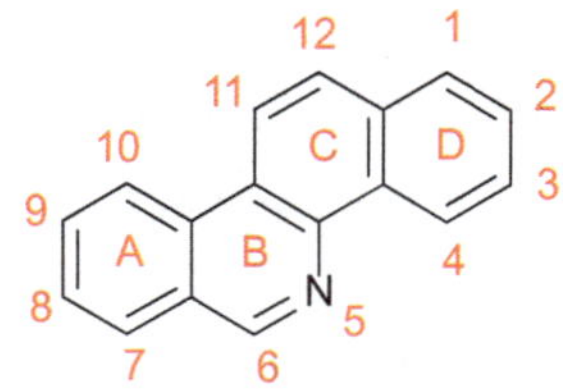

Figure 7.8 Benzo[*c*]phenanthridine scaffold.[174]

of the benzophenanthridine alkaloids are found within the three plant families: Papaveraceae, Fumariaceae, and Rutaceae.[32] The main sources of quarternary benzophenanthridines are plants belonging to the *Chelidonium majus* L., *Sanguinaria canadensis* L., *Dicranostigma lactucoides* Hook. f. et Thoms., *Macleaya* and *Bocconia* species from the *Papaveraceae* family, and some members of *Zanthoxylum* (Rutaceae).[176] The benzophenan further nathridine class also includes the following natural compounds and their synthetic derivatives, such as chelirubine (**38**), macarpine (**39**), fagaridine[177] (**40**), *O*-methylnorfagaronine[178] (**41**), nornitidine (**13**), zanthoxyline (**42**), terihanine (**43**), isoterihanine (**44**), 12-methoxychelerythrine[179] (**45**), norchelerythrine (**46**), chelilutine (**47**), avicine (**48**), noravicine (**49**), NK109 (**50**), and NK314 (**51**) (Figure 7.9).

Benzophenanthridines are pH- and redox-sensitive. Under physiological conditions, compounds in this class can be neutral, positively or negatively charged, which facilitates the bioavailability (Figure 7.10).

Sanguinarine (**5**) is the most prominent representative of the benzophenanthridine group.[7,10] The appearance of the sanguinarine chloride salt (**5**) is orange red, it is soluble in water and has a melting point of 277–280 °C.[141,145] Early investigations on the toxicity of sanguinarine (**5**) published in 1953 reported toxic effects on albino rats and no morbidity in rhesus monkeys.[3] The LD_{50} value in mice is 19.4 mg kg^{-1}.[145] Under alkaline conditions, the iminium form of sanguinarine (**5**) (charged), with a pK_a of 7.4, is transformed into the alkanolamine form (neutral) by hydroxylation of the C6 carbon atom.[140,145]

Sanguinarine is one of the most pH sensitive quarternary benzo[*c*]phenantridine alkaloids.[7] In the pH range 1–6 the iminium form of sanguinarine (**5**) dominates almost exclusively while in the pH range 8.5–11 the alkanolamine form is prevalent (Figure 7.11).[7]

Sanguinarine (**5**) showed antimicrobial, anti-inflammatory, antioxidant, and antitumour activity.[7,181] The use of sanguinarine (**5**) in toothpaste was approved by the FDA due to its antibacterial and antiplaque properties.[7] Cala *et al.* reported the inhibition of ouabain-sensitive K–Na pumps influencing the cation transport through the lipid bilayer of human red blood cells.[182] A screening of several isoquinoline alkaloids on a *Caenorhabditis elegans* model system revealed sanguinarine (**5**) as the most potent one with the ability to reduce lipid accumulation by activating the AMP-activated protein kinase.[183] Sanguinarine (**5**) also showed strong cytotoxicity against certain

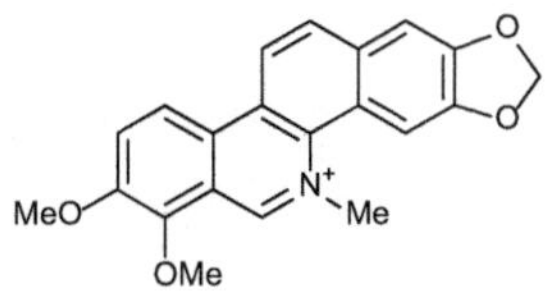

16 Chelerythrine

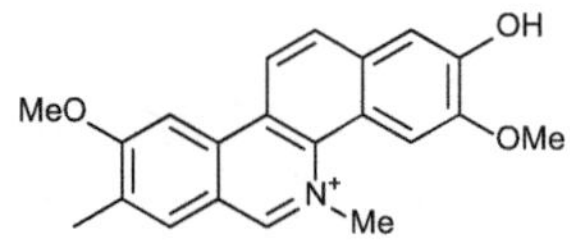

19 Fagaronine

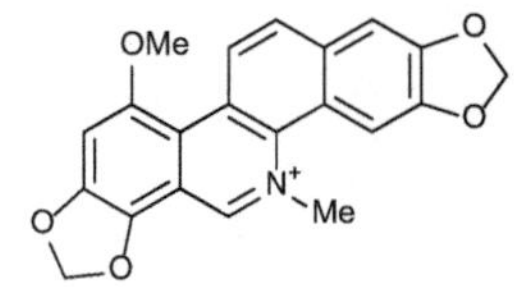

38 Chelirubine

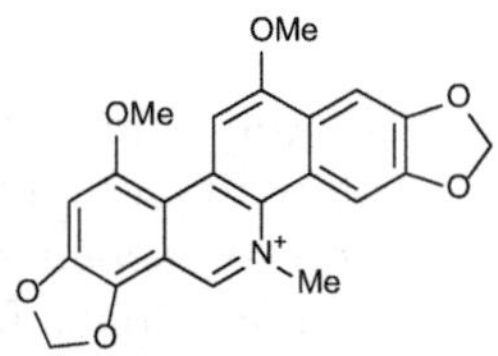

39 Macarpine

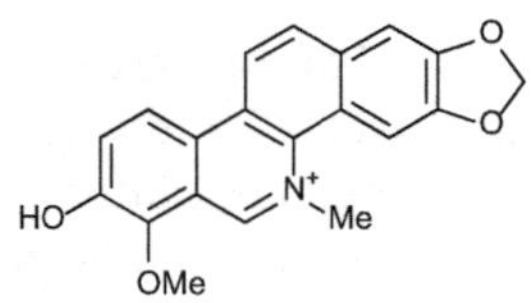

40 Fagaridine

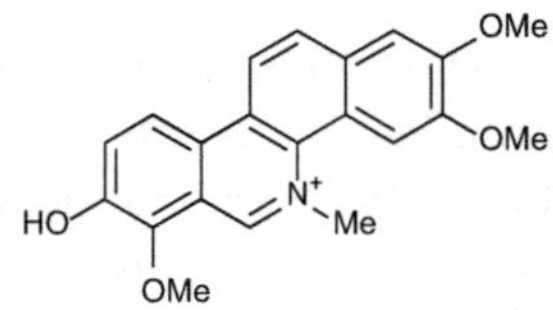

41 *O*-Methylnorfagaronine

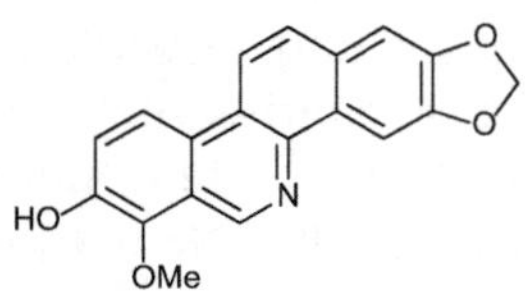

42 Zanthoxyline

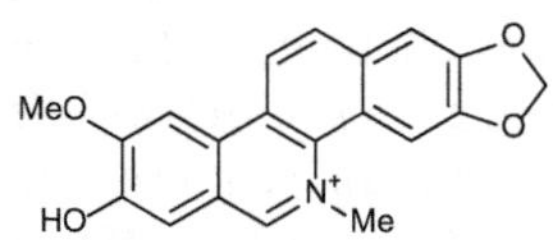

43 Terihanine

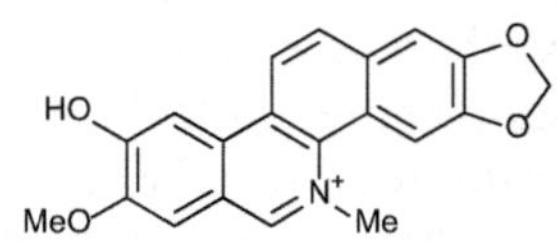

44 Isoterihanine

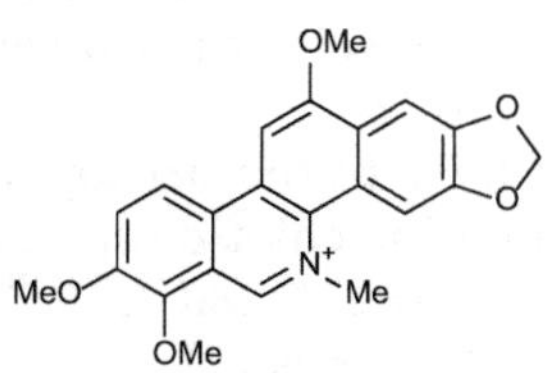

45 12-Methoxychelerythrine

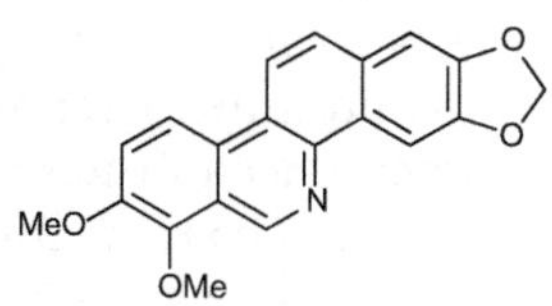

46 Norchelerythrine

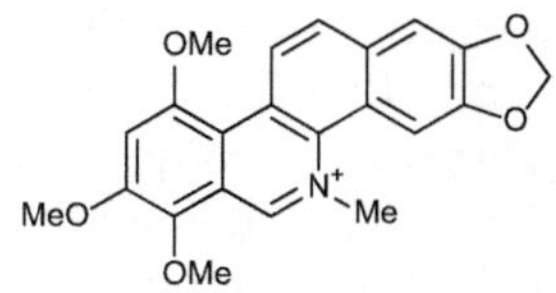

47 Chelilutine

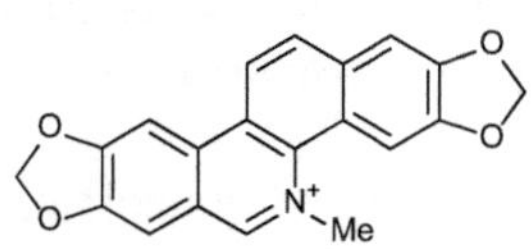

48 Avicine

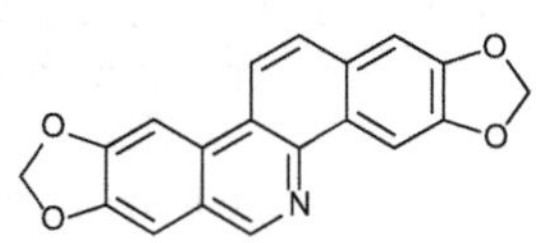

49 Noravicine

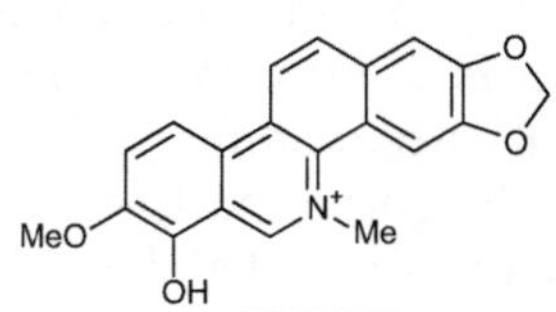

50 NK109

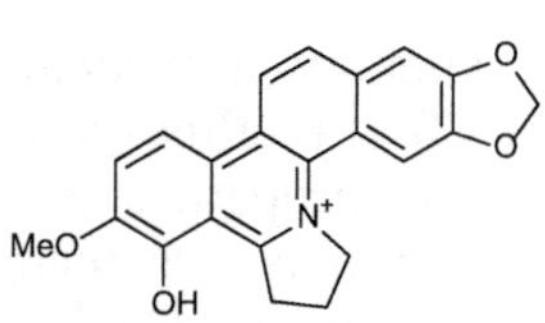

51 NK314

Figure 7.9 Structures of benzophenantridine derivatives.

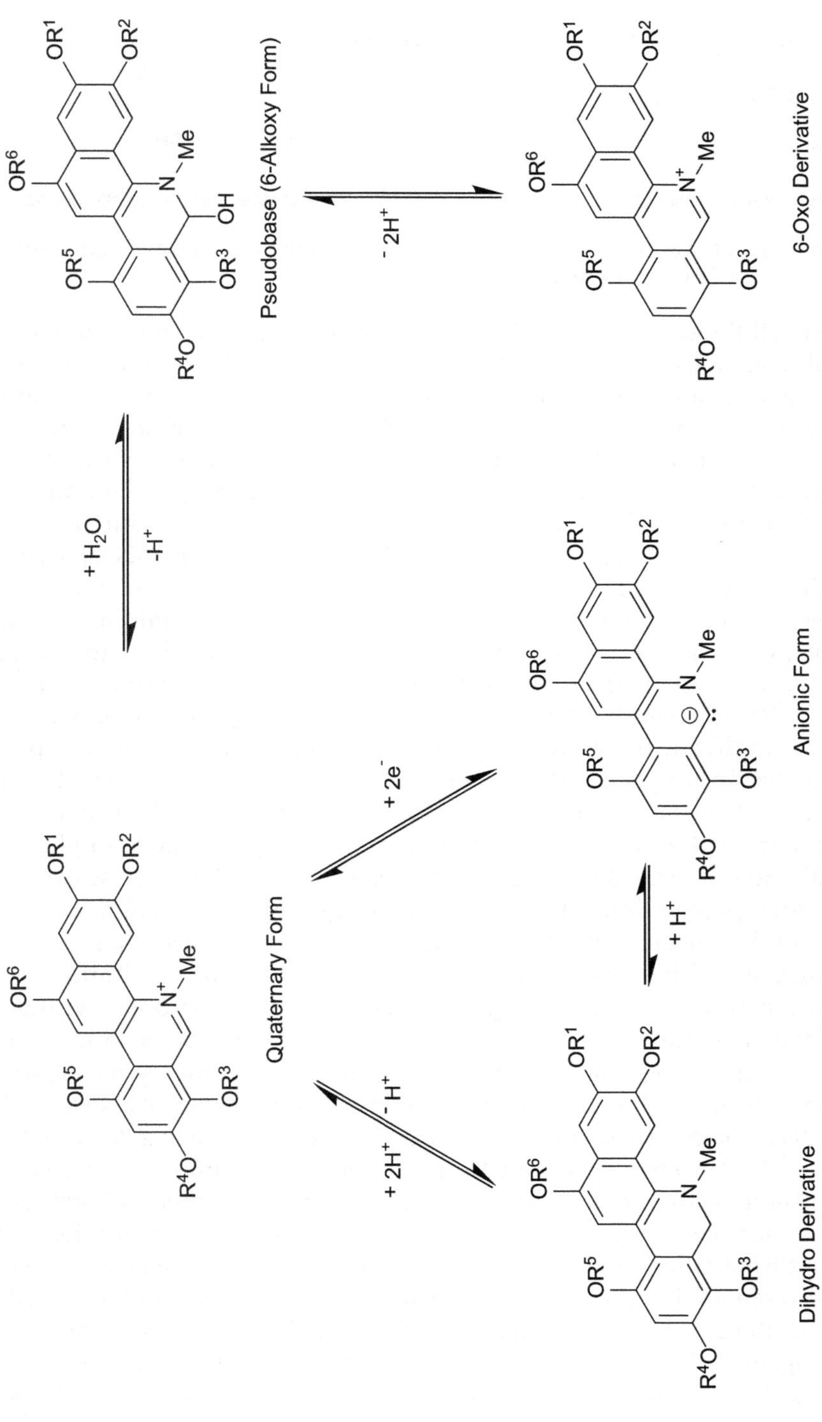

Figure 7.10 Benzo[*c*]phenanthridine under acidic and basic conditions *and* its redox conversion.[180]

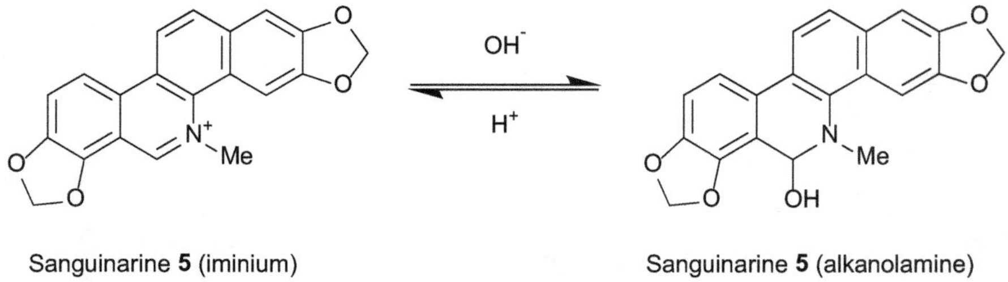

Figure 7.11 pH dependent equilibrium between iminium and alkanolamine forms of sanguinarine (5).[7]

cancer cell lines.[184] Sanguinarine (5) is reported to perturb microtubule assembly dynamics by binding to tubulin. Sanguinarine (5) also intercalates in DNA and can cause strand breaks among other damages to DNA.[7] Similar to berberine, sanguinarine (5) was found to bind to human telomeric G-quadruplex DNA.[185] Ferraroni *et al.* resolved a crystal structure of sanguinarine (5) intercalated in a $d(CGTACG)_2$ DNA sequence, which indicated the interaction of the iminium form of sanguinarine (5) with the DNA strand.[186] Furthermore, sanguinarine (5) can induce apoptosis in various cell lines.[7] A word of caution should be added at this point: due to the necrotic effect of some topically administered pastes containing sanguinarine (5), often referred to as Black Salve, supporters of alternative cures have recommended this unlicensed paste for skin cancer treatment, among others. Several reports of patient cases have recently been published in order to highlight the corrosive effect of this self-treatment which repeatedly ended in long-term plastic surgery therapy. For this reason, the importing of Black Salve is prohibited in some countries.[187–189] Furthermore, sanguinarine (5) is a known contaminant in cooking oil, and is implicated in outbreaks of heart disease, glaucoma, cancer, and other illnesses among the Indian population.[32] Traditional chemical synthesis of complex plant alkaloids has been a challenge, especially on a large scale, *e.g.* for the pharmaceutical industry. Fossati *et al.* describe the reconstitution of a 10-gene pathway in *Sachharomyces cerevisae* for the microbial synthesis of dihydrosanguinarine from which sanguinarine (5) can be easily obtained by oxidation.[9] Sanguinarine (5) exhibited anticancer efficacy on prostate cancer cells, thus being a promising therapy to overcome taxol resistance.[65]

In 1953, chelerythrine (**16**) was reported for the first time in a *Zanthoxylum* species.[190,191] It should be noted that the botanical name *Zanthoxylum* is synonymous with *Xanthoxylum* (see nitidine (**12**)).[32] Synthetic efforts have been undertaken early on. In 1956, Bailey *et al.* reported the synthesis of chelerythrine chloride (**16**).[192] Photolytic preparation of chelerythrine (**16**) was also reported in the 1980s.[117,193] Some years later, Ishii *et al.* reported a caesium fluoride-mediated total synthesis of chelerythrine (**16**) *via* Claisen rearrangement.[194] Chelerythrine (**16**) was obtained as crystalline yellow needles with a melting point of 199–200 °C.[176] The natural product showed anti-inflammatory activity when administered orally[195,196] and low acute toxicity of 95 mg kg^{-1} along with a low chronic toxicity.[195] It is also known to be a protein kinase C (PKC) inhibitor.[50,197] In contrast to sanguinarine (**5**) and nitidine (**12**), which had higher binding affinities, chelerythrine (**16**) stood out because of the sharply enhanced steady state fluorescence upon binding to DNA.[7] Chelerythrine (**16**) has been in development as a treatment for bipolar disorder and the cognitive deficits of schizophrenia based on its PKC inhibition.[198] Chelerythrine (**16**) was shown to bind to sequence-defined double-strand DNA. Docking studies with the crystal structure of AChE (*Torpedo californica*) as a model receptor suggested that chelerythrine (**16**) covers the gorge site, showing hydrogen bond interaction with TYR130 as well as π-stacking interactions with TYR121 and TYR334 peripheral anionic site (PAS) residues. Chelerythrine (**16**) showed activity against AChE and BChE, an ability to prevent aggregation of Aβ peptide into fibrils, and the ability to disintegrate preformed Aβ aggregates.[172] Chelerythrine (**16**) was shown to possess excellent bioactivity against phytopathogenic fungi[175,199] and induced apoptosis by generating reactive oxygen species in cardiac myocytes.[200] Specific tests were conducted to examine the inhibition of the antiapoptotic human Bcl-2-family protein by chelerythrine, compared to other non-isoquinoline natural products and synthetic compounds in pharmaceutical development, which revealed better inhibition activities.[201] Chelerythrine (**16**) is in preclinical trials as a Bcl-2 and Bcl-xL targeting small molecule as apoptosis-inducing cancer agent.[66] Derivatives such as 12-methoxychelerythrine (**45**) have been synthesised by Watanabe *et al.*[202]

The melting point of fagaronine chloride (**19**) was reported to be 202 °C.

Fagaronine (**19**) was synthesized[118,203] and showed microbiocide effects.[204] Studies suggested that fagaronine (**19**) is a strong intercalator in dublex DNA and a stoichiometry of one fagaronine (**19**) molecule per two DNA base pairs.[7] It was also reported that fagaronine (**19**) showed antileukemic activity and inhibits the topoisomerase enzyme.[32,77,205] Fagaronine (**19**) and nitidine (**12**) underwent development as anticancer drugs, but this was continued due to their low potency and incompatibility with biological fluids.[50] NK314 (**51**) is a fagaronine (**19**) derivative optimized for high activity and water-solubility.

Nitidine (**12**) is a natural alkaloid that was first isolated from *Xanthoxylum nitidum* collected in Hong Kong.[191,206] It has a melting point of 220 °C as chloride salt.[32] Ishii *et al.* described the preparation of nitidine (**12**) and the appearance as pale yellow needles with a melting point of 196–200 °C (decomposition).[174] The compound was described as rather unstable, with a tendency to disproportionate into dihydro- and oxy-derivatives.[32] Nitidine (**12**) showed antileukemic activity against L1210 in mice.[32,207,208] DNA-binding studies showed a great quenching effect of the steady state fluorescence of nitidine (**12**) (and sanguinarine (**5**))[7] and has been reported to be a topomerase I inhibitor.[77,205] Recently, Liao *et al.* elucidated the molecular basis for the antitumour activity of nitidine chloride (**12**), showing results of cancer cell apoptosis induction and down-regulation of cyclin D1, cyclin dependent kinase 4 (CDK4) and B-cell lymphoma 2 (Bcl-2) expression together with increased levels of p21 and Bcl-2 associated X protein (Bax) in hepatocellular carcinoma (HCC).[209]

NK314 (**51**) is a synthetic alkaloid compound derived from the fagaronine (**19**) structure that entered clinical trials as an antitumour back-up for NK109 (**50**). Metabolic reduction of NK109 (**50**) to 5,6-dihydro NK109 (**50**) made high doses necessary to maintain antitumour activity due to the inactivity of the formed metabolite.[82] In order to overcome this obstacle, derivatives were synthesized that were aimed at sustaining strong cytotoxicity and suppressing biological reduction.[82] Structure-activity relationship investigations revealed that substituents on position 6 made the molecule stable against biological reduction. At the same time, this resulted in weak cell growth activity. Additionally, substituents on position 8 also suppressed biological reduction, when not bulky and hydrophobic, but preserved cytotoxicity. The 8-*O*-Hydroxyethyl derivatives of NK109 (**50**) exhibited the most promising combination of favoured properties, thus serving as a lead compound.[82] About seven years later, NK314 (**51**) was published as the back-up compound tested in clinical trials. The mode of action was reported as topoisomerase IIα inhibition.[210] A great advantage is its potency against tumours that are resistant to other topoisomerase II inhibitors.[50] Toyoda *et al.* showed that the α-isoform of topomerase II is responsible for the NK314 (**51**) cytotoxicity. Due to the high selectivity towards the α-isoform, this compound appeared to be highly valuable for further clinical studies, because the involvement of the β-isoform by established drugs (etoposide, doxorubine, mitoxantrone) was shown to cause more serious side effects, such as secondary malignancies.[211] In 2011, Guo *et al.* described the intercalating properties of NK314 (**51**) and a potential DNA repair mechanism that was supposed to cause resistance to NK314 (**51**). DNA-dependent protein kinase (DNA-PK) and ataxia telangiectasia mutated (ATM) were suggested to contribute to cell survival in response to NK314 (**51**) and could therefore be potential targets to overcome resistance.[212] Based on their results, Hisatomi *et al.* suggested that NK314 (**51**) is a dual inhibitor of topoisomerase IIα and DNA-PKcs (DNA-dependent protein kinase catalytic subunit). Because adult T-cell leukemia-lymphoma (ATL) is incurable, and known to express high amounts of DNA-PKcs, NK314 (**51**) would be a promising treatment.[213] Clinical studies are currently underway.[79]

7.4.3 Pyridocarbazoles

Pyridocarbazoles (Figure 7.12) have been of great interest due to their pronounced antitumour activity.[214] Since then chemists elaborated synthetic routes for these natural products and their derivatives.

This alkaloid class includes compounds and derivatives such as ellipticine (**52**), 5,11-dimethylellipticine, 2*N*-methyl-9-hydroxyellipticine (MHE) (**53**), pazelliptine (**54**), olivacine (**55**), datelliptium (**56**), elliprabin (**57**), retelliptin (**58**), ditercalinium (**59**), and S 16020-2 (**60**), among many others (Figure 7.13).

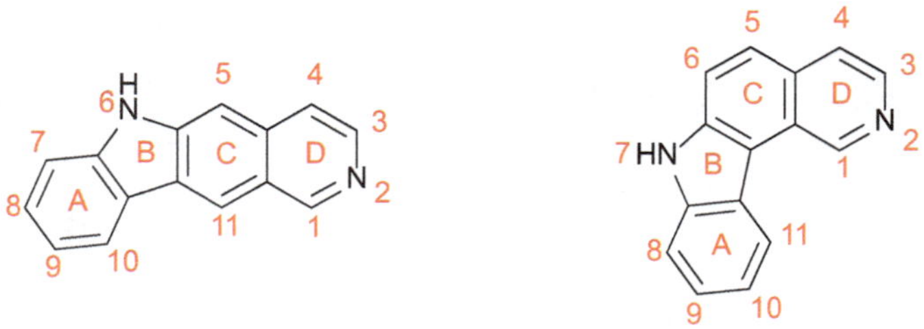

6*H*-pyrido[4,3-*b*]carbazole 7*H*-pyrido[4,3-*c*]carbazole

Figure 7.12 Pyridocarbazole scaffolds.[214]

Ellipticine (**52**) is a simple, naturally occurring alkaloid with a planar structure. In 1959, it was first isolated from the leaves of *Ochrosia elliptica* Labill (Apocynaceae).[68] Eight years later it was described as a natural antitumour alkaloid.[215,216] Ellipticine (**52**) and 9-methoxyellipticine (**61**) are highly cytotoxic to malignant cultured cells.[217] Since large quantities were unavailable from natural sources, effective synthetic routes were developed early on to obtain sufficient amounts for further testing.[216] The interest in ellipticines (**52**) derived from its limited toxicity and complete lack of haematological toxicity.[217] Ellipticine (**52**) and most of its derivatives have been found to intercalate in DNA and target topoisomerase II.[63] Topoisomerase II inhibitors are of particular interest because this ubiquitous enzyme plays a critical role in DNA metabolism and is therefore a target in cancer treatment.[63] Ellipticine (**52**) belongs to the category of topoisomerase poisons, in contrast to catalytic inhibitors.[63] Topoisomerase poisons target specific DNA sequences and the stabilization of the cleavable topoisomerase II-DNA complex that causes tumour cell death.[63,161] Ellipticines were also found to be mutagenic in *Salmonella* Ames tester strains, Chinese hamster ovary (CHO) cells and yeast.[217] Recent studies contributed to the elucidation of the mode of action by studying the ellipticine (**52**) metabolism by peroxidases and human cytochrome P450 (CYP), leading to DNA adducts.[218]

Ellipticine (**52**) was also found to inhibit RNA polymerase I (Pol-I), an enzyme linked to high rates of proliferation in cancer.[67] In general, ellipticine (**52**) showed promising properties that have been exploited for further lead optimization. Further development activities of ellipticine (**52**) as a lead strucutre have been discontinued due to cardiovascular toxicity and haemolysis that were observed in preclinical studies.[68] Nevertheless, ellipticine (**52**) was recently identified as a CK2 inhibitor providing structural insights by a CK2-inhibitor crystal structure that can be exploited for further lead optimization studies.[219]

Celiptium (**62**) (2*N*-methyl-9-hydroxyellipticinium, 9-OH-NME, elliptinium acetate, 9-hydroxy-2-methylellipticinium acetate, NMHE),[40,41,68,220] a fully synthetic 9-substituted derivative of ellipticine (**52**),[41] has been used for cancer treatment by taking advantage of the cytotoxic activity.[220] Dalton *et al.*

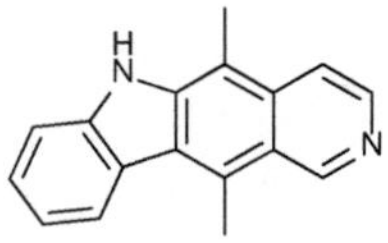

52 Ellipticine

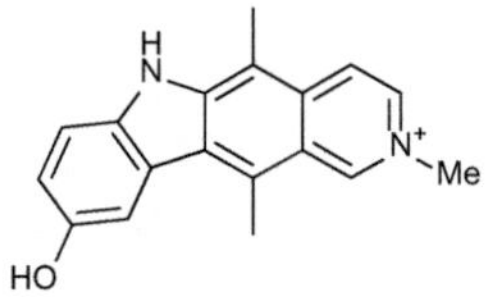

53 MHE

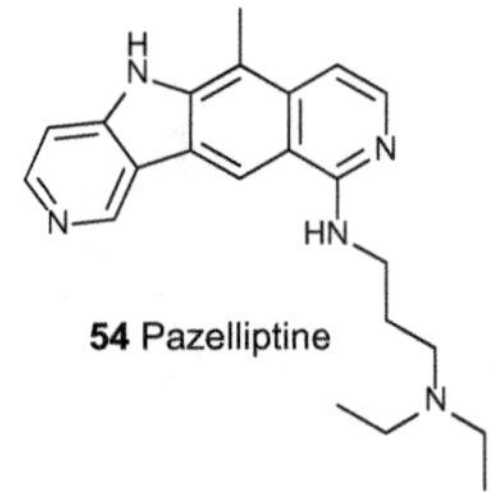

54 Pazelliptine

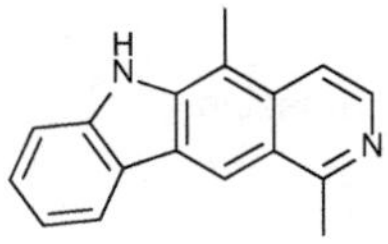

55 Olivacine

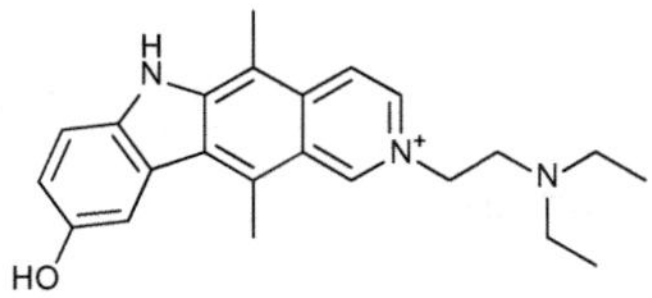

56 Datelliptium

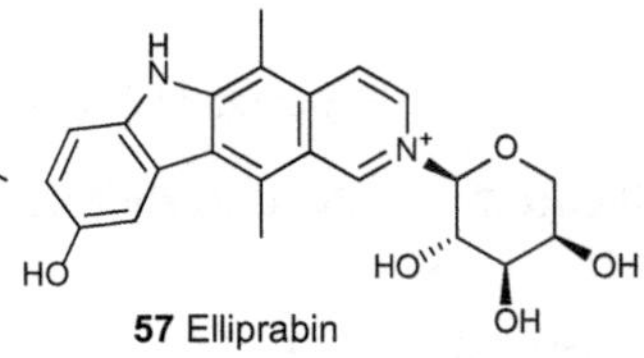

57 Elliprabin

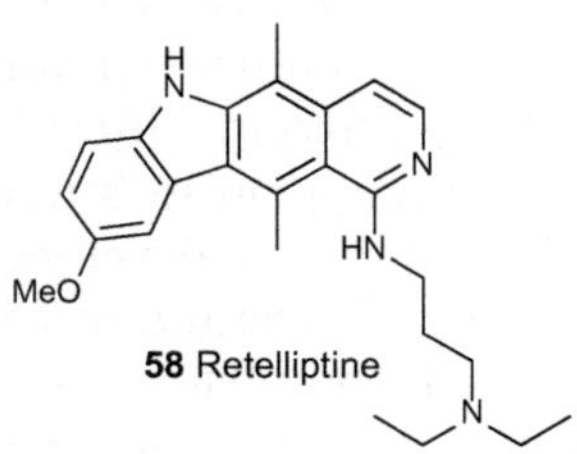

58 Retelliptine

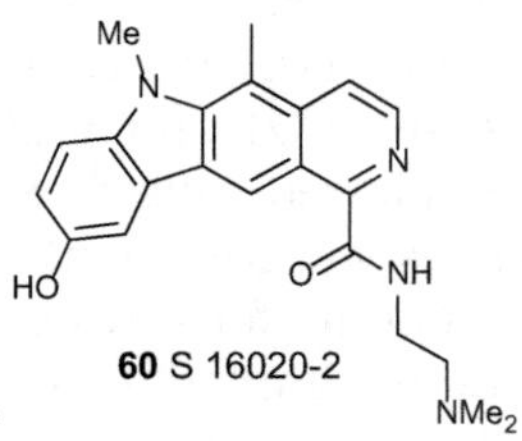

60 S 16020-2

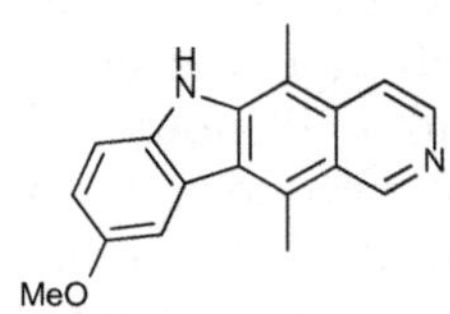

61 9-Methoxyellipticine

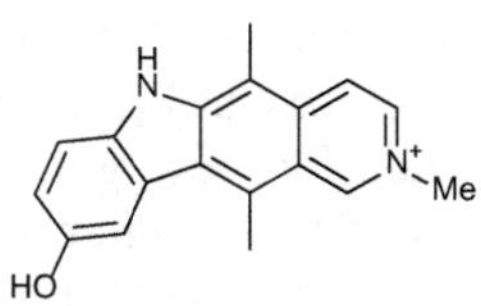

62 Elliptinium

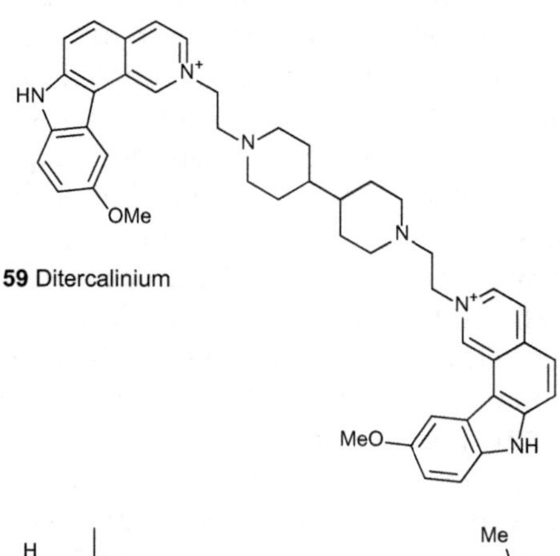

59 Ditercalinium

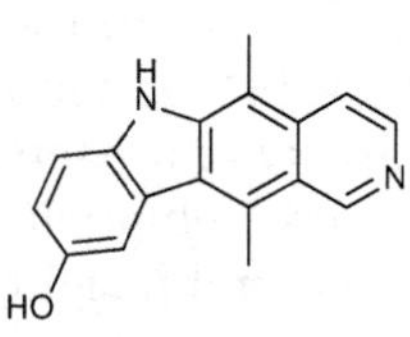

63 9-Hydroxyellipticine

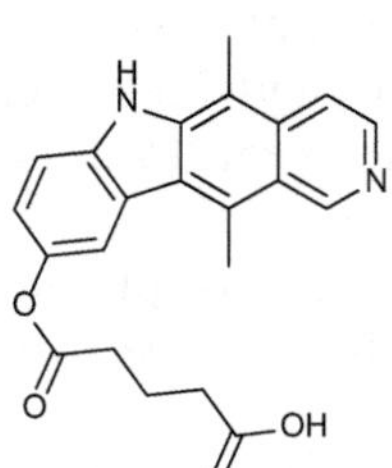

64 T-215

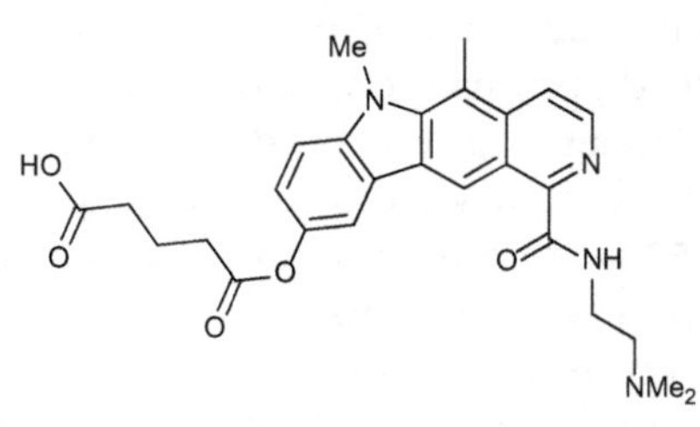

65 S 30972-1

Figure 7.13 Chemical structures of pyridocarbazole derivatives.

described the synthesis of ellipticine (**52**) and 9-Methoxyellipticine (**61**).[216] The main metabolite in rat and human urine was identified to be 9-(*O*)-glucuronide.[220] Generally, all ellipticine (**52**) derivatives and the natural product ellipticine (**52**) itself showed high *in vitro* cytotoxicity.[217] Two properties made ellipticines initially promising clinical candidates: their limited toxic side effects and the complete lack of haematological toxicity.[217] Celiptium (**62**) was tested in a phase II clinical study as salvage treatment in breast cancer. In addition to other studies, a weekly schedule revealed toxicities such as xerostomia and immune-mediated haemolytic reactions that were related to the development of *anti*-elliptinium IgM antibodies.[41] Renal[89] and gastrointestinal[35] toxicities were also observed. A modified three-week schedule showed reduced toxicity and no *anti*-elliptinium antibodies were observed.[41] Although elliptinium (**62**) is marketed in France for the treatment of breast cancer, little information is available.[72,73] Other 9-substituted derivatives such as 9-methoxyellipticine (**61**) and 9-hydroxyellipticine (**63**) revealed only limited activity in clinical trials.[68] Based on the experience of elliptinium (**62**), further experimental studies of ellipticine (**52**) derivatives were carried out to improve both activity and tolerability.[39] Numerous structure–activity relationship studies have been reported and many promising compounds were investigated in preclinical and clinical studies.

T-215 (9-pentanediolate-ellipticine) (**64**) was developed by Tanabe Seiyaku Co. Ltd. and patented in 1993 as an apoptosis-inducing agent of cancer cells *via* inhibition of phosphorylation of mutant p53 for clinical trials. Since then, no further information on its clinical success could be accessed. The chemical structure has been disclosed in early publications.[34]

Compared to celiptium (**62**), datelliptium (**56**) (DHE, 2-(diethylamino-2-ethyl)-9-hydroxyellipticinium-chloride) has a basic diethylaminoethyl chain at nitrogen 2 resulting in a significant increase in lipophilicity, decrease in acute toxicity and increase in cytotoxicity.[221] As all ellipticine analogues, datelliptium (**56**) also intercalates with DNA and shows inhibition of topoisomerase II.[227,328] It also revealed a high activity in decatenation.[34,221] Although datelliptium (**56**) was shown to be 100 times more cytotoxic than ellipticine (**52**), ellipticine (**52**) caused more DNA strand breaks, possibly because of limited access to the cell nucleus.[34] Datelliptium (**56**) has been studied for the treatment of advanced breast cancer and was active in metastatic breast cancer.[34] In a phase II study of datelliptium (**56**) for metastatic malignant melanoma, patients showed no response.[34] Toxic effects were mainly hepatic and it induced rare and mild leukopenia and fatigue.[34]

Due to promising results in *in vitro* human cell lines, ditercalinium (**59**) was put forward to clinical trials.[34] After a phase I study with irreversible hepatotoxicity as dose-limiting side-effect, a rat study revealed selective accumulation in hepatocyte mitochondria including its damage.[87] Ditercalinium (**59**) has been withdrawn from clinical trials due to these findings.[34]

Pazelliptine (**54**)[64,71] (10-[3-diethylaminopropylamino]-6-methyl-5*H*-pyrido|3′,4′:4,5]pyrrolo[2,3-*g*]isoquinoline, PZE) has been investigated in the context of interaction with DNA as antitumour drug. Strong interactions could be demonstrated, suggesting at least two different binding sites for the PZE/poly(dG-dC)–poly(dG-dC) complex.[222]

In the 1990s, results of retelliptine (SR 95325 B) (**58**) phase I studies showed reversible visual and EKG anomalies with a dose-limiting cardiac toxicity associated with a mild hypotension.[34,39] No further clinical reports could be traced.

Olivacine (**55**) was isolated from *Aspidosperma olivaceum* Müll. Arg. in 1958.[223,224] Since it was discovered, until the early 1980s, the total synthesis of olivacine (**55**) has been the subject of numerous studies due to the promising antitumour activities.[225] Olivacine (**55**) has served as a lead structure for many analogues, with improved antitumour activities that made the synthetic access crucial.[224] The total synthesis of olivacine (**55**) was reported by Kutney *et al.* in 1987.[226]

1-Diethylaminoethylolivacine or S-16020 (**60**) is an olivacine derivative that stimulates ATP-dependent topoisomerase II-mediated DNA cleavage.[63] Due to favourable outcomes in experimental antitumour activity models, pharmacokinetic characteristics, and its acceptable toxicity, S-16020 (**60**) was put forward to clinical trials. The main observed side effects in clinical phase I studies were asthenia and skin toxicity.[89] Neither haemolysis nor *anti*-S-16020 (**60**) antibodies were detected.[89] S30972-1 (**65**),[62,63] another synthetic derivative of olivacine (**55**), exhibited an improved therapeutic index compared to S-16020 (**60**). Based on those results, further pyridocarbazole modifications have been synthesized to optimize the physicochemical properties.[214,224,227]

7.4.4 Phenanthridine

The syntheses of phenanthridines (Figure 7.14) as chemotherapeutics against trypanosome infections were reported by Browning *et al.* early in 1938.[228]

Around 60 years later, in 1997, Geerts *et al.* described experiments with cows using isometamidium (**66**) and ethidium (**67**) as prophylactic agents against the parasite *Trypanosoma congolense*, which can for instance be transferred by tsetse flies.[230]

Phenanthridines (Figure 7.15) gained further interest through findings that certain compounds can intercalate with DNA, attracting attention as a

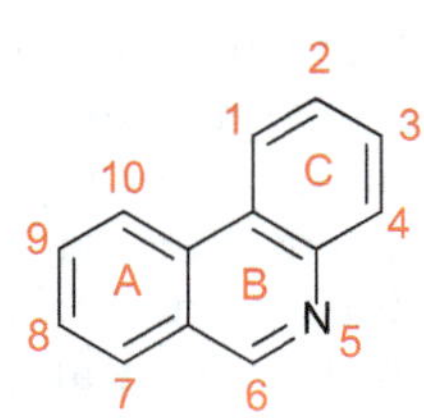

Figure 7.14 Phenanthridine scaffold.[229]

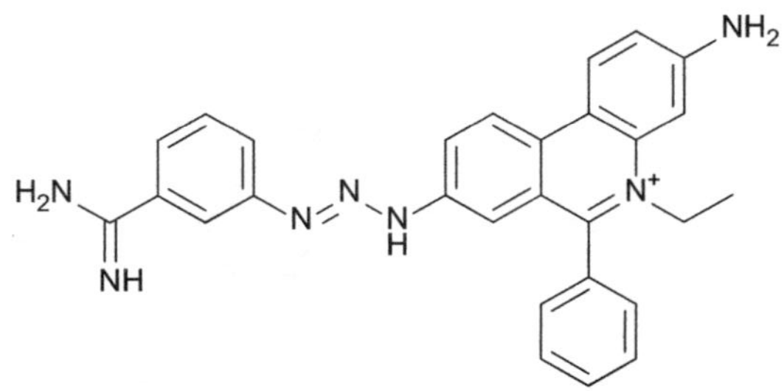

66 Isometamidium

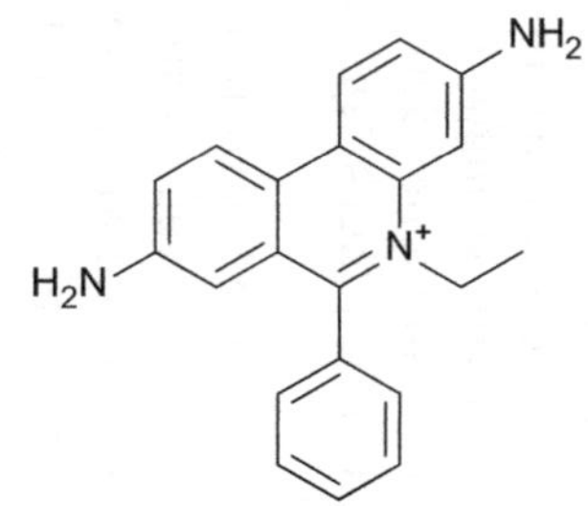

67 Ethidium

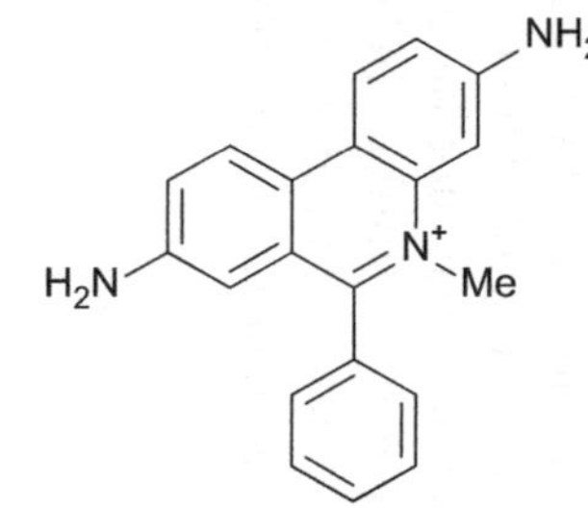

68 Dimidium

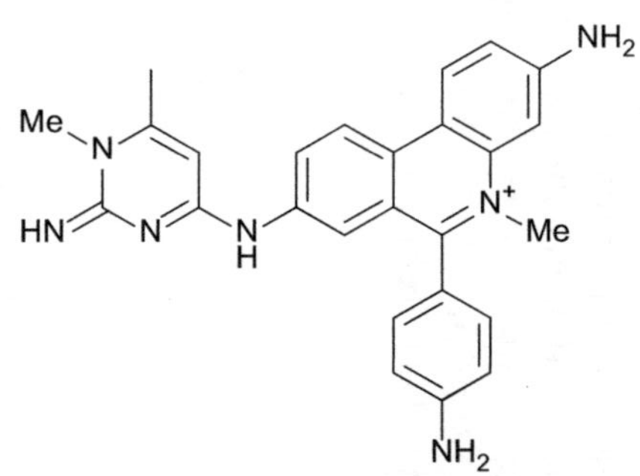

69 Prothidium

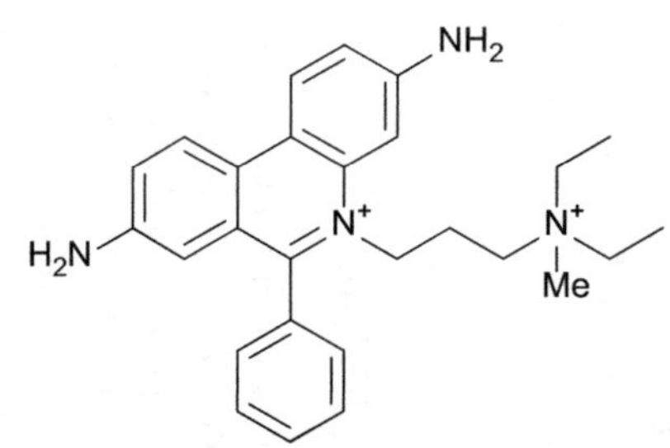

70 Propidium

Figure 7.15 Chemical structures of phenanthridine derivatives.

potential anti-cancer agent. In 1979, Hogan *et al.* described studies of the structure of short DNA sequences with ethidium bromide (**67**) as intercalating[231] drug after Jain *et al.* and Tsai *et al.* elucidated the drug-nucleic acid interaction with a crystalline complex.[232,233] Despite the early focus on phenanthridines as veterinary drugs such as ethidium (**67**), dimidium (**68**), or prothidium (**69**),[234] human preclinical development has not been set forth. Today, phenanthridines such as propidium (**70**) and ethidium (**67**) are used as fluorescence indicators for DNA probes.[235] Recent studies investigated propidium (**70**) in the context of Alzheimer disease as a control compound.[100,172]

7.4.5 Aspergillitines

Scientists are keen to find new sources for potential drug candidates. Despite the manifold natural plant compounds, the right combination of clinically relevant properties is difficult to find. Chemists are interested in lead compounds for lead optimization. In the past, novel marine sponges caught the attention of researchers. In 2003, a novel chromone derivative (Figure 7.16) with an isoquinoline motif from the marine fungus *Aspergillus versicolor*, namely aspergillitine (**71**), was reported.[236]

Simonetti *et al.* indicated that the formerly presented structure for aspergillitine (**71**) was similar to a natural product that showed ^{1}H and ^{13}C NMR resonances that matched with TMC-120B (**72**), leading to the conclusion that the originally described compound was not aspergillitine (**71**), but TMC-120B (**72**) (Figure 7.17). Therefore, the initially proposed structure of aspergillitine (**71**) remains unobserved among natural products.[237]

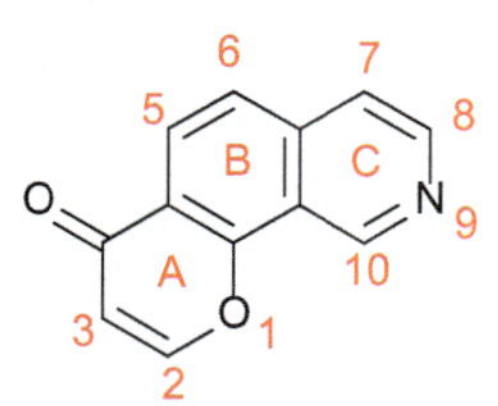

Figure 7.16 Aspergillitine scaffold.[43]

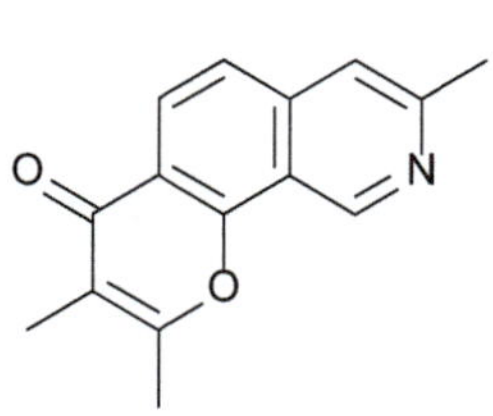

71 Aspergillitine

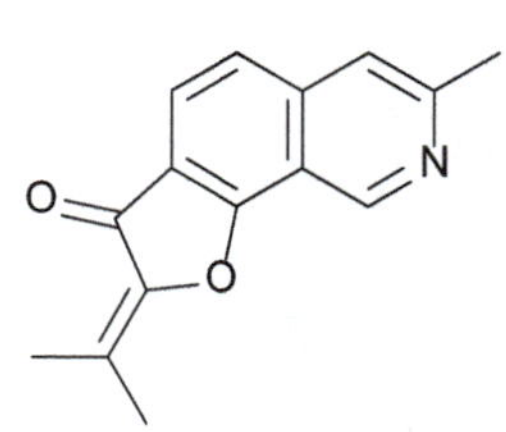

72 TMC-120B

Figure 7.17 Chemical structures of aspergillitine derivatives.

Nevertheless, these findings resulted in the total synthesis of both molecules (TMC-120B (**72**),[43,179,238–251] aspergillitine (**71**)[237,241,249,252]). Cases like this show impressively that, even today with the full range of analytical instrumentation, structure elucidation remains from time to time a tedious endeavour.

7.4.6 Benzylisoquinolines

There are approximately 2500 known structures of naturally occurring plant benzylisoquinolines (Figure 7.18).[10]

Papaverine (**3**), ethaverine (**73**), daphnine (**74**), neolitacumonine (**75**), anocherine A-C (**76-78**), *O*-methyl-neolitacumonine (**79**), *N*, *O*-dimethylneolitacumonine (**80**), hypecoumine (**81**) are some examples. Papaverine (**3**) is the most prominent representative of the benzylisoquinoline group. It can be extracted from the opium poppy (*Papaver somniferum*)[10,253] and has been used as a non-specific vasodilator due to its direct action on smooth muscle,[10] although it has not been approved by the FDA.[86,254] Papaverinol ((*S*)-enantiomer (**82**) and (*R*)-enantiomer (**83**)) and papaveraldine (**84**) (Figure 7.19) are known degradation products of papaverine (**3**).[255]

In principal, three groups of derivatives can be distinguished, namely those derived from papaverine (**3**) (Figure 7.20), papaverinol (**82, 83**) (Figure 7.21) and papaveraldine (**84**) (Figure 7.22).

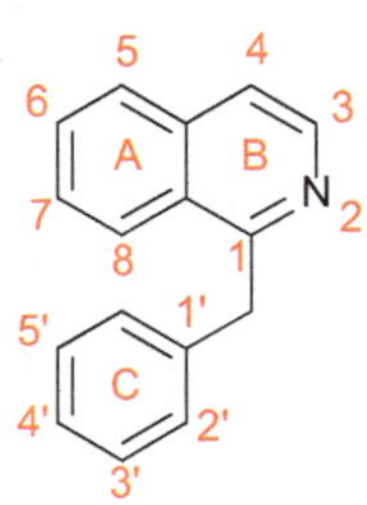

Figure 7.18 Benzylisoquinoline scaffold.[10]

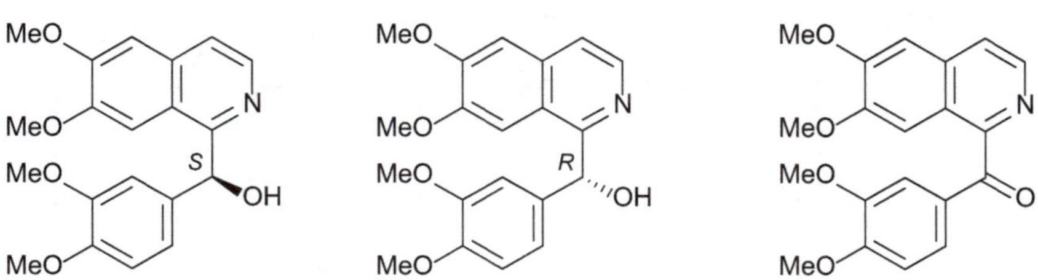

82 (S)-Papaverinol

83 (R)-Papaverinol

84 Papaveraldine

Figure 7.19 Degradation products of papaverine (**3**).

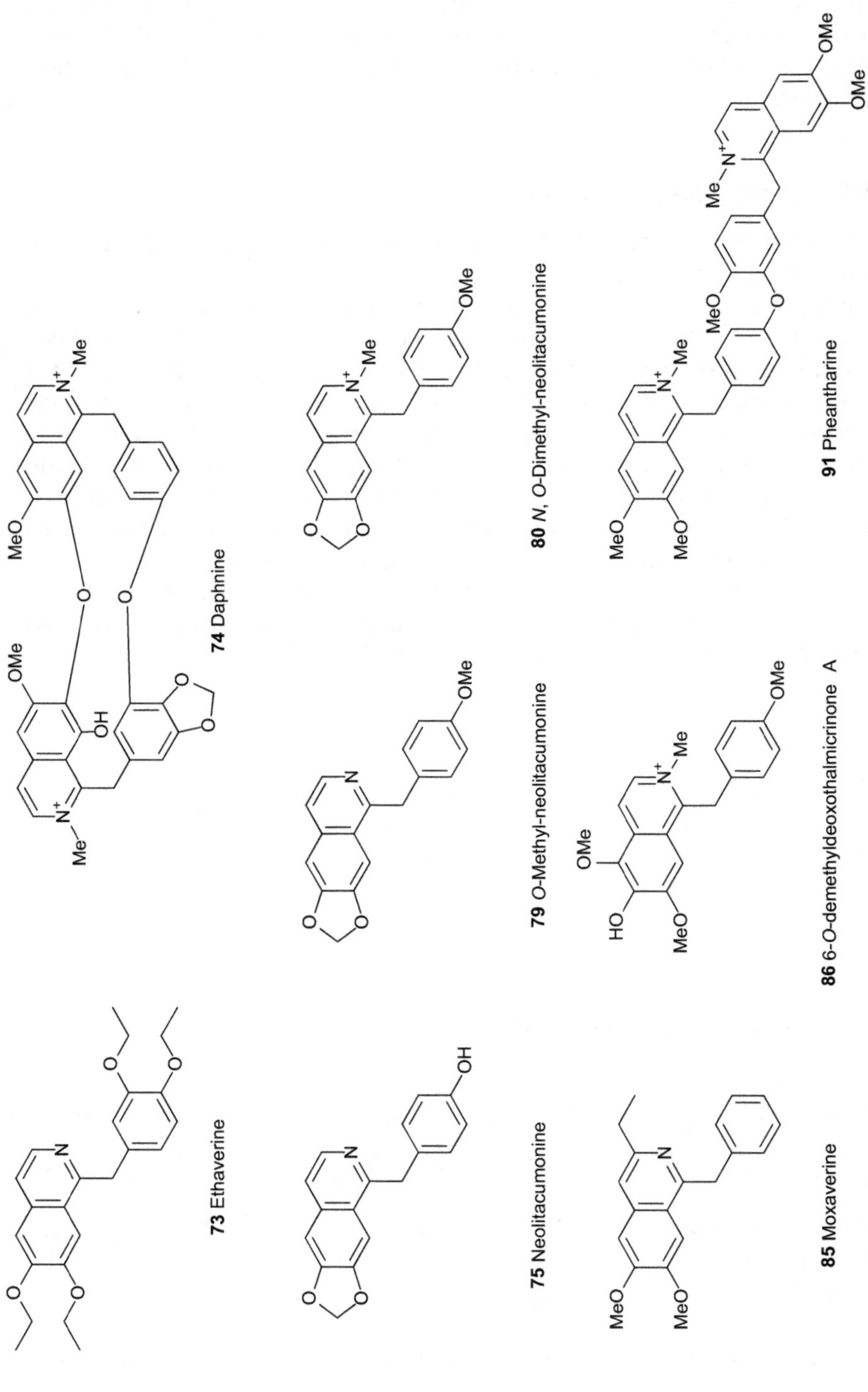

Figure 7.20 Papaverine derivatives.

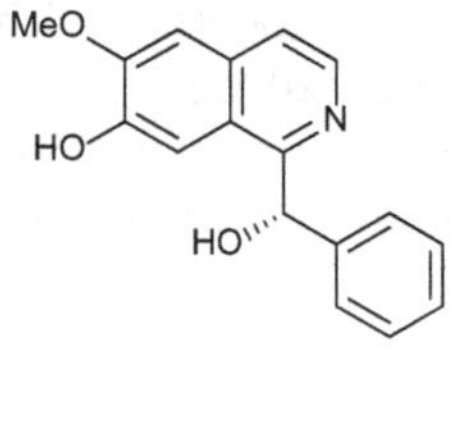

76 Anocherine A

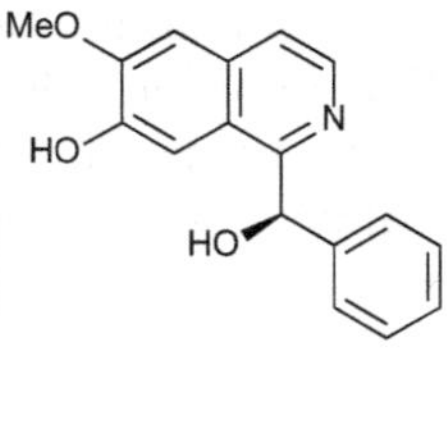

78 Anocherine C

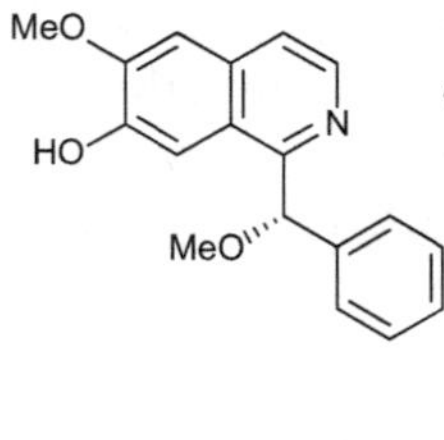

77 Anocherine B

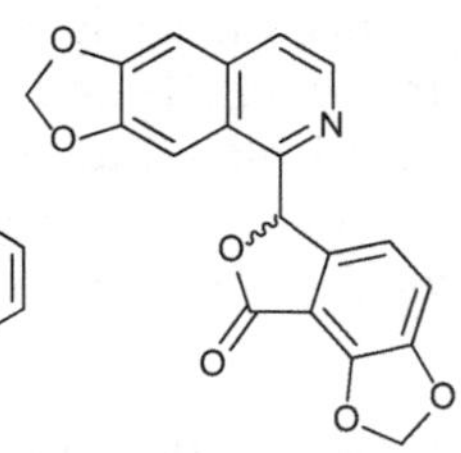

81 Hypecoumine

Figure 7.21 Papaverinol derivatives.

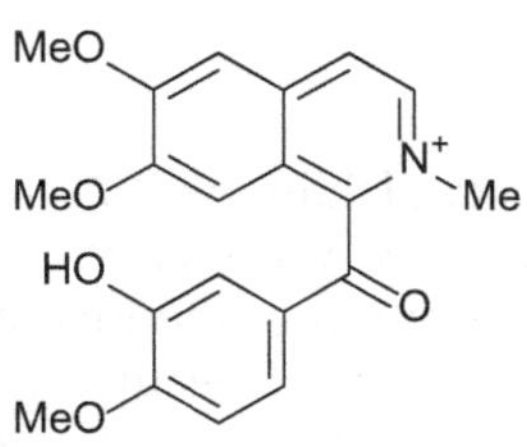

87 Thalprzewalskiinone

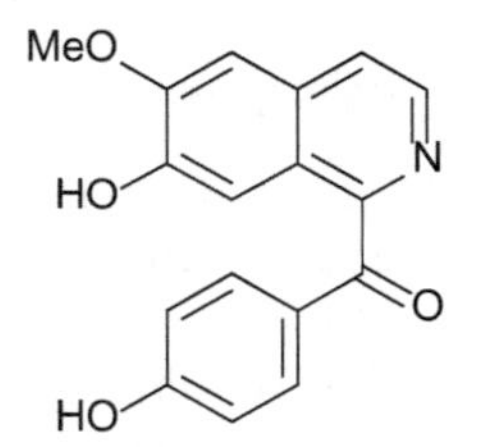

88 Oxodeoxyannocherine A

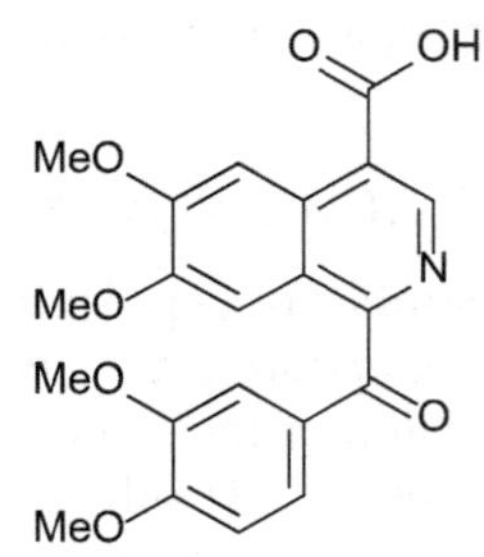

89 RO0509347A

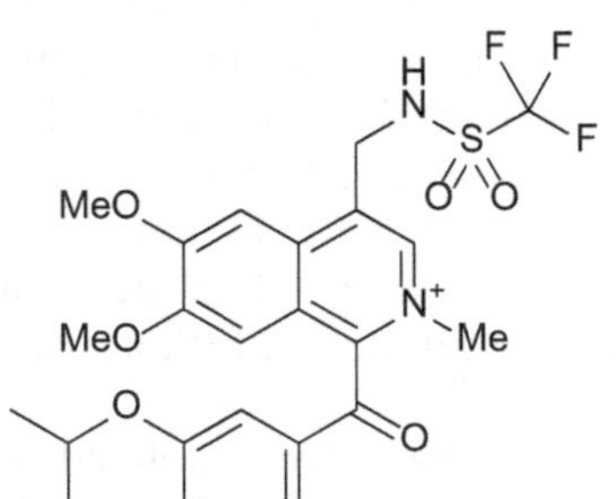

90

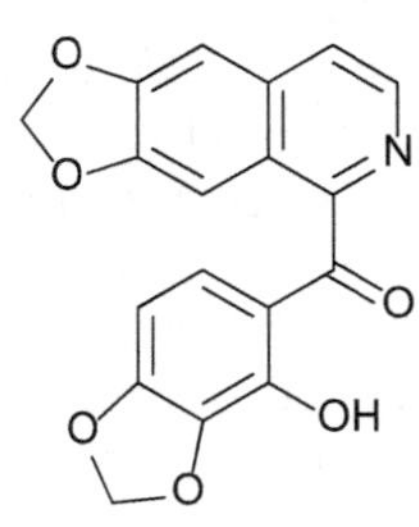

92 Sauvagnine

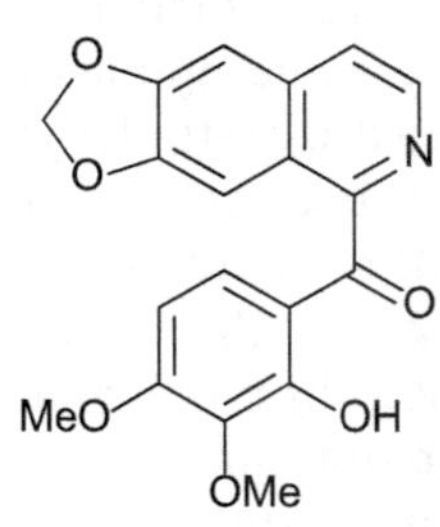

93 Linaresine/Rugosinone

Figure 7.22 Papaveraldine derivatives.

7.4.6.1 *Papaverine-analogues*

Moxaverine (**85**) is a phosphodiester (PDE) inhibitor currently tested on ocular blood flow.[256] Ethaverine (**73**), the tetraethoxy homologue of papaverine (**3**), was also reported to be a smooth muscle relaxant.[83] Ethaverine (**73**) additionally showed effects in PC12 cells by decreasing dopamine levels *via* tyrosine hydroxylase inhibition.[257] Daphnine (**74**) has been isolated from *Daphnandra dielsii* and has a rare bisbenzylisoquinoline scaffold that has not yet been investigated in more detail.[178] Neolitacumonine (**75**) together with its *O*-methyl- and *N*, *O*-dimethyl-neolitacumonine (**79** and **80**) derivatives were isolated about a century ago, but no further investigations could be found.[258] 6-*O*-demethyl-de-oxo-thalmicrinone (**86**) was isolated from *Thalictrum delavayi*.[259]

7.4.6.2 Papaverinol-analogues

Papaverinol-analogues (Figure 7.21) are not very common. Anocherine A–C[177,260] (**76–78**), isolated from *Annona cherimola*[177,260] and the lactone hypecoumine (**81**), which can be regarded as a masked papaverinol-derivative, are four examples.[261]

7.4.6.3 Papaveraldine-analogues

Similar to aspergillitine (**71**), the isolation of thalprzewalskiinone (**87**) from *Thalictrum przewalskii* was reported in 1999.[262] Only two years later, a revision of the structure was published regarding the position of the methoxy group.[263] Oxodeoxyannocherine A (**88**) is a natural product occurring in *Menispremum dauricum*.[259] RO0509347 (**89**) is a product of a classical chemical series based on an initial high throughput screening with subsequent hit identification and optimization.[54] The keto group was found to be necessary for the high potency for glutamine fructose-6-phosphate amidotransferase (GFAT) inhibition.[54] Compound **90** showed the most promising properties, suggesting its suitability for further studies.[54] Pheantharine (**91**) was isolated by Santos in 1932.[264] The structure was revised in 1983.[265] The same applies for sauvagnine (**92**), which was originally assigned the α-hydroxytetradehydrocularine structure, and linaresine (**93**). It turned out that both compounds have a benzoyl isoquinoline scaffold, with linaresine (**93**) being the known compound rugosinone (**93**).[266,267]

Although about 2500 known benzylisoquinolines have been isolated to date, little information on systematic screening was found. Most probably the amount isolated is often hardly sufficient for the analytical characterization and access of larger quantities for high throughput screenings, or subsequent preclinical characterization most probably requires a high-yielding total synthesis or semi-synthetic approach. The halo effect of the total syntheses of natural products for the absolute determination of the structure is that researchers gain access to a route for larger quantities.

7.4.7 Aporphines/Oxoaporphines

Aporphines and oxoaporphines (Figure 7.23) comprise a large group of compounds isolated from various plant species. The compounds of this class showed antimalarial, antitrypanosomal, cytotoxic, antioxidant, and larvicidal activities.[268]

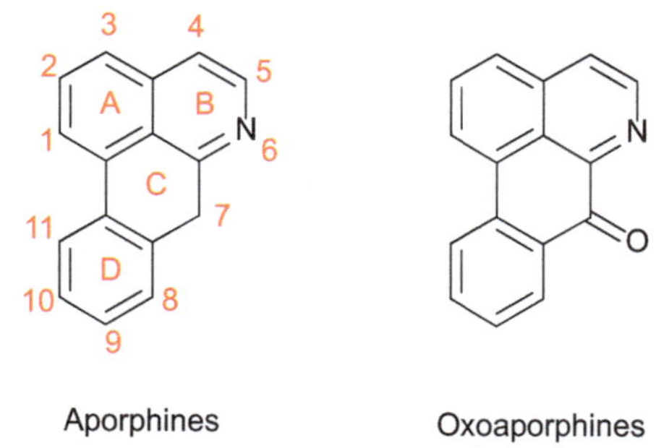

Figure 7.23 Aporphine and oxoaporphine scaffolds.[269]

Many novel compounds such as oxobuxifoline (**94**),[270,271] eletefine (**95**) and oxoeletefine (**96**),[271] oxoglaucidaline (**97**),[31] corydaturtshine A (**98**),[31] daurioxoisoporphine A-D (**99–102**),[260] artabonatine C (**103**) and D (**104**),[260] oxo-*O*-methylbulbocapnine (**105**),[260] dauriporphinoline (**106**),[178] sinofranine (**107**),[259] duguevalline (**108**),[259] fissiceine (**109**),[272] subsessiline (**110**),[273] 8-hydroxy-5-methoxyliriodenine (**111**),[274] and peruvianin (**112**)[275,276] have been isolated and characterized, but not investigated in greater detail (Figures 7.24 and 7.25). Cytotoxicity assays were conducted with 8-hydroxy-artabonatine C (**113**).[277] Lysicamine (**114**) exhibited strong anti-plasmodial activity along with low cytotoxicity.[156]

Dauriporphine (**115**), bianfugecine (**116**), dauriporphinoline (**106**), and menisporphine (**117**) were examined for their P-gp mediated multidrug resistance (MDR) reversal activity in human cancer cells, where dauriporphine (**115**) showed the most potent P-gp MDR inhibition activity.[278]

Zanthoxoaporphine A (**118**) was tested against *Anopheles gambiae* larvae and showed moderate larvicidal activity after 24 h.[268] *O*-methylmoschatoline (**119**) showed strong trypanocidal effects against the clinical relevant trypomastigote forms of *T. cruzi*.[279]

Synthetic oxoisoaporphine derivatives were tested for antileishmanial activities. The oxoisoaporphines with dihydroisoquinoline motif showed better effects than the oxoisoaporphine derivatives with isoquinoline scaffold.[280]

Oxoglaucine (**120**) showed marginal cytotoxicity against HepG2,[281] but indicated strong anticancer activity against HCT-8 and KB.[282] It exhibited moderate antiplasmodial activity with no observable cytotoxicity.[269] Oxoglaucine (**120**) has attracted attention as metal ligand and results indicated that oxoglaucine-metal complexes were selectively active against certain cell lines.[282]

Liriodenine (**121**) is mutagenic, induces DNA damage,[268] was found to have potent antifungal activity,[283] and results indicated cytotoxic activity.[260,284] It has been found to significantly inhibit platelet aggregation.[285] Liriodenine (**121**) was also used as a metal ligand, forming cytotoxic gold(III) complexes.[286] It is a potent topoisomerase II inhibitor.[280] Protais *et al.* tested three aporphines among other isoquinoline alkaloids, namely melosmine (**122**), liriodenine (**121**), and lysicamine (**114**), for their activity as dopamine uptake inhibitors, but the compounds did not display significant effects.[287]

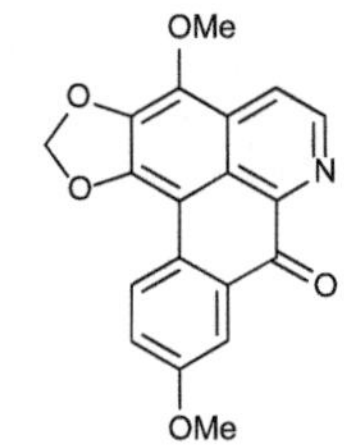

94 Oxobuxifoline

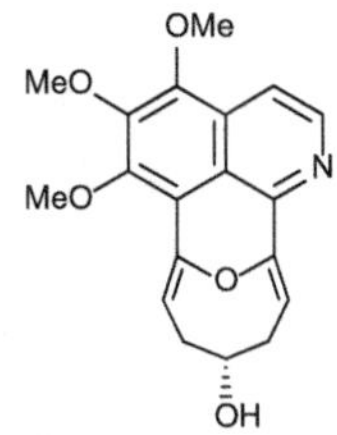

95 Eletefine

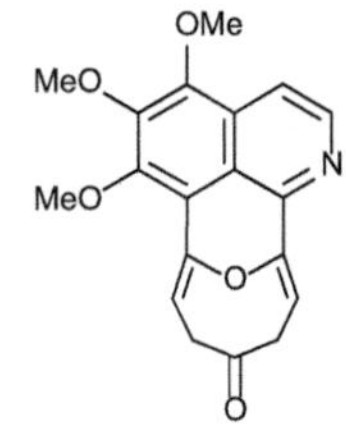

96 Oxoeletefine

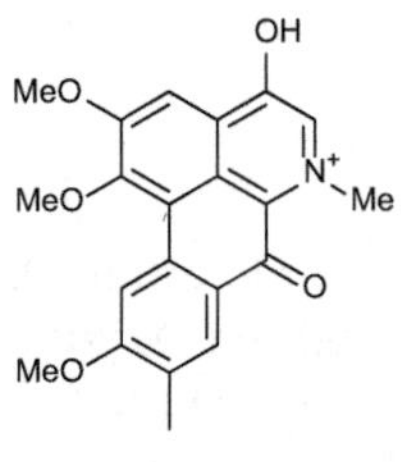

97 Oxoglaucindaline

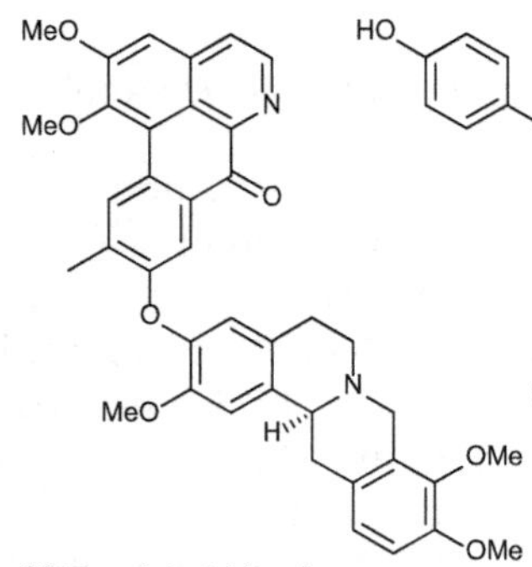

98 Corydaturtshine A

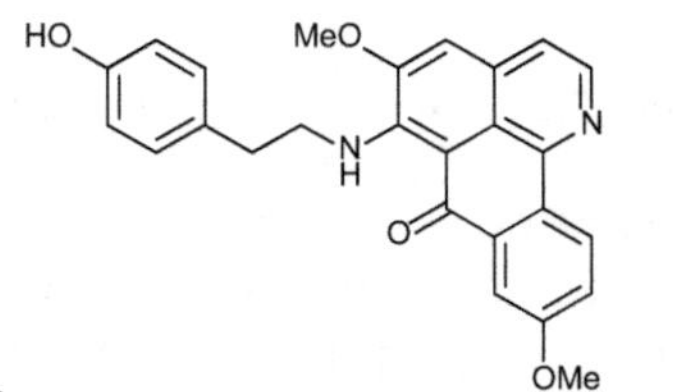

99 Daurioxoisoaporphine A

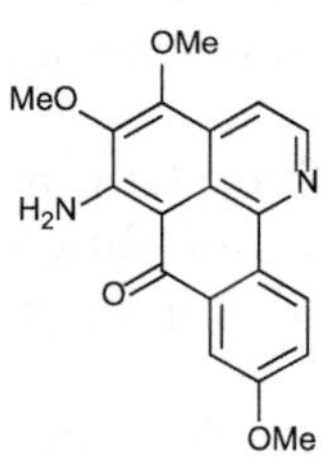

100 Daurioxoisoaporphine B

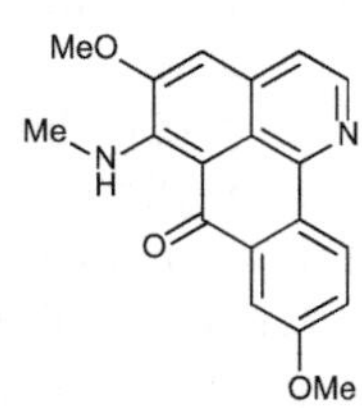

101 Daurioxoisoaporphine C

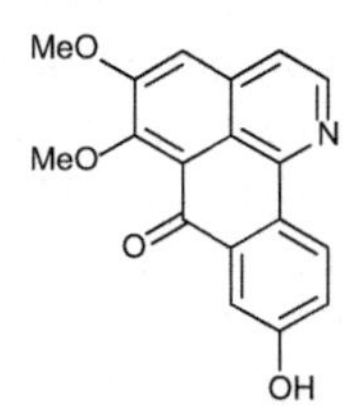

102 Daurioxoisoaporphine D

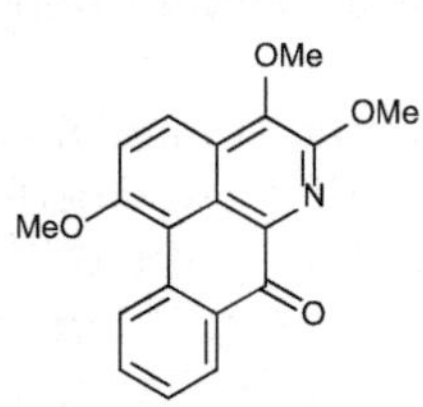

103 Artabonatine C

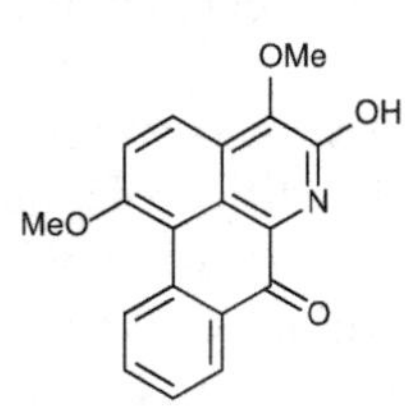

104 Artabonatine D

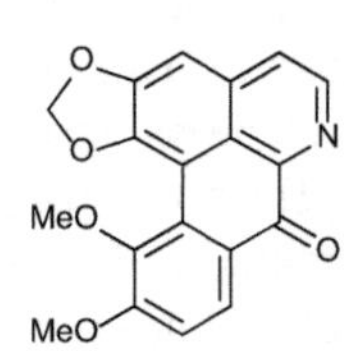

105 Oxo-*O*-methylbulbocapnine

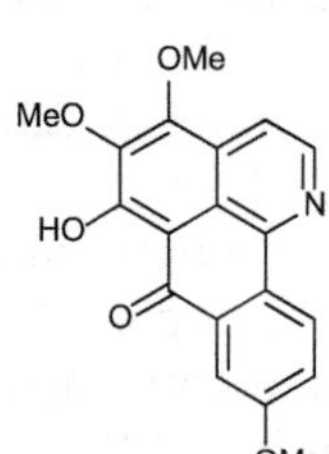

106 Dauriporphinoline

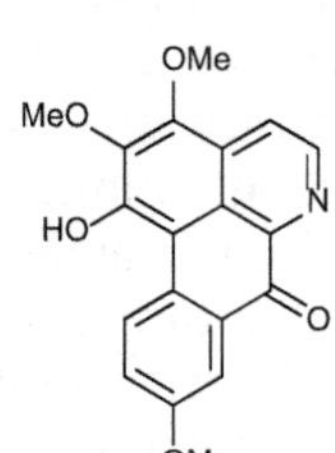

107 Sinofranine

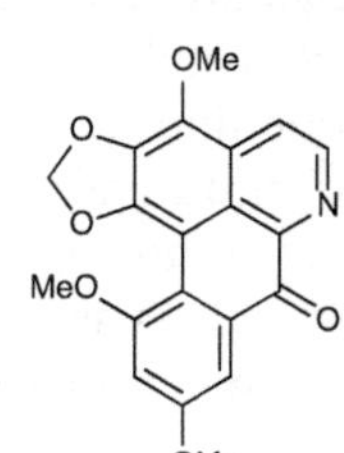

108 Duguevalline

Figure 7.24 Chemical structures of selected aporphine/oxoaporphine derivatives, Part 1.

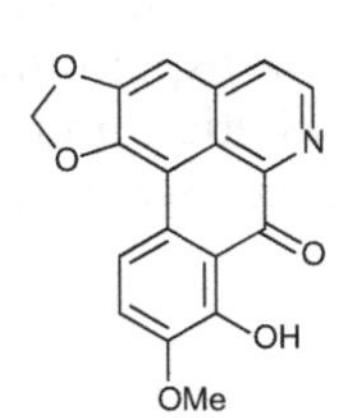

109 Fissiceine

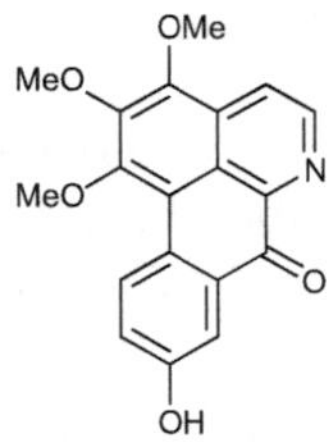

110 Subsessiline

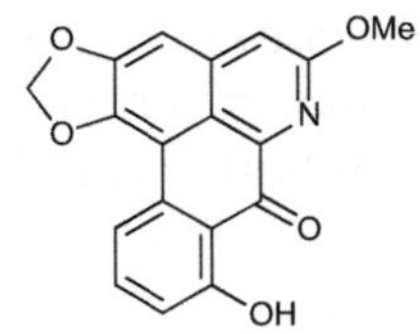

111 8-Hydroxy-5-methoxyliriodenine

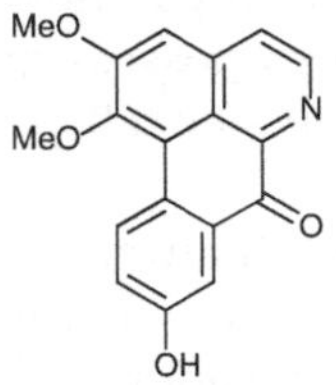

112 Peruvianine

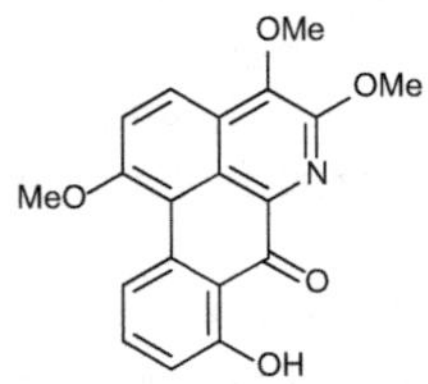

113 8-Hydroxyartabonatine

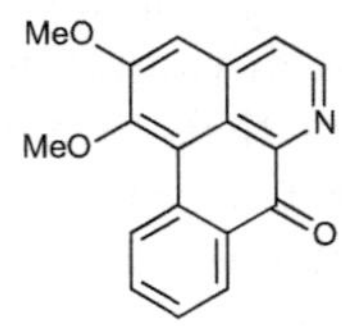

114 Lysicamine

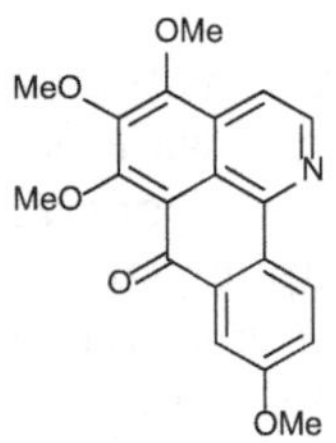

115 Dauriporphine

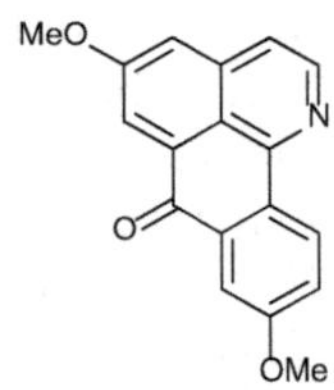

116 Bianfungecine

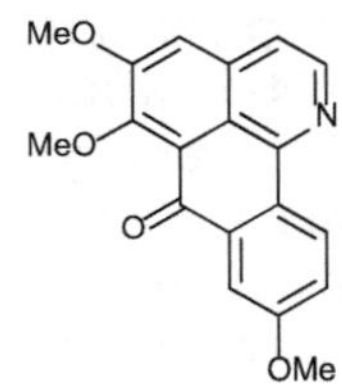

117 Menisporphine

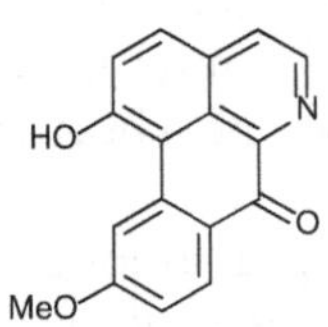

118 Zanthoxoaporphine A

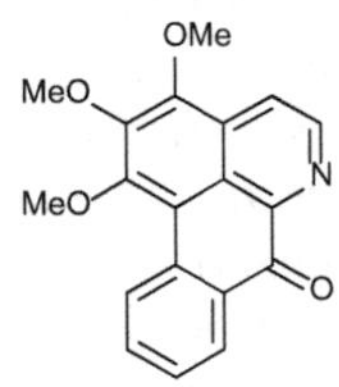

119 *O*-Methylmoschatolie

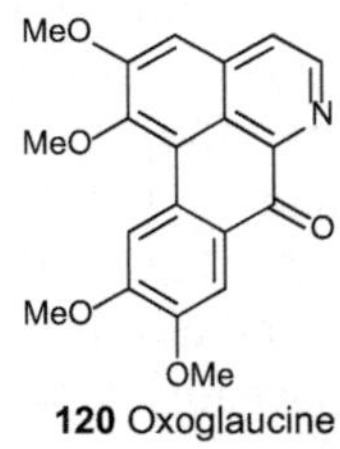

120 Oxoglaucine

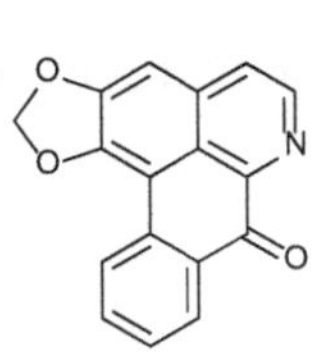

121 Liriodenine

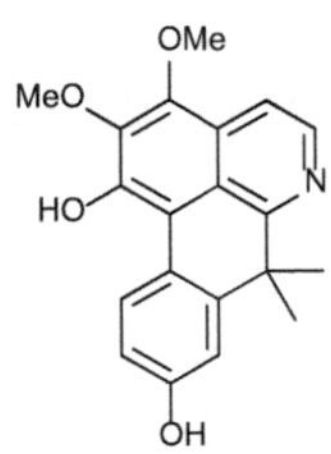

122 Melosmine

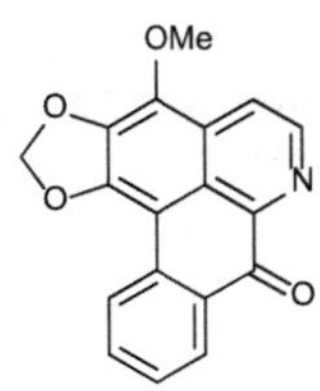

123 Atherospermidine

Figure 7.25 Chemical structures of selected aporphine/oxoaporphine derivatives, Part 2.

Bick *et al.* isolated atherospermidine (**123**) in 1956 from the bark *Atherosperma moschatum* Labill.[288] In 1965, the synthesis of atherospermidine (**123**)[289] was reported, followed by a photochemical synthetic route in 1983.[290] Atherospermidine (**123**) exhibited cytotoxicity against hepatocarcinoma cancer cell lines[291] and induced DNA damage.[268,284]

7.4.8 Azafluoranthenes and Related Tropolones

Much fewer azafluoranthene (Figure 7.26) natural products are known compared to aporphines.[292] Compounds of this class have been patented as constituents of wound-healing agents, have been reported to possess antidepressant activity, are demonstrated to be active against P-388 cells, and acted as 5-HT_3 agonists.[293]

Beside the biological activity, azafluoranthenes (Figure 7.27) attracted attention for their interesting spectral properties and have been therefore studied as agents in luminescent and electroluminescent applications.[292,294] Scherowsky *et al.*, for instance, synthesized an azafluoranthene derivative to develop discotic liquid crystals.[293,295]

Numerous preparation efforts have been published. The syntheses of rufescine (**124**) and imeluteine (**125**) have been reported by Cava *et al.* in 1972[296] and the total syntheses by Boger *et al.* in 1984.[297] Fu *et al.* and Zhao *et al.* reported a synthetic route using metalation – cross coupling strategies for the preparation of imeluteine (**125**).[298,299] In 1995, Boger *et al.* reported the total syntheses granditropone (**126**), grandirubine (**127**), imerubine (**128**), and isoimerubine (**129**).[300] Studies towards the total synthesis of grandirubine (**127**) and imerubine (**128**) have been reported by Banwell *et al.*[301] A new route to azafluoranthenes was reported by Ponnala *et al. via* direct arylation, resulting in the successful preparation of rufescine (**124**) and triclisine (**130**).[292,302] Ramana *et al.* introduced a new and short synthesis of triclisine (**130**) *via* photocyclization.[303]

Pareitropone (**131**), the first tropoisoquinoline discovered among the isoquinoline alkaloids,[304] was isolated in 1995 from *Cissampelos pareira* (Menispermaceae)[305] and was found to be a potent antitumour agent *in vitro*.[177,305] The total synthesis of pareitropone (**131**) was reported by Feldman *et al.*[177,306] and Hong *et al.*[307]

Pareirubrine A (**132**) and B (**133**) exhibited antileukemic activities against P-388 cells.[308]

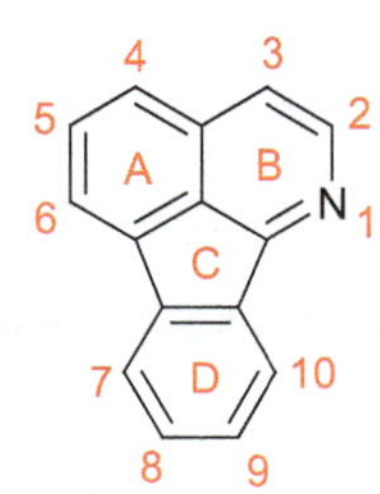

Figure 7.26 Azafluoranthene scaffold.[300,309]

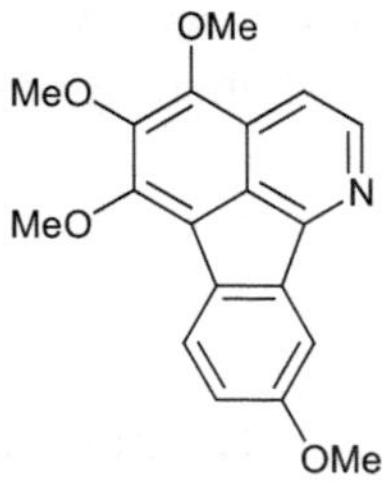

124 Rufescine

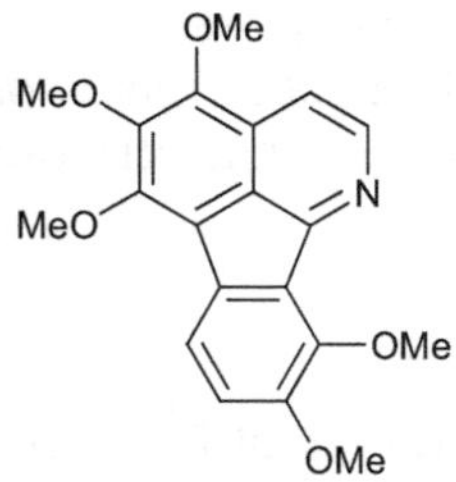

125 Imeluteine

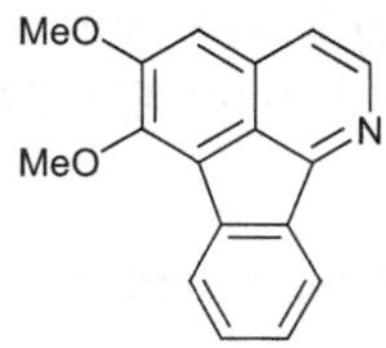

130 Triclisine

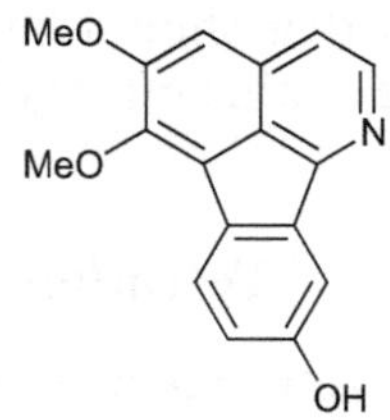

134 Telitoxine

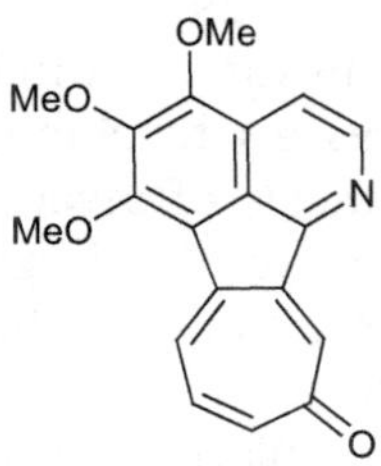

126 Granditropone

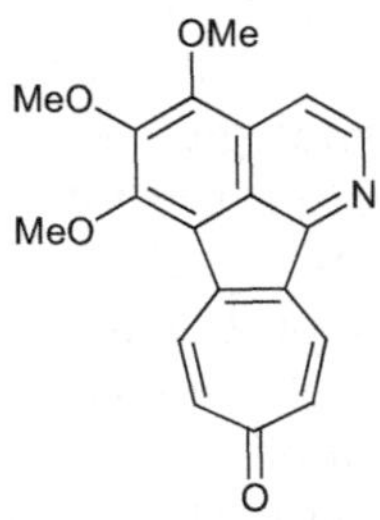

127 Grandirubine

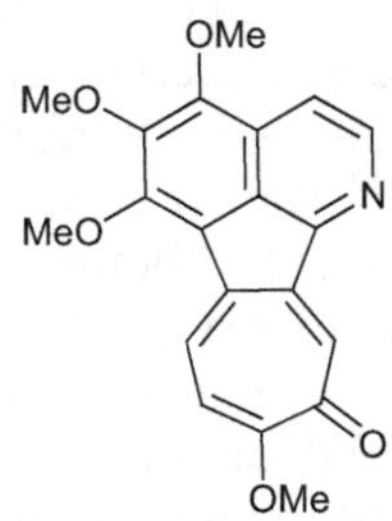

128 Imerubine

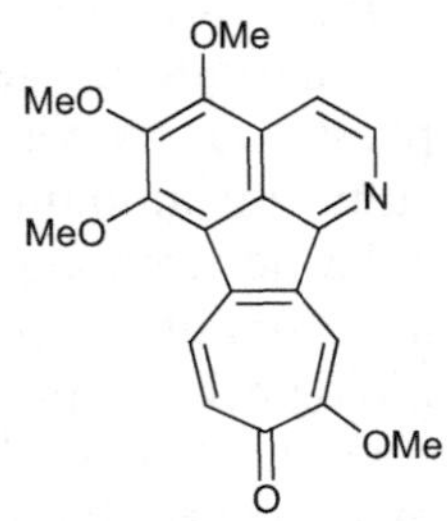

129 Isoimerubine

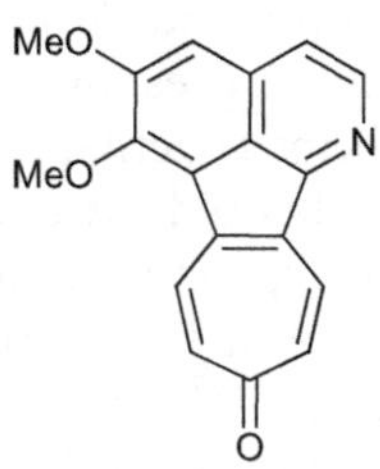

131 Pareitropone

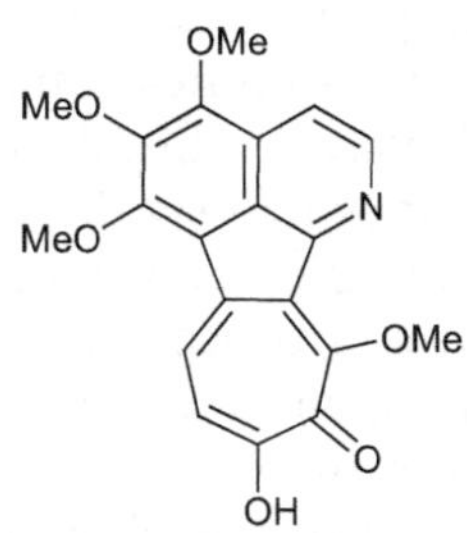

132 Pareirubrine A

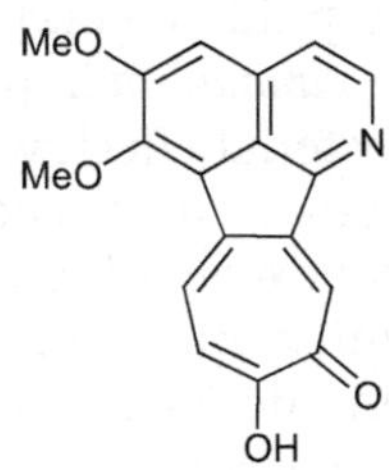

133 Pareirubrine B

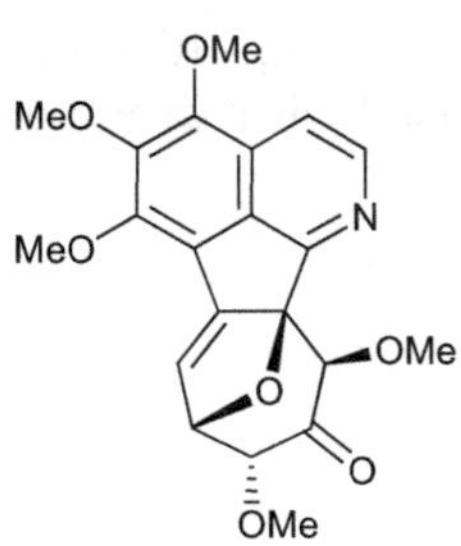

135

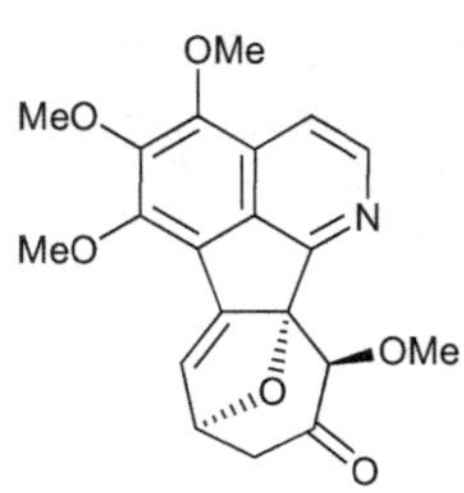

136

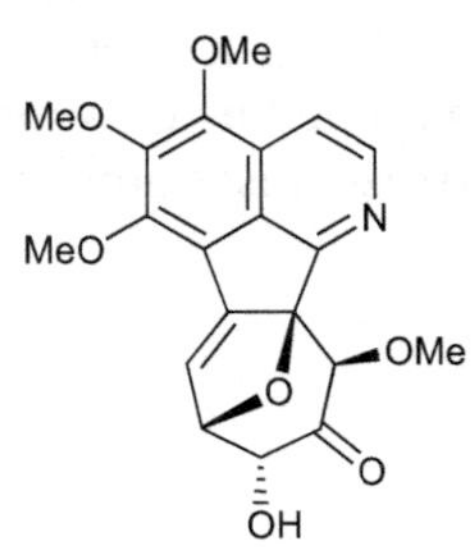

137

Figure 7.27 Chemical structures of azafluoranthene derivatives.

Telitoxine (**134**) was first isolated in 1981 by Menachery *et al.*[275] and synthesized in 1987.[309] Compounds **135, 136**, and **137** are of synthetic origin and have been prepared by Lee *et al.*[310]

This small class of azafluoranthenes comprises an underprivileged scaffold that should be addressed and investigated more comprehensively. Its side chains offer good variability for further derivatization.

7.4.9 Tetradehydrocularines

Cularines are a class with a dibenzoxepine motif (Figure 7.28). The natural compounds oxocularine (**138**),[178] oxocularicine (**139**),[178] oxosarcocapnine (**140**),[311] oxosarcocapnidine (**141**),[118] oxosarcophylline (**142**),[117] gouregine (**143**),[312] and oxocularicine (**138**)[178] have been isolated (Figure 7.29).

Cularines are generally present only in small quantities in their natural plant sources, which is why synthetic strategies were necessary to obtain sufficient amounts.[313] Garcia *et al.* reported the total synthesis of several cularine compounds *via* dibenzoxepines in 1995.[313] Despite the large amounts of synthesis reports, only few reports have been traceable on the biological and pharmacological activities of these compounds.

Orhan *et al.* evaluated among 33 isoquinoline alkaloids oxocularine (**138**), oxosarcocapnine (**140**), oxosarcocapnidine (**141**) for their *in vitro* antiviral and antimicrobial activities. The cularines exhibited higher activity against Gram-negative (*E. coli*, *P. aeruginosa*, *P. mirabilis*, *K. pneumonia*, and *A. baumannii*) than Gram-positive bacteria (*S. aureus*, *B. subtilis*).[314] Oxosarcocapnidine (**141**) showed significant inhibition towards *K. pneumoniae* and *A. baumannii* in particular. [314] The alkaloids did not present any notable antibacterial effect, while they had significant antifungal activity.[314]

Protais *et al.* tested among other isoquinoline alkaloids two cularines, namely oxocularine (**138**) and oxosarcophylline (**142**), for their activity as dopamine uptake inhibitors, but like the tested aporphine compounds, the cularine derivatives did not display significant effects.[287] Gouregine (**143**) was tested for the ability to displace ^{3}H-SCH 23 390 and ^{3}H-raclopride from the striatal binding site of the dopamine receptors D1 and D2, but was found to be hardly effective.[315]

The structures of sauvagnine (**92**) and linaresine (**93**), which were initially assigned as cularine derivatives, have been revised with strong evidence of benzoyl isoquinoline scaffolds.[266,267,316]

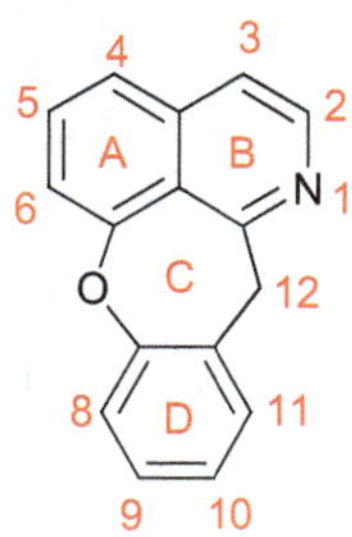

Figure 7.28 Tetradehydrocularine scaffold.[313]

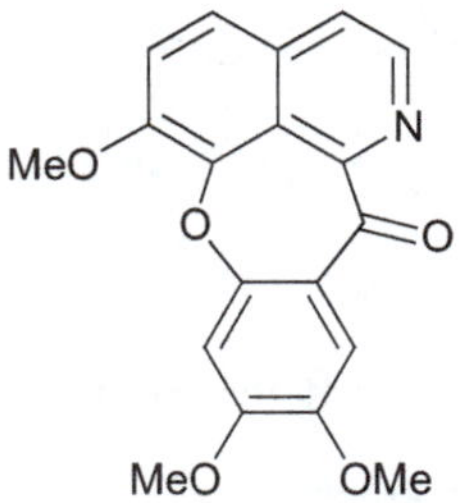

138 Oxocularine

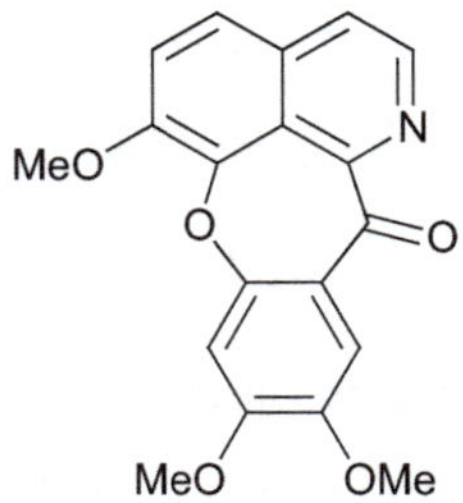

139 Oxocularicine

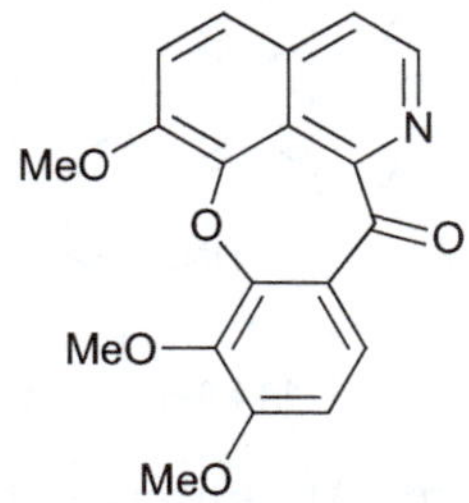

140 Oxosarcocapnine

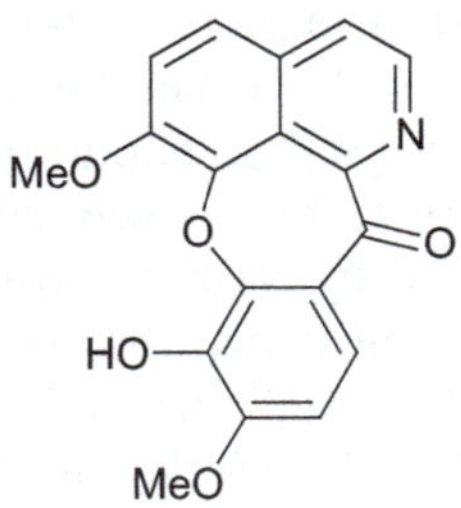

141 Oxosarcocapnidine

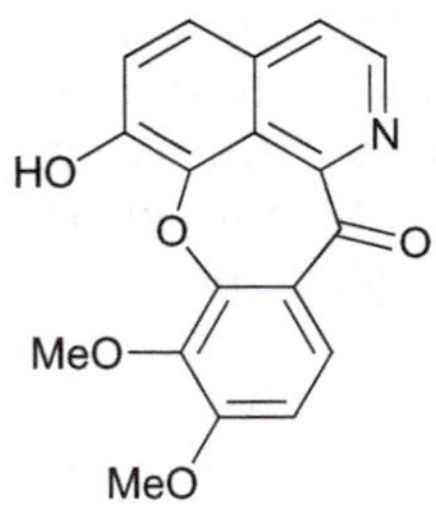

142 Oxosarcophylline

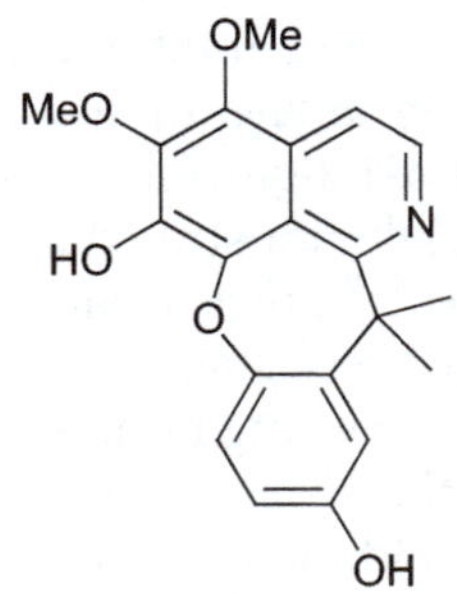

143 Gouregine

Figure 7.29 Chemical structures of tetradehydrocularine derivatives.

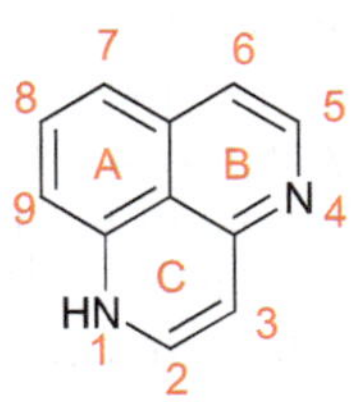

Figure 7.30 Aaptamine scaffold.[319]

7.4.10 Aaptamines

Marine sources have attracted increasing attention. Aaptamine (**144**), the compound this substance class is named after (Figure 7.30), can be isolated from the marine sponge *Aaptos suberitoides*[317] and *Aaptos aaptos*.[318]

Since the isolation of aaptamine (**144**) in 1982[320] from natural sources, numerous synthetic routes have been reported.[319,321,322] Only few aaptamines with true isoquinoline moieties are known (Figure 7.31).

The interest in aaptamines rose when the inhibitory effect on cancer cell growth was revealed.[56] The publication of Pettit *et al.* illustrates why the effort of total syntheses is generally justified. It required the extraction of

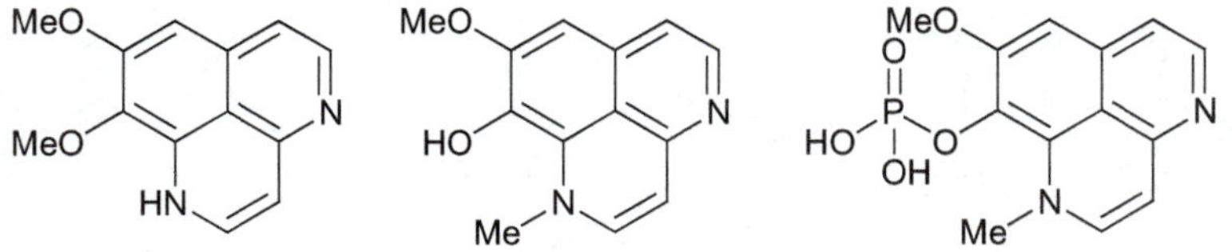

Figure 7.31 Chemical structures of benzylisoquinoline derivatives.

500 kg of the marine organism in order to obtain gram quantities of aaptamine (**144**) and isoaaptamine (**145**).[56] Aaptamine (**144**) has been reported to act as a multi-target agent, showing antiviral and anticancer activities, strong *in vivo* radical scavenging capacity, α-adrenoceptor action blocking, and α-1,3-glucanase and monoamine oxidase inhibition.[322] Isoaaptamine (**145**) demonstrated higher potency in a variety of cell assays than aaptamine (**144**).[322] Since isoaaptamine (**145**) turned out to be unstable, it was converted to a stable sodium phosphate prodrug designated hystatin 1 (**146**).[56] An increasing number of publications are investigating the structure-activity relationships of aaptamine derivatives.[322] Despite the small size of the scaffold, aaptamines exhibit great potential as a privileged scaffold for medicinal chemistry and offer a variety of derivatization options to improve preclinical and clinical properties.[322]

7.4.11 Simple Isoquinolines

Natural plant sources are known for the diversity of complex molecular structures. Nevertheless, there are also simple isoquinolines (Figure 7.32), usually bicyclic sometimes tricyclic (*e.g.* crispine B (**147**)),[323] with multiple substituents.

Some examples are dehydrohydrastinine (**148**),[177] crispine B (**147**), C (**149**), and D (**150**),[177] backebergine (**151**),[117,324] isobackebergine (**152**),[117,324] isosalsolidine (**153**),[117] isonortehuanine (**154**),[117] isonorweberine (**155**),[117] isopachycereine (**156**),[117] nigellimine-*N*-oxide (**157**),[323] and 2-methyl-6,7-dimethoxyisoquinolinium chloride (**158**) (Figure 7.33).[323] Of crispine B (**147**), C (**149**), and D (**150**) only crispine B (**147**) showed significant antitumour activity against SKOV3, KB and Hela human cancer cell lines.[325,326] Total syntheses of crispine B (**147**) and crispine C (**149**) have been reported by Yasuhara *et al.* in 2009,[327] and by Blair *et al.* in 2012, respectively.[325]

Backebergine (**151**) induced vasorelaxation in a time-dependent and dose-dependent manner in rat aorta pre-contracted with phenylephrine. The

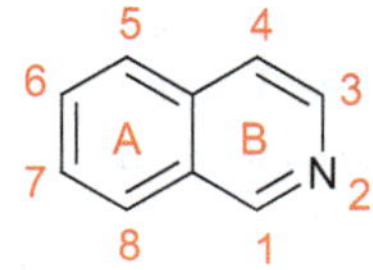

Figure 7.32 Isoquinoline scaffold.[10]

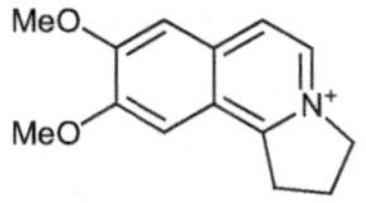

147 Crispine B

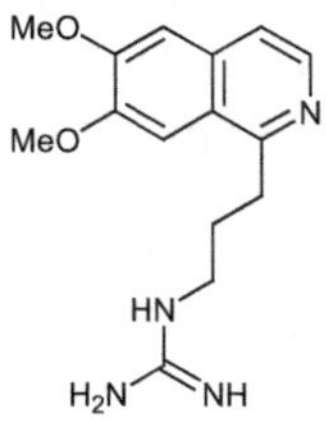

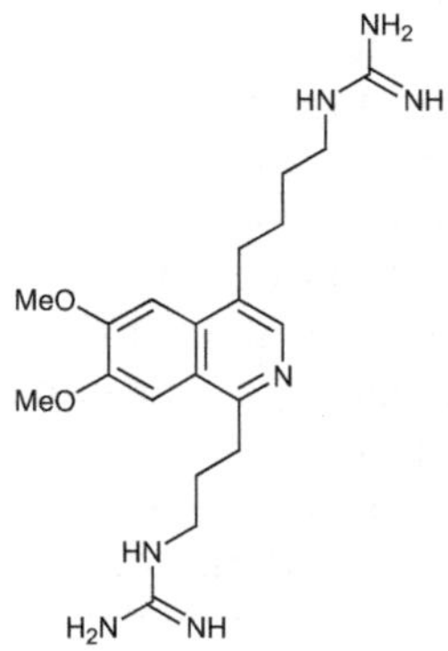

148 Dehydrohydrastanine

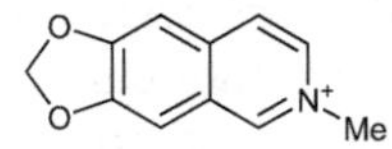

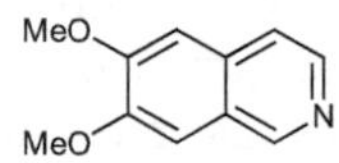

151 Backebergine

149 Crispine C

150 Crispine D

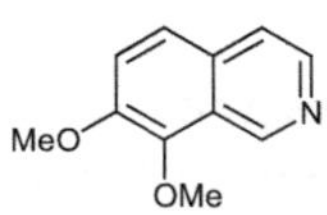

152 Isobackebergine

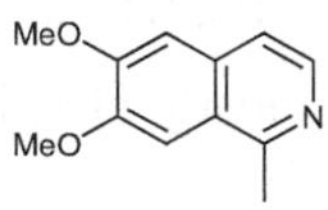

153 Isosalsolidine

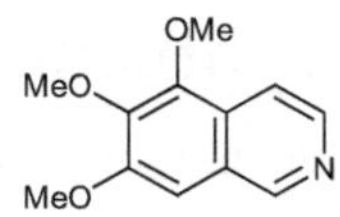

154 Isonortehuanine

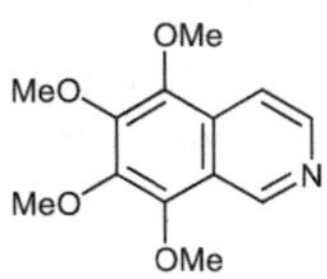

155 Isonorweberine

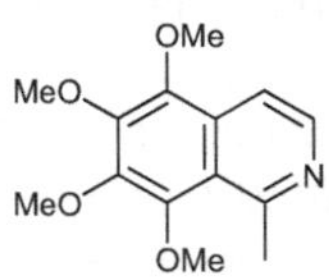

156 Isopachycereine

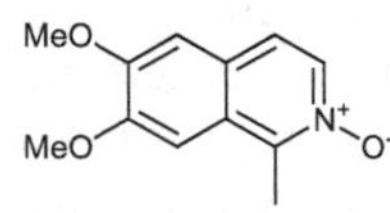

157 Nigellimine-*N*-oxide

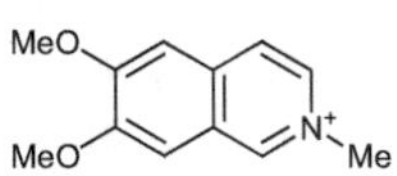

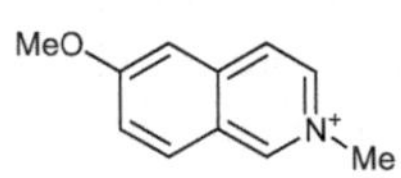

159 *N*-methyl-6-methoxy-isoquinolinium

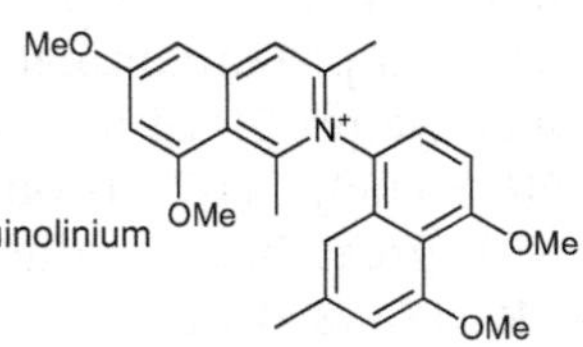

158 6,7-Dimethoxy-2-methylisoquinolinium

160 Ancisheynine

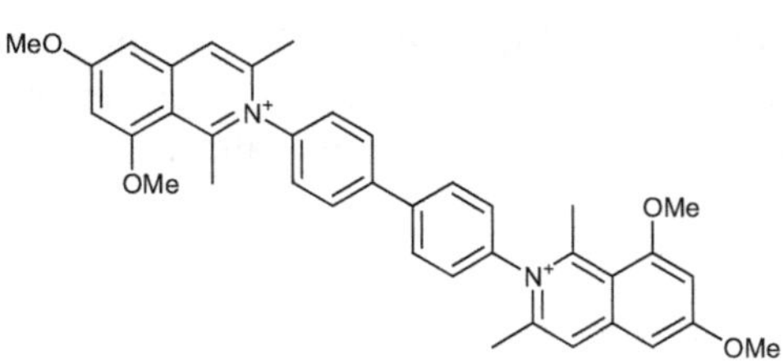

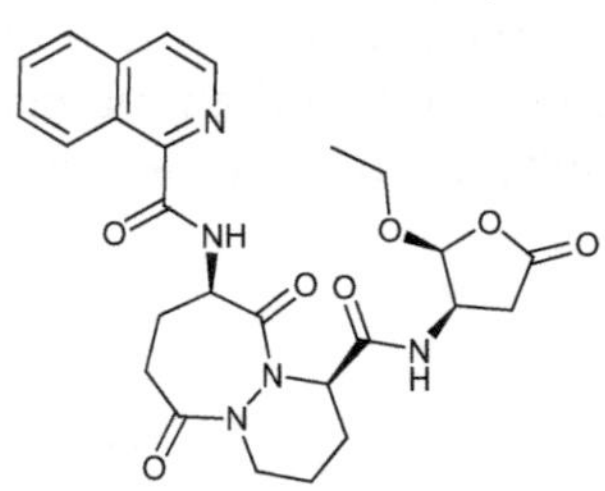

161 IQ-143

162 Pralnacasan

Figure 7.33 Chemical structures of simple isoquinoline derivatives.

action of backebergine (**151**) may be attributed to its inhibition of the voltage-dependent Ca^{2+} channel and receptor-operated Ca^{2+} channel.[328] Several mono- to trisubstituted isoquinolines were tested as mosquito repellents, but none of the compounds was superior to diethyltoluolamide.[329]

Thull *et al.* tested a series of substituted isoquinolines and isoquinolinium salts as monoamine oxidase (MAO) inhibitor. Of the series, *N*-methyl-6-methoxy-isoquinolinium (**159**) emerged as the most active inhibitor. The *N*-methylisoquinolinium ions were, as a group, the most selective MAO-A inhibitors, whereas the more lipophilic and uncharged isoquinolines showed weak MAO-A inhibitory effects, displaying poor or no selectivity towards MAO-B.[330] Naphthylisoquinoline alkaloids are a growing group of secondary metabolites. Despite more than 120 known natural naphthylisoquinolines, few have an aromatic isoquinoline scaffold.[331] Based on this novel *N,C*-coupled naphthyl-isoquinoline alkaloid ancisheynine (**160**), a synthetic compound named IQ-143 (**161**) was synthesized.[332] IQ-143 (**161**) was found to be a promising lead compound for antibiotic therapy against staphylococci.[332]

7.5 Synthetic Isoquinoline Derivatives

Synthetic isoquinolines are mostly simple isoquinolines that are usually mono- to tetra-substituted and use simple natural product isoquinolines as lead structures for the synthetic scaffolds.[259] The benzene ring is more electron rich than the heteroaromatic ring, which yields the 5-substituted isoquinoline as the major isomer in electrophilic aromatic substitution reactions.[333] Nucleophilic attack occurs on the heteroaromatic ring and results mainly in position 1-substituted isomers (Figure 7.34).[334] Many small molecule drug candidates derive from this class.

In the 1990s, Vertex Pharmaceuticals patented a series of inhibitors of interleukin-1β converting enzymes (ICE or caspase 1[335]).[336–338] Interleukin (IL) is a pro-inflammatory cytokine. ICE is a key enzyme that converts the inactive pro-IL-1β into the biologically active form.[339] VX-740 (pralnacasan, HMR3480)[51] (**162**) is one of those ICE inhibitors and has been investigated with great interest due to its oral bioavailability, and is the prodrug of VRT-18858 (RU36384) (**163**) (Figure 7.38).[339,340] Because of the promising inhibitory activities, it was announced that Vertex collaborated with Aventis Pharma AG in the development of pralnacasan (**162**) for the potential treatment of inflammatory diseases.[341] The companies reported positive results from a phase IIa study designed to evaluate safety, efficacy, and

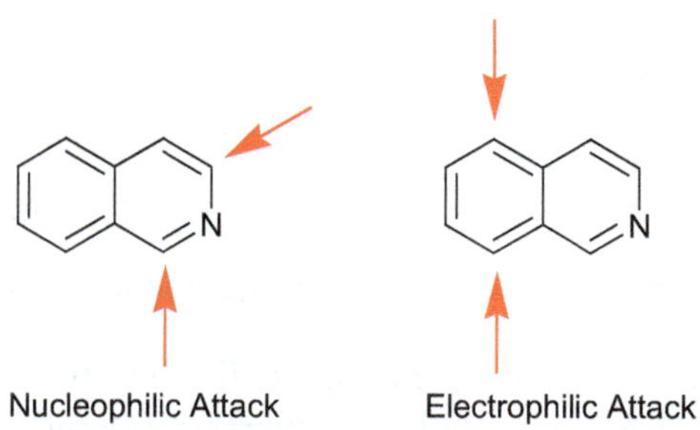

Figure 7.34 Favoured positions of nucleophilic and electrophilic attack on isoquinolines.[334]

tolerability in subjects with rheumatoid arthritis (RA).[51] Osteoarthritis (OA) is the most common joint disease for middle-aged and elderly people.[342,343] Results exhibited strong evidence of the contribution of proinflammatory cytokines to cartilage degradation.[343,344] Pralnacasan (**162**) showed effects in two murine models of osteoarthritis by reducing joint damage.[340] VRT-018858 (**163**) was found to markedly reduce ischaemic injury in rats. VRT-018858 (**163**) exhibited neuroprotective effects confirming that caspase-1 is an important target for intervention in acute CNS injury. This new class of caspase-1 inhibitors revealed great potential as highly effective neuroprotective agents.[335]

Substituted isoquinolinesulfonyl compounds such as HA-1004 (**164**) and HA-1077 (**2**) (Fasudil, AT877, Eril) were disclosed by Asahi Chemical Industry in the 1980s, suggesting drug applications in the area of smooth muscle of mammals, in particular as a vasodilator, cerebral circulation ameliorator, antihypertensive agent, for the prevention and treatment of angina, cerebrovascular thrombosis, hypertonia, arteriosclerosis, cerebral apoplexy and other circulatory organ diseases.[345,346] To date, numerous reports have been published on fasudil (**2**) and derivatives. Fasudil (**2**) appears to be involved in the inhibition of Rho GTPase, an enzyme necessary for the activation of the interleukin-1 inflammation pathway.[343,347,348] Rho kinases (ROCKs), the downstream targets of the small GTP-binding proteins belonging to the Rho family,[349] play an important role as abnormal activation of the Rho/ROCK pathway has been observed in various central nervous system (CNS) disorders. Activated ROCKs occur in cases of injury of the adult vertebrate brain or spinal cord and inhibit neurite growth and sprouting.[350] Results from studies suggest that ROCKs could also be involved in hypertension[81] and atherosclerosis.[349] *In vivo* and *in vitro* studies proved the vasodilation effect of fasudil mesylate (**2**) acting as a Rho kinase inhibitor.[351] Dimethylfasudil (diMF, H-1152P, BRD4911) (**165**) was found to selectively increase polyploidization, mature cell-surface marker expression and the apoptosis of malignant megakaryocytic leukemia cells.[352] Despite the structural similarity of dimethylfasudil (**165**) and fasudil (**2**), the small differences can have noteworthy effects. Dimethylfasudil (**165**) strongly inhibited the Aurora kinase family (A, B, and C) whereas fasudil (**2**) showed no effect at all.[352] Hydroxyfasudil (**166**) is a major active metabolite of fasudil (**2**) and showed more specific inhibitory effects of Rho-kinase after oral administration.[353,354] Intracoronary administration of fasudil (**2**) and hydroxyfasudil (**166**) inhibited coronary spasm in pigs *in vivo*.[355] Ono-Saito *et al.* reviewed the isoquinolinesulfonamide H-series protein kinase inhibitors (CKI-6 (**167**), CKI-7 (**168**), H-7 (**169**), H-8 (**170**), H-9 (**171**), HA-1077 (**2**), H-89 (**172**), CKA-1306 (**173**) and KN-62 (**174**)) (Figure 7.35).[356]

H-7 (**169**) is a compound that competitively inhibits protein kinase C (PKC) with respect to ATP.[357] H-8 (**170**) is a more potent cyclic nucleotide-dependent protein kinase inhibitor, freely reversible and also competitive with respect to ATP.[357] H-9 (**171**) showed rather non-selective protein kinase inhibition with weak inhibitory effects on CKI and CKII.[356] CKI-6 (**167**) exhibited better inhibitory potential of CKI and CKII.[356] The chlorinated derivative CKI-7 (**168**) revealed selective competitive inhibition for CKI with

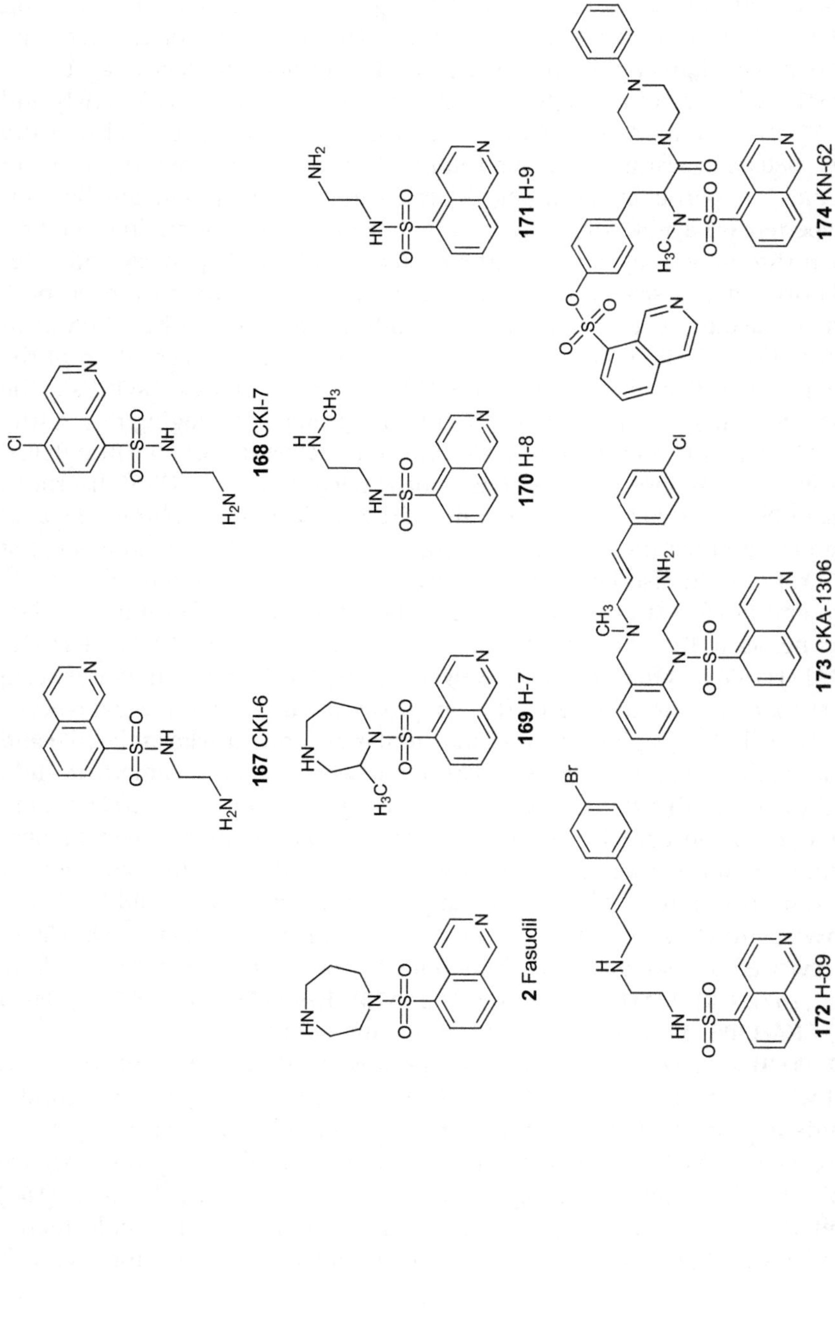

167 CKI-6, **168** CKI-7, **2** Fasudil, **169** H-7, **170** H-8, **171** H-9, **172** H-89, **173** CKA-1306, **174** KN-62

Figure 7.35 Isoquinolinesulfonamide H-series.

respect to ATP.[356] KN-62 (**174**) has been reported as the first selective calmoduline-competitive and ATP-noncompetitive inhibitor of calmodulin kinase II, and later also as calmodulin IV inhibitor.[356] H-89 (**172**) was shown to bind directly to the ATP-binding site of protein kinase A, being distinct from the other compounds by interacting with an additional site with the glycine flap of PKA. This might explain the selectivity and highest inhibitory effect of H-89 (**172**) among the H-series inhibitors.[356] CKA-1306 (**173**) showed competitive inhibitory effects of calmodulin kinase I with respect to ATP.[356]

Fasudil (**2**), an enhancement of the H-series, is a Rho-associated coiled-coil forming protein kinase inhibitor (Rho-kinase), which has been the first protein kinase inhibitor approved in Japan for clinical treatment of cerebral vasospasm.[357] HA-1077 (**2**) also exhibited significant vasodilatory activity and was reported to undergo clinical trials for the treatment of angina pectoris.[357] H-1152P (**165**) is a modification of HA-1077 (**2**) with a better inhibitory profile than HA-1077 (**2**).[357] Compound **175** was found to be a highly specific Rho-kinase inhibitor and exhibited reduction of the intraocular pressure in rabbit ocular hypertensive models.[357]

AR-12286 (**176**) (Aerie Pharmaceuticals) is currently under investigation in phase II clinical trials on ocular hypotensive effects (glaucoma, ocular hypertension, and elevated intraocular pressure).[49,358] The compound is a selective ROCK inhibitor with the ability to lower intraocular pressure by improving outflow of fluid *via* the trabecular pathway.[48] The chemical structure has not been disclosed and, to the best of the author's knowledge,[359] therefore there has been no reasonable means of checking the accuracy of the information.

Lavogina *et al.* conjugated oligo-D-arginine to fasudil (Figure 7.36). Compound ARC-3002 (**177**) resulted in increased affinity and selectivity towards Rho kinase.[360] Compounds **178**, **179**, and **180** also displayed ROCK1 inhibitory effects.[351,361]

Big Pharma has shown great interest in the transient receptor potential vanilloid type 1 (TRPV1), a promising target for chronic pain treatment.[58] Several pharmaceutical companies have developed lead compounds, such as A-425619 (**181**) (Abbott)[58] and GRC-6211 (**182**) (Lilly/Glenmark).[58] For GRC-6211 (**182**) clinical trials have been initiated and in one case suspended.[362] In rats A-425619 (**181**) showed after oral administration a transient period of hyperthermia followed by a period of hypothermia, representing a unique profile among the reported TRPV1 antagonists.[363] A general issue of TRPV1 antagonists is the body temperature increase, which has hampered the development of inflammatory and neuropathic pain agents for this target.[363]

Reilly *et al.* reported that a series of isoquinoline-scaffold based TRPV1 antagonists (Figure 7.37) has been tested (A-1165442 (**183**), A-1105512 (**184**), A-1165746 (**185**), A-1208747 (**186**), A-1233371 (**187**), A-1233372 (**188**), A-1241797 (**189**), A-1241407 (**190**)). It was concluded that acid-sparing antagonists do not significantly increase body temperature. A-1165442 (**183**) was different in a structure-activity relationship (SAR) study of TRPV1 antagonists. A-1165442 (**183**) completely inhibited capsaicin and NADA

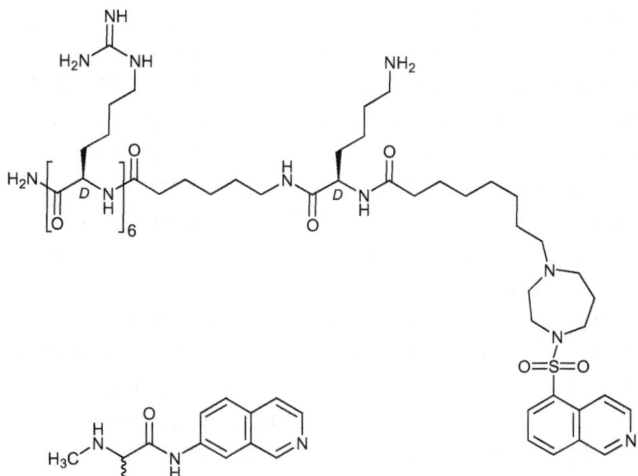

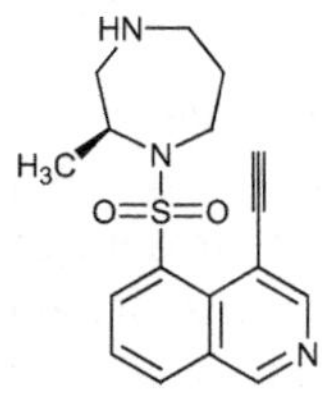

175

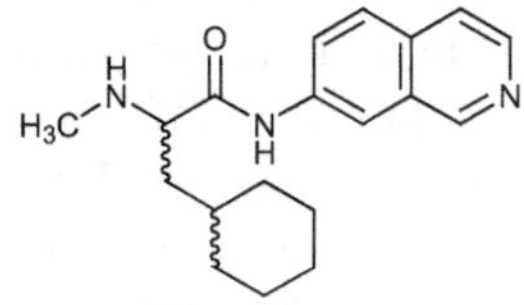

176 AR-12286

177 ARC-3002

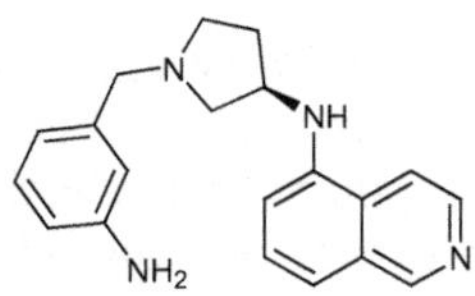

178

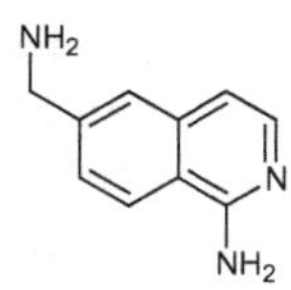

179

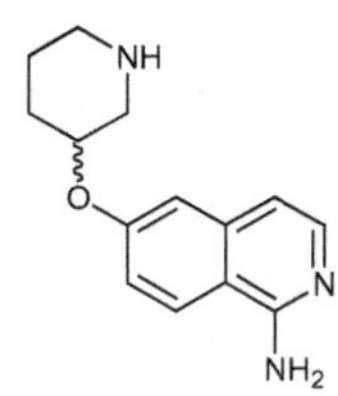

180

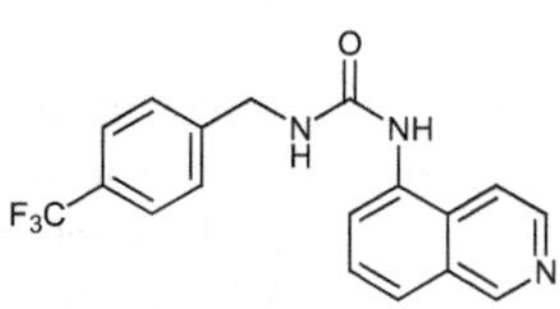

181 A-425619

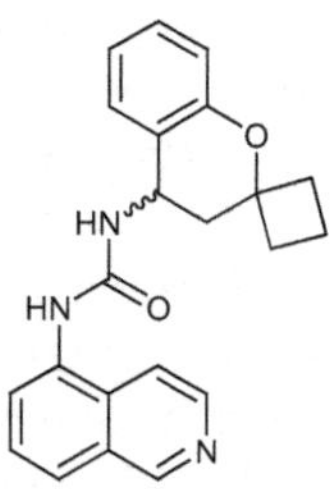

182 GRC-6211

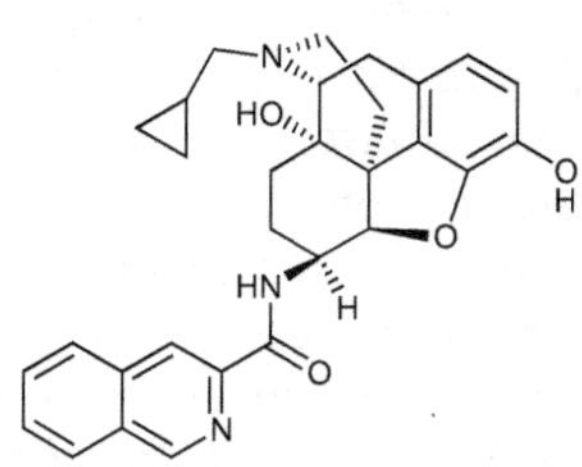

191 NAQ

Figure 7.36 Selected research compounds (stereochemical information has not been disclosed for all substances).

activation of rat and human TRPV1 channels. The reported data supports that the antagonists are able to block the acid activation of TRPV1, which is critical to predicting TRPV1 antagonist-induced hyperthermia.[364]

NAQ (**191**) is a naltrexone derivative.[365] Naltrexone has been used for opioid addiction and alcoholism treatment by mainly blocking the mu opioid receptor (MOR) exhibiting hepatotoxicity.[365] NAQ derivatives have been synthesized in a structure-activity relationship study. Several compounds with various substituents on the isoquinoline moiety displayed an

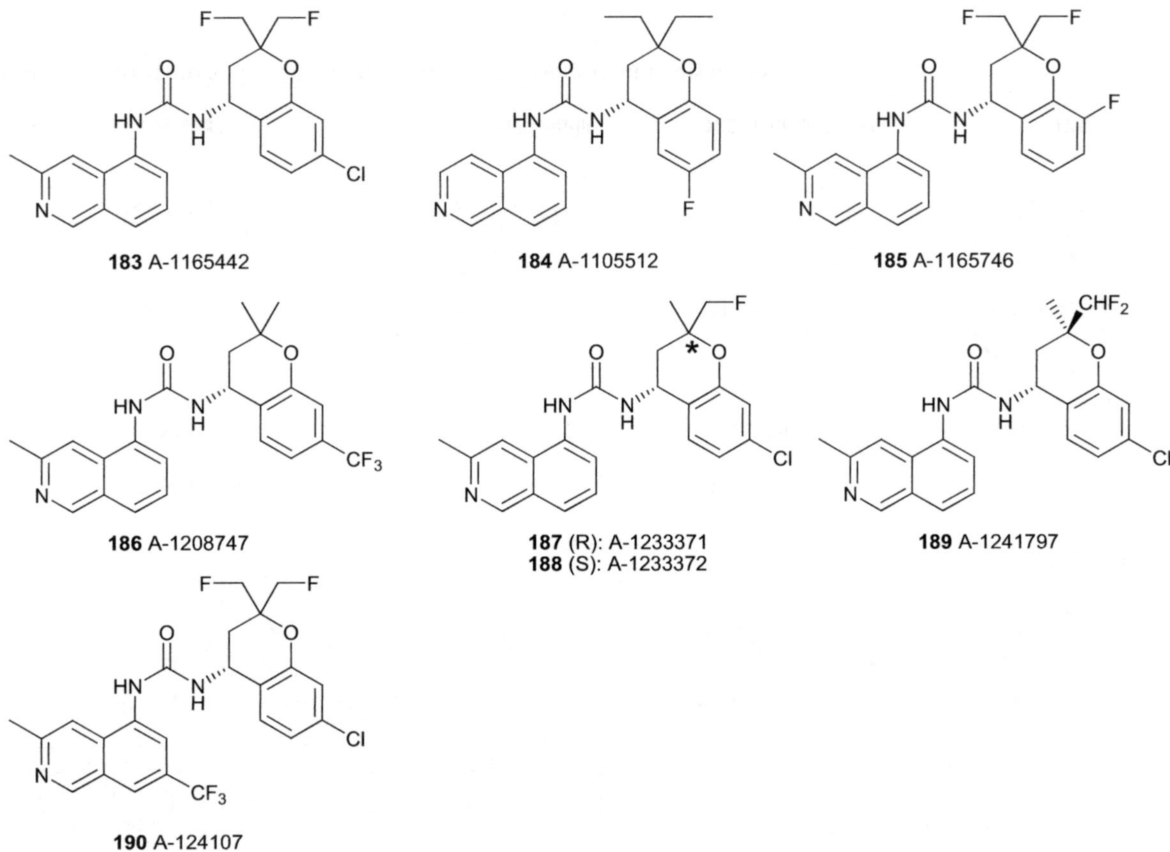

Figure 7.37 Series of isoquinoline-scaffold based TRPV1 antagonists (Abbott).

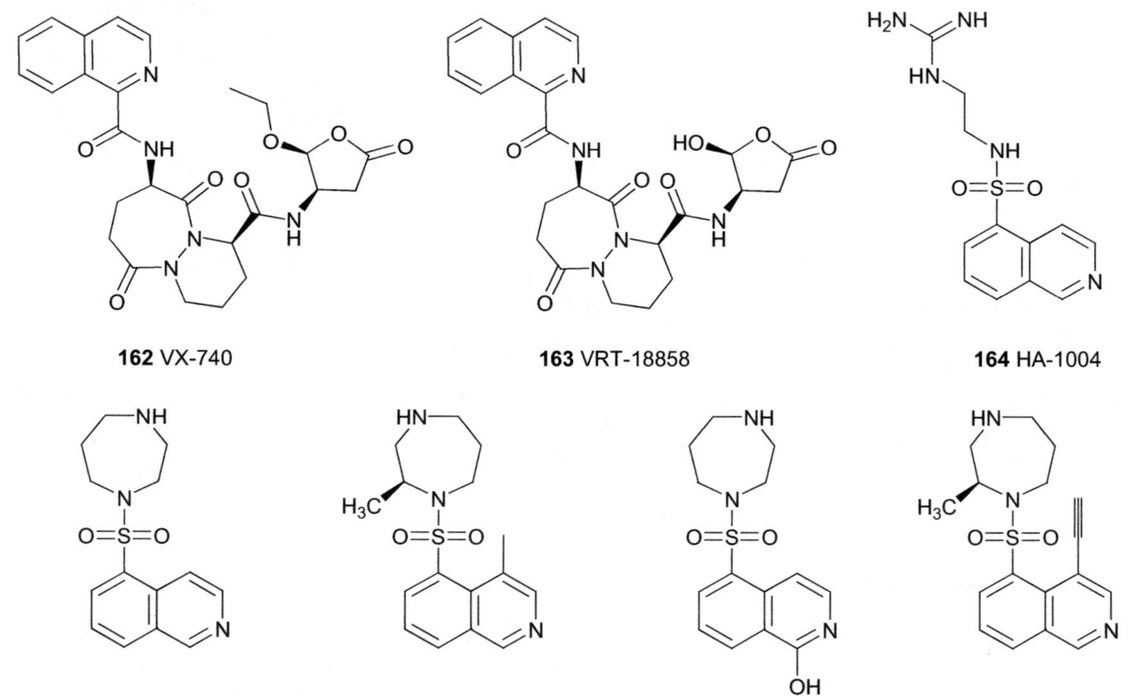

162 VX-740

163 VRT-18858

164 HA-1004

2 Fasudil

165 Dimethylfasudil

166 Hydroxyfasudil

175

Figure 7.38 Chemical structures of mono- and disubstituted isoquinoline derivatives.

approximately 10-fold higher selectivity over KOR and DOR and greatly improved selectivity for MOR.[365]

PK-11195 (**192**) (Figure 7.39) is a specific 18 kDa translocator protein (TSPO) ligand[366] that can be exploited as a positron emission tomography (PET) tracer by carbon-11 labelling (^{11}C-[R]-PK-11195).[367]

K-115 (**193**) is a ROCK inhibitor for the treatment of glaucoma and ocular hypertension.[47] It has been developed by D. Western Therapeutics Institute and out-licensed to Kowa Pharmaceutical. In 2013, the companies reported the successful filing of the NDA (new drug application), bringing K-115 (**193**) for the treatment of glaucoma to market (non-proprietary name: Ripasudil hydrochloride hydrate (**193**)).[368] Again, the chemical structure could not be verified by a reliable source.[369]

Asunaprevir (BMS-605339) (**194**) is a tripeptidic acylsulfonamide inhibitor of the NS3/4A enzyme and currently in phase III clinical trials for the treatment of hepatitis C virus (HCV) infection.[60,61]

A series of tipifarnib isoquinoline-based analogues, a clinical cancer drug candidate, has been synthesized. The results showed that these isoquinoline compounds have the potential to kill *Trypanosoma cruzi* amastigotes grown in mammalian host cells, therefore offering promising lead structures for the treatment of Chagas disease, caused by infection with the protozoan parasite.[370]

Dimethisoquin hydrochloride (**1**) has been used as an active surface anaesthetic and topically for the relief of itching, irritation, burning, or pain.[74]

FG-4592 (**195**) is an anaemia compound developed by FibroGen (Figure 7.40). AstraZeneca has committed to co-develop and

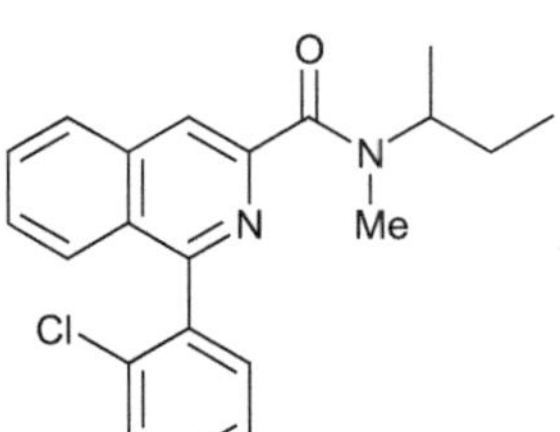

192 PK-11195

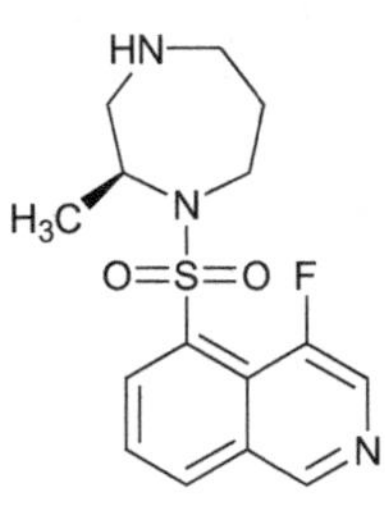

193 K-115

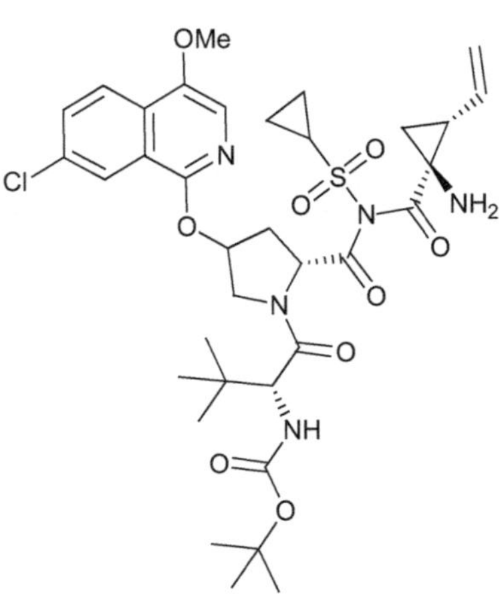

194 Asunaprevir

Figure 7.39 Isoquinoline derivatives.

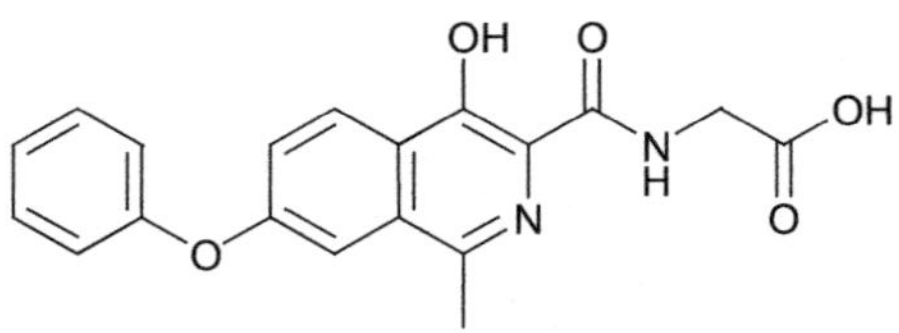

195 FG-4592

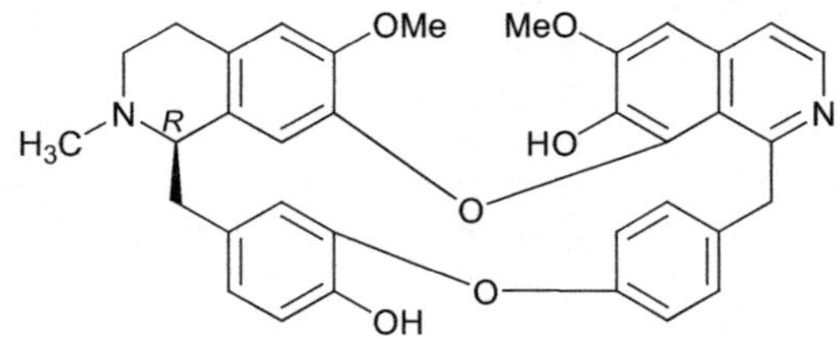

196 Racemosinine B

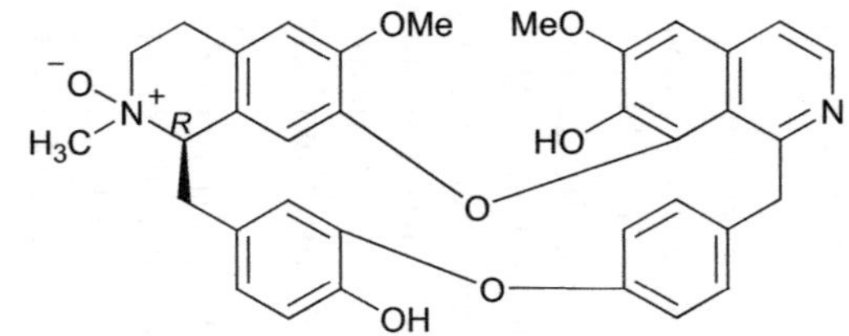

197 Racemosinine C

Figure 7.40 Chemical structures of polysubstituted isoquinoline derivatives.

co-commercialize the compound, which is currently in phase III trials. FG-4592 (**195**) is a small molecule inhibitor of hypoxia-induced factor prolyl hydroxylase (HIF-PH) for the treatment of anaemia in patients with chronic kidney disease.[371]

Racemonisin B (**196**) and C (**197**) are two natural products of the rare class of bisbenzylisoquinoline alkaloids and exhibited cytotoxic activities against HCT-8 and Bel-7402 cancer cell lines.[372]

7.6 Conclusion

There is relatively little publicly accessible data addressing the fundamental questions of how current drugs work and what the mode of action of drug candidates is.[373] Reasons are speculative, but the complexity of drug interactions in living organisms together with the often only small amounts of available substances might play a role. Nevertheless, many targets have been carved out which are addressed by isoquinoline scaffolds. This property makes isoquinoline a privileged scaffold, applying the concept of the ability to bind multiple targets.[374] This review shows that only few drugs succeeded in being approved. Even though generally only a fraction of research compounds enter the clinical phase, it appears that isoquinolines have not yet been fully exploited. A holistic and systematic research approach would help shed more light on this compound class. Although natural products are a great source for inspiration and a rich source of compounds, drug development remains a tedious endeavour.

References

1. E. Pászthory, *Chem. Unserer Zeit*, 1996, **30**, 96–102.
2. S. N. Kramer, *Am. J. Pharm. (1937–1978)*, 1954, **126**, 76–84.
3. S. B. Lal, A. C. Roy and G. Achari, *Curr. Sci.*, 1953, **22**, 374–375.
4. S. M. Sarkar, *Nature*, 1948, **162**, 265.
5. A. M. Polson, N. H. Stoller, P. J. Hanes, C. L. Bandt, S. Garrett and G. L. Southard, *J. Clin. Periodontol.*, 1996, **23**, 782–788.
6. A. Meyer and P. Imming, *J. Nat. Prod.*, 2011, **74**, 2482–2487.
7. K. Bhadra and G. S. Kumar, *Med. Res. Rev.*, 2011, **31**, 821–862.
8. B. Patwardhan and A. D. Vaidya, *Indian J. Exp. Biol.*, 2010, **48**, 220–227.
9. E. Fossati, A. Ekins, L. Narcross, Y. Zhu, J.-P. Falgueyret, G. A. W. Beaudoin, P. J. Facchini and V. J. J. Martin, *Nat. Commun.*, 2014, **5**, 4283/1–4283/11.
10. J. M. Hagel and P. J. Facchini, *Plant Cell Physiol.*, 2013, **54**, 647–672.
11. P. Villuendas and E. P. Urriolabeitia, *J. Org. Chem.*, 2013, **78**, 5254–5263.
12. C. Pomeranz, *Monatsh. Chem.*, 1893, **14**, 116–119.
13. P. Fritsch, *Ber. Dtsch. Chem. Ges.*, 1893, **26**, 419–422.
14. W. J. Gensler, *Organic Reactions*, John Wiley & Sons, Inc., 2004.
15. T. J. Donohoe, B. S. Pilgrim, G. R. Jones and J. A. Bassuto, *Proc. Natl. Acad. Sci. U. S. A.*, 2012, **109**, 11605–11608, S11605/11601-S11605/11164.

16. A. Pictet and A. Gams, *Ber. Dtsch. Chem. Ges.*, 1910, **43**, 2384–2391.
17. A. Pictet and T. Spengler, *Ber. Dtsch. Chem. Ges.*, 1911, **44**, 2030–2036.
18. W. M. Whaley and T. R. Govindachari, *Org. React.*, 1951, **VI**, 151–190.
19. A. Bischler and B. Napieralski, *Ber. Dtsch. Chem. Ges.*, 1893, **26**, 1903–1908.
20. W. M. Whaley and T. R. Govindachari, *Org. React.*, 1951, **VI**, 74–150.
21. E. Schlittler and J. Müller, *Helv. Chim. Acta*, 1948, **31**, 914–924.
22. D. L. Boger, C. E. Brotherton and M. D. Kelley, *Tetrahedron*, 1981, **37**, 3977–3980.
23. B. Wang, B. Lu, Y. Jiang, Y. Zhang and D. Ma, *Org. Lett.*, 2008, **10**, 2761–2763.
24. P.-C. Too, Y.-F. Wang and S. Chiba, *Org. Lett.*, 2010, **12**, 5688–5691.
25. T. Konno, J. Chae, T. Miyabe and T. Ishihara, *J. Org. Chem.*, 2005, **70**, 10172–10174.
26. T. Sakamoto, A. Numata and Y. Kondo, *Chem. Pharm. Bull.*, 2000, **48**, 669–672.
27. B. K. Ghorai, D. Jiang and J. W. Herndon, *Org. Lett.*, 2003, **5**, 4261–4263.
28. R. Alonso, P. J. Campos, B. Garcia and M. A. Rodriguez, *Org. Lett.*, 2006, **8**, 3521–3523.
29. B. S. Pilgrim, A. E. Gatland, C. T. McTernan, P. A. Procopiou and T. J. Donohoe, *Org. Lett.*, 2013, **15**, 6190–6193.
30. H. Jiang, Y. Cheng, R. Wang, Y. Zhang and S. Yu, *Chem. Commun.*, 2014, **50**, 6164–6167.
31. M. Iranshahy, R. J. Quinn and M. Iranshahi, *RSC Adv.*, 2014, **4**, 15900–15913.
32. B. D. Krane, M. O. Fagbule, M. Shamma and B. Gozler, *J. Nat. Prod.*, 1984, **47**, 1–43.
33. P. Zajdel, K. Marciniec, A. Maslankiewicz, K. Grychowska, G. Satala, B. Duszynska, T. Lenda, A. Siwek, G. Nowak, A. Partyka, D. Wrobel, M. Jastrzebska-Wiesek, A. J. Bojarski, A. Wesolowska and M. Pawlowski, *Eur. J. Med. Chem.*, 2013, **60**, 42–50.
34. M. Ohashi and T. Oki, *Expert Opin. Ther. Pat.*, 1996, **6**, 1285–1294.
35. D. Khayat, C. Borel, M. Azab, D. Paraisot, E. Malaurie, C. Bouloux and M. Weil, *Cancer Chemother. Pharmacol.*, 1992, **30**, 226–228.
36. Y. Z. Shu, *J. Nat. Prod.*, 1998, **61**, 1053–1071.
37. K. Jan, J. Bernardus, M. J. Jasper, V. Rogier and W. Ellen, *EP* 2628482A1, 2013, p. 28.
38. A. Alt, R. D. Hilgers, A. Tura, K. Nassar, T. Schneider, A. Hueber, K. Januschowski, S. Grisanti, J. Luke and M. Luke, *Cell. Physiol. Biochem.*, 2013, **32**, 218–234.
39. J. Kattan, M. Durand, J. P. Droz, M. Mahjoubi, J. P. Marino and M. Azab, *Am. J. Clin. Oncol.*, 1994, **17**, 242–245.
40. J. Treat, A. Greenspan, A. Rahman, M. S. McCabe and P. J. Byrne, *Invest. New Drugs*, 1989, **7**, 231–234.
41. J. Rouesse, M. Spielmann, F. Turpin, T. Le Chevalier, M. Azab and J. M. Mondesir, *Eur. J. Cancer*, 1993, **29A**, 856–859.

42. C. Paoletti, J. B. Le Pecq, N. Dat-Xuong, P. Juret, H. Garnier, J. L. Amiel and J. Rouesse, *Recent Results Cancer Res.*, 1980, **74**, 107–123.
43. S. O. Simonetti, E. L. Larghi, A. B. Bracca and T. S. Kaufman, *Org. Biomol. Chem.*, 2012, **10**, 4124–4134.
44. J. Kohno, M. Sakurai, N. Kameda, M. Nishio, K. Kawano, N. Kishi, T. Okuda and S. Komatsubara, *J. Antibiot.*, 1999, **52**, 913–916.
45. http://clinicaltrials.gov, Accessed March 2014.
46. http://www.chemspider.com/Chemical-Structure.25069704.html?rid = 4f024952-9341-447e-ac4e-a301dd4bd735, Accessed March 2014.
47. J. Chen, S. A. Runyan and M. R. Robinson, *Clin. Ophthalmol.*, 2011, **5**, 667–677.
48. http://www.businesswire.com/news/home/20121101006004/en/Aerie-Pharmaceuticals-Announces-Positive-Phase-2a-Clinical#.UyR3lYWDmk2, Accessed March 2014.
49. R. D. Williams, G. D. Novack, T. van Haarlem and C. Kopczynski, *Am. J. Ophthalmol.*, 2011, **152**, 834–841, e831.
50. T. Onda, E. Toyoda, O. Miyazaki, C. Seno, S. Kagaya, K. Okamoto and K. Nishikawa, *Cancer Lett.*, 2008, **259**, 99–110.
51. http://www.centerwatch.com/clinical-trials/results/new-therapies/nmt-details.aspx?CatID = 132, Accessed March 2014.
52. G. Mathe, M. Hallard and P. Pontiggia, *Biomed. Pharmacother.*, 1996, **50**, 510–511.
53. G. Mathe, P. Pontiggia, S. OrbachArbouys, K. Triana, N. Ambetima, C. Morette, M. Hallard and D. Blanquet, *Biomed. Pharmacother.*, 1996, **50**, 220–227.
54. Y. M. Qian, M. Ahmad, S. Q. Chen, P. Gillespie, N. Le, F. Mennona, S. Mischke, S. S. So, H. Wang, C. Burghardt, S. Tannu, K. Conde-Knape, J. Kochan and D. Bolin, *Bioorg. Med. Chem. Lett.*, 2011, **21**, 6264–6269.
55. E. L. Larghi, M. L. Bohn and T. S. Kaufman, *Tetrahedron*, 2009, **65**, 4257–4282.
56. G. R. Pettit, H. Hoffmann, J. McNulty, K. C. Higgs, A. Murphy, D. J. Molloy, D. L. Herald, M. D. Williams, R. K. Pettit, D. L. Doubek, J. N. Hooper, L. Albright, J. M. Schmidt, J. C. Chapuis and L. P. Tackett, *J. Nat. Prod.*, 2004, **67**, 506–509.
57. T. Asano, I. Ikegaki, S.-I. Satoh, M. Seto and Y. Sasaki, *Cardiovasc. Drug Rev.*, 1998, **16**, 76–87.
58. P. R. Kym, M. E. Kort and C. W. Hutchins, *Biochem. Pharmacol.*, 2009, **78**, 211–216.
59. R. El Kouhen, C. S. Surowy, B. R. Bianchi, T. R. Neelands, H. A. McDonald, W. Niforatos, A. Gomtsyan, C. H. Lee, P. Honore, J. P. Sullivan, M. F. Jarvis and C. R. Faltynek, *J. Pharmacol. Exp. Ther.*, 2005, **314**, 400–409.
60. P. M. Scola, L. Q. Sun, A. X. Wang, J. Chen, N. Sin, B. L. Venables, S. Y. Sit, Y. Chen, A. Cocuzza, D. M. Bilder, S. V. D'Andrea, B. Zheng, P. Hewawasam, Y. Tu, J. Friborg, P. Falk, D. Hernandez, S. Levine, C. Chen, F. Yu, A. K. Sheaffer, G. Zhai, D. Barry, J. O. Knipe, Y. H. Han,

R. Schartman, M. Donoso, K. Mosure, M. W. Sinz, T. Zvyaga, A. C. Good, R. Rajamani, K. Kish, J. Tredup, H. E. Klei, Q. Gao, L. Mueller, R. J. Colonno, D. M. Grasela, S. P. Adams, J. Loy, P. C. Levesque, H. Sun, H. Shi, L. Sun, W. Warner, D. Li, J. Zhu, N. A. Meanwell and F. McPhee, *J. Med. Chem.*, 2014, **57**, 1730–1752.

61. P. M. Scola, A. X. Wang, A. C. Good, L. Q. Sun, K. D. Combrink, J. A. Campbell, J. Chen, Y. Tu, N. Sin, B. L. Venables, S. Y. Sit, Y. Chen, A. Cocuzza, D. M. Bilder, S. D'Andrea, B. Zheng, P. Hewawasam, M. Ding, J. Thuring, J. Li, D. Hernandez, F. Yu, P. Falk, G. Zhai, A. K. Sheaffer, C. Chen, M. S. Lee, D. Barry, J. O. Knipe, W. Li, Y. H. Han, S. Jenkins, C. Gesenberg, Q. Gao, M. W. Sinz, K. S. Santone, T. Zvyaga, R. Rajamani, H. E. Klei, R. J. Colonno, D. M. Grasela, E. Hughes, C. Chien, S. Adams, P. C. Levesque, D. Li, J. Zhu, N. A. Meanwell and F. McPhee, *J. Med. Chem.*, 2014, **57**, 1708–1729.
62. L. Kraus-Berthier, N. Guilbaud, S. Leonce, T. Parker, P. Genissel, C. Guillonneau, S. Goldstein, G. Atassi and A. Pierre, *Cancer Chemother. Pharmacol.*, 2002, **50**, 95–103.
63. H. Malonne, S. Farinelle, C. Decaestecker, L. Gordower, J. Fontaine, F. Chaminade, J. M. Saucier, G. Atassi and R. Kiss, *Clin. Cancer Res.*, 2000, **6**, 3774–3782.
64. V. Pierson, A. Pierre, Y. Pommier and P. Gros, *Cancer Res.*, 1988, **48**, 1404–1409.
65. M. Sun, W. Lou, J. Y. Chun, D. S. Cho, N. Nadiminty, C. P. Evans, J. Chen, J. Yue, Q. Zhou and A. C. Gao, *Genes Cancer*, 2010, **1**, 283–292.
66. K. Taylor, D. Micha, M. Ranson and C. Dive, *Expert Opin. Invest. Drugs*, 2006, **15**, 669–690.
67. W. J. Andrews, T. Panova, C. Normand, O. Gadal, I. G. Tikhonova and K. I. Panov, *J. Biol. Chem.*, 2013, **288**, 4567–4582.
68. L. M. Shi, T. G. Myers, Y. Fan, P. M. O'Connor, K. D. Paull, S. H. Friend and J. N. Weinstein, *Mol. Pharmacol.*, 1998, **53**, 241–251.
69. H. Hiroyoshi, N. Masahiro and S. Kengo, *US* 2008/0064681 A1, 2008, p. 21.
70. http://www.evaluategroup.com/Universal/View.aspx?type = Story& id = 297059, Accessed August 2014.
71. S. P. Langdon, H. R. Hendriks, B. J. Braakhuis, G. Pratesi, D. P. Berger, O. Fodstad, H. H. Fiebig and E. Boven, *Ann. Oncol.*, 1994, **5**, 415–422.
72. G. M. Cragg and D. J. Newman, *J. Ethnopharmacol.*, 2005, **100**, 72–79.
73. A. B. Oliveira, M. F. Dolabela, F. C. Braga, R. L. Jacome, F. P. Varotti and M. M. Povoa, *An. Acad. Bras. Cienc.*, 2009, **81**, 715–740.
74. K. F. L. Smith, *J. Am. Med. Assoc.*, 1955, **159**, 114–115.
75. T. G. Administration, *TGA Approved Terminology for Medicines Section 1 – Chemical Substances*, Australian Government, Department of Health and Ageing, 1999.
76. http://www.fda.gov/downloads/aboutfda/centersoffices/officeofmedicalproductsandtobacco/cder/ucm135688.pdf, Accessed March 2014.

77. L. Guo, X. Liu, K. Nishikawa and W. Plunkett, *Mol. Cancer Ther.*, 2007, **6**, 1501–1508.
78. D. Davidson, L. Amrein, L. Panasci and R. Aloyz, *Front. Pharmacol.*, 2013, **4**, 5.
79. C. Jekimovs, E. Bolderson, A. Suraweera, M. Adams, K. J. O'Byrne and D. J. Richard, *Front. Oncol.*, 2014, **4**, 86.
80. P. Cohen, *Nat. Rev. Drug Discovery*, 2002, **1**, 309–315.
81. B. Kojonazarov, A. Myrzaakhmatova, T. Sooronbaev, T. Ishizaki and A. Aldashev, *Eur. Respir. J.*, 2012, **39**, 496–498.
82. T. Nakanishi, A. Masuda, M. Suwa, Y. Akiyama, N. Hoshino-Abe and M. Suzuki, *Bioorg. Med. Chem. Lett.*, 2000, **10**, 2321–2323.
83. W. J. Oswald and D. H. Baeder, *South. Med. J.*, 1975, **68**, 1481–1484.
84. F. S. Trainor, R. E. Phillips, D. D. Michie, S. R. Zellner, L. Hogan, Jr. and J. M. Chubb, *Angiology*, 1986, **37**, 343–351.
85. http://www.fda.gov/Drugs/DrugSafety/DrugShortages/ucm314742.htm#papaverine, Accessed March 2014.
86. optumrx.com, vol. 2014.
87. R. Fellous, D. Coulaud, I. el Abed, B. P. Roques, J. B. Le Pecq, E. Delain and A. Gouyette, *Cancer Res.*, 1988, **48**, 6542–6549.
88. G. L. Southard, R. J. Harkrader and J. A. Greene, *Compend. Contin. Educ. Dent.*, 1986, **7**, 189–192.
89. A. Awada, S. Giacchetti, B. Gerard, P. Eftekhary, C. Lucas, D. De Valeriola, M. G. Poullain, J. Soudon, C. Dosquet, M. H. Brillanceau, B. Giroux, M. Marty, H. Bleiberg, F. Calvo and M. Piccart, *Ann. Oncol.*, 2002, **13**, 1925–1934.
90. M. F. Roberts, M. Wink and Editors, *Alkaloids: Biochemistry, Ecology, and Medicinal Applications*, Plenum, 1998.
91. K. Shaari, K. C. Ling, Z. M. Rashid, T. P. Jean, F. Abas, S. M. Raof, Z. Zainal, N. H. Lajis, H. Mohamad and A. M. Ali, *Mar. Drugs*, 2009, **7**, 1–8.
92. L. Grycova, J. Dostal and R. Marek, *Phytochemistry (Elsevier)*, 2007, **68**, 150–175.
93. M. Imanshahidi and H. Hosseinzadeh, *Phytother. Res.*, 2008, **22**, 999–1012.
94. I. Thorne Research, *Altern. Med. Rev.*, 2000, **5**, 175–177.
95. F. Wang, G. Zhao, L. Cheng, H.-Y. Zhou, L.-Y. Fu and W.-X. Yao, *Brain Res.*, 2004, **999**, 91–97.
96. W. H. Peng, C. R. Wu, C. S. Chen, C. F. Chen, Z. C. Leu and M. T. Hsieh, *Life Sci.*, 2004, **75**, 2451–2462.
97. S. K. Kulkarni and A. Dhir, *Phytother. Res.*, 2010, **24**, 317–324.
98. G. Derosa and P. Maffioli, *Curr. Top. Med. Chem.*, 2014, **14**, 200–206.
99. R. Li, J. Wu, Y. He, L. Hai and Y. Wu, *Bioorg. Med. Chem. Lett.*, 2014, **24**, 1762–1765.
100. L. Huang, A. D. Shi, F. He and X. S. Li, *Bioorg. Med. Chem.*, 2010, **18**, 1244–1251.
101. L. Huang, Z. H. Luo, F. He, J. Lu and X. S. Li, *Bioorg. Med. Chem.*, 2010, **18**, 4475–4484.

102. A. Shi, L. Huang, C. Lu, F. He and X. Li, *Bioorg. Med. Chem.*, 2011, **19**, 2298–2305.
103. L. Huang, Z. H. Luo, F. He, A. D. Shi, F. F. Qin and X. S. Li, *Bioorg. Med. Chem. Lett.*, 2010, **20**, 6649–6652.
104. W. J. Shan, L. Huang, Q. Zhou, F. C. Meng and X. S. Li, *Eur. J. Med. Chem.*, 2011, **46**, 5885–5893.
105. H. Jiang, X. Wang, L. Huang, Z. Luo, T. Su, K. Ding and X. Li, *Bioorg. Med. Chem.*, 2011, **19**, 7228–7235.
106. T. Su, S. Xie, H. Wei, J. Yan, L. Huang and X. Li, *Bioorg. Med. Chem.*, 2013, **21**, 5830–5840.
107. L. Huang, T. Su, W. Shan, Z. Luo, Y. Sun, F. He and X. Li, *Bioorg. Med. Chem.*, 2012, **20**, 3038–3048.
108. C. Y. Lo, L. C. Hsu, M. S. Chen, Y. J. Lin, L. G. Chen, C. D. Kuo and J. Y. Wu, *Bioorg. Med. Chem. Lett.*, 2013, **23**, 305–309.
109. C. Liu, S. Y. Liu, Y. C. Wang, S. X. Wang, J. C. Zhang, S. H. Li, X. Y. Qin, X. L. Li, K. R. Wang and Q. Q. Zhou, *Med. Chem. Res.*, 2014, **23**, 1899–1907.
110. Y. H. Li, P. Yang, W. J. Kong, Y. X. Wang, C. Q. Hu, Z. Y. Zuo, Y. M. Wang, H. Gao, L. M. Gao, Y. C. Feng, N. N. Du, Y. Liu, D. Q. Song and J. D. Jiang, *J. Med. Chem.*, 2009, **52**, 492–501.
111. P. Yang, D. Q. Song, Y. H. Li, W. J. Kong, Y. X. Wang, L. M. Gao, S. Y. Liu, R. Q. Cao and J. D. Jiang, *Bioorg. Med. Chem. Lett.*, 2008, **18**, 4675–4677.
112. Y.-H. Li, H.-G. Fu, F. Su, L.-M. Gao, S. Tang, C.-W. Bi, Y.-H. Li, Y.-X. Wang and D.-Q. Song, *Chem. Cent. J.*, 2013, **7**(117), 118.
113. K. D. Park, J. H. Lee, S. H. Kim, T. H. Kang, J. S. Moon and S. U. Kim, *Bioorg. Med. Chem. Lett.*, 2006, **16**, 3913–3916.
114. P. V. Karamchandani, *Lancet*, 1930, **215**, 78.
115. G. Derosa, A. D'Angelo, A. Bonaventura, L. Bianchi, D. Romano and P. Maffioli, *Expert Opin. Biol. Ther.*, 2013, **13**, 475–482.
116. R. Izzo, G. de Simone, R. Giudice, M. Chinali, V. Trimarco, N. De Luca and B. Trimarco, *J. Hypertens.*, 2010, **28**, 1482–1487.
117. K. W. Bentley, *Nat. Prod. Rep.*, 1986, **3**, 153–169.
118. K. W. Bentley, *Nat. Prod. Rep.*, 1985, **2**, 81–96.
119. J. F. Pan, C. Yu, D. Y. Zhu, H. Zhang, J. F. Zeng, S. H. Jiang and J. Y. Ren, *Acta Pharmacol. Sin.*, 2002, **23**, 77–82.
120. F. Zuo, N. Nakamura, T. Akao and M. Hattori, *Drug Metab. Dispos.*, 2006, **34**, 2064–2072.
121. F. Qiu, Z. Zhu, N. Kang, S. Piao, G. Qin and X. Yao, *Drug Metab. Dispos.*, 2008, **36**, 2159–2165.
122. M. Ye, S. Fu, R. Pi and F. He, *J. Pharm. Pharmacol.*, 2009, **61**, 831–837.
123. Y. Li, G. Ren, Y.-X. Wang, W.-J. Kong, P. Yang, Y.-M. Wang, Y.-H. Li, H. Yi, Z.-R. Li, D.-Q. Song and J.-D. Jiang, *J. Transl. Med.*, 2011, **9**, 62.
124. J. Y. Ma, R. Feng, X. S. Tan, C. Ma, J. W. Shou, J. Fu, M. Huang, C. Y. He, S. N. Chen, Z. X. Zhao, W. Y. He, Y. Wang and J. D. Jiang, *J. Pharm. Sci.*, 2013, **102**, 4181–4192.

125. X. S. Tan, J. Y. Ma, R. Feng, C. Ma, W. J. Chen, Y. P. Sun, J. Fu, M. Huang, C. Y. He, J. W. Shou, W. Y. He, Y. Wang and J. D. Jiang, *PLoS One*, 2013, **8**, e77969.
126. S. Spinozzi, C. Colliva, C. Camborata, M. Roberti, C. Ianni, F. Neri, C. Calvarese, A. Lisotti, G. Mazzella and A. Roda, *J. Nat. Prod.*, 2014, 77(4), 766–772.
127. Y. X. Wang, Y. P. Wang, H. Zhang, W. J. Kong, Y. H. Li, F. Liu, R. M. Gao, T. Liu, J. D. Jiang and D. Q. Song, *Bioorg. Med. Chem. Lett.*, 2009, **19**, 6004–6008.
128. P. Yang, D. Q. Song, Y. H. Li, W. J. Kong, Y. X. Wang, L. M. Gao, S. Y. Liu, R. Q. Cao and J. D. Jiang, *Bioorg. Med. Chem. Lett.*, 2008, **18**, 4675–4677.
129. Y. Zhou, S. Cao, Y. Wang, P. Xu, J. Yan, W. Bin, F. Qiu and N. Kang, *Fitoterapia*, 2014, **92**, 230–237.
130. M. Zhong, Y. X. Ma, J. X. Liu and D. L. Di, *Nat. Prod. Res.*, 2014, **28**, 507–510.
131. B. Das and K. V. N. S. Srinivas, *Synth. Commun.*, 2002, **32**, 3027–3029.
132. S. Hayasaka, T. Kodama and A. Ohira, *Am. J. Chin. Med.*, 2012, **40**, 887–904.
133. H. S. Cui, S. Hayasaka, X. Y. Zhang, Y. Hayasaka, Z. L. Chi and L. S. Zheng, *Life Sci.*, 2006, **79**, 949–956.
134. K. C. Chen, S. S. Chang, H. J. Huang, T. L. Lin, Y. J. Wu and C. Y. Chen, *J. Biomol. Struct. Dyn.*, 2012, **30**, 662–683.
135. H.-L. Li, T. Han, R.-H. Liu, C. Zhang, H.-S. Chen and W.-D. Zhang, *Chem. Biodiversity*, 2008, **5**, 777–783.
136. N. Takao, Y. Yasumoto and K. Iwasa, *Yakugaku Zasshi*, 1973, **93**, 242–244.
137. F.-L. Zeng, Y.-F. Xiang, Z.-R. Liang, X. Wang, D.-e. Huang, S.-N. Zhu, M.-M. Li, D.-P. Yang, D.-M. Wang and Y.-F. Wang, *Am. J. Chin. Med.*, 2013, **41**, 119–130.
138. Y. Kan, C. Jiang, Q. Xi, X. Wang, L. Peng, J. Jiang and R. Yu, *Anal. Sci.*, 2014, **30**, 561–568.
139. J. Zhou, D. A. Sayre, Y. Zheng, H. Szmacinski and H. O. Sintim, *Anal. Chem.*, 2014, **86**, 2412–2420.
140. G. S. Kumar, *J. Biosci.*, 2012, **37**, 539–552.
141. M. Maiti and G. S. Kumar, *J. Nucleic Acids*, 2009, **2010**, 1–23.
142. J. H. Lin and W. L. Tseng, *Analyst*, 2014, **139**, 1436–1441.
143. S. Hazra, M. Hossain and G. S. Kumar, *Mol. Biosyst.*, 2013, **9**, 143–153.
144. K. Y. Zee-Cheng and C. C. Cheng, *J. Pharm. Sci.*, 1972, **61**, 969–971.
145. M. Maiti and G. S. Kumar, *Med. Res. Rev.*, 2007, **27**, 649–695.
146. K. Jewers, A. H. Manchanda and P. N. Jenkins, *J. Chem. Soc., Perkin Trans. 2*, 1972, 1393–1396.
147. H. A. Jung, B. S. Min, T. Yokozawa, J. H. Lee, Y. S. Kim and J. S. Choi, *Biol Pharm. Bull.*, 2009, **32**, 1433–1438.
148. Z. T. Ma, X. W. Yang, Y. Zhang and J. X. Liu, *J. Asian Nat. Prod. Res.*, 2014, **16**, 483–496.
149. H. J. Hong, P. Y. Chen, T. C. Shih, C. Y. Ou, M. D. Jhuo, Y. Y. Huang, C. H. Cheng, Y. C. Wu and J. G. Chung, *Mol. Med. Rep.*, 2012, **5**, 142–147.

150. Y. Deng, Q. Liao, S. Li, K. Bi, B. Pan and Z. Xie, *J. Chromatogr. B: Anal. Technol. Biomed. Life Sci.*, 2008, **863**, 195–205.
151. Z. Ye, Y. Li, J. Wen, K. Li and B. Ye, *Talanta*, 2014, **126**, 38–45.
152. H. Wu, K. He, Y. Wang, D. Xue, N. Ning, Z. Zou, X. Ye, X. Li, D. Wang and J. Pang, *Phytomedicine*, 2014, **21**(11), 1373–1381.
153. L. D. Kong, C. H. Cheng and R. X. Tan, *Planta Med.*, 2001, **67**, 74–76.
154. G. Q. Liu, Y. Jiang, G. Peng, Ibrahim and L. Q. Zhou, *Yaoxue Xuebao*, 1985, **20**, 566–570.
155. M. Cernakova, D. Kost'alova, V. Kettmann, M. Plodova, J. Toth and J. Drimal, *BMC Complementary Altern. Med.*, 2002, **2**, 2.
156. H. M. Malebo, T. Wenzler, M. Cal, S. M. Swaleh, M. O. Omolo, A. Hassanali, U. Sequin, D. Haussinger, P. Dalsgaard, M. Hamburger, R. Brun and I. O. Ndiege, *BMC Complementary Altern. Med.*, 2013, **13**, 48.
157. L. Zhu, B. Huang, X. Ban, J. He, Y. Chen, L. Han and Y. Wang, *Food Chem. Toxicol.*, 2012, **50**, 2584–2588.
158. L. Rackova, M. Majekova, D. Kost'alova and M. Stefek, *Bioorg. Med. Chem.*, 2004, **12**, 4709–4715.
159. V. Fajardo, C. Carcamo and B. Moreno, *Heterocycles*, 1996, **43**, 949–952.
160. Y. H. Long, L. P. Bai, Y. Qin, J. Y. Pang, W. H. Chen, Z. Cai, Z. L. Xu and Z. H. Jiang, *Bioorg. Med. Chem.*, 2006, **14**, 4670–4676.
161. R. R. K. Gupta and M. T. H. Khan, *Bioactive Heterocycles IV*, Springer, Berlin Heidelberg, 2007.
162. V. Preininger and F. Santavy, *Pharmazie*, 1970, **25**, 356–360.
163. W. Awe, *Arzneim.-Forsch.*, 1951, **1**, 287–288.
164. M. L. Colombo, C. Bugatti, A. Mossa, N. Pescalli, L. Piazzoni, G. Pezzoni, E. Menta, S. Spinelli, F. Johnson, R. C. Gupta and L. Dasaradhi, *Farmaco*, 2001, **56**, 403–409.
165. H. Hirano, E. Osawa, Y. Yamaoka and T. Yokoi, *Biol. Pharm. Bull.*, 2001, **24**, 1277–1281.
166. C. C. Lin, L. T. Ng, F. F. Hsu, D. E. Shieh and L. C. Chiang, *Clin. Exp. Pharmacol. Physiol.*, 2004, **31**, 65–69.
167. H. Tanabe, H. Suzuki, A. Nagatsu, H. Mizukami, Y. Ogihara and M. Inoue, *Phytomedicine*, 2006, **13**, 334–342.
168. J. Yi, X. Ye, D. Wang, K. He, Y. Yang, X. Liu and X. Li, *J. Ethnopharmacol.*, 2013, **145**, 303–310.
169. Q. Zhou, Y. Liu, X. Wang and X. Di, *J. Mass Spectrom.*, 2013, **48**, 111–118.
170. ekolab, 2014, vol. 2014, http://ekolab.free.fr/cadreGb.htm, Accessed March 2014.
171. P. Virtanen, V. Lassila, T. Njimi and D. E. Mengata, *Cells Tissues Organs*, 1988, **131**, 166–170.
172. G. Brunhofer, A. Fallarero, D. Karlsson, A. Batista-Gonzalez, P. Shinde, C. Gopi Mohan and P. Vuorela, *Bioorg. Med. Chem.*, 2012, **20**, 6669–6679.
173. K. W. Bentley, *Nat. Prod. Rep.*, 1990, **7**, 245–268.
174. H. Ishii, Y. Ichikawa, E. Kawanabe, M. Ishikawa, T. Ishikawa, K. Kuretani, M. Inomata and A. Hoshi, *Chem. Pharm. Bull.*, 1985, **33**, 4139–4151.

175. X. J. Yang, F. Miao, Y. Yao, F. J. Cao, R. Yang, Y. N. Ma, B. F. Qin and L. Zhou, *Molecules*, 2012, **17**, 13026–13035.
176. F. Miao, X.-J. Yang, L. Zhou, H.-J. Hu, F. Zheng, X.-D. Ding, D.-M. Sun, C.-D. Zhou and W. Sun, *Nat. Prod. Res.*, 2011, **25**, 863–875.
177. K. W. Bentley, *Nat. Prod. Rep.*, 2004, **21**, 395–424.
178. K. W. Bentley, *Nat. Prod. Rep.*, 1992, **9**, 365–391.
179. K. W. Bentley, *Nat. Prod. Rep.*, 2005, **22**, 249–268.
180. I. Slaninova, K. Pencikova, J. Urbanova, J. Slanina and E. Taborska, *Phytochem. Rev.*, 2014, **13**, 51–68.
181. X. Z. Dong, M. Zhang, K. Wang, P. Liu, D. H. Guo, X. L. Zheng and X. Y. Ge, *Biomed. Res. Int.*, 2013, **2013**, 517698.
182. P. M. Cala, J. G. Norby and D. C. Tosteson, *J. Membr. Biol.*, 1982, **64**, 23–31.
183. Y. L. Chow and F. Sato, *Biosci. Biotechnol. Biochem.*, 2013, 77, 2405–2412.
184. S. Y. Eid, M. Z. El-Readi and M. Wink, *Phytomedicine*, 2012, **19**, 1307–1314.
185. I. Bessi, C. Bazzicalupi, C. Richter, H. R. A. Jonker, K. Saxena, C. Sissi, M. Chioccioli, S. Bianco, A. R. Bilia, H. Schwalbe and P. Gratteri, *ACS Chem. Biol.*, 2012, 7, 1109–1119.
186. M. Ferraroni, C. Bazzicalupi, A. R. Bilia and P. Gratteri, *Chem. Commun.*, 2011, **47**, 4917–4919.
187. N. C. Ong, E. Sham and B. M. Adams, *Med. J. Aust.*, 2014, **200**, 314.
188. L. Ma, J. W. Dharamsi and T. Vandergriff, *Dermatitis*, 2012, **23**, 239–240.
189. K. L. Eastman, L. V. McFarland and G. J. Raugi, *J. Altern. Complementary Med.*, 2014, **20**, 284–289.
190. J. R. Cannon, G. K. Hughes, E. Ritchie and W. C. Taylor, *Aust. J. Chem.*, 1953, **6**, 86–89.
191. H. R. Arthur, W. H. Hui and Y. L. Ng, *J. Chem. Soc.*, 1959, 1840–1845.
192. A. S. Bailey and C. R. Worthing, *J. Chem. Soc.*, 1956, 4535–4543.
193. J. Smidrkal, *Collect. Czech. Chem. Commun.*, 1984, **49**, 1412–1420.
194. H. Ishii, T. Ishikawa, S. Takeda, S. Ohta, S. Ueki, M. Suzuki and T. Harayama, *Tennen Yuki Kagobutsu Toronkai Koen Yoshishu*, 1991, **33rd**, 164–171.
195. J. Lenfeld, M. Kroutil, E. Marsalek, J. Slavik, V. Preininger and V. Simanek, *Planta Med.*, 1981, **43**, 161–165.
196. K. M. Cho, I. D. Yoo and W. G. Kim, *Biol. Pharm. Bull.*, 2006, **29**, 2317–2320.
197. J. M. Herbert, J. M. Augereau, J. Gleye and J. P. Maffrand, *Biochem. Biophys. Res. Commun.*, 1990, **172**, 993–999.
198. B. Leblond, E. Beausoleil, C. Chauvignac, T. Taverne, V. Picard, O. C. De and F. Schweighoffer, *WO* 2009/007457A2, 2009.
199. Z. Hou, R. Yang, C. Zhang, L. F. Zhu, F. Miao, X. J. Yang and L. Zhou, *Molecules*, 2013, **18**, 10413–10424.
200. S. Yamamoto, K. Seta, C. Morisco, S. F. Vatner and J. Sadoshima, *J. Mol. Cell. Cardiol.*, 2001, **33**, 1829–1848.

201. D. Zhai, C. Jin, A. C. Satterthwait and J. C. Reed, *Cell Death Differ.*, 2006, **13**, 1419–1421.
202. T. Watanabe, Y. Ohashi, R. Yoshino, N. Komano, M. Eguchi, S. Maruyama and T. Ishikawa, *Org. Biomol. Chem.*, 2003, **1**, 3024–3032.
203. H. Ishii, I. S. Chen and T. Ishikawa, *Chem. Pharm. Bull.*, 1983, **31**, 2963–2966.
204. K. W. Bentley, *Nat. Prod. Rep.*, 1994, **11**, 555–576.
205. L. K. Wang, R. K. Johnson and S. M. Hecht, *Chem. Res. Toxicol.*, 1993, **6**, 813–818.
206. V. F. Kosteletzky, *Allg. Med. Pharm. Fl.*, 1831, **1**, 1782–1786.
207. S. D. Phillips and R. N. Castle, *J. Heterocycl. Chem.*, 1981, **18**, 223–232.
208. M. E. Wall, M. C. Wani and H. L. Taylor, 162nd National Meeting of American Chemical Society, Washington DC, 1971, MEDI 34.
209. J. Liao, T. Xu, J. X. Zheng, J. M. Lin, Q. Y. Cai, D. B. Yu and J. Peng, *Int. J. Mol. Med.*, 2013, **32**, 79–84.
210. L. Guo, X. Liu, K. Nishikawa and W. Plunkett, *Mol. Cancer Ther.*, 2007, **6**, 1501–1508.
211. E. Toyoda, S. Kagaya, I. G. Cowell, A. Kurosawa, K. Kamoshita, K. Nishikawa, S. Iiizumi, H. Koyama, C. A. Austin and N. Adachi, *J. Biol. Chem.*, 2008, **283**, 23711–23720.
212. L. Guo, X. Liu, Y. Jiang, K. Nishikawa and W. Plunkett, *Mol. Pharmacol.*, 2011, **80**, 321–327.
213. T. Hisatomi, N. Sueoka-Aragane, A. Sato, R. Tomimasu, M. Ide, A. Kurimasa, K. Okamoto, S. Kimura and E. Sueoka, *Blood*, 2011, **117**, 3575–3584.
214. Y. Erguen, S. Patir and G. Okay, *J. Heterocycl. Chem.*, 2003, **40**, 1005–1010.
215. J. Bernadou, B. Meunier, G. Meunier, C. Auclair and C. Paoletti, *Proc. Natl. Acad. Sci. U. S. A.*, 1984, **81**, 1297–1301.
216. L. K. Dalton, S. Demerac, B. C. Elmes, J. W. Loder, J. M. Swan and T. Teitei, *Aust. J. Chem.*, 1967, **20**, 2715–2727.
217. C. Auclair, *Arch. Biochem. Biophys.*, 1987, **259**, 1–14.
218. M. Stiborova, M. Rupertova, H. H. Schmeiser and E. Frei, *Biomed. Pap. Med. Fac. Univ. Palacky Olomouc Czech. Repub.*, 2006, **150**, 13–23.
219. R. Prudent, V. Moucadel, C. H. Nguyen, C. Barette, F. Schmidt, J. C. Florent, L. Lafanechere, C. F. Sautel, E. Duchemin-Pelletier, E. Spreux, O. Filhol, J. B. Reiser and C. Cochet, *Cancer Res.*, 2010, **70**, 9865–9874.
220. B. Monsarrat, M. Maftouh, G. Meunier, B. Dugue, J. Bernadou, J. P. Armand, C. Picard-Fraire, B. Meunier and C. Paoletti, *Biochem. Pharmacol.*, 1983, **32**, 3887–3890.
221. C. Auclair, A. Pierre, E. Voisin, O. Pepin, S. Cros, C. Colas, J. M. Saucier, B. Verschuere, P. Gros and C. Paoletti, *Cancer Res.*, 1987, **47**, 6254–6261.
222. M. Enescu, M.-P. Fontaine-Aupart, B. Soep and F. Tfibel, *J. Phys. Chem. B*, 1998, **102**, 3631–3636.
223. J. Schmutz and F. Hunziker, *Pharm. Acta Helv.*, 1958, **33**, 341–347.

224. R. Jasztold-Howorko, B. Tylinska, B. Biadun, T. Gebarowski and K. Gasiorowski, *Acta Pol. Pharm.*, 2013, **70**, 823–832.
225. J. P. Kutney, M. Noda, N. G. Lewis, B. Monteiro, D. Mostowicz and B. R. Worth, *Can. J. Chem.*, 1982, **60**, 2426–2430.
226. J. P. Kutney, M. Noda, N. G. Lewis, B. Monteiro, D. Mostowicz and B. R. Worth, *Heterocycles*, 1981, **16**, 1469–1472.
227. C. Guillonneau, A. Nault, E. Raimbaud, S. Leonce, L. Kraus-Berthier, A. Pierre and S. Goldstein, *Bioorg. Med. Chem.*, 2005, **13**, 175–184.
228. C. H. Browning, G. T. Morgan, J. V. M. Robb and L. P. Walls, *J. Pathol. Bacteriol.*, 1938, **46**, 203–204.
229. IUPAC, *Nomenclature of Organic Chemistry (Blue Book)*, The Royal Society of Chemistry, 2014, pp. 73–164.
230. S. Geerts, P. Kageruka, R. De Deken, J. R. Brandt, J. M. Kazadi, B. Diarra, M. C. Eisler, Y. Lemmouchi, E. Schacht and P. H. Holmes, *Acta Trop.*, 1997, **65**, 23–31.
231. M. Hogan, N. Dattagupta and D. M. Crothers, *Biochemistry*, 1979, **18**, 280–288.
232. S. C. Jain, C. C. Tsai and H. M. Sobell, *J. Mol. Biol.*, 1977, **114**, 317–331.
233. C. C. Tsai, S. C. Jain and H. M. Sobell, *J. Mol. Biol.*, 1977, **114**, 301–315.
234. G. Dougherty, *Int. J. Biochem.*, 1982, **14**, 493–504.
235. J. B. Le Pecq and C. Paoletti, *Anal. Biochem.*, 1966, **17**, 100–107.
236. W. Lin, G. Brauers, R. Ebel, V. Wray, A. Berg, Sudarsono and P. Proksch, *J. Nat. Prod.*, 2003, **66**, 57–61.
237. S. O. Simonetti, E. L. Larghi, A. B. Bracca and T. S. Kaufman, *Nat. Prod. Rep.*, 2013, **30**, 941–969.
238. T. Kumemura, T. Choshi, A. Hirata, M. Sera, Y. Takahashi, J. Nobuhiro and S. Hibino, *Chem. Pharm. Bull.*, 2005, **53**, 393–397.
239. T. Kumemura, T. Choshi, A. Hirata, M. Sera, Y. Takahashi, J. Nobuhiro and S. Hibino, *Heterocycles*, 2003, **61**, 13–17.
240. T. Choshi, T. Kumemura, H. Fujioka, Y. Hieda and S. Hibino, *Heterocycles*, 2012, **84**, 587–595.
241. S. O. Simonetti, E. L. Larghi, A. B. J. Bracca and T. S. Kaufman, *Org. Biomol. Chem.*, 2012, **10**, 4124–4134.
242. T. Choshi, T. Kumemura, H. Fujioka, Y. Hieda and S. Hibino, *Heterocycles*, 2012, **84**, 587–595.
243. J. Kohno, H. Hiramatsu, M. Nishio, M. Sakurai, T. Okuda and S. Komatsubara, *Tetrahedron*, 1999, **55**, 11247–11252.
244. J. Kohno, M. Sakurai, N. Kameda, M. Nishio, K. Kawano, N. Kishi, T. Okuda and S. Komatsubara, *J. Antibiot.*, 1999, **52**, 913–916.
245. T. Kumemura, T. Choshi, A. Hirata, M. Sera, Y. Takahashi, J. Nobuhiro and S. Hibino, *Chem. Pharm. Bull.*, 2005, **53**, 393–397.
246. T. Kumemura, T. Choshi, J. Nobuhiro and S. Hibino, *Fukuyama Daigaku Yakugakubu Kenkyu Nenpo*, 2007, **25**, 35–62.
247. A. Ogawa, C. Murakami, S. Kamisuki, I. Kuriyama, H. Yoshida, F. Sugawara and Y. Mizushina, *Bioorg. Med. Chem. Lett.*, 2004, **14**, 3539–3543.

248. A. M. Sarotti, *Org. Biomol. Chem.*, 2013, **11**, 4847–4859.
249. S. O. Simonetti, E. L. Larghi, A. B. J. Bracca and T. S. Kaufman, *Nat. Prod. Rep.*, 2013, **30**, 941–969.
250. G. J. Slack, E. Puniani, J. C. Frisvad, R. A. Samson and J. D. Miller, *Mycol. Res.*, 2009, **113**, 480–490.
251. K. Trisuwan, V. Rukachaisirikul, Y. Sukpondma, S. Phongpaichit, S. Preedanon and J. Sakayaroj, *Tetrahedron*, 2010, **66**, 4484–4489.
252. W. Lin, G. Brauers, R. Ebel, V. Wray, A. Berg, Sudarsono and P. Proksch, *J. Nat. Prod.*, 2003, **66**, 57–61.
253. G. A. Beaudoin and P. J. Facchini, *Planta*, 2014, **240**, 19–32.
254. http://dailymed.nlm.nih.gov/dailymed/lookup.cfm?setid = 9e5e2ce4-7a57-4c61-a826-64c8d11d038e, Accessed March 2014.
255. A. Colautti, F. Fontani and V. Maurich, *J. Pharm. Biomed. Anal.*, 1987, **5**, 493–499.
256. D. Schmidl, B. Pemp, M. Lasta, A. Boltz, S. Kaya, S. Palkovits, F. Prager, L. Schmetterer and G. Garhofer, *Graefe's Arch. Clin. Exp. Ophthalmol.*, 2013, **251**, 515–520.
257. J. S. Shin, J. J. Lee, Y. Kim, C. K. Lee, Y. P. Yun and M. K. Lee, *Biol. Pharm. Bull.*, 2001, **24**, 103–105.
258. F. R. Chang, T. J. Hsieh, T. L. Huang, C. Y. Chen, R. Y. Kuo, Y. C. Chang, H. F. Chiu and Y. C. Wu, *J. Nat. Prod.*, 2002, **65**, 255–258.
259. K. W. Bentley, *Nat. Prod. Rep.*, 2006, **23**, 444–463.
260. K. W. Bentley, *Nat. Prod. Rep.*, 2003, **20**, 342–365.
261. K. W. Bentley, *Nat. Prod. Rep.*, 1987, **4**, 677–702.
262. A. J. Al-Rehaily, M. H. M. Sharaf, M. A. Zemaitis, C. Gao, G. E. Martin, C. E. Hadden, T. J. Thamann, F. Lin and P. L. Schiff, Jr., *J. Nat. Prod.*, 1999, **62**, 146–148.
263. T. A. Al-Howiriny, M. A. Zemaitis, C. Y. Gao, C. E. Hadden, G. E. Martin, F. T. Lin and P. L. Schiff, Jr., *J. Nat. Prod.*, 2001, **64**, 823–826.
264. A. C. Santos, *Ber. Dtsch. Chem. Ges. B*, 1932, **65B**, 472–474.
265. T. A. Van Beek, R. Verpoorte, A. Baerheim Svendsen, A. C. Santos and L. P. Olay, *J. Nat. Prod.*, 1983, **46**, 226–231.
266. A. Garcia, L. Castedo and D. Dominguez, *J. Nat. Prod.*, 1996, **59**, 806–807.
267. A. Garcia, L. Castedo and D. Dominguez, *Tetrahedron*, 1996, **52**, 5929–5932.
268. F. N. Samita, L. P. Sandjo, I. O. Ndiege, A. Hassanali and W. Lwande, *Beilstein J. Org. Chem.*, 2013, **9**, 447–452.
269. R. Graziose, T. Rathinasabapathy, C. Lategan, A. Poulev, P. J. Smith, M. Grace, M. A. Lila and I. Raskin, *J. Ethnopharmacol.*, 2011, **133**, 26–30.
270. K. W. Bentley, *Nat. Prod. Rep.*, 1998, **15**, 341–362.
271. D. K. Semwal, R. B. Semwal, I. Vermaak and A. Viljoen, *J. Ethnopharmacol.*, 2014, **155**(2), 1011–1028.
272. K. W. Bentley, *Nat. Prod. Rep.*, 2002, **19**, 332–356.
273. D. S. Swaffar, C. J. Holley, R. W. Fitch, K. R. Elkin, C. Zhang, J. P. Sturgill and M. D. Menachery, *Planta Med.*, 2012, **78**, 230–232.

274. K. W. Bentley, *Nat. Prod. Rep.*, 2001, **18**, 148–170.
275. M. D. Menachery and M. P. Cava, *J. Nat. Prod.*, 1981, **44**, 320–323.
276. K. T. Buck, D. L. Edgren, G. W. Blake and M. D. Menachery, *Heterocycles*, 1993, **36**, 2489–2495.
277. W. Sesang, S. Punyanitya, S. Pitchuanchom, P. Udomputtimekakul, N. Nuntasaen, R. Banjerdpongchai, B. Wudtiwai and W. Pompimon, *Molecules*, 2014, **19**, 8762–8772.
278. Y. D. Min, S. U. Choi and K. R. Lee, *Arch. Pharmacal Res.*, 2006, **29**, 627–632.
279. E. V. Costa, M. L. Pinheiro, A. D. de Souza, A. Barison, F. R. Campos, R. H. Valdez, T. Ueda-Nakamura, B. P. Filho and C. V. Nakamura, *Molecules*, 2011, **16**, 9714–9720.
280. E. Sobarzo-Sanchez, P. Bilbao-Ramos, M. Dea-Ayuela, H. Gonzalez-Diaz, M. Yanez, E. Uriarte, L. Santana, V. Martinez-Sernandez, F. Bolas-Fernandez and F. M. Ubeira, *PLoS One*, 2013, **8**, e77560.
281. E. M. Hassan, R. A. Hassan, J. Y. Salib, S. M. Mohamed and S. A. El-Toumy, *J. Appl. Sci. Res.*, 2013, **9**, 4884–4888.
282. Z.-F. Chen, Y.-F. Shi, Y.-C. Liu, X. Hong, B. Geng, Y. Peng and H. Liang, *Inorg. Chem.*, 2012, **51**, 1998–2009.
283. A. M. Clark, E. S. Watson, M. K. Ashfaq and C. D. Hufford, *Pharm. Res.*, 1987, **4**, 495–498.
284. A. C. Goren, B. N. Zhou and D. G. Kingston, *Planta Med.*, 2003, **69**, 867–868.
285. K.-S. Chen, Y.-C. Wu, C.-M. Teng, F.-N. Ko and T.-S. Wu, *J. Nat. Prod.*, 1997, **60**, 645–647.
286. Z. F. Chen, Y. C. Liu, Y. Peng, X. Hong, H. H. Wang, M. M. Zhang and H. Liang, *J. Biol. Inorg. Chem.*, 2012, **17**, 247–261.
287. P. Protais, J. Arbaoui, E.-H. Bakkali, A. Bermejo and D. Cortes, *J. Nat. Prod.*, 1995, **58**, 1475–1484.
288. I. R. C. Bick, P. S. Clezy and W. D. Crow, *Aust. J. Chem.*, 1956, **9**, 111–119.
289. B. R. Pai and G. Shanmugasundaram, *Tetrahedron*, 1965, **21**, 2579–2584.
290. S. Nimgirawath and W. C. Taylor, *J. Sci. Soc. Thailand*, 1983, **9**, 73–79.
291. T. J. Hsieh, F. R. Chang, Y. C. Chia, C. Y. Chen, H. C. Lin, H. F. Chiu and Y. C. Wu, *J. Nat. Prod.*, 2001, **64**, 1157–1161.
292. S. Ponnala and W. W. Harding, *Eur. J. Org. Chem.*, 2013, **3013**, 1107–1115.
293. C. C. Silveira, E. L. Larghi, S. R. Mendes, A. B. J. Bracca, F. Rinaldi and T. S. Kaufman, *Eur. J. Org. Chem.*, 2009, 4637–4645, S4637/4631-S4637/4627.
294. P. Gasiorski, K. S. Danel, M. Matusiewicz, T. Uchacz, W. Kuznik and A. V. Kityk, *J. Fluoresc.*, 2012, **22**, 81–91.
295. G. Scherowsky, E. Frackowiak and D. Adam, *Acta Crystallogr., Sect. C: Cryst. Struct. Commun.*, 1997, **C53**, xi, IUC9700005.
296. M. P. Cava, K. T. Buck and A. I. DaRocha, *J. Am. Chem. Soc.*, 1972, **94**, 5931.

297. D. L. Boger and C. E. Brotherton, *J. Org. Chem.*, 1984, **49**, 4050–4055.
298. B. Zhao and V. Snieckus, *Tetrahedron Lett.*, 1991, **32**, 5277–5278.
299. J. M. Fu, B. P. Zhao, M. J. Sharp and V. Snieckus, *Can. J. Chem.*, 1994, **72**, 227–236.
300. D. L. Boger and K. Takahashi, *J. Am. Chem. Soc.*, 1995, **117**, 12452–12459.
301. M. G. Banwell, E. Hamel, N. K. Ireland, M. F. Mackay and A. K. Serelis, *J. Chem. Soc., Perkin Trans. 1*, 1993, 1905–1911.
302. S. Ponnala and W. W. Harding, *Eur. J. Org. Chem.*, 2013, **2013**, 1107–1115.
303. M. M. V. Ramana, R. H. Sharma and J. A. Parihar, *Tetrahedron Lett.*, 2005, **46**, 4385–4386.
304. K. W. Bentley, *Nat. Prod. Rep.*, 1997, **14**, 387–411.
305. H. Morita, K. Takeya and H. Itokawa, *Bioorg. Med. Chem. Lett.*, 1995, **5**, 597–598.
306. K. S. Feldman, T. D. Cutarelli and R. Di Florio, *J. Org. Chem.*, 2002, **67**, 8528–8537.
307. S.-K. Hong, H. Kim, Y. Seo, S. H. Lee, J. K. Cha and Y. G. Kim, *Org. Lett.*, 2010, **12**, 3954–3956.
308. H. Morita, K. Matsumoto, K. Takeya, H. Itokawa and Y. Iitaka, *Chem. Pharm. Bull.*, 1993, **41**, 1418–1422.
309. M. D. Menachery and C. D. Muthler, *J. Nat. Prod.*, 1987, **50**, 726–729.
310. J. C. Lee and J. K. Cha, *J. Am. Chem. Soc.*, 2001, **123**, 3243–3246.
311. J. D. Phillipson, M. F. Roberts, M. H. Zenk and Editors, *The Chemistry and Biology of Isoquinoline Alkaloids*, Springer-Verlag, 1985.
312. D. P. Allais and H. Guinaudeau, *J. Nat. Prod.*, 1983, **46**, 881–883.
313. A. Garcia, L. Castedo and D. Dominguez, *Tetrahedron*, 1995, **51**, 8585–8598.
314. I. Orhan, B. Ozcelik, T. Karaoglu and B. Sener, *Z. Naturforsch., C: J. Biosci.*, 2007, **62**, 19–26.
315. P. Protais, D. Cortes, J. L. Pons, S. Lopez, M. C. Villaverde and L. Castedo, *Experientia*, 1992, **48**, 27–30.
316. A. Garcia, L. Castedo and D. Dominguez, *Tetrahedron Lett.*, 1993, **34**, 1797–1798.
317. S. Tsukamoto, R. Yamanokuchi, M. Yoshitomi, K. Sato, T. Ikeda, H. Rotinsulu, R. E. Mangindaan, N. J. de Voogd, R. W. van Soest and H. Yokosawa, *Bioorg. Med. Chem. Lett.*, 2010, **20**, 3341–3343.
318. K. W. Bentley, K. W. Bentley and K. W. Bentley, *Nat. Prod. Rep.*, 1999, **16**, 367–388.
319. E. L. Larghi, B. V. Obrist and T. S. Kaufman, *Tetrahedron*, 2008, **64**, 5236–5245.
320. H. Nakamura, J. Kobayashi, Y. Ohizumi and Y. Hirata, *Tetrahedron Lett.*, 1982, **23**, 5555–5558.
321. E. Sugino, T. Choshi and S. Hibino, *Heterocycles*, 1999, **50**, 543–559.
322. J. J. Bowling, H. K. Pennaka, K. Ivey, S. Wahyuono, M. Kelly, R. F. Schinazi, F. A. Valeriote, D. E. Graves and M. T. Hamann, *Chem. Biol. Drug Des.*, 2008, **71**, 205–215.

323. M. D. Menachery, G. L. Lavanier, M. L. Wetherly, H. Guinaudeau and M. Shamma, *J. Nat. Prod.*, 1986, **49**, 745–778.
324. N. R. Ferrigni, S. A. Sweetana, J. L. McLaughlin, K. E. Singleton and R. G. Cooks, *J. Nat. Prod.*, 1984, **47**, 839–845.
325. A. Blair, L. Stevenson and A. Sutherland, *Tetrahedron Lett.*, 2012, **53**, 4084–4086.
326. Q. Zhang, G. Tu, Y. Zhao and T. Cheng, *Tetrahedron*, 2002, **58**, 6795–6798.
327. T. Yasuhara, N. Zaima, S. Hashimoto, M. Yamazaki and O. Muraoka, *Heterocycles*, 2009, **77**, 1397–1402.
328. K. Zaima, Y. Takeyama, I. Koga, A. Saito, H. Tamamoto, S. S. Syed Abd Azziz, M. R. Mukhtar, K. Awang, A. H. A. Hadi and H. Morita, *J. Nat. Med.*, 2012, **66**, 421–427.
329. W. A. Skinner, H. T. Crawford, H. Tong, D. Skidmore and H. I. Maibach, *J. Pharm. Sci.*, 1976, **65**, 1404–1407.
330. U. Thull, S. Kneubuehler, P. Gaillard, P.-A. Carrupt, B. Testa, C. Altomare, A. Carotti, P. Jenner and K. S. T. P. McNaught, *Biochem. Pharmacol.*, 1995, **50**, 869–877.
331. G. Bringmann, T. Gulder, M. Reichert and F. Meyer, *Org. Lett.*, 2006, **8**, 1037–1040.
332. A. Cecil, C. Rikanovic, K. Ohlsen, C. Liang, J. Bernhardt, T. A. Oelschlaeger, T. Gulder, G. Bringmann, U. Holzgrabe, M. Unger and T. Dandekar, *Genome Biol.*, 2011, **12**, R24.
333. W. D. Brown and A.-H. Gouliaev, *Synthesis*, 2002, 83–86.
334. W. Adam and A. Grimison, *Tetrahedron*, 1965, **21**, 3417–3422.
335. J. Ross, D. Brough, R. M. Gibson, S. A. Loddick and N. J. Rothwell, *Neuropharmacology*, 2007, **53**, 638–642.
336. G. W. Bemis, J. P. Duffy, W. H. Fridman, J. M. C. Golec, D. J. Livingston, M. D. Mullican, M. A. Murcko and R. E. Zelle, *WO* 9722618A1, 1997, p. 103.
337. J. M. C. Golec, D. J. Lauffer, D. J. Livingston, M. D. Mullican, M. A. Murcko, P. L. Nyce, A. L. C. Robidoux and M. W. Wannamaker, *WO* 9824805A1, 1998, p. 135.
338. J. M. C. Golec, D. J. Lauffer, D. J. Livingston, M. D. Mullican, P. L. Nyce, A. L. C. Robidoux and M. W. Wannamaker, *WO* 9824804A2, 1998, p. 107.
339. P. Galatsis, B. Caprathe, J. Gilmore, A. Thomas, K. Linn, S. Sheehan, W. Harter, C. Kostlan, E. Lunney, C. Stankovic, J. Rubin, K. Brady, H. Allen and R. Talanian, *Bioorg. Med. Chem. Lett.*, 2010, **20**, 5184–5190.
340. K. Rudolphi, N. Gerwin, N. Verzijl, P. van der Kraan and W. van den Berg, *Osteoarthritis Cartilage*, 2003, **11**, 738–746.
341. B. Siegmund and M. Zeitz, *IDrugs*, 2003, **6**, 154–158.
342. P. M. Brooks, *Curr. Opin. Rheumatol.*, 2002, **14**, 573–577.
343. H. Namazi, *Med. Hypotheses*, 2007, **68**, 1425.
344. S. R. Goldring and M. B. Goldring, *Clin. Orthop. Relat. Res.*, 2004, S27–S36.

345. H. Hidaka and R. Kobayashi, *Annu. Rev. Pharmacol. Toxicol.*, 1992, **32**, 377–397.
346. H. Hidaka and T. Sone, *EP* 187371A2, 1986, p. 85.
347. R. Singh, B. Wang, A. Shirvaikar, S. Khan, S. Kamat, J. R. Schelling, M. Konieczkowski and J. R. Sedor, *J. Clin. Invest.*, 1999, **103**, 1561–1570.
348. K. Ishikura, N. Yamada, M. Ito, S. Ota, M. Nakamura, N. Isaka and T. Nakano, *Circ. J.*, 2006, **70**, 174–178.
349. J. K. Liao, M. Seto and K. Noma, *J. Cardiovasc. Pharmacol.*, 2007, **50**, 17–24.
350. B. K. Mueller, H. Mack and N. Teusch, *Nat. Rev. Drug Discovery*, 2005, **4**, 387–398.
351. R. Guan, X. Xu, M. Chen, H. Hu, H. Ge, S. Wen, S. Zhou and R. Pi, *Eur. J. Med. Chem.*, 2013, **70**, 613–622.
352. Q. Wen, B. Goldenson, S. J. Silver, M. Schenone, V. Dancik, Z. Huang, L. Z. Wang, T. A. Lewis, W. F. An, X. Li, M. A. Bray, C. Thiollier, L. Diebold, L. Gilles, M. S. Vokes, C. B. Moore, M. Bliss-Moreau, L. Verplank, N. J. Tolliday, R. Mishra, S. Vemula, J. Shi, L. Wei, R. Kapur, C. K. Lopez, B. Gerby, P. Ballerini, F. Pflumio, D. G. Gilliland, L. Goldberg, Y. Birger, S. Izraeli, A. S. Gamis, F. O. Smith, W. G. Woods, J. Taub, C. A. Scherer, J. E. Bradner, B. C. Goh, T. Mercher, A. E. Carpenter, R. J. Gould, P. A. Clemons, S. A. Carr, D. E. Root, S. L. Schreiber, A. M. Stern and J. D. Crispino, *Cell*, 2012, **150**, 575–589.
353. H. Shimokawa, M. Seto, N. Katsumata, M. Amano, T. Kozai, T. Yamawaki, K. Kuwata, T. Kandabashi, K. Egashira, I. Ikegaki, T. Asano, K. Kaibuchi and A. Takeshita, *Cardiovasc. Res.*, 1999, **43**, 1029–1039.
354. H. Shimokawa, *J. Cardiovasc. Pharmacol.*, 2002, **39**, 319–327.
355. K. Satoh, Y. Fukumoto and H. Shimokawa, *Am. J. Physiol.*, 2011, **301**, H287–H296.
356. N. Ono-Saito, I. Niki and H. Hidaka, *Pharmacol. Ther.*, 1999, **82**, 123–131.
357. M. Tamura, H. Nakao, H. Yoshizaki, M. Shiratsuchi, H. Shigyo, H. Yamada, T. Ozawa, J. Totsuka and H. Hidaka, *Biochim. Biophys. Acta*, 2005, **1754**, 245–252.
358. http://www.clinicaltrial.gov/ct2/results?term = AR-12286, Accessed August 2014.
359. http://www.chemspider.com/Chemical-Structure.25069704.html?rid = 4f024952-9341-447e-ac4e-a301dd4bd735, Accessed March 2014.
360. D. Lavogina, K. Kalind, J. Bredihhina, M. Hurt, A. Vaasa, M. Kasari, E. Enkvist, G. Raidaru and A. Uri, *Bioorg. Med. Chem. Lett.*, 2012, **22**, 3425–3430.
361. P. Ray, J. Wright, J. Adam, J. Bennett, S. Boucharens, D. Black, A. Cook, A. R. Brown, O. Epemolu, D. Fletcher, A. Haunso, M. Huggett, P. Jones, S. Laats, A. Lyons, J. Mestres, J. de Man, R. Morphy, Z. Rankovic, B. Sherborne, L. Sherry, N. van Straten, P. Westwood and G. Z. R. Zaman, *Bioorg. Med. Chem. Lett.*, 2011, **21**, 97–101.

362. N. Khairatkar-Joshi and A. Szallasi, *Trends Mol. Med.*, 2009, **15**, 14–22.
363. C. Mills, M. McMackin, R. Jaffe, J. Yu, E. Zininberg, D. Slee, K. Gogas and M. Bradbury, *Neuroscience*, 2008, **156**, 165–174.
364. R. M. Reilly, H. A. McDonald, P. S. Puttfarcken, S. K. Joshi, L. Lewis, M. Pai, P. H. Franklin, J. A. Segreti, T. R. Neelands, P. Han, J. Chen, P. W. Mantyh, J. R. Ghilardi, T. M. Turner, E. A. Voight, J. F. Daanen, R. G. Schmidt, A. Gomtsyan, M. E. Kort, C. R. Faltynek and P. R. Kym, *J. Pharmacol. Exp. Ther.*, 2012, **342**, 416–428.
365. Y. Yuan, O. Elbegdorj, I. O. Beletskaya, D. E. Selley and Y. Zhang, *Bioorg. Med. Chem. Lett.*, 2013, **23**, 5045–5048.
366. C. Gemelli, B. M. Dongmo, F. Ferrarini, A. Grande and L. Corsi, *Life Sci.*, 2014, **107**, 14–20.
367. A. Kumar, O. Muzik, V. Shandal, D. Chugani, P. Chakraborty and H. T. Chugani, *J. Neuroinflammation*, 2012, **9**, 232.
368. http://www.evaluategroup.com/Universal/View.aspx?type = Story& id = 477384, Accessed March 2014.
369. H. Gracies, http://www.glaucome.net/spip.php?article77, Accessed March 2014.
370. N. K. Chennamaneni, J. Arif, F. S. Buckner and M. H. Gelb, *Bioorg. Med. Chem. Lett.*, 2009, **19**, 6582–6584.
371. M. H. Flight, *Nat. Rev. Drug Discovery*, 2013, **12**, 730.
372. J. Z. Wang, Q. H. Chen and F. P. Wang, *J. Nat. Prod.*, 2010, **73**, 1288–1293.
373. J. P. Overington, B. Al-Lazikani and A. L. Hopkins, *Nat. Rev. Drug Discovery*, 2006, **5**, 993–996.
374. M. E. Welsch, S. A. Snyder and B. R. Stockwell, *Curr. Opin. Chem. Biol.*, 2010, **14**, 347–361.

CHAPTER 8

Rhodanine

TIHOMIR TOMAŠIČ* AND LUCIJA PETERLIN MAŠIČ

University of Ljubljana, Faculty of Pharmacy, Aškerčeva 7, 1000 Ljubljana, Slovenia
*Email: tihomir.tomasic@ffa.uni-lj.si

8.1 Chemistry and Reactivity of Rhodanines

Rhodanine (Figure 8.1) is a five-membered heterocycle containing thioether and amino groups at positions 1 and 3, respectively. It is structurally related to thiazolidine-2,4-dione and 2-iminothiazolidine-4-one that include an oxo or imino group, respectively, instead of the thioxo group at position 2. It is also related to 4-thioxothiazolidin-2-one, which bears oxo and thioxo groups at positions opposite to those in rhodanine (Figure 8.1). Although these heterocycles appear to be very similar at first glance, analogous compounds based on these scaffolds usually differ in their biological activities.

The rhodanine ring enables the formation of several types of interaction with the amino acid residues in ligand binding sites of proteins: (i) hydrogen bonds, with rhodanine acting as a hydrogen bond acceptor or donor, (ii) hydrophobic interactions, (iii) π–π and cation–π interactions with amino acids with aromatic or charged side chains in the case of aromatic 5-benzylidenerhodanines, and (iv) interaction with metal ions, such as Zn^{2+}.[1] The rhodanine moiety has been utilised as a uracil mimic[2–6] and a (di)phosphate isostere.[7] All these possibilities give the rhodanine scaffold unique properties and, consequently, high potential for incorporation in compounds possessing biological activity.

RSC Drug Discovery Series No. 50
Privileged Scaffolds in Medicinal Chemistry: Design, Synthesis, Evaluation
Edited by Stefan Bräse

Published by the Royal Society of Chemistry, www.rsc.org

2-thioxothiazolidin-4-one **rhodanine** thiazolidine-2,4-dione 2-iminothiazolidin-4-one 4-thioxothiazolidin-2-one

Figure 8.1 Chemical structures of rhodanine and its analogues.

Analyses of crystal structures of proteins in complex with a rhodanine-based inhibitor, available in the Protein Data Bank, show that the rhodanine heterocycle can take part in a significant number of spatially defined interactions.[8] Moreover, the density of interactions is very high compared with that in the remaining parts of the ligands. Rhodanines possess a marked propensity for formation of polar interactions that can be attributed to the exocyclic, double bonded sulphur. Density functional theory calculations have shown that the HOMO orbital and the negative electrostatic potential are strongly localised at the exocyclic sulphur in rhodanines, giving them special biomolecule binding properties.[8] In addition, this sulphur is geometrically a more tolerant group than its carbonyl counterpart, and is thus able to form a larger number of hydrogen bonds. The sulphur in the thioxo group supports a more diffuse lone pair electron density distribution and is able to interact with up to four hydrogen bond donors.[9] Although hydrogen bonds to the thioxo group are, energetically, weaker than those to the carbonyl group, the desolvation penalty of breaking a stronger hydrogen bond between water and the carbonyl group can result in more favourable overall binding of a ligand containing the thioxo, rather than the carbonyl, group.[8]

The rhodanine scaffold offers several possibilities for chemical modification (Scheme 8.1). 5-Arylmethylidenerhodanines[10] are the most commonly synthesised derivatives and are usually obtained by base-catalysed Knoevenagel condensation between rhodanines or *N*-substituted rhodanines and aromatic aldehydes, using either conventional heating or microwave-assisted synthesis. Following introduction of a 5-benzylidene or 5-arylmethylidene moiety, the rhodanine ring becomes aromatic. The reaction usually gives the *Z* isomer, as confirmed by several crystal structures[8] and NMR.[11] The wide variety of commercially available aldehydes and the high reaction yields make large compound libraries of rhodanines easy to obtain.

However, the exocyclic double bond, which is conjugated to the carbonyl group at position 4 of the rhodanine moiety, is a potentially reactive site. It can react, as an electrophilic Michael acceptor, with nucleophilic amino acid side chains of the target proteins, such as cysteine with the reactive thiol group, to form a covalent adduct (Figure 8.2a). Indeed, this has already been observed in the crystal structure of the rhodanine-based inhibitor covalently bound to the Cys366 side chain in the allosteric binding site of the hepatitis C virus (HCV) RNA polymerase non-structural protein 5B (NS5B) (PDB entry:

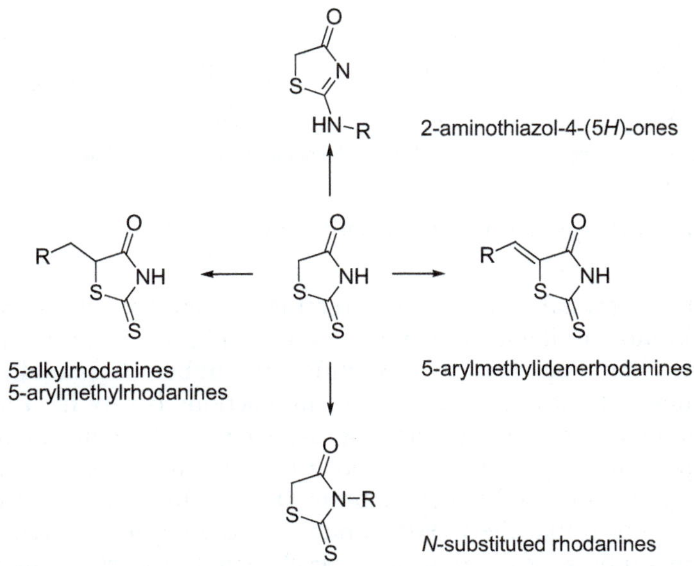

Scheme 8.1 Possible routes of derivatisation of the rhodanine ring.

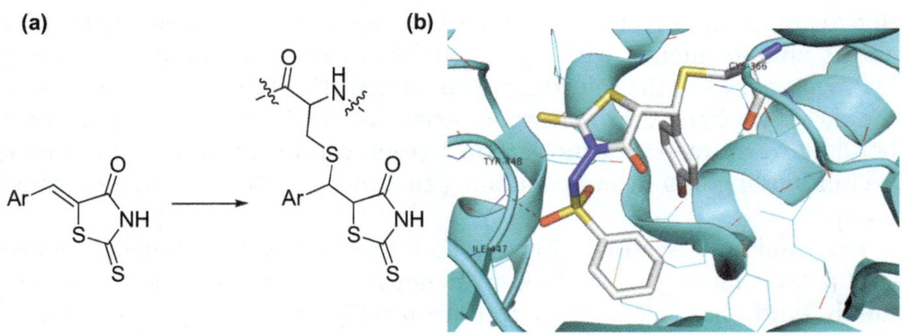

Figure 8.2 (a) Addition of a reactive cysteine thiol group to the exocyclic double bond of 5-arylmethylidenerhodanines; (b) Crystal structure of a rhodanine-based inhibitor (in grey sticks) covalently bound to the Cys366 side chain in the allosteric binding site of HCV RNA polymerase NS5B (in cyan, PDB entry: 2AWZ).
The figure was prepared by PyMOL.[12]

2AWZ)[13] (Figure 8.2b). However, covalent binding to the exocyclic double bond of the 5-benzylidenerhodanine-based inhibitor of the HCV RNA polymerase NS5B was found to be reversible,[13] as is the conjugate addition of dithiothreitol to the exocyclic double bond of the UDP-galactopyranose mutase inhibitors possessing the rhodanine scaffold.[14] Potential reactivity of the 5-arylmethylidenerhodanines in the 1,4-conjugate addition has been confirmed using ALARM NMR.[15,16] Recently, the reactivity of 5-benzylidene-barbiturates, -rhodanines, -hydantoins, -thiohydantoins, and -thiazolidine-2,4-diones was studied, using cysteamine as an exemplary biological nucleophile, with Avonto[17] NMR spectroscopy method.[18] In this study, 5-benzylidenerhodanines were found to be slightly reactive for addition of the cysteamine; however, their electrophilicity was significantly lower than those of the 5-benzylidenebarbiturates. Similarly, the lack of reactivity of the 5-benzylidenerhodanines was also observed when using glutathione as a nucleophile, indicating that the electrophilicity of such a Michael system is insignificant and the possible reaction highly unfavourable, since it would destroy the aromatic system of the ring.[8] Since reactivity of the cysteine thiol group is importantly dependent on the surrounding amino acids in a protein – which cannot be satisfactorily mimicked by smaller molecules like cysteamine or glutathione – reactivity of rhodanines possessing the exocyclic double bond cannot be completely excluded.[15]

A systematic study evaluating the biological activity of a large array of rhodanines, thiazolidine-2,4-diones, hydantoins and thiohydantoins against four targets (bacterial transferase MurA from *Escherichia coli*, serine proteases thrombin from bovine plasma, NS2B-NS3 protease of Dengue virus and metalloprotease methionine aminopeptidase from *E. coli*) was carried out by Mendgen and co-workers.[8] It was suggested that the distinct binding profile of such compounds is not related to common mechanisms of non-specific binding, such as aggregation or reactivity, but rather to the special electronic and hydrogen-bonding properties of the exocyclic sulphur atom (as described above), particularly in the case of the aromatic 5-benzylidenerhodanines.[8]

In order to avoid reactions that could result in off-target binding, the exocyclic double bond can be reduced[6] or saturated analogues synthesised by various ring closure reaction pathways.[19,20] However, when the double bond is reduced, the compounds become more flexible, with a new chiral centre, and their electronic properties are changed, since there is a loss of the conjugation between the rhodanine moiety and the aromatic ring at position 5 that is present in the unsaturated analogues. All these differences can lead to the activity of the saturated analogues being weaker than that of their unsaturated counterparts, a phenomenon that appears to be related to the loss of aromaticity of the heterocycle.[2]

8.2 Biological Activities of Rhodanines

Numerous scientific publications and patents describing a wide variety of biological activities of rhodanine-based compounds have been reported over the past 15 years.[10,21] A SciFinder search using the rhodanine heterocycle as query in a substructure search retrieved more than 116 550 compounds reported in nearly 8900 publications up to 17th July 2014. Approximately half are concerned with the evaluation of rhodanines in biological studies, reaching a peak in 2012, with 178 out of 374 publications reporting biological studies (Figure 8.3). A small decline in number occurred in 2011, compared to 2009 and 2010, which is probably related to the publications of

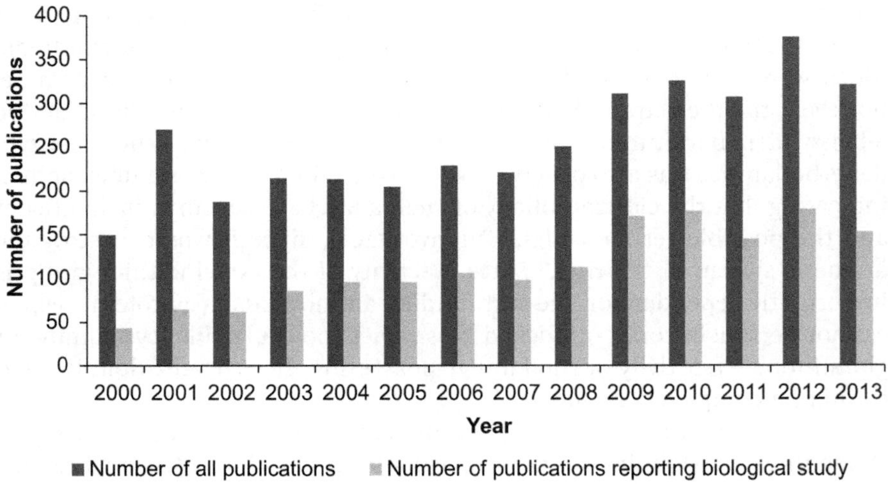

Figure 8.3 SciFinder search using the rhodanine heterocycle as a template for substructure search.

Baell and co-workers, who described rhodanine-based compounds as pan assay interference compounds (PAINS) and frequent hitters in biological screening campaigns.[22,23] They presented rhodanines as hits that are useless in the drug discovery process because of insufficient selectivity, and published a set of new substructure filters for removal of rhodanines and other PAINS from compound libraries prior to biological screening.[22]

While the more negative perspective is often associated with hits emerging from high-throughput screening campaigns, in which the same compounds appear as hits against several unrelated targets,[23] rhodanines can be viewed, from a more positive perspective, as hits with tuneable target affinity and selectivity that can be achieved in optimisation steps.[8] Moreover, their propensity for crystallisation in protein–ligand complexes makes them interesting chemical tools for studying proteins that could aid further optimisation and, possibly, replacement of the rhodanine heterocycle by a more acceptable core in subsequent steps. Nevertheless, biological studies have to be designed in a way that shows unambiguously that the investigated compound possesses a specific effect on the target macromolecule. There are several issues that need to be adequately addressed, when highlighting a rhodanine-based compound as a promising hit or lead compound. These include:

(i) **selectivity** against related and non-related targets – important for avoiding unwanted side effects resulting from off-target activity;
(ii) **specificity** of binding, since on-target activity can also result from non-specific binding, such as formation of aggregates, reactivity in the Michael addition, interaction with transition metals and signal interference in biological assays, since rhodanine-based compounds are highly conjugated and hence often coloured;

(iii) **proof of mode of action** when interpreting results from cell-based assays, since effects that are observed can also result from non-specific mechanisms, such as membrane damage, or from non-selective binding to cell proteins.

Although rhodanine-based compounds possess a variety of biological activities, there is only one rhodanine-based compound that is currently used in therapy (see Section 8.2.5). Rhodanines have been investigated most intensively as anticancer, antibacterial, antiviral, antifungal, antimalarial, anti-inflammatory agents and as compounds for the treatment of type 2 diabetes mellitus and associated complications.[10,21]

8.2.1 Antibacterial Activity

Of the many activities of rhodanine derivatives, their antibacterial activity has been investigated extensively. Studies are still regularly reporting potent antibacterial activities of rhodanines against a variety of Gram positive and Gram negative bacteria and their resistant strains. These activities were often obtained by random screening of in-house and commercial compound libraries or exhibited by the design of analogues of already published compounds. Antibacterial activities found in these whole cell-based assays should be interpreted with a certain degree of caution since, in the majority of cases, the molecular target responsible for the observed effect on bacterial growth was not investigated. Moreover, non-specific cytotoxicity and/or membrane damaging effects of these usually lipophilic and planar molecules were not considered in the studies. Since, as mentioned above, the effects observed from cell-based assays could also result from non-selective off-target binding, such antibacterially active compounds are of limited use for further optimisation and development.[10,23]

In contrast, by structure- and/or ligand-based design of bacterial enzyme inhibitors, several potent rhodanine-based compounds have been identified, but unfortunately development of a potent inhibitor to an antibacterially active compound still remains a very demanding challenge.[24] Rhodanine-based compounds are so far known to be involved in inhibiting enzymes in three main groups of activities. The first group is the biosynthesis of peptidoglycan and other components of the cell wall (Figure 8.4), involving MurC-F (**1** and **2**),[2–7,25,26] MurG (**3**),[27] penicillin-binding proteins (**4**),[28] UDP-galactopyranose mutase (**5**),[29] L,L-diaminopimelate aminotransferases[30] and class C β-lactamase.[31] The second is DNA replication (Figure 8.5), involving DNA gyrase (**6**)[32,33] and DNA helicase (**7**).[34] The third group (Figure 8.6) includes various enzymes, such as sortase A (**8**),[35] FtsZ (**9**)[36] and deoxyxylulose phosphate reductoisomerase (**10**).[37] A common feature of these inhibitors is good inhibitory activity, as measured on the isolated enzymes, but weak, if any, antibacterial activity.

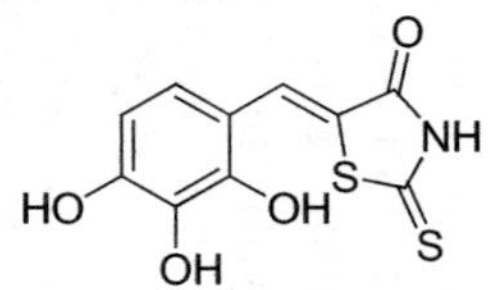

1, MurC-F inhibitor

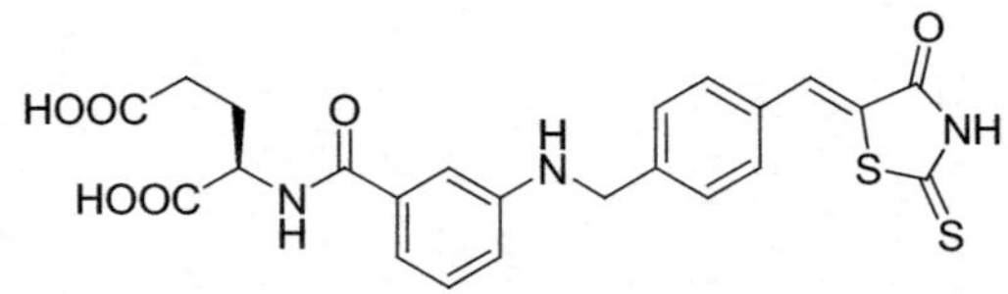

2, dual MurD and MurE ligase inhibitor

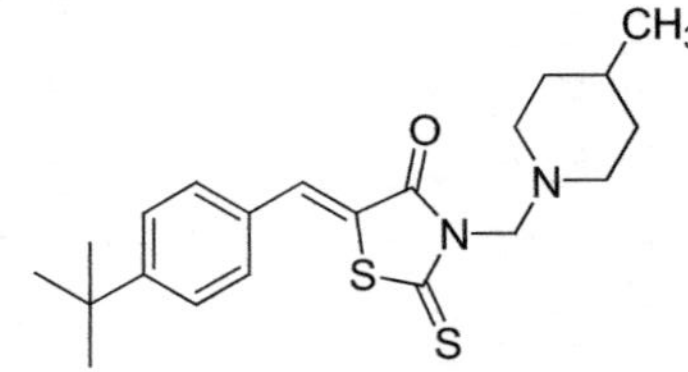

3, MurG inhibitor

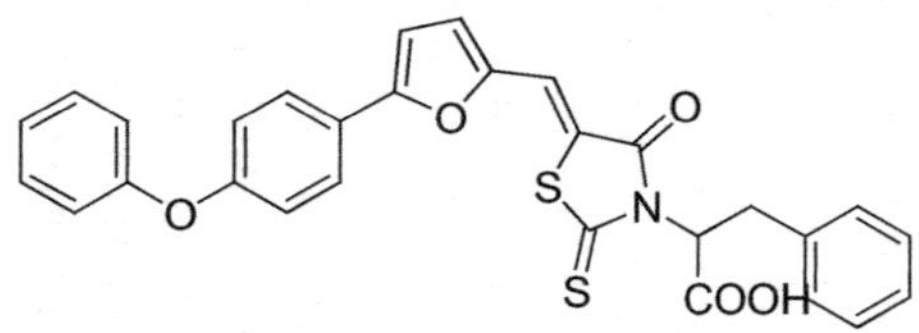

4, PBP inhibitor

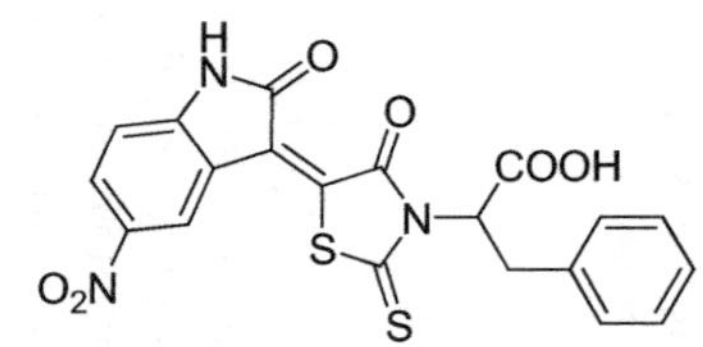

5, UDP-galactopyranose mutase inhibitor

Figure 8.4 Rhodanine derivatives that inhibit enzymes involved in bacterial cell wall formation.

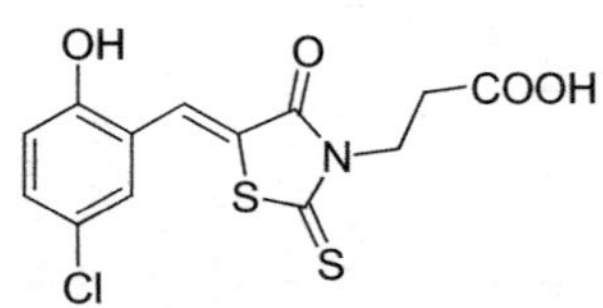

6, DNA gyrase B inhibitor

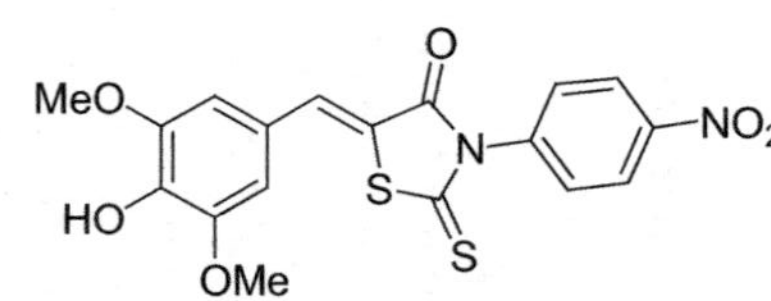

7, DNA helicase inhibitor

Figure 8.5 Rhodanine derivatives inhibiting DNA gyrase and DNA helicase in DNA replication.

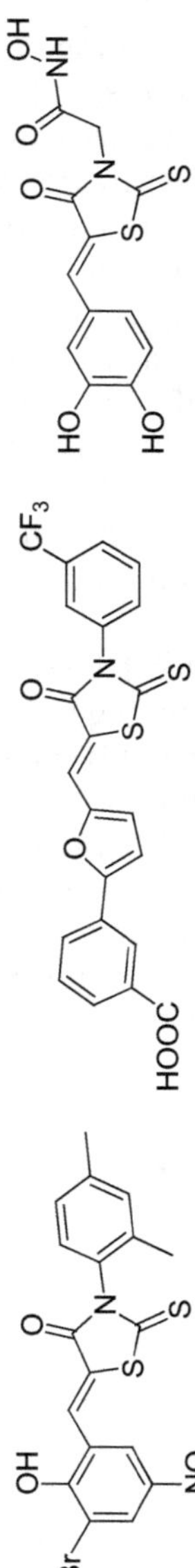

8, sortase A inhibitor

9, FtsZ inhibitor

10, deoxyxylulose phosphate reductoisomerase inhibitor

Figure 8.6 Inhibitors of sortase A, FtsZ and deoxyxylulose phosphate reductoisomerase.

8.2.2 Antiviral Activity

The activity of rhodanine-based compounds against hepatitis C virus (HCV), human immunodeficiency virus (HIV) and Dengue virus proteins has recently been investigated.

Non-structural protein 3 (NS3) from HCV is a serine protease that is vital for HCV replication. Micromolar rhodanine-based inhibitors of NS3 have been identified by screening (**11**, Figure 8.7), but were not selective against related proteases such as chymotrypsin, trypsin, plasmin and elastase.[38] In contrast, rhodanines containing bulkier hydrophobic groups (**12**, Figure 8.7) also inhibited NS3 in the micromolar range but showed selectivity against chymotrypsin.[39] Another HCV target protein inhibited by rhodanines is the non-structural protein 5B (NS5B) polymerase that functions as a catalytic subunit of the viral replicase. Low micromolar inhibitors of NS5B were identified by high-throughput[13] and virtual screening (**14**, Figure 8.8).[40] Subsequent optimisation of an HTS hit resulted in compound **13** (Figure 8.8) that inhibits NS5B with an IC_{50} value of 200 nM. Crystal structures of this series of inhibitors in complex with NS5B show covalent binding of the exocyclic double bond to the Cys366 thiol, as discussed above (Figure 8.2).[13]

Rhodanines are known that inhibit HIV-1 integrase, that catalyses the integration of viral cDNA into the human genome.[41] The most potent HIV-1 integrase inhibitors described in this study contain rhodanine and salicylic acid moieties and display enzyme inhibition and antiviral activities in the low micromolar range (**15**, Figure 8.9). Low micromolar inhibition by rhodanines of HIV-1 replication in MT-2 cells, which results from targeting the HIV-1 envelope glycoprotein transmembrane subunit gp41, has also been described (**16**, Figure 8.9).[42–44] Moreover, rhodanines have been discovered that inhibit Dengue virus protease NS2B-NS3.[8,45]

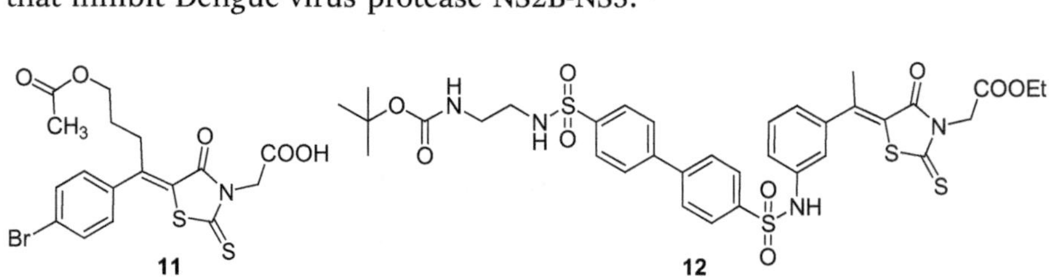

11 **12**

Figure 8.7 Hepatitis C virus NS3 inhibitors.

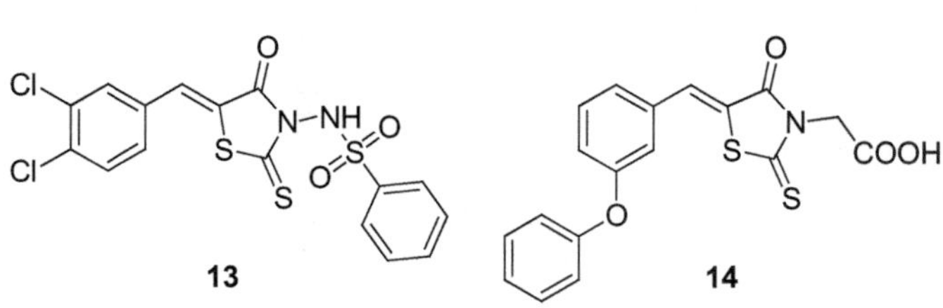

13 **14**

Figure 8.8 Hepatitis C virus NS5B inhibitors.

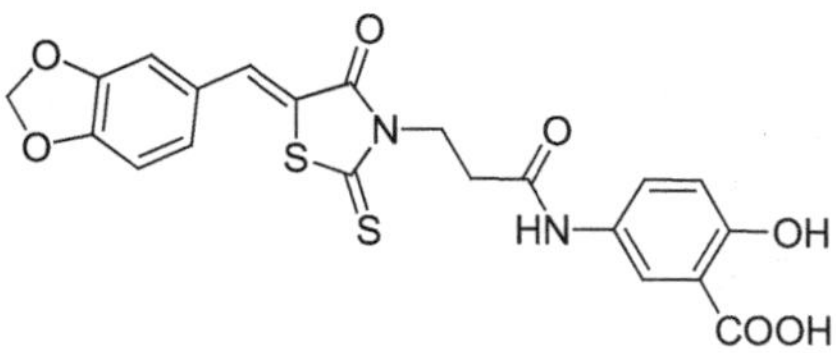

15, HIV-1 integrase inhibitor

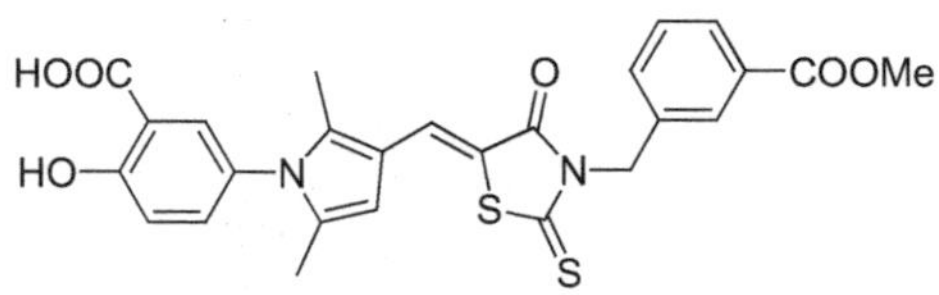

16, HIV-1 gp41 inhibitor

Figure 8.9 Representative inhibitors of HIV-1 integrase and gp41.

8.2.3 Anticancer Activity

Rhodanines have often emerged as hits from whole-cell screening on a variety of cancer cell lines. As already noted for whole cell-based assays for antibacterial activity, proof is often lacking that the compounds cause cell death *via* a specific mechanism and not by a non-specific effect such as membrane damage or interaction with cellular proteins important for cell viability.[46] However, rhodanine-based compounds often exhibit selective toxicity against selected normal cell lines.[47,48]

Rhodanines have been investigated as inhibitors of several enzymes involved in cancer pathogenesis. The latter include JNK-stimulating phosphatase-1 (JSP-1) (**17**),[49] sphingosine kinase (**18**),[50] phosphatase of regenerating liver 3 (**19**),[51] DNA polymerase λ (**20**)[52] and/or β,[53] 17β-hydroxysteroid dehydrogenase type 3 (**21**),[54] and inhibitors of the interaction between BH3 domain and Bcl-X_L (**22**) (Figure 8.10).[55]

8.2.4 Rhodanine-based Hits as Clinical Candidates

8.2.4.1 Phosphoinositide 3-kinase (PI3K) Inhibitors

Compound **23** (PI3Kγ $IC_{50} = 0.92$ μM, PI3Kα $IC_{50} > 20$ μM)[56] was discovered as a hit in high-throughput screening and optimised at Merck Serono to compound **24** (PI3Kγ $IC_{50} = 8$ nM, PI3Kα $IC_{50} = 60$ nM) (Figure 8.11) as a potent and selective inhibitor of phosphoinositide 3-kinase-γ (PI3Kγ). Non-covalent binding was revealed in the crystal structure of the PI3Kγ-**24** complex.[57] PI3Kγ inhibitor **24** progressed to Phase II clinical trials as an anti-inflammatory agent, until being terminated due to toxicity.[57,58] It served, however, as a starting point for the discovery at GlaxoSmithKline of **25** (PI3Kα $IC_{50} = 2$ nM) and **26** (PI3Kγ $IC_{50} = 60$ pM, PI3Kα $IC_{50} = 19$ pM) (Figure 8.11), which progressed to clinical trials as anticancer agents.[59]

8.2.5 Marketed Drugs Containing the Rhodanine Scaffold

8.2.5.1 Aldose Reductase Inhibitors

Aldose reductase catalyses the reduction of glucose to sorbitol, the first step in the polyol pathway of glucose metabolism. Since the glucose flux through

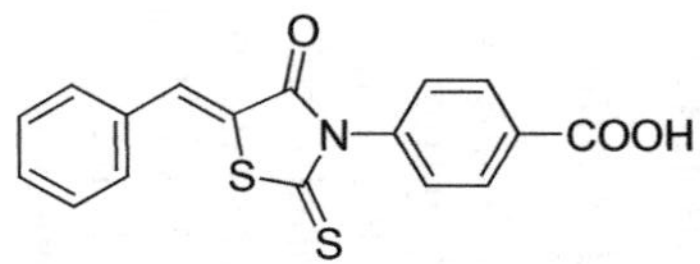

17, JSP-1 inhibitor

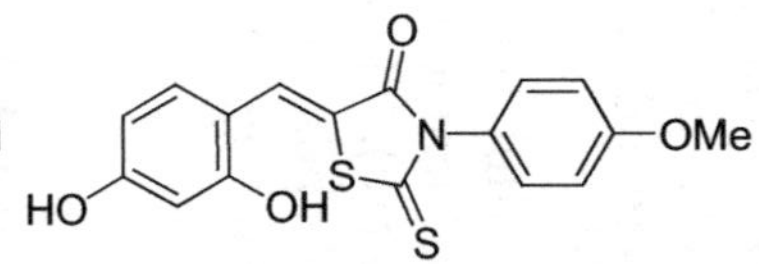

18, sphingosine kinase inhibitor

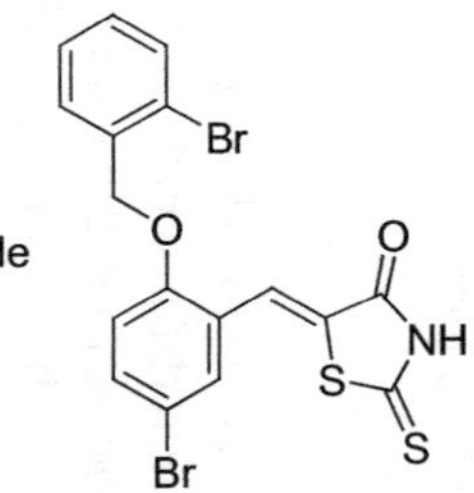

19, PRL-3 inhibitor

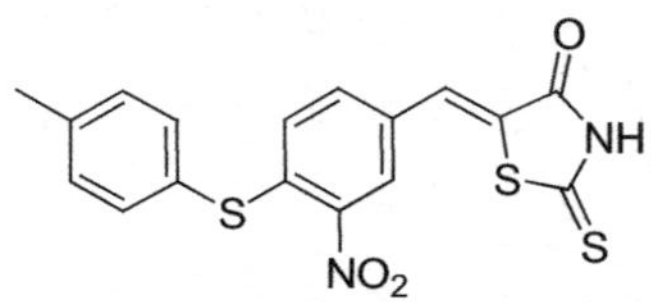

20, DNA polymerase λ

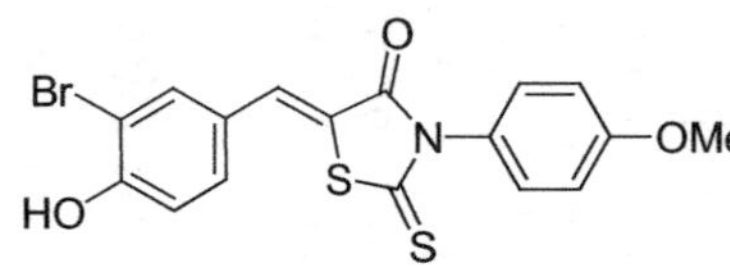

21, 17β-HSD3 inhibitor

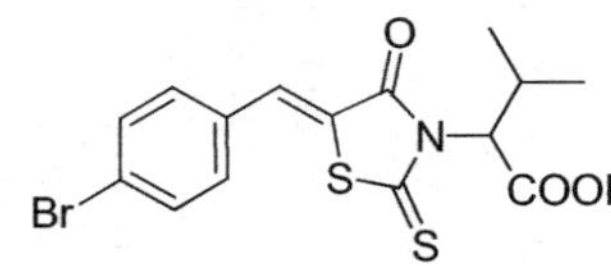

22, inhibitor of BH3 - Bcl-X_L interaction

Figure 8.10 Rhodanines targeting proteins involved in cancer pathogenesis.

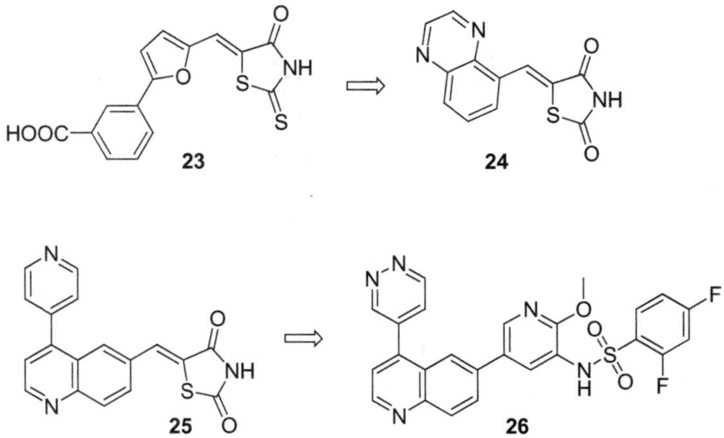

Figure 8.11 Representative phosphoinositide 3-kinase (PI3K) inhibitors **23–26.**

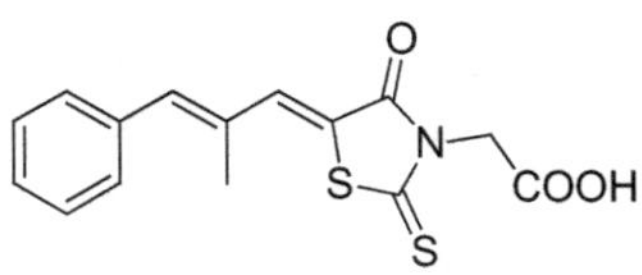

epalrestat

Figure 8.12 Epalrestat (ONO Pharmaceuticals).

the polyol pathway is significantly increased under chronic hyperglycaemic conditions such as in diabetes mellitus, aldose reductase is believed to be responsible for several complications in diabetes. **Epalrestat** (Figure 8.12) is an aldose reductase inhibitor used in the treatment of diabetic complications such as neuropathy, nephropathy and cataract. It was discovered by ONO Pharmaceuticals in 1982 in Japan,[60] and is approved there for the improvement of subjective neuropathy symptoms, abnormality of vibration sense, and abnormal changes in heart beat associated with diabetic peripheral neuropathy.[61]

8.2.5.2 PPAR-γ Agonists

While rhodanines are usually weakly active as agonists of peroxisome proliferator-activated receptor (PPAR)-γ, their closely related oxo analogues, thiazolidine-2,4-diones, are present in several drugs of this class.[21] PPAR-γ agonists are used in the treatment of insulin resistance in type 2 diabetes mellitus and of hyperlipidaemia in atherosclerosis. Structures of glitazones, incorporating the thiazolidine-2,4-dione ring, are presented in Figure 8.13.

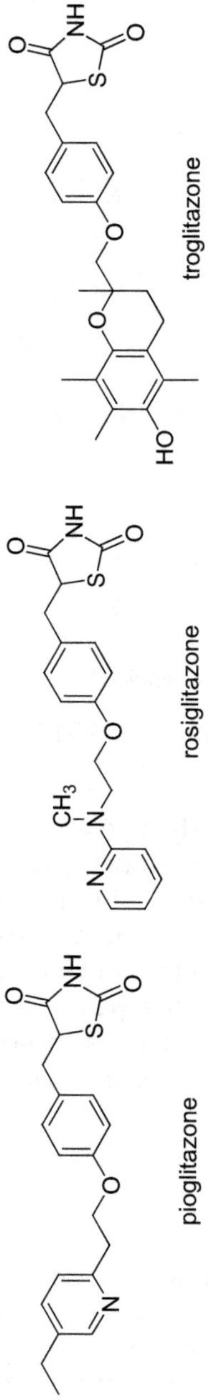

Figure 8.13 Pioglitazone (Takeda Pharmaceuticals), rosiglitazone (GlaxoSmithKline) and troglitazone (Daiichi Sankyo Co.).

References

1. M. Forino, S. Johnson, T. Y. Wong, D. V. Rozanov, A. Y. Savinov, W. Li, R. Fattorusso, B. Becattini, A. J. Orry, D. Jung, R. A. Abagyan, J. W. Smith, K. Alibek, R. C. Liddington, A. Y. Strongin and M. Pellecchia, *Proc. Natl. Acad. Sci. U. S. A.*, 2005, **102**, 9499–9504.
2. N. Zidar, T. Tomašič, R. Sink, V. Rupnik, A. Kovač, S. Turk, D. Patin, D. Blanot, C. Contreras Martel, A. Dessen, M. Müller Premru, A. Zega, S. Gobec, L. Peterlin Mašič and D. Kikelj, *J. Med. Chem.*, 2010, **53**, 6584–6594.
3. N. Zidar, T. Tomašič, R. Šink, A. Kovač, D. Patin, D. Blanot, C. Contreras-Martel, A. Dessen, M. M. Premru, A. Zega, S. Gobec, L. P. Mašič and D. Kikelj, *Eur. J. Med. Chem.*, 2011, **46**, 5512–5523.
4. T. Tomašič, N. Zidar, R. Šink, A. Kovač, D. Blanot, C. Contreras-Martel, A. Dessen, M. Müller-Premru, A. Zega, S. Gobec, D. Kikelj and L. P. Mašič, *J. Med. Chem.*, 2011, **54**, 4600–4610.
5. T. Tomašič, R. Šink, N. Zidar, A. Fic, C. Contreras-Martel, A. Dessen, D. Patin, D. Blanot, M. Müller-Premru, S. Gobec, A. Zega, D. Kikelj and L. P. Mašič, *ACS Med. Chem. Lett.*, 2012, **3**, 626–630.
6. T. Tomašič, N. Zidar, A. Kovač, S. Turk, M. Simčič, D. Blanot, M. Müller-Premru, M. Filipič, S. G. Grdadolnik, A. Zega, M. Anderluh, S. Gobec, D. Kikelj and L. Peterlin Mašič, *ChemMedChem*, 2010, **5**, 286–295.
7. T. Tomašič, A. Kovač, M. Simčič, D. Blanot, S. G. Grdadolnik, S. Gobec, D. Kikelj and L. Peterlin Mašič, *Eur. J. Med. Chem.*, 2011, **46**, 3964–3975.
8. T. Mendgen, C. Steuer and C. D. Klein, *J. Med. Chem.*, 2012, **55**, 743–753.
9. P. A. Wood, E. Pidcock and F. H. Allen, *Acta Crystallogr., Sect. B: Struct. Sci.*, 2008, **64**, 491–496.
10. T. Tomašič and L. Peterlin Mašič, *Expert Opin. Drug Discovery*, 2012, **7**, 549–560.
11. N. Zidar and D. Kikelj, *Acta Chim. Slov.*, 2011, **58**, 151–157.
12. PyMOL, Delano Scientific LLC, San Francisco, CA, http://pymol.sourceforge.net.
13. J. P. Powers, D. E. Piper, Y. Li, V. Mayorga, J. Anzola, J. M. Chen, J. C. Jaen, G. Lee, J. Liu, M. G. Peterson, G. R. Tonn, Q. Ye, N. P. Walker and Z. Wang, *J. Med. Chem.*, 2006, **49**, 1034–1046.
14. E. E. Carlson, J. F. May and L. L. Kiessling, *Chem. Biol.*, 2006, **13**, 825–837.
15. J. R. Huth, R. Mendoza, E. T. Olejniczak, R. W. Johnson, D. A. Cothron, Y. Liu, C. G. Lerner, J. Chen and P. J. Hajduk, *J. Am. Chem. Soc.*, 2005, **127**, 217–224.
16. J. R. Huth, D. Song, R. R. Mendoza, C. L. Black-Schaefer, J. C. Mack, S. A. Dorwin, U. S. Ladror, J. M. Severin, K. A. Walter, D. M. Bartley and P. J. Hajduk, *Chem. Res. Toxicol.*, 2007, **20**, 1752–1759.
17. C. Avonto, O. Taglialatela-Scafati, F. Pollastro, A. Minassi, V. Di Marzo, L. De Petrocellis and G. Appendino, *Angew. Chem. Int. Ed.*, 2011, **50**, 467–471.

18. E. Arsovska, J. Trontelj, N. Zidar, T. Tomašić, L. Peterlin Mašič, D. Kikelj, J. Plavec and A. Zega, *Acta Chim. Slov.*, 2014, **61**, 637–644.
19. A. C. Tripathi, S. J. Gupta, G. N. Fatima, P. K. Sonar, A. Verma and S. K. Saraf, *Eur. J. Med. Chem.*, 2014, **72**, 52–77.
20. R. B. Lesyk and B. S. Zimenkovsky, *Curr. Org. Chem.*, 2004, **8**, 1547–1577.
21. T. Tomašič and L. P. Mašič, *Curr. Med. Chem.*, 2009, **16**, 1596–1629.
22. J. B. Baell and G. A. Holloway, *J. Med. Chem.*, 2010, **53**, 2719–2740.
23. J. B. Baell, *Future Med. Chem.*, 2010, **2**, 1529–1546.
24. L. L. Silver, *Clin. Microbiol. Rev.*, 2011, **24**, 71–109.
25. M. M. Sim, S. B. Ng, A. D. Buss, S. C. Crasta, K. L. Goh and S. K. Lee, *Bioorg. Med. Chem. Lett.*, 2002, **12**, 697–699.
26. M. Simčič, K. Pureber, K. Kristan, U. Urleb, D. Kocjan and S. G. Grdadolnik, *Eur. J. Med. Chem.*, 2014, **83C**, 92–101.
27. Y. Hu, J. S. Helm, L. Chen, C. Ginsberg, B. Gross, B. Kraybill, K. Tiyanont, X. Fang, T. Wu and S. Walker, *Chem. Biol.*, 2004, **11**, 703–711.
28. A. Zervosen, W. P. Lu, Z. Chen, R. E. White, T. P. Demuth, Jr. and J. M. Frere, *Antimicrob. Agents Chemother.*, 2004, **48**, 961–969.
29. M. Soltero-Higgin, E. E. Carlson, J. H. Phillips and L. L. Kiessling, *J. Am. Chem. Soc.*, 2004, **126**, 10532–10533.
30. S. M. McKinnie, E. M. Rodriguez-Lopez, J. C. Vederas, J. M. Crowther, H. Suzuki, R. C. Dobson, T. Leustek, A. J. Triassi, M. S. Wheatley and A. O. Hudson, *Bioorg. Med. Chem.*, 2014, **22**, 523–530.
31. E. B. Grant, D. Guiadeen, E. Z. Baum, B. D. Foleno, H. Jin, D. A. Montenegro, E. A. Nelson, K. Bush and D. J. Hlasta, *Bioorg. Med. Chem. Lett.*, 2000, **10**, 2179–2182.
32. M. Brvar, A. Perdih, V. Hodnik, M. Renko, G. Anderluh, R. Jerala and T. Solmajer, *Bioorg. Med. Chem.*, 2012, **20**, 2572–2580.
33. M. M. Werner, Z. Li and R. J. Zauhar, *Bioorg. Med. Chem.*, 2014, **22**, 2176–2187.
34. D. Aiello, M. H. Barnes, E. E. Biswas, S. B. Biswas, S. Gu, J. D. Williams, T. L. Bowlin and D. T. Moir, *Bioorg. Med. Chem.*, 2009, **17**, 4466–4476.
35. N. Suree, S. W. Yi, W. Thieu, M. Marohn, R. Damoiseaux, A. Chan, M. E. Jung and R. T. Clubb, *Bioorg. Med. Chem.*, 2009, **17**, 7174–7185.
36. T. K. Beuria, P. Singh, A. Surolia and D. Panda, *Biochem. J.*, 2009, **423**, 61–69.
37. C. Zingle, D. Tritsch, C. Grosdemange-Billiard and M. Rohmer, *Bioorg. Med. Chem.*, 2014, **22**, 3713–3719.
38. K. Sudo, Y. Matsumoto, M. Matsushima, M. Fujiwara, K. Konno, K. Shimotohno, S. Shigeta and T. Yokota, *Biochem. Biophys. Res. Commun.*, 1997, **238**, 643–647.
39. W. T. Sing, C. L. Lee, S. L. Yeo, S. P. Lim and M. M. Sim, *Bioorg. Med. Chem. Lett.*, 2001, **11**, 91–94.
40. T. T. Talele, P. Arora, S. S. Kulkarni, M. R. Patel, S. Singh, M. Chudayeu and N. Kaushik-Basu, *Bioorg. Med. Chem.*, 2010, **18**, 4630–4638.
41. R. Dayam, T. Sanchez, O. Clement, R. Shoemaker, S. Sei and N. Neamati, *J. Med. Chem.*, 2005, **48**, 111–120.

42. A. R. Katritzky, S. R. Tala, H. Lu, A. V. Vakulenko, Q. Y. Chen, J. Sivapackiam, K. Pandya, S. Jiang and A. K. Debnath, *J. Med. Chem.*, 2009, **52**, 7631–7639.
43. X. Y. He, P. Zou, J. Qiu, L. Hou, S. Jiang, S. Liu and L. Xie, *Bioorg. Med. Chem.*, 2011, **19**, 6726–6734.
44. X. Y. He, L. Lu, J. Qiu, P. Zou, F. Yu, X. K. Jiang, L. Li, S. Jiang, S. Liu and L. Xie, *Bioorg. Med. Chem.*, 2013, **21**, 7539–7548.
45. C. Nitsche, V. N. Schreier, M. A. Behnam, A. Kumar, R. Bartenschlager and C. D. Klein, *J. Med. Chem.*, 2013, **56**, 8389–8403.
46. J. B. Baell, L. Ferrins, H. Falk and G. Nikolakopoulos, *Aust. J. Chem.*, 2013, **66**, 1483–1494.
47. B. T. Moorthy, S. Ravi, M. Srivastava, K. K. Chiruvella, H. Hemlal, O. Joy and S. C. Raghavan, *Bioorg. Med. Chem. Lett.*, 2010, **20**, 6297–6301.
48. D. Havrylyuk, L. Mosula, B. Zimenkovsky, O. Vasylenko, A. Gzella and R. Lesyk, *Eur. J. Med. Chem.*, 2010, **45**, 5012–5021.
49. N. S. Cutshall, C. O'Day and M. Prezhdo, *Bioorg. Med. Chem. Lett.*, 2005, **15**, 3374–3379.
50. K. J. French, R. S. Schrecengost, B. D. Lee, Y. Zhuang, S. N. Smith, J. L. Eberly, J. K. Yun and C. D. Smith, *Cancer Res.*, 2003, **63**, 5962–5969.
51. J. H. Ahn, S. J. Kim, W. S. Park, S. Y. Cho, J. D. Ha, S. S. Kim, S. K. Kang, D. G. Jeong, S. K. Jung, S. H. Lee, H. M. Kim, S. K. Park, K. H. Lee, C. W. Lee, S. E. Ryu and J. K. Choi, *Bioorg. Med. Chem. Lett.*, 2006, **16**, 2996–2999.
52. T. Strittmatter, B. Bareth, T. A. Immel, T. Huhn, T. U. Mayer and A. Marx, *ACS Chem. Biol.*, 2011, **6**, 314–319.
53. T. Strittmatter, A. Brockmann, M. Pott, A. Hantusch, T. Brunner and A. Marx, *ACS Chem. Biol.*, 2014, **9**, 282–290.
54. K. Harada, H. Kubo, A. Tanaka and K. Nishioka, *Bioorg. Med. Chem. Lett.*, 2012, **22**, 504–507.
55. A. Degterev, A. Lugovskoy, M. Cardone, B. Mulley, G. Wagner, T. Mitchison and J. Yuan, *Nat. Cell Biol.*, 2001, **3**, 173–182.
56. V. Pomel, J. Klicic, D. Covini, D. D. Church, J. P. Shaw, K. Roulin, F. Burgat-Charvillon, D. Valognes, M. Camps, C. Chabert, C. Gillieron, B. Francon, D. Perrin, D. Leroy, D. Gretener, A. Nichols, P. A. Vitte, S. Carboni, C. Rommel, M. K. Schwarz and T. Ruckle, *J. Med. Chem.*, 2006, **49**, 3857–3871.
57. M. Camps, T. Ruckle, H. Ji, V. Ardissone, F. Rintelen, J. Shaw, C. Ferrandi, C. Chabert, C. Gillieron, B. Francon, T. Martin, D. Gretener, D. Perrin, D. Leroy, P. A. Vitte, E. Hirsch, M. P. Wymann, R. Cirillo, M. K. Schwarz and C. Rommel, *Nat. Med*, 2005, **11**, 936–943.
58. D. F. Barber, A. Bartolome, C. Hernandez, J. M. Flores, C. Redondo, C. Fernandez-Arias, M. Camps, T. Ruckle, M. K. Schwarz, S. Rodriguez, A. C. Martinez, D. Balomenos, C. Rommel and A. C. Carrera, *Nat. Med.*, 2005, **11**, 933–935.
59. S. D. Knight, N. D. Adams, J. L. Burgess, A. M. Chaudhari, M. G. Darcy, C. A. Donatelli, J. I. Luengo, K. A. Newlander, C. A. Parrish, L. H. Ridgers,

M. A. Sarpong, S. J. Schmidt, G. S. Van Aller, J. D. Carson, M. A. Diamond, P. A. Elkins, C. M. Gardiner, E. Garver, S. A. Gilbert, R. R. Gontarek, J. R. Jackson, K. L. Kershner, L. Luo, K. Raha, C. S. Sherk, C. M. Sung, D. Sutton, P. J. Tummino, R. J. Wegrzyn, K. R. Auger and D. Dhanak, *ACS Med. Chem. Lett.*, 2010, **1**, 39–43.
60. H. Terashima, K. Hama, R. Yamamoto, M. Tsuboshima, R. Kikkawa, I. Hatanaka and Y. Shigeta, *J. Pharmacol. Exp. Ther.*, 1984, **229**, 226–230.
61. M. A. Ramirez and N. L. Borja, *Pharmacotherapy*, 2008, **28**, 646–655.

CHAPTER 9

Heterocycles Containing Nitrogen and Sulfur as Potent Biologically Active Scaffolds[†]

ANA MARTINEZ AND CARMEN GIL*

Centro de Investigaciones Biológicas (CSIC), Ramiro de Maeztu 9, 28040 Madrid, Spain
*Email: carmen.gil@csic.es

9.1 Introduction

Because one of the main objectives of organic and medicinal chemistry is the design, synthesis, and production of molecules having value as human therapeutic agents, privileged structures play an important role in the discovery of novel biologically active compounds.[1] Among these preferred molecular scaffolds that possess inherent biological activity, aromatic heterocycles have a central position. More than half of the known drugs that have been approved for the treatment of human diseases contain at least one heterocyclic component in their structure.[2] Also, nature makes many pharmacologically active compounds containing heteroaromatics rings. These privileged structures have been identified in many random screening assays and have been used as leads, not only to design new molecules that can mimic and enhance their pharmacological activity, but also as new hits for developing leads.[3]

[†]Dedicated to our dear Prof. José Elguero on the occasion of his anniversary.

RSC Drug Discovery Series No. 50
Privileged Scaffolds in Medicinal Chemistry: Design, Synthesis, Evaluation
Edited by Stefan Bräse

Published by the Royal Society of Chemistry, www.rsc.org

Heteroaromatic rings have been extensively used by medicinal chemists to design new drug candidates. They can mimic natural conformations of peptides or transition states in biologically relevant enzymatic reactions, increasing the activity of natural substrates.[4] Alternatively, the presence of a heteroatom in the ring has been shown to increase interactions with the receptor or the enzyme, and thus render greater pharmacological activity or specificity to resulting compounds.[5] Introduction of heteroaromatic rings into drug molecules also affects their physicochemical properties, which in turn can alter their absorption, distribution, metabolism, and excretion (ADME) profiles. For these reasons, attempts to correlate the effect of the change in the physicochemical properties such as log P/log D, acidity/basicity (pK_a), hydrogen bonding capability of molecules, and polar surface area (PSA) on their ADME properties have been done.[6]

Looking at the overall drug space listed in the Food and Drug Administration (FDA) orange book, only 351 diverse rings are present in drugs that came onto the market before 2013, as has recently been reported.[2] Among them, 325 frameworks (more than 92% of the total) included one or more heteroatom in their chemical structure, oxygen, nitrogen, and sulfur being the most common ones. These elements, ever present in the composition of living organisms, are suitable atoms to mimic biological activities being extensively used in drug design. The goal of this chapter is to summarize some relevant data about sulfur and nitrogen containing heterocycles as privileged structures for drug design.

The 10% of the top one hundred ring systems more frequently used in drugs approved by the FDA contain six- and five-membered rings with sulfur and nitrogen in their cyclic core.[2] Moreover, thirty more heterocycles containing sulfur and nitrogen atoms are also found in the overall 351 diverse rings present in drugs approved for the use in pharmacological treatments recently. All these privileged rings, which very often produce biologically active analogues in a target family, are depicted in Figure 9.1, and are the main focus of this chapter.

9.2 β-Lactams

One of the most important discoveries from the 20th century and, no doubt, the one with most impact on the health and life quality of human beings, is the discovery of Penicillin (Figure 9.2). This was attributed to the Scottish scientist and Nobel laureate Alexander Fleming in 1928. Penicillins, natural products isolated from different *Penicillium* spp., began the modern era of drug discovery. Their chemical structure was determined in 1945 and they are the most widely used antibiotics to date, and are still used for many Gram-positive bacterial infections.[7] Penicillin antibiotics were among the first drugs to be effective against many previously serious diseases, such as bacterial infections caused by staphylococci and streptococci, though misuse has now made many types of bacteria resistant.[8] All Penicillins are β-lactam antibiotics that means to have in their chemical structure a [4 + 5]

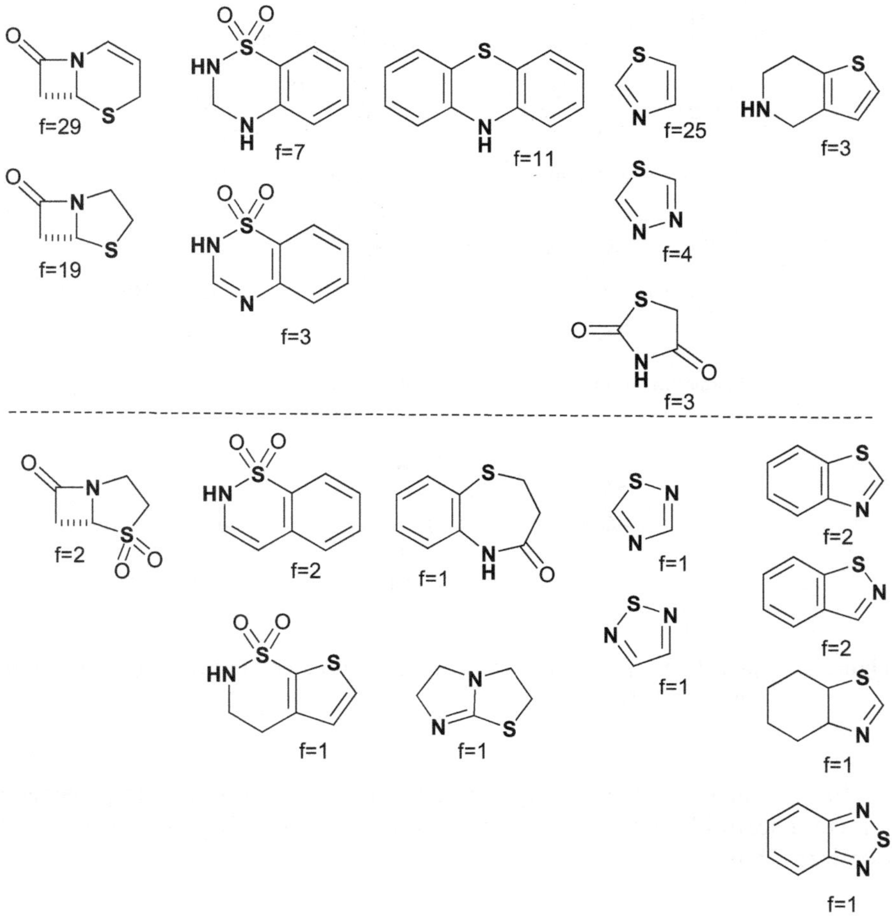

Figure 9.1 Sulfur and nitrogen containing heterocycles present in marketed drugs before 2013. Frequency of the same heterocycle in different drugs is noted.

condensed ring system with at least one nitrogen atom, containing in the cyclic amide bond, and not by chance but privileged by nature, one sulfur atom also.[9] In addition to Penicillins, the discovery of cephalosporines, isolated from *Cephalosporium acremonium* in 1948 as bactericidals less susceptible to β-lactamases than Penicillins, prompted the use of this privileged new [4 + 6] β-lactam scaffold in many antibiotic molecules.[10] The cephalosporin nucleus, 7-aminocephalosporanic acid (7-ACA), was derived from cephalosporin C and proved to be analogous to the Penicillin nucleus, 6-aminopenicillanic acid (6-APA), but it was not sufficiently potent for clinical use. Modification of the 7-ACA side chains resulted in the development of useful antibiotic agents, and the first agent, Cefalotin (Cephalothin), was launched by Eli Lilly and Company in 1964 and is actually in clinical use (Figure 9.2).[11]

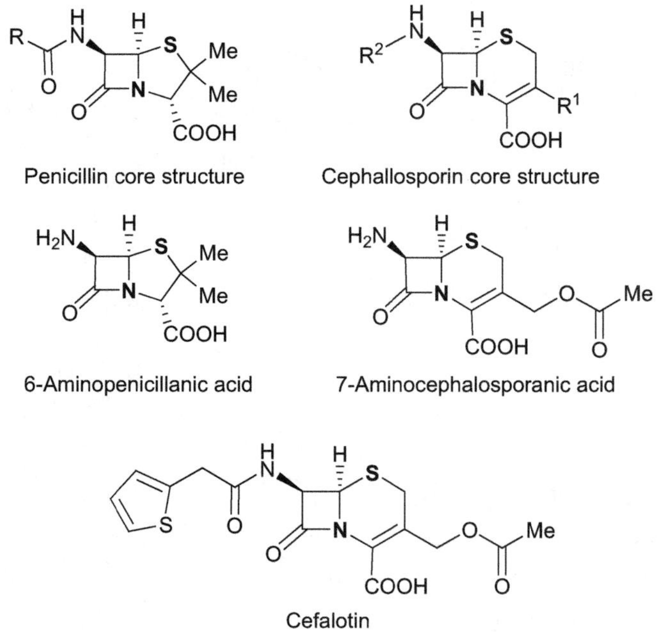

Figure 9.2 β-Lactams: privileged nitrogen and sulfur containing rings with antibiotic activity.

Since then, more than thirty new antibiotics containing this privileged β-lactam scaffold have been approved for human use. They are described in detail in Chapter 3.

9.3 Dioxides of Benzothiazines and Benzothiadiazines

Benzothiazines and benzothiadiazines are [6 + 6] sulfur and nitrogen containing heterocyclic compounds. Both families are known to represent a class of medicinally important heterocyclic compounds which are extensively used in drug design. They have wide biological properties which qualify them as excellent scaffolds in therapeutic and medicinal research. Thus, many derivatives of these compounds have been synthesized as target structures in novel drug development.

Although different isomers of these fused heterocycles, such as 1,4-benzothiazine[12,13] or 2,1,3-benzothiadiazine[14,15] derivatives have been reported to exhibit a wide range of pharmacological properties including antifungal, immunostimulating, anti-rheumatic, anti-allergic, vasorelaxant, anti-arrhythmic, antiviral, neuroprotective, and cytotoxic activities, none of these compounds have been approved for human use yet. However, 1,2-benzothiazines and 1,2,4-benzothiadiazine scaffolds are present in one of the main classes of non-steroidal anti-inflammatory drugs called "oxicams"[16] and in the thiazide class of diuretics.[17] Both families of drugs are extensively used in clinical pharmacology to treat inflammation, hypertension, and oedema.

9.3.1 Synthesis of Benzothiazines and Benzothiadiazines

1,2,4-Benzothiazines are typically prepared by reaction of amines with thioethers or sulfone derivatives (Scheme 9.1a and b). Because of the crucial importance of this heterocyclic skeleton, cyclization strategies by using solid phase synthesis have been reported. In one example, the heterocycle was obtained *via* a cyclative cleavage after reaction of a supported iminophosphorane with isocyanates (Scheme 9.1c).[18] Otherwise, the cyclization was achieved on solid support by reaction of previously synthesized sulfonamide resin with thiocarbonyldiimidazole (Scheme 9.1d).[19]

Regarding 1,2-benzothiazine, efforts to synthesize this ring system in a one-step approach (Scheme 9.1e),[20] instead of earlier multistep procedures, are remarkable.[21,22]

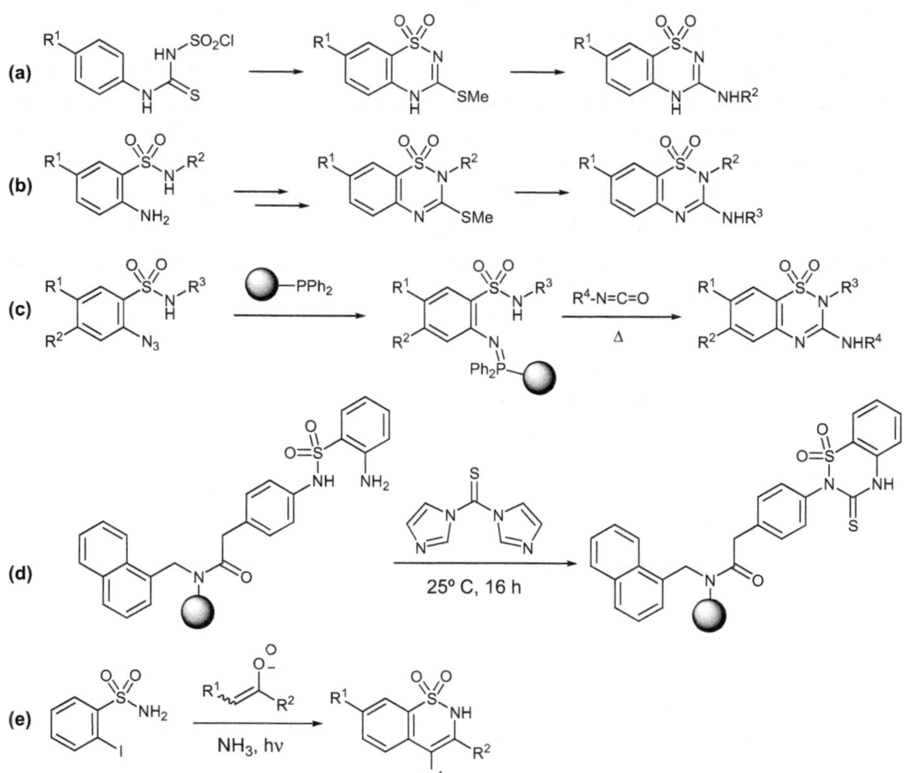

Scheme 9.1 General synthetic routes to obtain 1,2-benzothiazines and 1,2,4-benzothiazines.

9.3.2 Biological Activity of Benzothiazines and Benzothiadiazines

The 1,2-benzothiazine privileged scaffold is present in the widely used non-steroidal anti-inflammatory drugs (NSAIDs), called "oxicams".[16] They are unselective inhibitors of COX enzymes, being only meloxicam a slight (10 : 1) COX-2 inhibitor.[23] Only recently, the molecular basis of the interaction with their target enzymes has been reported.[24] There are several similar benzothiazines drugs that are clinically relevant: Piroxicam, Tenoxicam, Lornoxicam, and Meloxicam, which are manufactured by different pharmaceutical companies (Figure 9.3).

These drugs fall into the enolic acid group of NSAIDs. They contain a vinylogous carboxylic acid that exhibits a form of keto–enol tautomerism, given acidic properties to the whole molecule without being a carboxylic acid. They are used to relieve the symptoms of painful, inflammatory conditions such as arthritis by inhibition of cyclooxygenase (COX), the enzyme responsible for converting arachidonic acid into prostaglandin H_2, which is the first step in the synthesis of prostaglandins. Although selective COX-2 inhibitors are associated with a moderately increased risk of vascular events,[25] population exposure to systemic Piroxicam treatment remained unaffected by this health safety warning, but declined sharply after the introduction of prior authorization.[26] 1,2,4-Benzothiadiazine dioxide is the main privileged scaffold present in the so-called "thiazide" class of diuretics. They control hypertension by inhibiting reabsorption of sodium (Na^+) and chloride (Cl^-) ions from the distal convoluted tubules in the kidneys. They also target ATP-sensitive potassium channels[17] and increase calcium reabsorption at the distal tubule. The term "thiazide" is also often used for drugs with a similar action that do not have the thiazide chemical structure, such as Chlortalidone and Metolazone. These agents are more properly

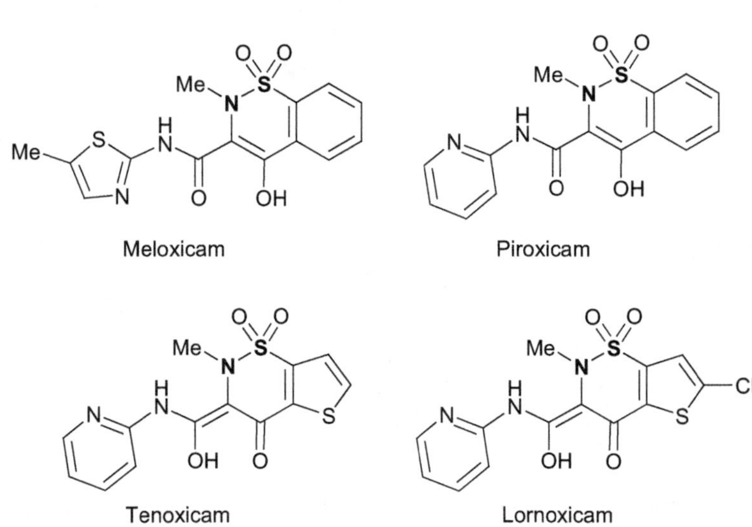

Figure 9.3 1,2-Benzothiazine anti-inflammatory drugs.

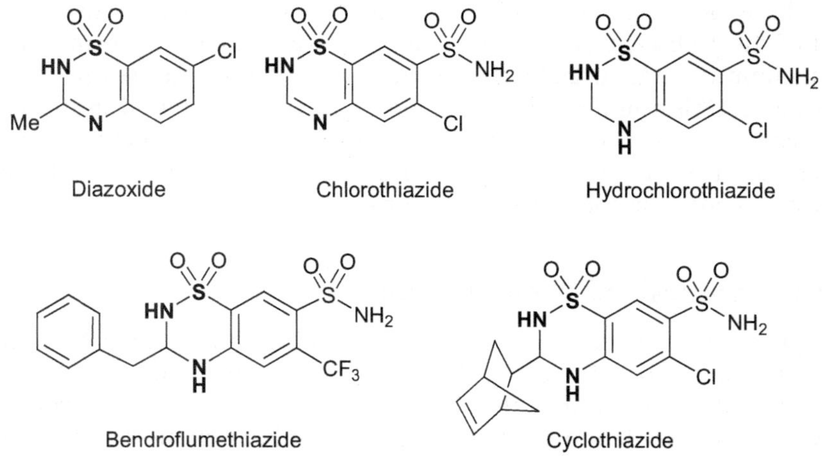

Figure 9.4 1,2,4-Benzothiadiazine containing drugs approved for human treatments.

termed thiazide-like diuretics. 1,2,4-Benzothioadiazine dioxides drugs such as Hydrochlorothiazide, Bendroflumethiazide, Chlorothiazide, Cyclothiazide, and Diazoxide, reduce the risk of death, stroke, heart attack, and heart failure due to hypertension, and are the cheapest antihypertensive drugs (Figure 9.4). In fact, Hydrochlorothiazide is on the World Health Organization's (WHO's) list of essential medicines, a list of the most important medication needed in a basic health system. It is frequently used for the treatment of hypertension, congestive heart failure, symptomatic oedema, diabetes insipidus, renal tubular acidosis, and the prevention of kidney stones.

Diazoxide is a potassium channel activator, which causes local relaxation in smooth muscle by increasing membrane permeability to potassium ions and it is used as a vasodilator in the treatment of acute hypertension or malignant hypertension.[27] As a potassium channel opener, Diazoxide may have a beneficial effect in diabetes treatment,[28] as it has been shown to control glycaemic levels in the first clinical trials with type I diabetic patients.[29] Diazoxide acts also as a positive allosteric modulator of the AMPA (α-amino-3-hydroxy-5-methyl-4-isoxazolepropionic acid) receptor, suggesting potential application as a cognitive enhancer.[30]

More recently, it was discovered that Cyclothiazide is a positive allosteric modulator of the AMPA receptor, and acts as a $GABA_A$ (γ-aminobutyric acid) receptor negative allosteric modulator. In animals, it is a powerful convulsant, robustly enhancing epileptiform activity and inducing seizures, but without producing any apparent neuronal death.[31]

9.4 Phenothiazines

Phenothiazine is a tricyclic fused [6 + 6 + 6] aromatic system that contains one sulfur and nitrogen atom, respectively, in the central ring. This privileged structure and its derivatives have held a prominent place in pharmacology and biomedicine, being the first family of antipsychotic agents.[32]

They have their origin in the development of the German dye industry, at the end of the 19th century.[33] Up to 1940, they were employed as antiseptics, antihelmintics, and antimalarials. Clinical use of *N*-substituted phenothiazines as antihistaminics, sedatives, and antipsychotics started in the 1940s and continues to this day. Recently, interest in these structures has re-emerged for a variety of diverse activities in relation to neurodegenerative diseases, and/or mycobacterial infections.

9.4.1 Synthesis of Phenothiazines

The phenothiazine core was first synthesized by Bernthsen in 1883 through the treatment of diphenylaniline with elemental sulfur at 250 °C (Scheme 9.2a).[34] Since then, the chemistry of phenothiazines has evolved in several directions due to the interest of the dye industry first and pharmaceutical companies later.[35,36] Almost all the phenothiazine syntheses depends upon the formation of the heterocyclic ring either from diaryl amines by thionation, or from two arenes that together bear the two heteroatoms. From the Methylene blue synthesis by vigorous heating of diphenylamine with elemental sulfur,[37] till the palladium-catalyzed three-component approach to Promazine (Scheme 9.2b),[38] stimulated research in this field has been done. Noteworthy is the fact that not only have ring closure reactions been investigated, but so have different reactions involving functional group insertion, removal or modifications.

9.4.2 Biological Activity of Phenothiazines

Phenothiazines belong to the oldest, synthetic antipsychotic drugs, which do not have their precursor in the world of natural compounds.[39] They are used to treat serious mental and emotional disorders, including schizophrenia and other psychotic disorders. Apart from their fundamental neuroleptic

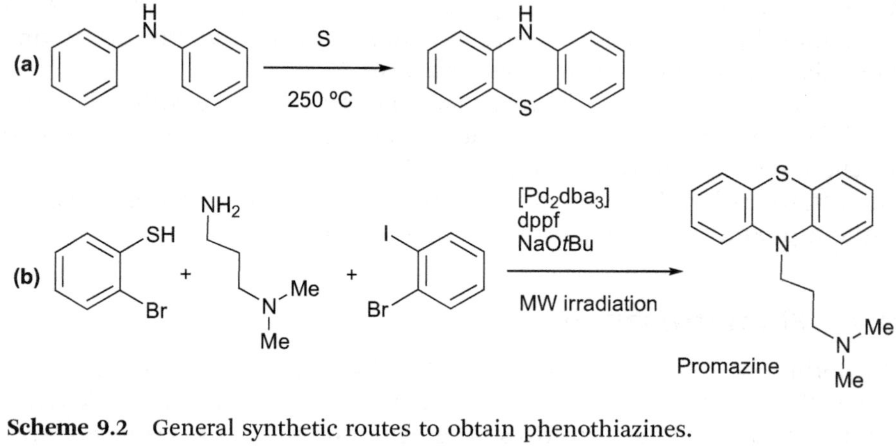

Scheme 9.2 General synthetic routes to obtain phenothiazines.

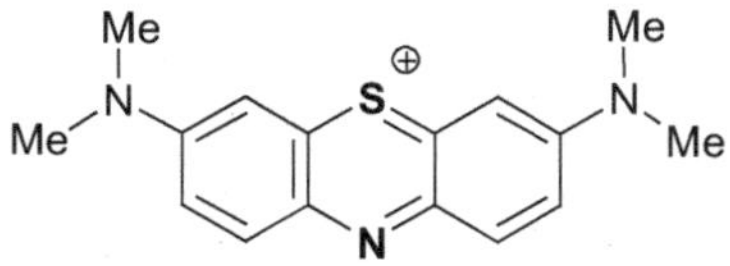

Methylene blue

Figure 9.5 Phenothiazine derivatives in clinical trials for Alzheimer's disease.

action connected with the dopaminergic receptors blockade, phenothiazine derivatives also exert diverse biological activities,[40] such as antihistaminic efficacy, calmodulin- and protein kinase C inhibitory-actions, anti-proliferative effect, inhibition of P-glycoprotein transport function and reversion of multidrug resistance.[41] They are also used as anti-emetics and to treat migraine and others moderate to severe pains in some hospitalized patients.[42] These activities were the result of the actions of phenothiazines on biological systems by the interaction of the multicyclic ring system (π–π interaction, intercalation in DNA) and the lipophilic character of the pharmacophoric substituent, in some cases of strict length, allowing penetration through the biological membranes.[43] The role of phenotiazines as inhibitors of type II NADH dehydrogenase, a key component of the respiratory chain of *Mycobacterium tuberculosis*, seems to be a rational option for the improvement of tuberculosis chemotherapy, including the recently emerged multi-, extensively, and totally drug resistant strains.[44] One of the most interesting new biological activities discovered for phenothiazines derivatives is their protein antiaggregation ability by intercalating between different peptidic monomers.[45] That is the case for methylthioninium chloride (Methylene blue) that has shown efficacy in preclinical models of Alzheimer's disease[46] and is being clinically developed for such neurodegenerative disorder[47] (Figure 9.5).

The discovery of the antipsychotic properties of Chlorpromazine in the 1950s was a fundamental event for the practice of psychiatry and for the genesis of the so-called "psychopharmacological revolution". Phenothiazine antipsychotics comprise more than 40 compounds grouped under three subtypes. Their biological activity is affected by substitutions at position 2 or 10 of the tricyclic scaffold. Based on the nature of side chains joined at the nitrogen atom of the middle ring, they are categorized into three subclasses: aliphatic, piperidine, and piperazine phenothiazines. The most relevant used drugs are depicted in Figure 9.6. Since phenothiazines are considerably less expensive than newer antipsychotics, they remain a valuable option in the treatment of psychotic disorders. Chlorpromazine is on the WHO's list of essential medicines.

9.5 Thiazoles and Thiazolidinones

Thiazole is a five-membered ring containing one sulfur and one nitrogen atom. Regarding the relative position of the heteroatoms in the cycle, there

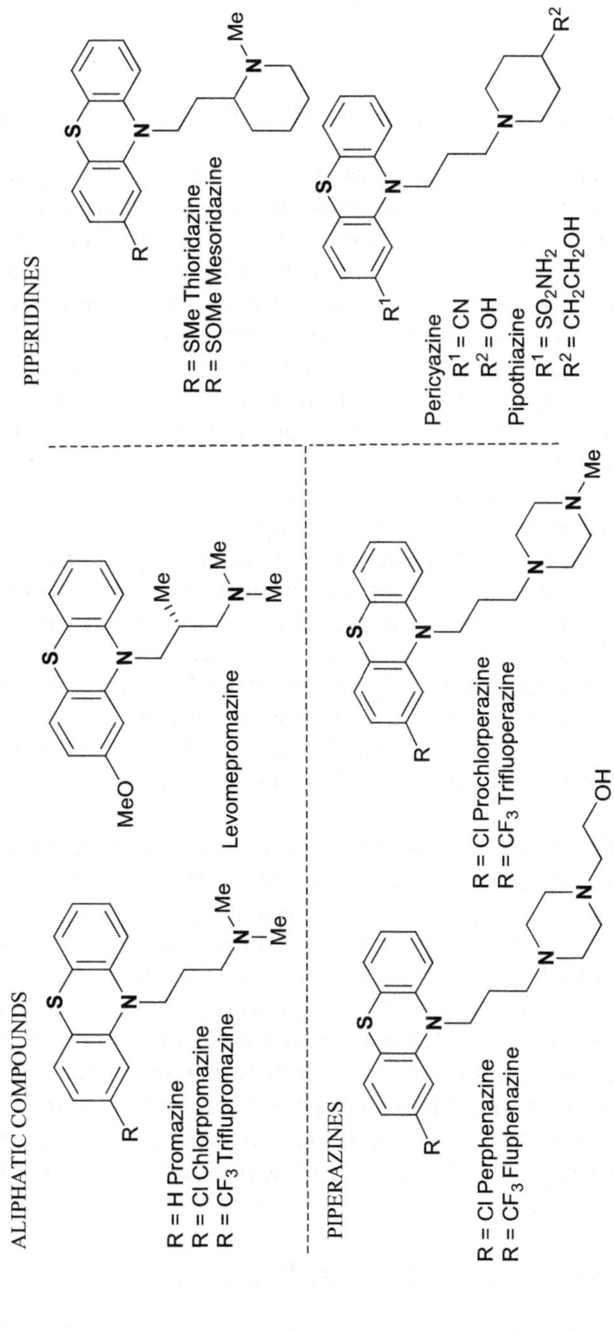

Figure 9.6 Phenothiazines: more common drugs.

are two isomers called 1,2- and 1,3-thiazoles. Thiazole can also be considered a functional group that is present in many organic molecules, 1,3-thiazole being a well-known isostere of pyridine. Thiazole is a planar, aromatic ring. Among their saturated forms, thiazolidinones, commonly named glitazones, are the most biologically relevant heterocycles.[48] Thiazoles are found in a variety of specialized products, being well represented in biomolecules such as vitamin B_1 and epothilone. Numerous natural products containing this heterocycle have been isolated and exhibit significant biological activities such as cytotoxic, immunosuppressive, antifungal, and enzyme inhibitory activity. Moreover, among the different aromatic heterocycles, thiadiazole occupies a prominent position in the drug discovery process and this ring structure is found in several marketed drugs.

9.5.1 Synthesis of Thiazoles

The most common method for the generation of 1,3-thiazoles was developed by Hantzsch in 1888 and is based on the condensation of α-haloketones and thioamides (Scheme 9.3a).[49] Other methods include the reaction of α-aminonitriles with carbon disulphide, such as the Cook–Heilbron synthesis (Scheme 9.3b)[50] or the reaction of α-acylaminoketones with the Lawesson's

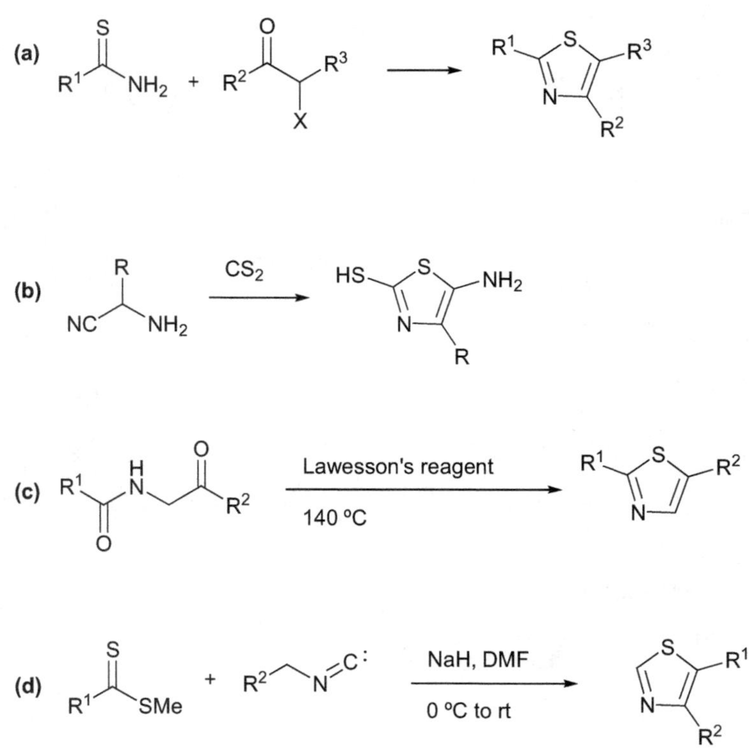

Scheme 9.3 General synthetic routes to obtain 1,3-thiazoles.

reagent or phosphorus pentasulfide in an adaptation of the Robinson-Gabriel synthesis (Scheme 9.3c).[51]

Furthermore, the biological interest of this heterocycle has prompted the development of efficient procedures as a base-induced click reaction of active methylene isocyanide with methyl dithiocarboxylates (Scheme 9.3d),[52] or the adaptation to fluorous synthesis in order to be able to obtain a large number of chemical libraries.[53]

9.5.2 Biological Activity of Thiazoles and Thiazolidinones

Thiazole and thiazolidinone heterocycles are the core structure in a variety of pharmaceuticals with a broad spectrum of biological activity including antioxidants,[54] analgesic, anti-inflammatory, antimicrobial, antifungal, antiviral, diuretic, anticonvulsant, neuroprotective, and antitumor or cytotoxic properties with fewer side effects. Commercial significant thiazoles include the fungicides Thiabendazole and Thifluzamide (Figure 9.7) or the

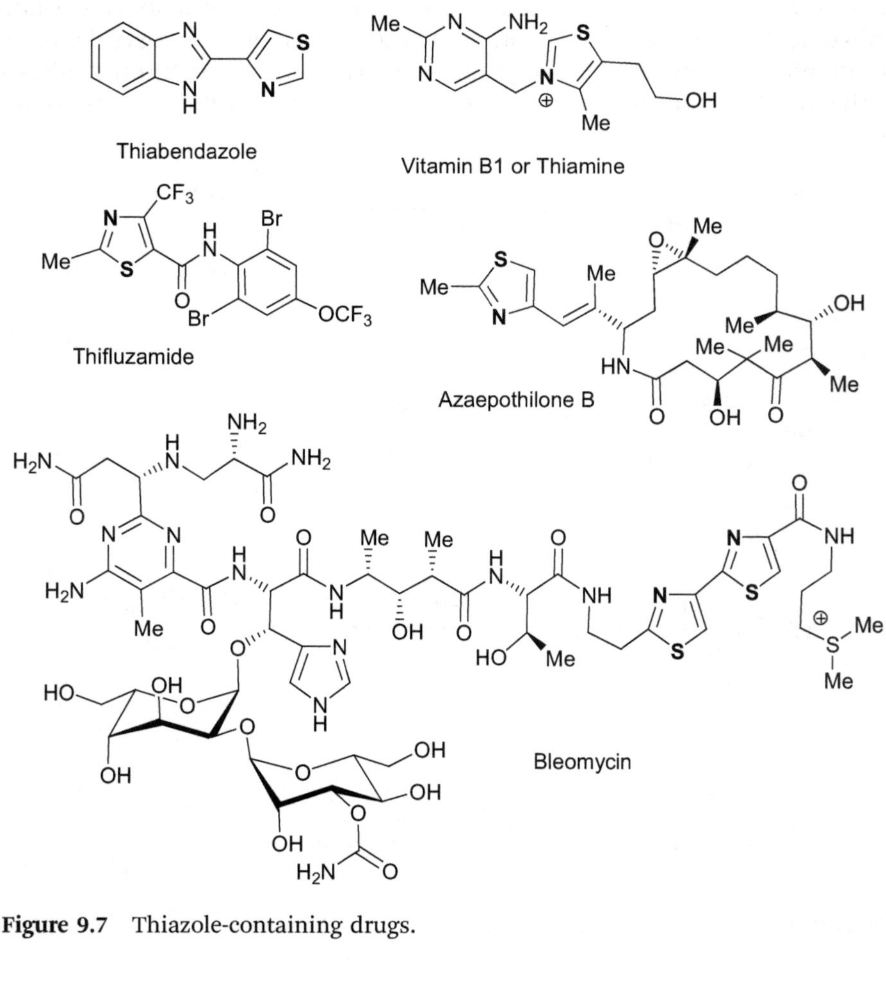

Figure 9.7 Thiazole-containing drugs.

widely used non-steroidal anti-inflammatory drug Meloxicam, described previously in this chapter (Figure 9.3). Moreover, some thiazole-containing molecules are present in natural products, such Bleomycin, a glycopeptide antibiotic produced by the bacterium *Streptomyces verticillus,* and Aza-epothilone B, produced by *Sorangium cellulosum* that induces apoptosis in cancer cells.[3] These compounds are two commercial drugs approved by the FDA for the treatment of Hodgkin's lymphoma, testicular cancer, and some carcinomas. Bleomycin acts by induction of DNA strand breaks and it is included on the WHO's list of essential medicines that are needed for basic health systems.

Moreover, in the last five years, some new drugs containing thiazole moiety have been approved by the FDA such as the protein kinase inhibitors Dasatinib and Dabrafenib, the proteases inhibitors Ritonavir and Simeprevir, and the β-adrenergic agonist, Mirabegron (Figure 9.8). Dasatinib and Dabrafenib are anti-cancer drugs that target different tyrosine kinases, such Abl/Src and B-Raf, respectively, being used as first line treatment for chronic myelogeneous leukaemia[55] or certain types of melanomas.[56] Ritonavir and Simeprevir are also thiazole-containing molecules approved by the FDA. While Ritonavir is an antiretroviral drug to treat HIV infections from the protease inhibitor class[57] with anti-inflammatory and anti-tubercular activity included on the WHO's list of essential medicines, Simeprevir is a drug for the treatment and cure of hepatitis C, which was approved by the FDA in 2013.[58] Mirabregon is a new drug for the treatment of overactive bladder acting on the β3 adrenergic-receptor in the detrusor muscle in the bladder.[59] These last examples show how thiazole derivatives are currently present in the discovery and development of effective drugs for human use, being a biologically privileged scaffold that continues to have great potential in chemical pharmaceutical research.[60,61]

Thiazolidinone is considered a biologically important active scaffold that possesses almost all types of biological activities. Successful introduction of Ralitoline as a potent anti-convulsant, Etozoline as an antihypertensive, and Pioglitazone as a hypoglycaemic agent proved the potential of thiazolidinone moiety (Figure 9.9). The thiazolidinedions is one of the most used scaffolds for the design of antidiabetic drugs. They are also known as glitazones and were introduced in the late 1990s to treat diabetes type II. Thiazolidinediones act by activating peroxisome proliferator-activated receptors, a group of nuclear receptors, with greatest specificity for the gamma isoform.[62] Chemically, the members of this class are derivatives of the parent compound thiazolidinedione, and include Rositiglitazone, Pioglitazone and Troglitazone, which was withdrawn from the market due to an increased incidence of drug-induced hepatitis. Replacing one oxygen atom in the carbonyl group attached to the thiazolidinedione heterocycle by an atom of sulfur gives Rhodanine, another privileged biological scaffold containing nitrogen and sulfur in its main core, which is discussed in detail in Chapter 8.[63] The new therapeutic activities found in these compounds open

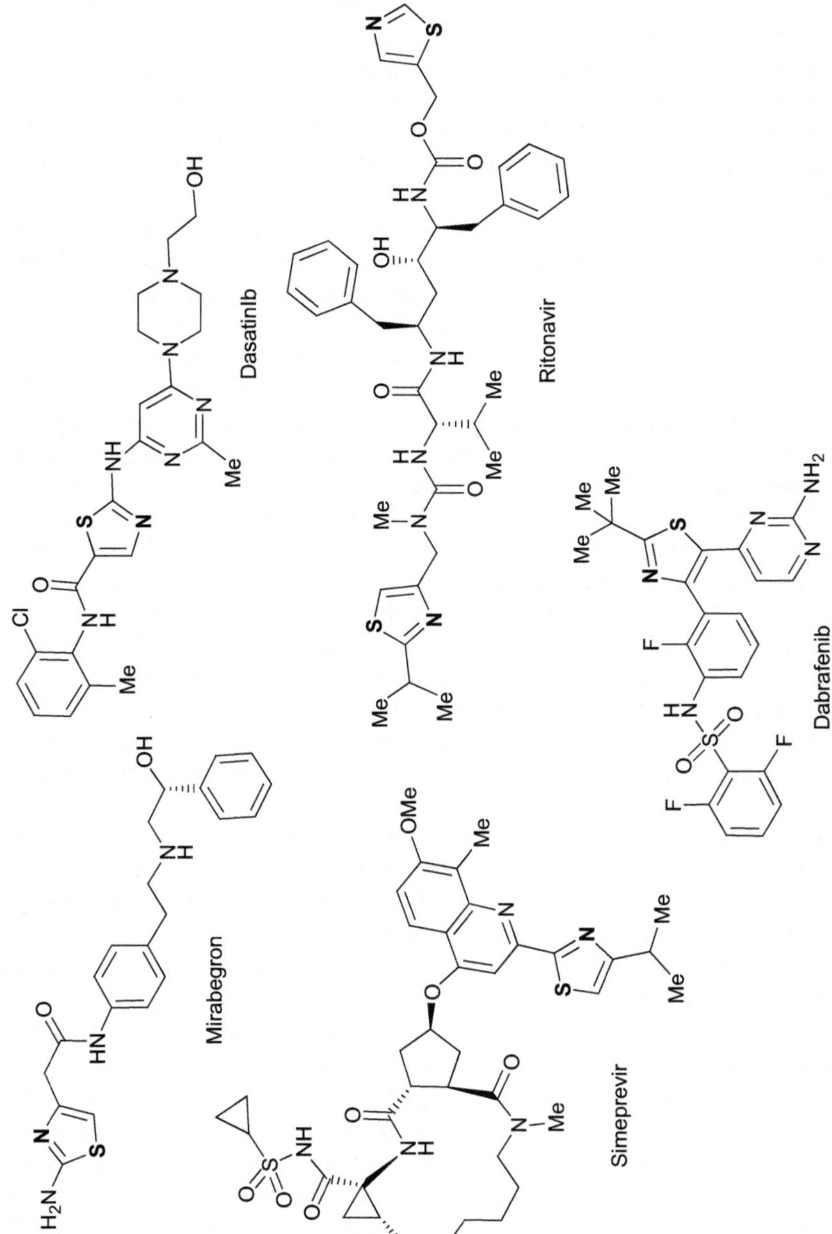

Figure 9.8 Recent thiazole-containing drugs approved by the FDA.

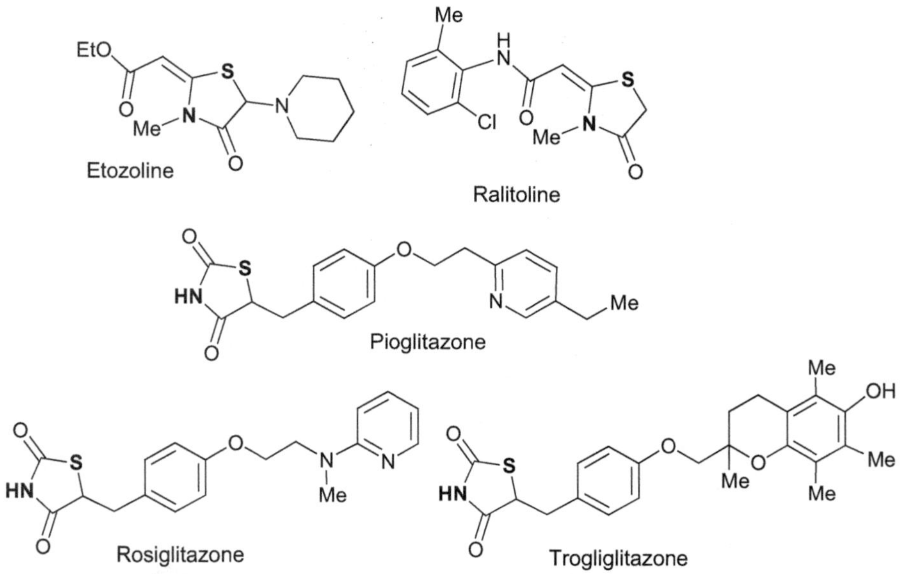

Figure 9.9 Drugs containing thiazolidinone scaffold.

new perspectives for the use of these drugs in severe unmet diseases as cancer[64] or Alzheimer's disease,[65] and the further development of other related derivatives.[66]

9.6 Benzothiazoles

In the family of heterocyclic compounds, benzothiazole ring has assumed special significance in synthetic chemistry and pharmaceutical chemistry, as well as in clinical applications. Benzothiazole, a group of compounds containing a benzene ring fused with a thiazole ring, are used worldwide for a variety of therapeutic applications.[67]

9.6.1 Synthesis of Benzothiadiazoles

Several methods for the synthesis and cyclization of benzothiazoles and its derivatives have been reported, such as the Hofmann method, Jacobson synthesis or oxidation by bromide, among others.[68] However, due to the importance of this chemical entity to the medicinal chemistry field, the development of new procedures for the rapid construction of libraries with high degree of structural diversity is desirable. In this sense, it is important to mention some attempts to synthesize this heterocycle on solid phase. Combinatorial synthesis through condensation of an aldehyde with the immobilized *o*-substituted aniline (Scheme 9.4a),[69] use of alkoxyamine linker (Scheme 9.4b)[70] or even a traceless solid supported protocol have been described (Scheme 9.4c).[71]

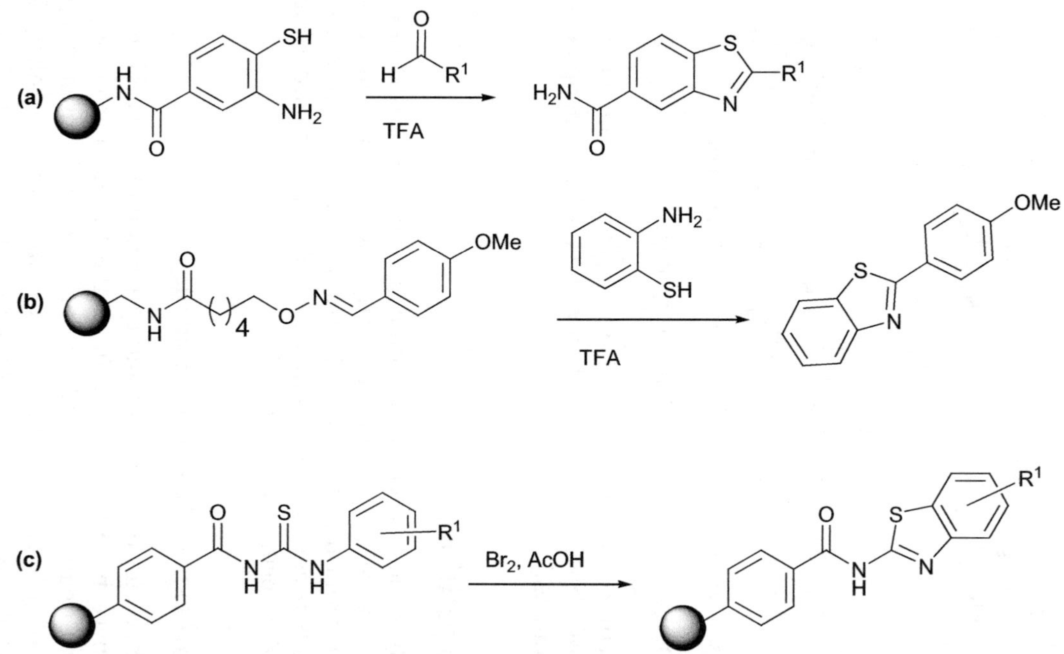

Scheme 9.4 General synthetic routes to obtain benzothiazoles on solid supports.

9.6.2 Biological Activity of Benzothiazoles

Although benzothiazole derivatives are relatively rare in nature, there are some important examples in the literature.[72] The well-known firefly compound Luciferin, which was isolated in the late 1940s,[73] possess a benzothiazole moiety, as do the natural products isolated from bacterial sources as Rifamycins P and Q,[74] the Thiazinotrienomycin F and G[75] or Erythrazoles A and B[76] (Figure 9.10). As far as we know, the only member of the benzothiazole class FDA-approved for human use, is Riluzole. Riluzole has antiglutamatergic action and blocks Tetrodotoxin-sensitive sodium channels, which are associated with damaged neurons. It is approved for the palliative treatment of amyotrophic lateral sclerosis (Figure 9.11).[77] Recently, the privileged biological activity of benzothiazoles has been used in the development of positron emission tomography (PET) radioligands for the non-invasive imaging of amyloid-β plaque burden being several radiolabelled molecules in later stages of drug development (phase II/III clinical trial studies) (Figure 9.12). These molecules will be provided not only a suitable diagnostic imaging agent but also a means to evaluate potential therapies for Alzheimer's disease.[78]

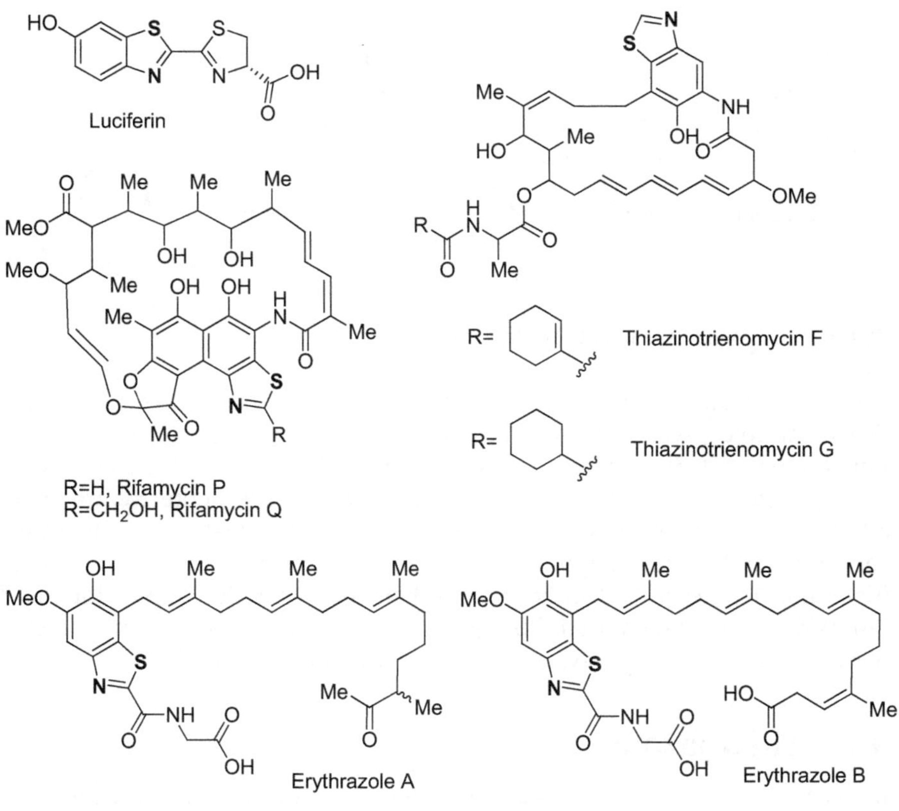

Figure 9.10 Benzothiazole-containing natural products.

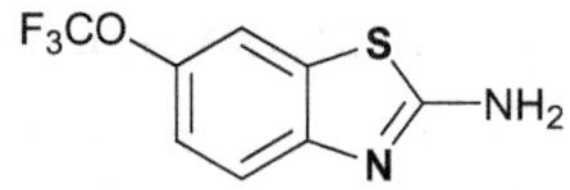

Figure 9.11 Chemical structure of Riluzole.

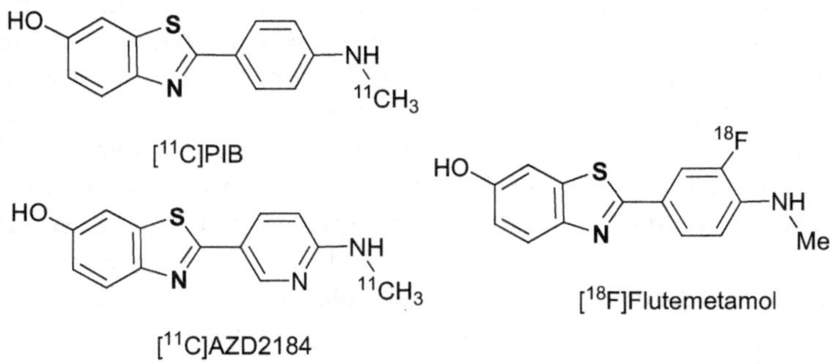

Figure 9.12 Benzothiazoles as PET radioligands.

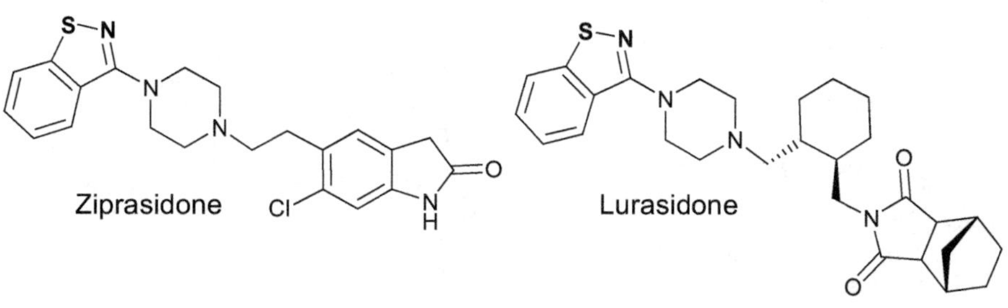

Figure 9.13 Benzoisothiazole containing drugs.

Finally, benzoisothiazole, the main benzothiazole isomer, is also a privileged scaffold present in two drugs acting on the central nervous system: Lurasidone and Ziprasidone (Figure 9.13). Both sulfur and nitrogen containing heterocycles are approved by the FDA for the acute treatment of adults with schizophrenia.[79] They are agonists for the most of the dopamine, serotonin and α_1-adrenergic receptors, having lower binding affinities for histamine H_1 receptor. Unlike many other antipsychotics, Lurasidone and Ziprasidone lack any anticholinergic side effects, improving their effectiveness in treating affective symptomatology and cognitive deficits.

9.7 Thiadiazoles

In recent decades, research has indicated that the thiadiazole ring is an important framework with broad-spectrum biological activity.[80] There are

1,2,4-thiadiazole 1,3,4-thiadiazole 1,2,3-thiadiazole 1,2,5-thiadiazole

Figure 9.14 Thiadiazole rings.

four types of thiadiazole: 1,3,4-, 1,2,4-, 1,2,5- and 1,2,3-thiadiazole (Figure 9.14). Among them, the most fully investigated of them are the 1,2,4- and 1,3,4-thiadiazoles. Although there is a large number of thiadiazole derivatives known for their pharmacological properties, there are only a few examples of thiadiazole-containing drugs currently on the market. Until now, none of the currently available drugs contain the 1,2,3-thiadiazole moiety.

9.7.1 Synthesis of 1,2,4- and 1,3,4-Thiadiazoles

Chemical synthetic procedures for these versatile frameworks have been reviewed in the last few years. A number of methods have been developed for the synthesis of 1,2,4- or 1,3,4-thiadiazole[81,82] and also synthetic strategies for the generation of these privileged scaffolds using resin-bound substrates have been described.[83]

For example, the main synthetic procedure to obtain 1,2,4-thiadiazoles usually includes an oxidation step of compounds containing a thioamide group by using oxidizing agents (Scheme 9.5a). Besides oxidation, thermolysis and rearrangements have allowed the synthesis of this ring. In fact, starting from oxathiadiazoles,[84] tetrazoles[85] or furoxans,[86] 1,2,4-thiadiazoles with different degree of substitution have been generated (Scheme 9.5b–d). Regarding solid-phase synthesis, cyclization of carboxamide thiourea resins achieved by using *p*-toluenesulfonyl chloride (TsCl) as activating agent yielded the 1,2,4-thiadiazole (Scheme 9.5e).[87]

Different synthetic methods have also been employed for the preparation of 1,3,4-thiadiazole. However, *S*-heterocycle synthesis suffers from drawbacks as moderate to low yields, by-product formation, long reaction times at elevated temperatures, and/or use of a large excess of the sulfurization reagent among others. For these reasons, one-pot reported procedures could represent a facile approach to overcome these drawbacks. In this sense, starting from acylhydrazines, a one-pot synthesis from carboxylic acids using propylphosphonic anhydride (T3P) as an efficient reagent was described (Scheme 9.6a).[88] Another way to simplify the generation of this ring is the synthesis under microwave irradiation mediated by Lawesson's reagent (Scheme 9.6b),[89] or by the interconversion of the corresponding 1,3,4-oxadiazoles with thiourea. (Scheme 9.6c).[90] 1,3,4-Thiadiazoles has been obtained on solid support *via* cyclization of acyldithiocarbazates (Scheme 9.6d).[91]

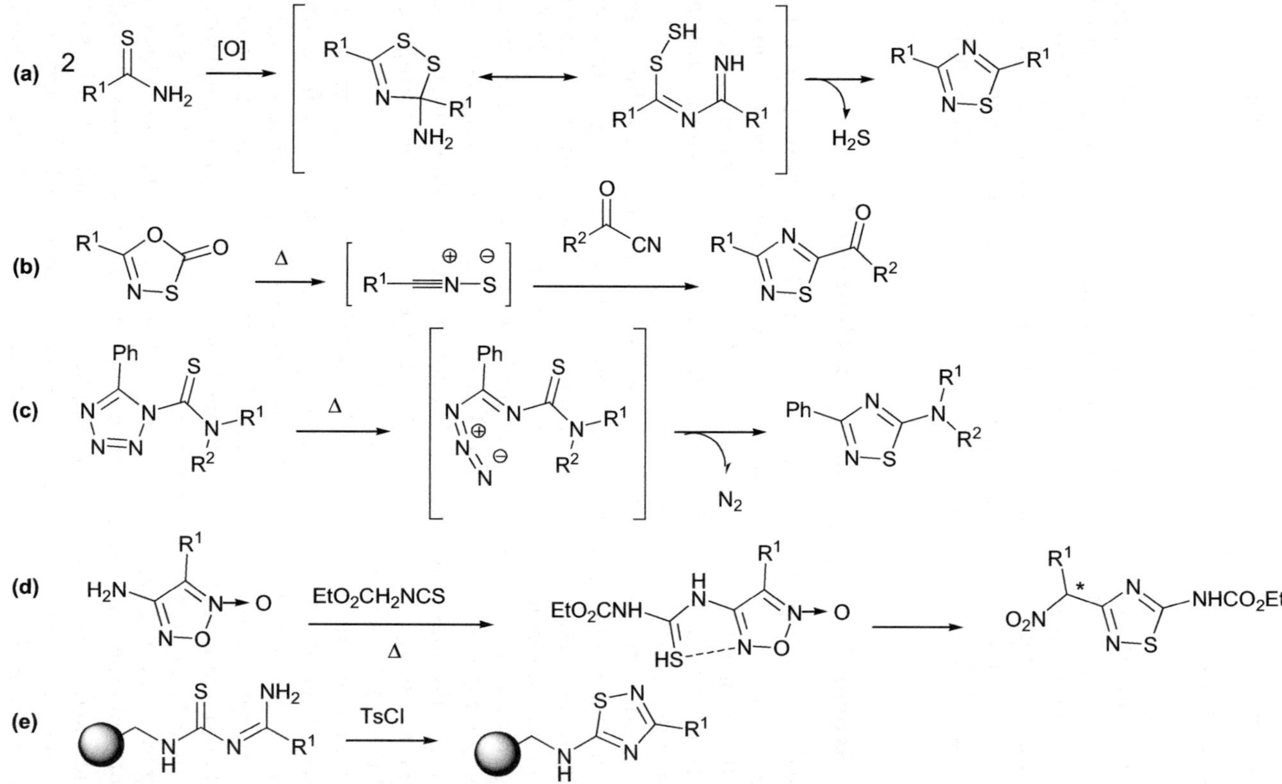

Scheme 9.5 General synthetic routes to obtain 1,2,4-thiadiazole.

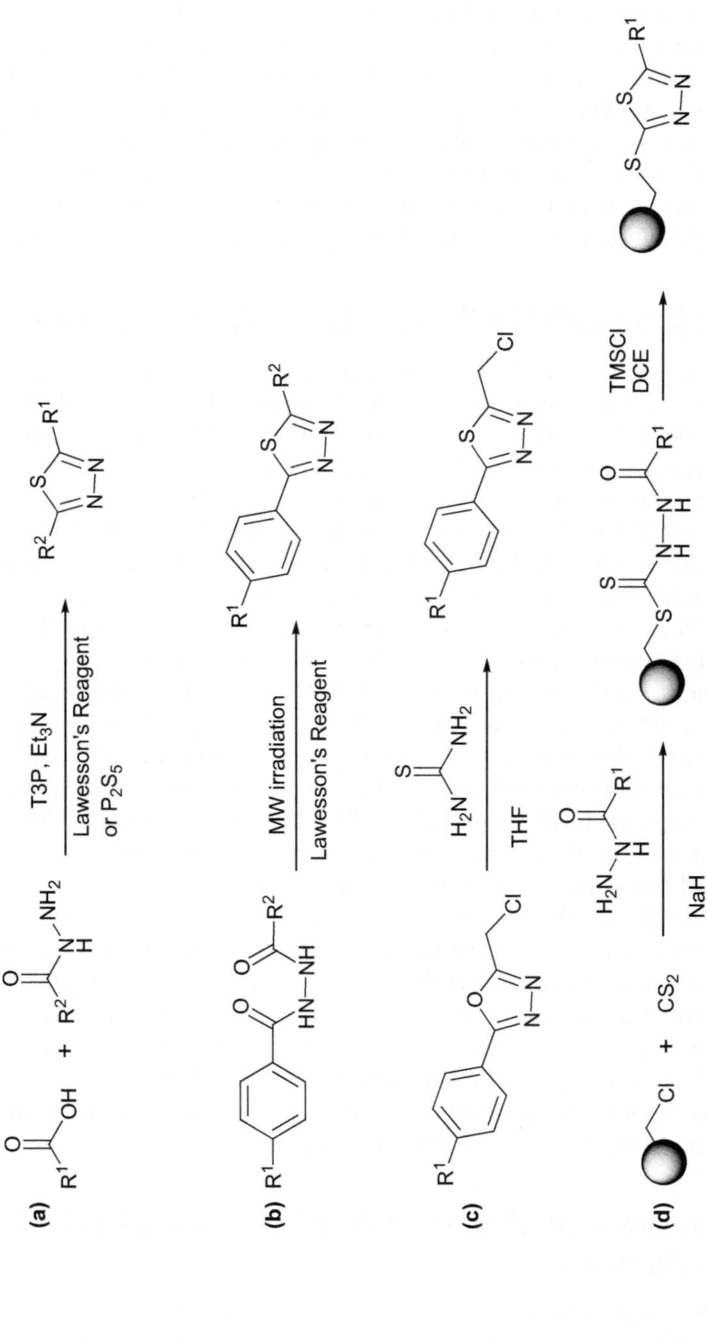

Scheme 9.6 General synthetic routes to obtain 1,3,4-thiadiazole rings.

9.7.2 Biological Activity of 1,2,4- and 1,3,4-Thiadiazoles

Thiadiazole-based structural scaffolds form an essential constituent of some synthetic drugs exhibiting a wide spectrum of biological activities as anti-cancer, anti-inflammatory, antibacterial, antifungal, antiviral, anticonvulsant, and antiparasitic activities.[92] The β-lactams antibiotics Cefozopram, Cefazolin sodium, and Cefazedone incorporate in their chemical structure one of these privileged heterocycles. Moreover, 1,3,4-thiadiazole is a privileged structure to inhibit carbonic anhydrases being Acetazolamide or Methazolamide used as diuretics to treat hypertension and/or glaucoma.[93,94] Other important drugs such the antiparasitic agent Megazol,[95] or the non-selective β-adrenergic blocker Timolol[96] are also approved for human use (Figure 9.15).

9.8 Thiadiazolidindiones (TDZDs): A Case Study

Finally, as a case study, it is worth mentioning the role of thiadiazolidindione heterocyclic ring as a privileged scaffold for the inhibition of glycogen synthase kinase 3 (GSK-3), a unique kinase involved in many unmet severe human diseases such as diabetes type II, bipolar disorders, cancer, and neurodegenerative diseases.[97] The small heterocyclic thiadiazolidindiones (TDZDs) were the first ATP non-competitive GSK-3 inhibitors reported in the literature. Two members of this family, named TDZD-8 and Tideglusib, have achieved particular relevance in the field. TDZD-8, commercially available from different sources, has been one of the most useful pharmacological tools in the chemical genetic approach followed by many scientists to explore GSK-3 functions. On the other hand, Tideglusib is on clinical trials for different neurodegenerative disorders where tau phosphorylation plays a key role, such as Alzheimer's disease or progressive supranuclear palsy.[98]

The kinase inhibitory activity of TDZDs was discovered in a GSK-3 target program initiated in the late 1990's. The previously characterized protein kinase C inhibitor Ro31-8220 was reported to inhibit GSK-3 in December 1999,[99] and based on similarities between the chemical structure of TDZDs and Ro31-8220, mainly the 1,3-dicarbonyl moiety in a five-member ring with a nitrogen atom between both carbonyls groups (Figure 9.16), several side products obtained from the synthesis of biological active compounds, such as potassium channel openers,[100] or acetylcholinesterase inhibitors,[101] were included in the screening program. Three out of the four tested compounds had an IC_{50} (the compound concentration that inhibits 50% of the enzyme activity) in the low micromolar range. Kinetic studies revealed the ATP-non competition of TDZDs in their GSK-3 inhibition.[102]

9.8.1 Synthesis of Thiadiazolidindiones and Hit-to-lead Optimization

The TDZD synthesis pathway is based on the reactivity of *N*-alkyl-*S*-[*N*′-(chlorocarbonyl)amino]isothiocarbamoyl chlorides with isocyanates (Scheme 9.7).[103]

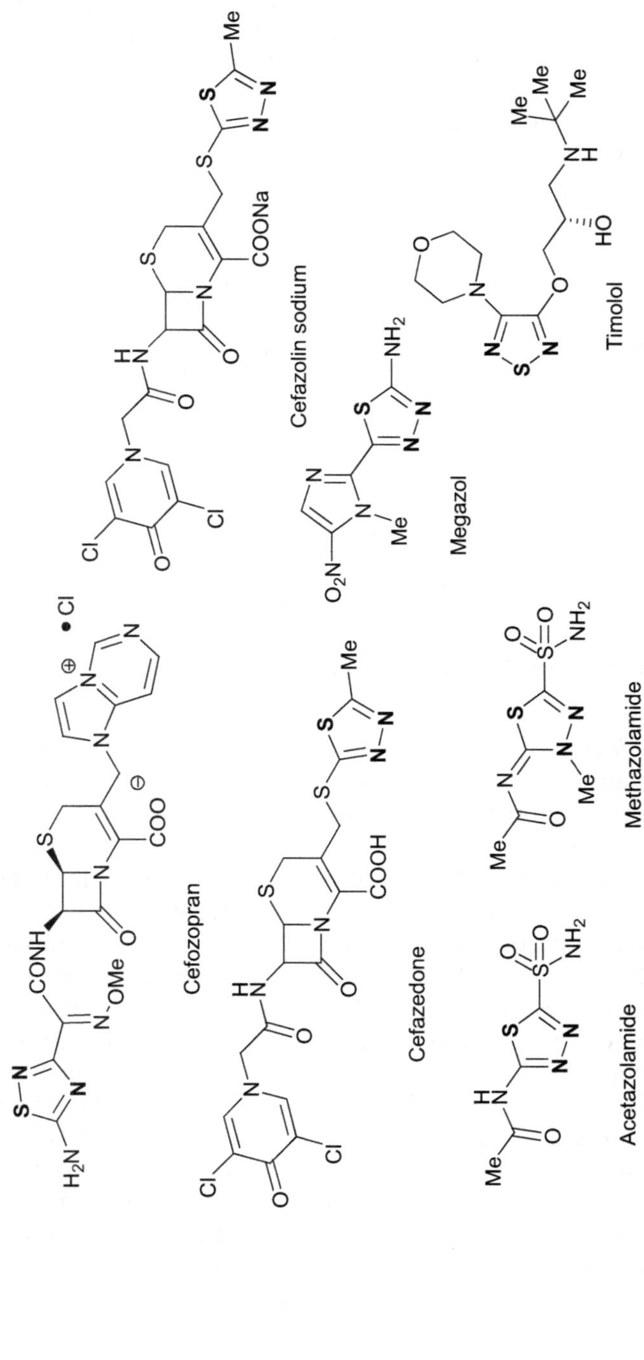

Figure 9.15 Thiadiazole-containing drugs available in the market.

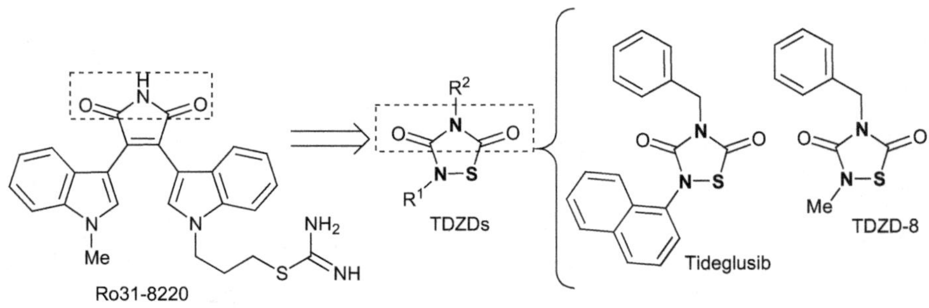

Figure 9.16 From Ro31-8220 to thiadiazolidindiones (TDZDs).

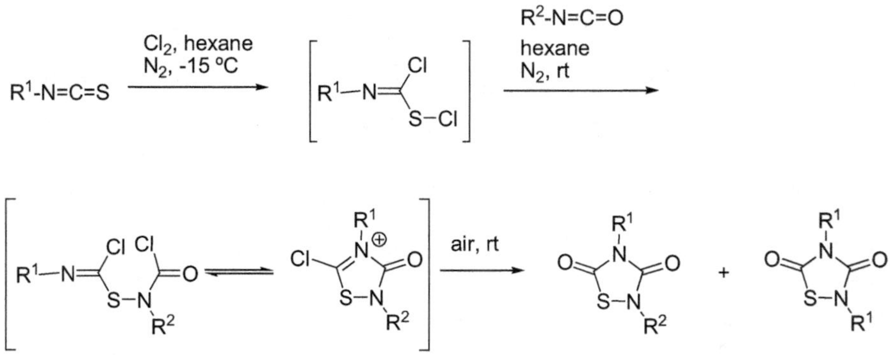

Scheme 9.7 General synthetic route to obtain TDZDs.

These intermediate heterocyclic salts are exceptionally reactive, and in the presence of moist air and *via* hydrolysis, it was possible to obtain the 1,2,4-thiadiazolidine-3,5-diones (TDZDs) as white crystalline solids after evolution of hydrogen chloride.

After the discovery of TDZDs as ATP non-competitive inhibitors of GSK-3, the structure activity relationships (SAR) studies were initiated to define crucial chemical features required for inhibition and, more importantly, to identify the best candidate for further therapeutic development. In the first approach, the nature of alkyl and aryl moieties attached to nitrogen atoms at positions 2 and 4 in the TDZD framework and the influence of the two carbonyl groups were studied.[102] Secondly, different structural modifications were introduced in the heterocyclic TDZD ring with the aim to test the influence of each heteroatom on biological activity.[104] The only compounds containing nitrogen and sulfur atoms were GSK-3 inhibitors and a crucial role for the sulfur atom in modulating the inhibitory activity against GSK-3 in this framework was confirmed. From the results of the GSK-3 inhibitory activity assessment and the SAR study performed with the heterocyclic families, TDZDs were identified as a privilege scaffold for the selective inhibition of GSK-3.

The selection of Tideglusib as a clinical candidate was based not only on its specific activity, but also in its pharmacokinetic properties. As Alzheimer's disease (AD) was the planned therapeutic target, the selected candidate should not only be able to cross the blood brain barrier (BBB), but should also be orally bioavailable for convenient administration to patients. These properties, together with a half-life time compatible with a single daily administration in humans were the guide from lead-to-candidate selection.[105]

9.8.2 Biological Activity of Thidiazolidindiones

TDZDs were discovered as ATP-non competitive GSK-3 inhibitors. To determine the target selectivity of this heterocyclic family, their effects on other kinases were studied.

Cyclin-dependent kinases (CDKs), in particular, CDK5 and CDK2, are the closest homologous kinases to GSK-3 (overall 33% amino acid identity). Consequently, several of the described synthetic small molecules that inhibit GSK-3 are also used to inhibit CDKs.[106] TDZD-8 did not significantly affect the activities of CDK-2 and other protein kinases such as CDK-1/cyclin B, PKA, CK-II and PKC.[102] Moreover, to determine the selectivity of kinase inhibition of this series of compounds, more than twenty TDZDs were tested against a wide panel of related kinases with no resulting significant inhibitory effect,[104] indicating a remarkable specificity of TDZDs with respect to GSK-3 inhibition.

In preclinical studies, Tideglusib has abolished the AD-phenotype in a double transgenic mice model.[107] Three months of chronic oral treatment with this thiadiazolidinone compound resulted in lower levels of tau phosphorylation, decreased amyloid deposition and plaque-associated astrocytic proliferation, protection of neurons in the entorhinal cortex and CA1 hippocampal subfield against cell death, and prevention of memory deficits in APPxtau transgenic mouse model. Moreover, Tideglusib has shown an important neuroprotective effect using a kainate excitotoxicity model,[108] and it is able to increase the potent neurotrophic peptide, insulin growth factor 1, on mice brains, both on wild type and APPxPS1 mice, after oral treatment.[109]

Recent clinical studies have shown that Tideglusib is a safe compound, being able to decrease the brain atrophy in a dose-dependent manner[110] and to improve cognition in mild-to-moderate AD patients.[111]

Moreover, as GSK-3 is involved in the regulation of several signal transduction pathways, whose deregulation has been implicated in several severe diseases, the therapeutical potential of TDZDs have been explored in different cellular and animal models being effective agents in important diseases as Parkinson's disease,[112] multiple sclerosis,[113] spinal cord injury,[114] arthritis,[115] or fragile X.[116] All these data support the potential therapeutic role for TDZDs and in particular for Tideglusib as pharmacologic treatment of mood disorders, and inflammatory or neurodegenerative diseases.

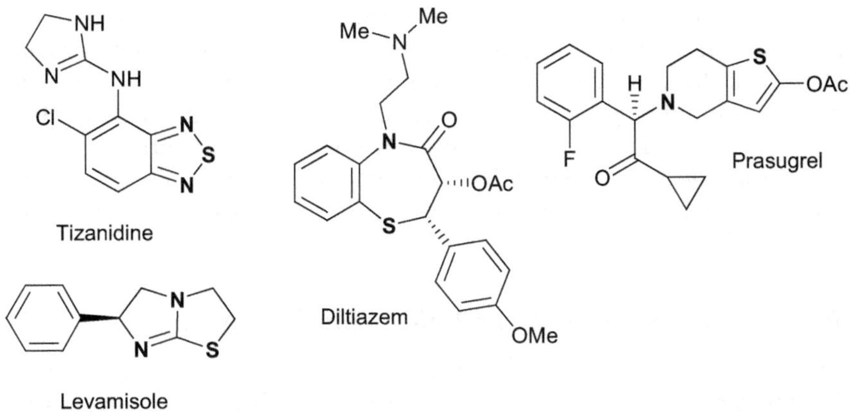

Figure 9.17 Chemical structures of Tizanidine, Levamisole, Diltiazem and Prasugrel.

9.9 Miscellaneous

Finally, it is worthwhile to mention that several diverse sulfur and nitrogen containing heterocycles are present in marketed drugs. That it is the case for the muscle relaxant Tizanidine, the antihelmintic Levamisole, the calcium channel antagonist Diltiazem or the platelet antiaggregant Prasugrel (Figure 9.17).

Tizanidine[117] is a central adrenergic agonist used as a muscle relaxant to treat the spasms and tightness of muscles caused by medical problems such as multiple sclerosis or certain other injuries to the spine or central nervous system. It is also prescribed off-label for migraine headaches and fibromyalgia. Levamisole[118] was discovered in 1966 and was used to treat infections with parasitic worms. Currently, evamisole remains in use in veterinary medicine only, as a dewormer for livestock. Prasugrel,[119] used to treat acute coronary syndromes, is a member of the thienopyridine class of adenosine diphosphate receptor inhibitors. These agents reduce the aggregation of platelets by irreversibly binding to $P2Y_{12}$ receptors. In fact, Prasugrel is a prodrug and its ester group is rapidly metabolized to the pharmacologically acidic active metabolite. Finally, Diltiazem[120] is a benzothiazepine used in the treatment of hypertension, angina pectoris, and some types of arrhythmia. It is a calcium channel antagonist with potent coronary and peripheral vessel vasodilator properties, and it is also an effective preventive medication for migraine.

9.10 Conclusion

Among different privileged scaffolds for drug discovery, five- and six-membered rings containing sulfur and nitrogen atoms have a prominent role. They are usually found in naturally occurring compounds and in many synthetic approved drugs for the treatment of human diseases. Although many therapeutic activities have been reported for them, the action of these compounds on central nervous system deserves a special mention. Phenothiazines and several five-membered rings such as thiazoles, thiadiazoles, thiazolidinones, or thiadiazolidinones are present in different antipsychotic drugs and currently in development for severe neurodegenerative diseases such as Alzheimer's disease or amyotrophic lateral sclerosis, offering hope for the future.

Acknowledgements

Financial support from MINECO (projects nos. SAF2012-37979-CO3-01 and SAF2012-33600) is acknowledged.

References

1. M. E. Welsch, S. A. Snyder and B. R. Stockwell, *Curr. Opin. Chem. Biol.*, 2010, **14**, 347.
2. R. D. Taylor, M. MacCoss and A. D. Lawson, *J. Med. Chem.*, 2014, **57**, 5845.
3. Q. Zhang, H. Zhou, S. Zhai and B. Yan, *Curr. Pharm. Des.*, 2010, **16**, 1826.
4. R. W. DeSimone, K. S. Currie, S. A. Mitchell, J. W. Darrow and D. A. Pippin, *Comb. Chem. High Throughput Screening*, 2004, **7**, 473.
5. J. B. Sperry and D. L. Wright, *Curr. Opin. Drug Discovery Dev.*, 2005, **8**, 723.
6. D. Dalvie, P. Kang, C.-M. Loi, L. Goulet and S. Nair, in *Metabolism, Pharmacokinetics and Toxicity of Functional Groups*, ed. D. A. Smith, Royal Society of Chemistry, Cambridge, 2010, p. 328.
7. J. P. Hou and J. W. Poole, *J. Pharm. Sci.*, 1971, **60**, 503.
8. R. R. Watkins, K. M. Papp-Wallace, S. M. Drawz and R. A. Bonomo, *Front. Microbiol.*, 2013, **4**, 392.
9. D. A. Nichols, A. R. Renslo and Y. Chen, *Future Med. Chem.*, 2014, **6**, 413.
10. J. M. Hamilton-Miller and W. Brumfitt, *Infection*, 1975, **3**, 183.
11. T. E. Long and J. T. Williams, *Expert Opin. Invest. Drugs*, 2014, 1.
12. O. O. Ajani, *Arch. Pharm.*, 2012, **345**, 841.
13. R. Fringuelli, L. Milanese and F. Schiaffella, *Mini-Rev. Med. Chem.*, 2005, **5**, 1061.
14. A. Martinez, A. Castro, C. Gil, M. Miralpeix, V. Segarra, T. Domenech, J. Beleta, J. M. Palacios, H. Ryder, X. Miro, C. Bonet, J. M. Casacuberta, F. Azorin, B. Piña and P. Puigdomenech, *J. Med. Chem.*, 2000, **43**, 683.
15. A. Martinez, C. Gil, C. Perez, A. Castro, C. Prieto, J. Otero, G. Andrei, R. Snoeck, J. Balzarini and E. De Clercq, *J. Med. Chem.*, 2000, **43**, 3267.
16. E. H. Wiseman and J. G. Lombardino, *Eur. J. Rheumatol. Inflammation*, 1981, **4**, 280.
17. B. Pirotte, P. de Tullio, S. Boverie, C. Michaux and P. Lebrun, *J. Med. Chem.*, 2011, **54**, 3188.

18. C. Blackburn, A. Achab, A. Elder, S. Ghosh, J. Guo, G. Harriman and M. Jones, *J. Org. Chem.*, 2005, **70**, 10206.
19. S. Makino, T. Okuzumi, E. Nakanishi and T. Tsuji, *Tetrahedron Lett.*, 2002, **43**, 8401.
20. W. J. Layman Jr., T. D. Greenwood, A. L. Downey and J. F. Wolfe, *J. Org. Chem.*, 2005, **70**, 9147.
21. E. Sianesi, R. Redaelli, M. J. Magistretti and E. Massarani, *J. Med. Chem.*, 1973, **16**, 1133.
22. J. G. Lombardino and E. H. Wiseman, *J. Med. Chem.*, 1971, **14**, 973.
23. N. M. Davies and N. M. Skjodt, *Clin. Pharmacokinet.*, 1999, **36**, 115.
24. S. Xu, D. J. Hermanson, S. Banerjee, K. Ghebreselasie, G. M. Clayton, R. M. Garavito and L. J. Marnett, *J. Biol. Chem.*, 2014, **289**, 6799.
25. P. M. Kearney, C. Baigent, J. Godwin, H. Halls, J. R. Emberson and C. Patrono, *Br. Med. J.*, 2006, **332**, 1302.
26. E. Carracedo-Martinez, A. Pia-Morandeira and A. Figueiras, *Pharmacoepidemiol. Drug Saf.*, 2012, **21**, 281.
27. W. A. Coetzee, *Pharmacol. Ther.*, 2013, **140**, 167.
28. V. Grill, M. Radtke, E. Qvigstad, M. Kollind and A. Bjorklund, *Diabetes, Obes. Metab.*, 2009, **11**(Suppl 4), 143.
29. M. A. Radtke, I. Nermoen, M. Kollind, S. Skeie, J. I. Sorheim, J. Svartberg, I. Hals, T. Moen, G. H. Dorflinger and V. Grill, *Diabetes Care*, 2010, **33**, 589.
30. B. Pirotte, P. Francotte, E. Goffin, P. Fraikin, L. Danober, B. Lesur, I. Botez, D. H. Caignard, P. Lestage and P. de Tullio, *Curr. Med. Chem.*, 2010, **17**, 3575.
31. S. Kong, B. Qian, J. Liu, M. Fan, G. Chen and Y. Wang, *Brain Res.*, 2010, **1355**, 207.
32. M. J. Ohlow and B. Moosmann, *Drug Discovery Today*, 2011, **16**, 119.
33. F. Lopez-Muñoz, C. Alamo, E. Cuenca, W. W. Shen, P. Clervoy and G. Rubio, *Ann. Clin. Psychiatry*, 2005, **17**, 113.
34. A. Bernthsen, *Ber. Dtsch. Chem. Ges.*, 1883, **16**, 2896.
35. S. M. Massie, *Chem. Rev.*, 1954, **54**, 797.
36. I. A. Silberg, G. Cormos and D. C. Oniciu, in *Advances in Heterocycles Chemistry*, ed. A. R. Katritzky, Academic Press, 2006, vol. 90, p. 205.
37. A. Bernthsen, *Ber. Dtsch. Chem. Ges.*, 1883, **16**, 2896.
38. D. Troels, C. W. Tornøe, B. Bang-Andersen, P. Nielsen and M. Jørgensen, *Angew. Chem., Int. Ed.*, 2008, **47**, 1726.
39. A. Jaszczyszyn, K. Gasiorowski, P. Swiatek, W. Malinka, K. Cieslik-Boczula, J. Petrus and B. Czarnik-Matusewicz, *Pharmacol. Rep.*, 2012, **64**, 16.
40. G. Sudeshna and K. Parimal, *Eur. J. Pharmacol.*, 2010, **648**, 6.
41. I. Tsakovska and I. Pajeva, *Curr. Drug Targets*, 2006, **7**, 1123.
42. A. Saguil and J. W. Lax, *Am. Fam. Physician*, 2014, **89**, 742.
43. K. Pluta, B. Morak-Mlodawska and M. Jelen, *Eur. J. Med. Chem.*, 2011, **46**, 3179.
44. S. Sharma and A. Singh, *Expert Opin. Invest. Drugs*, 2011, **20**, 1665.
45. E. Akoury, M. Pickhardt, M. Gajda, J. Biernat, E. Mandelkow and M. Zweckstetter, *Angew. Chem., Int. Ed.*, 2013, **52**, 3511.

46. V. Paban, C. Manrique, M. Filali, S. Maunoir-Regimbal, F. Fauvelle and B. Alescio-Lautier, *Neuropharmacology*, 2014, **76 Pt A**, 68.
47. C. Wischik and R. Staff, *J. Nutr., Health Aging*, 2009, **13**, 367.
48. A. K. Jain, A. Vaidya, V. Ravichandran, S. K. Kashaw and R. K. Agrawal, *Bioorg. Med. Chem.*, 2012, **20**, 3378.
49. A. Hantzsch, *Ber. Dtsch. Chem. Ges.*, 1888, **21**, 942.
50. A. H. Cook, I. Heilbron, S. F. MacDonald and A. P. Mahadevan, *J. Chem. Soc.*, 1949, 1064.
51. I. Thomsen, U. Pedersen, P. B. Rasmussen, B. Yde, T. P. Andersen and S. O. Lawesson, *Chem. Lett.*, 1983, **6**, 809.
52. G. S. Lingaraju, T. R. Swaroop, A. C. Vinayaka, K. S. S. Kumar, M. P. Sadashiva and K. S. Rangappa, *Synthesis*, 2012, **44**, 1373.
53. J. F. Sanz-Cervera, R. Blasco, J. Piera, M. Cynamon, I. Ibanez, M. Murguia and S. Fustero, *J. Org. Chem.*, 2009, **74**, 8988.
54. A. A. Geronikaki, E. P. Pitta and K. S. Liaras, *Curr. Med. Chem.*, 2013, **20**, 4460.
55. M. Lindauer and A. Hochhaus, *Recent Results Cancer Res.*, 2014, **201**, 27.
56. P. Rutkowski and C. Blank, *Expert Opin. Drug Saf.*, 2014, **13**, 1249.
57. S. L. Becker, *Expert Opin. Invest. Drugs*, 2003, **12**, 401.
58. A. S. deLemos and R. T. Chung, *Trends Mol. Med.*, 2014, **20**, 315.
59. U. Leone Roberti Maggiore, L. Cardozo, S. Ferrero, F. Sileo, A. Cola, F. Del Deo, M. Torella, N. Colacurci, M. Candiani and S. Salvatore, *Expert Opin. Pharmacother.*, 2014, **15**, 873.
60. A. Leoni, A. Locatelli, R. Morigi and M. Rambaldi, *Expert Opin. Ther. Pat.*, 2014, **24**, 201.
61. A. Leoni, A. Locatelli, R. Morigi and M. Rambaldi, *Expert Opin. Ther. Pat.*, 2014, **24**, 759.
62. H. Hauner, *Diabetes/Metab. Res. Rev.*, 2002, **18**(Suppl 2), S10.
63. T. Tomasic and L. P. Masic, *Curr. Med. Chem.*, 2009, **16**, 1596.
64. H. Joshi, T. Pal and C. S. Ramaa, *Expert Opin. Invest. Drugs*, 2014, **23**, 501.
65. K. Alagiakrishnan, S. Sankaralingam, M. Ghosh, L. Mereu and P. Senior, *Discovery Med.*, 2013, **16**, 277.
66. V. S. Jain, D. K. Vora and C. S. Ramaa, *Bioorg. Med. Chem.*, 2013, **21**, 1599.
67. R. Ali and N. Siddiqui, *J. Chem.*, 2013, **2013**, 345198.
68. A. Gupta and S. Rawat, *J. Curr. Pharm. Res.*, 2010, **3**, 13.
69. C. L. Lee, Y. Lam and S.-Y. Lee, *Tetrahedron Lett.*, 2001, **42**, 109.
70. H. Hioki, K. Matsushita, T. Noda, K. Yamaguchi, M. Kubo, K. Harada and Y. Fukuyama, *Tetrahedron Lett.*, 2012, **53**, 4337.
71. F. Piscitelli, C. Ballatore and A. B. Smith III, *Bioorg. Med. Chem. Lett.*, 2010, **20**, 644.
72. L. Le Bozec and C. J. Moody, *Aust. J. Chem.*, 2009, 639.
73. E. H. White, F. McCapra and G. F. Field, *J. Am. Chem. Soc.*, 1963, **85**, 337.
74. R. Cricchio, P. Antonini, G. C. Lancini, G. Tamborini, R. J. White and E. Martinelli, *Tetrahedron*, 1980, **36**, 1415.

75. N. Hosokawa, H. Naganawa, M. Hamada, H. Iinuma, T. Takeuchi, K. S. Tsuchiya and M. Hori, *J. Antibiot.*, 2000, **53**, 886.
76. Y. Hu and J. B. MacMillan, *Org. Lett.*, 2011, **13**, 6580.
77. R. Sufit and D. Newman, *Neurology*, 1996, **47**, S117.
78. H. Ito, H. Shimada, H. Shinotoh, H. Takano, T. Sasaki, T. Nogami, M. Suzuki, T. Nagashima, K. Takahata, C. Seki, F. Kodaka, Y. Eguchi, H. Fujiwara, Y. Kimura, S. Hirano, Y. Ikoma, M. Higuchi, K. Kawamura, T. Fukumura, E. L. Boo, L. Farde and T. Suhara, *J. Nucl. Med.*, 2014, **55**, 932.
79. R. S. McIntyre, D. S. Cha, M. Alsuwaidan, D. McIntosh, A. M. Powell and J. M. Jerrell, *Expert Opin. Pharmacother.*, 2012, **13**, 1653.
80. Y. Li, J. Geng, Y. Liu, S. Yu and G. Zhao, *ChemMedChem*, 2013, **8**, 27.
81. A. Castro, T. Castaño, A. Encinas, W. Porcal and C. Gil, *Bioorg. Med. Chem.*, 2006, **14**, 1644.
82. Y. Hu, C. Y. Li, X. M. Wang, Y. H. Yang and H. L. Zhu, *Chem. Rev.*, 2014, **114**, 5572.
83. Y. D. Gong and T. Lee, *J. Comb. Chem.*, 2010, **12**, 393.
84. M. C. McKie and R. M. Paton, *ARKIVOC*, 2002, 15.
85. K. Rosenbaum, S. Goldenberg and G. Weber, *J. Prakt. Chem.*, 1992, **334**, 283.
86. S. I. Molotov, A. S. Kulikov, N. N. Makhova and K. A. Lyssenko, *Mendeleev Commun.*, 2003, 188.
87. I. A. Ryu, J. Y. Park, H. C. Han and Y. D. Gong, *Synlett*, 2009, 999.
88. J. K. Augustine, V. Vairaperumal, S. Narasimhan, P. Alagarsamy and A. Radhakrishnan, *Tetrahedron*, 2009, **65**, 9989.
89. A. A. Kiryanov, P. Sampson and A. J. Seed, *J. Org. Chem.*, 2001, **66**, 7925.
90. V. Padmavathi, G. Dinneswara Reddy, S. Nagi Reddy and K. Mahesh, *Eur. J. Med. Chem.*, 2011, **46**, 1367.
91. J. Y. Hwang, H.-S. Choi, D.-H. Lee and Y.-D. Gong, *J. Comb. Chem.*, 2005, **7**, 816.
92. A. K. Jain, S. Sharma, A. Vaidya, V. Ravichandran and R. K. Agrawal, *Chem. Biol. Drug Des.*, 2013, **81**, 557.
93. E. R. Swenson, *Curr. Hypertens. Rep.*, 2014, **16**, 467.
94. A. Scozzafava and C. T. Supuran, *Subcell. Biochem.*, 2014, **75**, 349.
95. J. Keiser, A. Stich and C. Burri, *Trends Parasitol.*, 2001, **17**, 42.
96. R. C. Heel, R. N. Brogden, T. M. Speight and G. S. Avery, *Drugs*, 1979, **17**, 38.
97. A. Martinez, A. Castro, I. Dorronsoro and M. Alonso, *Med. Res. Rev.*, 2002, **22**, 373.
98. E. Tolosa, I. Litvan, G. U. Hoglinger, D. Burn, A. Lees, M. V. Andres, B. Gomez-Carrillo, T. Leon and T. Del Ser, *Mov. Disord*, 2014, **29**, 470.
99. I. Hers, J. M. Tavare and R. M. Denton, *FEBS Lett.*, 1999, **460**, 433.
100. A. Martinez, A. Castro, I. Cardelus, J. Llenas and J. M. Palacios, *Bioorg. Med. Chem.*, 1997, **5**, 1275.
101. A. Martinez, E. Fernandez, A. Castro, S. Conde, I. Rodriguez-Franco, J. E. Baños and A. Badia, *Eur. J. Med. Chem.*, 2000, **35**, 913.

102. A. Martinez, M. Alonso, A. Castro, C. Perez and F. J. Moreno, *J. Med. Chem.*, 2002, **45**, 1292.
103. G. Ottmann and H. Hooks, *Angew. Chem., Int. Ed.*, 1966, **5**, 672.
104. A. Martinez, M. Alonso, A. Castro, I. Dorronsoro, J. L. Gelpi, F. J. Luque, C. Perez and F. J. Moreno, *J. Med. Chem.*, 2005, **48**, 7103.
105. A. Martinez, M. Alonso, A. Castro and I. Dorronsoro, in *Medicinal Chemistry of Alzheimer Disease*, ed. A. Martinez, Transworld Research, Kerala, 2008, p. 225.
106. S. Leclerc, M. Garnier, R. Hoessel, D. Marko, J. A. Bibb, G. L. Snyder, P. Greengard, J. Biernat, Y. Z. Wu, E. M. Mandelkow, G. Eisenbrand and L. Meijer, *J. Biol. Chem.*, 2001, **276**, 251.
107. L. Sereno, M. Coma, M. Rodriguez, P. Sanchez-Ferrer, M. B. Sanchez, I. Gich, J. M. Agullo, M. Perez, J. Avila, C. Guardia-Laguarta, J. Clarimon, A. Lleo and T. Gomez-Isla, *Neurobiol. Dis.*, 2009, **35**, 359.
108. R. Luna-Medina, M. Cortes-Canteli, S. Sanchez-Galiano, J. A. Morales-Garcia, A. Martinez, A. Santos and A. Perez-Castillo, *J. Neurosci.*, 2007, **27**, 5766.
109. M. Bolos, S. Fernandez and I. Torres-Aleman, *J. Biol. Chem.*, 2010, **285**, 17693.
110. G. U. Höglinger, H. J. Huppertz, S. Wagenpfeil, M. V. Andres, V. Belloch, T. Leon and T. Del Ser, *Mov. Disord*, 2014, **29**, 479.
111. T. del Ser, K. C. Steinwachs, H. J. Gertz, M. V. Andres, B. Gomez-Carrillo, M. Medina, J. A. Vericat, P. Redondo, D. Fleet and T. Leon, *J. Alzheimers Dis.*, 2013, **33**, 205.
112. J. A. Morales-Garcia, C. Susin, S. Alonso-Gil, D. I. Perez, V. Palomo, C. Perez, S. Conde, A. Santos, C. Gil, A. Martinez and A. Perez-Castillo, *ACS Chem. Neurosci.*, 2013, **4**, 350.
113. E. Beurel, O. Kaidanovich-Beilin, W. I. Yeh, L. Song, V. Palomo, S. M. Michalek, J. R. Woodgett, L. E. Harrington, H. Eldar-Finkelman, A. Martinez and R. S. Jope, *J. Immunol.*, 2013, **190**, 5000.
114. S. Cuzzocrea, T. Genovese, E. Mazzon, C. Crisafulli, R. Di Paola, C. Muia, M. Collin, E. Esposito, P. Bramanti and C. Thiemermann, *J. Pharmacol. Exp. Ther.*, 2006, **318**, 79.
115. S. Cuzzocrea, E. Mazzon, R. Di Paola, C. Muia, C. Crisafulli, L. Dugo, M. Collin, D. Britti, A. P. Caputi and C. Thiemermann, *Clin. Immunol.*, 2006, **120**, 57.
116. A. V. Franklin, M. K. King, V. Palomo, A. Martinez, L. L. McMahon and R. S. Jope, *Biol. Psychiatry*, 2014, **75**, 198.
117. D. C. Vitale, C. Piazza, T. Sinagra, V. Urso, F. Cardi, F. Drago and S. Salomone, *Clin. Drug Invest.*, 2013, **33**, 885.
118. F. Chiadmi and J. Schlatter, *Clin. Lab.*, 2013, **59**, 439.
119. T. Pilgrim and S. Windecker, *Heart*, 2014, **100**, 1750.
120. S. E. O'Connor, A. Grosset and P. Janiak, *Fundam. Clin. Pharmacol.*, 1999, **13**, 145.

CHAPTER 10

Thiirane Class of Gelatinase Inhibitors as a Privileged Template that Crosses the Blood–Brain Barrier

MAJOR GOOYIT, ZHIHONG PENG, SHAHRIAR MOBASHERY AND MAYLAND CHANG*

Department of Chemistry and Biochemistry, University of Notre Dame, Notre Dame, IN 46556, USA
*Email: mchang@nd.edu

10.1 Brief Overview of Matrix Metalloproteinases

Matrix metalloproteinases (MMPs) constitute a large group of zinc-dependent endopeptidases that function in the remodeling of the extracellular matrix (ECM) and proteolytic processing of cytokines necessary for tissue growth and repair.[1] On the basis of substrate specificity and domain structure, MMPs can be subdivided into six groups: collagenases, gelatinases, matrilysins, membrane-type MMPs (MT-MMPs), stromelysins and other MMPs (refer to published articles[2,3] for detailed reviews on the structural domains of MMPs). Regulation of MMPs occurs at multiple levels including mRNA transcription, compartmentalization, posttranslational activation of zymogens, and endogenous inhibition. The expression of MMPs under physiological and pathological conditions is modulated by a variety of regulatory factors including cytokines, growth factors, reactive oxygen species (ROS), hormones

RSC Drug Discovery Series No. 50
Privileged Scaffolds in Medicinal Chemistry: Design, Synthesis, Evaluation
Edited by Stefan Bräse

Published by the Royal Society of Chemistry, www.rsc.org

and ECM interactions.[4] MMPs are produced as inactive zymogens (pro-MMPs), which require removal of the pro-domain (through disruption of the pro-peptide cysteine-sulfhydryl and catalytic zinc ion interaction) for activation.[5,6] The processing of pro-MMPs to generate the active forms can be accomplished either *in vitro* by chemical agents (organomercurials and chaotropic compounds) or *in vivo*, by serine proteases, plasmin, trypsin, and other MMPs.[4] MMP function is tightly controlled by endogenous inhibitors including α_2-macroglobulin and tissue inhibitor of metalloproteinases (TIMPs).[7] TIMPs maintain the latency of MMPs by the virtue of the putative cysteine–Zn^{2+} bond that coordinates the *N*-conserved cysteine thiol to the active site zinc ion of MMPs.[5,6] As such, TIMPs counteract aberrant MMP activity to regulate ECM turnover, tissue remodeling and cellular behavior. Uncontrolled MMP proteolysis or loss of the MMP/TIMP balance contributes to pathological conditions, including tumor metastasis, rheumatoid arthritis, and cardiovascular and neurological diseases, among others.[8–12]

10.2 The Gelatinases and their Multiple Roles in Diseases of the Extracellular Matrix

Of the 26 MMPs, the gelatinases, comprising MMP-2 (gelatinase A, ~72 kDa) and MMP-9 (gelatinase B, ~92 kDa), have received much attention due to their involvement in multiple disorders. In common with all the MMPs, the gelatinases are secreted from cells as latent enzymes, requiring the cleavage of the prodomain (cysteine switch) to achieve catalytic activity. Physiologic activation of pro-MMP-2 is initiated by proteolytic cleavage by MT1-MMP (MMP-14), followed by autolysis to produce mature active MMP-2.[13,14] Other reported activators of pro-MMP-2 include activated protein C,[15] thrombin,[16] and most MT-MMPs.[17,18] Several proteases have been demonstrated to activate pro-MMP-9, including plasmin,[19] trypsin,[20] tissue kallikrein,[21] and other MMPs (with MMP-3 being the most efficient activator).[22,23] Gelatinases display a high degree of substrate promiscuity, although they primarily degrade denatured collagen (gelatin) and type IV collagen in basal membranes. Elastin,[24] laminin,[25] and vitronectin[26] are a few of the many ECM components that have been identified as substrates of gelatinases. In addition, MMP-9, and to a lesser extent MMP-2, can degrade a variety of non-collagenous substrates, including cytokines, angiogenic factors, cell surface receptors, and growth factors.[27–29] Proteolysis of these substrates influences the cellular phenotype, which impacts several pathophysiological conditions. Herein, we focus on the implications of MMP-2 and MMP-9 activity in cancer metastases, neurological diseases, and chronic wounds, which are areas of interest to our labs.

10.2.1 Cancer Metastases

Metastatic cancer is cancer that spreads from a primary organ to anatomically distant sites. Brain metastases commonly arise from tumors

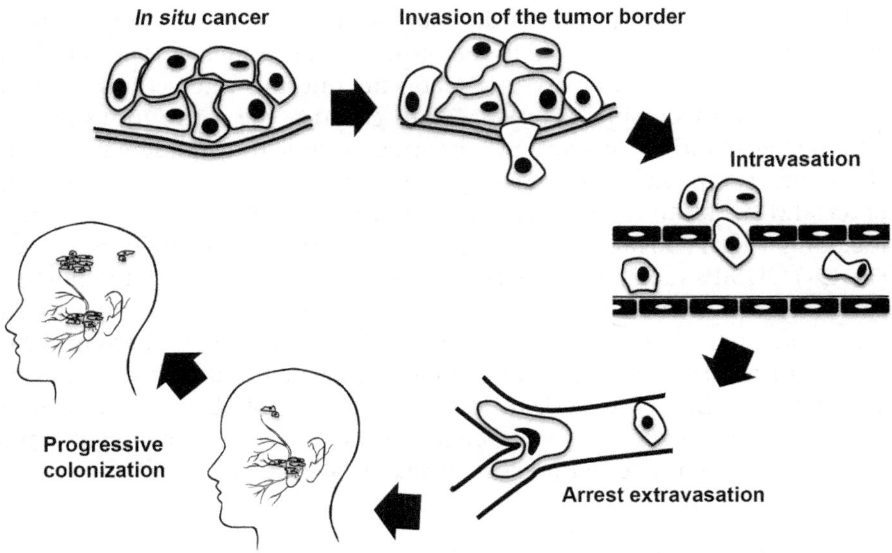

Figure 10.1 Stages of brain tumor metastasis.

originating in the lung, skin, breast, kidney, and colon. The metastatic cascade commences with invasion of the surrounding normal tissue and migration of cancer cells *via* the bloodstream or lymphatic system, followed by arrest in the vasculature at distant organ sites and extravasation into the parenchyma of the surrounding tissues.[30] Cancer cells that are able to cross the blood–brain barrier (BBB) and survive the brain microenvironment form micrometastases, and can proliferate to generate macroscopic neoplastic growths (Figure 10.1).[31] These pathological events involve, among others, breach of the basement membrane and proteolytic degradation of ECM components by MMPs, as well as production of angiogenic stimulators that contribute to tumor progression.[32]

The gelatinases are established mediators of tumor invasion and metastasis. Overexpression of MMP-2 and/or MMP-9 had been noted in a number of experimental and clinical studies, including colorectal,[33,34] lung,[35] ovarian,[26,36] prostate,[37,38] and pancreatic cancer metastases.[39] Several reports have suggested the essential role of MMP-2 in breast[40,41] and melanoma brain metastases.[42] A positive correlation of MMP-2 levels and angiogenesis was also documented in lung carcinoma metastasis to the central nervous system (CNS),[43] whereas MMP-9 was highly expressed in brain metastatic lung adenocarcinoma cells.[44] In a rat model of breast cancer metastasis to the brain, MMP-2, MMP-3, and MMP-9 protein expressions were significantly elevated in neoplastic brain tissue compared to normal brain.[45] Tumor metastases are responsible for ~90% of cancer-related deaths. Unfortunately, there is no anti-metastatic agent available in the market to prevent the dissemination of cancer cells.

10.2.2 Neurological Diseases

Stroke is a major cause of acute and chronic disability. Stroke occurs when the blood supply to the brain is blocked (also known as ischemic stroke, which accounts for ~87% of stroke cases), or when the blood vessel ruptures in the brain (hemorrhagic stroke). A series of cellular and molecular events ensues after ischemia, including potentiation of inflammatory responses, oxidative stress, and activation of cytotoxic agents such as nitric oxide (NO) and matrix degrading enzymes (Figure 10.2A).[46,47] MMPs can degrade a number of neurovascular matrix components and blood–brain barrier (BBB) tight junctions, thereby promoting BBB damage, edema, hemorrhage, and neuronal death.[48,49] Although the underlying mechanisms of these outcomes have not yet been fully understood, there is considerable evidence linking the gelatinases as major players in stroke pathology. MMP-9 activity is associated with brain injury, resulting from cerebral ischemia and hemorrhagic stroke in both humans and animals.[25,50,51] Additionally, smaller cerebral infarct size and reduced BBB injury were observed after focal cerebral ischemia in MMP-9-deficient mice.[52] In studies using baboons, significantly elevated MMP-2 levels correlated with neuronal injury following ischemia.[53] At present, tissue plasminogen activator (tPA) is the only FDA-approved therapy for ischemic stroke and is beneficial only within 4.5 h of stroke onset. However, tPA also has serious deficiencies as it promotes neurotoxicity and hemorrhage.[54] Moreover, the use of tPA was shown paradoxically to upregulate MMP-9 in the brain, which contributes to neuromatrix degradation and neuronal damage.[55,56]

Traumatic brain injury (TBI) is characterized by primary and secondary injury phases. Primary injury involves direct injury to brain cells caused by external mechanical forces, including cerebral contusion and skull fracture (Figure 10.2B). This leads to a cascade of biochemical processes that proceed to what is referred to as secondary injury, the critical outcome determinant of TBI. Secondary injury transpires as a result of the dynamic interplay of inflammatory and cytotoxic mechanisms.[57] Release of neurotransmitters, ROS, NO, and pro-inflammatory cytokines coupled with activation of proteases culminate in brain edema, BBB disruption, hemorrhage, and cell apoptosis.[58] Several reports have suggested the involvement of MMP-9 in the progression of secondary brain damage following trauma. Levels of MMP-9 were elevated in the cerebrospinal fluid (CSF) and blood of patients with severe TBI[59] and in murine brains following fluid percussion injury[60] and controlled cortical impact (CCI) injury.[61] In a CCI model of TBI, MMP-9-knockout mice manifested reduced morphological damage and motor deficits.[62] In another report, Shigemori and colleagues documented the upregulation of MMP-9, which contributed to BBB disruption and edema formation in rats following CCI injury.[63]

Gelatinase activity has also been associated with the pathology of spinal cord injury (SCI),[64] subarachnoid hemorrhage,[65] cerebral aneurysm,[66] as well as in neurodegenerative disorders including Alzheimer's disease,[67,68]

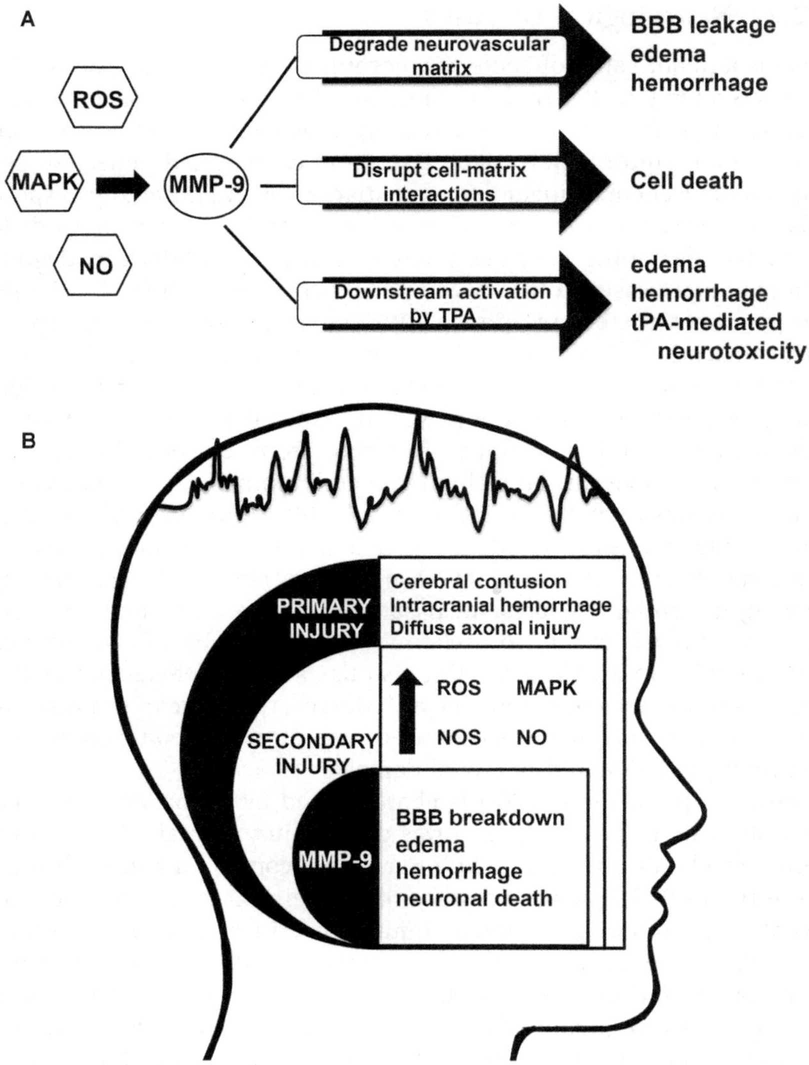

Figure 10.2 The pathophysiology of stroke and traumatic brain injury. (A) Role of MMP-9 after stroke. Activation of reactive-oxygen species (ROS), nitric oxide (NO) and mitogen-activated proteins kinases (MAPK) results in increased MMP-9 expression leading to blood–brain barrier (BBB) disruption and brain damage. (B) Primary and secondary events in TBI. Following primary mechanical insult, secondary injury biochemical processes occur including activation of pro-inflammatory mediators such as nitric oxide synthases (NOS, which produce NO), MAPK and ROS. As a result, MMP-9 is activated, which contributes to neurovascular damage and death.

Huntington's disease,[69,70] Parkinson's disease,[71] and amyotrophic lateral sclerosis (ALS, Lou Gehrig's disease).[71,72] MMP-9 was recently disclosed to induce fast motor neuron degeneration in ALS mice. This finding was

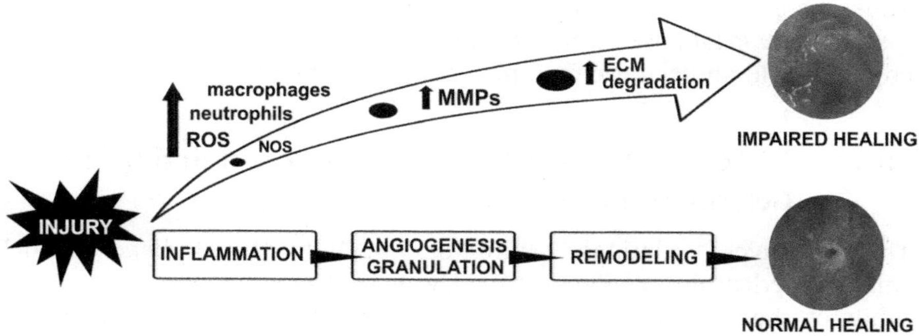

Figure 10.3 The process of wound healing involves inflammation, angiogenesis/granulation, and tissue remodeling. Following injury, a blood clot is formed to seal the wound, followed by recruitment of inflammatory cells and secretion of cytokines, growth factors, and proteinases. Angiogenesis is required to repair the injury and generate granulation tissue. MMPs are secreted to remodel the ECM and to convert granulation tissue into scar tissue. This process is dysregulated in diabetes, resulting in chronic wounds.

substantiated by genetic ablation, viral gene therapy and pharmacological inhibition studies, which remarkably delayed muscle denervation in mice,[72] thus implicating MMP-9 as a relevant therapeutic target in ALS.

10.2.3 Chronic Wounds

The process of wound healing involves inflammation, angiogenesis and granulation, followed by remodeling of the ECM (Figure 10.3). Hyperglycemia in diabetic patients can lead to vascular damage, resulting in ischemia, which contributes to the inability of wounds to heal.[73] Ischemia also triggers the production of ROS and MMPs, in particular MMP-9. Increased levels of MMP-2 and MMP-9 have been found in wound fluid from chronic leg ulcers[74] and in wound tissues from diabetic foot ulcers.[75] Higher levels of MMP-9 in wound fluid from patients with chronic wounds correlated with clinical severity of the ulcer.[76] In diabetic mice, MMP-9 is upregulated and makes the wounds refractory to healing.[77]

10.3 Pharmacological Intervention of Gelatinase-dependent Diseases

A number of reviews on the design and clinical evaluation of MMP inhibitors (MMPIs) have already been published.[78–80] The vast majority of known MMPIs are broad-spectrum inhibitors, inhibiting most, if not all MMPs, and in many cases, members of the related *A Disintegrin And Metalloproteinases* (ADAMs) family. In addition, the relative impenetrability of the BBB presents a major impediment in the development of neuropharmaceutics.[81,82]

Thus, there is an urgent need to design inhibitors that are not only able to discriminate gelatinases from other metalloproteases, but are also able to cross the BBB and achieve therapeutic levels in the brain.

10.3.1 SB-3CT, a Privileged Scaffold for Potent and Selective Gelatinase Inhibitors

The prototype gelatinase inhibitor, SB-3CT (**1**, 4-phenoxyphenylsulfonyl)methylthiirane, Figure 10.4A), was designed as a mechanism-based inhibitor that displays potency and selectivity towards the gelatinases.[83] Table 10.1 summarizes the inhibition profile of SB-3CT against several MMPs and ADAMs. SB-3CT functions as a slow-binding inhibitor of MMP-2 and MMP-9, with inhibitory constant (K_i) values of $K_i = 28 \pm 7$ nM and $K_i = 400 \pm 150$ nM, respectively.[84] Interestingly, these values are comparable to the slow-binding kinetic parameters observed for the endogenous MMP-inhibitors, TIMP-1 and TIMP-2, against the gelatinases.[83] SB-3CT also exhibits potency against MMP-14_{cat} ($K_i = 110$ nM), albeit the mode of inhibition is linear (reversible) competitive and not mechanism-based. The K_i values for the other MMPs and ADAMs tested were at best in the micromolar range, and occur either through a linear competitive (for MMP-1_{cat}, MMP-3_{cat}, and MMP-7) or a noncompetitive (for MMP-8_{cat} and ADAM-17 or TACE) mode of inhibition (Table 10.1).

10.3.2 Mechanism of Action

Forbes *et al.* shed light on the molecular basis of the selectivity of SB-3CT (**1**) against the gelatinases.[85] SB-3CT acts as a caged entity that undergoes an elimination reaction only in the active site of the gelatinases. Upon binding, the active site Glu404 initiates the slow-binding inhibition by deprotonation at the α-carbon adjacent to the sulfonyl moiety of SB-3CT, opening the thiirane to the corresponding thiolate, which then results in tight-binding coordination with the catalytic zinc ion, (Figure 10.4B).[85]

10.3.3 Metabolism, Pharmacokinetics, and Brain Distribution of SB-3CT

The biotransformation of SB-3CT was investigated in detail previously.[84,86] The major metabolic pathways of SB-3CT included hydroxylation at the *para*-position of the terminal phenyl ring to generate an active metabolite **2** and oxidation at the α-position to the sulfonyl group to give the inactive sulfinic acid **3** (Figure 10.5).[84] Incidentally, the *p*-hydroxy metabolite **2** is a more potent inhibitor of gelatinases with K_i values of 6 ± 3 nM and 160 ± 20 nM for human MMP-2 and MMP-9, respectively. Minor metabolites were the sulfoxide **4**, due to oxidation of the thiirane sulfur, and compound **5**, which was produced by ring-opening of the thiirane ring, followed by *S*-methylation

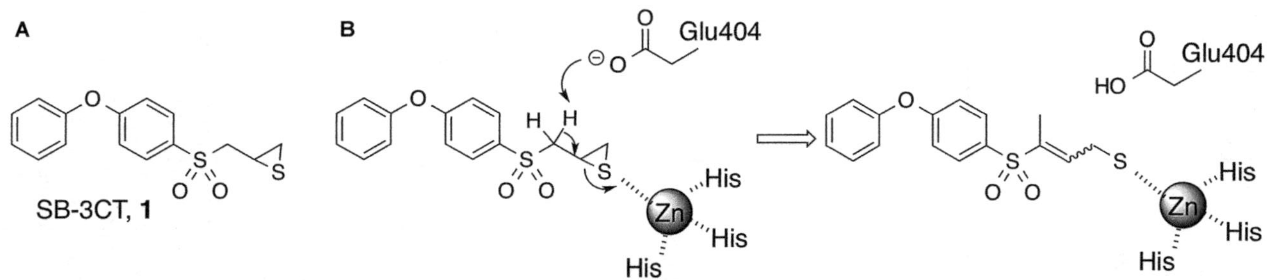

Figure 10.4 (A) Structure of SB-3CT and (B) its mechanism of action within the active site.

Table 10.1 MMP and ADAM kinetic profile of SB-3CT.

Enzyme	Mode of inhibition	$10^{-4}\ k_{on}$ ($M^{-1}\ s^{-1}$)	$10^3\ k_{off}$ (s^{-1})	K_i (μM)	Ref.
MMP-2	Slow-binding	2.0 ± 0.5	0.57 ± 0.03	0.028 ± 0.007	84
MMP-9	Slow-binding	2.8 ± 0.9	1.2 ± 0.1	0.40 ± 0.15	84
MMP-1_{cat}	Linear competitive	—	—	73 ± 5	84
MMP-3_{cat}	Linear competitive	—	—	4.0 ± 0.4	84
MMP-7	Linear competitive	—	—	67 ± 6	84
MMP-8_{cat}	Linear noncompetitive	—	—	2.1 ± 0.4	[a]
MMP-14_{cat}	Linear competitive	—	—	0.11 ± 0.01	84
MMP-19_{cat}	ND[b]	—	—	12%[c]	[a]
ADAM-9	ND[b]	—	—	36%[d]	[a]
ADAM-10	ND[b]	—	—	34%[d]	[a]
ADAM-17	Linear noncompetitive	—	—	2.3 ± 0.3	[a]

[a]Unpublished data.
[b]ND = not determined.
[c]Inhibition at 30 μM.
[d]Inhibition at 20 μM.

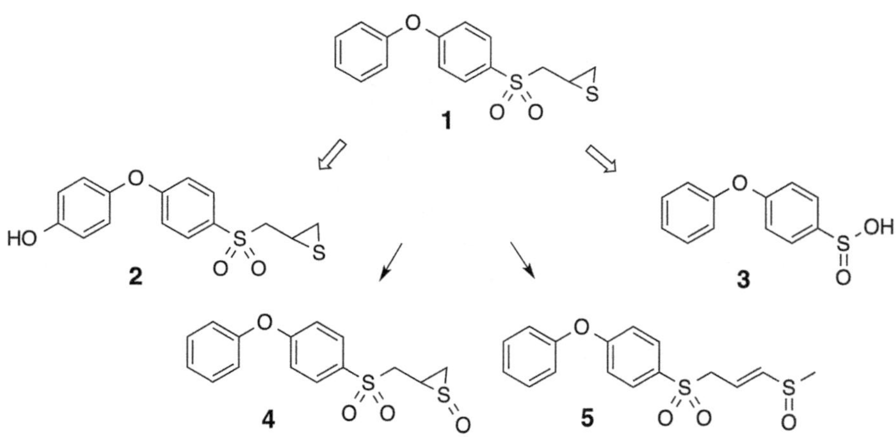

Figure 10.5 Metabolism of SB-3CT.

and oxidation (Figure 10.5); both metabolites are inactive against the gelatinases.

The pharmacokinetics (PK) and brain distribution of SB-3CT and its active metabolite **2** were reported earlier;[87] the PK parameters after intraperitoneal administration of SB-3CT and **2** to mice at 25 mg kg^{-1} are summarized in Table 10.2. SB-3CT was rapidly absorbed and distributed to the brain within 10 min, the first time point collected. Brain levels remained above the MMP-9 K_i of SB-3CT for 60 min and above that of SB-3CT thiolate for >3 hours. Systemic exposure, as measured by $AUC_{0-\infty}$ (*area under the curve*) of SB-3CT was 179 μM min in plasma and 122 pmol min mg^{-1} in brain. SB-3CT readily crossed the BBB with a brain to plasma *AUC* ratio of 0.68, whereas that of **2** after administration of SB-3CT was 2.0. Of note, the brain $AUC_{0-\infty}$ for SB-3CT

Table 10.2 Pharmacokinetic parameters of SB-3CT and compound **2** after a single intraperitoneal dose to mice at 25 mg kg^{-1}.[a]

	SB-3CT after dose of SB-3CT		Compound **2** after dose of SB-3CT		Compound **2** after dose of **2**	
Parameter	Brain	Plasma	Brain	Plasma	Brain	Plasma
$AUC_{0-\infty}$[b]	122	179	3.31	1.65	82.8	121
$t_{1/2\beta}$	46	22	94	78	29	31
$Brain_{AUC}/Plasma_{AUC}$	0.68		2.0		0.68	

[a]Data from reference.
[b]*AUC* in pmol · min per mg for brain and in μM · min for plasma.

was 37-fold higher than that for its metabolite **2**. The brain to plasma *AUC* ratio of compound **2** after a dose of **2** was 0.68, indicating that it crossed the BBB (Table 10.2). Both SB-3CT and compound **2** were found to distribute to all regions of the brain.[87] Moreover, SB-3CT does not accumulate and is eliminated from the brain after multiple intraperitoneal doses to mice, suggesting no neurotoxic effects would be observed.[61]

10.3.4 *In vitro* and *In vivo* Efficacy

Due to its unique mechanism of action that confers selectivity for the gelatinases, SB-3CT has been utilized in over 200 *in vitro* and *in vivo* studies to validate the roles of MMP-2 and MMP-9 in experimental models of diseases. It has become the gold standard in such studies. For the sake of brevity, experimental results dealing with cancer metastases and neurological diseases only will be communicated in this chapter.

10.3.4.1 Cancer Cell Invasion and Metastasis

In *in vitro* assays, SB-3CT suppressed the invasion and migration of human cancer lines including breast cancer MCF-7,[88] MDA-MB-231[89] and HB2(ErbB2) cells,[90] ovarian epithelial cancer CaOV-3 and SKOV-3 cells,[91,92] cervical cancer HeLa and CaSki cells,[93] testicular embryonal carcinoma NT2/D1 cells,[94] and squamous cell carcinoma PC1-37B cells.[95] The *in vivo* efficacy of SB-3CT has been demonstrated in a number of mouse models of cancer metastasis. Dong *et al.* previously reported net MMP-9 activity in PC3 tumor-bearing bones in the SCID-human model of prostate cancer.[96] SB-3CT-treatment resulted in inhibition of angiogenesis and intraosseous tumor growth within the marrow of human fetal bone implanted in SCID mice.[97] Induction of MMP-9 also correlated with the development of liver metastasis by L-CI.5s lymphoma cells in mice. Administration of SB-3CT caused a reduction in the number of liver metastases and increased survival in an aggressive murine model of T-cell lymphoma.[98] In another study, the Ras-BLT2 signaling pathway was shown to stimulate the production of MMP-9, resulting in enhanced metastasis *in vivo*.[99] Treatment with SB-3CT led to a dramatic reduction of metastatic lung nodules and tumor burden, and diminished mice mortality in Ras-induced pulmonary metastasis.[99] In a mouse model of breast cancer lung metastasis, pharmacological inhibition of host MMP-9 by SB-3CT attenuated tumor growth in the lungs of C57BL/6 mice.[100] The foregoing brain penetration and anti-metastatic effects indicate the considerable promise of SB-3CT as a treatment strategy for metastatic brain tumors.

10.3.4.2 Neurological Diseases

The efficacy of SB-3CT has been demonstrated in a number of neurological diseases. MMP-9 is markedly activated after transient focal cerebral[101] and hypoxic ischemia in mice.[102] SB-3CT abrogated MMP-9-induced laminin degradation and neuronal apoptosis, and provided significant protection against brain damage up to 6 h after ischemia.[25] SB-3CT also reduced the infarct size and significantly ameliorated neurological deficits after ischemic stroke.[25] In an embolic focal cerebral ischemia model, the more clinically relevant animal model of human ischemic stroke, SB-3CT-treated mice displayed reduced brain damage and notably improved neurobehavioral functions (Figure 10.6).[50] Pericyte loss and lumen contraction are distinctive features of ischemic brain, which contribute to diminished brain microcirculation, BBB disruption and ultimately brain vascular damage.[103] Cui *et al.* had shown in the same study that SB-3CT preserved the integrity of pericytes and significantly decreased intracranial hemorrhage volume from $>300\ \mu m^3$ to less than $100\ \mu m^3$ after embolic ischemia.[50]

In addition to neuroprotective effects, SB-3CT also prevents cerebral hemorrhage through interference of MMP-9 activity. Increased MMP-9 levels were observed on viral infection of immature dendritic cells in a hemorrhagic fever mouse model.[104] Gelatinolytic activity disrupts endothelial cell–cell adhesion molecules, resulting in vascular leakage, which was prevented after treatment with SB-3CT.[104] In a rat model of subarachnoid hemorrhage (SAH), SB-3CT prevented laminin degradation and neuronal apoptosis, thereby providing significant brain protection after SAH.[105,106] The mechanism of BBB-breakdown involves expression of apolipoprotein E4 (APOE4) in response to stroke or TBI,[107] inducing the downstream activation of the cyclophilin A (CypA)-nuclear factor-κB (NF-κB)-MMP-9 pathway, which leads to BBB degradation.[108] Bell *et al.* had shown that SB-3CT decreased APOE4-mediated vascular defects in mice through inhibition of MMP-9-mediated BBB disruption.[108]

In an electromagnetic-impactor-induced TBI mouse model, expression of active MMP-9 was observed up to 10 days post-trauma. Treatment with SB-3CT significantly diminished MMP-9 activity as well as reducing the cortical lesion volume in mouse brain (Figure 10.7).[61] Besides protecting the neurons from dendritic degeneration and attenuating glial activation, SB-3CT also conferred long-term protection from sensorimotor and cognitive deficits in TBI mice.[61] These results are consistent with those of Jia *et al.*, who

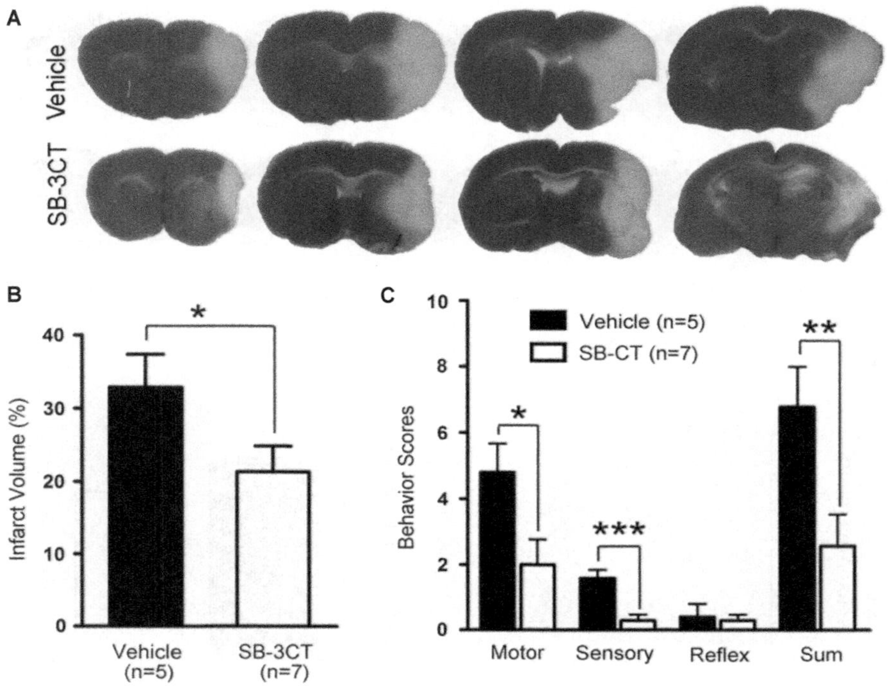

Figure 10.6 SB-3CT protects against brain damage and ameliorates neurobehavioral deficits after embolic middle cerebral artery occlusion in mice. Mice were injected ip (intraperitoneal) with SB-3CT (25 mg per kg body weight) or vehicle 2 and 4 hours after embolus-induced focal cerebral ischemia. Neurobehavioral tests were conducted 24 hours later and scored using the 14-point modified neurological severity scoring, followed by assessment of infarct volume by TTC staining. (A) Representative images of TTC staining. (B) Quantification of infarct volumes analyzed by TTC staining using ImageJ tools. *, $p < 0.05$ by one-tailed Student's t-test. (C) Improvement in motor and sensory function following post-ischemic SB-3CT treatment compared to the vehicle-treated controls; *, $p < 0.05$ for motor; ***, $p < 0.001$ for sensory; and **, $p < 0.01$ for the sum of the behavioral scores, respectively, by one-tailed Student's t-test. Number of animals (n) in each group is shown in parentheses, and data are expressed as means ± SEM, whereas reflex activity was unchanged.
This figure and its legend are reproduced from the paper by Cui *et al.*[50] with permission of the publisher.

confirmed that SB-3CT treatment preserves hippocampal neurons and attenuates behavioral deficits in rats subjected to fluid percussion injury.[60]

SB-3CT therapy also prevented BBB disruption and neuronal apoptosis after traumatic SCI in SOD1 rats.[109] In a separate study, MMP-9 was shown to control the proliferation of NG2 + glial cells in the damaged CNS following SCI in rats, and that blocking MMP-9 activity with SB-3CT stimulated oligodendrocyte maturation and remyelination, as well as facilitated functional recovery after SCI.[110] Neurodegenerative disorders, such as

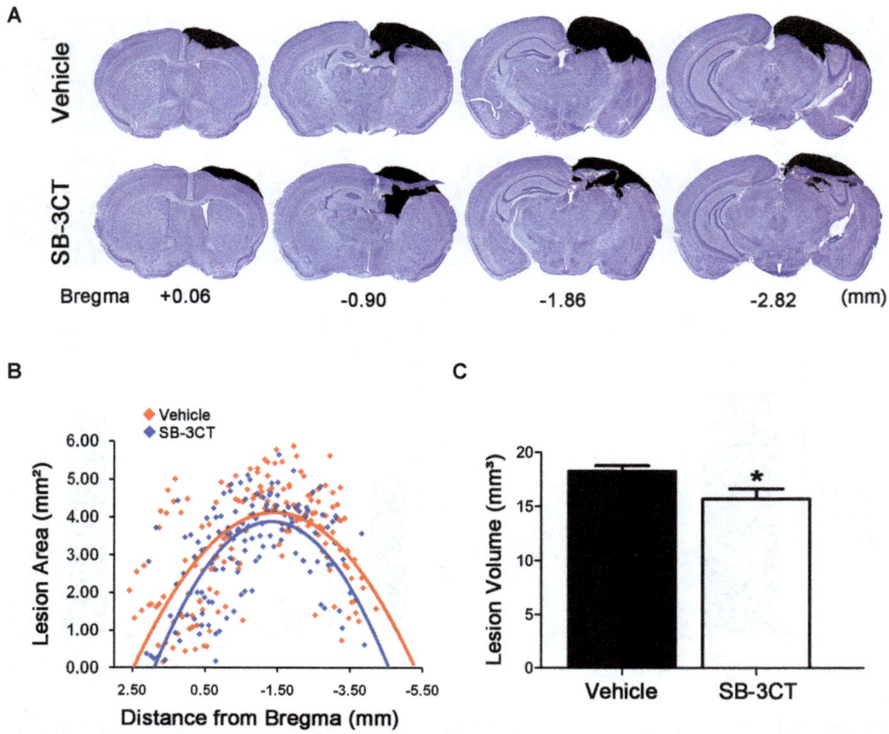

Figure 10.7 Histopathological quantification of lesion volumes in cresyl violet-stained brain sections at 7 days post-trauma. (A) Representative cresyl violet-stained coronal brain sections from vehicle and SB-3CT-treated mice marked with their coordinates to Bregma. The black area in each section shows the contralateral hemisphere superimposed on top of the lesioned hemisphere to visualize the brain damaged regions. (B) Stereological scatter-plot of lesion areas in the cresyl-violet stained sections of vehicle and SB-3CT-treated mice at 7 days post-trauma. Each data point represents the lesion area in one cresyl violet-stained brain section, and plotted according to the rostro-caudal axis of the brain coordinate to Bregma. A second-degree polynomial was generated to fit data points to visualize data trends. The graphs indicate a difference in lesion area between vehicle and SB-3CT-treated mice. (C) Quantification of cortical lesion volume at 7 days post-trauma in the SB-3CT-treated mice compared to the vehicle-treated mice. $n = 6$ in each group; *, $p < 0.05$ by one-tailed, unpaired Student's *t*-test. Data expressed as mean $\pm$ SEM.
This figure and its legend are reproduced from the paper by Hadass *et al.*[61] with permission of the publisher.

Parkinson's and Alzheimer's disease, are characterized by excessive activation of microglia and astrocytes.[111,112] The ability of SB-3CT to impede ATP-induced microglial migration[113] presents a strategy for intervention of neurodegenerative disorders. APOE4, the genetic risk factor for Alzheimer's disease, was earlier shown to compromise the BBB *via* activation of the cyclophilin A-NF-κB-MMP-9 signaling pathway.[108] SB-3CT treatment reversed BBB impairment in mice through inhibition of MMP-9.[108] On a related note, Schultz *et al.* demonstrated that MMP-9 mediates the oligomeric amyloid-β-induced shedding of the pericyte NG2, a proteoglycan that plays a crucial role in the maintenance of vascular integrity.[68] In the presence of SB-3CT, shed NG2 was significantly decreased, thereby preventing pericyte malfunction and vascular damage.

10.4 Second-generation Thiirane Inhibitors

The success of SB-3CT in experimental models of diseases, both *in vitro* and *in vivo*, is noteworthy. However, SB-3CT has limitations in that it is prone to metabolism[84] and has a poor water solubility of 2.3 μg mL^{-1}.[114] The structure-activity relationship (SAR) of SB-3CT was extensively studied in efforts to increase potency and metabolic stability.[115] The presence of the sulfonylmethylthiirane moiety and the phenoxyphenyl group was found to be critical for gelatinase inhibition; whereas, *para*- and *meta*-substitutions of the terminal phenyl ring were generally tolerated.[115] Table 10.3 presents

Table 10.3 Inhibition constants and metabolic stability of representative gelatinase inhibitors.

Compound		K_i (μM) MMP-2[a]	MMP-9[a]	$t_{1/2}$[b] (min)	Ref.
1	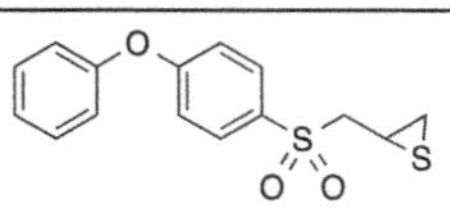	0.028	0.40	4.4	115
3	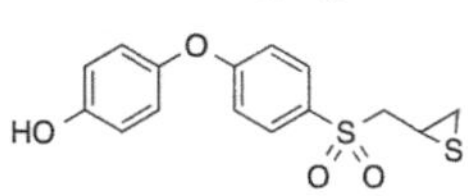	0.006	0.16	23	84
6	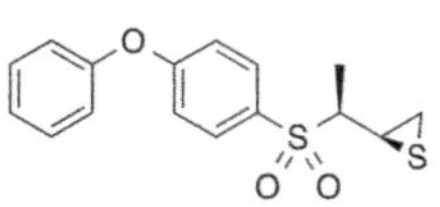	0.18	3.5	ND[c]	116
7	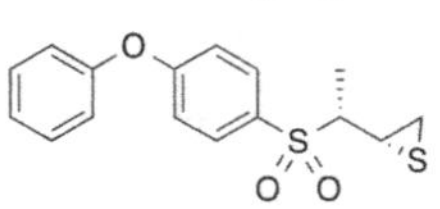	0.39	3.3	ND[c]	116
8	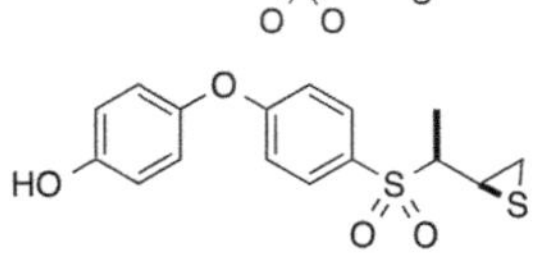	0.22	1.9	ND[c]	116
9	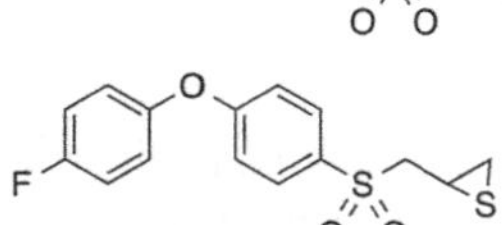	0.061	0.044[d]	3.7	115
10	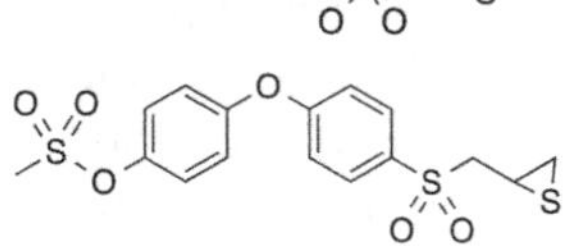	0.023	0.005	27	118

Table 10.3 (*Continued*)

Compound	K_i (μM)			
	MMP-2[a]	MMP-9[a]	$t_{1/2}$[b] (min)	Ref.
11 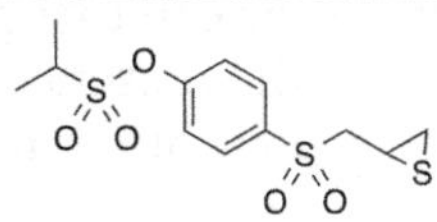	0.24	3.5	36	114
12 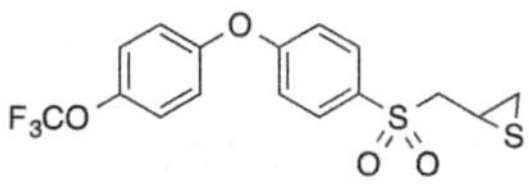	0.11	0.93	7.3	115
13 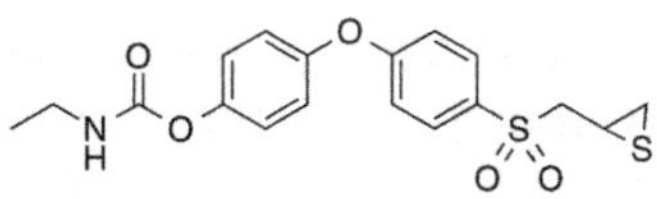	0.55	12	23	117
14 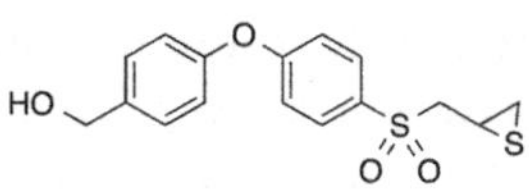	0.078	0.39	13	115
15 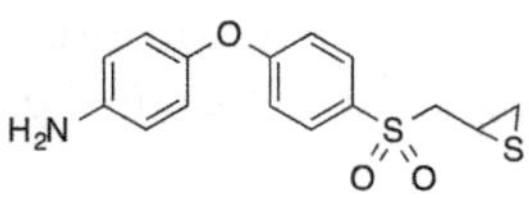	0.024	0.87	11; 50	119
16 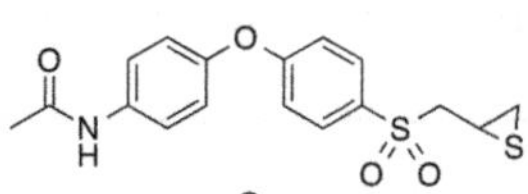	0.11	0.13[d]	ND[c]	120
17 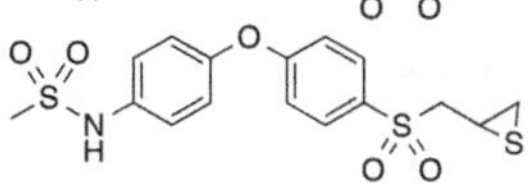	0.016	0.18	ND[c]	120
18 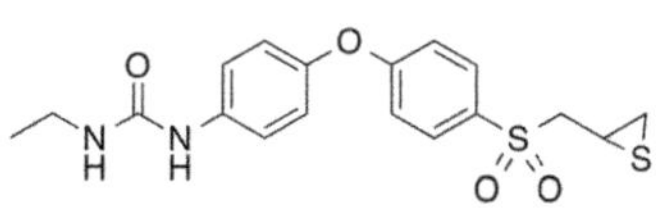	0.44	28%[e]	41	117
19 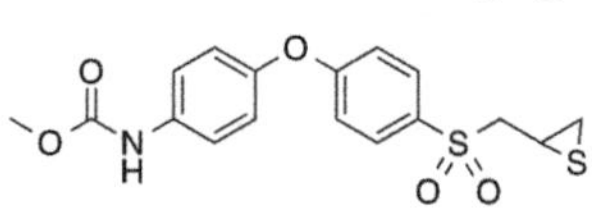	0.050	0.18	18	117
20 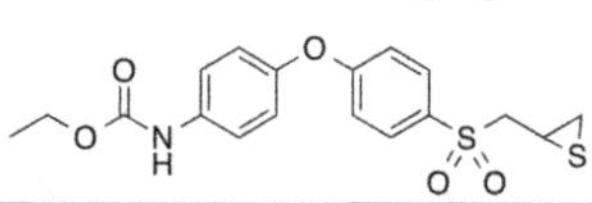	0.29	0.86	25	117

[a]K_i is calculated from the ratio of k_{off}/k_{on}.
[b]Half-life in rat liver S9.
[c]ND = not determined.
[d]K_i is calculated using Dixon plot for competitive inhibition.
[e]Inhibition at 50 μM.

some of the >500 synthesized analogs of SB-3CT, along with their inhibition constants against MMP-2 and MMP-9, and half-lives in rat liver S9. To block oxidation at the α-position to the sulfonyl group of SB-3CT, a methyl substituent was incorporated at that position to afford the four diastereomers.[116] Of these stereoisomers, compounds **6** and **7** were slow-binding inhibitors of MMP-2, MMP-9, and MMP-14. Moreover, a PK study of the α-methyl variant **6** in mice showed significant increase in systemic exposure and longer elimination half-lives in both plasma and brain.[87] Various functionalities were introduced at the terminal phenyl ring of SB-3CT to block metabolism, including halogens, sulfonates, urea, and carbamates.[115,117] These resulted in the identification of highly potent MMP-9 inhibitors (compounds **9**, **10**, and **16**) as well as highly selective MMP-2 inhibitors (*O*-phenyl carbamate **13** and urea **18**), with significantly improved metabolic stability. Sulfonate derivatives of SB-3CT increased the aqueous solubility (up to 100-fold for compound **11**) while retaining potency against MMP-2, however, decreased inhibition against MMP-9 was observed.[114]

The thiirane class of gelatinase inhibitors is a privileged scaffold in that it readily distributes to the brain. Figure 10.8 displays the brain and plasma levels of some thiirane inhibitors after administration to mice. In general, brain levels of the compounds were above their K_i values for MMP-2 for at least 2 hours, except for urea **18**, which had brain levels below the K_i for MMP-2 at all times.[117] At a higher dosage of 100 mg kg^{-1}, compound **18** would achieve therapeutic MMP-2 levels in the brain. The MMP-2-selective inhibitor **13** was not detected in either plasma or brain after a subcutaneous dose at 25 mg kg^{-1}, however, its metabolite (compound **2**) easily distributed to the brain with concentrations well above the K_i for both MMP-2 and MMP-9 (Figure 10.8E).[117] After a subcutaneous dose at 25 mg kg^{-1} to mice, brain levels of compound **15** (also referred to as ND-322) were below its K_i for MMP-9 of 0.87 μM (Figure 10.8C). ND-322 is metabolized by *N*-acetyltransferases to compound **16** (also referred to as ND-364), which distributed to the brain at concentrations above its K_i for MMP-9 of 0.13 μM for up to 4 hours.[121] When administered by itself, ND-364 also displayed sustained brain levels above its K_i for MMP-9 with brain to plasma *AUC* ratio of 0.39 (Figure 10.8D).[121]

In a mouse model of wound healing, MMP-8 and MMP-9 were identified in diabetic wounds.[77] Topical treatment of wounds with the MMP-9 inhibitor ND-322 accelerated wound healing, increased re-epithelialization and reduced apoptosis (Figure 10.9). In contrast, selective inhibition of MMP-8 delayed wound healing, decreased re-epithelialization and resulted in increased apoptosis.[77] MMP-9 was found to be detrimental to wound healing, whereas MMP-8 was beneficial. Thus, selective MMP-9 inhibition holds promise in the treatment of chronic wounds.

10.5 Water-soluble Gelatinase Inhibitor Prodrugs

In an attempt to address the poor aqueous solubility of SB-3CT, amino acids were used as pro-moieties to impart water solubility to the gelatinase inhibitor (>5000 fold over that of SB-3CT).[119] Whereas the ester prodrugs **21** were chemically unstable in water and hydrolyzed completely to **2** in plasma within 2 min, the amide prodrugs **22** displayed increased stability and released the active constituent **15** in human blood within 30 min

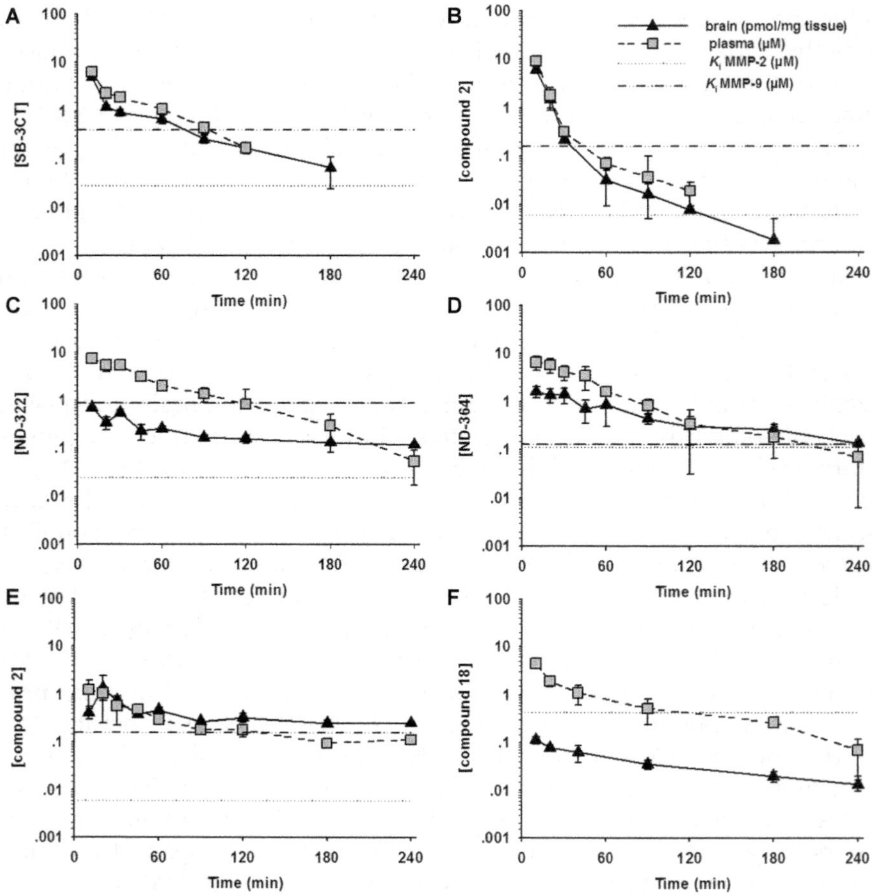

Figure 10.8 Plasma and brain concentration-time curves in mice after single dose administration of (A) SB-3CT at 25 mg kg^{-1} (intraperitoneal dose), (B) compound **2** at 25 mg kg^{-1} (intraperitoneal dose), (C) ND-322 (compound **15**) at 25 mg kg^{-1} (subcutaneous dose), (D) ND-364 (compound **16**) at 28 mg kg^{-1} (subcutaneous dose), (E) compound **13** at 25 mg kg^{-1} (subcutaneous dose), and (F) compound **18** at 25 mg kg^{-1} (subcutaneous dose).

(Figure 10.10). The metabolism and PK of the arginyl prodrug ND-478 (**22d**) were fully investigated.[119,121] The major metabolism pathway of ND-478 is hydrolysis to ND-322; the minor pathway is *N*-acetylation to the active metabolite ND-364 (Figure 10.11). In liver microsomes, ND-322 is hydroxylated at the terminal phenyl ring to give compound **23**. Oxidation of the amino group of ND-322 to give the potentially toxic hydroxylamine **24** was expressly looked for and not observed both *in vitro* and *in vivo*. Moreover, both ND-322 and ND-478 were non-mutagenic in the Ames II mutagenecity assay.[119] Following IV administration of ND-478 to mice, ND-478 was not detectable in the brain, whereas both its active metabolites ND-322 and ND-364 readily

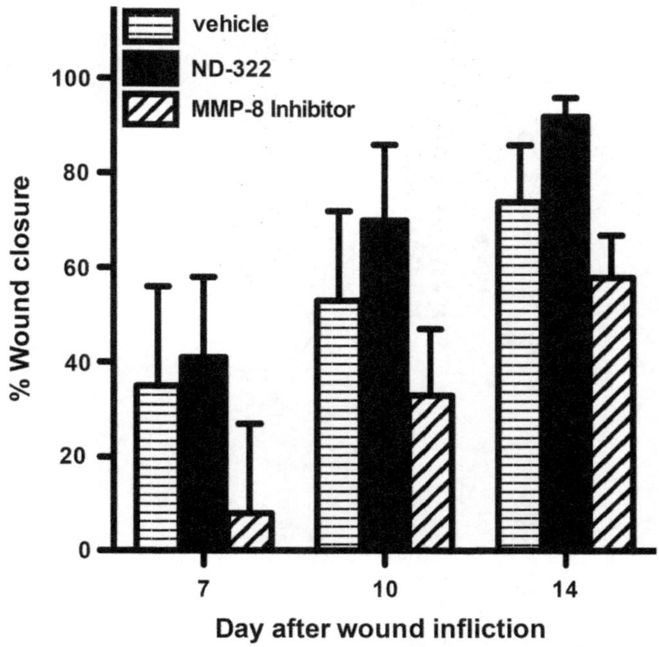

Figure 10.9 Selective inhibition of MMP-9 accelerates diabetic wound healing while selective inhibition of MMP-8 delays diabetic wound healing.

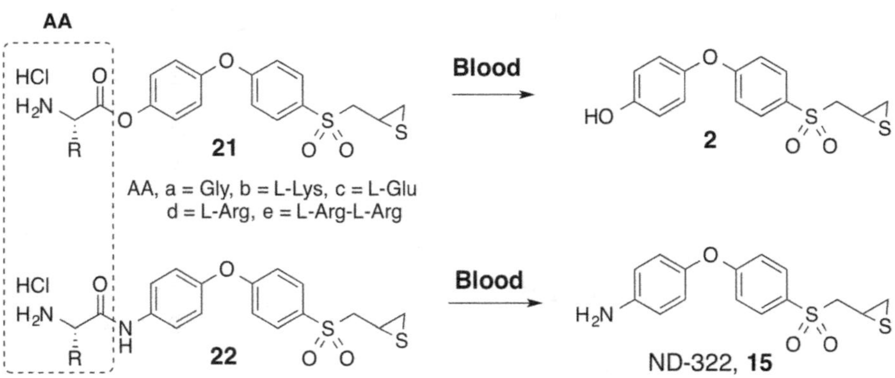

Figure 10.10 Prodrugs of compounds **2** and **10**.

crossed the BBB, with ND-364 preferentially distributing to the brain (5-fold higher than in plasma).[121]

10.6 Future of the Thiirane Class of Gelatinase Inhibitors

MMP-9 plays an important role in the pathology of many neurological diseases, as well a detrimental role in diabetic wound healing. While

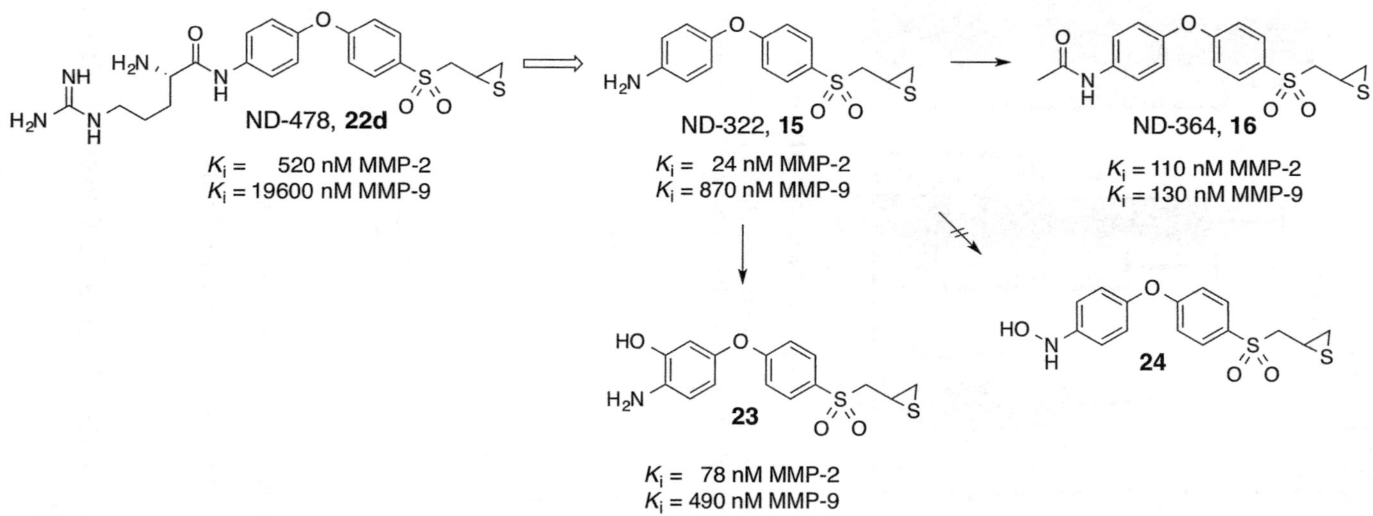

Figure 10.11 Metabolism pathway of the arginyl prodrug ND-478.

broad-spectrum MMPIs failed in clinical trials in patients with cancer, it is now recognized that one important aspect of the failure was due to broad inhibition of many MMPs and other related zinc-dependent enzymes. As MMPs exhibit a dichotomy of functions, involvement in the pathology of the disease and mediating repair and recovery, the use of selective MMP inhibitors is required. In addition, for neurological diseases, not only must the drug cross the BBB, it must achieve therapeutic concentrations in the brain, yet be cleared from the brain to avoid CNS toxicity. The thiirane class of gelatinase inhibitors is a privileged template, which meets these requirements. The compounds have a unique mechanism of action that involves a reaction catalyzed by the gelatinases, resulting in slow-binding and tight-binding inhibition. This mechanism is also at the root of selectivity enjoyed by the thiiranes. In addition, the thiiranes cross the BBB and achieve therapeutic concentrations in the brain. A major roadblock in the development of CNS drugs is the inability of >98% of small-molecule drugs to penetrate the brain. Furthermore, the thiiranes do not accumulate in the brain and are not toxic. These attributes indicate that the thiirane class of gelatinase inhibitors holds great promise for intervention of neurological diseases and chronic wounds.

References

1. A. Page-McCaw, A. J. Ewald and Z. Werb, *Nat. Rev. Mol. Cell Biol.*, 2007, **8**, 221.
2. I. Massova, L. P. Kotra, R. Fridman and S. Mobashery, *FASEB J.*, 1998, **12**, 1075.
3. H. Nagase, R. Visse and G. Murphy, *Cardiovasc. Res.*, 2006, **69**, 562.
4. H. J. Ra and W. C. Parks, *Matrix Biol.*, 2007, **26**, 587.
5. L. P. Kotra, J. B. Cross, Y. Shimura, R. Fridman, H. B. Schlegel and S. Mobashery, *J. Am. Chem. Soc.*, 2001, **123**, 3108.
6. G. Rosenblum, S. Meroueh, M. Toth, J. F. Fisher, R. Fridman, S. Mobashery and I. Sagi, *J. Am. Chem. Soc.*, 2007, **129**, 13566.
7. K. Brew, D. Dinakarpandian and H. Nagase, *Biochim. Biophys. Acta*, 2000, **1477**, 267.
8. A. R. Farina and A. R. Mackay, *Cancers*, 2014, **6**, 240.
9. M. Zhou, S. Qin, Y. Chu, F. Wang, L. Chen and Y. Lu, *Int. J. Clin. Exp. Pathol.*, 2014, **7**, 3048.
10. A. Azevedo, A. F. Prado, R. C. Antonio, J. P. Issa and R. F. Gerlach, *Basic Clin. Pharmacol. Toxicol.*, 2014, **115**, 301.
11. A. Mukherjee and S. Swarnakar, *Mol. Biol. Rep.*, 2015, **42**, 1.
12. E. Galliera, L. Tacchini and M. M. Corsi Romanelli, *Clin. Chem. Lab. Med.*, 2015, **53**, 349.
13. H. Will, S. J. Atkinson, G. S. Butler, B. Smith and G. Murphy, *J. Biol. Chem.*, 1996, **271**, 17119.

14. S. J. Atkinson, T. Crabbe, S. Cowell, R. V. Ward, M. J. Butler, H. Sato, M. Seiki, J. J. Reynolds and G. Murphy, *J. Biol. Chem.*, 1995, **270**, 30479.
15. M. Nguyen, J. Arkell and C. J. Jackson, *J. Biol. Chem.*, 2000, **275**, 9095.
16. Z. Wang, L. Kong, J. Kang, J. H. Morgan III, S. D. Shillcutt, J. S. Robinson Jr. and D. K. Nakayama, *Neurosci. Lett.*, 2009, **451**, 199.
17. C. J. Morrison, G. S. Butler, H. F. Bigg, C. R. Roberts, P. D. Soloway and C. M. Overall, *J. Biol. Chem.*, 2001, **276**, 47402.
18. Y. Itoh, A. Takamura, N. Ito, Y. Maru, H. Sato, N. Suenaga, T. Aoki and M. Seiki, *EMBO J.*, 2001, **20**, 4782.
19. Y. Gong, E. Hart, A. Shchurin and J. Hoover-Plow, *J. Clin. Invest.*, 2008, **118**, 3012.
20. S. T. Vilen, P. Nyberg, M. Hukkanen, M. Sutinen, M. Ylipalosaari, A. Bjartell, A. Paju, V. Haaparanta, U. H. Stenman, T. Sorsa and T. Salo, *Exp. Cell Res.*, 2008, **314**, 914.
21. S. Menashi, R. Fridman, S. Desrivieres, H. Lu, Y. Legrand and C. Soria, *Ann. N. Y. Acad. Sci.*, 1994, **732**, 466.
22. R. Fridman, M. Toth, D. Pena and S. Mobashery, *Cancer Res.*, 1995, **55**, 2548.
23. M. W. Olson, M. M. Bernardo, M. Pietila, D. C. Gervasi, M. Toth, L. P. Kotra, I. Massova, S. Mobashery and R. Fridman, *J. Biol. Chem.*, 2000, **275**, 2661.
24. A. C. Lau, T. T. Duong, S. Ito and R. S. Yeung, *Arthritis Rheum.*, 2008, **58**, 854.
25. Z. Gu, J. Cui, S. Brown, R. Fridman, S. Mobashery, A. Y. Strongin and S. Lipton, *J. Neurosci.*, 2005, **25**, 6401.
26. H. A. Kenny, S. Kaur, L. M. Coussens and E. Lengyel, *J. Clin. Invest.*, 2008, **118**, 1367.
27. D. Xu, N. Suenaga, M. J. Edelmann, R. Fridman, R. J. Muschel and B. M. Kessler, *Mol. Cell. Proteomics*, 2008, **7**, 2215.
28. K. J. Greenlee, D. B. Corry, D. A. Engler, R. K. Matsunami, P. Tessier, R. G. Cook, Z. Werb and F. Kheradmand, *J. Immunol.*, 2006, **177**, 7312.
29. P. E. Van den Steen, P. Proost, A. Wuyts, J. Van Damme and G. Opdenakker, *Blood*, 2000, **96**, 2673.
30. M. R. Fein and M. Egeblad, *Dis. Models & Mech.*, 2013, **6**, 580.
31. M. Valiente, A. C. Obenauf, X. Jin, Q. Chen, X. H.-F. Zhang, D. J. Lee, J. E. Chaft, M. G. Kris, J. T. Huse, E. Brogi and J. Massague, *Cell*, 2014, **156**, 1002.
32. S. S. McAllister and R. A. Weinberg, *Nat. Cell Biol.*, 2014, **16**, 717.
33. M. A. Lee, J. H. Park, S. Y. Rhyu, S. T. Oh, W. K. Kang and H. N. Kim, *BMC Cancer*, 2014, **14**, 125.
34. A. H. Said, J. P. Raufman and G. Xie, *Cancers*, 2014, **6**, 366.
35. M. K. El-Badrawy, A. M. Yousef, D. Shaalan and A. Z. Elsamanoudy, *J. Bronchology Interv. Pulmonol.*, 2014, **21**, 327.
36. E. M. Yousef, M. R. Tahir, Y. St-Pierre and L. A. Gaboury, *BMC Cancer*, 2014, **14**, 609.

37. T. V. Tian, N. Tomavo, L. Huot, A. Fluorens, E. Bonnelye, S. Flajollet, D. Hot, X. Leroy, Y. de Launoit and M. Duterque-Coquillaud, *Oncogene*, 2014, **33**, 2204.
38. Y. Gong, U. D. Chippada-Venkata and W. K. Oh, *Cancers*, 2014, **6**, 1298.
39. C. Xu, D. M. Hu and Q. Zhu, *Clin. Exp. Metastasis*, 2013, **30**, 933.
40. A. M. Tester, M. Waltham, S. J. Oh, S. N. Bae, M. M. Bills, E. C. Walker, F. G. Kern, W. G. Stetler-Stevenson, M. E. Lippman and E. W. Thompson, *Cancer Res.*, 2004, **64**, 652.
41. O. Mendes, H. T. Kim, G. Lungu and G. Stoica, *Clin. Exp. Metastasis*, 2007, **24**, 341.
42. T. X. Xie, F. J. Huang, K. D. Aldape, S. H. Kang, M. Liu, J. E. Gershenwald, K. Xie, R. Sawaya and S. Huang, *Cancer Res.*, 2006, **66**, 3188.
43. M. V. Rojiani, J. Alidina, N. Esposito and A. M. Rojiani, *Int. J. Clin. Exp. Pathol.*, 2010, **3**, 775.
44. L. Hu, J. Zhang, H. Zhu, J. Min, Y. Feng and H. Zhang, *Med. Oncol.*, 2010, **27**, 708.
45. O. Mendes, H. T. Kim and G. Stoica, *Clin. Exp. Metastasis*, 2005, **22**, 237.
46. A. Rosell and E. H. Lo, *Curr. Opin. Pharmacol.*, 2008, **8**, 82.
47. B. Q. Zhao, E. Tejima and E. H. Lo, *Stroke*, 2007, **38**, 748.
48. B. V. Zlokovic, *Nat. Med.*, 2006, **12**, 390.
49. S. E. Lakhan, A. Kirchgessner, D. Tepper and A. Leonard, *Front. Neurol.*, 2013, **4**, 32.
50. J. Cui, S. Chen, C. Zhang, F. Meng, W. Wu, R. Hu, O. Hadass, T. Lehmidi, G. J. Blair, M. Lee, M. Chang, S. Mobashery, G. Y. Sun and Z. Gu, *Mol. Neurodegener.*, 2012, **7**, 21.
51. A. Rosell, A. Ortega-Aznar, J. Alvarez-Sabin, I. Fernandez-Cadenas, M. Ribo, C. A. Molina, E. H. Lo and J. Montaner, *Stroke*, 2006, **37**, 1399.
52. M. Asahi, X. Wang, T. Mori, T. Sumii, J. C. Jung, M. A. Moskowitz, M. E. Fini and E. H. Lo, *J. Neurosci.*, 2001, **21**, 7724.
53. D. I. Chang, N. Hosomi, J. Lucero, J. H. Heo, T. Abumiya, A. P. Mazar and G. J. del Zoppo, *J. Cereb. Blood Flow Metab.*, 2003, **23**, 1408.
54. S. N. Chapman, P. Mehndiratta, M. C. Johansen, T. L. McMurry, K. C. Johnston and A. M. Southerland, *Vasc. Health Risk Manage.*, 2014, **10**, 75.
55. K. Tsuji, T. Aoki, E. Tejima, K. Arai, S. R. Lee, D. N. Atochin, P. L. Huang, X. Wang, J. Montaner and E. H. Lo, *Stroke*, 2005, **36**, 1954.
56. M. Ning, K. L. Furie, W. J. Koroshetz, H. Lee, M. Barron, M. Lederer, X. Wang, M. Zhu, A. G. Sorensen, E. H. Lo and P. J. Kelly, *Neurology*, 2006, **66**, 1550.
57. M. Prins, T. Greco, D. Alexander and C. C. Giza, *Dis. Models & Mech.*, 2013, **6**, 1307.
58. D. J. Loane and A. I. Faden, *Trends Pharmacol. Sci.*, 2010, **31**, 596.
59. M. Grossetete, J. Phelps, L. Arko, H. Yonas and G. A. Rosenberg, *Neurosurgery*, 2009, **65**, 702.

60. F. Jia, Y. H. Yin, G. Y. Gao, Y. Wang, L. Cen and J.-Y. Jiang, *J. Neurotrauma*, 2014, **31**, 1225.
61. O. Hadass, B. N. Tomlinson, M. Gooyit, S. Chen, J. J. Purdy, J. M. Walker, C. Zhang, A. B. Giritharan, W. Purnell, C. R. Robinson 2nd, D. Shin, V. A. Schroeder, M. A. Suckow, A. Simonyi, G. Y. Sun, S. Mobashery, J. Cui, M. Chang and Z. Gu, *PLoS One*, 2013, **8**, e76904.
62. X. Wang, J. Jung, M. Asahi, W. Chwang, L. Russo, M. A. Moskowitz, C. E. Dixon, M. E. Fini and E. H. Lo, *J. Neurosci.*, 2000, **20**, 7037.
63. Y. Shigemori, Y. Katayama, T. Mori, T. Maeda and T. Kawamata, *Acta Neurochir.*, 2006, **96**, 130.
64. C. N. Hansen, L. C. Fischer, R. J. Deibert, L. B. Jakeman, H. Zhang, L. Noble-Haeusslein, S. White and D. M. Basso, *J. Neurosci.*, 2013, **33**, 13101.
65. Z. Wang, Q. Fang, B. Q. Dang, X. M. Shen, Z. Shu, G. Zuo, W. C. He and G. Chen, *Ann. Clin. Lab. Sci.*, 2012, **42**, 14.
66. R. Serra, G. Volpentesta, L. Gallelli, R. Grande, G. Buffone, A. Lavano and S. de Franciscis, *Br. J. Neurosurg.*, 2015, 1.
67. X. X. Wang, M. S. Tan, J. T. Yu and L. Tan, *Biomed Res. Int.*, 2014, **2014**, 908636.
68. N. Schultz, H. M. Nielsen, L. Minthon and M. Wennstrom, *J. Neuropathol. Exp. Neurol.*, 2014, **73**, 684.
69. J. Duran-Vilaregut, J. del Valle, G. Manich, A. Camins, M. Pallas, J. Vilaplana and C. Pelegri, *Neuropathol. Appl. Neurobiol.*, 2011, **37**, 525.
70. K. H. Chang, Y. R. Wu, Y. C. Chen and C. M. Chen, *Brain, Behav., Immun.*, 2015, **44**, 121.
71. X. He, L. Zhang, X. Yao, J. Hu, L. Yu, H. Jia, R. An, Z. Liu and Y. Xu, *PLoS One*, 2013, **8**, e73777.
72. A. Kaplan, K. J. Spiller, C. Towne, K. C. Kanning, G. T. Choe, A. Geber, T. Akay, P. Aebischer and C. E. Henderson, *Neuron*, 2014, **81**, 333.
73. E. Tsourdi, A. Barthel, H. Rietzsch, A. Reichel and S. R. Bornstein, *Biomed Res. Int.*, 2013, **2013**, 385641.
74. A. B. Wysocki, L. Staiano-Coico and F. Grinnell, *J. Invest. Dermatol.*, 1993, **101**, 64.
75. Y. Liu, D. Min, T. Bolton, V. Nube, S. M. Twigg, D. K. Yue and S. V. McLennan, *Diabetes Care*, 2009, **32**, 117.
76. E. A. Rayment, Z. Upton and G. K. Shooter, *Br. J. Dermatol.*, 2008, **158**, 951.
77. M. Gooyit, Z. Peng, W. R. Wolter, H. Pi, D. Ding, D. Hesek, M. Lee, B. Boggess, M. M. Champion, M. A. Suckow, S. Mobashery and M. Chang, *ACS Chem. Biol.*, 2014, **9**, 105.
78. J. F. Fisher and S. Mobashery, *Cancer Metastasis Rev.*, 2006, **25**, 115.
79. N. G. Li, Y. P. Tang, J. A. Duan and Z. H. Shi, *Expert Opin. Ther. Pat.*, 2014, **24**, 1039.
80. M. Tauro, J. McGuire and C. C. Lynch, *Cancer Metastasis Rev.*, 2014, **33**, 1043.
81. W. M. Pardridge, *Drug Discovery Today*, 2007, **12**, 54.

82. J. F. Deeken and W. Loscher, *Clin. Cancer Res.*, 2007, **13**, 1663.
83. S. B. Brown, M. M. Bernardo, Z. H. Li, L. P. Kotra, Y. Tanaka, R. Fridman and S. Mobashery, *J. Am. Chem. Soc.*, 2000, **11**, 6799.
84. M. Lee, A. Villegas-Estrada, G. Celenza, B. Boggess, M. Toth, G. Kreitinger, C. Forbes, R. Fridman, S. Mobashery and M. Chang, *Chem. Biol. Drug Des.*, 2007, **70**, 371.
85. C. Forbes, Q. Shi, J. F. Fisher, M. Lee, D. Hesek, L. I. Llarrull, M. Toth, M. Gossing, R. Fridman and S. Mobashery, *Chem. Biol. Drug Des.*, 2009, **74**, 527.
86. G. Celenza, A. Villegas-Estrada, M. Lee, B. Boggess, C. Forbes, W. R. Wolter, M. A. Suckow, S. Mobashery and M. Chang, *Chem. Biol. Drug Des.*, 2008, **71**, 187.
87. M. Gooyit, M. A. Suckow, V. A. Schroeder, W. R. Wolter, S. Mobashery and M. Chang, *ACS Chem. Neurosci.*, 2012, **3**, 730.
88. C. W. Lin, S. C. Shen, W. C. Hou, L. Y. Yang and Y. C. Chen, *Mol. Cancer Ther.*, 2008, **7**, 1195.
89. H. Ling, H. Yang, S. H. Tan, W. K. Chui and E. H. Chew, *Br. J. Pharmacol.*, 2010, **161**, 1763.
90. Z. Ke, H. Lin, Z. Fan, T. Q. Cai, R. A. Kaplan, C. Ma, K. A. Bower, X. Shi and J. Luo, *Int. J. Cancer*, 2006, **119**, 8.
91. H. Y. Zhou, Y. L. Pon and A. S. Wong, *Endocrinology*, 2007, **148**, 5195.
92. J. H. Choi, K. C. Choi, N. Auersperg and P. C. Leung, *Cancer Res.*, 2006, **66**, 3912.
93. Y. Huang, J. Li, L. Xiang, D. Han, X. Shen and X. Wu, *Eur. J. Obstet. Gynecol. Reprod. Biol.*, 2012, **165**, 307.
94. P. L. Yao, Y. C. Lin and J. H. Richburg, *Biol. Reprod.*, 2012, **86**, 1.
95. N. Guo, F. Liu, L. Yang, J. Huang, X. Ding and C. Sun, *Oncol. Rep.*, 2014, **32**, 794.
96. Z. Dong, R. D. Bonfil, S. Chinni, X. Deng, J. C. Trindade Filho, M. M. Bernardo, U. Vaishampayan, M. Che, B. F. Sloane, S. Sheng, R. Fridman and M. L. Cher, *Am. J. Pathol.*, 2005, **166**, 1173.
97. R. D. Bonfil, A. Sabbota, S. Nabha, M. M. Bernardo, Z. Dong, H. Meng, H. Yamamoto, S. R. Chinni, I. T. Lim, M. Chang, L. C. Filetti, S. Mobashery, M. L. Cher and R. Fridman, *Int. J. Cancer*, 2006, **118**, 2721.
98. A. Kruger, M. J. Arlt, M. Gerg, C. Kopitz, M. M. Bernardo, M. Chang, S. Mobashery and R. Fridman, *Cancer Res.*, 2005, **65**, 3523.
99. E. Y. Kim, J. M. Seo, K. J. Cho and J. H. Kim, *Oncogene*, 2010, **29**, 1167.
100. M. D. Martin, K. J. Carter, S. R. Jean-Philippe, M. Chang, S. Mobashery, S. Thiolloy, C. C. Lynch, L. M. Matrisian and B. Fingleton, *Cancer Res.*, 2008, **68**, 6251.
101. Z. Gu, M. Kaul, B. Yan, S. J. Kridel, J. Cui, A. Strongin, J. W. Smith, R. C. Liddington and S. A. Lipton, *Science*, 2002, **297**, 1186.
102. H. S. Ranasinghe, A. Scheepens, E. Sirimanne, M. D. Mitchell, C. E. Williams and M. Fraser, *Dev. Neurosci.*, 2012, **34**, 417.

103. A. Armulik, G. Genove, M. Mae, M. H. Nisancioglu, E. Wallgard, C. Niaudet, L. He, J. Norlin, P. Lindblom, K. Strittmatter, B. R. Johansson and C. Betsholtz, *Nature*, 2010, **468**, 557.
104. N. Luplertlop, D. Misse, D. Bray, V. Deleuze, J. P. Gonzalez, V. Leardkamolkarn, H. Yssel and F. Veas, *EMBO Rep.*, 2006, **7**, 1176.
105. Z. Guo, X. Sun, Z. He, Y. Jiang, X. Zhang and J. H. Zhang, *Neurol. Res.*, 2010, **32**, 715.
106. Z. D. Guo, X. D. Zhang, H. T. Wu, B. Lin, X. C. Sun and J. H. Zhang, *Acta Neurochir.*, 2011, **110**, 81.
107. G. E. Barreto, X. Sun, L. Xu and R. G. Giffard, *PLoS One*, 2011, **6**, e27881.
108. R. D. Bell, E. A. Winkler, I. Singh, A. P. Sagare, R. Deane, Z. Wu, D. M. Holtzman, C. Betsholtz, A. Armulik, J. Sallstrom, B. C. Berk and B. V. Zlokovic, *Nature*, 2012, **485**, 512.
109. F. Yu, H. Kamada, K. Niizuma, H. Endo and P. H. Chan, *J. Neurotrauma*, 2008, **25**, 184.
110. H. Liu and V. I. Shubayev, *Exp. Neurol.*, 2011, **231**, 236.
111. M. L. Block, L. Zecca and J. S. Hong, *Nat. Rev. Neurosci.*, 2007, **8**, 57.
112. J. Rogers, D. Mastroeni, B. Leonard, J. Joyce and A. Grover, *Int. Rev. Neurobiol.*, 2007, **82**, 235.
113. M. S. Choi, K. S. Cho, S. M. Shin, H. M. Ko, K. J. Kwon, C. Y. Shin and K. H. Ko, *Arch. Pharmacal. Res.*, 2010, **33**, 257.
114. S. A. Testero, M. Lee, R. T. Staran, M. Espahbodi, L. I. Llarrull, M. Toth, S. Mobashery and M. Chang, *ACS Med. Chem. Lett.*, 2011, **2**, 177.
115. M. Lee, M. Ikejiri, D. Klimpel, M. Toth, M. Espahbodi, D. Hesek, C. Forbes, M. Kumarasiri, B. C. Noll, M. Chang and S. Mobashery, *ACS Med. Chem. Lett.*, 2012, **3**, 490.
116. M. Gooyit, M. Lee, D. Hesek, B. Boggess, A. G. Oliver, R. Fridman, S. Mobashery and M. Chang, *Chem. Biol. Drug Des.*, 2009, **74**, 535.
117. M. Gooyit, W. Song, K. V. Mahasenan, K. Lichtenwalter, M. A. Suckow, V. A. Schroeder, W. R. Wolter, S. Mobashery and M. Chang, *J. Med. Chem.*, 2013, **56**, 8139.
118. M. Lee, G. Celenza, B. Boggess, J. Blase, Q. Shi, M. Toth, M. M. Bernardo, W. R. Wolter, M. A. Suckow, D. Hesek, B. C. Noll, R. Fridman, S. Mobashery and M. Chang, *Chem. Biol. Drug Des.*, 2009, **73**, 189.
119. M. Gooyit, M. Lee, V. A. Schroeder, M. Ikejiri, M. A. Suckow, S. Mobashery and M. Chang, *J. Med. Chem.*, 2011, **54**, 6676.
120. M. Ikejiri, M. M. Bernardo, R. D. Bonfil, M. Toth, M. Chang, R. Fridman and S. Mobashery, *J. Biol. Chem.*, 2005, **280**, 33992.
121. W. Song, Z. Peng, M. Gooyit, M. A. Suckow, V. A. Schroeder, W. R. Wolter, M. Lee, M. Ikejiri, J. Cui, Z. Gu and M. Chang, *ACS Chem. Neurosci.*, 2013, **4**, 1168.

CHAPTER 11

Coumarins

STEFAN BRÄSE, FRANZISKA GLÄSER* AND THOMAS HURRLE

Karlsruhe Institute of Technology (KIT), 76131 Karlsruhe, Germany
*Email: glaeser.f@web.de

11.1 General Considerations of Coumarins

Coumarin or 1,2-benzopyrone (**1**) is the lactone of the corresponding *o*-hydroxy carboxylic acid, *i.e.* *o*-coumaric acid (**2**, Figure 11.1).[1] It is a natural compound that occurs in many plants, microorganisms and a few animals and was first extracted from the tonka bean.[2] The name for coumarin (**1**) is etymologically derived from the kumaru, the Tupi word for tonka bean tree.[3] It lends its name to a large class of natural products which are based on the benzopyrone motif.[4] Substituents on the coumarin core are labelled according to the numbering scheme shown in Figure 11.1.

11.1.1 Metabolic Aspects

The biosynthesis of coumarin (**1**, Scheme 11.1) in plants was investigated *via* the synthesis of ^{14}C labelled compounds and their use in metabolic studies.[5–7] Starting from phenylalanine (**3**), elimination of the amine leads to cinnamic acid (**4**), which is in turn hydroxylated to give *o*-coumaric acid (**2**). Coumarin (**1**) is then formed in a condensation reaction *via* their metabolic intermediates, the glucosides **5**.[5–8] The shown pathway (Scheme 11.1) is simplified for coumarin (**1**). Various enzymes are involved in the biosynthesis of the more complex coumarins, which are shown later on (see Section 11.1.2).[9]

RSC Drug Discovery Series No. 50
Privileged Scaffolds in Medicinal Chemistry: Design, Synthesis, Evaluation
Edited by Stefan Bräse

Published by the Royal Society of Chemistry, www.rsc.org

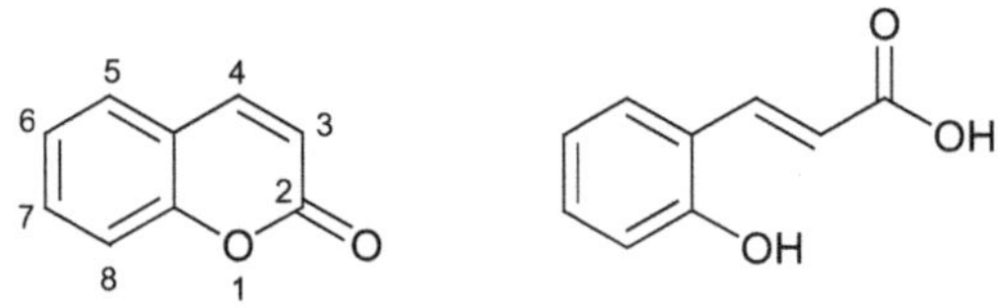

Figure 11.1 Coumarin (**1**) with numbering and *o*-coumaric acid (**2**).[1]

After oral absorption, coumarins (**1**) are immediately hydroxylated in an intense first-pass effect, which decreases the bioavailability of unmodified coumarins to less than 4%. The main metabolite, 7-hydroxy coumarin (**6**) is formed by enzyme CYP2A6 and quickly conjugates to 7-hydroxy coumarin glucuronide (**7**). The conjugate has a biological half-life period of 1–1.5 hours and is then excluded (Scheme 11.2).[1]

In principle, this mechanism can be assumed to be similar for most naturally occurring coumarins. The diversity in the natural products originating from coumarin can mostly be ascribed to the modification of the coumarin core, limiting the diversity in natural products to a certain extent. The next section bears a compilation of exemplary coumarins extracted from plants.

11.1.2 Coumarins – Natural Products

The parent compound coumarin (**1**) was initially found in the tonka bean (*dipteryx odorata* wild) and 150 different species are distributed over 30 different families, like *Rutaceae*, *Umbelliferae*, *Clusiaceae*, *Guttiferae*, *Caprifoliaceae*, *Oleaceae*, *Nyctaginaceae*, and *Apicaceae*, to name only the important ones. Coumarins occur in a large number of plants, particularly in grasses, clovers, and in carrots, and occur in the highest amounts in fruits, berries, seeds, roots, green tea and chicory.[10–13] They are found in substantial proportions in essential oils such as cassia oil, cinnamon bark oil and lavender oil.[9,14] Coumarin can act as a phytohormone[15–18] and allelochemical.[19,20] Other derivatives like psoralen (**16**) or bergapten (**19**) are reported as kairomones.[21,22] Beside their semiochemical properties,[23] naturally occurring coumarins have a rather limited chemical diversity. In this case, the coumarin derivatives are members of six main types: simple coumarins (Figure 11.2), furano coumarins, dihydrofurano coumarins (Figure 11.3), linear type pyrano coumarins, angular type pyrano coumarins, phenyl coumarins, and biscoumarins (Figure 11.4).[24]

11.1.3 Syntheses of Coumarins

This chapter gives a representative overview of synthetic methods to produce coumarins. Coumarins with substituents in 5-, 6-, 7- and 8-position can generally be prepared by reactions using readily available starting materials (*e.g.* salicylaldehydes) that provide the desired substitution pattern. Substitution in the 3 and 4 positions is often achieved in the condensation step or *via* post-condensation methods.

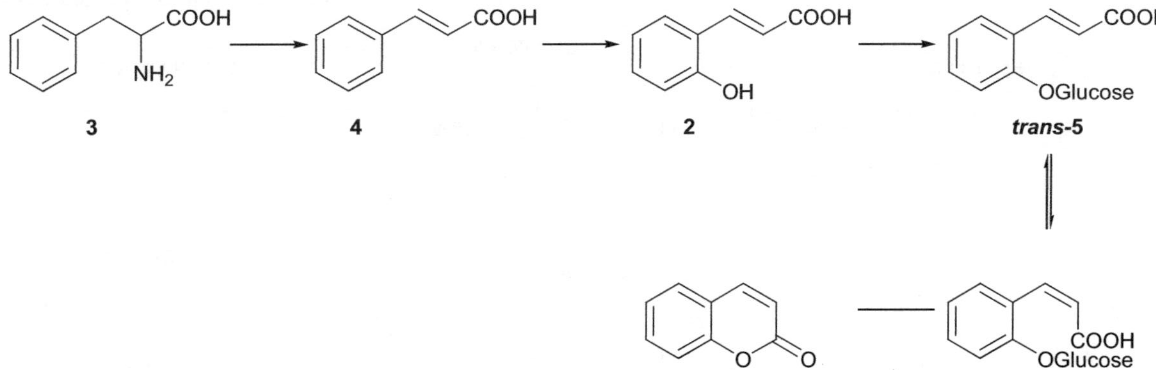

Scheme 11.1 Simplified biosynthesis of coumarin (**1**).[8]

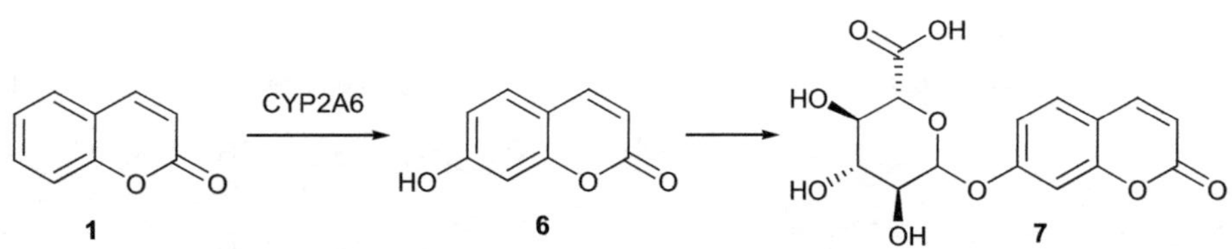

Scheme 11.2 Metabolism of coumarin (**1**).[1]

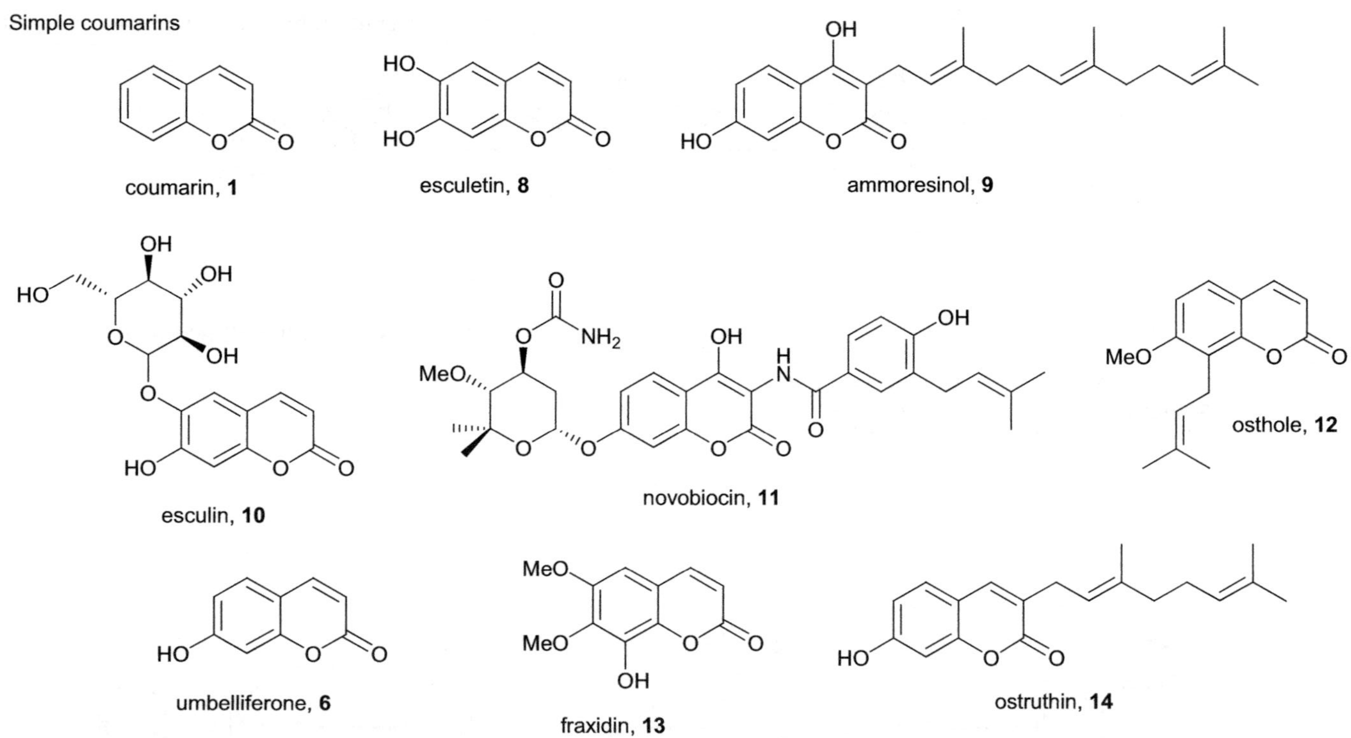

Figure 11.2 Examples of simple coumarins.

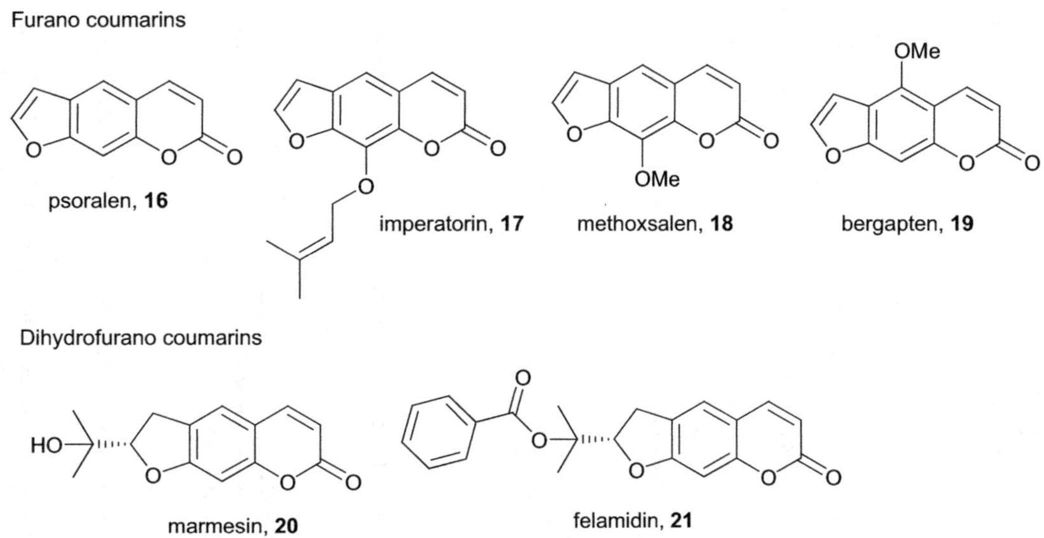

Figure 11.3 Examples of furano coumarins.

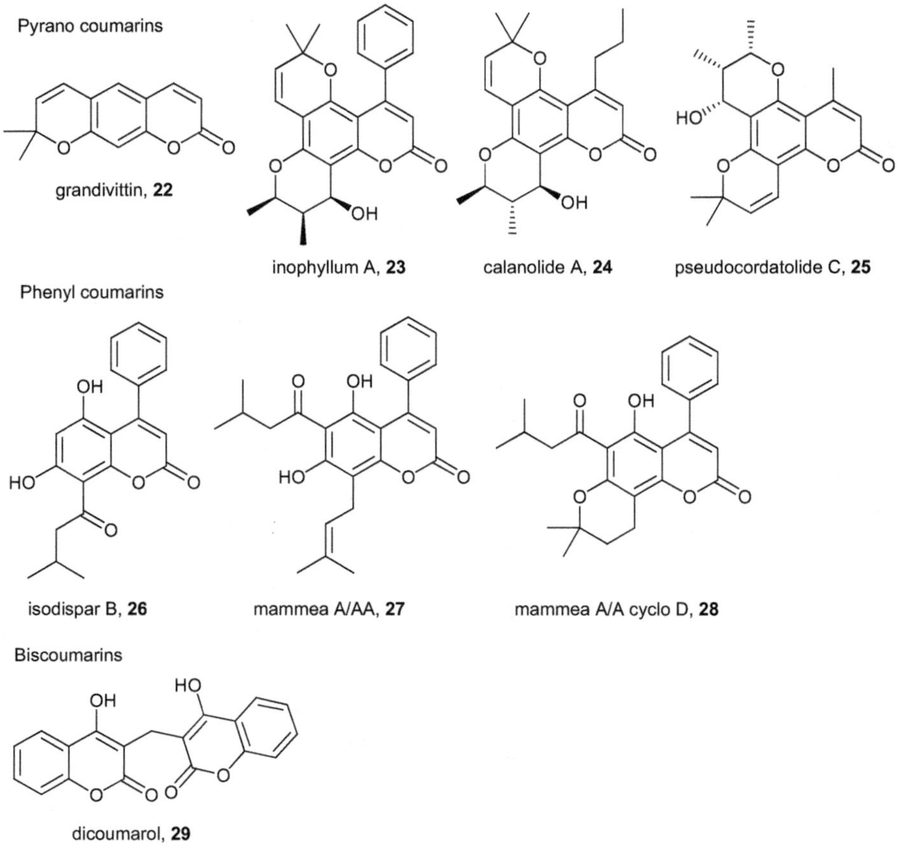

Figure 11.4 Examples of pyrano coumarins, phenyl coumarins and biscoumarins.

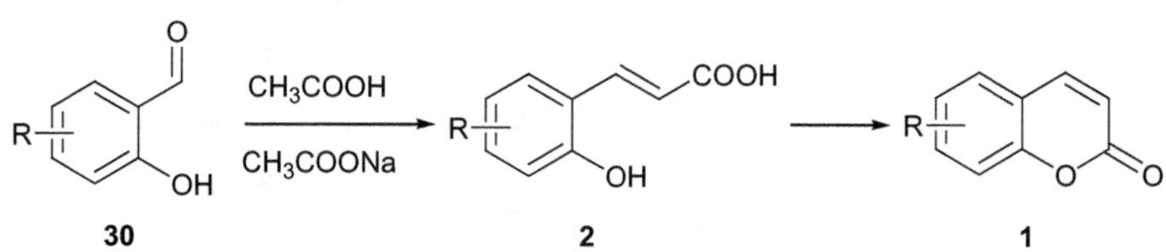

Scheme 11.3 Perkin reaction for the preparation of coumarin (**1**, R = H).

Coumarin (**1**) can be prepared from acetic acid and salicylaldehyde (**30**) *via* a modified Perkin reaction (Scheme 11.3).[25] Salicylaldehyde is treated with acetic acid and sodium acetate to yield *o*-coumaric acid (**2**) and an intramolecular esterification of acid **2** then forms coumarin (**1**).

A classical method for the synthesis of 3-carboxy coumarins **32** is the Knoevenagel condensation under microwave irradiation,[26] which allows reduced reaction times, simplified synthetic preparations and higher purities of the crude products, simplifying the purification (Scheme 11.4).[26]

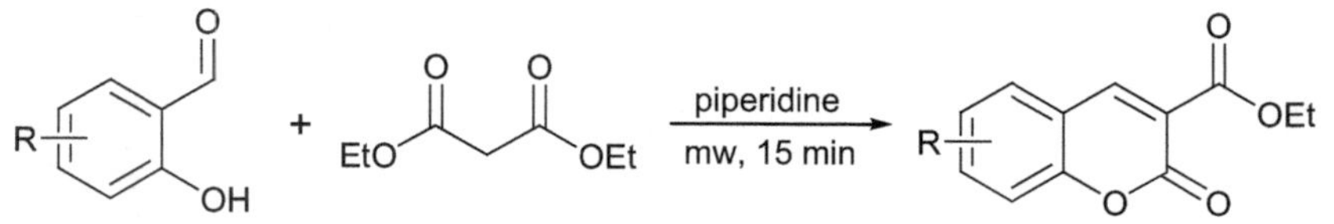

30 **31** **32**

Scheme 11.4 Knoevenagel reaction for the synthesis of 3-carboxy coumarins **32**.[26]

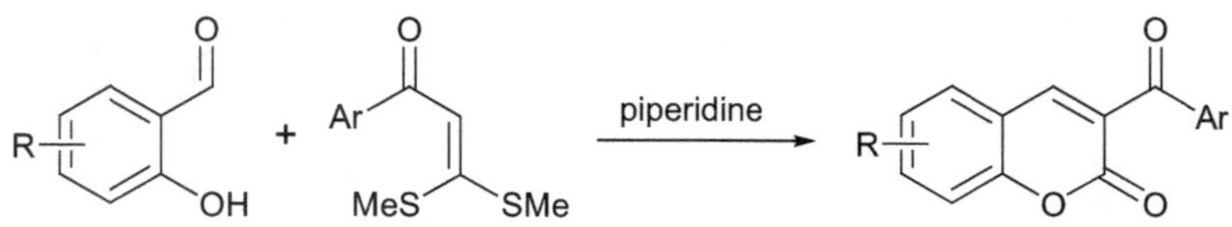

30 **33** **34**

Scheme 11.5 Synthesis of 3-aroylcoumarins (**34**).[27]

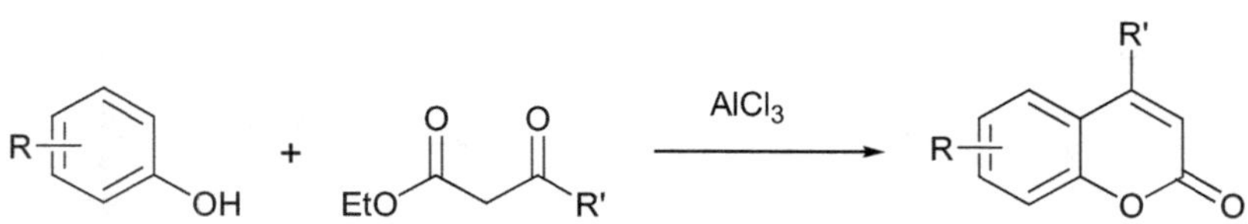

35 **36** **37**

Scheme 11.6 Pechmann condensation for 4-substituted coumarins **37**.

A facile and efficient synthesis of 3-aroylcoumarins **34** was developed by Rao *et al.* in 2006.[27] Here, a combinatorial library was synthesised by the condensation of α-aroylketene dithioacetals and 2-hydroxybenzaldehydes with catalytic amounts of piperidine in refluxing THF (Scheme 11.5).

The Pechmann condensation can be used for the synthesis of 4-substituted coumarins **37**. The reaction of phenols **35** with β-ketoesters **36** is often catalysed by the Lewis acid $AlCl_3$ (Scheme 11.6).[28,29]

Coumarins with substitutions in position 4 can also be obtained by hydroarylation of aryl propionic acid methyl esters **38** with various aryl boronic acids **39** under copper catalysis[30] or by palladium-catalysed oxidative cyclocarbonylation of 2-vinylphenols **40** with low pressure of carbon monoxide (Scheme 11.7).[31]

A very interesting approach is the carbamoyl rendition of the Baker–Venkataraman rearrangement, which allows a regiospecific route to substituted 4-hydroxy coumarins **44**.[32] The intermediate arylcarbamate **43** is prepared by directed *ortho*-metalation followed by $ZnCl_2$ transmetalation and cross-coupling with acid chlorides **42** under conditions described by Negishi.[33] Then, the Baker–Venkataraman rearrangement in the presence of NaH as base gives an easy access to 4-hydroxy coumarins (**44**, Scheme 11.8).

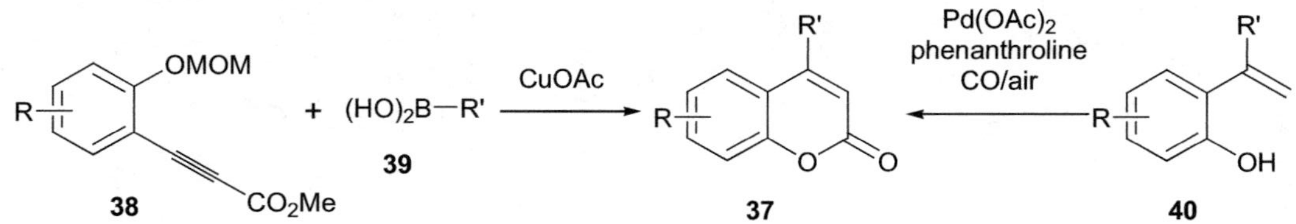

Scheme 11.7 Metal-catalysed coumarin syntheses.

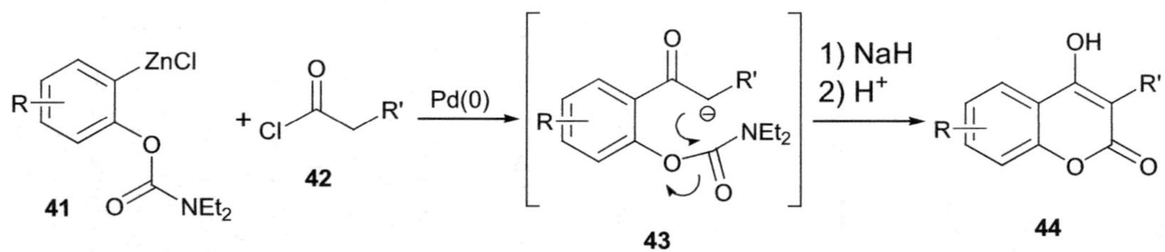

Scheme 11.8 Directed *ortho*-metalation – cross-coupling linkage followed by a Baker–Venkataraman rearrangement for the synthesis of 4-hydroxy coumarins **44**.

During the last decade, *N*-heterocyclic carbenes (NHC) have been effectively applied as organocatalysts, *e.g.* for the synthesis of 3-benzylcoumarins **47**.[34] The latter method is based on NHC-promoted Umpolung of α,β-unsaturated aldehydes **45** followed by annulation with salicylaldehydes **30** (Scheme 11.9). Various combinations of ionic liquid and base were tested and dimethyl 1,3-dimethylimidazolium phosphate **46** and potassium carbonate turned out to be the most effective ones.

This Umpolung reaction allowed the synthesis of a large library of 3-benzylcoumarins **47** under microwave irradiation by using a variety of salicylaldehydes **30** and cinnamaldehydes **45**. However, the largest library of synthesised coumarins so far has been achieved by post-condensation modification techniques.[23,35] Due to the successful application of various functional group transformations for the modification of the coumarin skeleton, the work of Bäuerle *et al.*,[36] resulting in a 151-membered library is up to now one of the most important contributions to combinatorial coumarin syntheses. Bromo-substituted coumarin precursors **49** were modified by addition of several alkenes, alkynes, and boronic acids (Scheme 11.10).

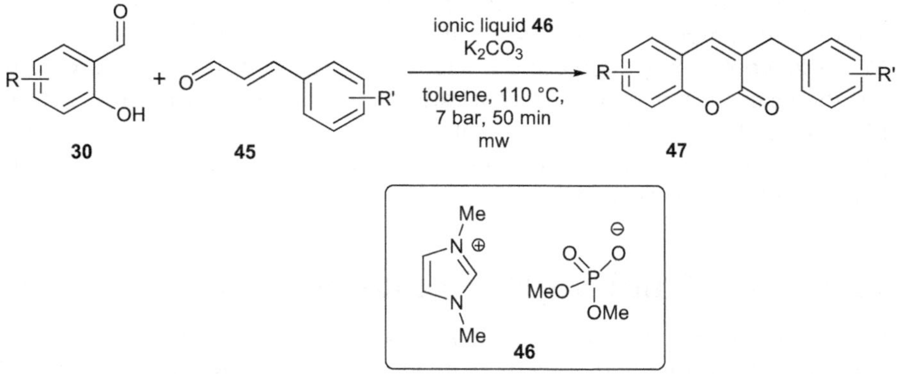

Scheme 11.9 NHC-promoted 3-benzylcoumarin synthesis.[34]

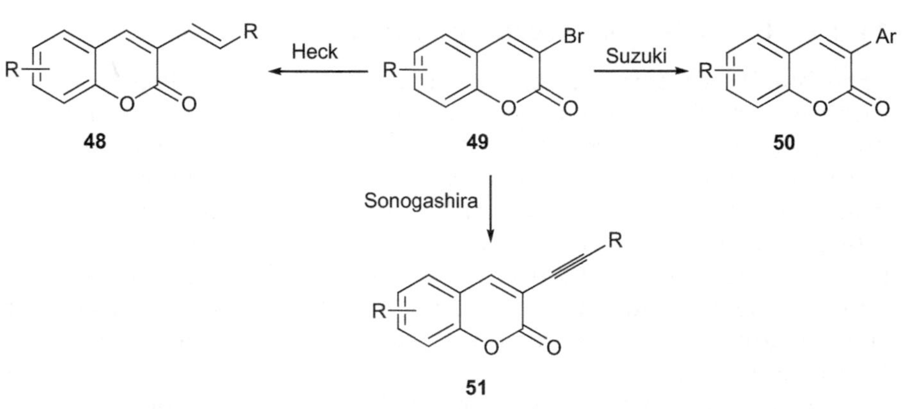

Scheme 11.10 Post-condensation modifications on 3-bromo coumarins **49**.

11.1.4 Coumarins and their Fluorescence

Besides their biological activity (see Section 11.2), coumarins are known for their optoelectronic properties. Coumarin and its derivatives play important roles in fluorescent metal ion sensors,[37] laser dyes,[38] and as organic sensitizers in high-efficiency dye-sensitized solar cells.[39–41] The different applications require diverse modifications. The optoelectronic properties can then be tuned through the attachment of substituents to the coumarin core.

Coumarins have been investigated to rationalise their UV-Vis absorption wavelengths, molar extinction coefficients, and fluorescence quantum efficiency.[42] The analysis of depicted coumarins **52–55** (Figure 11.5) showed that an electron-donating amino group limits nonemissive twisted intramolecular charge transfer (TICT) and results in increased fluorescence and lasing efficiencies of these dyes. Furthermore, electron-withdrawing groups (–COOH or –COOEt) situated in 3-position enhance ICT and shift the absorption and alter the fluorescence and lasing spectra of the respective coumarins towards longer wavelengths.[42]

These photoelectronic features of coumarins allowed their use as novel fluorescent probes for thiols, *e.g.* in biological systems. In this case, a fluorescent compound **56** is firstly transformed into a non-fluorescent probe **57**. Thiol-containing compounds, such as cysteine, homocysteine or glutathione can then react in a Michael addition to the α,β-unsaturated ketone **57**. The Michael addition is associated with the cleavage of the hemiketal group,[43] leading to the formation of a fluorescent 7-hydroxyl coumarin derivative **58** (Scheme 11.11).

11.2 Case Studies – New Leads and Marketed Drugs

In this section, we present a selected number of coumarin-derived drugs and new lead structures with a coumarin motif. Due to the large number of applications,[44] we aimed for a selected number of case studies.

11.2.1 Cannabinoid Receptor Agonists

Cannabinoid receptors belong to the endogenous cannabinoid system, which is part of the nervous system, involved in the various physiological processes. It consists of cannabinoid receptors, endocannabinoids as well as enzymes that synthesise the endocannabinoids.[45] Cannabinoid receptor ligands or extracts from *cannabis sativa* demonstrate high pharmacological activity and have been extensively evaluated in clinical trials.[46] For example, they were shown to be effective in the relief of pain, in the treatment of multiple sclerosis and neuropathic pain, as well in the treatment of Tourette's syndrome tics.[45,47] Furthermore, cannabinoids are approved for the treatment of nausea and vomiting for cancer patients in chemotherapy and as an appetite stimulant for AIDS patients.[45] There is also evidence that CB_1 antagonists can be used in the treatment of Parkinson's and Huntington's

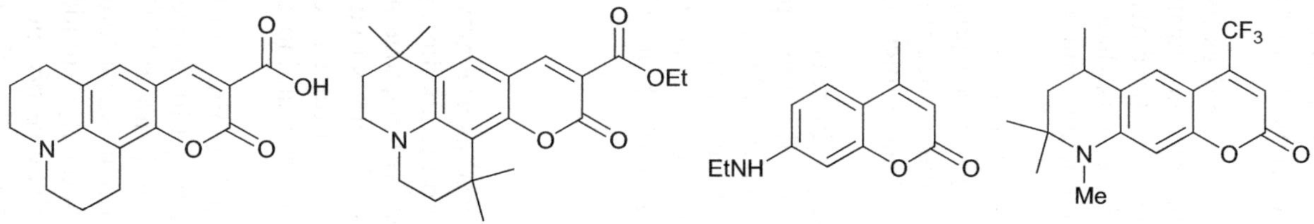

coumarin 343, **52** coumarin 314, **53** coumarin 445, **54** coumarin 522B, **55**

Figure 11.5 Fluorescent coumarins **52–55.**[42]

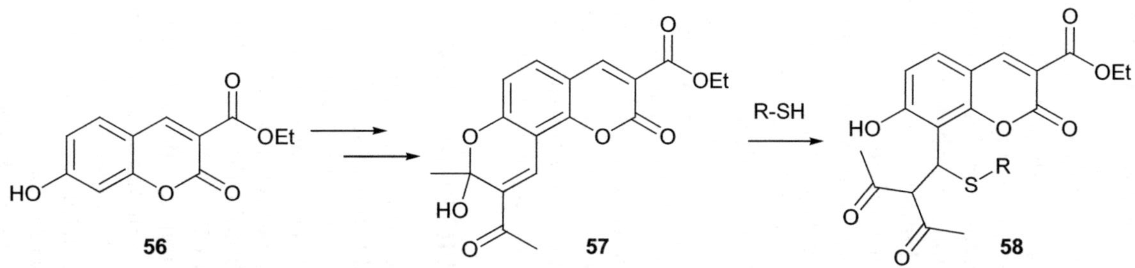

Scheme 11.11 Transformation of fluorescent compound **56** to non-fluorescent probe **57** and the transformation of probe **57** in presence of thiols to fluorescent compound **58.**[43]

diseases, probably due to an interaction between adenosine A_{2A} and cannabinoid CB_1 receptors in the brain striatum.[48,49]

Both of the up-to-now known subtypes of the cannabinoid receptors, the CB_1 and the CB_2 receptor, are part of the G protein-coupled receptor (GPCR) superfamily and are coupled *via* $G_{i/o}$ proteins, on the one hand inhibiting adenylate cyclase and on the other hand activating mitogen-activated protein kinase.[50]

In recent investigations, a novel series of coumarin derivatives has been reported, with a focus on their potency as cannabinoid receptor ligands.[51] A library of 3-benzylcoumarins **47** was synthesised by the abovementioned NHC-promoted Umpolung reaction (Scheme 11.9)[34] and was then investigated in radioligand binding assays to determine their activity towards both of the cannabinoid receptor subtypes CB_1 and CB_2.

Structure activity relationship studies revealed the influence of bulky or hydrophilic substituents on the affinity for the investigated receptor family and the selectivity between CB_1 and CB_2 receptors. Bulky substituents in 7-position led to CB_2 **47b** selectivity, whereas their absence increased the selectivity towards CB_1 **47a.**

However, the alkyl chains in 7-position were necessary to increase activity in the first place.[52] It is assumed that this side chain is the dominant pharmacophore in THC.[53,54] The structural comparison between tetrahydrocannabinol (THC, **48**) and coumarins **47** (Figure 11.6) suggests a similar activity.

In addition, functional properties were investigated in cAMP (cyclic adenosine monophosphate) assays using CHO cells stably expressing the human CB_1 or CB_2 receptor subtype, respectively.[52]

11.2.2 GPR55-antagonists

Recently, the orphan receptor GPR55 has been reported as a putative novel cannabinoid receptor subtype that is activated by lipid metabolites, such as lysophosphatidic acid and sphingosine 1-phosphate.[55] GPR55, together with the still relatively uninvestigated receptor GPR18, have been associated with the cannabinoid receptors. This association was based on their interaction with cannabinoid ligands rather than correlation of the amino acid sequence with $CB_{1/2}$.[55–57]

GPR55 is found especially in the cells of the heart,[58] brain, and tumour tissue[59] and is of particular interest as potential drug target in the treatment of diabetes, Parkinson's disease, neuropathic pain and cancer.[60–63] In an extensive structure-activity relationship study, 3-benzyl substituted coumarins, who are also agonists to the receptors CB_1, and CB_2 were identified as a novel antagonist class for GPR55.[60] The selectivity *versus* the related receptors CB_1, CB_2 and GPR18 was assessed. The study led to the identification of competitive GPR55 antagonists. Long aliphatic chains in 7-position led to potent, possibly allosteric GPR55 antagonists additionally showing $CB_{1/2}$ receptor affinity. In particular 7-(1,1-dimethyloctyl)-5-hydroxy-3-(2-hydroxybenzyl)-2H-chromen-2-one (**47d**) ($IC_{50} = 0.113$ μM, $K_B = 0.561$ μM)

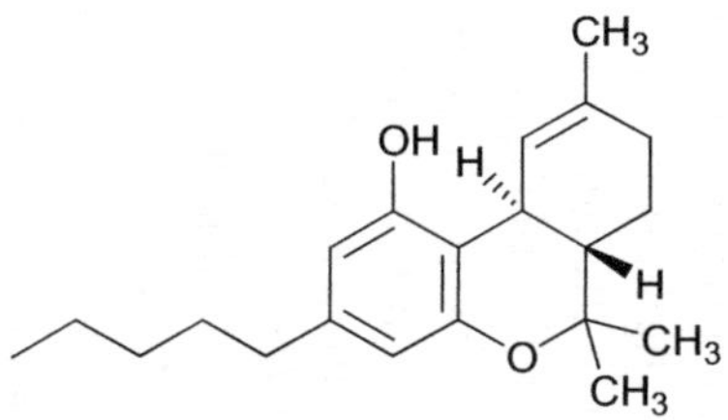

THC, 48
K_i (hCB_1) = 39.5 nM
K_i (hCB_2) = 40.0 nM

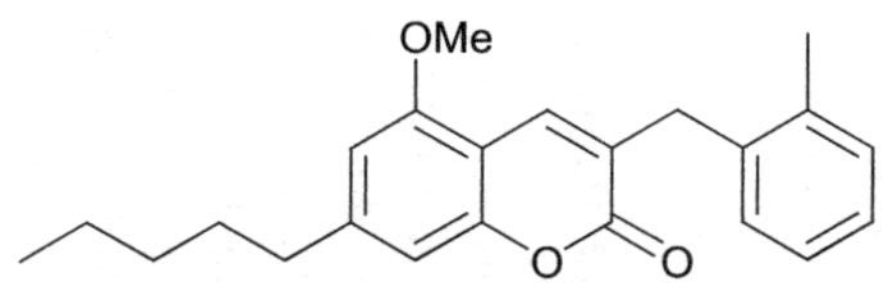

47a
K_i (hCB_1) = 22 nM
K_i (hCB_2) = 405 nM

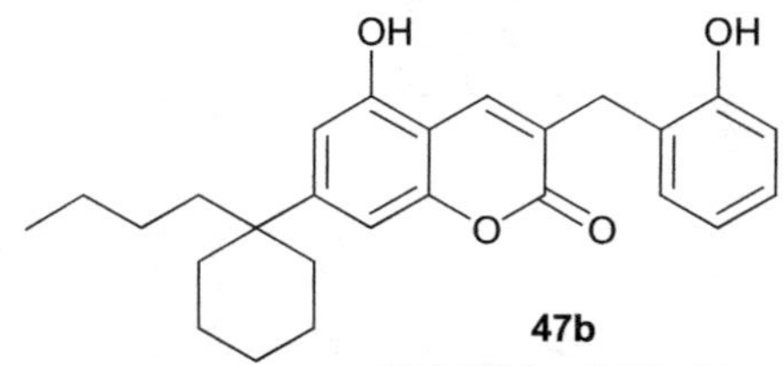

47b
K_i (hCB_1) = 4890 nM
K_i (hCB_2) = 49 nM

Figure 11.6 High affinity coumarins with selective activity for CB_1 (**47a**) and CB_2 (**47b**) and their comparison with THC (**48**).

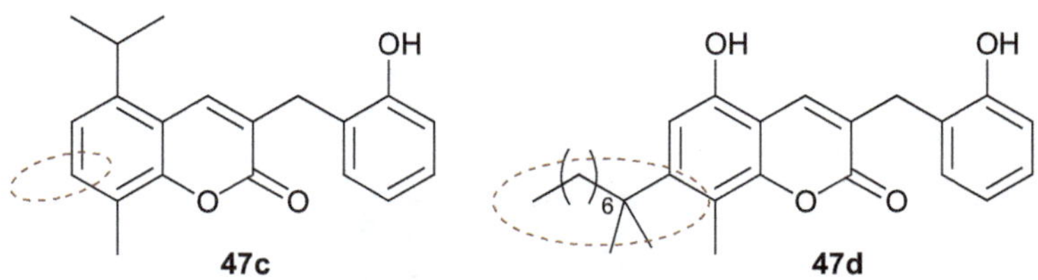

47c **47d**

human GPR55: pA_2 0.547 μM
human CB_1: K_i > 10 μM
human CB_1: K_i > 10 μM
human GPR18: IC_{50} > 10 μM

human GPR55: pA_2 0.483 μM
human CB_1: K_i 1.17μM
human CB_1: K_i 0.292 μM
human GPR18: IC_{50} > 10 μM

Figure 11.7 Two antagonists for GPR55.[60]

and 7-(1,1-dimethylheptyl)-5-hydroxy-3-(2-hydroxybenzyl)-2*H*-chromen-2-one ($IC_{50} = 0.261$ μM) were shown to be very potent GPR55 antagonists within the assorted samples (Figure 11.7).[60]

Computational quantitative structure activity relationship studies on these compounds led to models that can predict the activities of putative compounds and is therefore a possible tool for the design of novel GPR55 inhibitors.[64]

11.2.3 Vitamin-K-antagonists/Anticoagulants

A large class of coumarins based on the lead structure of warfarin (**49**) (coumadin),[65] a 4-hydroxy coumarin derivative, has been shown to act as vitamin K (**51**) antagonists and is therefore used as anticoagualants. Other well-known 4-hydroxy coumarins with the same activity are Phenprocoumon (**50**), Acenocumarol, Tromexan, Coumatetralyl, Bromadiolon, Brodifacoum (**53**) and Flocoumafen (**52**), which are FDA approved for use as anticoagulant drugs in humans (Figure 11.8).[66,67] Although most of these coumarins are chiral, racemic mixtures are used clinically (and as rodenocids).

Historically, the discovery of the anticoagulant activity of dicoumarol (**50**), a natural coumarin derivative occurring in sweet clover,[68] led the path to the systematic optimization and finally to the development of warfarin.[69]

The 4-hydroxy group is crucial for the biological activity, as these structures mimic vitamin K (**51**) and its benzoquinone moiety. All these 4-hydroxycoumarins inhibit the vitamin K reductase, resulting in depletion of the reduced form of vitamin K (vitamin KH_2).[70] As K vitamins are cofactors for the carboxylation of glutamate residues on the N-terminal regions of vitamin K-dependent proteins, this limits the γ-carboxylation and subsequent activation of the vitamin K-dependent coagulant proteins. The synthesis of vitamin K-dependent coagulation factors II, VII, IX, and X, as well as the anticoagulant proteins C and S is therefore inhibited. The depression of three of the four vitamin K-dependent coagulation factors (factors II, VII, and X) lead to decreased prothrombin levels and reduce the amount of thrombin generated and bound to fibrin. This reduces the

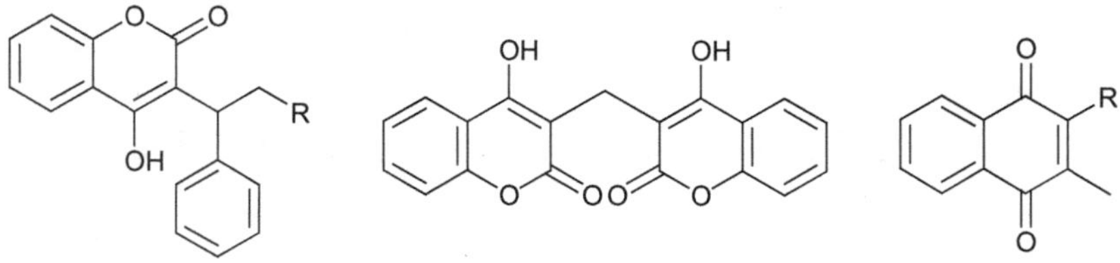

R = $COCH_3$ Warfarin **49**
R = CH_3 Phenprocoumon **50**

Dicoumarol **29**

Vitamin K **51**
K_1: R = $C_{20}H_{39}$
K_3: R = H

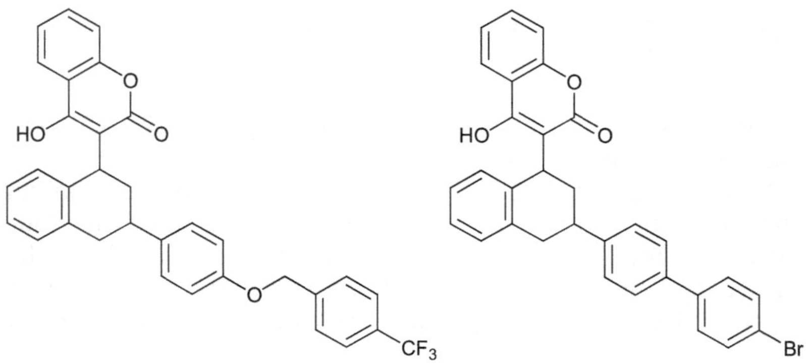

Flocoumafen **52**

Brodifacoum **53**

Figure 11.8 Anticoagulant coumarin derivatives. Brodifacoum (**53**) and Flocoumafen (**52**) are used as diasteremeric mixtures.

thrombogenicity of clots and is therefore used in the prophylactic and acute treatment of strokes and heart attacks.

11.2.4 Cytostatic Agents

Interestingly, the parent compound coumarin (**1**) itself possesses some anticancer activity.[71–73] However, it seems to be limited to certain cases[73] and it has been shown that large quantities of coumarin can also cause cancer.[74–76] This is also true for some simple analogues like 5,7-dimethoxycoumarin.[77,78] Moreover, the hepatotoxicity of coumarin prevents the broad use of the parent compound.[79] Likewise, there is a strong controversial debate over the use of coumarin as a food supplement.[2,80]

In contrast, the antitumor activity of a novel coumarin derivative, 5,7-dihydroxy-4-methyl-6-(3-methylbutanoyl)-coumarin (DMAC, **52**), on colorectal carcinoma was investigated. DMAC (**52**) was initially investigated due to its structural similarity to the cytostatic agent ochrocarpin B (**53**, Figure 11.9). Ochrocarpin B (**53**) was extracted from the bark of *Ochrocarpos punctatus,* along with several other coumarins and benzophenone derivatives, all exhibiting cytotoxicity against an ovarian cancer cell line.[81] Indeed, a large number of synthetic and natural coumarins act as cytotoxic agents.[82]

Treatment of colon cancer HCT116 and LoVo cells with DMAC (**52**) resulted in substantial proapoptotic activity.[83] The combination of DMAC (**52**)

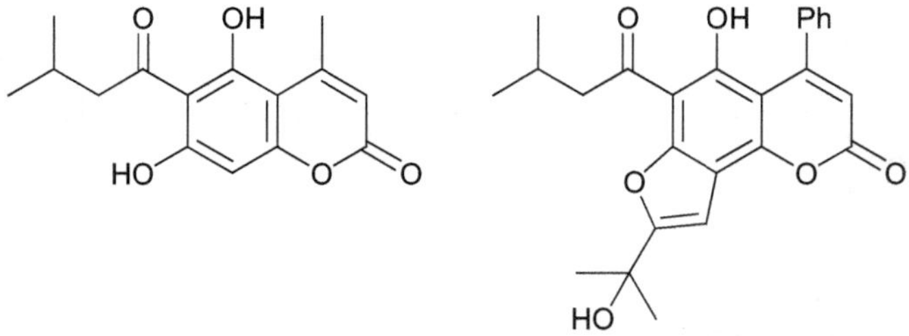

DMAC (**52**) Ochrocarpin B (**53**)

Figure 11.9 Biologically active coumarin derivatives DMAC (52) and Ochrocarpin B (53).

treatment with the established anticancer drugs 5-FU and CPT-11 enhanced their therapeutic efficacies.

Structure activity studies on DMAC (52) related compounds led to the conclusion that substitution at 6-position is crucial for inducing cell apoptosis and a phenyl group at 4-position presumably enhances the bioactivity.[83] New cytostatic agents based on the DMAC structure could therefore lead to interesting new therapeutic valuable compounds.

11.2.5 Neuroprotective Effects on the Central Nervous System

Neurodegenerative diseases have become more prevalent with the aging population.[84] Amongst various natural products that exhibit neuroprotective activity, coumarins have been identified as a new class.[85] Even though the mechanisms of action are not fully understood, several compounds were successfully shown to have neuroprotective effects. The *n*-butanol extract of the root of *Angelica gigas* contains coumarins and has been shown to protect cortical cell cultures against glutamate-induced toxicity (a model for several neurodegenerative diseases).[86] In the extracts, 9 coumarins were identified and structure activity relationship studies involving a range of a total 25 compounds was conducted. As a result, pyranocoumarins like decursinol (**54**), dihydrofuranocoumarins such as marmesin (**20**) and simple coumarins like umbelliferone (**6**)[86] and esculetin (**8**)[87] have been identified as neuroprotectives with their activity decreasing in this order (Figure 11.10).

Several other coumarins exhibit neuroprotective effects, *e.g.* prenyloxyoumarins **55**.[88,89] These compounds could be valuable lead structures in the development of neuroprotective drugs. One of the most prominent neurodegenerative diseases is Alzheimer's disease. A promising target for treatment of this ailment is acetylcholinesterase (AChE), which counteracts the sinking acetylcholine neurotransmitter levels. Several coumarins act as AChE inhibitors by binding to the peripheral anionic binding site or the catalytic binding site of the enzyme.[84,90,91]

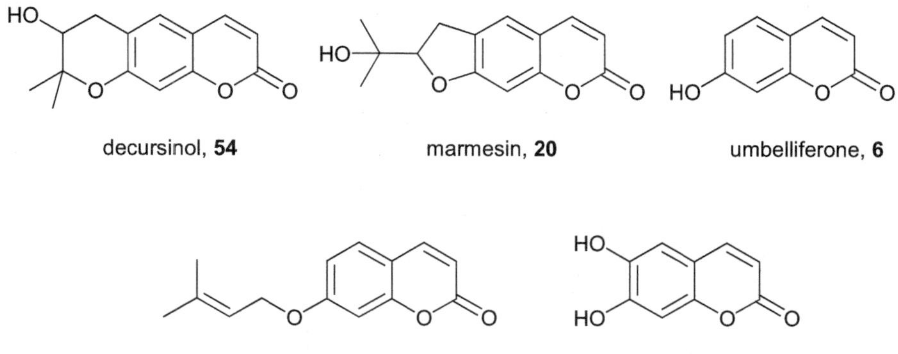

decursinol, **54** marmesin, **20** umbelliferone, **6**

7- isopentenyloxycoumarin, **55** esculetin, **8**

Figure 11.10 Selection of neuroprotective coumarins.

11.2.6 Anti-inflammatory Agents

A large number of coumarin derivatives exhibit anti-inflammatory activity.[92] The effects are tested *in vivo*, *e.g.* by the TNBS (2,4,6-trinitrobenzenesulfpnic acid)-model for chronic inflammation.[93] Hereby, TNBS is used to trigger acute dose-dependent colonic ulceration and inflammation in rats. Another method uses cotton pellets, which are implanted in small cuts in the skin of rats and lead to a subacute inflammation of the wound.[94] With these reproducible inflammations, compounds can be tested for their anti-inflammatory effects by measuring the spread and size of the inflammation of treated *versus* untreated animals. More target-oriented *in vitro* tests measure the inhibitory activity against the production of inflammatory mediators, such as nitric oxide in mouse RAW264.7 macrophages.[95,96]

Anti-inflammatory effects have been observed for coumarins with a wide structural range. For example, the simple coumarin Esculetin (**8**)[97] as well as 3-arylcoumarins **56** [96] and prenylated natural products such as Kayeassamin G (**57**)[98] show anti-inflammatory effects (Figure 11.11). The anti-inflammatory efficacy of these coumarins seems to be related to their antioxidant properties.[99,100]

Novel coumarins extracted from plants[99,101,102] and synthetic coumarins[92,94,96,103] acting as anti-inflammatory agents are continuously reported, having in common only their coumarin motif. Therefore, coumarin constitutes a rewarding core structure in the search for potent anti-inflammatory drugs. A more specific example in anti-inflammation is given in the next section.

11.2.7 Treatment of Asthma, Anti-leukotrienes

Asthma is a disease of the pulmonary system, where the airflow is restricted. A precise definition is not practical, since there are many variations in the disease pattern and causes.[104] However, several biological pathways related

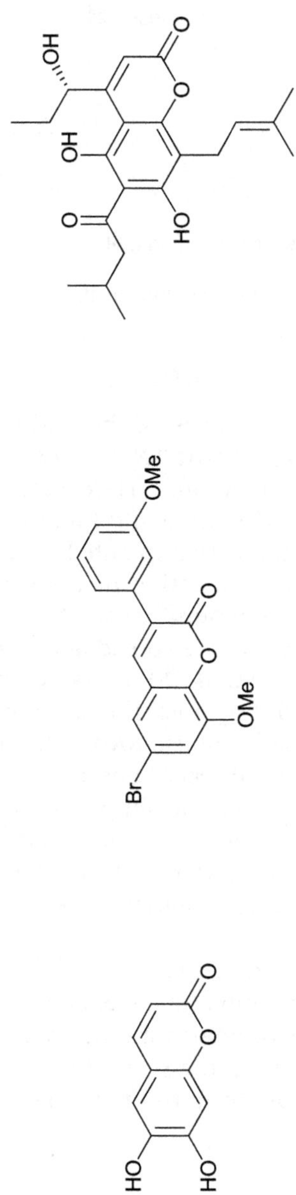

esculetin, **8**

6-bromo-8-methoxy-3-(3-methoxyphenyl)-2*H*-chromen-2-one, **56**

Kayeassamin G, **57**

Figure 11.11 Exemplary coumarins with anti-inflammatory effects.

to the disease and targets for the treatment of asthma have been identified. For example, airway inflammation is a central symptom in the pathogenesis of asthma. Leukotrienes are pro-inflammatory mediators[105] and play an important role in the 5-lipoxygenase pathway, representing biochemical products derived from arachidonic acid.[106,107] Inhibition of 5-lipoxygenase leads to a decreased biosynthesis of leukotrienes and consequently to reduced inflammation. Chronic, as well as spontaneous and responsive inflammation in the case of asthma can therefore be decreased by 5-lipoxygenase inhibitors. A potent example is the 5-lipoxygenase inhibitor **58** which is based on a coumarin core and leads to decreased leukotriene levels (Figure 11.12).[108]

11.2.8 HIV-reverse-transcriptase Inhibitors

The coumarin derivative BPRHIV001 (**59**, Figure 11.13) has shown to inhibit the replication of human immunodeficiency virus type 1 (HIV-1).[109] In contrast to the established HIV-1 reverse transcriptase inhibitors AZT (Retrovir®) and EFV (Sustiva®), the target of BPRHIV001 is the RNA-binding protein Tat. BPRHIV001 (**59**) strongly inhibits the Tat-mediated transactivity of HIV-1 by reducing the p300 protein levels, which represses the virus replication. Inhibition of HIV replication occurred with an EC_{50} of 1.3 nM,

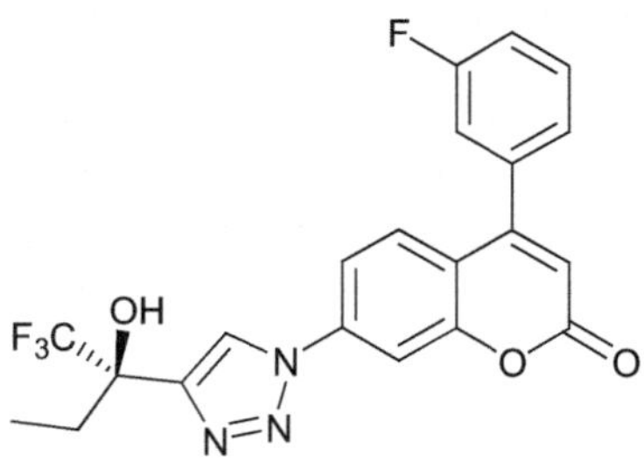

58

Figure 11.12 Example for a 5-lipoxygenase inhibitor.

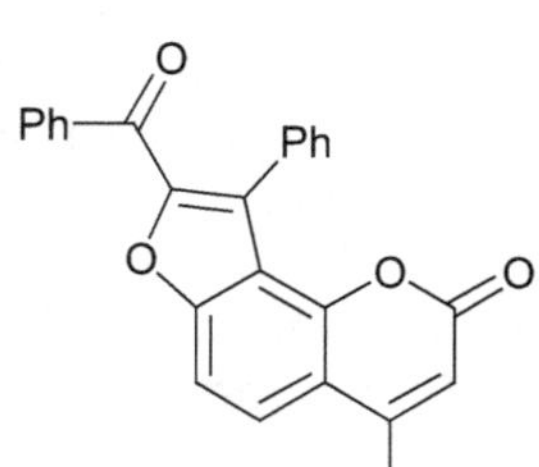

BPRHIV001, **59**

Figure 11.13 BPRHIV001 (**59**), inhibits Tat-mediated transactivity and leads to significant decrease in HIV-1 replication.[109]

while the 50% cytotoxicity concentration (CC_{50}) was 1.3 μM, which leads to a selective index (SI) of 1000 towards HIV-1.

Since BPRHIV001 (**59**) is addressing a different target than AZT[110] and EFV,[111] it is still effective against AZT and EFV resistant viruses with EC_{50}'s of 0.9 nM and 3.9 nM. It also shows strong synergistic effects with the reverse transcriptase inhibitors AZT and EFV. In conclusion, BPRHIV001 (**59**) is a potential lead compound for the development of a novel therapeutic agent against HIV-1 infection.[109]

11.3 Summary and Outlook

Coumarins constitute a diverse and vast class of natural and synthetic compounds. Their biological activities are not limited to the small selection of case studies presented in this chapter. New compounds are frequently reported and more libraries are continually synthesised to increase the number of identified compounds with the coumarin core.

Their asset of being broadly active on so many targets is also their weakness. Especially for simple coumarins, which often exhibit effects on multiple bimolecular targets, such as Esculetin (**33**), which is *e.g.* an inhibitor for lipoxygenase,[112] targets apoptosis-related proteins Bcl-2 and Bax[87] and exhibits antioxidant and anti-inflammatory activity.[97] This makes their pharmacological activity on complete organisms unpredictable. However, more complicated coumarins have been developed into specialised and marketed drugs.

Overall, the coumarins motif constitutes a highly interesting starting point towards manifold lead compounds. Development of novel drugs containing the coumarin core structure seems to be merely a question of time.

References

1. O. S. R. Hänsel, *Pharmakognosie - Phytopharmazie*, Springer Verlag, Berlin, Heidelberg, 2007.
2. B. G. Lake, *Food Chem. Toxicol.*, 1999, **37**, 423–453.
3. P. Mikaili, M. Sharifi, J. Shayegh and S. Sarahroodi, *J. Basic. Appl. Sci. Res.*, 2012, **2**(4), 3235–3241, http://www.researchgate.net/profile/Shadi_Sarahroodi/publication/231167865_A_Review_on_Pharmacognotic_and_Pharmaceutical_Terms_Originated_from_Islamic_Sources/links/09e415062d17b646da000000.pdf.
4. R. H. Murray, in *Fortschritte der Chemie organischer Naturstoffe/Progress in the Chemistry of Organic Natural Products*, ed. W. Herz, H. Falk, G. W. Kirby and R. E. Moore, Springer, Vienna, 2002, ch. 1, vol. 83, pp. 1–619.
5. F. Weygand and H. Wendt, *Z. Naturforsch., B: Anorg. Chem., Org. Chem., Biochem., Biophys., Biol.*, 1959, **14**, 421–427.
6. S. A. Brown, G. Towers and D. Wright, *Can. J. Biochem. Physiol.*, 1960, **38**, 143–156.

7. J. Stoker and D. Bellis, *J. Biol. Chem.*, 1962, **237**, 2303–2305.
8. K. G. Edwards and J. R. Stoker, *Phytochemistry*, 1967, **6**, 655–661.
9. F. Bourgaud, A. Hehn, R. Larbat, S. Doerper, E. Gontier, S. Kellner and U. Matern, *Phytochem. Rev.*, 2006, **5**, 293–308.
10. J. A. Mead, J. N. Smith and R. T. Williams, *Biochem. J.*, 1958, **68**, 67–74.
11. C. Spino, M. Dodier and S. Sotheeswaran, *Bioorg. Med. Chem. Lett.*, 1998, **8**, 3475–3478.
12. S. Rosselli, A. M. Maggio, N. Faraone, V. Spadaro, S. L. Morris-Natschke, K. F. Bastow, K.-H. Lee and M. Bruno, *Nat. Prod. Commun.*, 2009, **4**, 1701–1706.
13. A.-u. Rahman, M. Shabbir, S. S. Ziauddin, A. Jabbar and I. M. Choudhary, *Phytochemistry*, 1997, **44**, 683–685.
14. J. Choi, K.-T. Lee, H. Ka, W.-T. Jung, H.-J. Jung and H.-J. Park, *Arch. Pharmacal Res.*, 2001, **24**, 418–423.
15. E. V. Morozova and V. V. Kats, *Effect of Coumarin on the Initial Stage of Cucumber Seed Germination and Phytohormone Activity*, 1982.
16. J. S. Knypl, *Acta Soc. Bot. Pol.*, 1970, **39**, 321–332.
17. T. Hogetsu, H. Shibaoka and M. Shimokoriyama, *Plant Cell Physiol.*, 1974, **15**, 265–272.
18. G. F. Stallknecht and S. Farnsworth, *Am. Potato J.*, 1982, **59**, 17–32.
19. A. Lupini, A. Sorgona, A. J. Miller and M. R. Abenavoli, *Plant Signaling Behav.*, 2010, **5**, 1395–1400.
20. F. Ahrabi, S. Enteshari and A. Moradshahi, *J. Med. Plants Res.*, 2011, **5**, 5104–5109.
21. T. L. Harrison, A. R. Zangerl, M. A. Schuler and M. R. Berenbaum, *Arch. Insect Biochem. Physiol.*, 2001, **48**, 179–189.
22. I. Boulogne, P. Petit, H. Ozier-Lafontaine, L. Desfontaines and G. Loranger-Merciris, *Environ. Chem. Lett.*, 2012, **10**, 325–347.
23. N. Jung, F. Gläser and S. Bräse, *Phytochem. Rev.*, 2013, **12**, 603–651.
24. K. N. Venugopala, V. Rashmi and B. Odhav, *BioMed Res. Int.*, 2013, **2013**, 14.
25. W. H. Perkin, *J. Chem. Soc.*, 1875, **28**, 10–15.
26. I. S. D. Bogdal, J. Sanetra and E. Gondek, *Polimery*, 2003, **48**, 111–115.
27. H. S. P. Rao and S. Sivakumar, *J. Org. Chem.*, 2006, **71**, 8715–8723.
28. S. K. De and R. A. Gibbs, *Synthesis*, 2005, 1231–1233.
29. M. K. Potdar, S. S. Mohile and M. M. Salunkhe, *Tetrahedron Lett.*, 2001, **42**, 9285–9287.
30. Y. Yamamoto and N. Kirai, *Org. Lett.*, 2008, **10**, 5513–5516.
31. J. Ferguson, F. Zeng and H. Alper, *Org. Lett.*, 2012, **14**, 5602–5605.
32. A. V. Kalinin, A. J. M. da Silva, C. C. Lopes, R. S. C. Lopes and V. Snieckus, *Tetrahedron Lett.*, 1998, **39**, 4995–4998.
33. E.-i. Negishi, V. Bagheri, S. Chatterjee, F.-T. Luo, J. A. Miller and A. T. Stoll, *Tetrahedron Lett.*, 1983, **24**, 5181–5184.
34. J. Toräng, S. Vanderheiden, M. Nieger and S. Bräse, *Eur. J. Org. Chem.*, 2007, **2007**, 943–952.

35. I. Zhuravel, S. Kovalenko, S. Vlasov and V. Chernykh, *Molecules*, 2005, **10**, 444–456.
36. M.-S. Schiedel, C. A. Briehn and P. Bäuerle, *J. Organomet. Chem.*, 2002, **653**, 200–208.
37. Z. Xu, X. Liu, J. Pan and D. R. Spring, *Chem. Commun.*, 2012, **48**, 4764–4766.
38. F. J. D. L. W. Hillman, *Dye Laser Principles, with Applications*, Academic Press, San Diego, CA, 1990.
39. K. Hara, K. Sayama, Y. Ohga, A. Shinpo, S. Suga and H. Arakawa, *Chem. Commun.*, 2001, 569–570.
40. K. Hara, Z.-S. Wang, T. Sato, A. Furube, R. Katoh, H. Sugihara, Y. Dan-oh, C. Kasada, A. Shinpo and S. Suga, *J. Phys. Chem. B*, 2005, **109**, 15476–15482.
41. Z. S. Wang, Y. Cui, K. Hara, Y. Dan-oh, C. Kasada and A. Shinpo, *Adv. Mater.*, 2007, **19**, 1138–1141.
42. X. Liu, J. M. Cole, P. G. Waddell, T.-C. Lin and S. McKechnie, *J. Phys. Chem. C*, 2013, **117**, 14130–14141.
43. L. Long, L. Zhou, L. Wang, S. Meng, A. Gong, F. Du and C. Zhang, *Org. Biomol. Chem.*, 2013, **11**, 8214–8220.
44. D. Egan, R. O'Kennedy, E. Moran, D. Cox, E. Prosser and R. D. Thornes, *Drug Metab. Rev.*, 1990, **22**, 503–529.
45. K. Mackie, *Annu. Rev. Pharmacol. Toxicol.*, 2006, **46**, 101–122.
46. W. Notcutt, M. Price, R. Miller, S. Newport, C. Phillips, S. Simmons and C. Sansom, *Anaesthesia*, 2004, **59**, 440–452.
47. J. Zajicek, P. Fox, H. Sanders, D. Wright, J. Vickery, A. Nunn and A. Thompson, *Lancet*, 2003, **362**, 1517–1526.
48. J. Romero, I. Lastres-Becker, R. de Miguel, F. Berrendero, J. A. Ramos and J. Fernández-Ruiz, *Pharmacol. Ther.*, 2002, **95**, 137–152.
49. P. Carriba, O. Ortiz, K. Patkar, Z. Justinova, J. Stroik, A. Themann, C. Muller, A. S. Woods, B. T. Hope, F. Ciruela, V. Casado, E. I. Canela, C. Lluis, S. R. Goldberg, R. Moratalla, R. Franco and S. Ferre, *Neuropsychopharmacology*, 2007, **32**, 2249–2259.
50. A. C. Howlett, C. S. Breivogel, S. R. Childers, S. A. Deadwyler, R. E. Hampson and L. J. Porrino, *Neuropharmacology*, 2004, **47**(Suppl 1), 345–358.
51. A. Behrenswerth, N. Volz, J. Toräng, S. Hinz, S. Bräse and C. E. Müller, *Bioorg. Med. Chem.*, 2009, **17**, 2842–2851.
52. V. Rempel, N. Volz, S. Hinz, T. Karcz, I. Meliciani, M. Nieger, W. Wenzel, S. Brase and C. E. Muller, *J. Med. Chem.*, 2012, **55**, 7967–7977.
53. A. C. Howlett, F. Barth, T. I. Bonner, G. Cabral, P. Casellas, W. A. Devane, C. C. Felder, M. Herkenham, K. Mackie, B. R. Martin, R. Mechoulam and R. G. Pertwee, *Pharmacol. Rev.*, 2002, **54**, 161–202.
54. A. D. Khanolkar, S. L. Palmer and A. Makriyannis, *Chem. Phys. Lipids*, 2000, **108**, 37–52.
55. R. G. Pertwee, *Br. J. Pharmacol.*, 2007, **152**, 984–986.

56. D. McHugh, J. Page, E. Dunn and H. B. Bradshaw, *Br. J. Pharmacol.*, 2012, **165**, 2414–2424.
57. S. P. H. Alexander, *Br. J. Pharmacol.*, 2012, **165**, 2411–2413.
58. J. Yu, E. Deliu, X.-Q. Zhang, N. E. Hoffman, R. L. Carter, L. A. Grisanti, G. C. Brailoiu, M. Madesh, J. Y. Cheung, T. Force, M. E. Abood, W. J. Koch, D. G. Tilley and E. Brailoiu, *J. Biol. Chem.*, 2013, **288**, 22481–22492.
59. C. M. Henstridge, N. A. B. Balenga, J. Kargl, C. Andradas, A. J. Brown, A. Irving, C. Sanchez and M. Waldhoer, *Mol. Endocrinol.*, 2011, **25**, 1835–1848.
60. V. Rempel, N. Volz, F. Glaeser, M. Nieger, S. Brase and C. E. Mueller, *J. Med. Chem.*, 2013, **56**, 4798–4810.
61. H. Sharir and M. E. Abood, *Pharmacol. Ther.*, 2010, **126**, 301–313.
62. C. M. Henstridge, N. A. Balenga, R. Schröder, J. K. Kargl, W. Platzer, L. Martini, S. Arthur, J. Penman, J. L. Whistler and E. Kostenis, *Br. J. Pharmacol.*, 2010, **160**, 604–614.
63. C. M. Henstridge, N. A. Balenga, J. Kargl, C. Andradas, A. J. Brown, A. Irving, C. Sanchez and M. Waldhoer, *Mol. Endocrinol.*, 2011, **25**, 1835–1848.
64. F.-F. Deng, M.-H. Xie, P.-Z. Li, Y.-L. Tian, X.-Y. Zhang and H.-L. Zhai, *Chemom. Intell. Lab. Syst.*, 2014, **131**, 51–60.
65. I. Manolov and N. D. Danchev, *Eur. J. Med. Chem.*, 1995, **30**, 531–535.
66. J.-C. Jung, E. Lim, Y. Lee, D. Min, J. Ricci, O.-S. Park and M. Jung, *Molecules*, 2012, **17**, 2091–2102.
67. O. S. Park and B. S. Jang, *Arch. Pharmacal Res.*, 1995, **18**, 277–281.
68. H. A. Campbell and K. P. Link, *J. Biol. Chem.*, 1941, **138**, 21–33.
69. N. Kresge, R. D. Simoni and R. L. Hill, *J. Biol. Chem.*, 2005, **280**, e5.
70. J. K. Tie and D. W. Stafford, *Vitam. Horm.*, 2008, **78**, 103–130.
71. M. E. Marshall, K. Butler, J. Cantrell, C. Wiseman and L. Mendelsohn, *Cancer Chemother. Pharmacol.*, 1989, **24**, 65–66.
72. J. L. Mohler, L. G. Gomella, E. D. Crawford, L. M. Glode, C. D. Zippe, W. R. Fair and M. E. Marshall, *Prostate*, 1992, **20**, 123–131.
73. A. Maucher, M. Kager and E. von Angerer, *J. Cancer Res. Clin. Oncol.*, 1993, **119**, 150–154.
74. H. Sun, *J. Chem. Inf. Comput. Sci.*, 2004, **44**, 1506–1514.
75. J. F. Contrera, A. C. Jacobs and J. J. DeGeorge, *Regul. Toxicol. Pharmacol.*, 1997, **25**, 130–145.
76. J. F. Contrera, A. C. Jacobs, H. R. Prasanna, M. Mehta, W. J. Schmidt and J. De George, *J. Am. Coll. Toxicol.*, 1995, **14**, 1–10.
77. D. Alesiani, R. Cicconi, M. Mattei, C. Montesano, R. Bei and A. Canini, *Int. J. Oncol.*, 2008, **32**, 425–434.
78. M. A. Musa, J. S. Cooperwood and M. O. F. Khan, *Curr. Med. Chem.*, 2008, **15**, 2664–2679.
79. K. Abraham, F. Woehrlin, O. Lindtner, G. Heinemeyer and A. Lampen, *Mol. Nutr. Food Res.*, 2010, **54**, 228–239.
80. F. M. Johnson, *Environ. Mol. Mutagen.*, 2002, **39**, 69–80.

81. V. P. Chaturvedula, J. K. Schilling and D. G. Kingston, *J. Nat. Prod.*, 2002, **65**, 965–972.
82. I. Kostova, *Curr. Med. Chem.: Anti-Cancer Agents*, 2005, **5**, 29–46.
83. M.-H. Lin, C.-H. Cheng, K.-C. Chen, W.-T. Lee, Y.-F. Wang, C.-Q. Xiao and C.-W. Lin, *Chem.-Biol. Interact.*, 2014, **218**, 42–49.
84. P. Anand, B. Singh and N. Singh, *Bioorg. Med. Chem.*, 2012, **20**, 1175–1180.
85. F. Epifano, M. Curini, L. Menghini and S. Genovese, *Mini-Rev. Med. Chem.*, 2009, **9**, 1262–1271.
86. S. Y. Kang and Y. C. Kim, *Arch. Pharmacal Res.*, 2007, **30**, 1368–1373.
87. C. Wang, A. Pei, J. Chen, H. Yu, M.-L. Sun, C.-F. Liu and X. Xu, *J. Neurochem.*, 2012, **121**, 1007–1013.
88. F. Epifano, G. Molinaro, S. Genovese, R. T. Ngomba, F. Nicoletti and M. Curini, *Neurosci. Lett.*, 2008, **443**, 57–60.
89. F. Epifano, C. Pelucchini, M. Curini and S. Genovese, *Nat. Prod. Commun.*, 2009, **4**, 1755–1760.
90. M. Catto, L. Pisani, F. Leonetti, O. Nicolotti, P. Pesce, A. Stefanachi, S. Cellamare and A. Carotti, *Bioorg. Med. Chem.*, 2013, **21**, 146–152.
91. S. F. Razavi, M. Khoobi, H. Nadri, A. Sakhteman, A. Moradi, S. Emami, A. Foroumadi and A. Shafiee, *Eur. J. Med. Chem.*, 2013, **64**, 252–259.
92. K. C. Fylaktakidou, D. J. Hadjipavlou-Litina, K. E. Litinas and D. N. Nicolaides, *Curr. Pharm. Des.*, 2004, **10**, 3813–3833.
93. G. P. Morris, P. L. Beck, M. S. Herridge, W. T. Depew, M. R. Szewczuk and J. L. Wallace, *Gastroenterology*, 1989, 795–803.
94. M. Ghate, D. Manohar, V. Kulkarni, R. Shobha and S. Y. Kattimani, *Eur. J. Med. Chem.*, 2003, **38**, 297–302.
95. Y. Bansal, P. Sethi and G. Bansal, *Med. Chem. Res.*, 2013, **22**, 3049–3060.
96. W. Pu, Y. Lin, J. Zhang, F. Wang, C. Wang and G. Zhang, *Bioorg. Med. Chem. Lett.*, 2014, **24**, 5432–5434.
97. A. Witaicenis, L. N. Seito and L. C. Di Stasi, *Chem. - Biol. Interact.*, 2010, **186**, 211–218.
98. T. Morikawa, M. Sueyoshi, S. Chaipech, H. Matsuda, Y. Nomura, M. Yabe, T. Matsumoto, K. Ninomiya, M. Yoshikawa, Y. Pongpiriyadacha, T. Hayakawa and O. Muraoka, *Bioorg. Med. Chem.*, 2012, **20**, 4968–4977.
99. A. Witaicenis, L. N. Seito, A. da Silveira Chagas, L. D. de Almeida Jr, A. C. Luchini, P. Rodrigues-Orsi, S. H. Cestari and L. C. Di Stasi, *Phytomedicine*, 2014, **21**, 240–246.
100. R. K. Arora, N. Kaur, Y. Bansal and G. Bansal, *Acta Pharm. Sin. B*, 2014, **4**, 368–375.
101. J.-Y. Cho, T.-L. Hwang, T.-H. Chang, Y.-P. Lim, P.-J. Sung, T.-H. Lee and J.-J. Chen, *Food Chem.*, 2012, **135**, 17–23.
102. H.-N. Lv, S. Wang, K.-W. Zeng, J. Li, X.-Y. Guo, D. Ferreira, J. K. Zjawiony, P.-F. Tu and Y. Jiang, *J. Nat. Prod.*, 2015, **78**, 279–285.
103. C. Kontogiorgis and D. Hadjipavlou-Litina, *J. Enzyme Inhib. Med. Chem.*, 2003, **18**, 63–69.

104. M. Sears and P. M. R. Sears, *Allergy*, 1993, **48**, 12–16.
105. P. M. O'byrne, E. Israel and J. M. Drazen, *Ann. Intern. Med.*, 1997, **127**, 472–480.
106. B. Samuelsson, *Science*, 1983, **220**, 568–575.
107. S. P. G. Ouellet, D. Gauvreau, M. Cameron, S. Dolman, L.-C. Campeau, G. Hughes, P. D. O'Shea and I. W. Davies, *Org. Process Res. Dev.*, 2012, **16**, 214–219.
108. E. L. Grimm, Y. Ducharme, R. Frenette, R. Friesen, M. Gagnon, H. Juteau, S. Laliberte, B. Mackay and Y. Gareau, Google Patents, 2008.
109. P.-H. Lin, Y.-Y. Ke, C.-T. Su, H.-Y. Shiao, H.-P. Hsieh, Y.-K. Chao, C.-N. Lee, C.-L. Kao, Y.-S. Chao and S.-Y. Chang, *J. Virol.*, 2011, **85**, 9114–9126.
110. M. A. Fischl, D. D. Richman, M. H. Grieco, M. S. Gottlieb, P. A. Volberding, O. L. Laskin, J. M. Leedom, J. E. Groopman, D. Mildvan, R. T. Schooley, G. G. Jackson, D. T. Durack and D. King, *N. Engl. J. Med.*, 1987, **317**, 185–191.
111. S. D. Young, S. F. Britcher, L. O. Tran, L. S. Payne, W. C. Lumma, T. A. Lyle, J. R. Huff, P. S. Anderson, D. B. Olsen and S. S. Carroll, *Antimicrob. Agents Chemother.*, 1995, **39**, 2602–2605.
112. S.-J. Lee, U.-S. Lee, W.-J. Kim and S.-K. Moon, *Mol. Med. Rep.*, 2011, **4**, 337–341.

CHAPTER 12

Xanthones are Privileged Scaffolds in Medicinal Chemistry – but are they Over-privileged?

TIM WEZEMAN[a] AND KYE-SIMEON MASTERS*[b]

[a] Institute of Organic Chemistry (IOC), Karlsruhe Institute of Technology (KIT), Fritz-Haber-Weg 6, 76131 Karlsruhe, Germany; [b] School of Chemistry, Physics and Engineering, Queensland University of Technology, PO Box 2434, Brisbane, Queensland 4001, Australia
*Email: kye.masters@qut.edu.au

12.1 General Considerations

12.1.1 Physico-chemical Properties of Xanthones

The simple appearance of the tricyclic xanthone core belies the amazingly rich chemistry of this motif, a trend which is clearly visible from examining the physico-chemical and structural properties of these molecules. Despite this richness, only a few examples of xanthone-containing compounds have ever been marketed, or even trialled for use as active pharmaceutical ingredients. The apparent disparity of xanthones as drug leads compared to xanthones as drugs (or drug candidates) can be readily explained by examining the aforementioned structural features in a biological context (see Section 12.4) (Figure 12.1).

RSC Drug Discovery Series No. 50
Privileged Scaffolds in Medicinal Chemistry: Design, Synthesis, Evaluation
Edited by Stefan Bräse

Published by the Royal Society of Chemistry, www.rsc.org

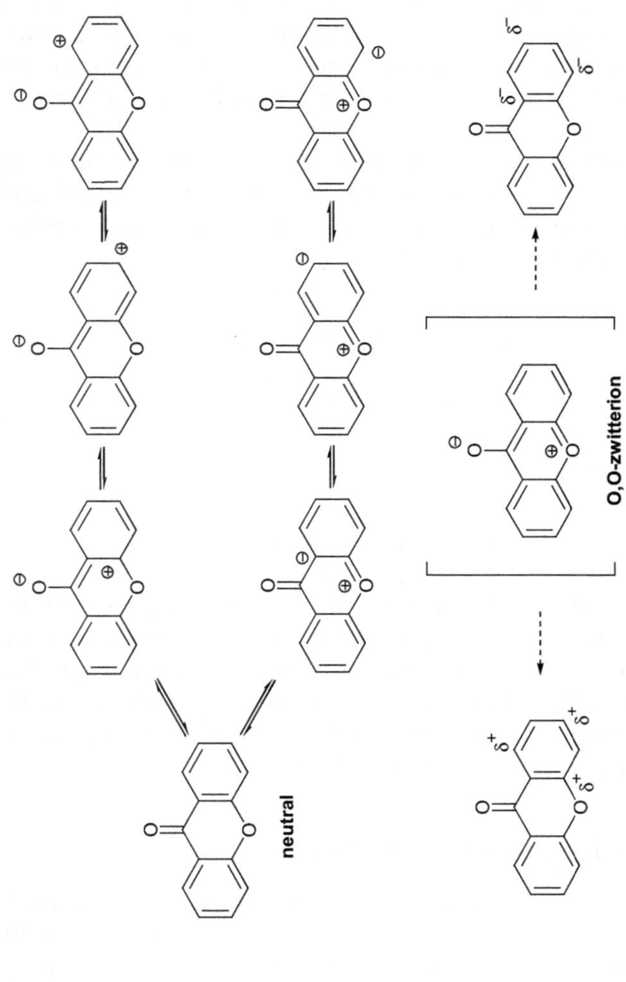

Figure 12.1 Xanthone and dominant resonance forms (without respect to substituents).

The anthracene-analogue core of xanthones features incorporation of two oxygen atoms; one internal to the ring system (biaryl ether) oxygen which donates electrons to the ring, and one external to the ring (carbonyl) oxygen that removes electrons from the ring system – the latter of these is often a more powerful effect. A zwitterionic form can thus be drawn for the xanthone nucleus; this characteristic, in conjunction with the aromatic planar nature of xanthones, leads to often poor solubility in the xanthones, wherein the core is less highly substituted with solubilizing moieties. The further oxygenation of specific positions of the xanthone core system is performed by specific enzymes. For example, Beerhues and co-workers have identified the presence of xanthone-6-hydroxylase, a form of cytochrome P450 monooxygenase, in cell cultures of *Centaurium erythraea* RAFN and *Hypericum androsaemum* L.[1]

Xanthones are found in all shapes and forms, from simple, monomeric core structures with perhaps a few functional groups here and there, to fully prenylated or caged structures. Dimeric, trimeric and glycosylated species have been reported as well.[2]

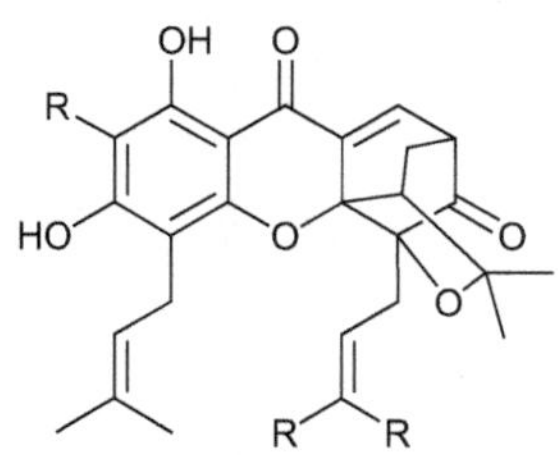

1 Prenylated and caged xanthone core

With the aid of new techniques, it is increasingly easy to identify and discover new xanthones. For example, recently, an effective qualitative method for rapidly profiling the caged xanthones in the resin of *G. hanburyi* was developed, employing multiple mass spectrometric scanning methods. This resulted in the characterization of 34 caged xanthones (see compound **1**) including 18 previously unreported ones.[3]

12.1.2 The Diversity of Xanthone Scaffolds

Xanthones have a wide range of biological and pharmacological properties, such as monoamine oxidase inhibition, and antioxidant, antimicrobial, cytotoxic, and hepatoprotective activity. The antifungal activity of many xanthones has also been well documented. Recent insights into biological activities of xanthones have been investigated by Pinto and co-workers,[4] showing some examples of the remarkable diversity of these compounds in terms of their biological activities.

Xanthones are active in an extremely varied array of disease states and upon a wide variety of structurally despondent biotargets. Several factors make it extremely difficult to link the bioactivities observed for these dimeric xanthones and the structural features that enable these bioactivities. First and foremost, no systematic study of the biological activity of xanthones has been conducted. Furthermore, ascribing an entire range of bioactivities, (*e.g.*, the antibacterial activity of chrysoxanthone to the presence of a biaryl ether linkage is overly simplistic). One caveat that should be taken into account is that it may be a fair criticism of xanthones as potential active pharmaceutical ingredients, or even as starting points for medicinal chemistry campaigns to attain a selective chemotherapeutic agent, that the xanthone core commonly features a polyhydroxylated and carbonyl-substituted polyarene core, a motif which is known to be indiscriminate in its binding to a plethora of target biomolecules (see Section 12.5).

12.1.3 Traditional Medicines Containing Xanthones

Kikuchi and co-workers have reported the identification of four xanthones from *Hypericum mysorense*, a plant from the Hypericum genus which has been traditionally used by peoples of India and Sri Lanka for medicinal purposes.[5] The plant *Garcinia hanburyi* Gamboges exudes a yellow gum that is used as a traditional Chinese medicine and as a folk medicine in India and nations nearby for the external treatment of scabies, tinea and maligna, and has been effectively used for the treatment of skin carcinoma. When later analysed by HPLC methods, gamboge was found to contain 12 cytotoxic caged xanthones, including forbesin and gambogic acid analogues.[6] Interesting studies from Theodoraki and co-workers on varying the ring-oxygenation pattern of gambogic acid and semi-synthesis on the gambogic acid framework (Pd-catalysed deprenylation and Diels-Alder reactions) was instructive in showing that the bioactivities of the caged Garcinia xanthones can be modulated *via* remote electronic effects.[7]

In Mongolia, a *Lomatogonium carinthiacum* (Wulfen) extract is used to treat liver and bile disease. Several xanthones (**2–5**) were isolated from the extract of this annual plant together with some flavones and iridoids. So far it is not determined which compounds in the extract are responsible for any biological effects.[8]

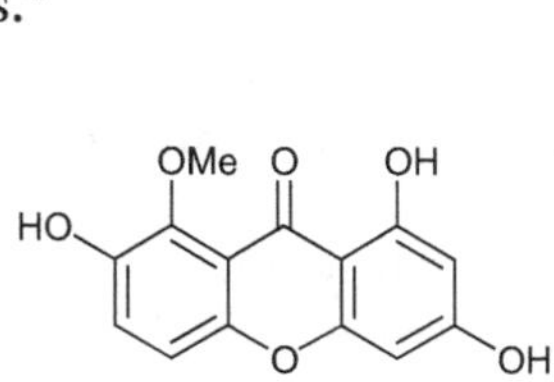

2

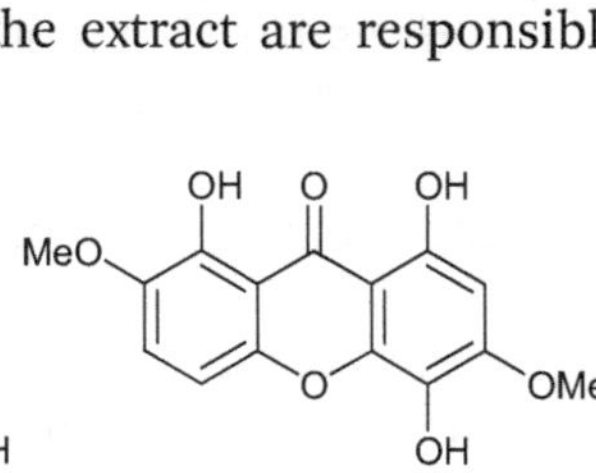

3

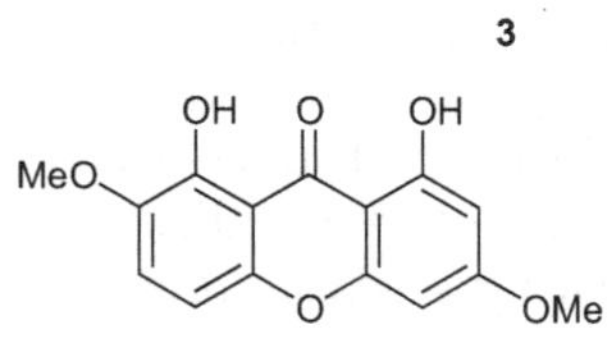

4

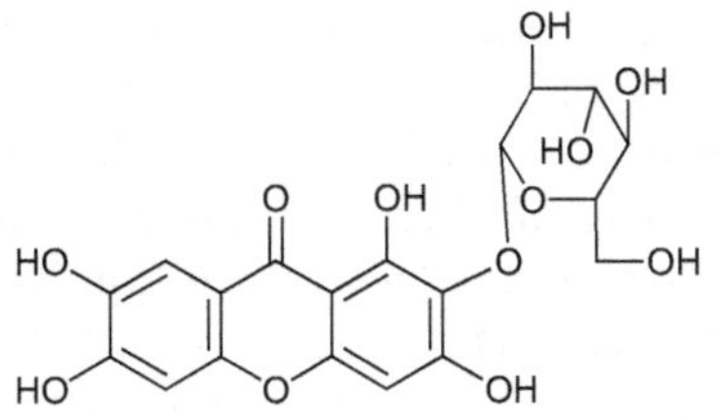

5

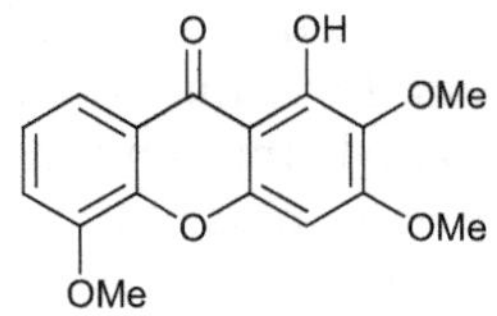

6 1-hydroxy-2,3,5-trimethoxyxanthone

The xanthone constituents from *Halenia elliptica*, a Tibetan medicinal plant used to treat liver and gall bladder diseases in China, were studied intensively by looking at their metabolic transformation in rat liver microsomes *in vitro.*[9] One of its main constituents, 1-hydroxy-2,3,5-trimethoxyxanthone (**6**), was investigated further, showing interesting substrate inhibition and metabolism-based drug–drug interactions.[10]

12.1.4 Crude Extracts and Neutraceuticals

In some cases, the extracts of plants and other biological sources of xanthones are known to possess biological activities without necessarily knowing which compounds give rise to them. For example, the crude methanolic extract (CME) from mangosteen pericarp including 25% α-mangostin (**29**, Section 12.2.4) as an active xanthone was used in this study. The inhibition on tumour cell proliferation of CME was preliminarily evaluated against the murine colon cancer cell line NL-17 with an IC_{50} value of 17 and 84 $\mu g\, mL^{-1}$.[11] Brazilian plants of the genus *Calophyllum* contain xanthones alongside coumarins, flavonoids, steroids and triterpenes, some of them with relevant biological activities.[12]

The chemistry of several plant-derived xanthones has been reported by Sultanbawa as early as 1980.[13] Prenylated caged xanthones are 'privileged structures' characterized by the presence of the unusual 4-oxo-tricyclo[4.3.1.0 3,7]dec-8-en-2-one scaffold. The natural sources of these compounds are confined mainly to the *Garcinia* genus in the family of *Guttiferae*.[14]

Interestingly, there is currently a healthy and growing consumer market for the crude extracts of mangosteen fruits as neutraceuticals. These are usually in the form of numerous health benefit juices which describe the contents as including xanthones, hence the entry of the term 'xanthone' into the popular lexicon. The products may have names such as 'Xanthone Juice' (http://xanthonecorp.com/index.php and http://www.xango.com/about/products/xango-juice). The true efficacy of such extracts and juices in combating diseases remains unclear to date, but sales are increasing.

12.2 Specific Bioactivities

12.2.1 Anti-algal

The dimeric xanthone Phomoxanthone A (7) has been found to exhibit anti-algal activity in an agar diffusion disk test at 0.2 cm at 10 μg per disk against *Chlorella fusca*.[15]

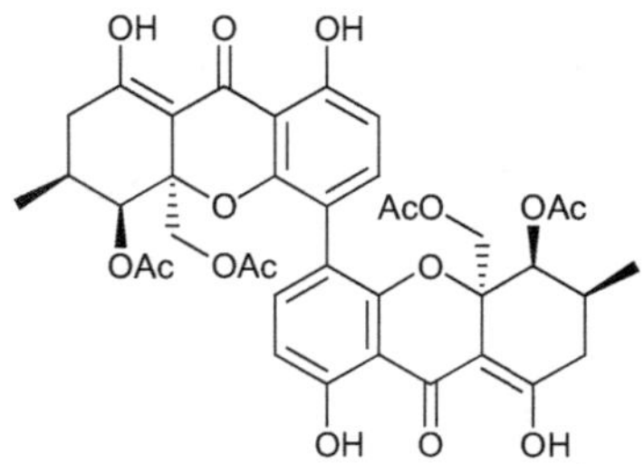

7 Phomoxanthone A
(absolute configuration)

12.2.2 Anti-allergic Properties

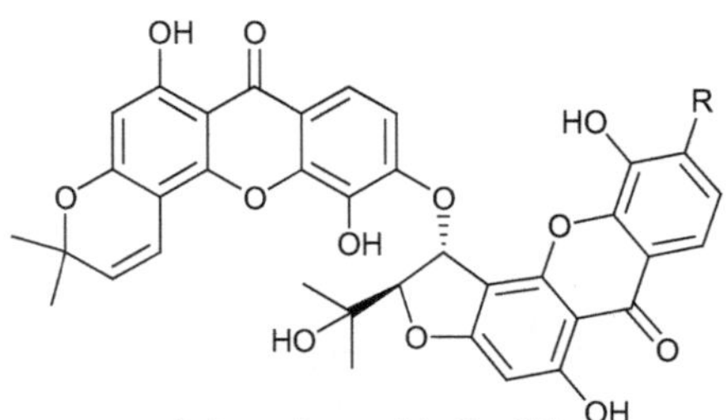

8 Jacarelhyperol A; R = OH
9 Jacarelhyperol B; R = H

Xanthones have been shown to be useful as anti-allergy and anti-asthma compounds, the latter effect being related to bronchodilatory activity.[16] Isolated from the plant *Hypericum japonicum,* the dimeric xanthones Jacarelhyperols A and B (**8, 9**) were found to significantly inhibit platelet-activating factor (PAF) induced hypotension at 10 mg kg^{-1} in mice without causing their blood pressure to rise. PAF inhibitors are considered to be potential drugs against allergic diseases, since PAF was found to act as primary initiator of HEL-induced anaphylactic hypotension.[17]

12.2.3 Antibacterial Xanthones

1,5-Dihydroxyxanthone selectively has been shown to inhibit Gram-positive bacteria.[18] Several dimeric xanthones that were isolated from different fungal sources were found to exhibit very interesting antibacterial features. Phomalevones A, B and C (**10–12**), all isolated from a Hawaiian isolate of *Phoma* species were found to actively inhibit the growth of *Bacillus subtilis* and *Staphylococcus aureus* using agar disk diffusion assays.[19]

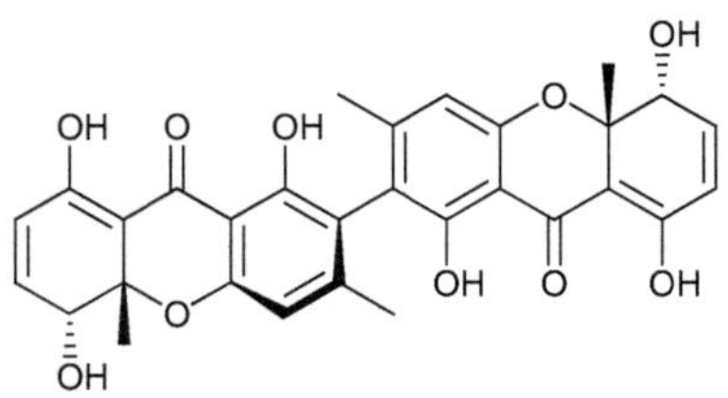

10 Phomalevone A

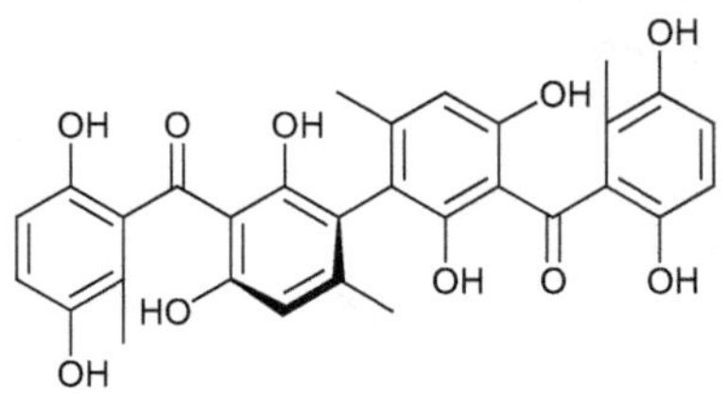

11 Phomalevone B

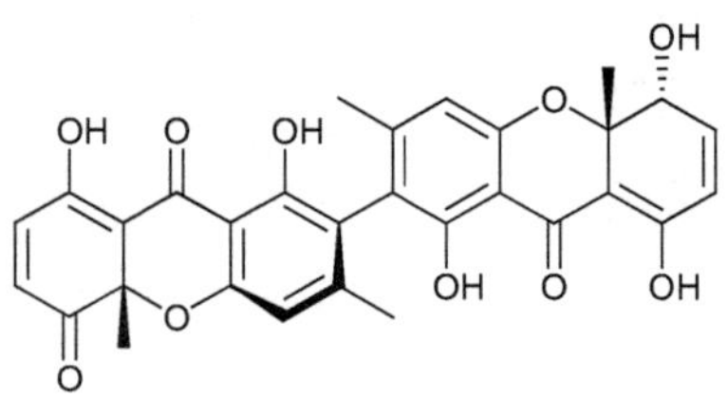

12 Phomalevone C

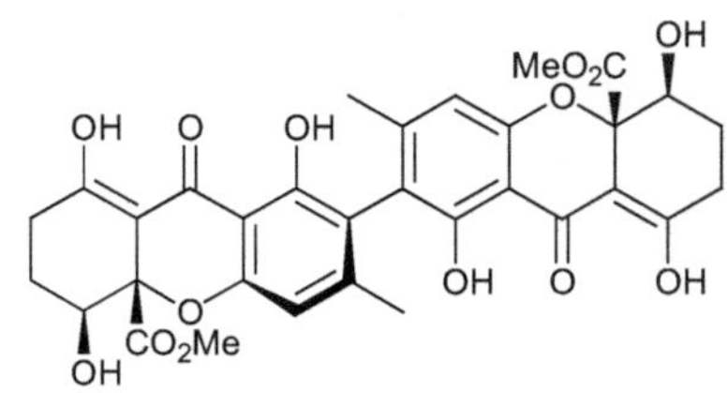

13 Rugulotrosin A
(relative configuration)

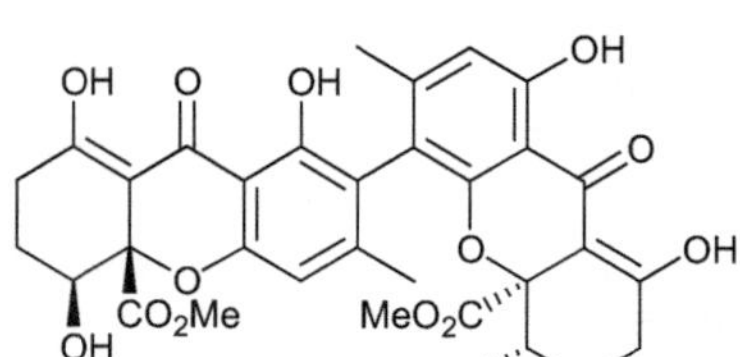

14 Rugulotrosin B
(relative configuration)

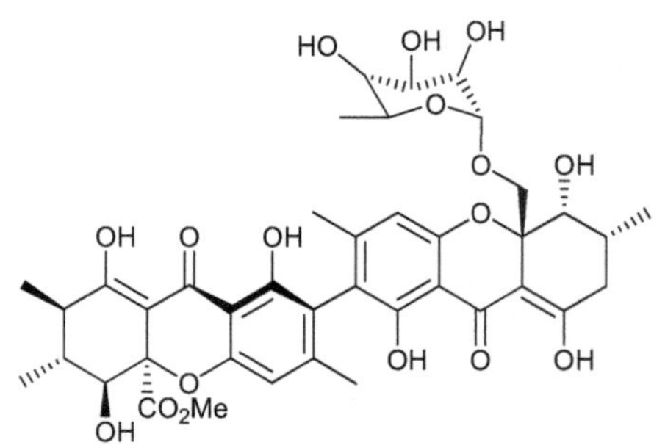

15 Hirtusneanoside A
(absolute stereochemistry)

Rugulotrosin A and B and Hirtusneanoside (**13–15**) have also been tested against several microbes. Rugulotrosin A showed a very strong activity against *Bacillus subtilis, Enterococcus faecalis* and *Bacillus cereus*, but performed rather badly against *Staphylococcus aureus* (LD_{99}: 1.6–5.5 μg mL^{-1} *versus* an LD_{99}: 200 μg mL^{-1}).[20] Rugulotrosin B and Hirtusneanoside[21] showed only moderate activities, although Hirtusneanoside did inhibit *Staphylococcus aureus with an* LD_{50} of 0.0034 μM. The dicerandrols (**105–107**, see Section 12.2.23), a family of dimeric xanthones, have been found to show antimicrobial activities (*Bacillus subtilis* and *Staphylococcus aureus*) with increased activities related to the decreased degree of acetylation of the molecules (*i.e.*, A > B > C).[22] The dicerandrols were tested extensively against several *Xanthomonas oryzae* strains that cause bacterial blight in rice. Dicerandrol A was also tested for antimicrobial activity against several Gram-positive and Gram-negative bacteria, a fungus and a yeast. It was found that dicerandrol A showed a relatively high activity in this broad spectrum of species, although it was often bested by commercial antibiotics.[23]

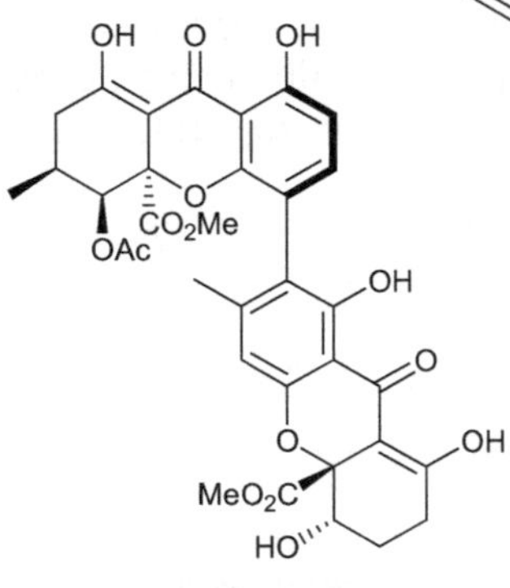

16 Neosartorin
(absolute configuration)

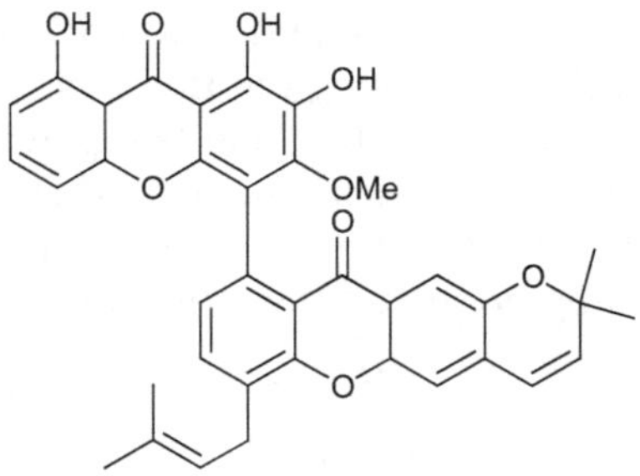

17 Globulixanthone E

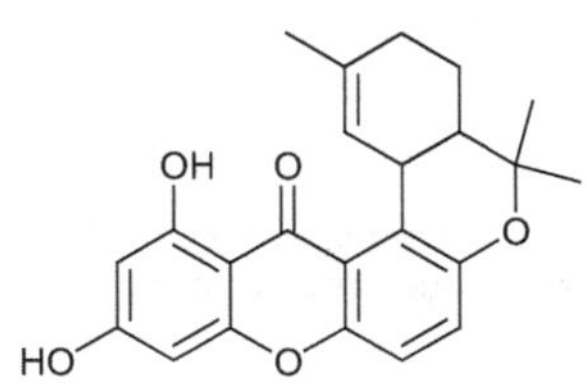

18 Calozeyloxanthone

Neosartorin (**16**) was first isolated in 1998 from a cultured mycelium of the soil mould *Neosartorya fischeri*[24] and was found to show a strong activity against Gram-positive bacteria such as *Staphylococcus aureus* (MIC: 8 $\mu g\,mL^{-1}$) and *Bacillus subtilis* (MIC: 4 $\mu g\,mL^{-1}$). However, it did not affect the tested Gram-negative bacteria. In cytotoxicity tests against several cancer cell lines it scored very badly, showing almost no notable cytotoxicity.[25]

The prenylated xanthones globulixanthone C–E were isolated from a tree that is widely used in Cameroon as a medicinal plant and a laxative for pregnant women. The three xanthones were tested for their *in vitro* antimicrobial activities against *Staphylococcus aureus*, *Bacillus subtilis*, *Vibrio anguillarium* and *Escherichia coli* in an agar well diffusion assay. The monomeric xanthones C and D showed only moderate activity to *S. aureus* (MIC: 8.05–14.05 $\mu g\,mL^{-1}$) and *B. subtilis* (MIC: 8.24–12.5 $\mu g\,mL^{-1}$), but globulixanthone E (**17**) showed a remarkably high activity against all tested organisms (MIC: 5.56–3.12 $\mu g\,mL^{-1}$) except *E. coli* and outperformed the positive control in the case of *S. aureus* (MIC: 4.51 $\mu g\,mL^{-1}$ *versus* 6.25 $\mu g\,mL^{-1}$ for streptomycin sulfate).[26] Calozeyloxanthone (**18**) is a known antibacterial agent.[27]

12.2.4 Anti-cancer

In 2008, Luo, Xu and co-workers reported polyprenylated xanthones (**19–24**) from the bark of *Garcinia lancilimba* which demonstrated apoptotic effects against HeLa-C3 Cells *in vitro*.[28] These compounds promoted apoptotic cell death, with similar results to paclitaxel. The caged garcinia xanthone cluvenone (**21**) was found by Theodorakis and co-workers to be a potent immune modulator and induces both apoptosis and cellular stress in great many cancer cell lines.[29] Among the findings, T-cell acute lymphoblastic leukemia cells ($EC_{50} = 0.25\ \mu mol\ L^{-1}$) and had potent growth-inhibitory activity against the NCI60 cell panel, including those that are multidrug-resistant, with a GI 50 range of 0.1 to 2.7 $\mu mol\ L^{-1}$. Isoalvaxanthone inhibits colon cancer cell proliferation, migration and invasion through inactivating Rac1 and AP-1.[30]

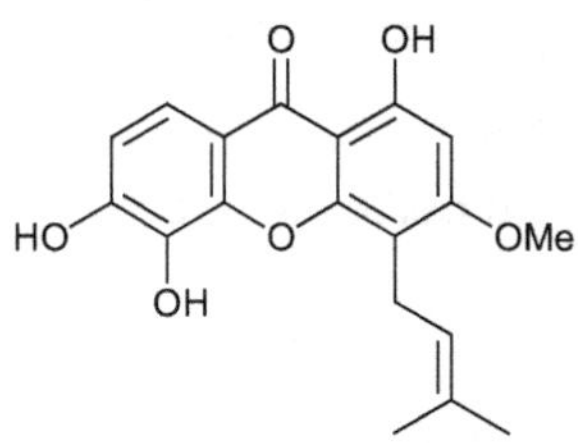

19 1,5,6-trihydroxy-3-methoxy-4-(3-methylbut-2-enyl)xanthone

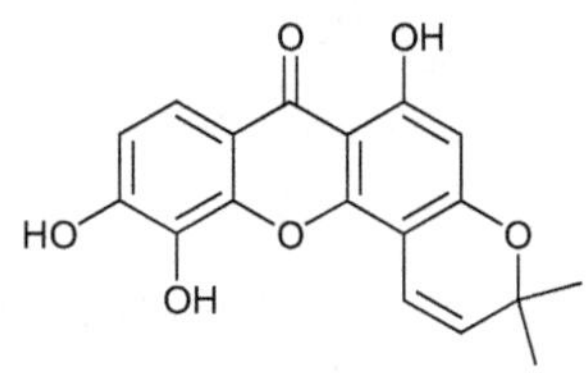

20 isojacareubin

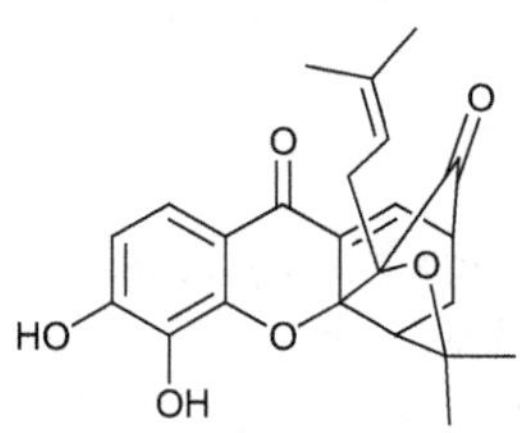

21 cluvenone

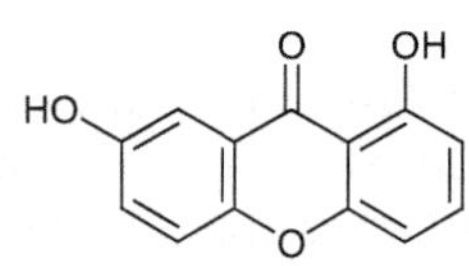

22 1,7-dihydroxyxanthone

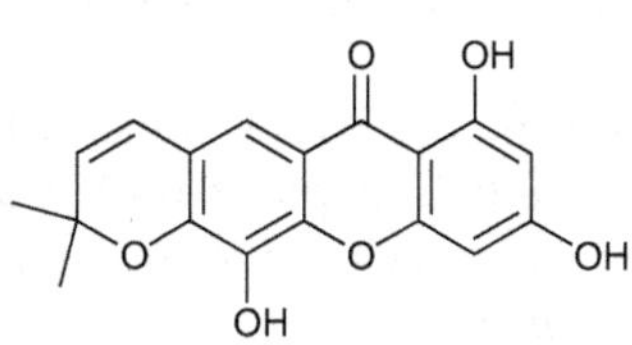

23 1,3,5-trihydroxy-13,13-dimethyl-2H-pyran [7,6-b] xanthone (TDP)

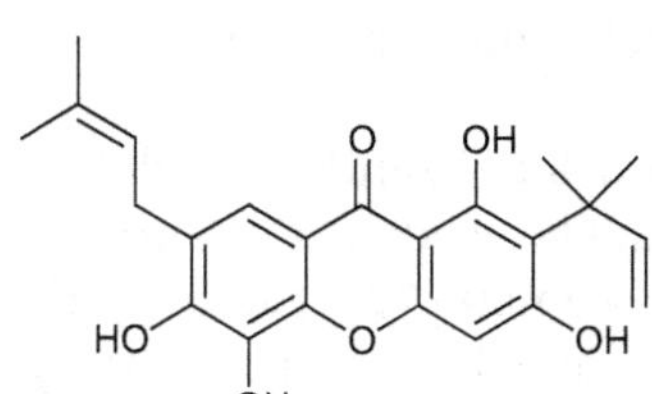

24 isoalvaxanthone

Hepatocellular carcinoma is a significant global killer with a very limited number of treatment options, but Xu, Kung and co-workers have reported that a xanthone isolated from *Garcinia oblongifolia*, 1,3,5-trihydroxy-13,13-dimethyl-2H-pyran [7,6-*b*] xanthone (TDP, **23**) has significantly inhibited growth of hepatocarcinoma cells and induced caspase-dependent mitochondrial apoptosis in them *via* strongly suppressing Hsp27 expression.[31]

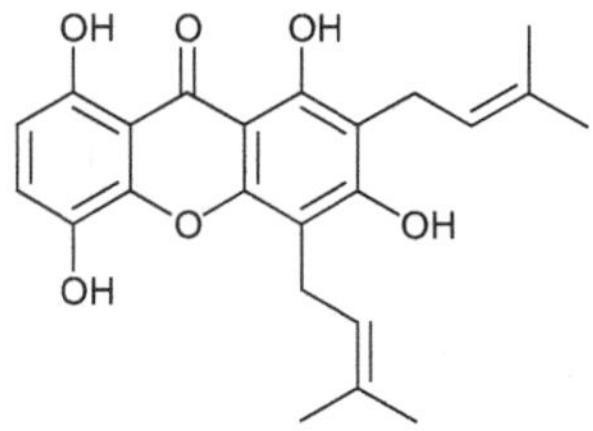

25 gartanin

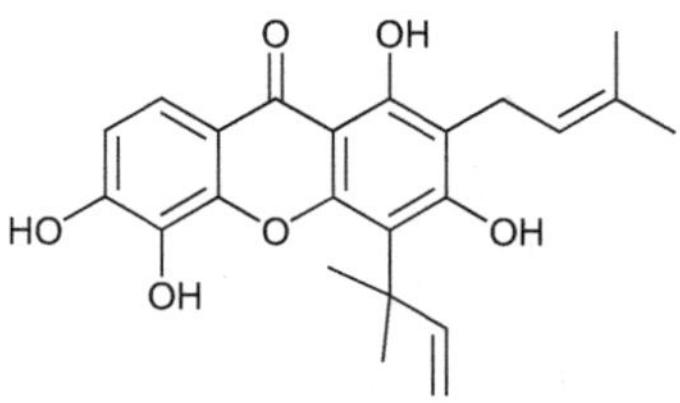

26 gerontoxanthone I

Sometimes a structural–activity relationship can be perceived based on common structural motifs with the same bioactivities. For example, gartanin (**25**) from mangosteen juice, which has reported effects on ailments ranging from skin infections and inflammation to urinary tract infections, was found by Li and co-workers to be a modulator of the mTOR pathway in human urinary bladder cancer cell lines, enhancing autophagy, apoptosis, and inhibiting the growth of cancer cells.[32] Boonnak and co-workers have isolated an impressive diversity (32 identified compounds) of prenylated xanthones from *Cratoxylum formosum* ssp. *pruniflorum*. Among these, several displayed antimicrobial activities. Some were also cytotoxic to cancer cell lines. In particular, it was found that the previously known compound gerontoxanthone I (**26**) was effective against MCF-7, HeLa, HT-29 and KB cancer cell lines with IC50's lower than 1 $\mu g\,mL^{-1}$.[33] 1,3,5-trihydroxy-13,13-dimethyl-2*H*-pyran [7,6-*b*] xanthone (**27**), isolated from the traditional Chinese medicinal herb, *Garcinia oblongifolia,* has been found to effectively inhibit tumour cell growth and induce caspase-dependent mitochondrial apoptosis in hepatocellular carcinoma. Nude mouse model testing confirmed the anti-tumour effect of the xanthone.[31] In a study of six natural products that are widely used for several human ailments, one prenylated xanthone (named xanthone V_1 (**28**)) was thoroughly investigated for its anti-cancer properties.[34]

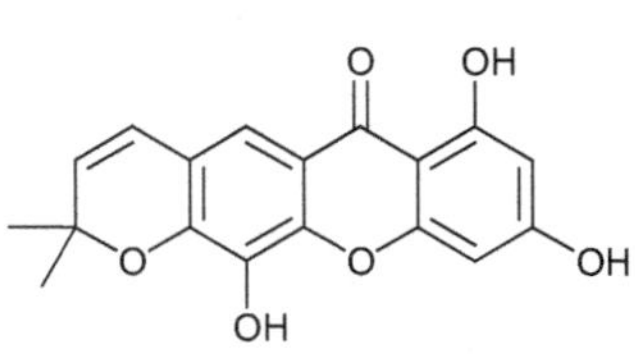

27 1,3,5-trihydroxy-13,13-dimethyl-2H-pyran [7,6-b] xanthone

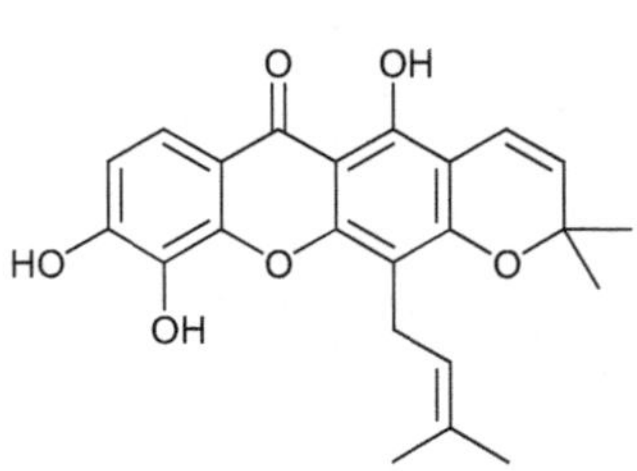

28 xanthone V_1

In order to overcome some inherent drawbacks of xanthones for use as anti-cancer medicine, such as its low aqueous solubility, Okonogi *et al.* prepared xanthone-loaded micelles and concluded that these can be very attractive delivery systems for treatment.[35]

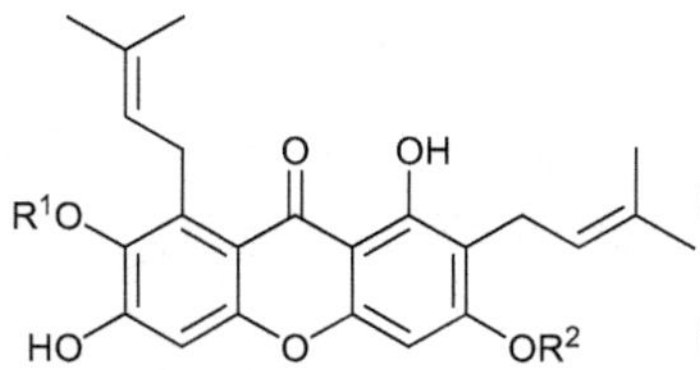

29 R^1 = Me, R^2 = H; α-mangostin
30 R^1 = Me, R^2 = Me; β-mangostin
31 R^1 = H, R^2 = H; γ-mangostin

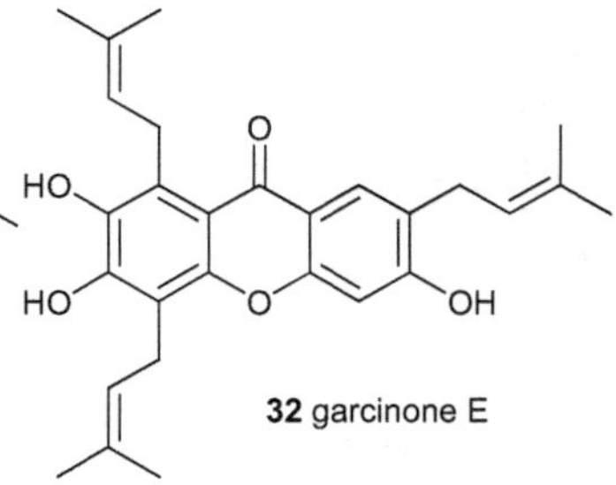

32 garcinone E

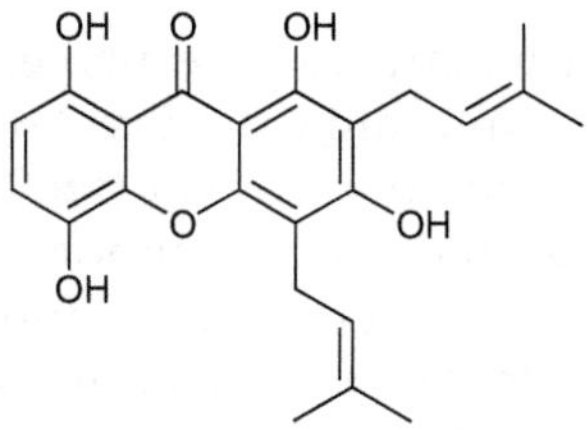

33 gartanine

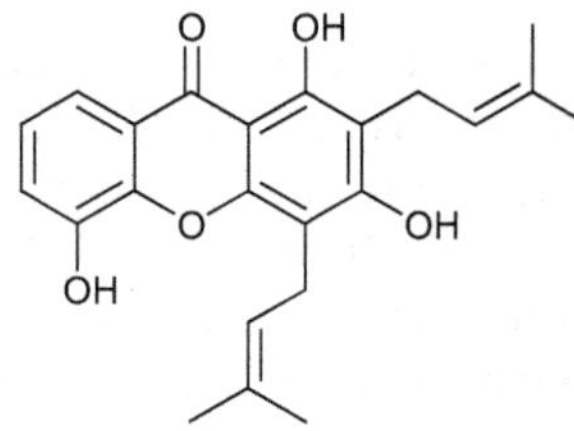

34 8-deoxygartanine

Several xanthones from the Mangosteen tree, including α-mangostin (**29**) were found to induce apoptosis in human melanoma cell lines. The apoptotic effect of α-mangostin was found to be *via* caspase activation and disruption of mitochondrial membrane pathways as evidenced by 25-fold increased caspase-3 activity and 9-fold decreased mitochondrial membrane potential when compared to untreated cells.[36] A review from 2008 covers the xanthones isolated from the mangosteen tropical tree. The most prominent xanthones discussed are shown below. In this review a wide variety of properties is briefly discussed, including antioxidant, anti-cancer and antimicrobial features.[37]

12.2.5 Anti-fungal

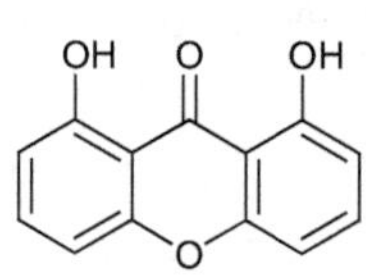

35 1,7-dihydroxyxanthone (euxanthone)

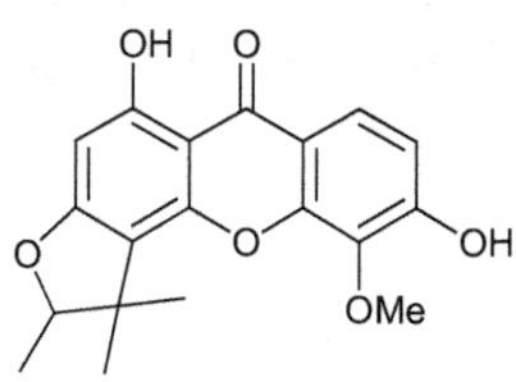

36 5-O-methyl-2-deprenyl rheediaxanthone B

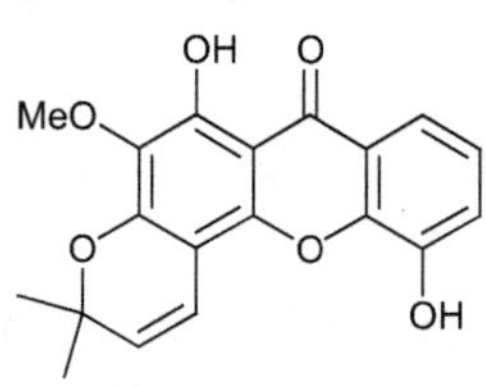

37 caledonixanthone E

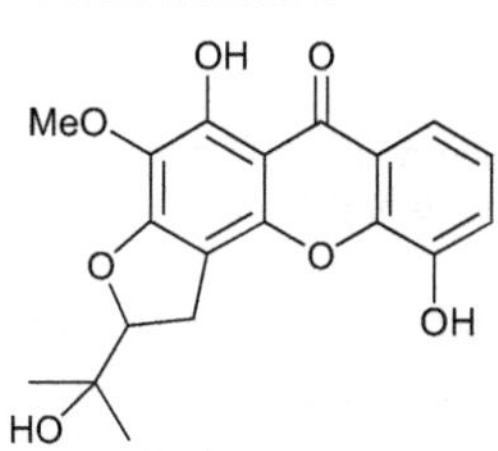

38 caledonixanthone F

Tocci, Simonetti and co-workers have reported that *Hypericum perforatum* subsp. *angustifolium* can be elicited with chitosan to produce a rich extract of xanthones with antifungal activity against a large collection of human fungal pathogen strains (30 *Candida* species, 12 *Cryptococcus* neoformans, and 16 *dermatophytes*).[38] The xanthones responsible for this antifungal activity were reported as 1,3,5,6-tetrahydroxyxanthone and 1,3,6,7-tetrahydroxyxanthone (mixture), 1,7-dihydroxyxanthone (euxanthone, **35**), cadensin G, toxyloxanthone B, paxanthone, 5-*O*-methyl-2-deprenylrheediaxanthone B (**36**). The caledonixanthones E and F (**37** and **38**) were isolated from CH_2Cl_2 extracts of *Calophyllum caledonicum* and found to be effective against pathogenic fungi *A. fumigatus* and *C. albicans*.[39] Further isolated from the same plant were the antifungal compounds caledol and dicaledol, particularly against *A. fumigatus*.[40]

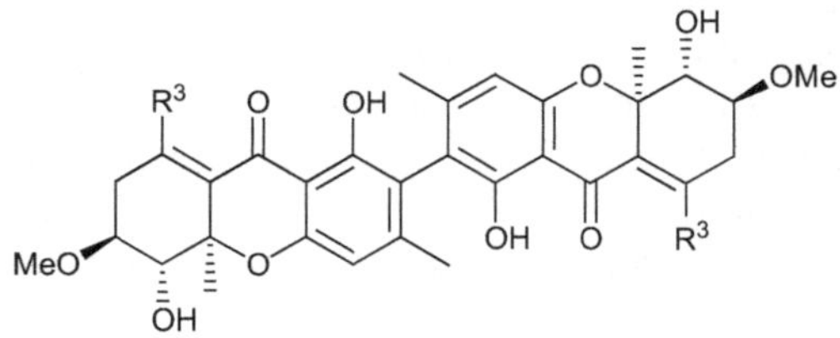

39 Ascherxanthone A; R^1 = OMe, R^2 = Me, R_3 = H (relative stereochem.)
40 Ascherxanthone B; R^1 = OMe, R^2 = Me, R_3 = OH (relative stereochem.)

Ascherxanthone B (**40**), a dimeric xanthone produced by *Aschersonia luteola* BCC 8774, was found to be active against *Magnaporthe grisea* with an IC_{90} value of 0.58 $\mu g\,mL^{-1}$. While searching for new drugs against this rice blast fungus, culture broth extracts from over 800 fungal strains were investigated. Interestingly, the closely related ascherxanthone A was found to be inactive in the same biological assay against *M. grisea*.[41]

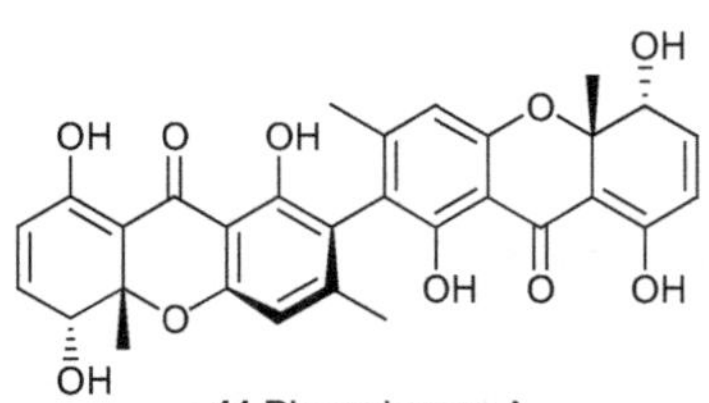

41 Phomalevone A

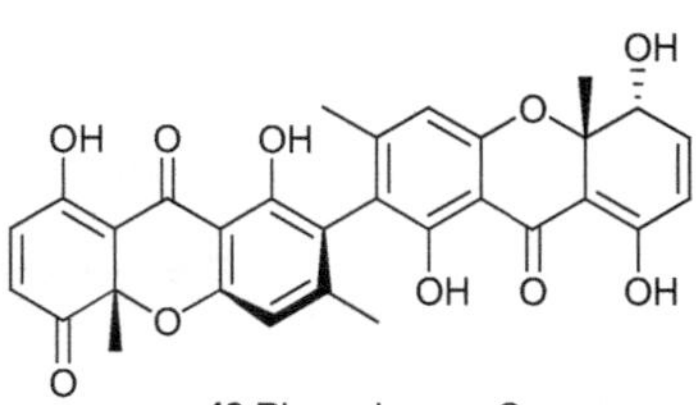

42 Phomalevone C

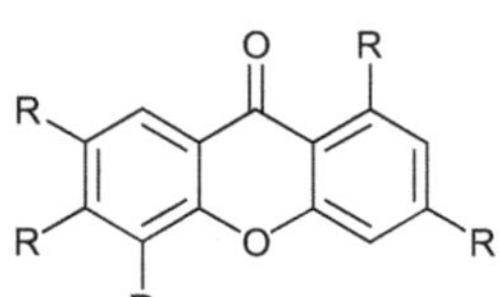

43 Calophyllum derivatives

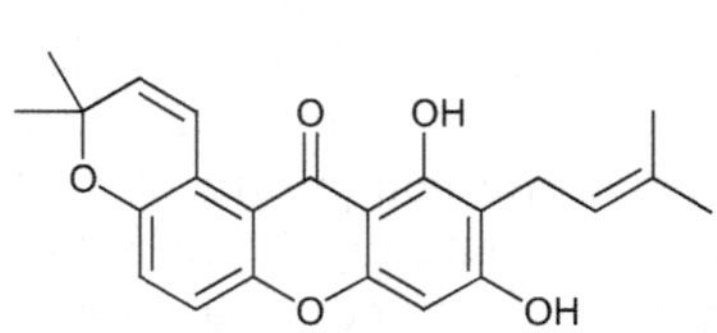

44 calothwaitesixanthone

Phomalevone C (**42**) showed anti-fungal activity against *Fusarium verticillioides* and *Aspergillus flavus* and Phomalevone B was found to inhibit the yeast *Candida albicans* in a agar disk diffusion assay at 18 mm at 100 μg per disk.[19] In another agar disk diffusion assay, Phomoxanthone A (**41**) inhibited *Ustilago violacea* at 0.5–0.8 cm at 10 μg per disk.[15] A small group of xanthones with the core structure (**43**) and calothwaitesixanthone (**44**) were isolated from the root bark of the Sri Lankan *Calophyllum* species. Very interesting anti-fungal and anti-oxidant properties were found, but not in all isolated xanthones.[42]

12.2.6 Anti-HIV

In a typical example of the highly-bioactive nature of xanthones which is also indicative of their 'over-privileged' nature, the novel xanthone 1,4,6-trihydroxy-3-methoxy-2-(3-methyl-2-butenyl)-5-(1,1-dimethyl-prop-2-enyl)xanthone (**45**) was isolated by Magadula from Tanzanian *Garcinia edulis* (Clusiaceae).[43] This compound was found to be active not only against HIV with an anti HIV-protease activity of IC_{50} value of 11.3 μg mL^{-1}, but also to possess potent cytotoxic character against brine shrimp larvae (2.36 μg mL^{-1} *in vitro*).

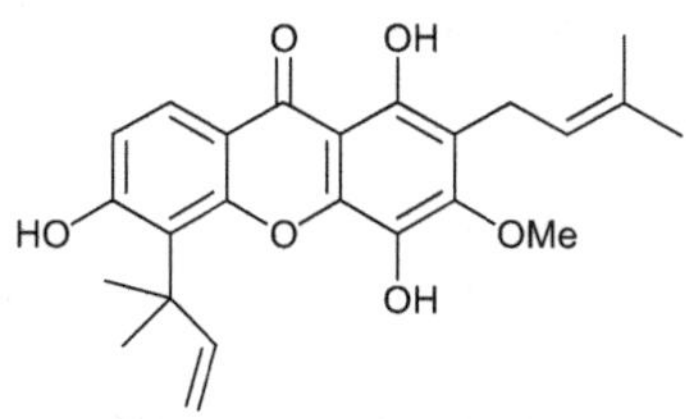

45 1,4,6-trihydroxy-3-methoxy-2-(3-methyl-2-butenyl)-5-(1,1-dimethyl-prop-2-enyl)xanthone

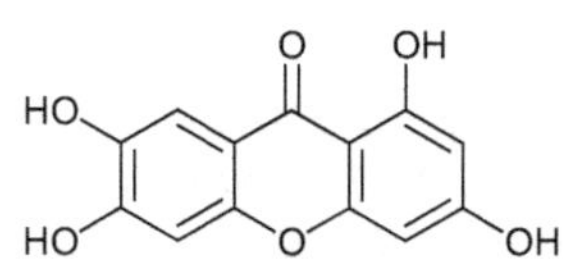

46 Norathyriol

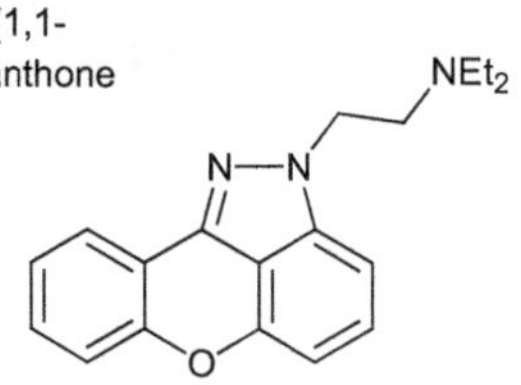

47 9-methoxypyrazoloacridine

Norathyriol (**46**) was also found to inhibit cancers generated by UV-light exposure by acting as an inhibitor of extracellular signal-regulated kinase (ERK)1/2 activity (as confirmed by cocrystal analysis) to attenuate UVB-induced phosphorylation in mitogen-activated protein kinases signalling cascades.[44] The xanthone moiety in norathyriol acted as an adenine mimetic to anchor the compound by hydrogen bonds to the hinge region of the protein ATP-binding site on ERK2.

Amino-substituted xanthones (for example **47**) synthesized from xanthones to possess structural analogy to the potent anti-cancer agent

9-methoxypyrazoloacridine (**47**) were shown to have significant cytotoxic activity against a panel of cancer cell lines.[45] These hybrid molecules were also found to retain activity against the multidrug resistant MES-SA/Dx5 subline.

12.2.7 Xanthones with Anti-inflammatory Properties

Crockett and co-workers isolated 1,6-dihydroxy-5-methoxy-4,5-dihydro-4,4,5-trimethylfurano-(2,3:3,4)-xanthone (48), a component of St John's wort, *Hypericum perforatum*, and found it to be anti-inflammatory.[46] In initial bioassay testing against enzymes involved in inflammation response (COX-1, COX-2 and 5-LOX product formation), 48 demonstrated a very high inhibition of LTB4 formation in the *in vitro* assay, tested initially at 50 $\mu g\,mL^{-1}$. This compound was also found to be anti-fungal against fungi pathogenic to plants (*Phomopsis obscurans* and *P. viticola*).

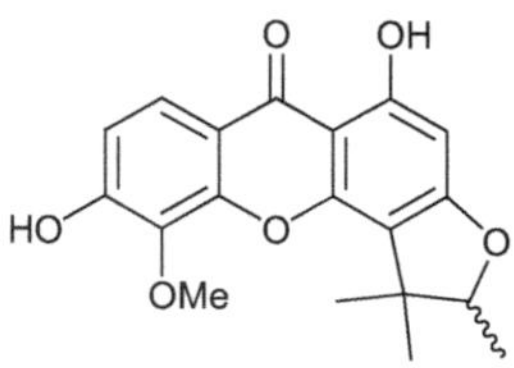

48 1,6-dihydroxy-5-methoxy-4,5-dihydro-4,4,5-trimethylfurano-(2,3:3,4)-xanthone

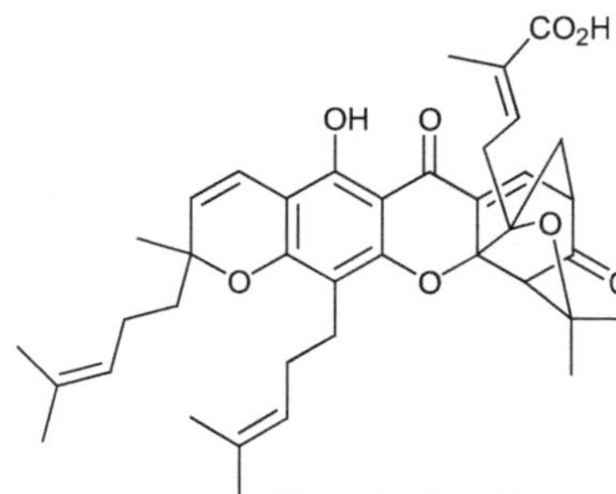

49 gambogic acid

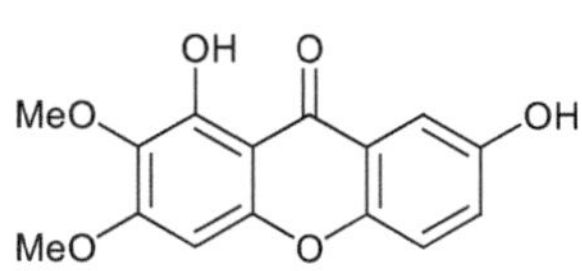

50 1,7-dihydroxy-2,3-dimethoxyxanthone

Gambogic acid (**49**) exhibits potent anti-inflammatory activity, which according to research from Prabhu and co-workers covalently modifies a cysteine residue in the Iκ B kinase-β subunit to mediate suppression of lipopolysaccharide-induced activation of NF-κ B in macrophages. 1,7-Dihydroxy-2,3-dimethoxyxanthone (**50**) was found by Calixto and co-workers to antagonise, in a non-competitive but reversible manner, the contractions induced by chemical inflammatory mediators in the guinea pig trachea *in vitro*.[47]

12.2.8 Anti-mutagenic

Isolated from *V. harveyi*, these five aminoalkanolic xanthones (**51**) were found to exhibit *in vitro* antimutagenic activity.[48]

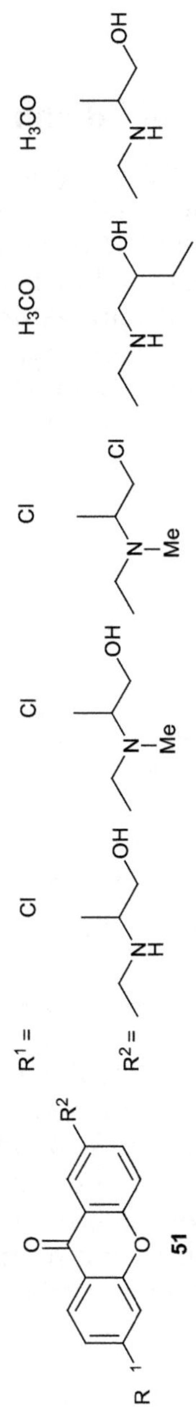

51

12.2.9 Anti-leukaemia

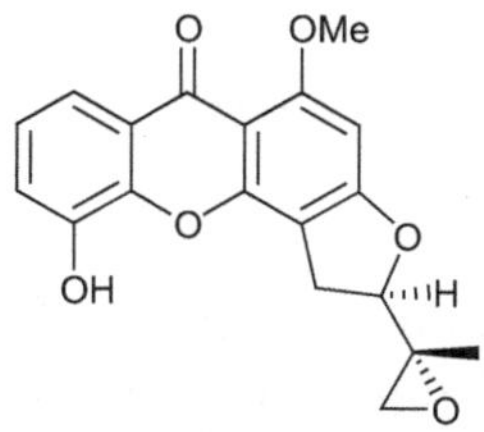

52 psorospermine

Xanthones isolated from Guttiferae family plants have been shown to act as anti-leukaemia agents.[49] Cassady and co-workers detailed chemical studies of the cytotoxic extract of the plant *Psorospermum febrifugum* (Guttiferae) have led to the re-isolation of the antileukaemic xanthone psorospermine (**52**) and the new discovery of several bioactive analogues, in a report in which they also ascertain the absolute stereochemistry of psorospermine.[50] The significant anti-leukemic activity established the importance of the configuration and functionality of the epoxydihydrofuran moiety.

12.2.10 Antimalarial

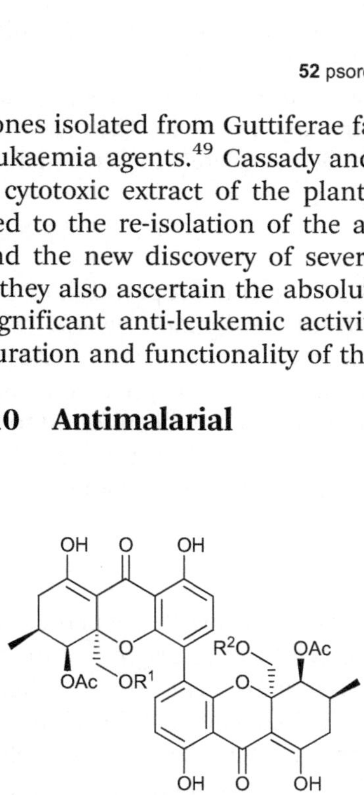

53 Phomoxanthone A
(absolute configuration); R^1 = Ac, R^2 = Ac

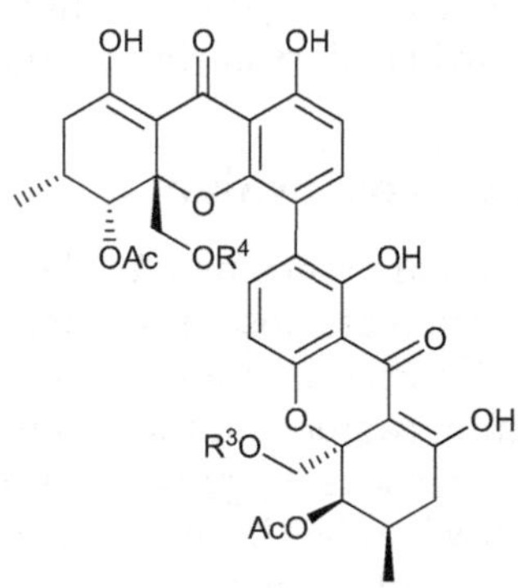

54 Phomoxanthone B; R^3,4 = Ac

Intrigued by the antimalarial activity of methanolic extracts of mycelia of the endophytic fungus *Phomopsis* sp. BCC 1323, researchers isolated and identified two dimeric xanthones named phomoxanthones A and B (**53–54**). The phomoxanthones A and B are the first reported examples of naturally occurring xanthone dimers with 4,4′ or 2,4 linkages. Phomoxanthones A and B showed antimalarial (*Plasmodium falciparum* K1, multi drug resistant strain, IC_{50} of 0.11 and 0.33 μg mL^{-1}, compared to IC_{50} of 0.16 and 0.0011 μg mL^{-1} for chloroquine diphosphate and artemisinin, respectively).[51] A few years later, the same research group found Ascherxanthone A (**39**), another dimeric xanthone, in the methanolic extracts of the mycelia of *Aschersonia* sp. BCC 8401, after they observed antimalarial activity in *Aschersonia* cultures. Ascherxanthone A has a significant activity against *Plasmodium falciparum* K1 with an IC_{50} of 0.20 μg mL^{-1}).[52]

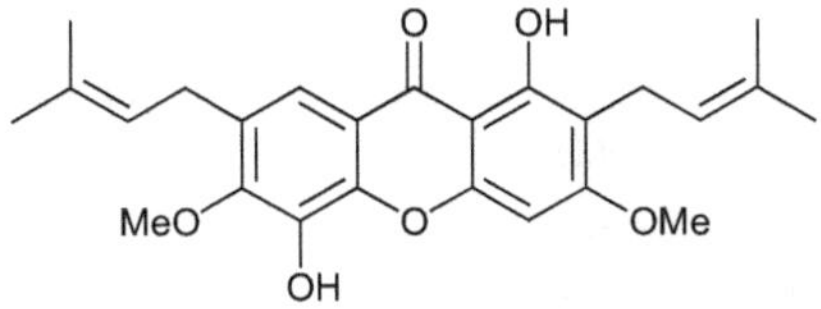

55

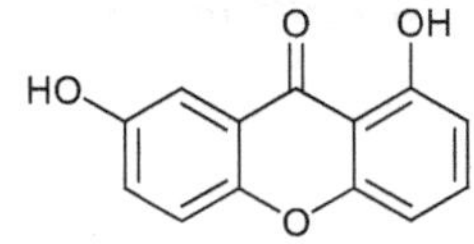

35

In a search for compounds with interesting biological properties 1,5-dihydroxy-3,6-dimethoxy-2,7-diprenylxanthone (**55**) and 1,7-dihydroxy-xanthone (**35**) were tested thoroughly. The prenylated xanthone (**55**) showed selective activity against *P. falciparum* with an IC_{50} of 7.25 μM, but all other tests revealed the molecule to be virtually inactive. The 1,7-dihydroxy-xanthone (euxanthone, **35**) was found to be completely inactive in all tests.[53]

12.2.11 Anti-nociception

Xanthones have been found to possess pain-killing properties. For example, 1,7-dihydroy-2,3-dimethoxyxanthone from *Polygala cyparissias* (Polygalaceae), which grows abundantly on Brazil's Atlantic coast.[54] Additionally, 1,5-dihydroxyxanthone displayed potent anti-nociceptive effects when evaluated.[55,56] This same dihydroxyxanthone has also been found to possess cytotoxic activity against KB-line cells and potential antimicrobial activity, which further hints at the multiple activities a single xanthone may have.[57]

12.2.12 Anti-oxidant

Xanthones are renowned for their anti-oxidant properties, whether naturally derived, such as those from mango,[58] or synthetic in origin,[59] and the more active of these xanthones are hydroxylated or bear the catechol motif. A preliminary study on the ability of polyhydroxylated 2,3-diarylxanthones to scavenge reactive oxygen and nitrogen species ($^{\bullet}O_2^{-}$, HOCl, $^{\bullet}NO^{-}$, NO_3^{-} and singlet oxygen) has been performed using chemical methods.[60] Additionally, Martinez and co-workers have studied the anti-oxidant features (single electron transfer) of twenty isolated xanthones which are present in the tropical fruit mangosteen, *Garcinia mangostana*, and some of their anionic forms, as many of these compounds exist under biological conditions as the conjugate bases.[61] They found that these xanthones react with $^{\bullet}OH$ at diffusion-limited rates, proving there is evidence for the often-promoted healthy antioxidant nature of mangosteen.

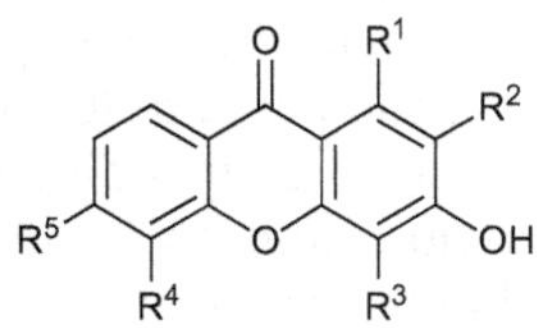

56 A. luzonensis xanthones

Anaxagorea luzonensis A. GRAY is a tree used traditionally in Thai medicine as a blood tonic antipyretic, and for muscular pain.[62] Analysis of the heartwood of the tree led to the isolation of five novel xanthones (generic formula **56**), which were found to have little antioxidant activity in comparison to a co-isolated flavone.

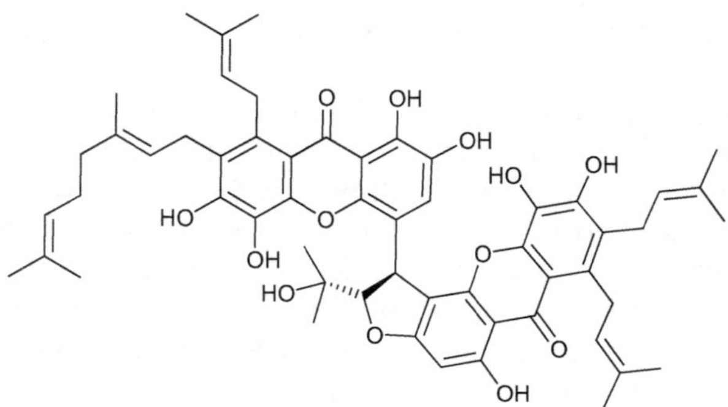

57 Bigarcinenone A
(relative stereochemistry)

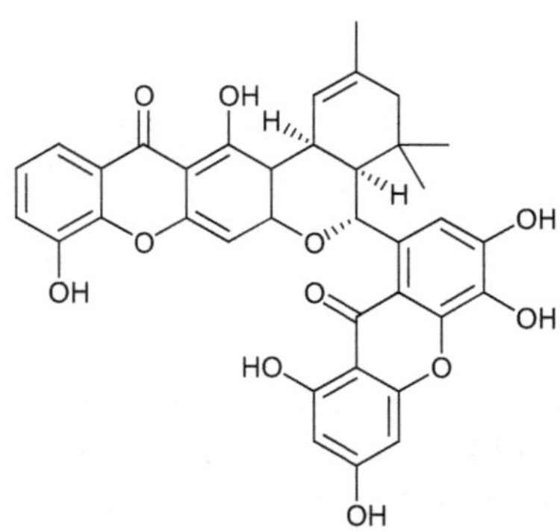

58 Bigarcinenone B
(relative stereochemistry)

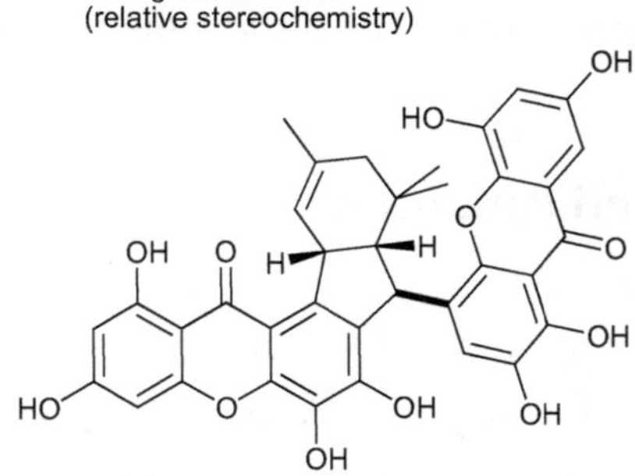

59 Griffipavixanthone

Bigarcinenone A (**57**) was isolated from the Chinese medicinal plant *Garcinia xanthochymus* after Yang *et al.* observed a strong antioxidant activity (IC_{50}: 4.6 μg mL^{-1}, as determined by a 1,1-diphenyl-2-picrylhydrazyl (DPPH) radical scavenging bioassay) in the ethyl acetate soluble fraction from the ethanol extracts. When subjected to a DPPH radical scavenging activity assay the dimeric xanthone Bigarcinenone A outperformed (IC_{50}: 9.2 μg mL^{-1}) all other isolated xanthones from this extract (IC_{50}: 16.3–250 μg mL^{-1}) as well as BHT (IC_{50}: 20.0 μg mL^{-1}), a well-known synthetic antioxidant.[63] In 2011, Bigarcinenone B (**58**) was isolated from the bark of the same plant and its antioxidant activity was also tested with a DPPH assay (IC_{50}: 20.14 μM versus 13.16 μM for ascorbic acid) and a H_2O_2 assay (IC_{50}: 2.85 μM versus 0.76 μM for ascorbic acid).[64] Griffipavixanthone (**59**), isolated from several species of the Malaysian plants *Garcinia pavifolia*, *G. griffithi*,[65] G. *oblongifolia*,[66] G. virgate[67] and *G. maingayii*[68] was the first bisxanthone found that was connected through a 5 and a 6 membered ring. In a DPPH free radical scavenging assay it was found that griffipavixanthone performed rather well, outperforming two known antioxidant references. (EC_{50}: Griffipavixanthone: 11.5 μg per 100 mL, 2,6-di-*tert*-butyl-4-hydroxy-anisol: 13.6 μg per 100 mL, α-tocopherol: 13.8 μg per 100 mL).[67] In that same study two monomeric prenylated xanthones, the virgataxanthone A and B (not shown) were also isolated, but they were found to perform rather poorly in tests.

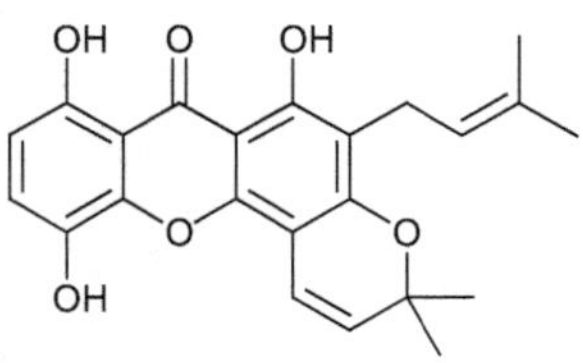

60

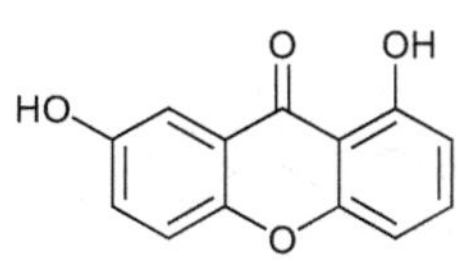

35

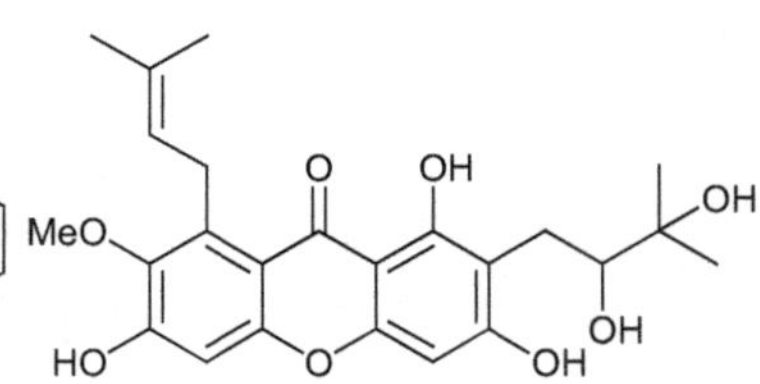

61

Not all xanthones that are found always exhibit any useful activities. In a search for new plant-based antioxidants and antibacterial agents, euxanthone (**35**), morusignin J (**60**) and the thus far unreported dulcisxanthone G (**61**) were isolated from *Garcinia dulcis* (Guttiferae). Sadly, none of these xanthones was found to show notable antibacterial or antioxidant activity.[69]

12.2.13 Anti-Parkinson's

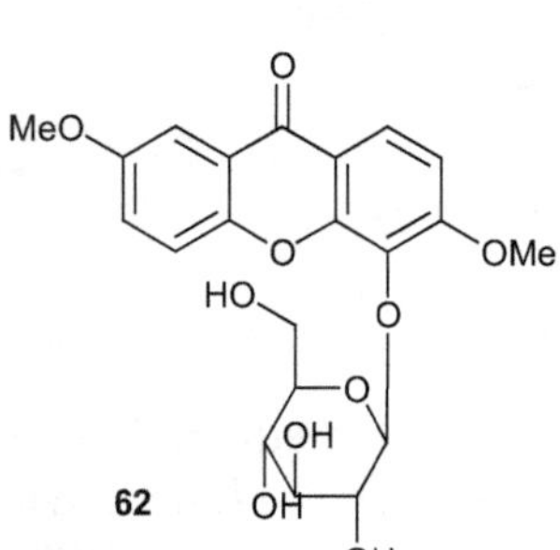

62

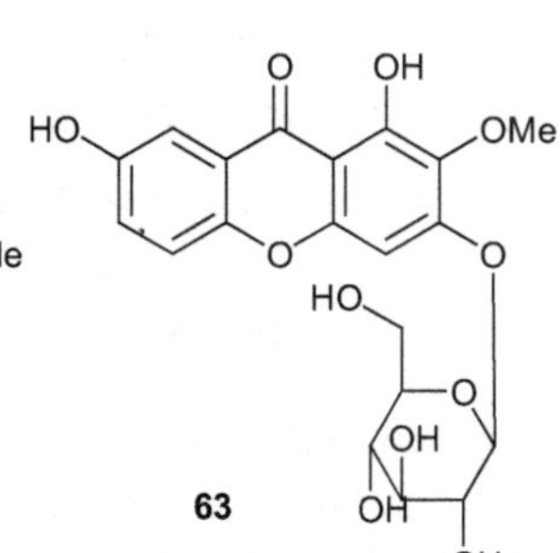

63

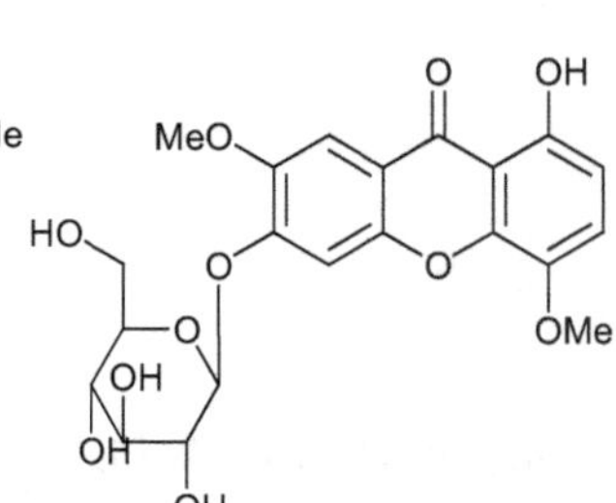

64

In the continued search for novel antidepressant and anti-Parkinson's disease drugs, three new xanthone glycosides named securixanside A (**62**), securixanside B (**63**), and securixanside C (**64**), were isolated from the stems of *Securidaca inappendiculata* Hassk.[70]

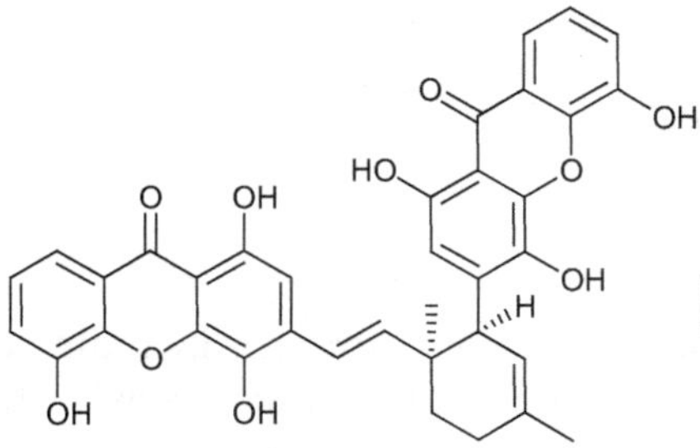

65 Garcilivin A

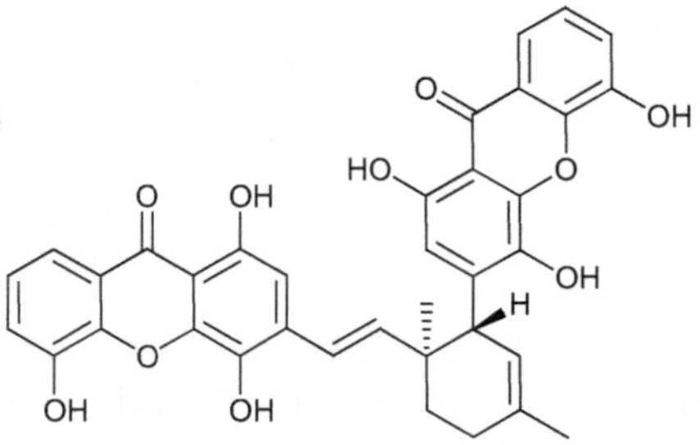

66 Garcilivin C

12.2.14 Anti-protozoal

Garcilivins A and C (**65–66**) are two dimeric and prenylated xanthones isolated from the South African plant *Garcinia livingstonei.*[71] Garcilivin A showed a very strong non-selective activity in the assays against *Trypanosoma brucei*, *T. cruzi* and *Plasmodium falciparum,* but was inactive against *Leishmania infantum,* whereas garcilivin C only showed significant activity against *T. brucei.*[72]

12.2.15 Anti-tubercular

Phomoxanthones A and B, (**53–54**) isolated from the methanolic extracts of mycelia of the endophytic fungus *Phomopsis* sp. BCC 1323 have been shown to exhibit anti-tubercular activity (*Mycobacterium tuberculosis* H37Ra strain MIC of 0.5 and 6.25 $\mu g\, mL^{-1}$, compared to MIC of 0.05 and 2.5 $\mu g\, mL^{-1}$ for isoniazid and kanamycin sulfate, respectively).[51]

12.2.16 Anti-viral

One plant in this genus dried stem bark of *Calophyllum brasiliense* allowed isolation of seven new xanthones (brasixanthones A–G), from which three (brasixanthones B to D, **67–69**) were shown to be active against Epstein–Barr virus.[73]

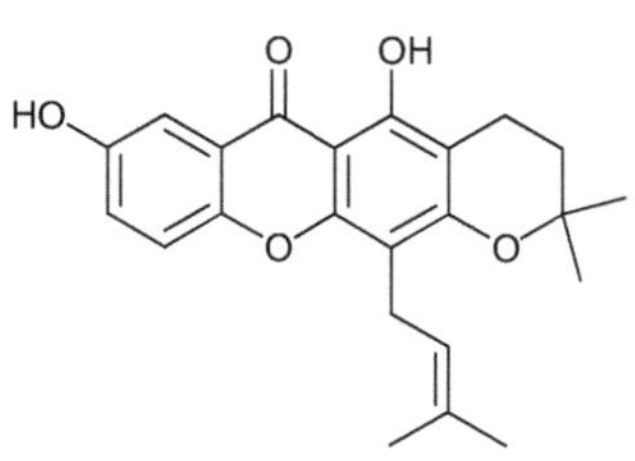

67 brasixanthone B

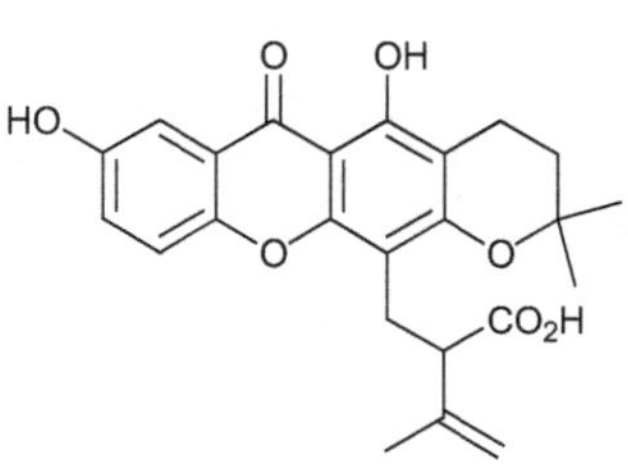

68 brasixanthone C

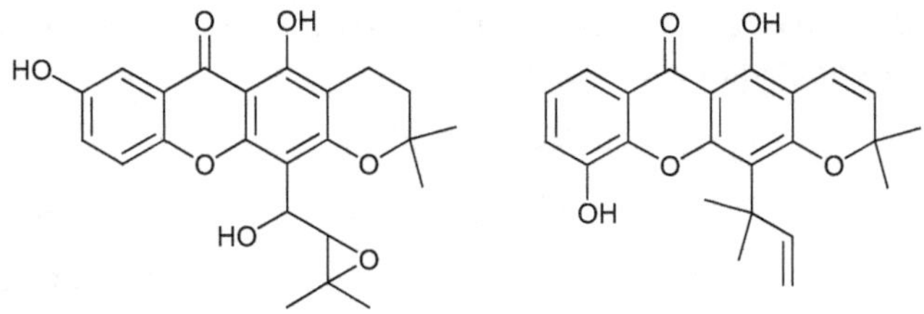

69 brasixanthone D **70** blancoxanthone

*Calophyllum blanco*i was found to contain several xanthones, including the structurally similar compound blancoxanthone (**70**), which likewise had considerable anti-viral activity, this time against coronavirus.[74]

12.2.17 Anthelmintic

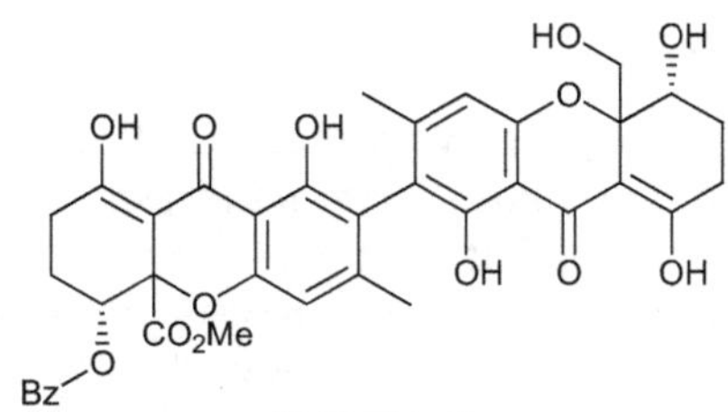

71 Xanthonol

A dimeric xanthone named xanthonol (**71**) was found to be active as anthelmintic, inhibiting larval growth of *Lucilia sericata, Aedes aegypti* and *Haemonchus contortus* with LD_{90} of 33 $\mu g\,mL^{-1}$, 8 $\mu g\,mL^{-1}$ and 50 $\mu g\,mL^{-1}$, respectively. Xanthonol is an unsymmetrical 2,2′-biaryl-linked dimeric xanthone that was isolated from the fermentation broth of a non-sporulating fungi found in the leaf litter of the plant *Manikara bidentata.*[75]

In 2012, the anthelminthic properties of α-mangostin (**29**) and a synthetic mangostin diacetate derivative were investigated. Both compounds showed very low activity against *A. ceylanicum* adults, but α-mangostin affected the tested trematodes in *in vitro* tests. However, these results did not transfer to *in vivo* tests.[76]

12.2.18 Enzyme Inhibition

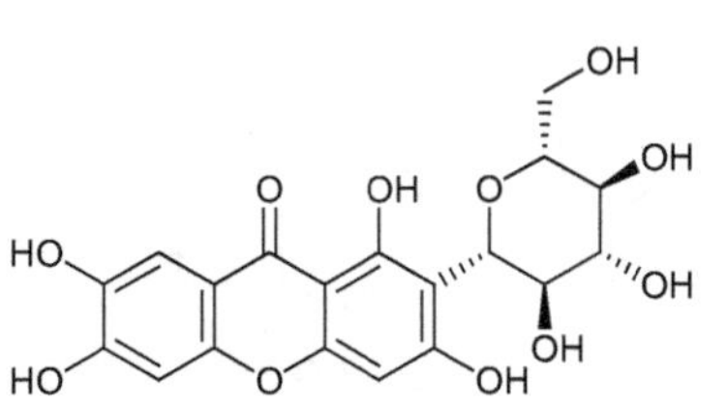

72 mangiferin

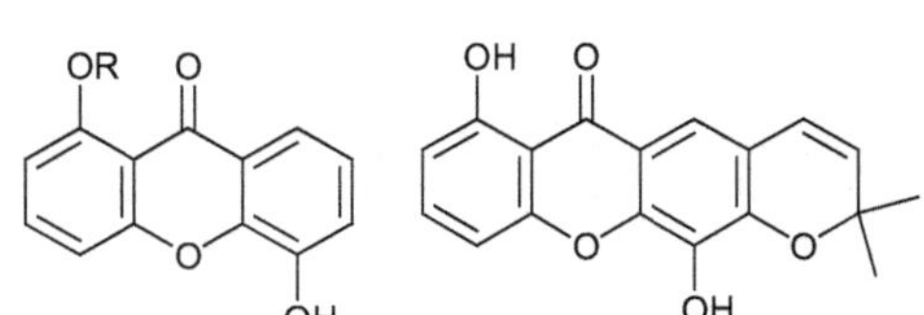

73: R = H
74: R = Me

75 6-deoxyjacareubin

Mangiferin (**72**), a natural bioactive xanthone C-glycoside, is widely present in medicinal plants like the leaf of *Mangifera indica* L. (Anacardiaceae). It has been reported that mangiferin possesses a variety of effects including: (1) antidiabetic, (2) hepatoprotective, (3) anti-inflammatory, (4) antioxidant, and (5) anticarcinogenic activities: further evidence for multiple bioactivities and activity at multiple sites of action for compounds within the xanthone class. The specific activity of mangiferin upon several kinases were investigated by Han and Chin.[77] They found that specific inhibition of anaplastic lymphoma kinase 0.81 μmoL, the insulin receptor >10 μmoL and epidermal growth factor receptor >10 μmoL. Taken together, these results are suggestive of a new mode of action of this compound pertinent to cancer intervention. Li and co-workers also observed a reducing effect of mangiferin on serum uric acid levels in mice.[78] It was found in a study by Nakatani and co-workers that γ-mangostin (**31**) directly inhibits cyclooxygenase (COX) activity as well as prostaglandin E2 synthesis.[79] *Hypericum brasiliense* contains compounds like 1,5-dihydroxyxanthone, 5-hydroxy-1-methoxyxanthone and 6-deoxyjacareubin (**74–75**) which had differing effects in terms of monoamine oxidase inhibition.[80] Jacareubin and several of its derivatives (**76–79**) were tested on the inhibition of gastric H^+, K^+-ATPases and performed with IC_{50} values ranging from 47 μM to 1.6 mM.[81]

Two xanthones are currently listed as experimental drugs, both active in enzyme inhibition. *N*-(1,4-dihydro-5*H*-tetrazole-5-ylidene)-9-oxo-9*H*-xanthene-2-sulfonamide (**80**) was found to inhibit 1,3-dehydroquinate dehydratase from a strain of *Helicobacter pylori*, although its true pharmacological action remains unknown. 1,8-Di-hydroxy-4-nitro-xanthen-9-one (**81**) was reported to interact with the human casein kinase II subunit alpha.[82]

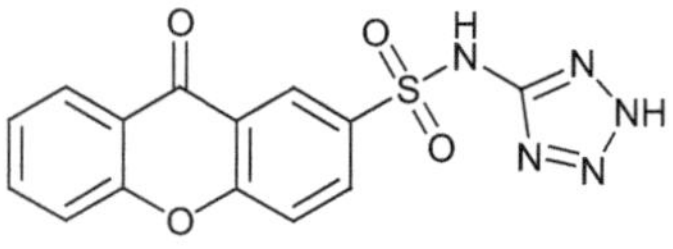

80

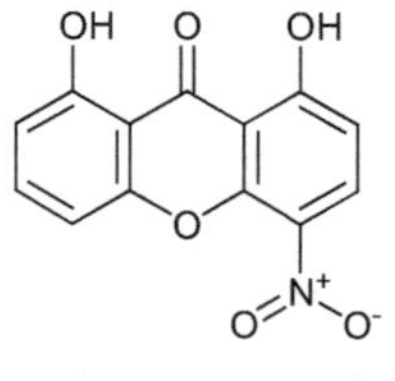

81

12.2.19 Hepatoprotection

Pseudonolin (**82**) was isolated alongside 13 other known constituents from a Chinese natural medicine, *Swertia pseudochinensis* HARA, *S. pseudochinenses* was traditionally used to treat acute or chronic hepatitis. Pseudonolin, as well as the other twelve compounds were found to be hepatoprotective when tested on hepatocyte toxicity induced by exposure to CCl_4.[83]

Wang and co-workers found the xanthone HM-1 from Tibetan herb *Halenia elliptica*, that is used in tablet form ('Yiganjian tablets') to treat liver and gall bladder diseases in China. It contains 1-hydroxy-2,3,5-trimethoxyxanthone (**83**) in their *in vitro* identification of which isoforms of cytochrome P_{450} are actively involved in the metabolism of this compound, providing evidence of substrate inhibition and metabolism-based drug–drug interaction for the medicinal preparations containing HM-1 used in clinic.[84]

This same compound has also been shown to be a vasorelaxor endothelium-independent mechanism by inhibiting Ca^{2+} influx through L-type voltage-operated Ca^{2+} channels.[85] The same authors studied the metabolite of that compound, HM-5 (**84**), and showed that the mechanisms of the vasorelaxant effects of HM-5 are distinctly different from those of its parent drug HM-1. The vasorelaxant effect of HM-5 was mediated through the opening of the potassium channel (4-AP) and the altering of intracellular calcium by partial inhibition of Ca^{2+} influx through L-type voltage-operated Ca^{2+} channels and intracellular Ca^{2+} stores.[86]

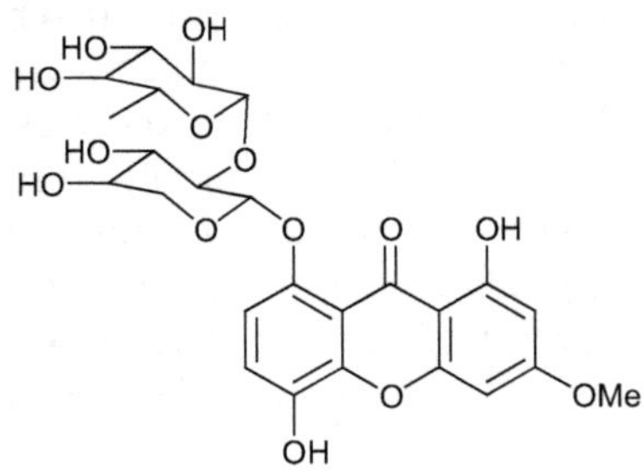

82 Pseudonolin

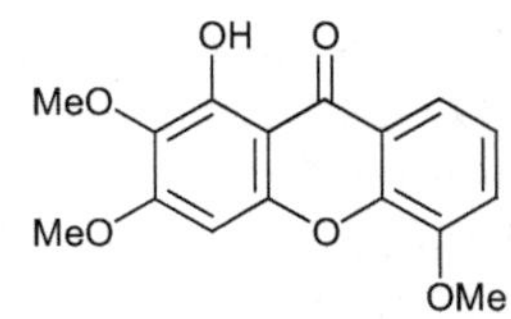

83 1-hydroxy-2,3,5-trimethoxyxanthone

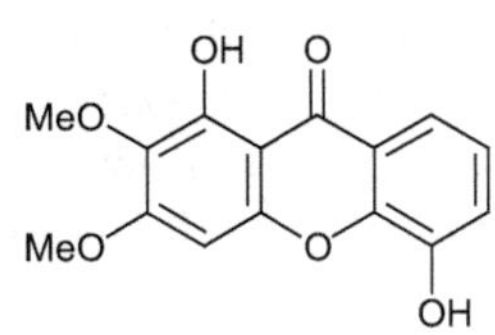

84 1,5-dihydroxy-2,3-dimethoxyxanthone

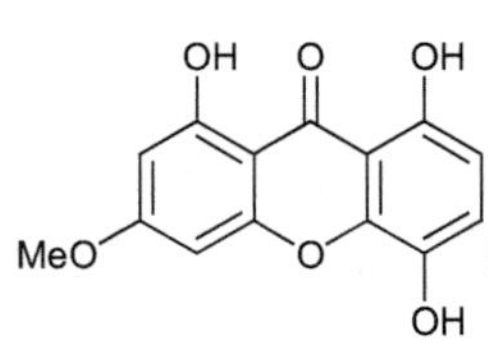

85

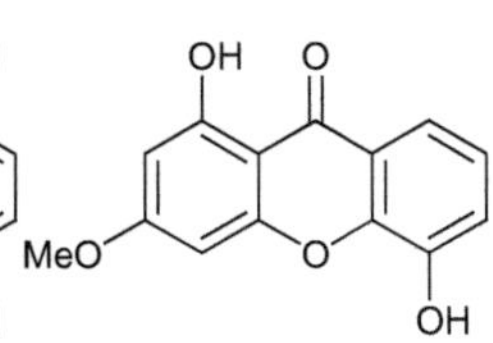

86

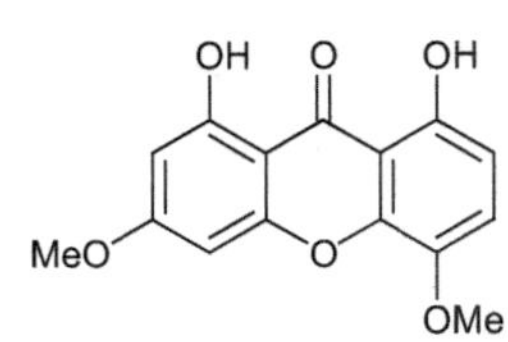

87

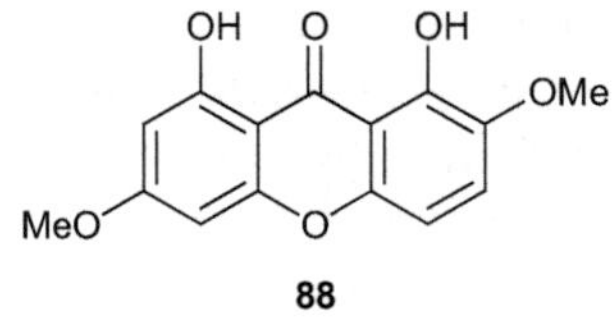

88

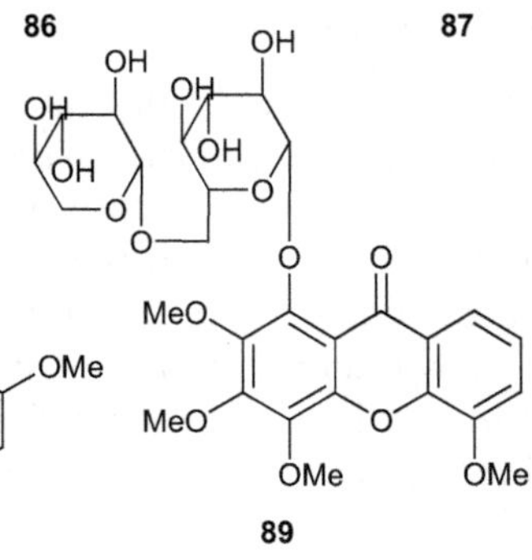

89

The total iridoid and xanthone extract from *Swertia mussotii* was found to exhibit significant hepatoprotecive effects. A closer examination of these extracts revealed that they contained significant amounts of 1,5,8-trihydroxy-3-methoxyl xanthone (**85**), 1-hydroxy-3,5-dimethoxy xanthone (**86**), 1,8-dihydroxy-3,7-dimethoxy xanthone (**87**), 1,8-dihydroxy-3,5-dimethoxy xanthone (**88**) and 2,3,4,5-tetramethoxy-1-*O*-primeverosyloxanthane (**89**). Whether these xanthones are actively contributing to these effects is still under investigation.[87]

12.2.20 Nerve-growth Factor Inducing Activity

Garcinia xanthochymus is a perennial medicinal plant (native to the north of Thailand and Myanmar), the wood of which was shown to contain three prenylated xanthones (**90–92**) which displays a markedly enhancing activity on nerve-growth factor (NGF)-mediated neurite outgrowth on PC12D cells at concentrations of 10–30 mM.[88]

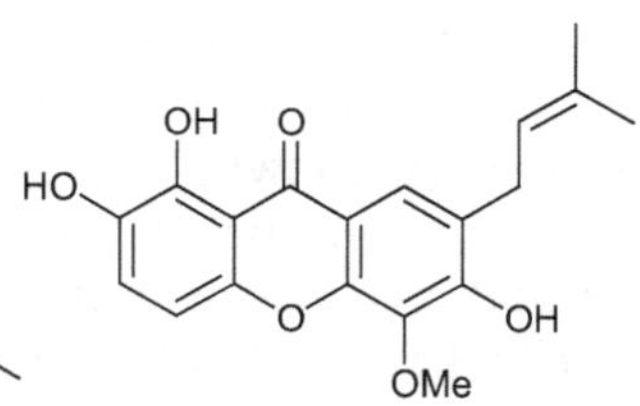

90 1,2,6-trihydroxy-5-methoxy-7-(3-methylbut-2-enyl)xanthone

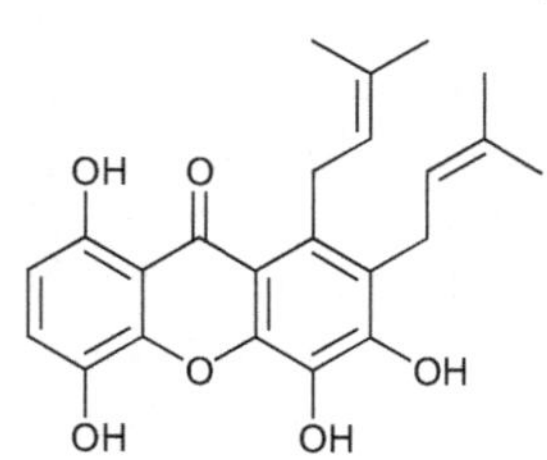

91 1,4,5,6-tetrahydroxy-7,8-di(3-methylbut-2-enyl)xanthone

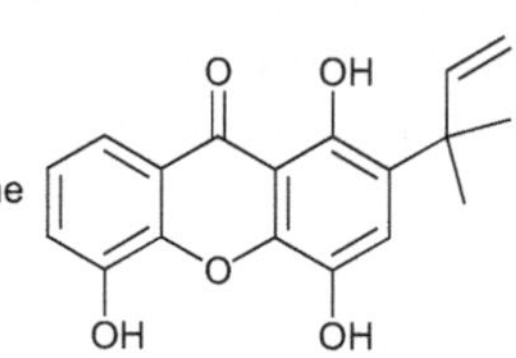

92 12b-hydroxy-des-d-garcigerrin A

12.2.21 Neurogenic Inflammation and Vasorelaxant Activity

Norathyriol (**93**), the aglycon of mangiferin, was isolated by Wang and coworkers from *Tripterospermum lanceolatum*, and used to inhibit plasma leakage elicited by the passive cutaneous anaphylactic reaction in normal as well as in adrenalectomized mice, an effect which the authors have shown is not related to the release of steroid hormones from the adrenal gland.[89] Additionally, 1-hydroxy-2,3,5-trimethoxy-xanthone (**94**) was found to show potent vasorelaxant activity.[90,91]

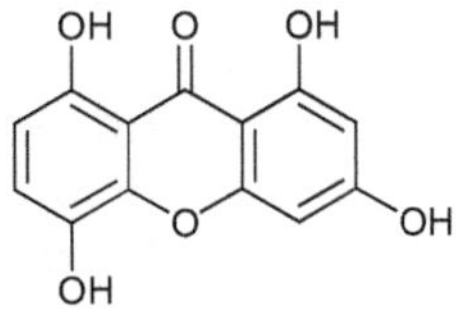

93 norathyriol

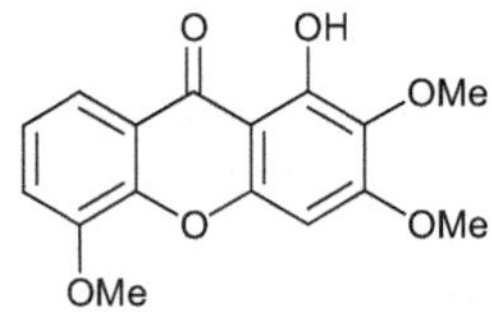

94 1-hydroxy-2,3,5-trimethoxy-xanthone

12.2.22 Neuroprotective

Several related dimeric xanthones have been found to exhibit potent neuroprotective capabilities. Xanthone treated and untreated PC12 cells were subjected to 1-methyl-4-phenylpyridiniumion (MPP+), rotenone or hydrogen peroxide. Using MTT cell death assays it was found that 3-*O*-demethyl-swertipunicoside can effectively protect the cells from injury. Additionally, it was found that this effect likely occurs by elevation of the TH and DJ-1 protein levels.[92] Swertia-bisxanthone I 8′-*O*-β-D-glucopyranoside (**95**), isolated from *Gentianella amarella* by Hostettmann *et al.*,[93] puniceasides A–E (puniceaside B **96**, others not shown), 3-*O*-demethyl-swertipunicoside (not shown) and swertipunicoside (not shown) were also tested for their neuroprotective activity against hydrogen peroxide-induced PC12 cell damage. Most notably was puniceaside B, which showed a cell viability of $98.1 \pm 6.8\%$ at a concentration of 25 $\mu g\,mL^{-1}$. Perhaps even more interesting, swertiabisxanthone I 8′-*O*-β-D-glucopyranoside and 3-*O*-demethylswertipunicoside were found to potently stimulate the damaged PC12 cells to grow, resulting in cell viability scores of 123% and 158%. In 2012, Guo *et al.* found 16 new xanthone compounds from extracts of *Swertia punicea* that yet have not been tested for biological activity.[94]

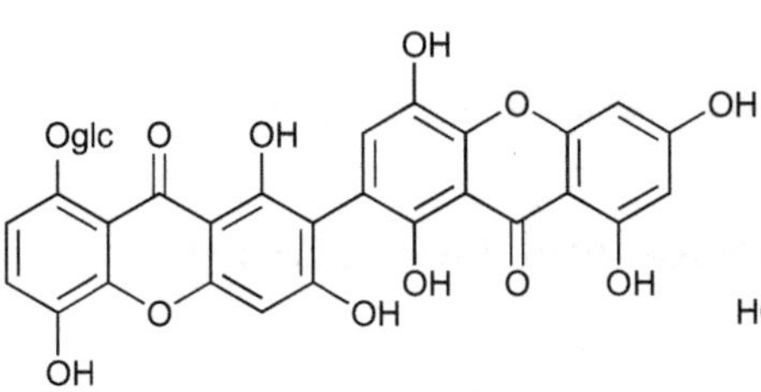

95 Swertiabisxanthone I 8′-*O*-β-D-glucopyranoside

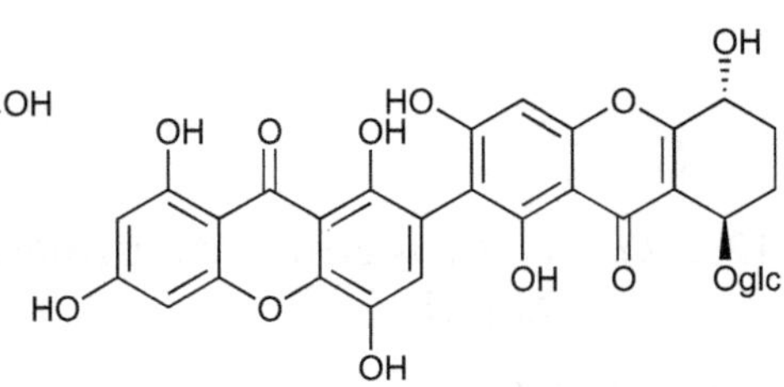

96 Puniceaside B

12.2.23 Novel Cytotoxicity

In the search for new cancer therapeutic agents, cell cytotoxicity tests are one of the most tested biological properties of xanthones. Many dimeric xanthones have been found to be cytotoxic. In 1957, Nakamura and co-workers isolated bikaverin (**97**), a deep-red coloured compound, from *Gibberella fujikori*.[95] The compound has been isolated from several other sources, and is known to have a vacuolation effect in fungi,[96] to have a specific anti-protozoal activity against *Leishmania brasiliensis*,[96] and to be cytotoxic to various tumour cell types.[97,98]

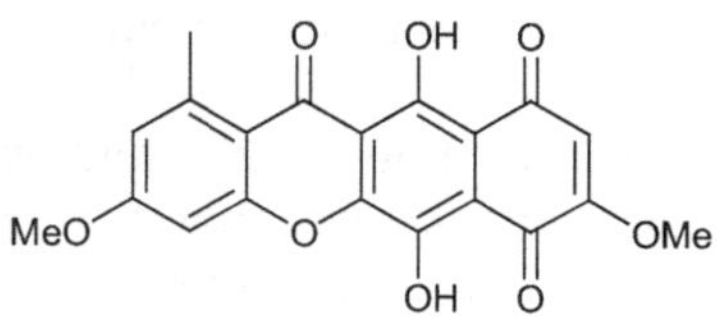

97 bikaverin

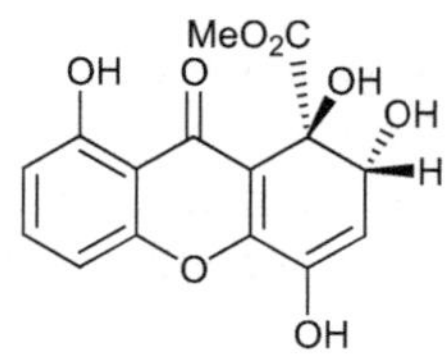

98 globosuxanthone A

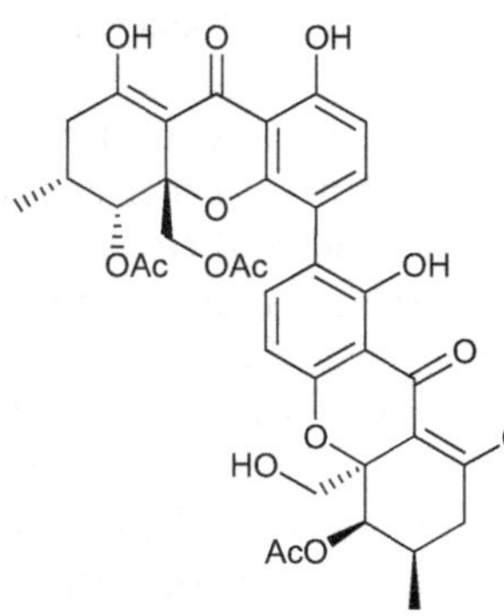

99 deacetylphomoxanthone B / penexanthone A

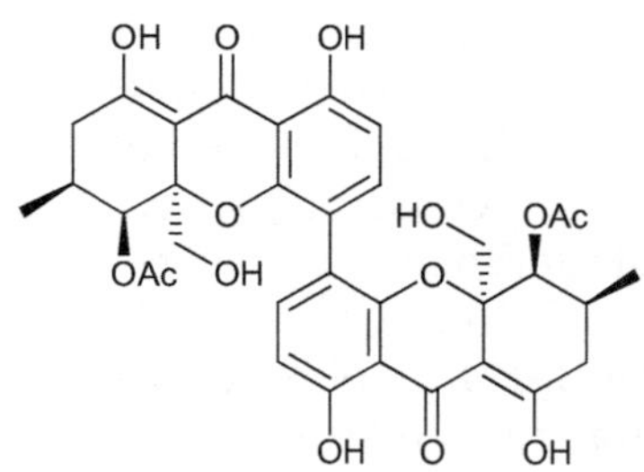

100 deacetylphomoxanthone C (relative configuration)

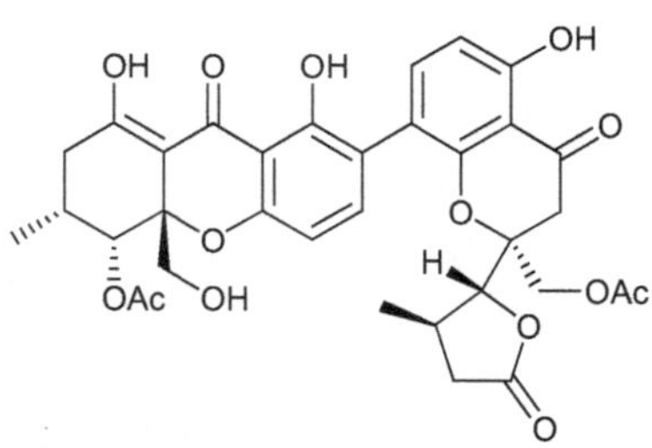

101 phomalactonexanthone A

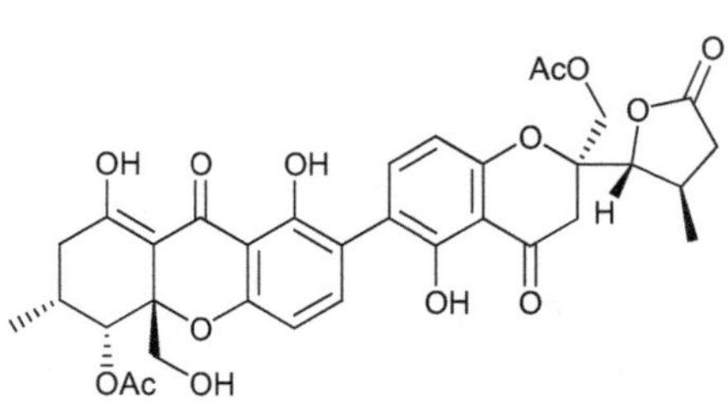

102 phomalactonexanthone B

Globosuxanthone A (**98**) was found to be particularly cytotoxic against multiple human cancer cell lines.[99] Phomoxanthones A and B (**53–54**) have been tested against cancer cell lines (KB cells, BC-1 cells and Vero cells) and were found to be cytotoxic (IC_{50} μg mL^{-1}; 0.51–1.4 for and IC_{50} μg mL^{-1} 0.7–4.1 for phomoxanthone B). However, it must be noted that the currently available standard drugs are more potent. A deacetyl derivative of phomoxanthone A was also investigated in these assays and was found to be mostly inactive in all the tests; it is speculated that this might be due to its lower lipophilicity.[51] The closely related deacetylphomoxanthone B and

C (**99–100**) and phomolactonexanthones A and B (**101–102**) were also tested in respect to their cytotoxicity against a series of human cancer cell lines, and it was found that while deacetylphomoxanthone B, also named penexanthone A, has a reasonable cytotoxicity against all tested cancer cell lines, deactylphomoxanthone C and the phomolactonexanthones were virtually inactive against all cell lines.[100] Phomoxanthone A and deacetylphomoxanthone A (not shown) were also found to be actively apoptosis-inducing.[101]

The plant derived dimeric xanthone griffipavixanthone (**59**), that was found to have very interesting anti-oxidant properties, also showed high activity against P388, LL/2 and Wehil64 cell lines.[65] A recent Chinese patent claims promising anti-cancer properties to a variety of cancer cell lines including human lung, breast, prostatic and intestinal cancer cells. Also, **59** showed no cytotoxicity to normal kidney epidermal cells. In more detail they found that for the lung cancer cells H520 the cell cycle was blocked in the S stage, thereby preventing the cancer from propagating.[102]

Ascherxanthone A (**39**), that also displays very interesting antimalarial properties, was found to be cytotoxic to *Vero* cells and three cancer cell lines (IC_{50} values between 1.7 and 0.16 $\mu g\,mL^{-1}$ against KB, BC and NCIH187 cell lines).[52] Chrysoxanthone (**103**), a 2-hydroxychrysophanol coupled to blennolide A (hemisecalonic acid B) through a biaryl ether linkage[103] is an unusual xanthone in the subclass of the 'xanthraquinones', that also includes the beticolins[104–107] and the xanthoquinodins.[108–111] Biological tests on Jurkat, L-1210, Colo-320 and HeLa-S3 cell lines showed a moderate to low cytotoxicity. Chrysoxanthone was also found to exhibit antibacterial and antifungal properties.[103]

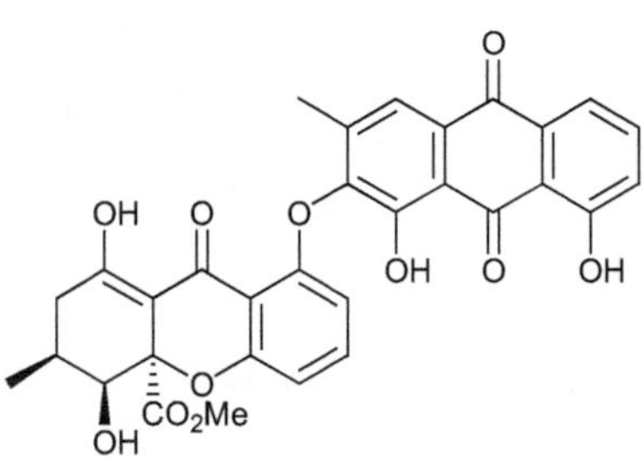

103 chrysoxanthone

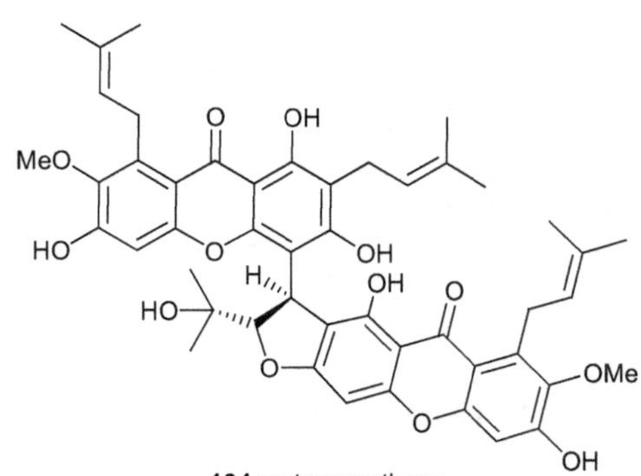

104 cratoxyxanthone

Cratoxyxanthone (**104**) was first isolated from the bark of *Cratoxylum cochinchinense* In 1995[112] and later from the chloroform soluble extracts of the stem bark of *Garcinia mangostana*.[113] This dimeric xanthone having a high degree of prenylation was found to perform very poor in cytotoxicity (HT-29 colon cancer cell line) and ELISA NF-κB (p65 and p50) assays.

Dicerandrols A–C (**105–107**) are dimeric xanthones with a 2,2′-linkage isolated from the fungus *Phomopsis longicolla*.[22] Besides their antibacterial properties, as discussed in Section 12.2.3, they were found to be active against two cancer cell lines, HCT-116 and A549 (colon and lung tumour, respectively).[22] In another study, the dicerandrols and the structurally similar penexanthone A were submitted in several biological activity tests versus a broad range of tumour cell lines.[114] In these tests the effect of the presence of non-malignant accessory cells, such as bone marrow stromal cells, was evaluated. It was found that the activities of potential drugs can be affected by microenvironment-dependent drug resistance or sensitization of the tumour cells.[115] Most notable among the tested compounds was dicerandrol B, which showed moderate activity against several cancer cell lines including myeloma, lymphoma, leukemia, breast and prostate cancers cell lines in the presence of stromal cells. Also, dicerandrol B was evaluated as extra promising as a novel drug, due to a relatively low toxicity against human immortalized non-malignant cells, such as HS-5 bone marrow stromal cells, HOBIT osteoblast-like cells, THLE-3 hepatocytes, and SVGp12 astrocytes.[114] The other dicerandrols and penexanthone A (**99**) showed also activities, but were in general less active. Ding *et al.* also evaluated the dicerandrols A–C on their cytotoxicity against human breast, colon, lung and liver cancer and breast epithelial cell lines. Dicerandrol A showed a broad anti-tumour activity, but was also cytotoxic to the breast epithelial cells. Dicerandrol B and penexanthone A were found to be quite selective and damage the breast epithelial cells less. Dicerandrol C was found to be not very cytotoxic, suggesting the free hydroxy groups to be important as key pharmacophore.[100] Dicerandrols B and C are both cytotoxic to murine lymphoma cancer cell lines (IC_{50} values of 10 and 1.1 μM, respectively) and dicerandrol C was found to be mildly pro-apoptotic.[101]

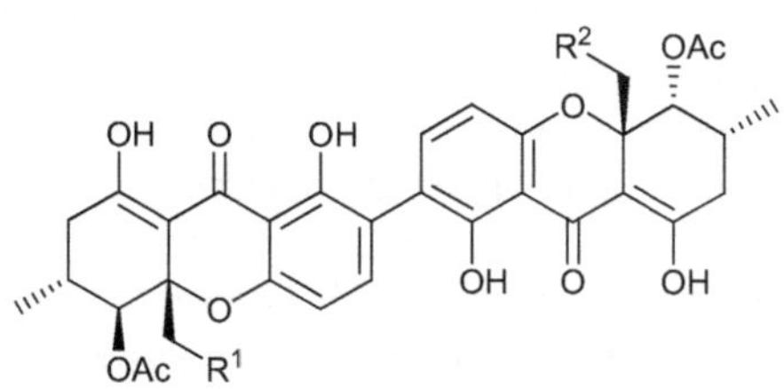

105 Dicerandrol A (relative configuration); R^1 = OH, R^2 = OH
106 Dicerandrol B (absolute configuration); R^1 = OAc, R^2= OH
107 Dicerandrol C (absolute configuration); R^1 = OAc, R^2 = OAc

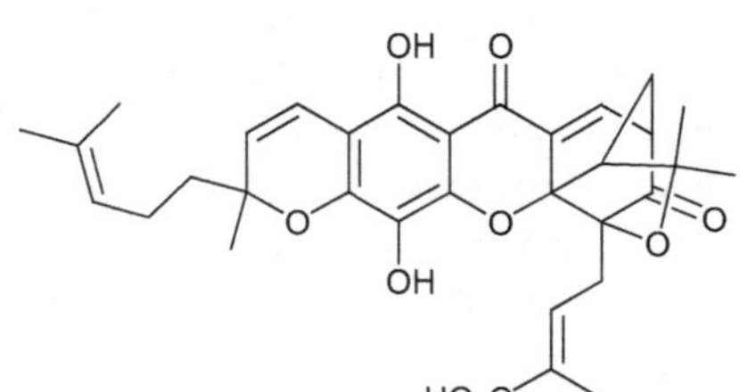

108 gambogenic acid

The dimeric garcilivin A and C (**65–66**) showed interesting differences in a cytotoxicity test against MRC-5 cells. With an IC_{50} of 2.0 μM garcilivin A proved to be 25 times stronger than Garcilivin C.[72]

The discovery of novel effective chemotherapy regimens for glioblastoma multiform therapy is urgent. The application of Gambogenic acid (**108**) to U251 glioblastoma cells induced time- and dose-dependent growth inhibition and apoptosis in those cells. The putative mechanism involves Akt pathway inactivation.[116]

12.3 Xanthone Drugs

Although no drugs are currently on the market with a xanthone structure as a core, two compounds in particular have shown extensive positive results in terms of anti-tumour activities: Gambogic acid and dimethylxanthoneacetic acid.

12.3.1 Gambogic Acid (GA)

Gambogic acid (GA, **49**) is the main component of gamboge,[117] which is a traditional medicine in South-East Asia. The caged xanthone-containing scaffold of GA features a 4-oxa-tricyclo [4.3.1.03,7]dec-2-one ring system and has been extensively studied[118] as a potent anti-tumour and anti-inflammatory agent. It has entered phase I clinical trials in China, after initial studies showed that GA inhibits cancer proliferation and metastasis[119] GA has been reported to interact with several biological target molecules; *e.g.*, to induce apoptosis by binding to the transferrin receptor[120] and interfering with pathways involving nuclear factor-κB (NF-κB).[121] GA also plays the role of a conjugate acceptor which irreversibly binds to a cysteine residue of IKKβ and thereby prevents NF-κB activation.[122] It is believed that GA can bind to heat shock protein 90 (HSP90) to decrease the activity of a number of biomolecules which are dependent on their interaction with this protein to create angiogenesis, cell growth and metastasis. Recently, Wang and co-workers have synthesized insight-providing derivatives.[123]

12.3.2 Dimethylxanthone-4-acetic Acid (DMXAA)

Dimethylxanthone acetic acid (DMXAA, **109**) and its salts are important compounds, which have promising anti-cancer activities due to their capacity to act as vascular disruption agents (VDAs);[124] they are the subject of several patents.[125] The anti-tumour activity has been demonstrated to be *via* the activation of cytokine tumour necrosis factor α (α-TNF),[126] which leads to decreased blood flow to the tumour. DMXAA were administered 1 hr after a range of doses of cisplatin, the tumour cell kill was 10–500-fold greater than that seen with chemotherapy alone.[127] Phase I clinical trials for the treatment of vascular cancers started in 2005; the results indicated that DMXAA has anti-tumour activity at well-tolerated doses.[128] DXMAA then entered phase II clinical trials in the United States for the treatment of prostate cancer in 2007.

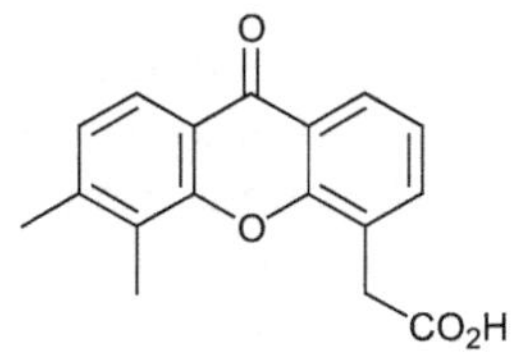

109 DMXAA

12.4 Are Xanthones 'Over-privileged'?

Xanthones are privileged scaffolds in medicinal chemistry. However, there is a significant caveat: some of the effects noted above may be due to their nature as poly-hydroxylated aromatics. This is significant, as phenols and polyphenols have been identified by medicinal chemists over time as giving rise to a number of non-specific modes of action. Andersen and co-workers have shown that phenolics from natural sources have membrane-destabilisation effects,[129] which can be imparted by the 'detergent-like' amphiphilic nature of these compounds, whereby the compounds gather at the membrane/aqueous fluid interface. These effects can go on to change the function of diverse membrane proteins. Baell and co-workers have noted that such compounds may fall under the category of 'PAINS'[130] – 'pan-assay interference compounds', and vigilance is advised when investigating, for example, catechols, quinones and hydroquinones.[131]

12.5 Conclusions

This chapter has given a necessarily concise entry-point onto the staggeringly diverse array of disease states in which xanthones are known to have an effect. The number of biological target molecules is also extremely diverse and, in some cases, as with gambogic acid, we have seen that multiple biotargets are affected, with the combined effect of down-regulating a single disease state. It may well be the case that some xanthones, particularly the more simple structures, can give a 'false positive' hit in assays. The medicinal chemist should be aware, but by no means ignore the potential for xanthones as lead compounds, drug candidates and even APIs.

References

1. W. Schmidt, S. Peters and L. Beerhues, *Phytochemistry*, 2000, **53**, 427–431.
2. K.-S. Masters and S. Bräse, *Chem. Rev.*, 2012, **112**, 3717–3776.
3. J. Yang, L. Ding, L. Hu, S. Jin, W. Liu, Q. You and Q. Guo, *J. Pharm. Biomed. Anal.*, 2012, **60**, 71–79.
4. M. M. M. Pinto, M. E. Sousa and M. S. Nascimento, *Curr. Med. Chem.*, 2005, **12**, 2517–2538.
5. T. Kikuchi, S. Kadota, S. Matsuda, K. Tanaka and T. Namba, *Chem. Pharm. Bull.*, 1985, **33**, 557–564.
6. J.-Z. S. S.-L. Li, Q.-B. Han, C.-F. Qiao and H.-X. Xu, *Biomed. Chromatogr.*, 2008, **22**, 637.
7. K. M. Elbel, G. Guizzunti, M. A. Theodoraki, J. X. Ayse Batova, M. Dakanali and E. A. Theodorakis, *Org. Biomol. Chem.*, 2013, **11**, 3341–3348.
8. L. Li, M. H. Li, N. Zhang and L. Q. Huang, *Biochem. Syst. Ecol.*, 2011, **39**, 766–768.

9. R. Feng, Y. Y. Zhang, X. Chen, Y. Wang, J. G. Shi, C. T. Che, J. H. Yeung, J. Y. Ma, X. S. Tan, C. Yang, Y. L. Deng and Y. K. Zhang, *J. Pharm. Biomed. Anal.*, 2012, **62**, 228–234.
10. R. Feng, X. Zhou, X. S. Tan, P. M. Or, T. Hu, J. Fu, J. Y. Ma, M. Huang, C. Y. He, J. G. Shi, C. T. Che, J. H. Yeung and Y. Wang, *Chem.–Biol. Interact.*, 2014, **210**, 12–19.
11. N. Kosem, K. Ichikawa, H. Utsumi and P. Moongkarnd, *J. Nat. Med.*, 2013, **67**, 255–263.
12. V. Cechinel Filho, C. Meyre-Silva and R. Niero, *Chem. Biodiversity*, 2009, 313.
13. M. U. S. Sultanbawa, *Tetrahedron*, 1980, **36**, 1465.
14. N. Anantachoke, P. Tuchinda, C. Kuhakarn, M. Pohmakotr and V. Reutrakul, *Pharm. Biol.*, 2012, **50**, 78–91.
15. B. Elsässer, K. Krohn, U. Flörke, N. Root, H.-J. Aust, S. Draeger, B. Schulz, S. Antus and T. Kurtán, *Eur. J. Org. Chem.*, 2005, 4563–4570.
16. W. D. Jones, W. I. Albrecht, N. L. Munro and K. T. Stewart, *J. Med. Chem.*, 1977, **20**, 594.
17. K. Ishiguro, S. Nagata, H. Oku and M. Yamaki, *Planta Med.*, 2002, **68**, 258–261.
18. J. B. Pretto, V. Cechinel-Filho, V. F. Noldin, M. R. K. Sartori, D. E. B. Isaias and A. B. Cruz, *Z. Naturforsch., C: J. Biosci.*, 2004, **59**, 657.
19. S. H. Shim, J. Baltrusaitis, J. B. Gloer and D. T. Wicklow, *J. Nat. Prod.*, 2011, **74**, 395–401.
20. M. Stewart, R. J. Capon, J. M. White, E. Lacey, S. Tennant, J. H. Gill and M. P. Shaddock, *J. Nat. Prod.*, 2004, **67**, 728–730.
21. T. Rezanka and K. Sigler, *J. Nat. Prod.*, 2007, **70**, 1487–1491.
22. M. M. Wagenaar and J. Clardy, *J. Nat. Prod.*, 2001, **64**, 1006–1009.
23. C. Lim, J. Kim, J. N. Choi, K. Ponnusamy, Y. Jeon, S.-U. Kim, J. G. Kim and C. H. Lee, *J. Microbiol. Biotechnol.*, 2010, **20**, 494–500.
24. B. Proksa, D. Uhrin, T. Liptaj and M. Sturdikova, *Phytochemistry*, 1998, **48**, 1161–1164.
25. A. R. B. Ola, A. Debbab, A. H. Aly, A. Mandi, I. Zerfass, A. Hamacher, M. U. Kassack, H. Brotz-Oesterhelt, T. Kurtan and P. Proksch, *Tetrahedron Lett.*, 2014, **55**, 1020–1023.
26. A. E. Nkengfack, P. Mkounga, M. Meyer, Z. T. Fomum and B. Bodo, *Phytochemistry*, 2002, **61**, 181–187.
27. Y. Sakagam, in ed. I. Ahmad, F. Aqil and M. Qwais, *Modern Phytomedicine: Turning Medicinal Plants into Drugs*, Wiley-VCH, Weinheim, 2006, pp. 138–153.
28. Q.-B. Han, H.-L. Tian, N.-Y. Yang, C.-F. Qiao, J.-Z. Song, D. C. Chang, K. Q. Luo and H.-X. Xu, *Chem. Biodiversity*, 2008, **5**, 2710–2717.
29. A. Batova, D. Altomare, O. Chantarasriwong, K. L. Ohlsen, K. E. Creek, Y.-C. Lin, A. Messersmith, A. L. Yu, J. Yu and E. A. Theodorakis, *Mol. Cancer Ther.*, 2010, **9**, 2869–2878.
30. L. K. L. Wang, X. Pan, J. Liu, Q. Wang, B. Du, D. Li, J. Luo, M. Liu, A. Hou and M. Qian, *Int. J. Cancer*, 2010, **127**, 1220–1229.

31. W.-m. Fu, J.-f. Zhang, H. Wang, Z.-c. Xi, W.-m. Wang, P. Zhuang, X. Zhu, S.-c. Chen, T.-m. Chan, K.-S. Leung, G. Lu, H.-X. Xu and H.-f. Kung, *J. Proteomics*, 2012, 4833–4843.
32. Z. Liu, M. Antalek, L. Nguyen, X. Li, X. Tian, A. Le and X. Zi, *Nutr. Cancer*, 2013, **65**(S1), 68–77.
33. N. Boonnak, C. Karalai, S. Chantrapromma, C. Ponglimanont, H.-K. Fun, A. Kanjana-Opas and S. Laphookhieo, *Tetrahedron*, 2006, **62**, 8850–8859.
34. V. Kuete, H. K. Wabo, K. O. Eyong, M. T. Feussi, B. Wiench, B. Krusche, P. Tane, G. N. Folefoc and T. Efferth, *PLoS One*, 2011, **6**, e21762.
35. R. Khonkarn, S. Mankhetkorn, M. Talelli, W. E. Hennink and S. Okonogi, *Colloids Surf., B*, 2012, **94**, 266–273.
36. J. J. Wang, B. J. Sanderson and W. Zhang, *Food Chem. Toxicol.*, 2011, **49**, 2385–2391.
37. J. Pedraza-Chaverri, N. Cardenas-Rodriguez, M. Orozco-Ibarra and J. M. Perez-Rojas, *Food Chem. Toxicol.*, 2008, **46**, 3227–3239.
38. N. Tocci, G. Simonetti, F. D. D'Auria, S. Panella, A. T. Palamara, A. Valletta and G. Pasqua, *Appl. Microbiol. Biotechnol.*, 2011, **91**, 977–987.
39. C. Morel, J. Séraphin, J.-M. Oger, M. Litaudon, T. Sévenet, P. Richomme and J. Bruneton, *J. Nat. Prod.*, 2000, **63**, 1471.
40. C. Ito, M. Itoigawa, Y. Miyamoto, K. S. Rao, J. Takayasu, T. Mukainaka, H. Tokuda, H. Nishino and H. Furukawa, *J. Nat. Prod.*, **62**, 1668–1671.
41. C. Chutrakul, T. Boonruangprapa, R. Suvannakad, M. Isaka, P. Sirithunya, T. Toojinda and K. Kirtikara, *J. Appl. Microbiol.*, 2009, **107**, 1624–1631.
42. H. R. Dharmaratne, M. T. Napagoda and S. B. Tennakoon, *Nat. Prod. Res.*, 2009, **23**, 539–545.
43. J. J. Magadula, *Fitoterapia*, 2010, **81**, 420–423.
44. J. Li, M. Malakhova, M. Mottamal, K. Reddy, I. Kurinov, A. Carper, A. Langfald, N. Oi, M. O. Kim, F. Zhu, C. P. Sosa, K. Zhou, A. M. Bode and Z. Dong, *Cancer Res.*, 2012, **72**, 260–270.
45. I. K. Kostakis, R. Tenta, N. Pouli, P. Marakos, A.-L. Skaltsounis, H. Pratsinisc and D. Kletsas, *Bioorg. Med. Chem. Let.*, 2005, **15**, 5057–5060.
46. S. L. Crockett, B. Poller, N. Tabanca, E.-M. Pferschy-Wenzig, O. Kunert, D. E. Wedge and F. Bucar, *J. Sci. Food Agric.*, 2011, **91**, 428–434.
47. M. E. Sayah, V. Cechinel-Filho, T. R. Pinheiro, R. A. Yunes and J. B. Calixto, *Inflamm. Res.*, 1999, **48**, 218–223.
48. K. Sloczynska, E. Pekala, A. Wajda, G. Wegrzyn and H. Marona, *Lett. Appl. Microbiol.*, 2010, **50**, 252–257.
49. S. M. Kupchan, D. R. Steelman and A. J. Snedan, *J. Nat. Prod.*, 1980, **43**, 296–301.
50. A. M. Habib, D. K. Ho, S. Masuda, T. McCloud, K. S. Reddy, M. Aboushoer, A. McKenzie, S. R. Byrn, C.-J. Chang and J. M. Cassady, *J. Org. Chem.*, 1987, **52**, 412–418.

51. M. Isaka, A. Jaturapat, K. Rukseree, K. Danwisetkanjana, M. Tanticharoen and Y. Thebtaranonth, *J. Nat. Prod.*, 2001, **64**, 1015–1018.
52. M. Isaka, S. Palasarn, K. Kocharin and J. Saenboonrueng, *J. Nat. Prod.*, 2005, **68**, 945–946.
53. E. Elfita, M. Muharni, M. Latief, D. Darwati, A. Widiyantoro, S. Supriyatna, H. H. Bahti, D. Dachriyanus, P. Cos, L. Maes, K. Foubert, S. Apers and L. Pieters, *Phytochemistry*, 2009, **70**, 907–912.
54. P. de Campos, A. R. S. Santos, Z. R. Vaz, T. R. Pinheiro, M. G. Pizzolatti, V. Cechinel Filho, F. Delle Monache, R. A. Yunes and J. B. Calixto, *Life Sci.*, 1997, **61**, 1619–1630.
55. D. E. B. Isaias, R. Niero, V. F. Noldin, F. d. Campos-Buzzi, R. A. Yunes, F. D. Monache and V. Cechinel-Filho, *Pharmazie*, 2004, **59**, 897.
56. K. L. Silva, A. R. S. Santos, P. E. O. Mattos, R. A. Yunes, F. Delle Monache and V. Cechinel-Filho, *Therapie*, 2001, **56**, 431.
57. M. C. Yimdjo, A. G. Azebaze, A. E. Nkengfack, A. M. Meyer, B. Bodo and Z. T. Fomum, *Phytochemistry*, 2004, **56**, 2789.
58. A. J. N. Sellès, H. T. V. Castro, J. Agüero-Agüero, J. González-González, F. Naddeo, F. de Simone and L. Rastelli, *J. Agric. Food Chem.*, 2002, **50**, 762–766.
59. C. M. M. Santos, A. M. S. Silva, P. Filipe, R. Santus, L. K. Patterson, J.-C. Mazière, J. A. S. Cavaleiro and P. Morlière, *Org. Biomol. Chem.*, 2011, **9**, 3965–3974.
60. C. M. M. Santos, M. Freitas, D. Ribeiro, A. Gomes, A. M. S. Silva, J. A. S. Cavaleiro and E. Fernandes, *Bioorg. Med. Chem.*, 2010, 6776–6784.
61. A. Martínez, E. Hernández-Marin and A. Galano, *Food Funct.*, 2012, **3**, 442–450.
62. R. Gonda, T. Takeda and T. Akiyama, *Chem. Pharm. Bull.*, 2000, **48**, 1219–1222.
63. F.-F. Zhong, Y. Chen and G.-Z. Yang, *Helv. Chim. Acta*, 2008, **91**, 1695–1703.
64. Y. Chen, H. Fan, G.-Z. Yang, Y. Jiang, F.-F. Zhong and H.-W. He, *Helv. Chim. Acta*, 2011, **94**, 662–668.
65. Y.-J. Xu, S.-G. Cao, X.-H. Wu, Y.-H. Lai, B. H. K. Tan, J. T. Pereira, S. H. Goh, G. Venkatraman, L. J. Harrison and K.-Y. Sim, *Tetrahedron Lett.*, 1998, **39**, 9103–9106.
66. S. Feng, Y. Jiang, J. Li, S. Qiu and T. Chen, *Nat. Prod. Res.*, 2014, **28**(2), 81–85.
67. J. Merza, M.-C. Aumond, D. Rondeau, V. Dumontet, R. A.-M. Le, D. Seraphin and P. Richomme, *Phytochemistry*, 2004, **65**, 2915–2920.
68. S. Hartati, H.-B. Wang, L. B. S. Kardono, S. Kosela and G.-W. Qin, *Zhongguo Tianran Yaowu*, 2007, **5**, 272–276.
69. S. Deachathai, S. Phongpaichit and W. Mahabusarakam, *Nat. Prod. Res.*, 2008, **22**, 1327–1332.

70. X. D. Yang, N. An, L. Z. Xu and S. L. Yang, *J. Asian Nat. Prod. Res.*, 2002, **4**, 141–145.
71. I. Sordat-Diserens, M. Toyota, M. Hamburger, B. Sordat and K. Hostettmann, *Bull. Liaison – Groupe Polyphenols*, 1990, **15**, 175–178.
72. Z. H. Mbwambo, M. C. Kapingu, M. J. Moshi, F. Machumi, S. Apers, P. Cos, D. Ferreira, J. P. J. Marais, B. D. Vanden, L. Maes, A. Vlietinck and L. Pieters, *J. Nat. Prod.*, 2006, **69**, 369–372.
73. C. Ito, M. Itoigawa, Y. Mishina, V. Cechinel-Filho, T. Mukainaka, H. Tokuda, H. Nishino and H. Furukawa, *J. Nat. Prod.*, 2002, **65**, 65–68.
74. Y.-C. Shen, L.-T. Wang, A. T. Khalil, L. C. Chiang and P.-W. Cheng, *Chem. Pharm. Bull.*, 2005, **53**, 244–247.
75. J. G. Ondeyka, A. W. Dombrowski, J. P. Polishook, T. Felcetto, W. L. Shoop, Z. Guan and S. B. Singh, *J. Antibiot.*, 2006, **59**, 288–292.
76. J. Keiser, M. Vargas and R. Winter, *Parasitol. Int.*, 2012, **61**, 369–371.
77. S.-Y. Han and Y.-W. Chin, *J. Enzyme Inhib. Med. Chem.*, 2011, **26**, 445–448.
78. Y. Niu, W. Lu, L. Gao, H. Lin, X. Liu and L. Li, *Pharm. Biol.*, 2012, **50**, 1177–1182.
79. K. Nakatani, N. Nakahata, T. Arakawa, H. Yasuda and Y. Ohizumi, *Biochem. Pharmacol.*, 2002, **63**, 73–79.
80. L. Rocha, A. Marston, M. Auxiliawra, C. Kaplan, H. Stoeckli-Evans, U. Thull, B. Testa and K. Hostettemann, *Phytochemistry*, 1994, **36**, 1381–1385.
81. R. Reyes-Chilpa, C. H. Baggio, D. Alavez-Solano, E. Estrada-Muniz, F. C. Kauffman, R. I. Sanchez and S. Mesia-Vela, *J. Ethnopharmacol.*, 2006, **105**, 167–172.
82. H. M. Berman, J. Westbrook, Z. Feng, G. Gilliland, T. N. Bhat, H. Weissig, I. N. Shindyalov and P. E. Bourne, *Nucleic Acids Res.*, 2000, **28**, 235–242.
83. Y. Z. Meng, Z. A. Liang, Y. X. Lu and A. S. Hay, *Polymer*, 2005, **46**, 11117–11124.
84. R. Feng, X. Zhou, X.-S. Tan, P. M. Y. Or, J. F. Tao Hu, J.-Y. Ma, M. Huang, C.-Y. He, J.-G. Shi, C.-T. Che, J. H. K. Yeung and Y. Wang, *Chem. -Biol. Interact.*, 2014, **210**, 12–19.
85. Y. Wang, J.-G. Shi, M.-Z. Wang, C.-T. Che and J. H. K. Yeung, *Life Sci.*, 2007, **81**, 1016–1023.
86. Y. Wang, J.-G. Shi, M.-Z. Wang, C.-T. Che and J. H. K. Yeung, *Life Sci.*, 2008, **82**, 91–98.
87. C. Tian, T. Zhang, L. Wang, Q. Shan and L. Jiang, *J. Ethnopharmacol.*, 2014, **154**, 259–266.
88. W. Chanmahasathien, Y. Li, M. Satake, Y. Oshima, N. Ruangrungsi and Y. Ohizumi, *Phytochemistry*, 2003, **64**, 981–986.
89. S.-L. R. J.-P. Wang, C.-N. Lin and C.-M. Teng, *Eur. J. Pharmacol.*, 1994, **251**, 35–42.
90. Y. Wang, J. G. Shi, M. Z. Wang, C. T. Che and J. H. Yeung, *Life Sci.*, 2007, **81**, 1016–1023.

91. Y. Wang, J. G. Shi, M. Z. Wang, C. T. Che and J. H. Yeung, *Life Sci.*, 2008, **82**, 91–98.
92. S. P. Zhang, X. G. Du and X. P. Pu, *Biol. Pharm. Bull.*, 2010, **33**, 1529–1533.
93. A. Urbain, A. Marston, L. S. Grilo, J. Bravo, O. Purev, B. Purevsuren, D. Batsuren, M. Reist, P. A. Carrupt and K. Hostettmann, *J. Nat. Prod.*, 2008, **71**, 895–897.
94. X.-G. Du, W. Wang, Q.-Y. Zhang, J. Cheng, B. Avula, I. A. Khan and D.-A. Guo, *Rapid Commun. Mass Spectrom.*, 2012, **26**, 2913–2923, S2913/2911–S2913/2920.
95. Y. Nakamura, T. Shinomura and J. Ona, *Nippon Nogei Kagaku*, 1957, **31**.
96. J. W. Cornforth, G. Ryback, P. M. Robinson and D. Park, *J. Chem. Soc., C*, 1971, 2786.
97. J. Fuska, L. Ivanitskaya, L. Makukho and L. Volkova, *Antibiotiki*, 1974, **19**, 890–893.
98. N. Katagiri, J. Nakano and T. Kato, *J. Chem. Soc., Perkin Trans. 1*, 1981, 2710–2716.
99. E. Wijeratne, T. Turbyville, A. Fritz, L. Whitesell and A. Gunatilaka, *Bioorg. Med. Chem.*, 2006, **14**, 7917–7923.
100. B. Ding, J. Yuan, X. Huang, W. Wen, X. Zhu, Y. Liu, H. Li, Y. Lu, L. He, H. Tan and Z. She, *Mar. Drugs*, 2013, **11**, 4961–4972.
101. D. Rönsberg, A. Debbab, A. Mándi, V. Vasylyeva, P. Böhler, B. Stork, L. Engelke, A. Hamacher, R. Sawadogo, M. Diederich, V. Wray, W. Lin, M. U. Kassack, C. Janiak, S. Scheu, S. Wesselborg, T. Kurtán, A. H. Aly and P. Proksch, *J. Org. Chem.*, 2013, **78**, 12409–12425.
102. S. Feng and T. Chen, *Shenzhen Fairy Lake Botanical Garden Management Office, Peop. Rep. China*, 2012, CN102440985A.
103. A. Schüffler, J. C. Liermann, H. Kolshorn, T. Opatz and H. Anke, *Tetrahedron Lett.*, 2009, **50**, 4813–4815.
104. M.-L. Milat and J.-P. Blein, *J. Chromatogr. A*, 1995, **699**, 277–283.
105. C. Goudet, M.-L. Milat, H. Sentenac and J.-B. Thibaud, *Mol. Plant-Microbe Interact.*, 2000, **13**, 203–209.
106. P.-H. Ducrot, J. Einhorn, L. Kerhoas, J.-Y. Lallemand, M.-L. Milat, J.-P. Blein, A. Neuman and T. Prangé, *Tetrahedron Lett.*, 1996, **37**, 3121–3124.
107. P.-H. Ducrot, J.-Y. Lallemand, M.-L. Milat and J.-P. Blein, *Tetrahedron Lett.*, 1994, **47**, 8797–8800.
108. J. Y. Ueda, M. Takagi and K. Shin-ya, *J. Antibiot.*, 2010, **63**, 615–618.
109. K. Matsuzaki, N. Tabata, H. Tomoda, Y. Iwai, H. Tanaka and S. Omura, *Tetrahedron Lett.*, 1993, **34**, 8251–8254.
110. N. Tabata, H. Tomoda, Y. Iwai and S. Omura, *J. Antibiot.*, 1995, **49**, 267–271.
111. N. Tabata, H. Tomoda, K. Matsuzaki and S. Omura, *J. Am. Chem. Soc.*, 1993, **115**, 8558–8564.
112. G.-L. Sia, G. J. Bennett, L. J. Harrison and K.-Y. Sim, *Phytochemistry*, 1995, **38**, 1521–1528.

113. A.-R. Han, J.-A. Kim, D. D. Lantvit, L. B. S. Kardono, S. Riswan, H. Chai, d. B. E. J. Carcache, N. R. Farnsworth, S. M. Swanson and A. D. Kinghorn, *J. Nat. Prod.*, 2009, **72**, 2028–2031.
114. S. Cao, D. W. McMillin, G. Tamayo, J. Delmore, C. S. Mitsiades and J. Clardy, *J. Nat. Prod.*, 2012, **75**, 793–797.
115. D. W. McMillin, J. Delmore, J. Negri, M. Ooi, S. Klippel, C. V. Miduturu, N. S. Gray, P. G. Richardson, K. C. Anderson, A. L. Kung and C. S. Mitsiades, *PLoS One*, 2011, **6**, e20226.
116. H.-B. Chen, L.-Z. Zhou, L. Mei, X.-J. Shi, X.-S. Wang, Q.-L. Li and L. Huang, *J. Nat. Med.*, 2012, **66**, 62–69.
117. H. Auterhoff, H. Frauendorf, W. Liesenklas and C. Schwandt, *Arch. Pharm.*, 1962, **295/67**, 833–846.
118. T. J. R. Weakley, S. K. Cai, H. Z. Zhang and J. F. W. Keana, *J. Chem. Crystallogr.*, 2001, **31**, 501–505.
119. Y. Yang, L. Yang, Q. D. You, F. F. Nie, H. Y. Gu, L. Zhao, X. T. Wang and Q. L. Guo, *Cancer Lett.*, 2007, **256**(2), 259–266.
120. S. Kasibhatla, K. A. Jessen, S. Maliartchouk, J. Y. Wang, N. M. English, J. Drewe, L. Qiu, S. P. Archer, A. E. Ponce, N. Sirisoma, S. C. Jiang, H. Z. Zhang, K. R. Gehlsen, S. X. Cai, D. R. Green and B. Tseng, *Proc. Natl. Acad. Sci. U. S. A.*, 2005, **102**, 12095–12100.
121. M. K. Pandey, B. Sung, K. S. Ahn, A. B. Kunnumakkara, M. M. Chaturvedi and B. B. Aggarwal, *Blood*, 2007, **110**, 3517–3525.
122. U. D. Palempalli, U. Gandhi, P. Kalantari, H. Vunta, R. J. Arner, V. Narayan, A. Ravindran and K. S. Prabhu, *Biochem. J.*, 2009, **419**, 401–409.
123. G. Miao, J. Ma, K. Yang, Z. Huang, Q. Gu, Y. Wang, Q. Guo, Q. You and J. Wang, *Aust. J. Chem.*, 2014, **68**(6), 872–880.
124. J. W. I. Lippert, *Bioorg. Mol. Chem.*, 2007, **15**, 605–615.
125. B. C. Baguley, L.-M. Ching and M. Philpott, *PCT WO 02/09700 A1*, 2002.
126. G. W. Rewcastle, G. J. Atwell, L. Zhuang, B. C. Baguley and W. A. Denny, *J. Med. Chem.*, 1991, **34**, 217–222.
127. D. W. Siemann, E. Mercer, S. Lepler and A. M. Rojiani, *Int. J. Cancer*, 2002, **99**, 1–6.
128. J. B. Jameson, P. I. Thompson, B. C. Baguley, B. D. Evans, V. J. Harvey, D. J. Porter, M. R. McCrystal, M. Small, K. Bellenger, L. Gumbrell, G. W. Halbert and P. Kestell, *Br. J. Cancer*, 2003, **88**, 1844–1850.
129. H. I. Ingolfsson, P. Thakur, K. F. Herold, E. A. Hobart, N. B. Ramsey, X. Periole, D. H. de Jong, M. Zwama, D. Yilmaz, K. Hall, T. Maretzky, H. C. Hemmings, C. Blobel, S. J. Marrink, A. Kocer, J. T. Sack and O. S. Andersen, *ACS Chem. Biol.*, 2014, **9**, 1788–1798.
130. J. B. Baell and G. A. Holloway, *J. Med. Chem.*, 2010, **53**, 2719–2740.
131. J. Baell and M. A. Walters, *Nature*, 2014, **513**, 481–483.

CHAPTER 13

Natural Product Scaffolds of Value in Medicinal Chemistry[†]

DAVID J. NEWMAN* AND GORDON M. CRAGG

Natural Products Branch, Developmental Therapeutics Program, DCTD, NCI, Frederick National Laboratory, P. O. Box B, Frederick, MD 21702, USA
*Email: dn22a@nih.gov

13.1 Introduction

The number of natural product scaffolds (*i.e.* the base structure of a natural product which, when utilized or modified by direct substitution and/or by use of isosteric modifications) that may lead to, or are being considered as, "leads to drugs" in a large number of pharmacologic areas, is very high, base scaffolds being isolated from all domains of life.[1] However, it has now become obvious that in a number of cases, the actual producer of the compound(s) is not the organism from which they were isolated and identified.

Good examples of the current situation are shown in recent articles from three different research groups. The first, from the Crews laboratory at the University of California, Santa Cruz, where marine-derived structures have been compared and contrasted with very similar structures obtained from non-marine sources.[2] The second, from the Schmidt group at Utah, demonstrates that cyclic peptides thought to be from non-ribosomal processing following initial ribosomal synthesis are actually not, and come from as yet uncultivated microbes.[3–5] The third, from the Piel group at the ETH in

[†]Note: The opinions expressed in this article are those of the authors, not necessarily those of the US Government.

RSC Drug Discovery Series No. 50
Privileged Scaffolds in Medicinal Chemistry: Design, Synthesis, Evaluation
Edited by Stefan Bräse

Published by the Royal Society of Chemistry, www.rsc.org

Zurich, shows the enormous versatility of another as yet uncultured microbe found in the sponge *Theonella swinhoei.*[6]

All of these scaffolds are useful as leads for synthesis as, in most cases, they have significant cytotoxicity against human tumor cell lines; but this activity may only be a function of the funding agencies (*i.e.* the National Cancer Institute in the USA for the first two), rather than a full representation of their potential.

As one might expect, synthetic and medicinal chemists have generated very large numbers of nitrogen-based heterocyclic compounds, even though analyses of natural product sources show that oxygen-related heterocycles predominate[7] and these initial findings are still valid 12 years later.[8] In addition to the well-known scaffolds, there are a very considerable number of "under-represented scaffolds"; though obviously not all are from natural sources, it is quite possible that in years to come, many more examples will be found first in synthetic chemistry laboratories and then from nature, just as histamine itself was first synthesized in the late 1800s but not identified in animals until the early 1900s.

In order to give an idea of what is still "out there", in 2009, chemists at UCB-Celltech in the UK reported that they could identify approximately 25 000 small aromatic ring systems by using only mono and bicyclic rings with 5 or 6 atoms in the ring(s) with C, H, N, O, and S, with all obeying Hückel's aromaticity rule. As of the publication date of that article, less than 1800 had been reported in the literature (research and patent).[9] A literature search in June 2014 showed 87 citations to this article, with one that was a "sub-citation", demonstrating the use of a synthetic biology platform based upon baker's yeast to yield previously unknown scaffolds of approximately the same size range that are suitable for further optimization.[10] Thus, there are very significant numbers of "not yet represented" scaffolds that are open for synthesis and utilization.

In the sections to follow, we will take selected scaffolds from nature and demonstrate how they have led, or are leading to, agents with medicinal potential. We will not constrain our discussion to only antitumor agents, but equally we will not attempt to give examples from all possible disease areas, as rather than a single chapter we would need multiple volumes to do justice to the topic. However, most of the examples that we will use are in fact based more on nitrogen hetero atoms rather than oxygen, in spite of the comments above.

13.2 Privileged Structures

All secondary metabolites, by which we mean those compounds produced by an organism, frequently in response to a stimulus of some type, which is often another chemical agent (see later) that are not required for the basic life of the organism, irrespective of nominal or actual source are "privileged structures". This term was first defined by Evans *et al.*[11,12] when discussing the biological activities of synthetic benzodiazepines based upon the known anxiolytic structures as potential cholecystokinin antagonists. To show the continued influence of this term/concept, a search conducted in April 2014 using Evans' original paper as the root of the search, listed over 740 citations, and of the recent papers that cited Evans, four are of interest. Three use the "privileged structure" concept as defined by Evans,[13–15] but with some variation. The fourth, from Ganesan's group at the University of East Anglia in the UK, has an interesting "view" on the PS concept that can be best described by quoting a part of their introduction; "A scaffold that leads to biologically active compounds will attract interest by medicinal chemists who will then produce more examples of the same and discover new active compounds that further confirm the hypothesis. There should then exist examples of '*underprivileged scaffolds*' that are intrinsically suitable for drug discovery applications but in practice are underrepresented or absent."[16] These aspects will be covered later in this chapter.

13.2.1 Modified Nucleosides, Privileged Structures giving Antitumor and Antiviral Agents that "Contradicted Dogma"

Until the work of the Bergmann group at Yale in the early to mid 1950s, it was dogma that for a nucleoside derivative to demonstrate biological activity in an assay, the sugar had to be either deoxyribose or ribose, though the base could vary significantly. With the identification of spongouridine (Figure 13.1, **1**) and spongothymidine (Figure 13.1, **2**) came the realignment of dogma to include arabinose as a sugar that did not remove biological activity. These two compounds can be considered to be the prototypes of all the nucleoside analogs that have crossed the antiviral and antitumor stages since that time, as chemists then began to use modifications of the bases used and once these demonstrated activity, modification of the sugar moieties were not far behind, even using acyclic variations rather than regular pentoses.

These early experiments led to a vast number of derivatives that were tested extensively as antiviral and antitumor agents over the next six decades. In a review in 1991, Suckling[17] demonstrated how such structures evolved in the (then) Wellcome laboratories, leading ultimately to molecules such as azidothymidine (Figure 13.1, **3**) or AZT, though no direct mention was made of the original "privileged structures from natural sources". Demonstrating an interesting temporal reversal where chemists synthesized a compound that was later found in nature, in 1960, Lee *et al.* reported the synthesis of arabinosyladenine (Ara-A or Vidarabine (Figure 13.1, **4**) as a potential antitumor agent.[18] A later report showed production by fermentation of *S. antibioticus*[19] and then it was isolated from the Mediterranean gorgonian *Eunicella cavolini* by the Cimino group in 1984.[20] To this list we can also add Ara-C (Figure 13.1, **5**), which was synthesized by Evans *et al.*[21] following the early discoveries above and covered in work by Cohen's

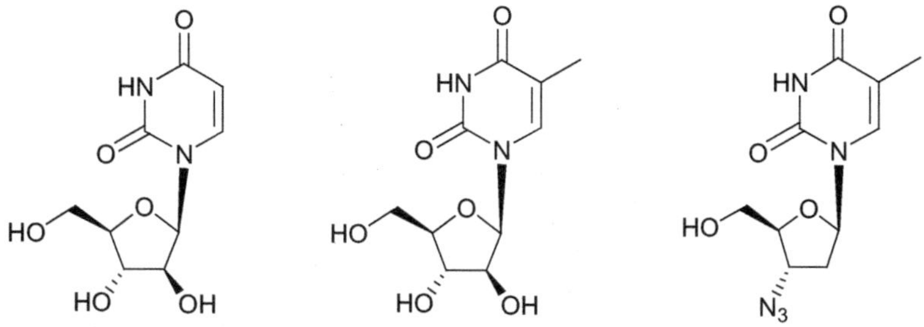

1 Spongouridine

2 Spongothymidine

3 AZT
HIV, 1987

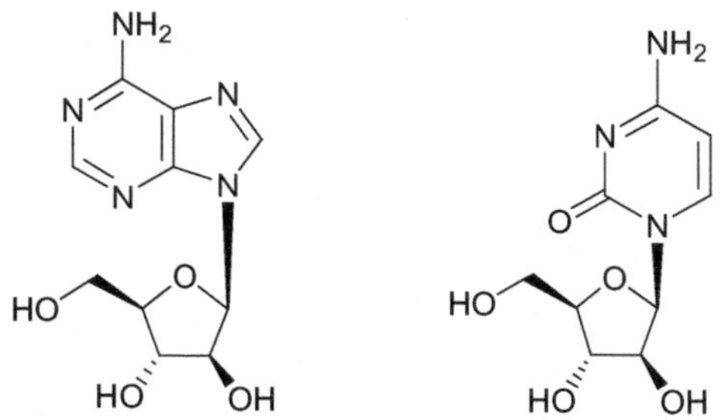

4 Ara A
Herpes, 1984

5 Ara C
Lymphoma, 1969

Figure 13.1 Arabinose-linked nucleosides.

group[22] and the report of Chu and Fischer[23] on the potential mechanism as an antileukemic agent.

As a result of these initial forays into different sugars and also modified bases, the following approved drugs are simply a soupcon of the many compounds derived by medicinal chemists since the identification of the base structures by Bergmann. In Figure 13.2, we have given some of the molecules synthesized as a result of these initial discoveries that are now drugs for diseases for which there were few or no agents prior to their introduction. We have simply given one of the diseases for which they are approved and the year of approval. Acyclovir (Figure 13.2, **6**), famiciclovir (Figure 13.2, **7**), valacyclovir (Figure 13.2, **8**), tenofovir (Figure 13.2, **9**), emtricitabine (Figure 13.2, **10**), entecavir (Figure 13.2, **11**), clofarabine (Figure 13.2, **12**) and nelarabine (Figure 13.2, **13**). However, the underlying thrust is that without the recognition that bioactivity was independent of the "sugar" moiety, none of these would have been designed.

13.3 Alkaloids

The number of naturally occurring molecules that fall under this generic heading is vast. We will begin with a story that is now over fifty

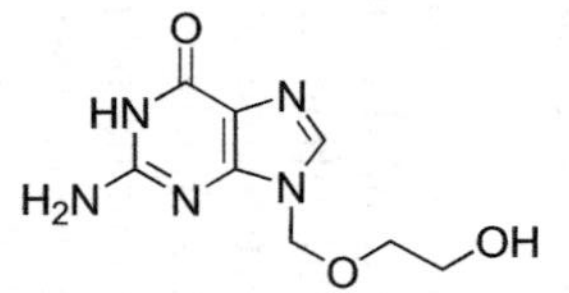

6 Acyclovir (Herpes,1981)

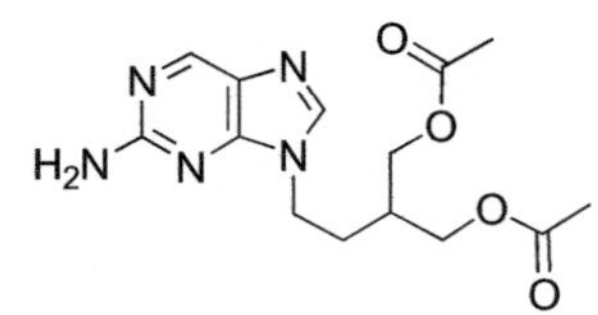

7 Famiciclovir (Herpes, 1994)

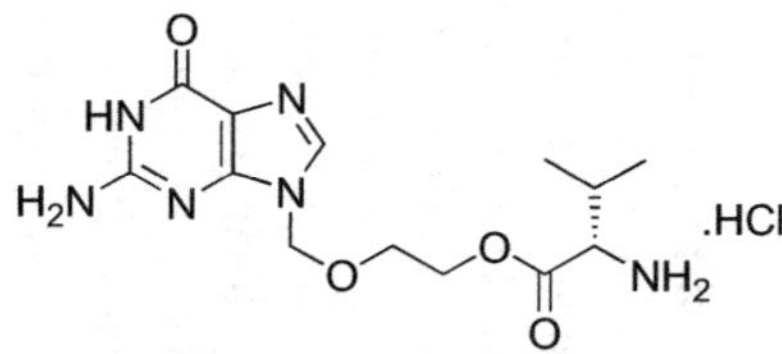

8 Valacyclovir HCl (Herpes, 1995)

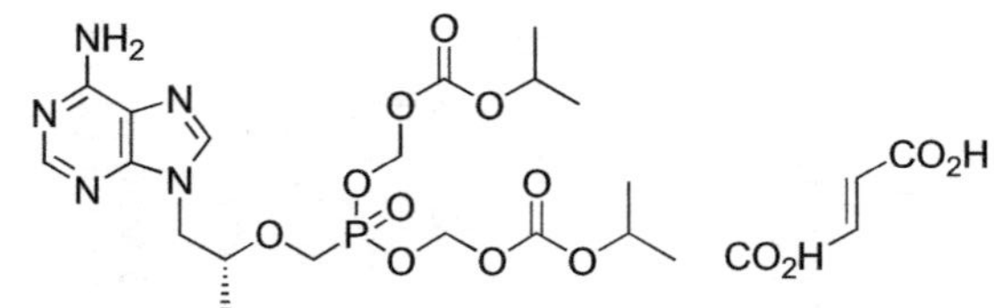

9 Tenofovir disoproxil fumarate (Hep B, 2001)

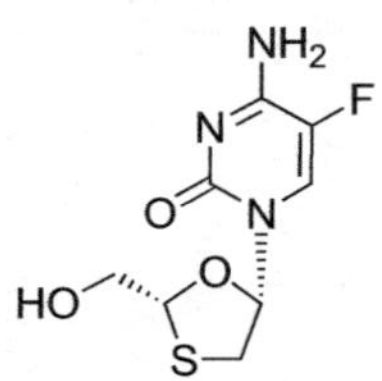

10 Emtricitabine (HIV, Hep-B, 2003)

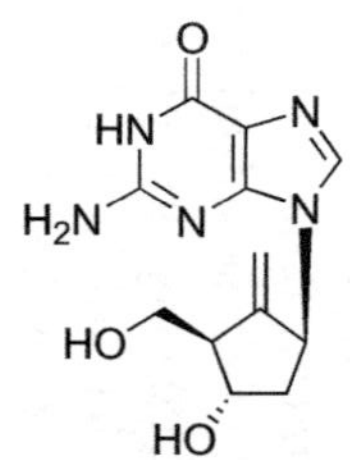

11 Entecavir (Hep B, 2005)

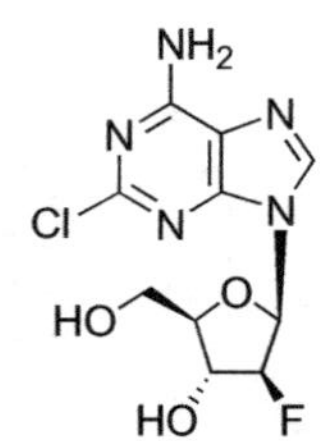

12 Clofarabine (ALL, 2005)

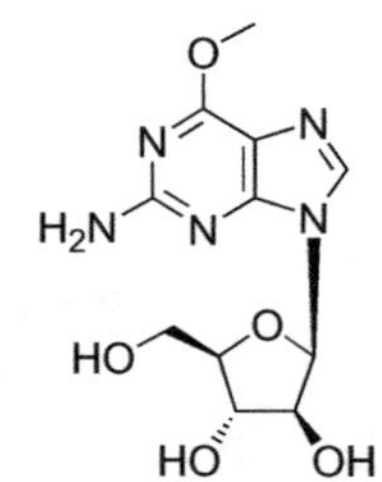

13 Nelarabine (Lymphoma, 2006)

Figure 13.2 Approved drugs based upon modification of the nucleoside sugar.

years old but still viable, then give a more recent example with unusual fused ring systems first reported from marine sources, and finally we will give a recent example of how medicinal chemists are still using this well-known but extremely diverse chemical class to produce novel molecules.

13.3.1 Vinca Alkaloids

The first example in this section deals with the discovery and subsequent utilization of the molecules known colloquially as the "Vinca Alkaloids". The story of these discoveries, which led to a drug series that converted childhood leukemia from a death sentence prior to their use, to a potentially treatable disease nowadays, with better than 90% survival, started in 1949 in Canada at the University of Western Ontario. There, researchers were intrigued by reports from Jamaica that extracts of the rosy periwinkle (then *Vinca rosea,* now *Catharanthus roseus*) were used as a tea to control diabetes. Following controlled experiments using diabetic/normal rats/rabbits and oral administration, there was no effect on blood sugar levels, nor was there any effect on the response to glucagon seen in the experiments. In contrast, following intravenous dosing, the rats succumbed within a week and on necropsy, evidence of septicemia was present but the injected fluid was sterile. On further investigation, it became obvious that the white blood cells were being significantly depressed.

In 1955, the laboratory commenced a more thorough investigation using a bioactivity-linked isolation process and isolated "vinleukoblastine" (VLB and now known as vinblastine, Figure 13.3, **14**) as its sulfate salt. The enriched crude fractions demonstrated activity (carcinostatic) against a transplantable sarcoma in rats and against a mammary carcinoma in DBA/JAX mice.[24]

Concomitantly, an independent study by Svoboda *et al.,* starting from ethnobotanical reports on the use of extracts of the plant in the then Dutch East Indies during World War II as a treatment for diabetes, led to studies by this group at Lilly that demonstrated the cytotoxic activity of the extract against lipocytes and then reported the alkaloids leurosine and vinleukoblastine (Figure 13.3, **14**) approximately a year after Noble *et al.*[25,26] These papers were closely followed by one in 1960 demonstrating the *in vivo* activity of both leurosine and VLB in a mouse model of acute lymphocytic leukemia in DBA/2 mice.[27]

Since these original publications, the number of approved (meaning launched following approval by the FDA or its equivalent in other countries) derivatives of vinblastine (launched in 1963 or 1965; sources differ) and vincristine (Figure 13.3, **15**, launched in 1963) have risen to three further discrete compounds with approval of vindesine (Figure 13.3, **16**) in 1979, vinorelbine (Figure 13.3, **17**) in 1989, vinflunine (Figure 13.3, **18**) in 2010 and in 2013 a liposomal formulation of vincristine, 50 years after the base compound was approved in the USA.

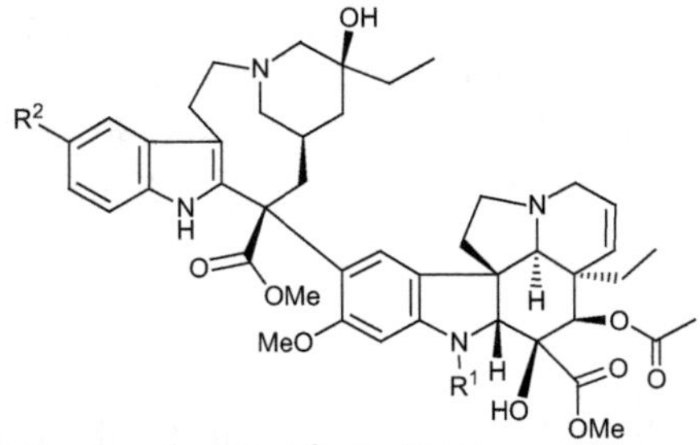

14 Vinblastine $R^1 = CH_3$; $R^2 = H$` (1963)
15 Vincristine R = CHO; $R^2 = H$ (1963)
19 12'-Methylthiovinblastine $R^1 = CH_3$; $R^2 = S\text{-}CH_3$ (Phase I, 2014)

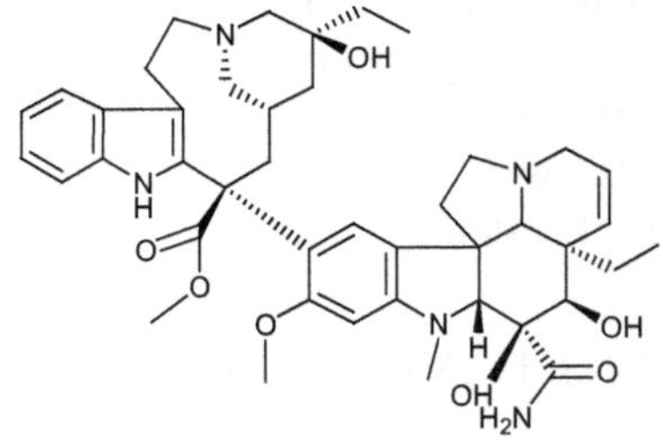

16 Vindesine (1979)

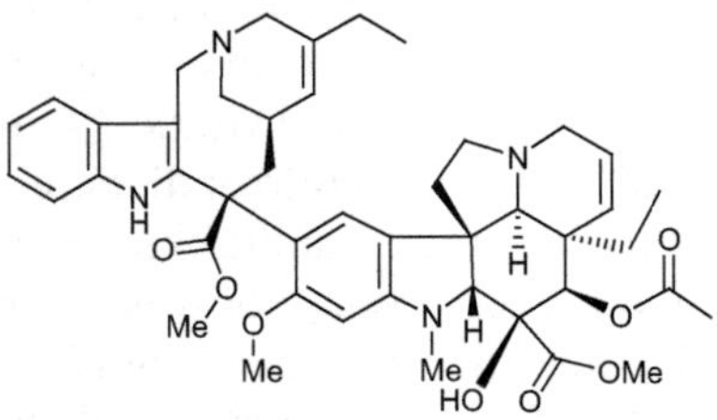

17 Vinorelbine (1989)

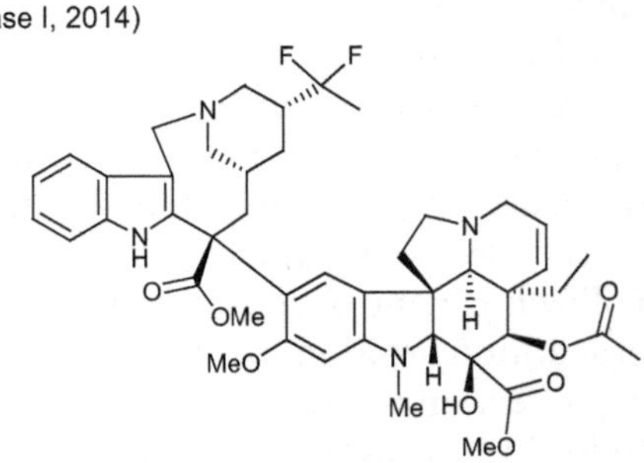

18 Vinflunine (2010)

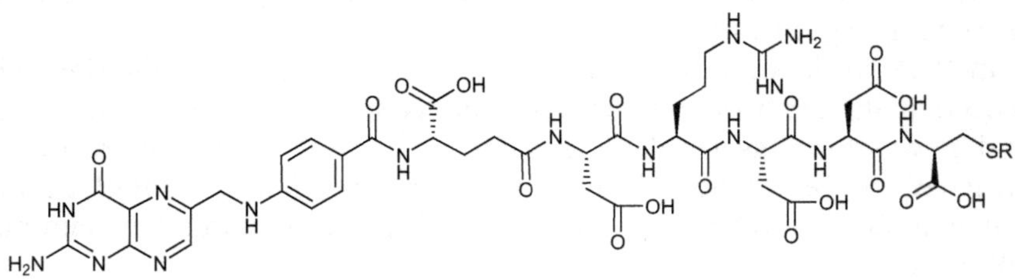

20 Vintafolide, R =
EU 2014?

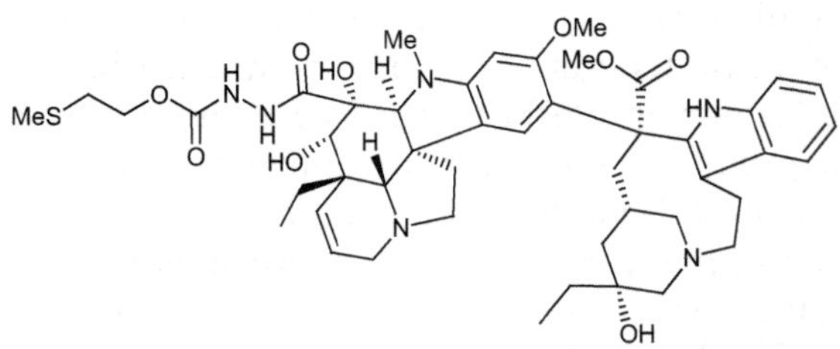

Figure 13.3 Vinca alkaloids and drugs based thereon.

Currently, there are three variations on the "vincas" in varying stages of clinical development, with the liposomal variation of vinorelbine being in Phase I with Tekmira in California; 12′-methylthiovinblastine (Figure 13.3, **19**) also in Phase I against solid tumors (Albany Molecular Research/Bessor Pharma in MA) and the most advanced being vintafolide, which is a conjugate of desacetylvinblastine hydrazide-folate (Figure 13.3, **20**) from Endocyte, now licensed to Merck. This molecule was recommended for approval by the European Union in March 2014, but withdrawn by Merck in the US two months later. Currently its status is uncertain.

Except for this last construct, where the folic acid derivative that is linked to vinblastine, all other modifications that have been published (and there have been many over the last 50 years) have been modifications of the two parts of the heterodimer, catharanthine (the left hand half as conventionally drawn) and vindoline (the right-hand half). Two recent review articles[28,29] have covered, albeit briefly in some aspects, the modifications that have been made around the monomers and dimers from a medicinal chemistry aspect, but though excellent chemistry was performed, with hundreds of modifications in the literature over the last 50 years, from a pharmaceutical aspect, it appears that Mother Nature might have produced close to the optimal structure initially. However, this vast amount of chemical knowledge is in no way wasted, as novel methods of substitution and ring opening techniques have been published.

For a thorough series of opinions on the subject, one can consult the 2012 review by Keglevich *et al.*[28] covering modifications to the basic skeletons of the vinca alkaloids, the excellent review by Roussi *et al.* the same year that covers a lot more of the mechanism of action of the various modifications that were made utilizing medicinal chemistry techniques.[29] Very recently, the Keglevich group demonstrated that cyclopropanation of the 14,15 double bond in both vincristine and vinblastine retained antitumor cell line activity, though the corresponding reduced forms of the parent molecules did not. This interesting finding tends to confirm the unusual properties of a cyclopropane ring over simple reduction to give a fully saturated molecule.[30]

13.3.2 Lamellarins

The first report of the lamellarin structures came from work reported by the Faulkner group at Scripps Institute of Oceanography in 1985 as lamellarins A–D (Figure 13.4, **21–24**) isolated from a Palauan prosobrach mollusk, *Lamellaria* sp.[31] Since then, well over 70 related compounds have been described with varying fused ring forms, but all with a central pyrrole that at times, particularly in the storniamides (Figure 13.4, **25**) and didemnimides (Figure 13.4, **26**), is not part of a fused system. The multiplicity of sources that these compounds have been isolated from definitely implies a microbial source, which is probably an as yet uncultivated microbe. That such sources are real was demonstrated by the recent publication from the Piel group.[6]

A number of reviews on the lamellarins from both a chemical and a biological aspect have been published in the last 10 years and, in addition, there was interest from PharmaMar in the base molecules as antitumor candidates. The early reviews on discovery and synthesis gave both an historical perspective on the discovery and the concomitant biological activities.[32,33] These ranged from antibiotic activity with the storniamides, to HIV integrase

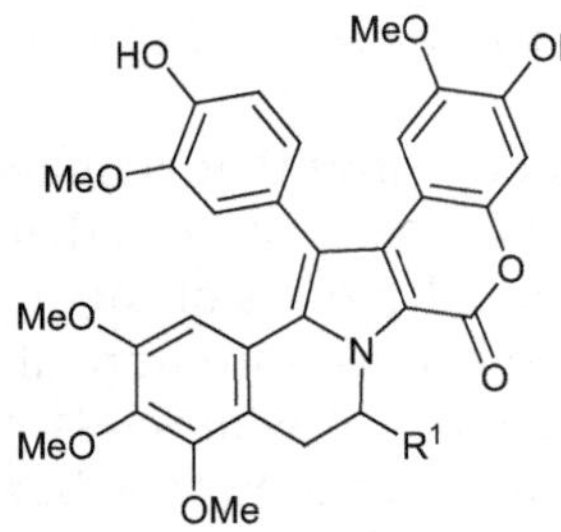

21 Lamellarin A; R^1 = OH
23 Lamellarin C; R^1 = H

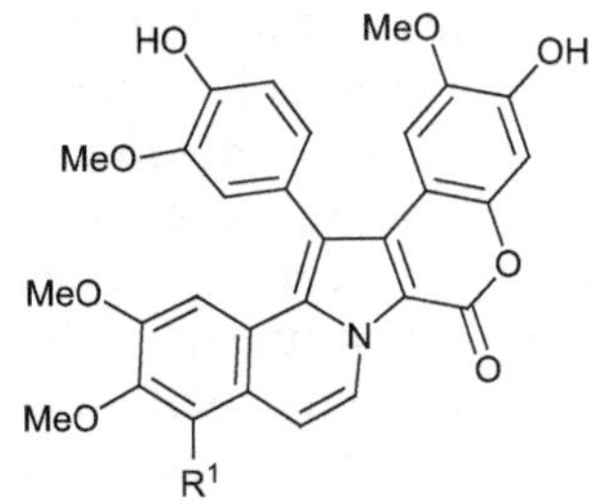

22 Lamellarin B; R^1 = OCH_3
24 Lamellarin D; R^1 = H

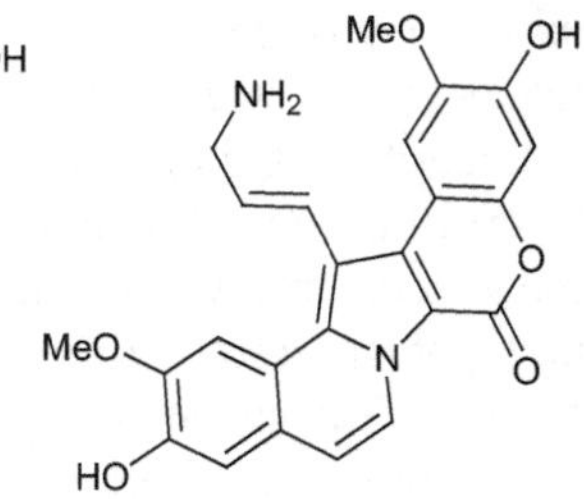

27 PM-031379

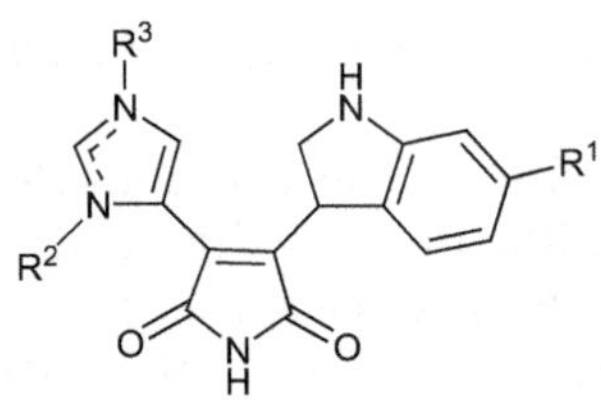

25 Base structure Didemnimide A-E

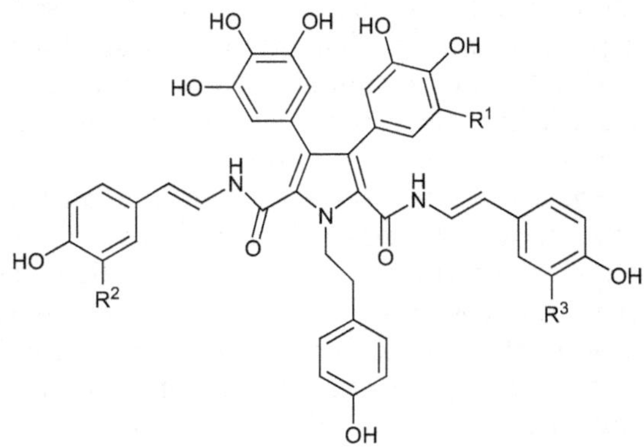

26 Base structure Storniamide A-D

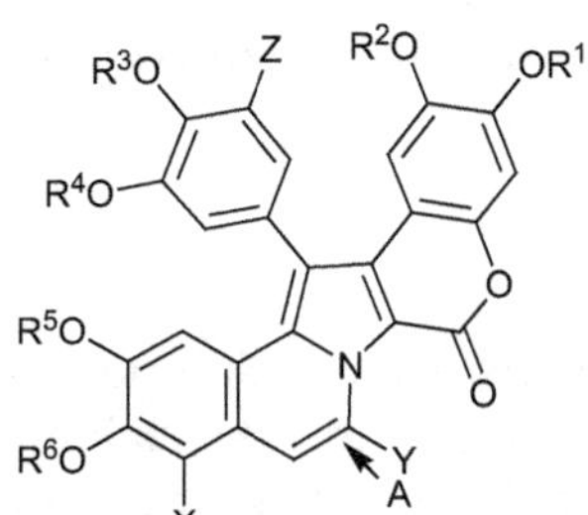

27 SAR Lamellarins

Required for Cytotoxicity

R^1 = OH
R^4 = OH or OCH_3
R^6 = OH
X = OH
Olefin at "A"

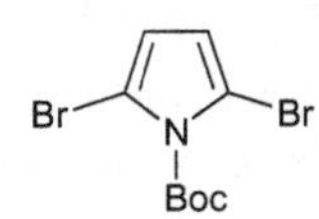

28 Iwao's dibromo pyrrole

Figure 13.4 Lamellarins.

activity with a variation, lamellarin α-20 as the sulfate salt demonstrating both *in vitro* and *in vivo* activities, to significant antitumor effects with lamellarin D (Figure 13.4, **24**) as a topoisomerase I inhibitor.[33] In his 2004 review, Bailly[32] pointed out the multidrug reversal (MDR) activity of a permethylated derivative of storniamide A synthesized by Boger *et al.*[34] without concomitant cytotoxicity. This work was based upon a report three years earlier from PharmaMar scientists demonstrating that lamellarin I was an order of magnitude more active than verapamil as an MDR-reversal agent at levels well below its cytotoxic concentrations, whereas other lamellarins were equipotent against wild-type and doxorubicin, daunomycin and vinblastine resistant cell line pairs.[35] Thus, these scaffolds could well have been developed as agents against MDR tumor types, but we can find no published evidence of this. More recently, in 2009, Bailly's group reported evidence that in addition to its effect on topoisomerase I, lamellarin D, and other molecules in the series such as the synthetic derivative PM-031379 (Figure 13.4, 27) synthesized by PharmaMar,[36] were involved in mitochondrial directed apoptosis *via* the intrinsic pathway.[37]

In the last few years, a significant number of investigators have published their efforts on synthesis and modification of the base scaffolds mainly as a result of the interesting biological activities of this class of alkaloids. The individual papers and review articles were published by investigators based in Europe, Thailand, Japan and the People's Republic of China. The list below is not meant in any way to be comprehensive, but to give an idea of the work that is still on-going with this privileged alkaloid class.

In 2008, Pla *et al.*[38] published a review, giving brief but quite reasonable coverage of the many methods used to synthesize a whole gamut of these molecules, and commented on their structure activity relationships on molecules based upon lamellarin D. They included commentary on the docking studies by the Bailly group, where lamellarin D was modelled into the camptothecin binding site of topoisomerase 1/DNA. A more thorough discussion of the syntheses around the basic structure was provided by Fan *et al.*, in their article in *Chemical Reviews* the same year,[33] and these should be read together for a more comprehensive overview.

In two papers, one in 2009[39] and the next a year later,[40] a group at the Chulabhorn Research Institute in Bangkok, Thailand, reported on the cytotoxicities and SAR of both natural and synthetic lamellarins, producing the SAR-related structure (Figure 13.4, 27). They followed up with analysis of the "drug likeness" of an enlarged set of both natural and synthetic lamellarins, demonstrating that using the logP parameter, these compounds do not show the expected relationships assumed in most modelling packages for the effect of substitutions on the value of a calculated logP. Thus, such analyses required experimental determinations of this parameter, and by such an experimental evaluation, they were then able to demonstrate that these molecules are just within the Lipinski values for "drug likeness" and intend to use their parameters in later synthetic schemes. This finding aptly demonstrates that reliance on the "calculated values" of such SAR parameters may well lead investigators astray.

In 2011, Li *et al.*,[41] in a paper in *Organic Letters*, demonstrated a very concise synthesis of lamellarins D and H in seven steps, together with lamellarin R and ningalin B in five steps, using three oxidative reactions as key steps that resembled the biosynthesis of these compounds. The basic method for lamellarins D and H are shown in Scheme 13.1. By subtle alteration of the subsequent steps from intermediate A, ningalin B could be obtained in 87% yield and alteration of the starting amine to

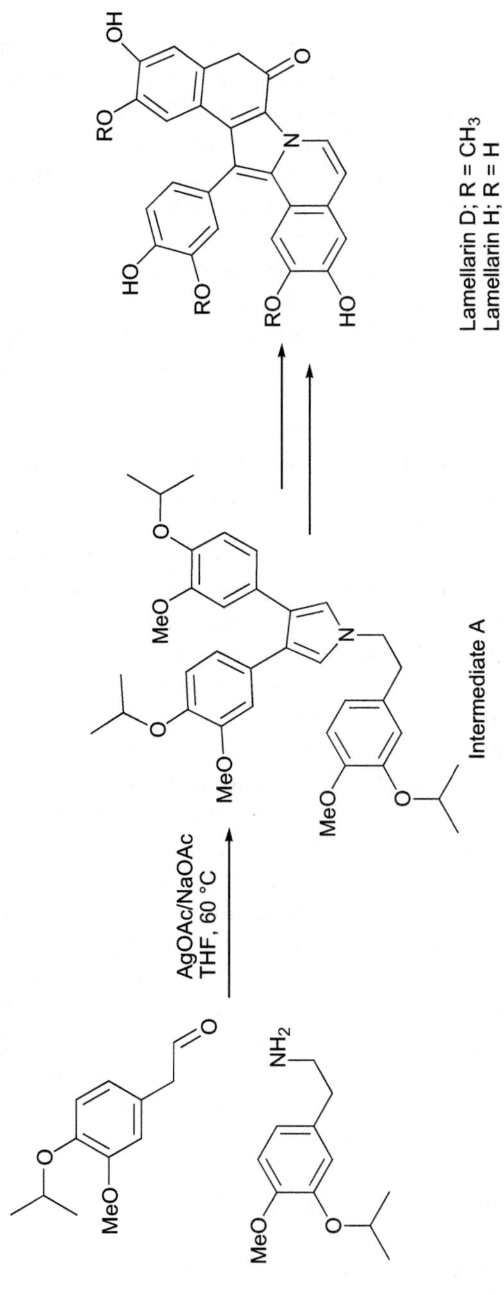

Scheme 13.1 Synthetic route to lamellarins.

p-methoxyaniline enabled the synthesis of the trisubstituted pyrrole derivative, lamellarin R (reaction not shown) with an overall yield from the starting aldehyde and amine of 44%. These reaction sequences can easily be modified to give other, non-natural lamellarin-type molecules in good yields.

The same year, Iwao's group published an excellent review covering the synthesis and biological activity of both the naturally occurring and synthetic lamellarins.[42] In this work, they covered both the general synthetic methods that led to many of this class of compounds and also discussed the various routes to the molecules from simple pyrroles or from more complex methodologies. As an introduction to what has been accomplished, together with an idea of the multiplicity of biological activities that these privileged structures have and may lead to, this particular paper is well worth the investment in time to read thoroughly. Then very recently, Iwao's group published another synthetic variation, starting with the very simple trisubstituted pyrrole, 2,5-dibromo-1-(*tert*-butoxycarbonyl)-1*H*-pyrrole (Figure 13.4, **28**) that permitted them to produce a variety of closely related lamellarins for further evaluation.[43]

13.3.3 Alkaloids as Chemical Probes

In a novel extension of the "privileged structure concept", Aube's group recently published a very interesting report on the use of natural product alkaloids that were reported to have multiple biological activities, rather than just depending upon one quoted biological target.[13] From these initial structures, a primary scaffold could be identified, which could then give rise to secondary scaffolds by variation around the rings with respect to size, connectivity or even presence. An example from their work with neostenine alkaloids is shown in Scheme 13.2, and though it appears formally equivalent to the type of analyses that the Waldmann group published on the derivation of their BIOS system,[44] and updated in a recent article in *Chemical Reviews* demonstrating applications to protein–protein interactions,[45] it differs in that the new structures can be considered to be as complex as the originals, rather than the more simplified compounds usually derived from the BIOS analyses.

A principal component analysis of the libraries produced by the Aube group using this technique, comparing the calculated physiochemical properties, though not bioactivity data at this time, of the 686 previously unknown structures from 55 separate scaffolds and 631 analogs, demonstrated that these compounds compared well in "expressing" the attributes often considered to be "more natural product like", such as high sp^3 counts.[46] This may well be due to the original choice of the scaffolds, but in addition, the libraries were found to be both Lipinski (with 100% meeting at least 3 of the 4 criteria) and Veber (100%) compliant.[47] It will be very interesting to see what activities will be found for this and similar libraries in the future.

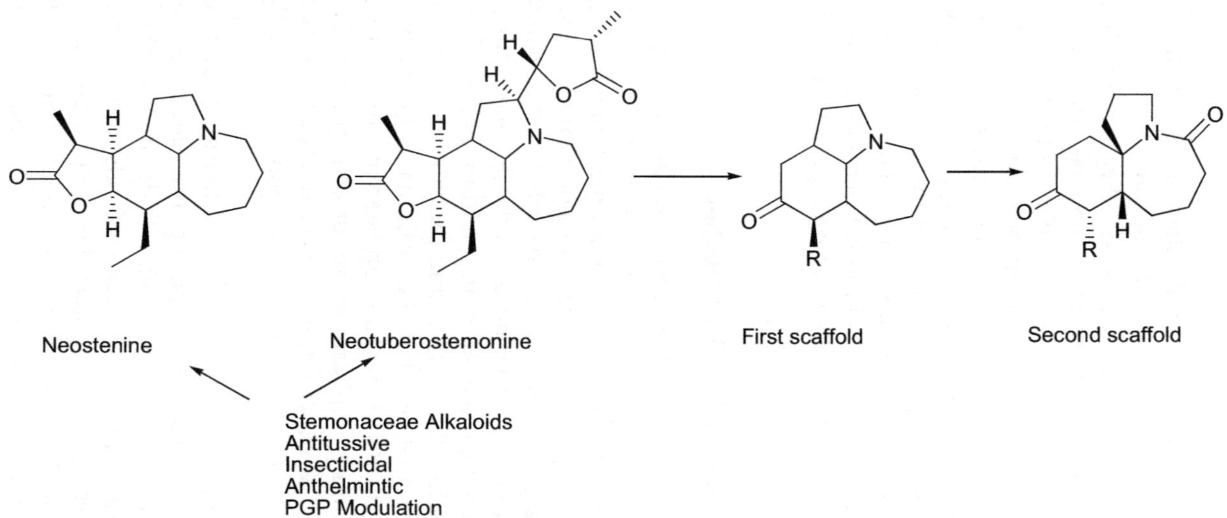

Scheme 13.2 Pharmacophore leads from stenine alkaloids.

13.4 Underprivileged Scaffolds; Diketopiperazines and Derivatives

Though these can also be considered as "alkaloidal in structural terms" we have elected to treat them separately. In 2010, Ganesan's group at the University of East Anglia published an article[16] that began to explore the potential of subtle modifications to the well-known class of heterocyclic natural products, the diketopiperazines (DKPs). These are formed by condensation of two similar or dissimilar amino acids, providing a class of natural products frequently found in marine habitats. As shown in the article by Zhao,[48] which appears in Chinese, the structures show a fair number of substituted DKPs, and microbial habitats.[49] In addition, there are many reports of sulfur-bridged DKPs, a well-known class of fungal metabolites isolated from terrestrial and marine sources, with a recent example being the novel thiodiketopiperazines known as phomazines A–C (Figure 13.5, **29–31**) from a marine *Phoma* species.[50]

The bioactivities of what may be considered "regular DKPs" are well known and cover a variety of drug targets.[51,52] In the last few years, the 2012 review by Borthwick has garnered a significant number of citations that have updated the earlier targets and cover a much larger area of pharmacology,[53] with activities shown in areas ranging from phosphodiesterase V inhibitors through hormone antagonists, tubulin depolymerizing agents to antiviral and antibiotic drug candidates.

The Ganesan group decided to investigate a simple modification of the DKP structure where another nitrogen atom would be placed into the basic diazadione system (Figure 13.5, **32**), generating a triazadione analog (Figure 13.5, **33**). Following some excellent chemistry based on solid-phase combinatorial methodologies, they reported the syntheses of 32 examples, with starting materials that included all variations on regular amino acids, variations on aldehydes and, in particular, a propargyl derivative that may well be amenable to a "click chemistry" link with potential targets. To date, no biological activities related to these compounds have yet been published, but with the previous record of DKPs, it is only a matter of time before biologically active compounds from this series will be identified.

In 2012, Gonzalez *et al.* published an excellent review demonstrating how these "privileged structures" (unlike Ganesan they did consider them to be "main line" structures) could act as building blocks to generate complex natural products, with just one example being the synthesis of ET743, Trabectedin from a simple tricyclic DKP (Scheme 13.3).[54] In addition to this synthesis they also made mention of the total synthesis of chaetocin A, another in the thio-bridged fungal metabolites with very interesting biological activities.

Finally, in 2013, the Zhu group in China published a patent review covering "attractive" DKP-based compounds published before 2012, with approximately 150 patents covered.[55] Of significant interest in this article were the 1000-plus structures shown and the pharmacological areas of the compounds when organized by the number of compounds claimed. These

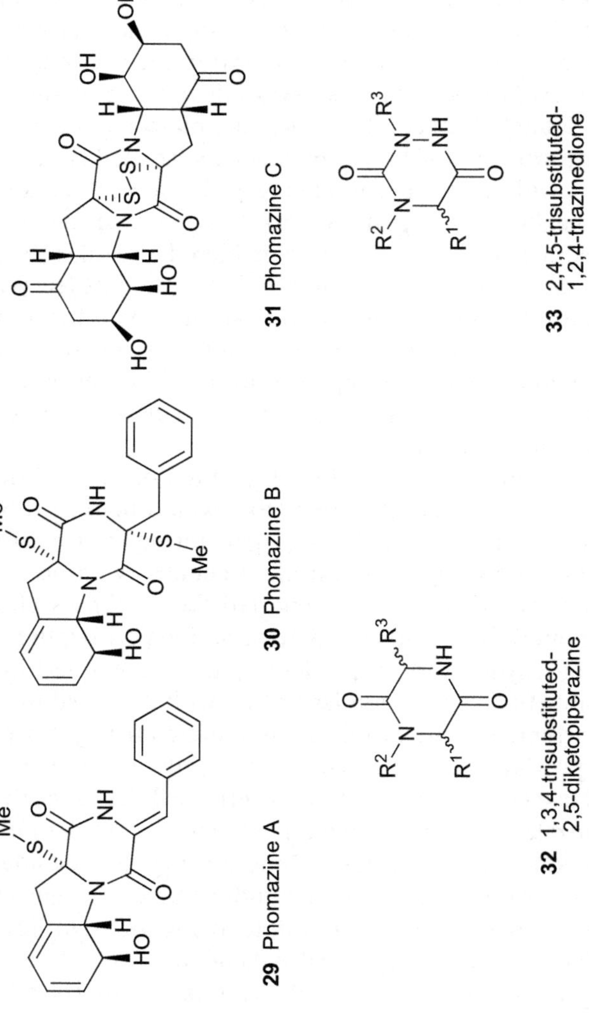

Figure 13.5 Diketopiperazine-based structures.

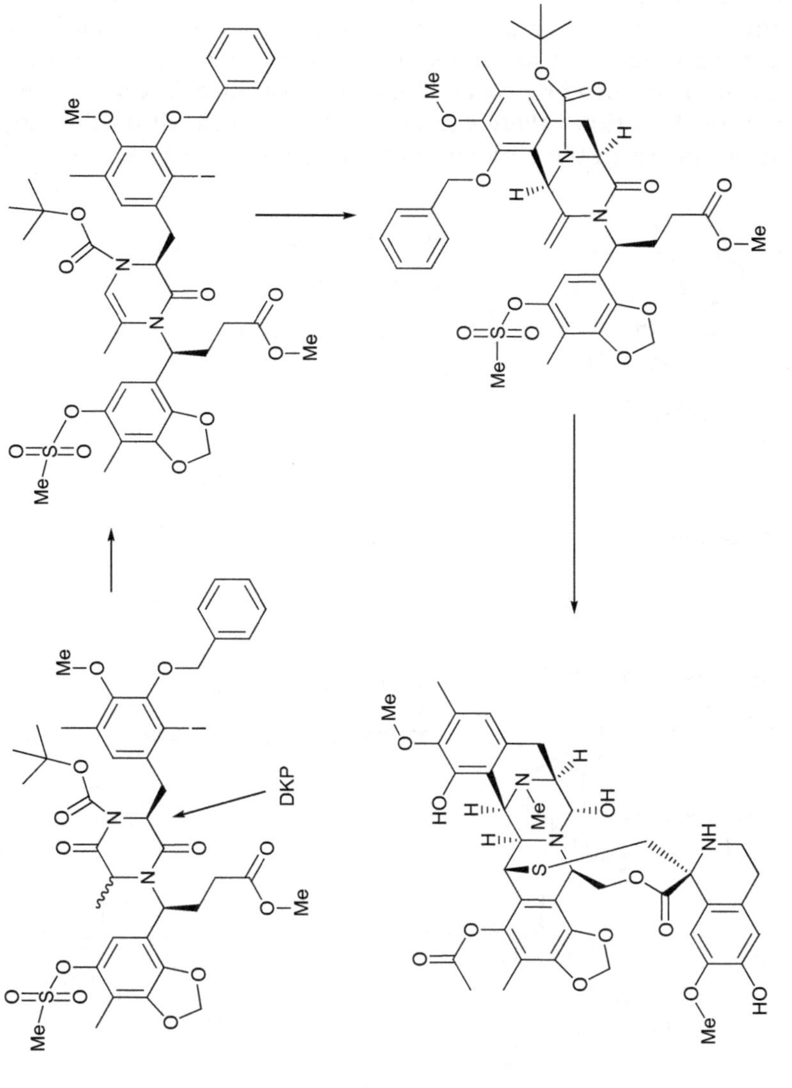

ET743; Trabectedin

Scheme 13.3 DKP to ET743 synthesis.

were 46% as oxytocin inhibitors, 19% as plasminogen activator inhibitors, 10% as antitumor compounds and 10% as PDE5 inhibitors, which included tadalafil, better known as Cialis®.

13.5 Ansamycins

The basic structures of these classes of biologically active compounds (containing materials that act as antimicrobial, antiparasitic and antitumor agents) can be thought of as comprising two interlocking ring systems, usually an aromatic or close to aromatic ring within a larger macrolide structure, though at times there can be two fused rings within the larger macrolide, as shown in the structure of rifamycin (Figure 13.6, **34**), as the

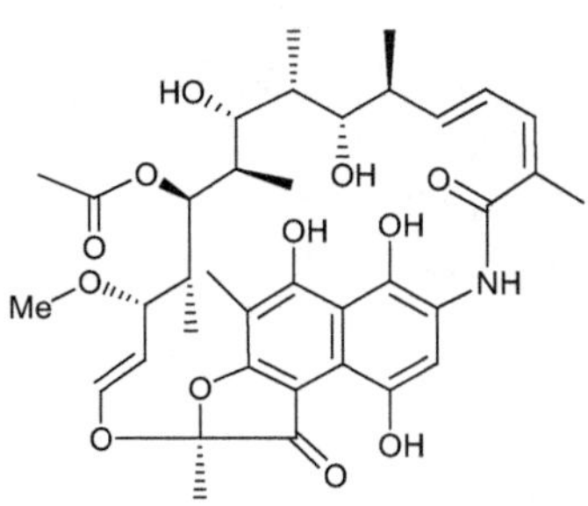

34 Rifamycin

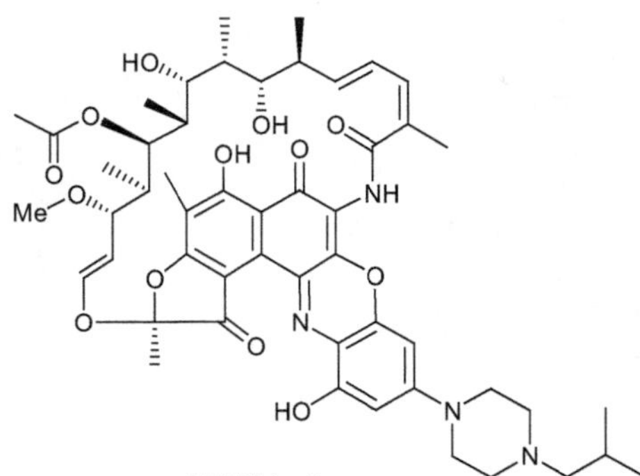

35 Rifalazil

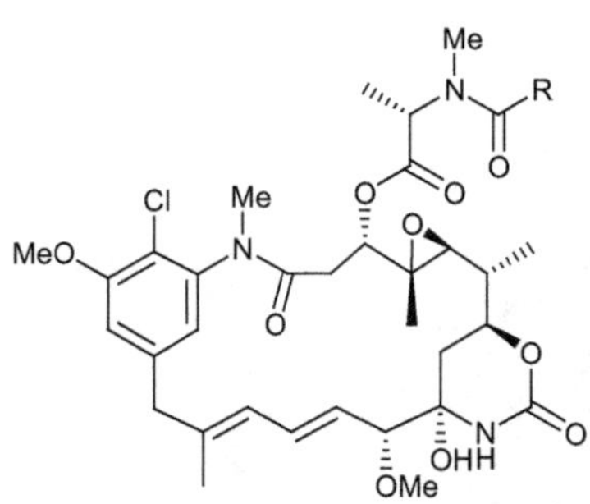

36 Maytansine, R = CH_3
37 DM1, R = CH_2CH_2SH
38 DM4, R = $CH_2CH_2C(CH_3)_2SH$

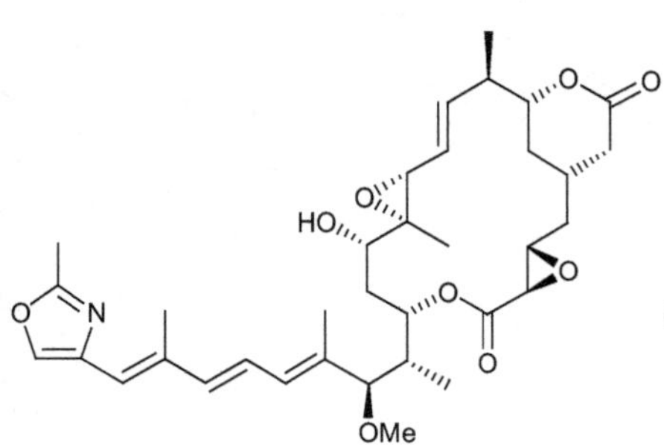

39 Rhizoxin

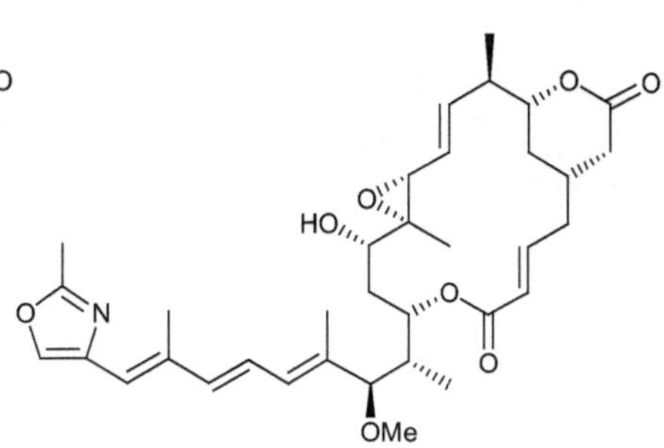

40 WF-1360F

Figure 13.6 Ansamycins I, rifamycin, ansamitocins and rhizoxin.

base molecule in this series and launched in the mid 1960s as a treatment for tuberculosis (as an antimycobacterial agent). In the intervening five decades, well over 300 variations on the structure have been reported as being in biological assessments ranging from *in vitro* testing through clinical trials to becoming approved drugs. A search of the Thomson-Reuters Integrity™ database in June 2014 showed 180 different compounds listed that were of similar structure at all stages from biological assessment to active clinical trials.

13.5.1 Rifamycins

A 2009 paper by Mariani and Maffioli[56] gives an excellent comparison of the various rifamycins and also covers other ansamycin molecules, including the geldanamycins and ansamitocins (both of which will be discussed later in this chapter).

In addition to rifamycin, four other chemical variations have been launched and approved by the FDA or equivalent organization in other countries. These drugs are: rifampicin in 1967, rifamixin in 1988, rifabutin in 1992 and rifapentine in 1998. All have the same basic nucleus but differ predominately in the South-East corner in terms of their substitution pattern. In contrast to the ansamycins of the geldanamycin class, no major substitution in the ansa-rings has led to a drug entity at this moment. At the time of writing (June 2014) one compound, rifalazil (Figure 13.6, **35**) is shown as being in Phase II clinical trials. However, this is now being tested against uncomplicated genital *Chlamydia trachomatis* infection in women, not as an antimycobacterial agent.

This is a compound with a very long history and was in fact part of the second-generation rifamycin-likes, with its earliest report in a full paper demonstrating bactericidal activity against *M. leprae* in 1992.[57] In 2012, Gill *et al.* reported on variations around rifalazil, effectively all modifications of the hydroxyl and piperazine substituents on the SE corner (Figure 13.6, **35**) that demonstrated increased anti *M. tuberculosis* activity and lack of P450 induction.[58] Part of the same group recently published the crystal structure of *E. coli* RNA Polymerase complexed with these novel agents, so it is probable that a new rifamycin-like compound will enter trials against *M. tuberculosis* in due course, based upon both chemistry and molecular modeling.[59] Currently, there appear to be no rifamycin-like molecules in clinical trials for mycobacterial infections, though a relatively recent publication in the infectious disease literature does imply that increased doses of these agents in conjunction with other antituberculosis drugs are still viable treatments.[60]

13.5.2 Ansamitocins (Tubulin Interactive Agents)

In 2012, one of the first "plant-derived" tubulin interactive compounds, maytansine (Figure 13.6, **36**) from the Ethiopian tree *Maytenus serrata*, was effectively granted a new lease of life as a slightly modified "warhead" on a monoclonal antibody, and was approved by the FDA.[61] From the initial determination of its structure (Figure 13.6, **36**) natural product chemists wondered if the compound was microbial in origin, due to its similarity to the "ansa" antibiotics such as the rifamycins. This was confirmed in 1977, when scientists at Takeda Chemical Industries reported the structures of the bacterial products, the ansamitocins, which very closely resembled the maytansenoids. Later work on compounds isolated from the bacterium *Actinosynnema pretiosum* demonstrated that they were in fact identical to those isolated from other plant genera. The work leading up to this determination was well covered in a review by Kirchning *et al.* in 2008.[62] In addition, the chapter by Yu *et al.* in 2012[63] should also be read, as together these cover the chemistry and biosynthesis of these microbial compounds, demonstrating the various modifications that have been made to the base molecule.

By utilizing microbial derived "precursors" of maytansine, specifically DM-1 (Figure 13.6, **37**) and DM-4 (Figure 13.6, **38**) in which suitable chemical linkages were added to maytansine, the resulting molecules were linked as "warheads" to specific monoclonal antibodies directed against tumor-linked epitopes. For discussions of the utility of such linked materials, the papers by Senter in 2009,[64] plus those of Alley[65] and Caravella[66] in 2010 should also be consulted.

The paper by Lambert in 2010 refers specifically to the DM1-linked conjugates from the aspect of their "construction" and their clinical efficacies.[67] We recommend that this article should be read in conjunction with the 2011 paper by Kümler *et al.*, where the story of trastuzumab emtansine,[68] which is the combination of the antitumor biologic agent Herceptin® with a specific linkage to DM1, is given in detail. Fairly recently, Barginear *et al.* published an update on the clinical trials with this combination, which is also well worth consulting, particularly for people interested in what is needed to progress such an entity through human trials.[69]

13.5.3 Rhizoxin (Tubulin Interactive Agents)

This particular ansamycin (Figure 13.6, **39**) has quite a past, as it entered clinical trials in Europe as a tubulin interactive agent,[70] but as a result of lack of activity in humans, it was discontinued.[71] For ten more years, it was a scientific curiosity, until a report in 2005 identified the producing organism as an endophytic bacterium, not the host fungus thought until then to be the producer.[72,73]

The stories of the genetic dissections and the biosynthesis of the rhizoxin complex, plus a full analysis of the symbiotic bacterium were published by the Hertweck group, who were responsible for the discovery of this unique biological interaction.[74–76] This group also reported the very surprising discovery that the oxidation of the double bond to produce the second epoxide in rhizoxin was probably performed by the host fungus, implying that the actual precursor of rhizoxin is the metabolite WF-1360F (Figure 13.6, **40**).[76]

A report showing the total synthesis of WF-1360F that confirmed the bioactivity of the molecule, was recently published by the Altmann group at the ETH,[77] and this paper should be read in conjunction with the reports of biosynthetic oxazole-nitrile conversions downstream of the biosynthesis of rhizoxin, as indications of where chemical and/or biochemical intervention may well produce novel compounds from this basic skeleton.[78]

13.5.4 Geldanamycin and Analog/HSP90 Inhibitors

The full story of the discovery and utilization of geldanamycin (Figure 13.7, **41**) and its derivatives as HSP90 inhibitors was presented by Snader in early 2011,[79,80] and readers interested in the early work should read those two reviews, as well as the 2013 review by Franke *et al.*,[81] as they show the routes of a number of geldanamycin-related antitumor agents that went into clinical trials up to Phase III in man, As of June 2014 only one trial is currently in progress with original clinical candidate 17-AAG or tanespimycin (Figure 13.7, **42**) at the Phase I/II level when either the US or EU clinical trials databases are accessed, though earlier studies had reached Phase III.

In this section, apart from an introductory paragraph paraphrasing the reviews by Snader, our aim is to show how, by using genetic modifications to the biosynthetic pathways, molecules that have HSP90 binding activity have been identified and produced biochemically, giving structures that have not yet been approached synthetically. Thus, our ultimate aim is to show structures to synthetic chemists that will "catch their fancy" in terms of structures to synthesize and test.

The benzoquinone ansamycin antibiotic geldanamycin (Figure 13.7, **41**) from *Streptomyces. hygroscopicus* var *geldanus* was initially reported by The Upjohn Company in 1970[82] and demonstrated antiparasitic activity. Later studies suggested good antitumor activity, which was thought to be due to inhibition of the tyrosine specific kinase (*v-Src*), which is involved in regulating growth and cell proliferation as well as several signal transduction pathways.[83,84] However, in 1994, Whitesell and co-workers[85] reported that it bound to heat shock protein 90 (HSP90), with the binding site being identified by Stebbins *et al.* as the ATP-binding site at the amino terminus end of HSP90, the first time that a molecule such as this was identified as an ATP-mimic,[86] leading indirectly to cell death as a result of altering the chaperone activity of HSP90.

By utilizing the information on biosynthetic pathways derived from total genome sequences, and the ability to "mix and match" genes within biosynthetic clusters and/or add exogenous genes, reports have been published in the last few years of potentially active structures from such efforts. Thus, analogs such as 19-hydroxy-4,5-dihydrogeldanamycin (Figure 13.7, **43**) and thiazinogeldanamycin (Figure 13.7, **44**) have been reported from engineered strains of *S. hygroscopicus* JCM4427 produced together with other known derivatives,[87] though their bioactivities have not been fully delineated. Even today, novel naturally occurring geldanamycin derivatives are still being

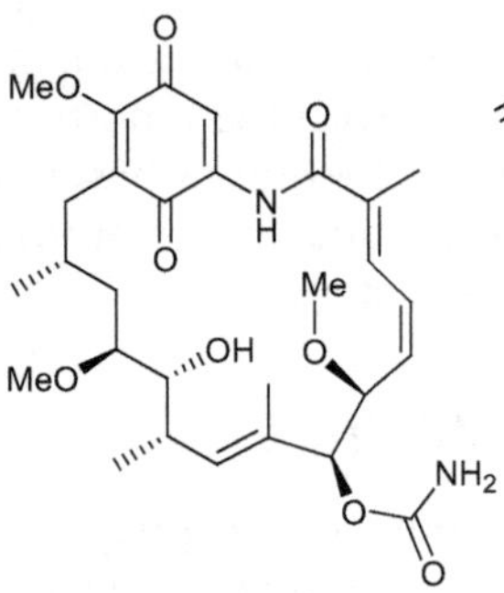

41 Geldanamycin

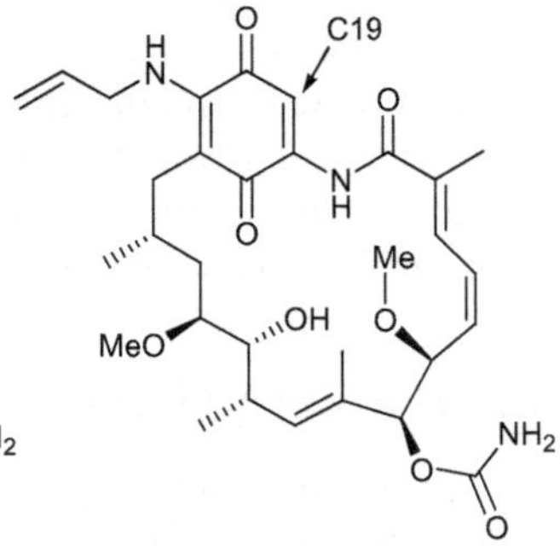

42 17-AAG
Tanespimycin

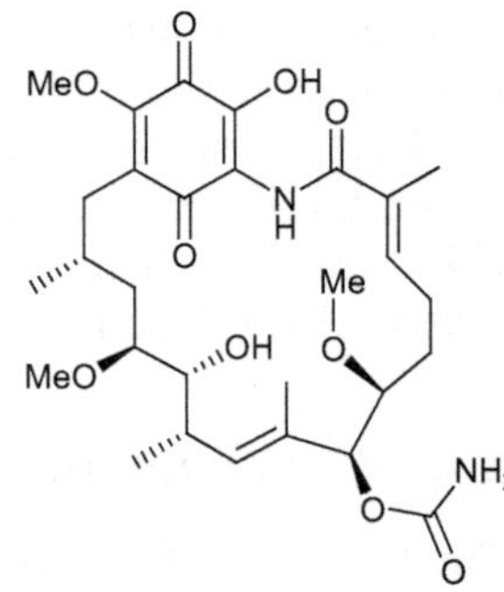

43 19-Hydroxy-4,5-dihydro-geldanamycin

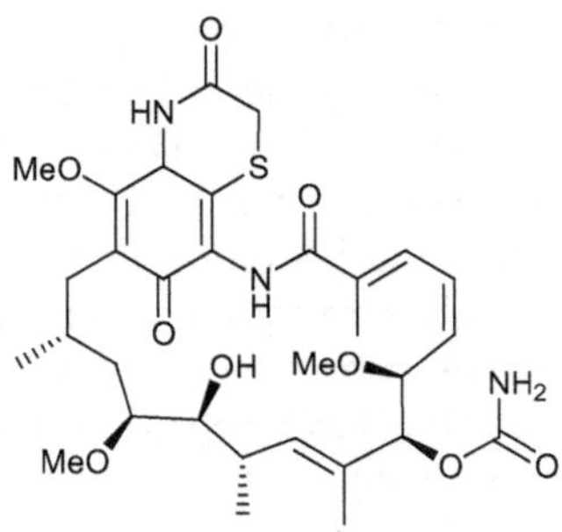

44 Thiazinogeldanamycin

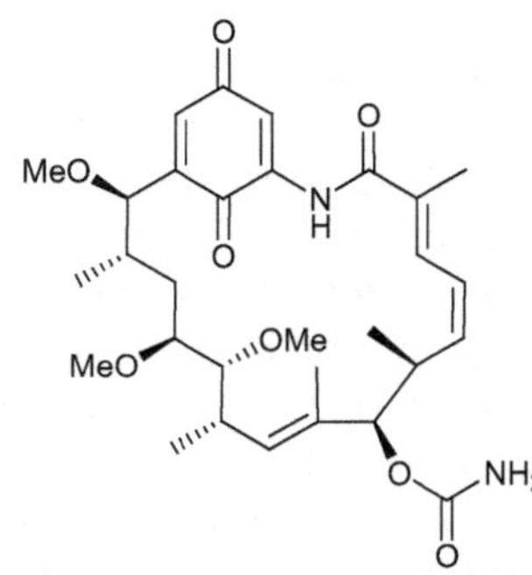

45 Macbecin I

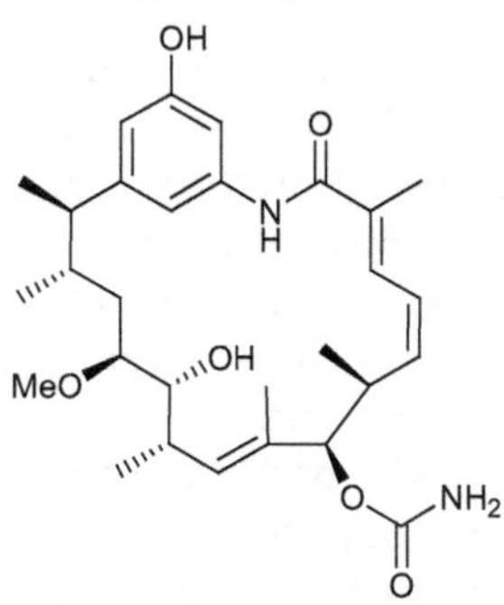

46 Macbecin-derived phenol

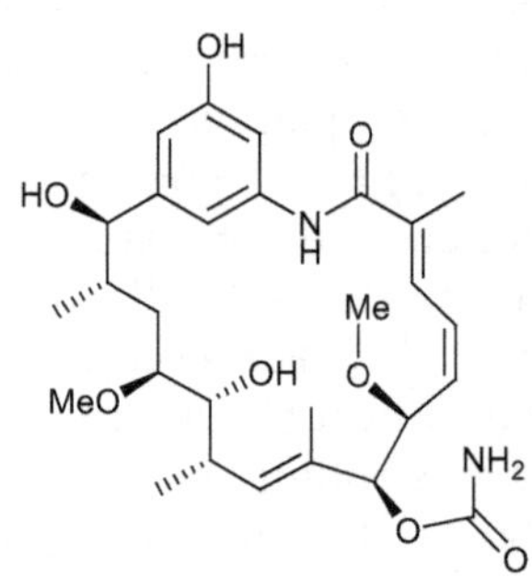

47 Geldanamycin-derived phenol

Figure 13.7 Ansamycins II, geldanamycin and macbecins.

reported, with the latest having novel substituents at the 19 position on the benzoquinone and its 4, 5 dihydro-derivative (structures not shown). Interestingly, though both were more water-soluble than geldanamycin, they showed lower activities against tumor cell lines.[88]

Very recently, chemical modifications at the 19 position in geldanamycin have yielded many new semi-synthetic geldanamycins,[89] including modifications of the clinical candidates 17-AAG and 17-DMAG that may have much lower "off-target effects" as drug candidates due to their lack of interactions with thiols such as glutathione, as a result of the blocked 19 positions (ortho to the quinone in the ring).[90] It will be interesting to see how these progress in time.

What is of definite interest, however, is that a close relative of geldanamycin, macbecin I (Figure 13.7, **45**) was reported in 2008 to be an HSP90 inhibitor by researchers from the small UK biotech company, Biotica. They reported that it had both *in vitro* and *in vivo* activity in mice and was more water-soluble and less toxic than the geldanamycin derivatives then in current trials.[91] Later that same year, they reported the optimization and production of macbecin-based molecules that were derived by genetic modification of the macbecin biosynthetic complex in *Actinosynnema pretiosum* subsp. *pretiosum*.[92] Of microbiological interest is that this is nominally the same genus and species that the ansamitocins were first isolated from.

Some of the "off-target" bioactivities of the base geldanamycin structure were often attributed to the quinone moieties undergoing redox cycling, so the Biotica team chose to make a molecule where the quinonoid ring was replaced by a phenol (Figure 13.7, **46**). The production of this compound was then further optimized to >200 mg L^{-1} by genetic manipulation in the same microbe. *In vitro* and *in vivo* experiments demonstrated an activity profile similar to that shown by 17-AAG, the initial geldanamycin-based molecule to go into human trials as an HSP-90 inhibitor,[79,80] but the compound was a tighter binder to HSP90 and active at a lower molar dose in both cellular and murine assays.[92] Somewhat similar modifications using the geldanamycin producer *S. hygroscopicus* were reported in 2011 by Wu *et al.*, producing a molecule similar (Figure 13.7, **47**) to that optimized by the Biotica group, effectively differing only two substituents from the phenol-containing macbecin analog.[93] Thus, different modified ansamycin macrocycles with HSP90 activity are available for future screening.

The utility of a mixed biosynthetic and chemical synthetic approach to these molecules was discussed with other examples, in addition to the geldanamycins, by Kirchning and Hahn in 2012.[94] This paper should be read by any synthetic chemist who wishes to utilize microbial products and their biosynthetic machinery as potential routes to other novel compounds with a high probability of being biologically active.

13.6 mTOR or FRAP1 Inhibitors

13.6.1 Rapamycin and Derivatives

Rapamycin and its close chemical relatives are almost "molecules for most diseases" since the rapamycins now cover compounds that have biological properties ranging from antifungal through immunomodulation to antitumor therapies and even comprise molecules to use in stents to avoid plaque formation. These molecules are formally too large and complex to meet Lipinski's rules, but natural product-based molecules were never quoted, even by Lipinski, as being subject to those rules.

In 1975, scientists at Wyeth-Ayerst laboratories reported the isolation of the 31-membered macrocyclic antibiotic rapamycin (Figure 13.8, **48a**) as an antifungal agent from a strain of *Streptomyces hygroscopicus*,[95–97] but due to its immunosuppressive effects, it was not successful as an antifungal agent. In the mid 1980s, its activity against syngeneic murine tumors was reported by Sehgal and co-workers,[98] but was not continued with. In 1991, by using yeast cells, the molecular target of rapamycin was identified (TOR or "target of rapamycin"),[99] and three years later, Brown *et al.* identified the mammalian homologue (mTOR),[100] and rapamycin was shown to be a "relatively" small molecule that effectively interfered with protein–protein interactions.

Initially, chemical modifications were made at the carbon atom at C^{43} on the rapamycin base structure (numeration as in Zech *et al.*[101] rather than the alternative numbering system of McAlpine *et al.*[102]) led to a total of four clinically approved drugs: rapamycin (sirolimus), everolimus, temsirolimus and zotarolimus (Figure 13.8, **48a–d**). In 1999, sirolimus (rapamycin) was approved as an immunosuppressive agent. Everolimus was first launched as an immunosuppressive agent in 2004 and then in 2009, 2010, 2011 and 2012, the compound was approved for the treatment of kidney, brain, pancreatic and breast cancers, respectively. In addition, in 2012, everolimus was released by Abbot to be used as a stent in the treatment of coronary and peripheral arterial diseases in the USA. In contrast, temsirolimus (CCI-779) was first approved for renal carcinoma treatment in the USA in 2007, then in 2010 it was approved in Japan. Zotarolimus was launched in the USA in 2005 for the treatment of arterial restenosis (as a component of a stent) and, recently, the EU approved a stent containing novolimus, which is a metabolite of rapamycin, where the methoxyl at C^7 has been demethylated giving an active C^7-hydroxy molecule.

Merck & Co. and Ariad Pharmaceuticals collaborated to develop another rapamycin derivative, ridaforolimus (Figure 13.8, **48e**), which went into Phase III clinical trials for the treatment of soft tissue carcinoma and bone cancer. Currently, it is in Phase II trials for different indications in cancer under each company as they dissolved the collaboration.

13.6.2 Rapalogs

A very interesting part of the rapamycin story has to do with the use of bacterial "genetic engineering". In such programs, rapamycin derivatives are produced that are composed of biosynthetic gene clusters that have been modified, either *via* genetic modifications or by feeding unusual substrates at some point in the biosynthetic pathway(s), or expressed in unusual environments, with the aim of producing analogs with different structures and perhaps biological activities.

One of the groups that spent many years studying the biosynthesis of rapamycin, following on from the pioneering work of the Demain group at

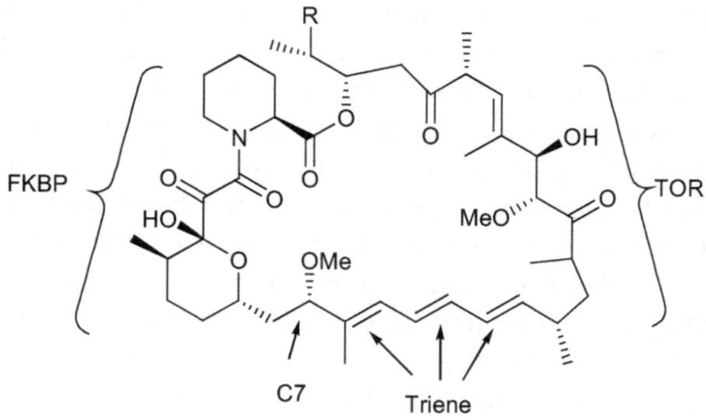

48 Base Rapamycin Macrolide

R =

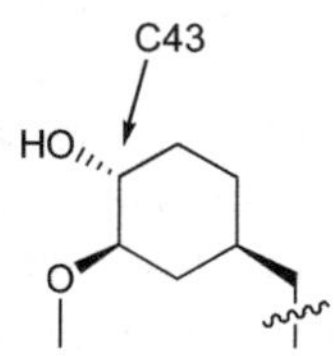

48a Sirolimus

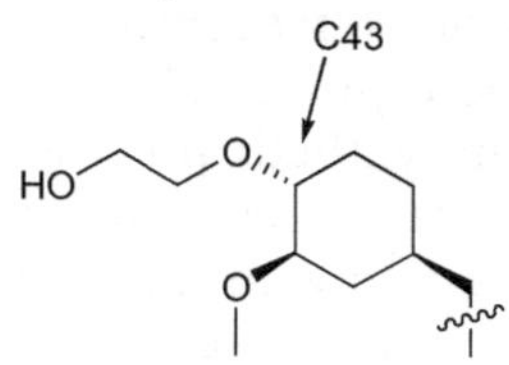

48b Everolimus

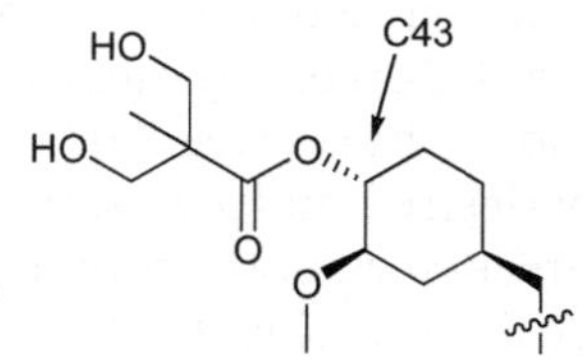

48c Temsirolimus

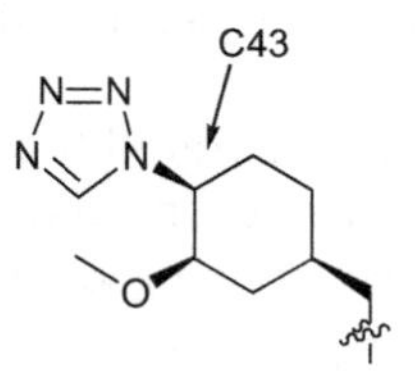

48d Zotarolimus

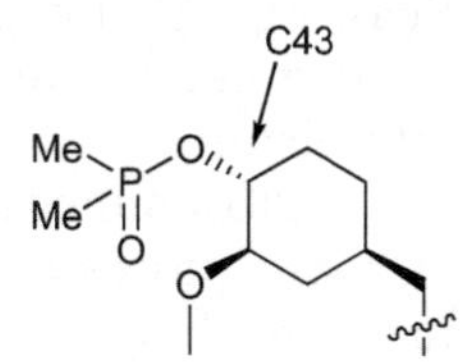

48e Ridaforolimus

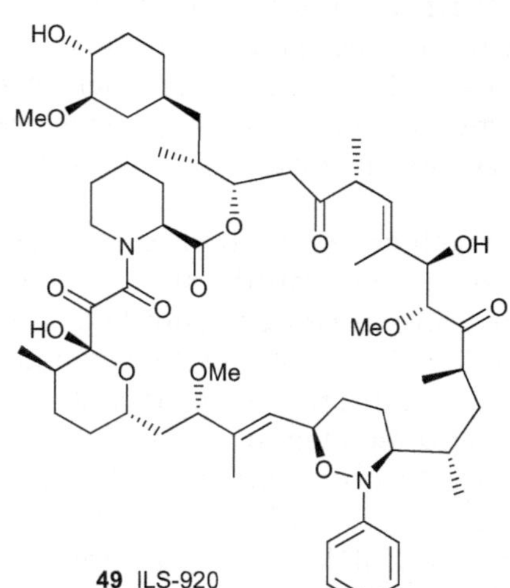

49 ILS-920

Figure 13.8 Rapamycin and derivatives.

MIT,[103,104] was the group at Cambridge University in the UK under Peter Leadley and James Staunton that spun off the biotech company known as Biotica, mentioned earlier in the discussion on macbecin derivatives. Though rapamycin was not the only target that this company investigated, they were instrumental in adapting the biosynthetic machinery of the rapamycin producer to generate agents that had not been seen previously.

The methodologies and some of the compounds developed since the 2004 review by Demain[104] have been presented by a number of authors in the last few years, beginning with a review by Graziani covering 2003 to 2008,[105] which should be read together with the 2010 review by Park *et al.*[106] These papers demonstrate the multiplicity of materials that can be produced by modification of these biosynthetic units.

In particular, these reviews recognize the problems involved in the regulation of any biosynthetic process, including what is now realized to be the "Achilles Heel" of mutasynthetic processes designed to increase yields of desired molecules, that is the provision in the microbe used of sufficient precursors both to maintain growth and also increase production of the desired molecule(s). An excellent example of this, though not from studies related to rapamycin biosynthesis, is given in the 2013 paper from Keasling's group discussing the production of artemisinic acid from a bio-engineered yeast strain as a precursor for the semi-synthesis of artemisinin.[107]

The 2013 publications from the Biotica group demonstrated the potential of these processes in producing engineered strains for producing novel rapalogs,[108] and how to utilize them in order to use their products for biosynthetic medicinal chemistry.[109] Both of these papers should be read in conjunction with two European patent applications by Biotica scientists,[110,111] since these together show the rage of possibilities with these techniques and their potential to produce unusual rapalogs. The research leaders from Biotica started another small biotech company, Isomerase Therapeutics, in the UK in 2013, having licensed some of Biotica's enabling technology, so the materials from Biotica may still be available in due course.

Using methodologies from the rapalog/Biotica collaboration, Wyeth Pharmaceuticals developed a rather interesting derivative of rapamycin with a modified ring structure (Figure 13.8, **49**), ILS-920, since by modifying the triene portion of the molecule, mTOR binding would be disrupted. However, ILS-920 has a different target since it is a nonimmunosuppressive neurotrophic analog reported to exhibit over a 900-fold higher binding affinity for FKBP52 over FKBP12 compared to that of rapamycin. It promotes neuronal survival and outgrowth *in vitro*, binding to the β1 subunit of L-type calcium channels (CACNB1).[112] ILS-920 was under development for treating stroke,[113] and a Phase I clinical trial for the treatment of acute ischemic stroke has been completed. A recent chapter by Graziani gives the story of ILS-920 in significant detail and should be consulted for the *in vivo* and clinical details.[114]

13.7 In Conclusion

There are many more very interesting structures from nature that could be covered in a chapter such as this, but the major problem is deciding when to stop. It might be that we have finished too early or perhaps have belabored some aspects more than others. However, we would certainly be remiss if we did not mention perhaps the most "chemically underserved group of all", peptides and their mimetics. These often are overlooked by mainstream synthetic chemists, but just to give a flavor of the area, we recommend that classical synthetic chemists read the excellent and very recent review by Avan *et al.* from the Katritzky laboratory at the University of Florida.[115]

If one couples this work to the vast number of potent cyclic peptides found in nature, particularly in the marine environment exemplified by the work in just the following two areas, cone snails,[116–121] and cyclic peptides from tunicates,[4] then the field is wide open for modifications.

On that final note/suggestion, we will close this chapter, hoping that we have shown enough areas for synthetic chemists to look at some unusual structures to modify.

References

1. C. R. Woese, O. Kandler and M. L. Wheelis, *Proc. Natl. Acad. Sci. U. S. A.*, 1990, **87**, 4576.
2. O. K. Radjasa, Y. M. Vaske, G. Navarro, H. C. Vervoort, K. Tenney, R. G. Linington and P. Crews, *Bioorg. Med. Chem.*, 2011, **19**, 6658.
3. J. C. Kwan, M. S. Donia, A. W. Han, E. Hirose, M. G. Haygood and E. W. Schmidt, *Proc. Natl. Acad. Sci. U. S. A.*, 2012, **109**, 20655.
4. E. W. Schmidt, M. S. Donia, J. A. McIntosh, W. F. Fricke and J. Ravel, *J. Nat. Prod.*, 2012, **75**, 295.
5. M. D. B. Tianero, M. S. Donia, T. S. Young, P. G. Schultz and E. W. Schmidt, *J. Am. Chem. Soc.*, 2012, **134**, 418.
6. M. C. Wilson, T. Mori, C. Rückert, A. R. Uria, M. J. Helf, K. Takada, C. Gernert, U. A. E. Steffens, N. Heycke, S. Schmitt, C. Rinke, E. J. N. Helfrich, A. O. Brachmann, C. Gurgui, T. Wakimoto, M. Kracht, M. Crüsemann, U. Hentschel, I. Abe, S. Matsunaga, J. Kalinowski, H. Takeyama and J. Piel, *Nature*, 2014, **506**, 58.
7. T. Henkel, R. M. Brunne, H. Muller and F. Reichel, *Angew. Chem., Int. Ed.*, 1999, **38**, 643.
8. A. B. Yongye, J. Waddell and J. L. Medina-Franco, *Chem. Biol. Drug Des.*, 2012, **80**, 717.
9. W. R. Pitt, D. M. Parry, B. G. Perry and C. R. Groom, *J. Med. Chem.*, 2009, **52**, 2952–2963.
10. J. Klein, J. R. Heal, W. D. O. Hamilton, T. Boussemghoune, T. O. Tange, F. Delagrange, G. Jaeschke, A. Hatsch and J. Heim, *ACS Synth. Biol.*, 2014, **3**, 314.

11. B. E. Evans, K. E. Rittle, M. G. Bock, R. M. DiPardo, R. M. Freidinger, W. L. Whitter, G. F. Lundell, D. F. Veber, P. S. Anderson, R. S. L. Chang, V. J. Lotti, D. J. Cerino, T. B. Chen, P. J. Kling, K. A. Kunkel, J. P. Springer and J. Hirshfield, *J. Med. Chem.*, 1988, **31**, 2235.
12. B. E. Evans, *Drugs Future*, 1989, **14**, 970.
13. M. C. McLeod, G. Singh, J. N. Plampin III, D. Rane, J. L. Wang, V. W. Day and J. Aubé, *Nat. Chem.*, 2014, **6**, 133.
14. J. Polanski, A. Kurczyk, A. Bak and R. Musiol, *Curr. Med. Chem.*, 2012, **19**, 1921.
15. J. J. Sahn and S. F. Martin, *ACS Comb. Sci.*, 2012, **14**, 496.
16. D. Bonnet, J.-F. Argathe, S. Radford, E. Pflimlin, S. Riché, P. Doman, M. Hibert and A. Ganesan, *ACS Comb. Sci.*, 2012, **14**, 323.
17. C. J. Suckling, *Sci. Prog.*, 1991, **75**, 323.
18. W. W. Lee, A. Benitez, L. Goodman and B. R. Baker, *J. Am. Chem. Soc.*, 1960, **82**, 2648.
19. P. Davis, *UK* 1159290, 23 Jul 1969.
20. G. Cimino, S. De Rosa and S. De Stefano, *Experientia*, 1984, **40**, 339.
21. J. S. Evans, E. A. Musser, G. D. Mengel, K. R. Forsblad and J. H. Hunter, *Proc. Soc. Exp. Biol. Med.*, 1961, **106**, 350.
22. L. I. Pizer and S. S. Cohen, *J. Biol. Chem.*, 1960, **235**, 2387.
23. M. Y. Chu and G. A. Fischer, *Biochem. Pharmacol.*, 1962, **11**, 423.
24. R. L. Noble, C. T. Beer and J. H. Curtis, *Ann. N. Y. Acad. Sci.*, 1958, **76**, 882.
25. N. Neuss, M. Gorman, G. H. Svoboda, C. T. Maciak and C. T. Beer, *J. Am. Chem. Soc.*, 1959, **81**, 4754.
26. M. Gorman, N. Neuss and G. H. Svoboda, *J. Am. Chem. Soc.*, 1959, **81**, 4745.
27. I. S. Johnson, H. F. Wright, G. H. Svoboda and J. Vlantis, *Cancer Res.*, 1960, **20**, 1016.
28. P. Keglevich, L. Hazai, G. Kalaus and C. Szántay, *Molecules*, 2012, **17**, 5893.
29. F. Roussi, F. Gueritte and J. Fahy, in *Anticancer Agents from Natural Products*, ed. G. M. Cragg, D. G. I. Kingston and D. J. Newman, Taylor and Francis, Boca Raton, FL, 2nd edn, 2012, p. 177.
30. P. Keglevich, L. Hazai, Z. Dubrovay, M. Dekany, C. Szántay Jr, G. Kalaus and C. Szantay, *Heterocycles*, 2014, **89**, 653.
31. R. J. Andersen, D. J. Faulkner, C. H. He, G. D. Van Duyne and J. Clardy, *J. Am. Chem. Soc.*, 1985, **107**, 5492.
32. C. Bailly, *Curr. Med. Chem.: Anti-Cancer Agents*, 2004, **4**, 363.
33. H. Fan, J. Peng, M. T. Hamann and J.-F. Hu, *Chem. Rev.*, 2008, **108**, 264.
34. D. L. Boger, C. W. Boyce, M. A. Labroli, C. A. Sehon and Q. Jin, *J. Am. Chem. Soc.*, 1999, **121**, 54.
35. A. R. Quesada, M. D. García Grávalos and J. L. Fernández Puentes, *Br. J. Cancer*, 1996, **74**, 677.
36. M. A. Gallego, C. Ballot, J. Kluza, N. Hajji, A. Martoriati, L. Castéra, C. Cuevas, P. Formstecher, B. Joseph, G. Kroemer, C. Bailly and P. Marchetti, *Oncogene*, 2008, **27**, 1981.

37. C. Ballot, J. Kluza, A. Martoriati, U. Nyman, P. Formstecher, B. Joseph, C. Bailly and P. Marchetti, *Mol. Cancer Ther.*, 2009, **8**, 3307.
38. D. Pla, F. Albericio and M. Alvarez, *Anticancer Agents Med. Chem.*, 2008, **8**, 746.
39. M. Chittchang, P. Batsomboon, S. Ruchirawat and P. Ploypradith, *ChemMedChem*, 2009, **4**, 457.
40. M. Chittchang, M. P. Gleeson, P. Ploypradith and S. Ruchirawat, *Eur. J. Med. Chem.*, 2010, **45**, 2165.
41. Q. Li, J. Jiang, A. Fan, Y. Cui and Y. Jia, *Org. Lett.*, 2011, **13**, 312.
42. T. Fukuda, F. Ishibashi and M. Iwao, *Heterocycles*, 2011, **83**, 491.
43. M. Komatsubara, T. Umeki, T. Fukuda and M. Iwao, *J. Org. Chem.*, 2014, **79**, 529.
44. H. Lachance, S. Wetzel, K. Kumar and H. Waldmann, *J. Med. Chem.*, 2012, **55**, 5989.
45. S. Rizzo and H. Waldmann, *Chem. Rev.*, 2014, **114**, 4621.
46. R. A. Bauer, J. M. Wurst and D. S. Tan, *Curr. Opin. Chem. Biol.*, 2010, **14**, 308.
47. D. F. Veber, S. R. Johnson, H.-Y. Cheng, B. R. Smith, K. W. Ward and K. D. Kopple, *J. Med. Chem.*, 2002, **45**, 2615.
48. C. Zhao, T. Zhu and W. Zhu, *Chin. J. Org. Chem.*, 2013, **33**, 1195.
49. T. T. Bladt, J. C. Frisvad, P. B. Knudsen and T. O. Larsen, *Molecules*, 2013, **18**, 11338.
50. F. Kong, Y. Wang, P. Liu, T. Dong and W. Zhu, *J. Nat. Prod.*, 2014, 77, 132.
51. M. Imamura and C. Prasad, *Peptides*, 2003, **24**, 445–448.
52. P. G. Wyatt, M. J. Allen, A. D. Borthwick, D. E. Davies, A. M. Exall, R. J. D. Hatley, W. R. Irving, D. G. Livermore, N. D. Miller, F. Nerozzi, S. L. Sollis and A. K. Szardenings, *Bioorg. Med. Chem. Lett.*, 2005, **15**, 2579–2582.
53. A. D. Borthwick, *Chem. Rev.*, 2012, **112**, 3641.
54. J. F. González, I. Ortín, E. De La Cuesta and J. C. Menéndez, *Chem. Soc. Rev.*, 2012, **41**, 6902.
55. Y. Wang, P. Wang, H. Ma and W. Zhu, *Expert Opin. Ther. Pat.*, 2013, **23**, 1415.
56. R. Mariani and S. I. Maffioli, *Curr. Med. Chem.*, 2009, **16**, 430.
57. M. Gidoh, S. Tsutsumi, T. Yamane, K. Yamashita, K. Hosoe and T. Hidaka, *Lepr. Rev.*, 1992, **63**, 319.
58. S. K. Gill, H. Xu, P. D. Kirchhoff, T. Cierpicki, A. J. Turbiak, B. Wan, N. Zhang, K.-W. Peng, S. G. Franzblau, G. A. Garcia and H. D. H. Showalter, *J. Med. Chem.*, 2012, **55**, 3814.
59. V. Molodtsov, I. N. Nawarathne, N. T. Scharf, P. D. Kirchhoff, H. D. H. Showalter, G. A. Garcia and K. S. Murakami, *J. Med. Chem.*, 2013, **56**, 4758.
60. D. Mitchison and G. Davies, *Int. J. Tuberc. Lung Dis.*, 2012, **16**, 724.
61. I. Krop and E. P. Winer, *Clin. Cancer Res.*, 2014, **20**, 15.
62. A. Kirschning, K. Harmrolfs and T. Knobloch, *C. R. Chim.*, 2008, **11**, 1523.

63. J.-W. Yu, H. G. Floss, G. M. Cragg and D. J. Newman, in *Anticancer Agents from Natural Products*, ed. G. M. Cragg, D. G. I. Kingston and D. J. Newman, Taylor and Francis, Boca Raton, FL, 2nd edn, 2012, p. 407.
64. P. D. Senter, *Curr. Opin. Chem. Biol.*, 2009, **13**, 235.
65. S. C. Alley, N. M. Okeley and P. D. Senter, *Curr. Opin. Chem. Biol.*, 2010, **14**, 529.
66. J. Caravella and A. Lugovskoy, *Curr. Opin. Chem. Biol.*, 2010, **14**, 520.
67. J. M. Lambert, *Drugs Future*, 2010, **35**, 471.
68. I. Kümler, C. Ehlers Mortensen and D. L. Nielsen, *Drugs Future*, 2011, **36**, 825.
69. M. F. Barginear, V. John and D. R. Budman, *Mol. Med.*, 2013, **18**, 1473.
70. P. Lelieveld and H. R. Hendriks, *Cancer Treat. Rev.*, 1990, **17**, 119.
71. A. R. Hanauske, G. Catimel, S. Aamdal, W. ten Bokkel Huinink, R. Paridaens, N. Pavlidis, S. B. Kaye, A. te Velde, J. Wanders and J. Verweij, *Br. J. Cancer*, 1996, **73**, 397.
72. L. P. Partida-Martinez and C. Hertweck, *Nature*, 2005, **437**, 884.
73. L. P. Partida-Martinez, I. Groth, I. Schmitt, W. Richter, M. Roth and C. Hertweck, *Int. J. Syst. Evol. Microbiol.*, 2007, **57**, 2583.
74. G. Lackner, N. Moebius, L. P. Partida-Martinez, S. Boland and C. Hertweck, *BMC Genomics*, 2011, **12**, 210.
75. G. Lackner, N. Moebius, L. P. Partida-Martinez and C. Hertweck, *J. Bacteriol.*, 2011, **193**, 783.
76. K. Scherlach, B. Busch, G. Lackner, U. Paszkowski and C. Hertweck, *Angew. Chem., Int. Ed.*, 2012, **51**, 9615.
77. C. M. Neuhaus, M. Liniger, M. Stieger and K.-H. Altmann, *Angew. Chem., Int. Ed.*, 2013, **52**, 5866.
78. K. Scherlach, N. Brendel, K. Ishida, H.-M. Dahse and C. Hertweck, *Org. Biomol. Chem.*, 2012, **10**, 5756.
79. K. M. Snader, in *Anticancer Agents from Natural Products*, ed. G. M. Cragg, D. G. I. Kingston and D. J. Newman, Taylor and Francis, Boca Raton, FL, 1st edn, 2005, p. 339.
80. K. M. Snader, in *Anticancer Agents from Natural Products*, ed. G. M. Cragg, D. G. I. Kingston and D. J. Newman, Taylor and Francis, Boca Raton, FL, 2nd edn, 2012, p. 429.
81. J. Franke, S. Eichner, C. Zeilinger and A. Kirschning, *Nat. Prod. Rep.*, 2013, **30**, 1299.
82. C. DeBoer, P. A. Meulman, R. J. Wnuk and D. H. Peterson, *J. Antibiot.*, 1970, **23**, 442.
83. Y. Uehara, M. Hori, T. Takeuchi and H. Umezawa, *Mol. Cell. Biol.*, 1986, **6**, 2198.
84. Y. Uehara, Y. Murakami, S. Mizuno and S. Kawalt, *Virology*, 1988, **164**, 294.
85. L. Whitesell, E. G. Mimnaugh, B. De Costa, C. E. Myers and L. M. Neckers, *Proc. Natl. Acad. Sci. U. S. A.*, 1994, **91**, 8324.

86. C. E. Stebbins, A. A. Russo, C. Schneider, N. Rosen, F. U. Hartl and N. P. Pavletich, *Cell*, 1997, **89**, 239.
87. C.-Z. Wu, J.-H. Jang, J.-S. Ahn and Y.-S. Hong, *J. Microbiol. Biotechnol.*, 2012, **22**, 1478.
88. S. Li, S. Ni, L. Wu, L. Li, B. Jiang, H. Wang, G. Sun, M. Gan, J. Li, W. He, L. Lin, Y. Wang, S. Bai and S. Si, *J. Nat. Prod.*, 2014, **76**, 969.
89. R. R. A. Kitson, C.-H. Chang, R. Xiong, H. E. L. Williams, A. L. Davis, W. Lewis, D. L. Dehn, D. Siegel, S. M. Roe, C. Prodromou, D. Ross and C. J. Moody, *Nat. Chem.*, 2013, **5**, 307.
90. C. H. Chang, D. A. Drechsel, R. R. Kitson, D. Siegel, Q. You, D. S. Backos, C. Ju, C. J. Moody and D. Ross, *Mol. Pharmacol.*, 2014, **85**, 849.
91. C. J. Martin, S. Gaisser, I. R. Challis, I. Carletti, B. Wilkinson, M. Gregory, C. Prodromou, S. M. Roe, L. H. Pearl, S. M. Boyd and M.-Q. Zhang, *J. Med. Chem.*, 2008, **51**, 2853.
92. M.-Q. Zhang, S. Gaisser, M. Nur-E-Alam, L. S. Sheehan, W. A. Vousden, N. Gaitatzis, G. Peck, N. J. Coates, S. J. Moss, M. Radzom, T. A. Foster, R. M. Sheridan, M. A. Gregory, S. M. Roe, C. Prodromou, L. Pearl, S. M. Boyd, B. Wilkinson and C. J. Martin, *J. Med. Chem.*, 2008, **51**, 5494.
93. C.-Z. Wu, A. N. Moon, J.-H. Jang, D. Lee, S.-Y. Kang, J.-T. Park, J. S. Ahn, B. Y. Hwang, Y. H. Kim, H.-S. Lee and Y.-S. Hong, *J. Antibiot.*, 2011, **64**, 461.
94. A. Kirschning and F. Hahn, *Angew. Chem., Int. Ed.*, 2012, **51**, 4012.
95. S. N. Sehgal, H. Baker and C. Vezina, *J. Antibiot.*, 1975, **28**, 727.
96. C. Vezina, A. Kudelski and S. N. Sehgal, *J. Antibiot.*, 1975, **28**, 721.
97. H. Baker, A. Sidorowicz, S. N. Sehgal and C. Vezina, *J. Antibiot.*, 1978, **31**, 539.
98. C. P. Eng, S. N. Sehgal and C. Vezina, *J. Antibiot.*, 1984, **37**, 1231.
99. J. Heitman, N. R. Movva and M. N. Hall, *Science*, 1991, **253**, 905.
100. E. J. Brown, M. W. Albers, T. Bum Shin, K. Ichikawa, C. T. Keith, W. S. Lane and S. L. Schreiber, *Nature*, 1994, **369**, 756.
101. S. G. Zech, M. Car, Q. K. Mohemmad, N. I. Narasimhan, C. Murray, L. W. Rozamus and D. C. Dalgarno, *J. Antibiot.*, 2011, **64**, 649.
102. J. B. McAlpine, S. J. Swanson, M. Jackson and D. N. Whittern, *J. Antibiot.*, 1991, **44**, 688.
103. N. L. Pavia, A. L. Demain and M. F. Roberts, *J. Nat. Prod.*, 1991, **54**, 167.
104. A. L. Demain, *Annu. Rev. Microbiol.*, 2004, **58**, 1.
105. E. I. Graziani, *Nat. Prod. Rep.*, 2009, **26**, 602.
106. S. R. Park, Y. J. Yoo, Y.-H. Ban and Y. J. Yoon, *J. Antibiot.*, 2010, **63**, 434.
107. C. J. Paddon, P. J. Westfall, D. J. Pitera, K. Benjamin, K. Fisher, D. McPhee, M. D. Leavell, A. Tai, A. Main, D. Eng, D. R. Polichuk, K. H. Teoh, D. W. Reed, T. Treynor, J. Lenihan, M. Fleck, S. Bajad, G. Dang, D. Dengrove, D. Diola, G. Dorin, K. W. Ellens, S. Fickes, J. Galazzo, S. P. Gaucher, T. Geistlinger, R. Henry, M. Hepp, T. Horning, T. Iqbal, H. Jiang, L. Kizer, B. Lieu, D. Melis, N. Moss, R. Regentin, S. Secrest, H. Tsuruta, R. Vazquez, L. F. Westblade, L. Xu, M. Yu,

Y. Zhang, L. Zhao, J. Lievense, P. S. Covello, J. D. Keasling, K. K. Reiling, N. S. Renninger and J. D. Newman, *Nature*, 2013, **496**, 528.

108. S. G. Kendrew, H. Petkovic, S. Gaisser, S. J. Ready, M. A. Gregory, N. J. Coates, M. Nur-e-Alam, T. Warneck, D. Suthar, T. A. Foster, L. McDonald, G. Schlingman, F. E. Koehn, J. S. Skotnicki, G. T. Carter, S. J. Moss, M.-Q. Zhang, C. J. Martin, R. M. Sheridan and B. Wilkinson, *Metab. Eng.*, 2013, **15**, 167.
109. M. A. Gregory, A. L. Kaja, S. G. Kendrew, N. J. Coates, T. Warneck, M. Nur-e-Alam, R. E. Lill, L. S. Sheehan, L. Chudley, S. J. Moss, R. M. Sheridan, M. Quimpere, M.-Q. Zhang, C. J. Martina and B. Wilkinson, *Chem. Sci.*, 2013, **4**, 1046.
110. M. A. Gregory, S. Gaisser, H. Petkovic and S. J. Moss, EP 2 277 898 A2, 24 Sep 2011.
111. M. A. Gregory, S. J. Moss and B. Wilkinson, WO 2013093493 A1, 27 Jun 2013.
112. B. Ruan, K. Pong, F. Jow, M. Bowlby, R. A. Crozier, D. Liu, S. Liang, Y. Chen, M. L. Mercado, X. Feng, F. Bennett, D. von Schack, L. McDonald, M. M. Zaleska, A. Wood, P. H. Reinhart, R. L. Magolda, J. Skotnicki, M. N. Pangalos, F. E. Koehn, G. T. Carter, M. Abou-Gharbia and E. I. Graziani, *Proc. Natl. Acad. Sci. U. S. A.*, 2008, **105**, 33.
113. M. Abou-Gharbia and W. Childers, *Pure Appl. Chem.*, 2012, **84**, 1543.
114. E. I. Graziani, in *Natural Products: Discourse, Diversity, and Design*, ed. A. Osbourn, R. J. Goss and G. T. Carter, Wiley Blackwell, Oxford, UK, 2014, p. 469.
115. I. Avan, C. D. Hall and A. R. Katritzky, *Chem. Soc. Rev.*, 2014, **43**, 3575.
116. B. M. Olivera, L. J. Cruz, V. De Santos, G. W. LeCheminant, D. Griffin, R. Zeikus, J. M. McIntosh, R. Galyean and J. Varga, *Biochemistry*, 1987, **26**, 2086.
117. R. J. Clark, J. Jensen, S. T. Nevin, B. P. Callaghan, D. J. Adams and D. J. Craik, *Angew. Chem., Int. Ed.*, 2010, **49**, 6545.
118. N. L. Daly and D. J. Craik, *Drugs Future*, 2011, **36**, 25.
119. A. E. Fedosov, S. A. Moshkovskii, K. G. Kuznetsova and B. M. Olivera, *Biochemistry (Moscow) Supp. Ser. B: Biomed. Chem.*, 2012, **6**, 107.
120. R. J. Clark, M. Akcan, Q. Kaas, N. L. Daly and D. J. Craik, *Toxicon*, 2012, **59**, 446.
121. C. I. Schroeder and D. J. Craik, *Future Med. Chem.*, 2012, **4**, 1243.

CHAPTER 14

Ergot Alkaloids

DOROTA JAKUBCZYK AND SARAH O'CONNOR*

The John Innes Centre, Department of Biological Chemistry,
Norwich NR4 7UH, Norwich, United Kingdom
*Email: Sarah.o'connor@jic.ac.uk

14.1 History of Ergot Alkaloids

Ergot alkaloids are produced in sclerotia of grass symbionts, namely fungi of the genus *Claviceps,* along with other filamentus fungi in the genus *Aspergillus*, *Neotyphodium*, *Arthroderma*, *Penicillium*, *Epichloe*, *Balansia* and the recently described *Periglandula*.[1,2] The EA have been referenced in ancient history. Abnormally infected grain was noticed as early as 1900–1700 BC, in Mesopotamia,[3] and by 600 BC the Assyrians were able to differentiate between different diseases affecting grain. References to grain diseases have also been found in the Bible, in the Old Testament (850–550 BC). Ergots were used in 1100 BC in China for the treatment of various obstetric conditions. The Eleusinian Mysteries of ancient Greece were linked to hallucinations caused by EA. In 550 BC, the Egyptians recommended a mixture of ergot, oil and honey as a treatment for hair growth. Moreover, in about 350 BC, the Parsi wrote about the "noxious grasses that cause pregnant women to drop the womb and die in childbed'".

In the Middle Ages, the first reported ergotism epidemic was in 944–1000 AD, when about half the population of the Aquitane region of France (about 60 000 people) died of ergot poisoning.[4,5] Other epidemics have occurred in Germany in 1581, 1587 and 1596, largely due to consumption of contaminated rye flour. In 1764, Von Munchhausen finally recognized the causative agent of ergotism as a fungus that parasitizes grain crops. The

RSC Drug Discovery Series No. 50
Privileged Scaffolds in Medicinal Chemistry: Design, Synthesis, Evaluation
Edited by Stefan Bräse

Published by the Royal Society of Chemistry, www.rsc.org

gangrenous form of the disease (*Ergotismus gangraenosus*) was commonly known as "ergotism", "holy fire", "infernal fire" or "St. Anthony's fire". Symptoms include delirium and hallucinations, muscle spasms, convulsions and gangrene of the limbs. Livestock is also subject to similar symptoms upon poisoning by EA. Ergotism was associated with the Salem Witch Trials and the Great Fear of the French Revolution.[1,4] At the end of the 17th century, people finally associated ergotism with the consumption of infected rye and general awareness and knowledge reduced these mass poisonings. The early medicinal uses of ergots were documented first in 1582, for a "quickening childbirth". However, after the number of stillborn neonates increased, the Medical Society of New York initiated an investigation, which resulted in a reduction of the use of ergots only to control postpartum haemorrhage. The history of medicinal applications of ergot alkaloids is very rich, due to their high biological activity (see Section 14.7) and, undoubtedly, further applications remain to be discovered.

14.2 Ergot Alkaloid Classes

All naturally occurring ergot alkaloids share a common tetracyclic scaffold, the so-called "ergoline scaffold" (Figure 14.1A). EA are divided into three major classes based on the substituents that decorate this scaffold: clavines (festuclavine and agroclavine derived), simple lysergic acid derivatives and ergopeptides.[5,6] The clavines include partially or fully saturated ring species D such as agroclavine **1** or festuclavine **2** (Figure 14.1B). Simple lysergic acid derivatives consist of the basic D-lysergic acid structure as an alkyl amide (Figure 14.1C) and ergopeptides based also on D-lysergic acid and a cyclic tripeptide moiety (Figure 14.1D).

14.3 Production of Ergot Alkaloids in Nature

Ergot alkaloids are produced by fungi belonging to the family *Clavicipitaceae*; *Claviceps purpurea* and *Neotyphodium lolii* from the order *Eurotiales*, which are parasitic or mutualistic plant symbionts, are well-known examples. Another known EA producer, *Aspergillus fumigatus* from the order *Eurotiales*, is an opportunistic pathogen of mammals.[5,7–9] Notably, this diverse group of fungi produces surprisingly similar alkaloid profiles. Derivatives of lysergic acid and ergopeptides (Figure 14.1, C, D) are produced by *Clavicipitaceous* fungi *Claviceps purpurea* and *Neotyphodium lolii*, and are believed to protect the fungi from predation by mammals and insects. Clavine-type ergot alkaloids (Figure 14.1B) are produced by *Aspergillus fumigatus* and *Aspergillus japonicus* during conidiation.[10,11] However, the biological role of EA in the survival of conidia during invasive aspergillosis is not completely understood.[12] Recently, the *Arthrodermataceae* family of fungi has been studied in the context of EA production.[13] It was demonstrated that *Arthroderma benhamiae* produces chanoclavine-I aldehyde **14** (Figure 14.2), the common biosynthetic intermediate of all EA biosynthesis. Finally,

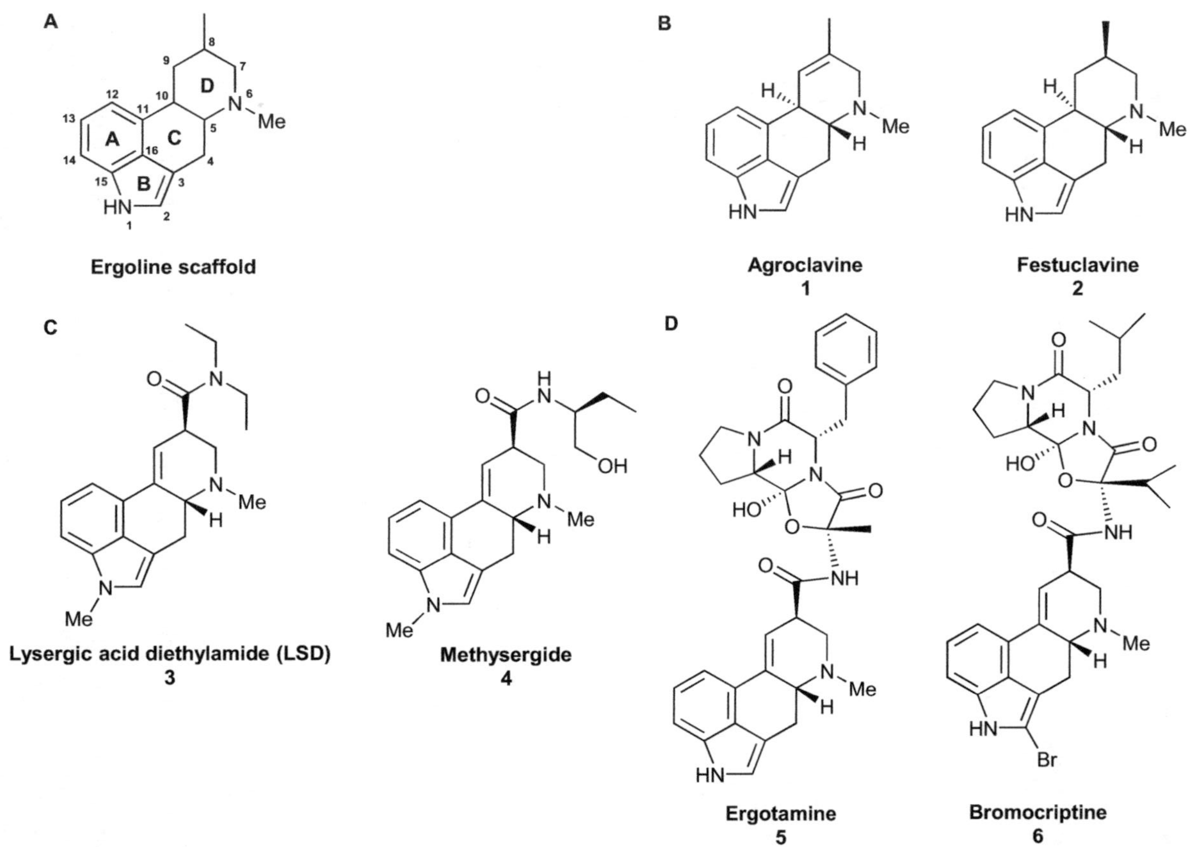

Figure 14.1 **A.** Tetracyclic ergoline ring structure with conventional numbering and lettering. **B.** Examples of clavines. **C.** Simple lysergic acid derivatives. **D.** Ergopeptides consist of D-lysergic acid with a cyclic tripeptide moiety.

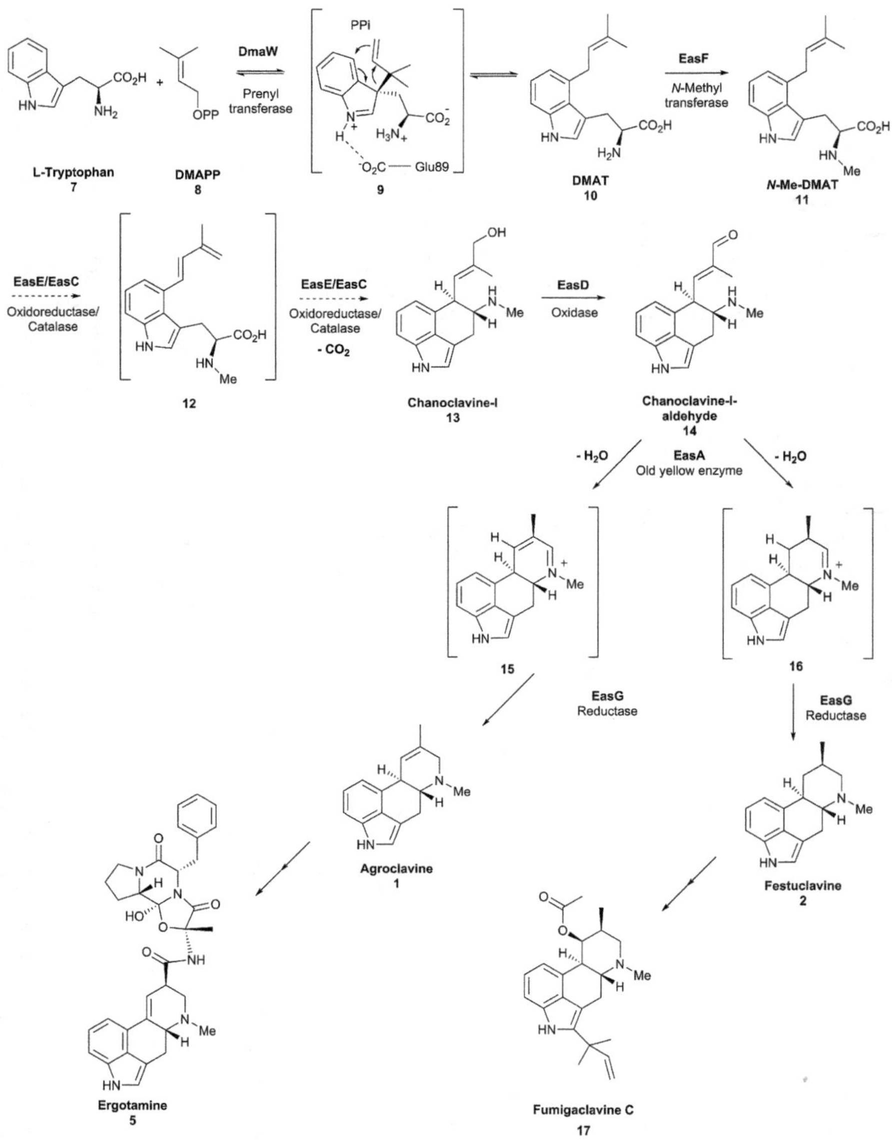

Figure 14.2 Proposed scheme of ergot alkaloids biosynthetic pathway.

Pleurobranchus forskalii, a species of marine gastropod mollusc is responsible for production of an ergot alkaloid peptide ergosinine, indicating that ergots may also be produced in marine organisms.[14]

Ergot alkaloids have also been found in plant taxa *Convolvulaceae* (Solanales), which is known to be associated with *Clavicipitaceous* fungi.[15,16] Recently, one of the unresolved questions why these alkaloids are present in such diverse taxa as the fungal *Clavicipitaceae* and a higher plant family as *Convolvulaceae*, has been answered. While horizontal gene transfer or a repeated origin of the EA biosynthetic pathway has been proposed,[17] recent work has shown that the morning glory family (*Convolvulaceae*) is colonized by an ergot alkaloid-producing *Clavicipitaceous* fungus and is seed-transmitted.[15,18] It has been demonstrated that treatment of the colonized host leaves with fungicides led to elimination of leaf-associated fungus and concomitant loss of alkaloids from the plant.[19] It turned out that these endophytic fungi live in a mutualistic symbiosis with plants and cause no symptoms of infection. The defensive mutualism relies on production of these alkaloids by plants for a protection from herbivores. In turn, fungi benefit from being in a protected position and receiving nutrition from the plant. Therefore, the ecological role of ergot alkaloids is to support environmental tolerance of plants, their fitness, resistance from drought and feeding deterrence from mammals and insects.[5,17,20–26] It has been proven that the fungal symbionts are vertically transmitted through the seed of the narrow range of the host plant.[27] However, the mechanism of how the fungi spread in the host plant remains cryptic. There are no signs of penetration of the plant epidermis by an epibiotic fungus. It has been observed that fungal hyphae are in close contact with the oil secretory glands of the plant cuticle, which may play a major role in the metabolic interaction fungus–host plants.[28]

14.4 Biosynthesis of Ergot Alkaloids

14.4.1 Biosynthetic Pathway

The EA biosynthetic pathway was initially investigated through feeding of isotopically labelled substrates to cultures of *C. purpurea*,[17] which led to a proposed biosynthetic pathway for these alkaloids (Figure 14.2). In the first step of EA biosynthesis L-tryptophan **7** is prenylated by dimethylallyl pyrophosphate (DMAPP) **8**, to yield 4-(γ, γ-dimethylallyl)tryptophan (DMAT) **10**.[29,30] In the next step DMAT **10** is *N*-methylated to yield 4-dimethyl-L-abrine (*N*-Me-DMAT) **11**.[31] *N*-Me-DMAT **11** is in turn converted into chanoclavine-I **13** through series of successive oxidation steps that catalyze the intramolecular cyclization of the prenyl and indole moieties to form ring C (Figure 14.2).[32–36] Subsequently, chanoclavine-I **13** is oxidized to form chanoclavine-I-aldehyde (**14**), which is the last common precursor of all classes of ergot alkaloids. At this crucial branch point, chanoclavine-I-aldehyde **14** can undergo intramolecular cyclization to form either fully saturated ring D of tetracyclic festuclavine **2** (*A. Fumigatus*) or the unsaturated ring D of agroclavine **1** (*C. purpurea, N. lolii*). Festuclavine **2** and argoclavine **1** then branch into lysergic acid amides/peptides and fumigaclavine type EA, respectively, as described in Section 14.4.4 (Figure 14.2).

14.4.2 Gene Clusters

Ergot alkaloid biosynthetic genes have been demonstrated to be clustered on the genome of *A. fumigatus*[9] (Figure 14.3A) and *Clavicipitaceous* fungi

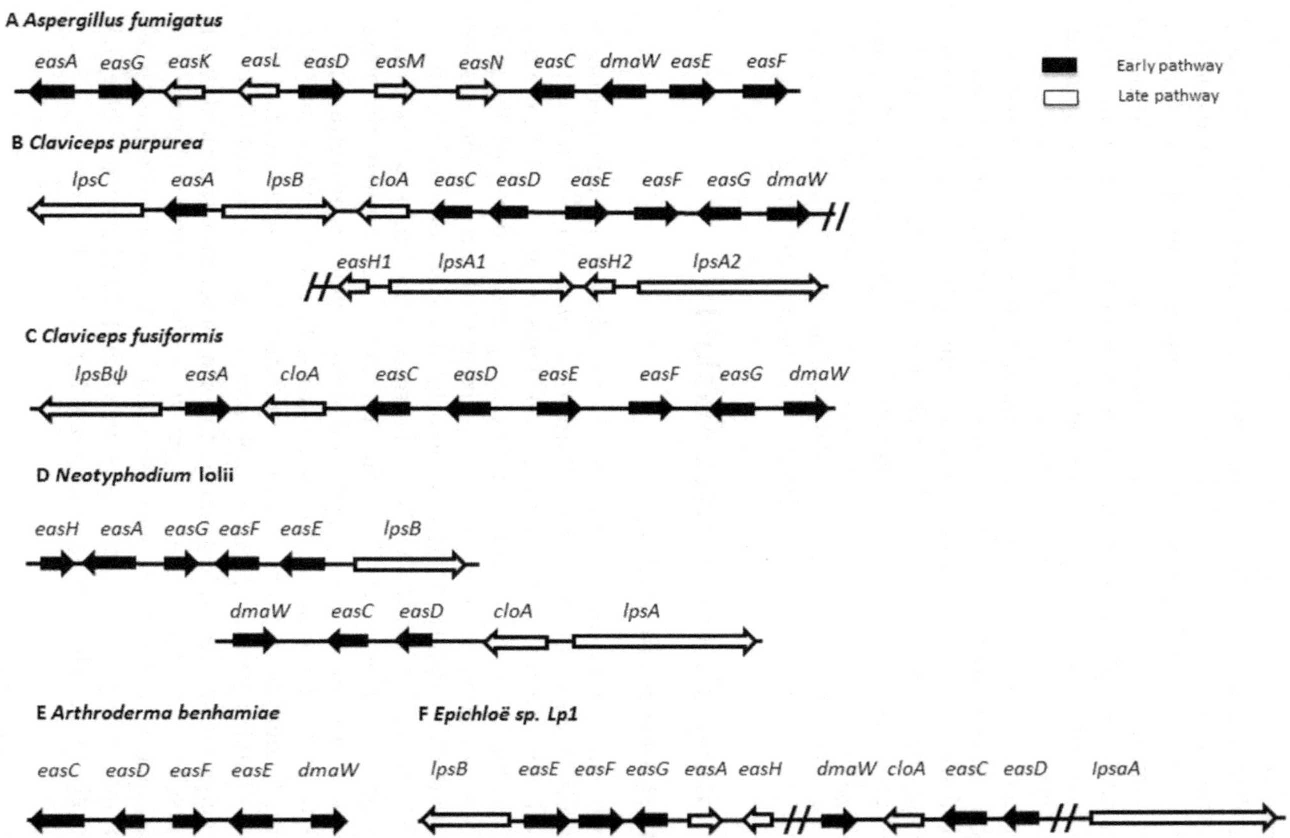

Figure 14.3 Representative ergot alkaloid gene clusters. **A.** *A. fumigatus*. **B.** *C. purpurea*. **C.** *C. fusiformis*. **D.** *N. lolli*. **E.** *A. benhamiae*. **F.** *Epichloë* sp. *Lp1*.

C. purpurea[37,38] (Figure 14.3B), *C. fusiformis*[39] (Figure 14.3C), *N. lolii*[40] (Figure 14.3D), *Arthroderma benhamiae*[13] (Figure 14.3E) and *Epichloë* sp.[41] Homologues common among these species participate in the early steps of ergot biosynthesis. Species-unique genes are most likely responsible for further downstream modifications that result in the production of specific ergot alkaloid classes unique to each individual species, as discussed further in Section 14.4.4 (Figure 14.3). Tsai *et al.* have successfully identified and cloned the gene coding for L-tryptophan dimethylallyl prenyl transferase (DmaW) from *C. purpurea.*[42] This discovery allowed the identification of the ergotamine biosynthesis cluster (68.5 kb) from *C. purpurea* – the first ergot gene cluster to be discovered – *via* chromosome walking (Figure 14.3).[37] This gene cluster included open reading frames encoding non-ribosomal peptide synthetase (NRPS) modules (Lps1 and Lps2) that would be expected to be involved with the later biosynthetic pathway formation of ergopeptides.[43–45] Moreover, it was observed that comparison of cluster sequences within *C. purpurea* strain P1 (ergotamine producer) with strain *C. purpurea* ECC93 (ergocristine producer) displayed conservation of most genes associated with formation of the ergoline ring, yet displayed high variation in genes associated with the NRPS production of the peptide ergot moiety.

A recent review by Schardl *et al.* compares ergot alkaloid profiles, biosynthetic genes and genomic arrangements of those genes among 15 Clavicipitaceae.[2,46] The dramatic differences in ergot alkaloid profiles are caused by the presence of specific mid-pathway or late-pathway genes, as well as differences in substrate or product specificity due to gene sequence variations. The authors correlated chemotypes of *Claviceps* species with the presence or absence of the genes *lpsA, lpsB, lpsC, easH, easO and easP.* This comprehensive work exhibits association of particular fungi with particular metabolites, which in turn reveals evolutionary changes in this pathway. A gene cluster for EA biosynthesis that was subsequently found in *Neotyphodium* sp. *Lp1* (a natural hybrid *Neotyphodium lolii*×*Epichloe typhina*), studied by Panaccione *et al.*,[47] allowed the experimental validation that disruption of the NRPS Lps1 homologue (*LpsA*) involved in ergopeptide biosynthesis causes the loss of downstream alkaloid ergovaline. Fleetwood *et al.* later identified part of the ergot alkaloid cluster for ergovaline biosynthesis (~19 kb) in *N. lolli* using both chromosome walking and southern blot (Figure 14.3D).[40] It was unambiguously demonstrated that the LpsB gene in *N. lolli*, a homologue of the *C. purpurea Lps2*, was associated with ergovaline production.[40]

The discovery of *A. fumigatus* biosynthetic gene cluster (22 kb) was facilitated by the published genome sequence of *A. fumigatus*, which was associated with the production of fumigaclavines A, B, C, (**21**, **20**, **17**, respectively) and festuclavine **2.**[9] Further analysis of gene function in this cluster led to the characterization of easF and easD gene products, which are responsible for catalysing early steps in the ergot pathway.[48,49] A recent survey of various isolates of *A. fumigatus* were shown to have variable production of ergot alkaloids, which could be linked to changes in the ergot gene cluster.[50]

Genome sequence analysis of fungi of the *Arthrodermataceae* revealed the presence of a gene cluster with high sequence similarity to those involved in the early common steps of ergot alkaloid biosynthesis in *Aspergillus fumigatus* and *Claviceps purpurea.*[13] However, this system has not been studied in depth.

14.4.3 Early Ergot Alkaloid Biosynthetic Enzymes

The enzymology of EA biosynthesis is fascinating, and a number of gene products from these ergot alkaloid biosynthetic clusters have been biochemically characterized *in vitro*. The first step of the EA biosynthetic pathway is catalyzed by dimethylallyl prenyltransferase (DmaW)[51] from cultures of ergot alkaloid producing *C. fusiformis.*[29,30] DmaW prenylates L-tryptophan *via* an electrophilic aromatic substitution reaction.[30,52] Recent work suggests the mechanism involves substitution on C-3 (instead of substitution on weakly nucleophilic C-4, as previously suggested) followed by a Cope rearrangement (Figure 14.2).[36] Furthermore, two lysine amino acids have been implicated in the mechanism.[36] DmaW homologues from *A. fumigatus*, *C. purpurea* and *N. lolli* have also been characterized.[42,53,54] The structure of this enzyme has been solved recently, which facilitates an understanding of this enzyme's specificity for the substrate and regioselectivity.[55] Recently, it was demonstrated that alternate substrates, 4-methyltryptophan, 4-methoxytryptophan and 4-aminotryptophan, can also be prenylated by DmaW.[56]

The next enzyme in the early pathway, EasF belongs to the *N*-methyltransferases enzyme family and is responsible for *N*-methylation of DMAT **10**. EasF was first purified by Otsuka *et al.* from cell free cultures of *C. purpurea.*[31] This enzyme methylates the amine nitrogen of dimethylallyl tryptophan using the *S*-adenosyl methionine (SAM) co-factor. After the identification of the ergot biosynthetic gene cluster in *A. fumigatus*, the *easF* gene was successfully cloned and heterologously expressed. The expressed EasF could also methylate DMAT **10** to yield *N*-Me-DMAT **11** (dimethylallyl L-abrine).[48] Following methylation by EasF, two oxidation reactions are proposed to transform *N*-Me-DMAT **11** to chanoclavine-I **13**, thus forming ergoline ring C. Kozikowski *et al.* predicted these two oxidation steps of the pathway based on isotopic feeding studies.[34] They observed that a proposed diene intermediate **12** was incorporated into downstream ergot alkaloids upon feeding to *C. purpurea*[34] and that oxygen from hydroxyl group of chanoclavine-I **13** was incorporated from molecular oxygen.[33] Enzyme candidates for carrying out oxidation reactions were proposed to be EasC and EasE. These enzymes display protein sequence similarity to catalases and FAD oxygenases, respectively. The role of the EasE and EasC in the oxidations of *N*-Me-DMAT **11** to chanoclavine-I **13** in *C. purpurea* has also been demonstrated by gene disruption experiments.[57] The disruption of *easE* and *easC* genes in *A. fumigatus* indicate that EasC and EasE are both required for ring C formation.[58] Heterologous expression of EasC revealed a catalase activity of this protein.[58] However, demonstration of *in vitro* activity for EasE has remained elusive.

EasD is an NAD^+ binding oxidase that is responsible for the oxidation of the hydroxyl group of chanoclavine-I **13** to carbonyl group of chanoclavine-I-aldehyde **14**. EasD was initially cloned and characterized from *A. fumigatus* by Wallwey *et al.*[59] An *easD* homologue from *Arthroderma benhamiae* was heterologously expressed and also oxidized chanoclavine-I **13** in the presence of NAD^+ to form chanoclavine-I aldehyde **14**.[13]

The formation of ergoline ring D involves two enzymes, EasA and EasG, which are responsible for the cyclization of chanoclavine-I-aldehyde **14** into either festuclavine **2** (*A. fumigatus*) or agroclavine **1** derived alkaloids (*C. purpurea/N. lolii*). Homologues of EasA in the ergot cluster show protein sequence similarity to enzymes of the Old Yellow Enzyme (OYE) family. OYEs are responsible for the reduction of alpha beta unsaturated ketones and aldehydes,[60] which initially suggested that these enzymes would be capable of reducing the alpha beta unsaturated carbons of chanoclavine-I-aldehyde (**14**) to give the cyclized iminium intermediates **15, 16** in ring D formation (Figure 14.2).[49,61,62]

A crucial difference between the ergot alkaloid classes is the fully saturated D ring of the clavine type alkaloids compared to the unsaturated ergoline D ring of the ergotamine type alkaloids (Figure 14.1A).[63] As opposed to the EasA homologue from *A. fumigatus*, which forms festuclavine **2**, an EasA from *N. lolli* is involved in production of agroclavine **1** and acts as an isomerase.[64] A mutant of EasA which is capable of producing both festuclavine **2** and agroclavine **1** products confirms this critical branch point in ergot alkaloids biosynyhesis.[64]

The EasG protein encoded by the cluster displays similarity to Rossman fold NADPH reductases and its function is to reduce the proposed cyclized iminium products **15, 16** of EasA to form festuclavine **2** (*A. fumigatus*) or agroclavine **1** (*C. purpurea/N. lolii*).[64,65]

14.4.4 Late Ergot Alkaloid Biosynthetic Enzymes

The early pathway of ergoline ring biosynthesis is defined by agroclavine **1** or festuclavine **2** production. Late pathway enzymes, which vary among different fungi species, are associated with biosynthesis of diverse alkaloid profiles. Transformation of agroclavine **1** into ergopeptides is observed in the *Clavicipitaceous* fungi *C. purpurea* and *N. lolli*. The late-pathway biosynthetic genes in these organisms encode non-ribosomal peptide synthase (NRPS) domains. These genes have been studied by both gene disruption and *in vitro* characterization (Figure 14.4). It has been demonstrated that ergopeptide formation occurs *via* an enzyme complex composed of NRPS subunits D-lysergyl peptidyl synthetase (Lps2) that activates lysergic acid, and (Lps1) that forms the tripeptide moiety.[43–45,47,66–70] The enzyme CloA was also demonstrated to be crucial for the oxidation of elymoclavine **18** to paspalic acid **19**. This enzyme oxidizes paspalic acid **19**, which in turn forms

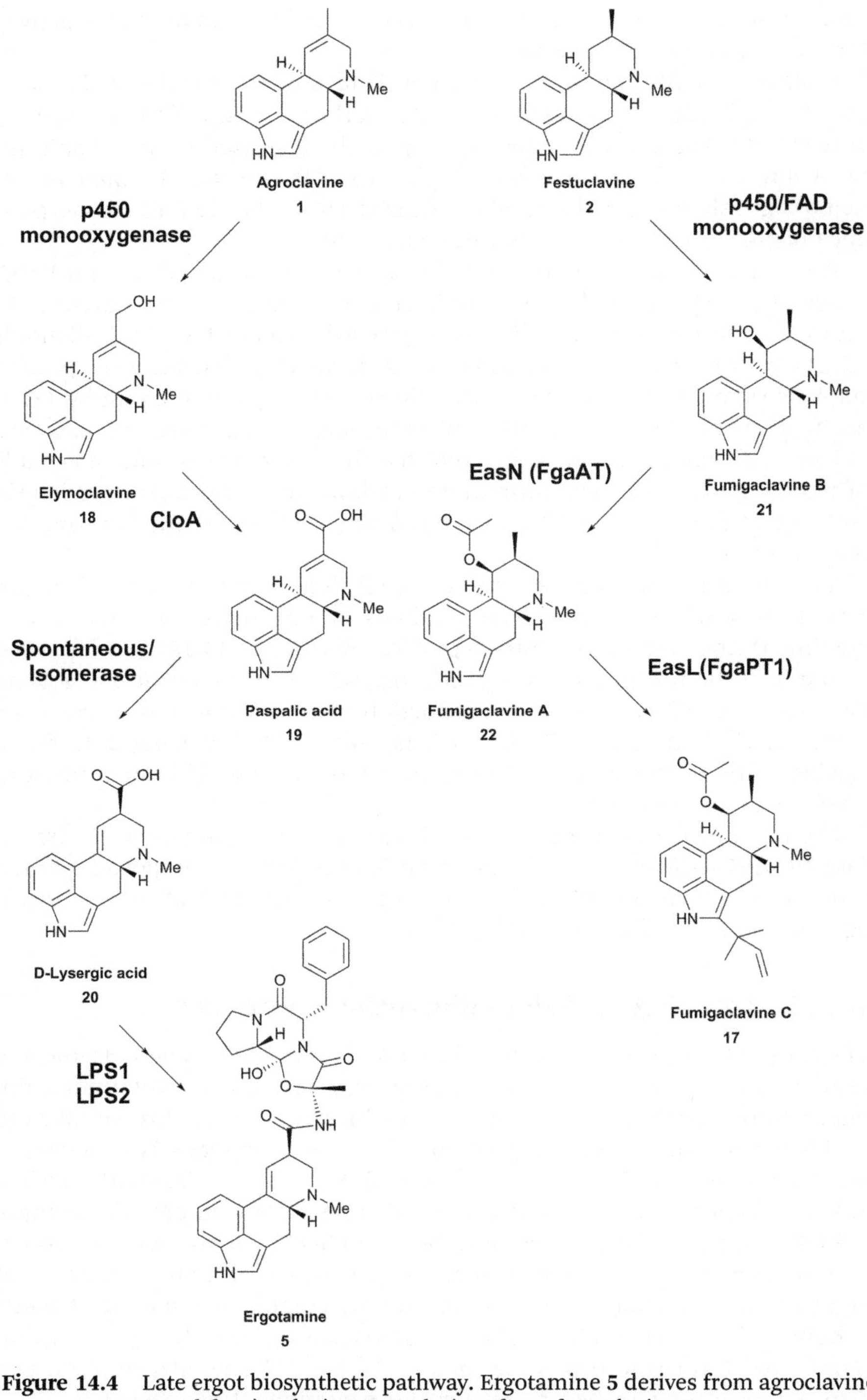

Figure 14.4 Late ergot biosynthetic pathway. Ergotamine **5** derives from agroclavine **1** and fumigaclavine C **17** derives from festuclavine **2**.

lysergic acid **20** either spontaneously or *via* an isomerase rearrangement (Figure 14.4).[71] Recently, *easH* from *C. purpurea* was heterologously expressed and characterized by Havemann *et al.* This enzyme is annotated as a nonheme-iron dioxygenase, which cyclizes dihydrolysergyl-Ala-Phe-Pro-lactam to dihydroergotamine.[72]

In contrast, conversion of festuclavine **2** into fumigaclavine A **22**, B **21** and C **17**, is carried out by the *A. fumigatus* biosynthetic gene cluster. The late ergot pathway genes of this cluster have been demonstrated to show acetylation and reverse prenyl transferase activities.[73,74] *A. fumigatus* does not appear to harbour any genes that encode NRPS domains such as the ones observed in ergot biosynthetic clusters of *N. lolii* and *C. purpurea* (Figure 14.4). However, the nonribosomal peptide synthetases PesL and Pes1, previously believed to be involved in biosynthesis of fungal quinazoline derived natural products, have been shown to be essential for fumigaclavine C (**17**) biosynthesis in *A. fumigatus* by gene deletion experiments.[75] Notably, these synthetases are not found in the core ergot cluster. *A. fumigatus* also produces fumitremorgin B, which requires an additional *N*-prenylation step in addition to the one catalyzed by DmaW.[76]

14.5 Production of Ergot Alkaloids *De novo*

Production of ergot alkaloids in *A. fumigatus* is limited to conidiating cultures.[77] Cultures typically accumulate several pathway intermediates, with most of the alkaloid content associated with the fungal colonies, and are not exported to the media. Therefore, the native hosts are not always amenable for large-scale production of these compounds. A two-stage culture process including shake culture and static culture was shown to increase the production of fumigaclavine C (**17**) to 60 mg L^{-1}.[78,79] Hulvova *et al.* have recently described the challenges and progress in the use of *Claviceps* as a source for biotechnological production of ergot alkaloids.[80] Very recently, heterologous reconstitution of biosynthetic pathways reveals another option for expression of the ergot alkaloids.

Genes of the early steps of this pathway – *dmaW, easF, easE, easC* – have been reconstituted in *Aspergillus nidulans* (a non-producer of ergots)[81] and in *Saccharomyces cerevisea*, resulting in *de novo* production of the chanoclavine-I **13**.[11] Finally, a recent review of Wallwey *et al.* highlights the methods of production, detection and purification of clavine-type ergot alkaloids.[82]

14.6 Chemical Synthesis of Ergot Alkaloids

Ergot alkaloids are also highly interesting and challenging targets for the organic chemistry community. A number of synthetic studies relied on a stepwise approach for construction of the C/D ring system, particularly of lysergic acid **20** – a crucial intermediate in the EA biosynthetic pathway.[83–90] Oppolzer and co-workers reported another strategy based on simultaneous construction of the C and D rings *via* an intramolecular imino-Diels-Alder

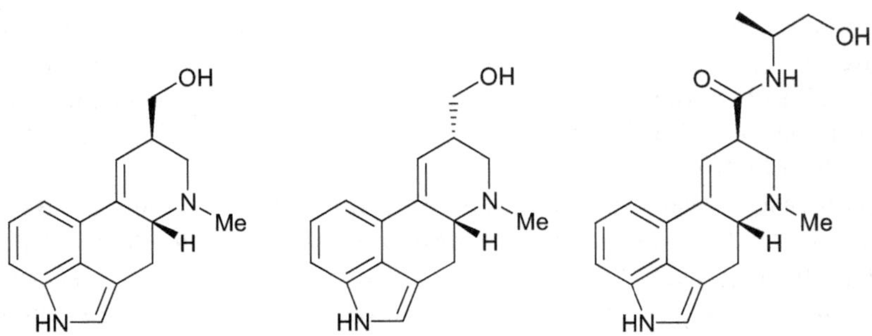

Lysergol **23** **Isolysergol** **24** **Ergonovine** **25**

Figure 14.5 Derivatives of lysergic acid: Lysergol (**23**), isolysergol (**24**) and ergonovine (**25**).

reaction.[85] Recently, Kalinin *et al.* have reported C/D ring synthesis by intramolecular Heck and ring-closing metathesis reactions.[91] The total synthesis of lysergic acid **20** and its derivatives, lysergol **23**, isolysergol **24** and ergonovine **25** has been reported (Figure 14.5).[84,92,93] Based on palladium-catalyzed domino cyclization of amino allenes bearing a bromoindolyl group, both racemic[92] and enantioselective[93] approaches to obtain lysergic acid **20**, lysergol **23** and isolysergol **24**, have been accomplished.

Very recently, the synthesis of cycloclavine – an unusual ergot alkaloid containing cyclopropyl ring – has been reported.[94–96] The formal synthesis of (±)-cycloclavine (**27**) was carried out in seven steps and 27% overall yield from the known 2-(4-bromo-1-tosyl-1*H*-indol-3-yl)acetaldehyde (**26**). Key steps include an iron(III)-catalyzed aza-Cope – Mannich cyclization and an intramolecular Heck reaction or a self-terminating 6-*exo-trig* aryl radical–alkene cyclization (Figure 14.6).[95]

The total synthesis of cycloclavine **27** was achieved in 14 steps, with a 1.2% overall yield. The crucial features of this synthesis include rapid construction of the heterocyclic core segments by two Diels-Alder reactions. An indole annulation was achieved by a late-stage intramolecular Diels-Alder furan cycloaddition, and a methylenecyclopropane dienophile was used for a stereoselective intramolecular [4 + 2] cycloaddition to give the cyclopropa[*c*]indoline building block.[94]

An important aspect of the chemical synthesis of these alkaloids is the synthesis of mechanistic probes, including the synthesis of isotope labelled building blocks for feeding experiments. For example, pioneering studies by Floss and co-workers[33–35,97] showed that the origin of oxygen atoms in chanoclavine-I **13** and elymoclavine **18** was molecular oxygen.[33] In addition, the mechanistic basis of ring C formation was investigated using synthetic probes by introducing a tritium label on α-carbon of L-tryptophan **7a**, $[^{13}C^{2}H_3]$methionine and an isotopic label on C-2 of DL-mevalonic acid. These studies also concluded that the oxygen atoms in **13a** and **18a** (Figure 14.7)

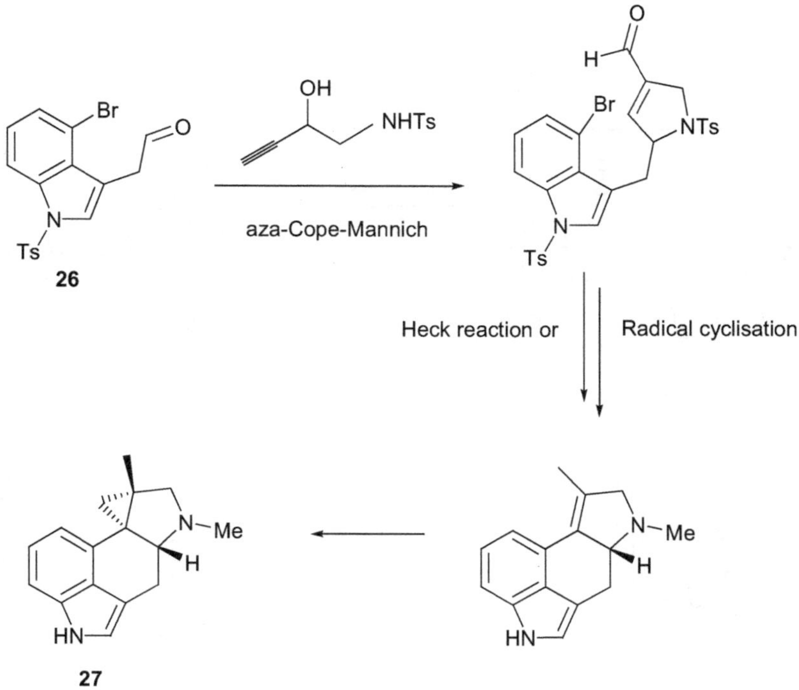

Figure 14.6 Scheme of the formal synthesis of cycloclavine **27** starting from 2-(4-bromo-1-tosyl-1*H*-indol-3-yl)acetaldehyde **26**.

derive from molecular oxygen. Moreover, this work suggested that formation of ring C proceeds *via* carbocation formation at the benzylic position and formation of a carbanion at the α-carbon of the alanine side chain. The tritium label at the α-carbon of L-tryptophan has confirmed the hypothesis that the decarboxylation step has to occur prior to or simultaneous with ring C closure as this α-hydrogen has retained (Figure 14.7).

The chemical synthesis of natural products such as ergot alkaloids is challenging and expensive due to the complex structures of these molecules, which contain multiple stereogenic centres. The yields from total synthesis are relatively low, making biosynthesis and semi-synthesis a promising approach to obtain high yields of these bioactive molecules.

14.7 Application of Ergoline Scaffold in Medicinal Chemistry

Historically, the abusive uses of ergot alkaloids have overshadowed the beneficial medicinal properties of these compounds. The first clinical applications of EA were mentioned in 1100 BC in China, followed by a recurrence in medicinal usage in the early 19th century (see Section 14.1).[3] Ergot-derived medicines were used to facilitate obstetric deliveries or to treat

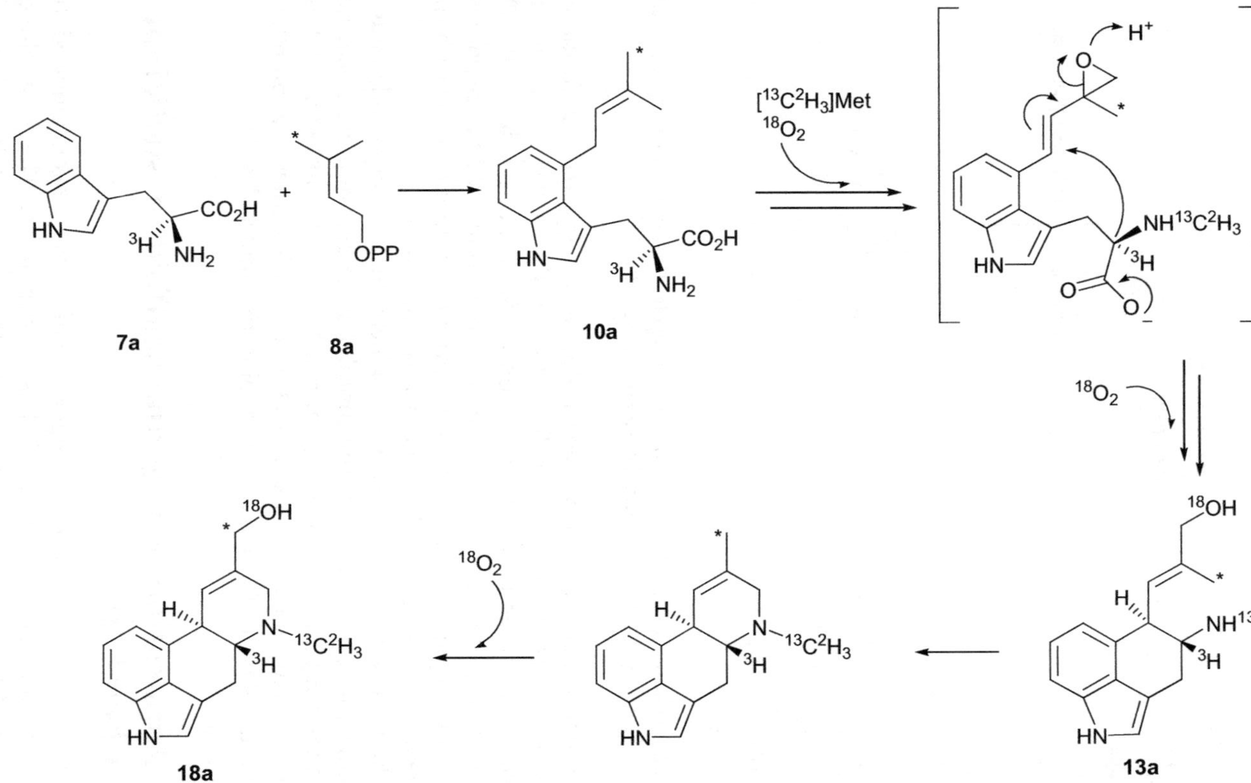

Figure 14.7 Isotopic labelling studies that provide insight into the origin of oxygen atoms of chanoclavine-I **13** and elymoclavine **18**. Additional ^{13}C, ^{3}H, and ^{2}H labelling enabled a mechanistic hypothesis for ring C formation to be proposed.

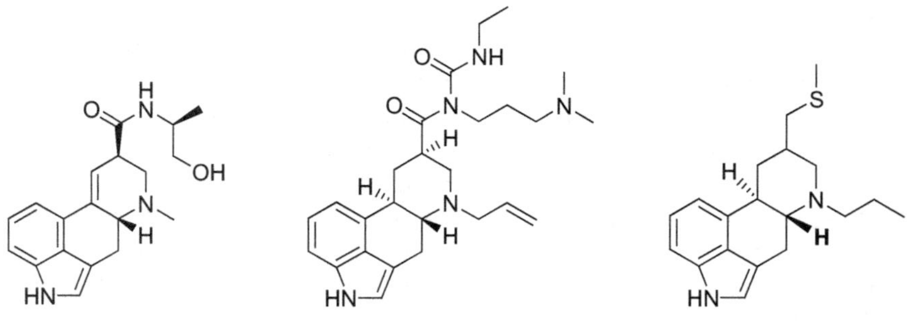

Figure 14.8 Ergot alkaloid inspired pharmaceuticals: Natural – Ergometrine **28**; and Semi-synthetic derivatives: Cabergoline **29** and pergolide **30**.

postpartum haemorrhage (ergometrine). Intensive research on the oxytocic activity of ergots resulted in the synthesis of lysergic acid diethylamide (LSD) **3** in 1938 (Figure 14.1), the most hallucinogenic compound yet discovered.[3] LSD **3** has become infamous for its use as an illicit recreational drug. However, ergot alkaloids are also the inspiration for a wide range of semi-synthetic derivatives that find wide-ranging medicinal application as treatments for migraine (methysergide **4**, ergotamine **5**), parkinsonism (bromocriptine **6**, cabergoline, pergolide), tumour (ergotamine **5**) or restless leg syndrome (cabergoline, pergolide) (Figures 14.1 and 14.8).

The high bioactivity of ergot alkaloids is correlated with the ability of these compounds to act as agonists or antagonists toward neuroreceptors for dopamine, serotonin and adrenaline.[17,98,99] In 2010, the production of these alkaloids was approximately 20 000 kg, of which field cultivation contributed about 50%.[80] Semi-synthetic derivatives of ergot alkaloids aim to tailor their potent bioactivity toward specific receptors, reducing adverse side effects. Therefore, the ergoline scaffold is one of the most important in terms of its application in medicinal chemistry. An ability to harness the biosynthetic pathways of these compounds will only enhance our ability to produce greater numbers of EA analogues that may have new and improved bioactivities.

References

1. U. Steiner, S. Leibner, C. L. Schardl, A. Leuchtmann and E. Leistner, *Mycologia*, 2011, **103**, 1133–1145.
2. C. L. Schardl, C. A. Young, U. Hesse, S. G. Amyotte, K. Andreeva, P. J. Calie, D. J. Fleetwood, D. C. Haws, N. Moore, B. Oeser, D. G. Panaccione, K. K. Schweri, C. R. Voisey, M. L. Farman, J. W. Jaromczyk, B. A. Roe, D. M. O'Sullivan, B. Scott, P. Tudzynski, Z. An, E. G. Arnaoudova, C. T. Bullock, N. D. Charlton, L. Chen, M. Cox, R. D. Dinkins, S. Florea, A. E. Glenn, A. Gordon, U. Guldener,

D. R. Harris, W. Hollin, J. Jaromczyk, R. D. Johnson, A. K. Khan, E. Leistner, A. Leuchtmann, C. Li, J. Liu, J. Liu, M. Liu, W. Mace, C. Machado, P. Nagabhyru, J. Pan, J. Schmid, K. Sugawara, U. Steiner, J. E. Takach, E. Tanaka, J. S. Webb, E. V. Wilson, J. L. Wiseman, R. Yoshida and Z. Zeng, *PLoS Genet.*, 2013, **9**, e1003323.

3. P. L. Schiff, *Am. J. Pharm. Educ.*, 2006, **70**, 1–10.
4. P. M. Dewick, *Medicinal Natural Products*, John Wiley & Sons, Ltd., Chichester, West Sussex, England, 2nd edn, 2002, 368–376.
5. C. L. Schardl, D. G. Panaccione and P. Tudzynski, *Alkaloids Chem. Biol.*, 2006, **63**, 45–86.
6. J. Mukherjee and M. Menge, *Adv. Biochem. Eng./Biotechnol.*, 2000, **68**, 1–20.
7. N. P. Keller, G. Turner and J. W. Bennett, *Nat. Rev. Microbiol.*, 2005, **3**, 937–947.
8. J. L. Brookman and D. W. Denning, *Curr. Opin. Microbiol.*, 2000, **3**, 468–474.
9. C. M. Coyle and D. G. Panaccione, *Appl. Environ. Microbiol.*, 2005, **71**, 3112–3118.
10. L. Pařenicová, P. Skouboe, J. Frisvad, R. A. Samson, L. Rossen, M. ten Hoor-Suykerbuyk and J. Visser, *Appl. Environ. Microbiol.*, 2001, **67**, 521–527.
11. C. Nielsen, C. Folly, A. Hatsch, A. Molt, H. Schroder, O. C. SE and M. Naesby, *Microb. Cell Fact.*, 2014, **13**, 1–11.
12. C. M. Coyle, S. C. Kenaley, W. R. Rittenour and D. G. Panaccione, *Mycologia*, 2007, **99**, 804–811.
13. C. Wallwey, C. Heddergott, X. Xie, A. A. Brakhage and S. M. Li, *Microbiology*, 2012, **158**, 1634–1644.
14. T. Wakimoto, K. C. Tan and I. Abe, *Toxicon*, 2013, **72**, 1–4.
15. M. A. Ahimsa-Muller, A. Markert, S. Hellwig, V. Knoop, U. Steiner, C. Drewke and E. Leistner, *J. Nat. Prod.*, 2007, **70**, 1955–1960.
16. A. Hofmann, *Planta Med.*, 1961, **9**, 354–367.
17. D. Groger and H. G. Floss, *Alkaloids (Academic Press)*, 1998, **50**, 171–218.
18. U. Steiner, M. A. Ahimsa-Muller, A. Markert, S. Kucht, J. Gross, N. Kauf, M. Kuzma, M. Zych, M. Lamshoft, M. Furmanowa, V. Knoop, C. Drewke and E. Leistner, *Planta*, 2006, **224**, 533–544.
19. S. Kucht, J. Gross, Y. Hussein, T. Grothe, U. Keller, S. Basar, W. A. Konig, U. Steiner and E. Leistner, *Planta*, 2004, **219**, 619–625.
20. H. A. Schultes and R. Evans, *Plants of the Gods: Their Sacred, Healing and Hallucinogenic Powers*, Healing Arts Press, Rochester, 1992.
21. J. F. White Jr, C. W. Bacon, N. L. Hywel-Jones and J. W. Spatafora, *Clavicipitalean Fungi, Evolutionary Biology, Chemistry, Biocontrol, and Cultural Impacts*, Marcel Dekker, New York, 2003.
22. C. Boyle, B. Schulz and T. Sieber, *Microbial Root Endophytes*, Springer, Berlin, Heidelberg, New York, 2006.
23. A. Varma, L. Abbott, D. Werner and R. Hampp, *Plant Surface Microbiology*, Springer, Berlin, Heidelberg, 2008.

24. P. Tudzynski and J. Scheffer, *Mol. Plant Pathol.*, 2004, **5**, 377–388.
25. C. W. Bacon and P. Lyons, Ecological fitness factors for fungi within the Balansieae and Clavicipitaceae, in *The Fungal Community, Its Organisation and Role in the Ecosystem*, ed. J. Dighton, J. F. White and P. Oudemans, CRC Taylor & Francis, Boca Raton, FL, 3rd edn, pp. 519–532.
26. C. Schardl and A. Leuchtmann, The Epichloe endophytes of grasses and the symbiotic continuum, in *The Fungal Community, its Organization and Role in the Ecosystem*, ed. J. Dighton, J. F. White and P. Oudemans, CRC Taylor and Francis, Boca Raton, 2005, pp. 475–503.
27. D. G. Panaccione, W. T. Beaulieu and D. Cook, *Funct. Ecol.*, 2014, **28**, 299–314.
28. U. Steiner and E. Leistner, *Fungal Ecol.*, 2012, **5**, 316–321.
29. J. C. Gebler and C. D. Poulter, *Arch. Biochem. Biophys.*, 1992, **296**, 308–313.
30. J. C. Gebler, A. B. Woodside and C. D. Poulter, *J. Am. Chem. Soc.*, 1992, **114**, 7354–7360.
31. H. Otsuka, F. R. Quigley, D. Groeger, J. A. Anderson and H. G. Floss, *Planta Med.*, 1980, **40**, 109–119.
32. H. G. Floss, M. Tcheng-Lin, C.-J. Chang, B. Naidoo, G. E. Blair, C. I. Abou-Chaar and J. M. Cassady, *J. Am. Chem. Soc.*, 1974, **96**, 1898–1909.
33. M. Kobayashi and H. G. Floss, *J. Org. Chem.*, 1987, **52**, 4350–4352.
34. A. P. Kozikowski, C. Chen, J. P. Wu, M. Shibuya, C. G. Kim and H. G. Floss, *J. Am. Chem. Soc.*, 1993, **115**, 2482–2488.
35. A. P. Kozikowski, J. P. Wu, M. Shibuya and H. G. Floss, *J. Am. Chem. Soc.*, 1988, **110**, 1970–1971.
36. L. Y. Luk, Q. Qian and M. E. Tanner, *J. Am. Chem. Soc.*, 2011, **133**, 12342–12345.
37. T. Haarmann, C. Machado, Y. Lubbe, T. Correia, C. L. Schardl, D. G. Panaccione and P. Tudzynski, *Phytochemistry*, 2005, **66**, 1312–1320.
38. P. Tudzynski, K. Holter, T. Correia, C. Arntz, N. Grammel and U. Keller, *Mol. Gen. Genet.*, 1999, **261**, 133–141.
39. N. Lorenz, E. V. Wilson, C. Machado, C. L. Schardl and P. Tudzynski, *Appl. Environ. Microbiol.*, 2007, **73**, 7185–7191.
40. D. J. Fleetwood, B. Scott, G. A. Lane, A. Tanaka and R. D. Johnson, *Appl. Environ. Microbiol.*, 2007, **73**, 2571–2579.
41. S. L. Robinson and D. G. Panaccione, *Appl. Environ. Microbiol.*, 2014, **90**, 6465–6472.
42. H.-F. Tsai, H. Wang, J. C. Gebler, C. D. Poulter and C. L. Schardl, *Biochem. Biophys. Res. Commun.*, 1995, **216**, 119–125.
43. T. Correia, N. Grammel, I. Ortel, U. Keller and P. Tudzynski, *Chem. Biol.*, 2003, **10**, 1281–1292.
44. B. Riederer, M. Han and U. Keller, *J. Biol. Chem.*, 1996, **271**, 27524–27530.
45. B. Walzel, B. Riederer and U. Keller, *Chem. Biol.*, 1997, **4**, 223–230.
46. C. L. Schardl, S. Florea, J. Pan, P. Nagabhyru, S. Bec and P. J. Calie, *Curr. Opin. Plant Biol.*, 2013, **16**, 480–488.

47. D. G. Panaccione, R. D. Johnson, J. Wang, C. A. Young, P. Damrongkool, B. Scott and C. L. Schardl, *Proc. Natl. Acad. Sci. U. S. A.*, 2001, **98**, 12820–12825.
48. O. Rigbers and S. M. Li, *J. Biol. Chem.*, 2008, **283**, 26859–26868.
49. C. Wallwey, M. Matuschek, X. L. Xie and S. M. Li, *Org. Biomol. Chem.*, 2010, **8**, 3500–3508.
50. S. L. Robinson and D. G. Panaccione, *Mycologia*, 2012, **104**, 804–812.
51. S.-L. Lee, H. G. Floss and P. Heinstein, *Arch. Biochem. Biophys.*, 1976, **177**, 84–94.
52. L. Y. P. Luk and M. E. Tanner, *J. Am. Chem. Soc.*, 2009, **131**, 13932–13933.
53. J. Wang, C. Machado, D. G. Panaccione, H.-F. Tsai and C. L. Schardl, *Fungal Genet. Biol.*, 2004, **41**, 189–198.
54. I. A. Unsold and S. M. Li, *Microbiology*, 2005, **151**, 1499–1505.
55. U. Metzger, C. Schall, G. Zocher, I. Unsold, E. Stec, S. M. Li, L. Heide and T. Stehle, *Proc. Natl. Acad. Sci. U. S. A.*, 2009, **106**, 14309–14314.
56. J. D. Rudolf, H. Wang and C. D. Poulter, *J. Am. Chem. Soc.*, 2013, **135**, 1895–1902.
57. N. Lorenz, J. Olsovska, M. Sulc and P. Tudzynski, *Appl. Environ. Microbiol.*, 2010, **76**, 1822–1830.
58. K. E. Goetz, C. M. Coyle, J. Z. Cheng, S. E. O'Connor and D. G. Panaccione, *Curr. Genet.*, 2011, **57**, 201–211.
59. C. Wallwey, M. Matuschek and S. M. Li, *Arch. Microbiol.*, 2010, **192**, 127–134.
60. H. S. Toogood, J. M. Gardiner and N. S. Scrutton, *ChemCatChem*, 2010, **2**, 892–914.
61. J. Z. Cheng, C. M. Coyle, D. G. Panaccione and S. E. O'Connor, *J. Am. Chem. Soc.*, 2010, **132**, 1776–1777.
62. X. Xie, C. Wallwey, M. Matuschek, K. Steinbach and S. M. Li, *Magn. Reson. Chem.*, 2011, **49**, 678–681.
63. C. M. Coyle, J. Z. Cheng, S. E. O'Connor and D. G. Panaccione, *Appl. Environ. Microbiol.*, 2010, **76**, 3898–3903.
64. J. Z. Cheng, C. M. Coyle, D. G. Panaccione and S. E. O'Connor, *J. Am. Chem. Soc.*, 2010, **132**, 12835–12837.
65. M. Matuschek, C. Wallwey, X. Xie and S. M. Li, *Org. Biomol. Chem.*, 2011, **9**, 4328–4335.
66. D. G. Panaccione, B. A. Tapper, G. A. Lane, E. Davies and K. Fraser, *J. Agric. Food Chem.*, 2003, **51**, 6429–6437.
67. S. L. Annis and D. G. Panaccione, *Can. J. Microbiol.*, 1998, **44**, 80–86.
68. P. Damrongkool, A. B. Sedlock, C. A. Young, R. D. Johnson, K. E. Goetz, B. Scott, C. L. Schardl and D. G. Panaccione, *DNA Sequence*, 2005, **16**, 379–385.
69. T. Haarmann, N. Lorenz and P. Tudzynski, *Fungal Genet. Biol.*, 2008, **45**, 35–44.
70. I. Ortel and U. Keller, *J. Biol. Chem.*, 2009, **284**, 6650–6660.
71. T. Haarmann, I. Ortel, P. Tudzynski and U. Keller, *ChemBioChem*, 2006, **7**, 645–652.

72. J. Havemann, D. Vogel, B. Loll and U. Keller, *Chem. Biol.*, 2014, **21**, 146–155.
73. X. Liu, L. Wang, N. Steffan, W.-B. Yin and S.-M. Li, *ChemBioChem*, 2009, **10**, 2325–2328.
74. I. A. Unsold and S. M. Li, *ChemBioChem*, 2006, **7**, 158–164.
75. K. A. O'Hanlon, L. Gallagher, M. Schrettl, C. Jochl, K. Kavanagh, T. O. Larsen and S. Doyle, *Appl. Environ. Microbiol.*, 2012, **78**, 3166–3176.
76. A. Grundmann, T. Kuznetsova, S. Afiyatullov and S. M. Li, *ChemBioChem*, 2008, **9**, 2059–2063.
77. P. Mulinti, N. A. Allen, C. M. Coyle, F. N. Gravelat, D. C. Sheppard and D. G. Panaccione, *Curr. Microbiol.*, 2014, **68**, 1–5.
78. L.-Y. Yao, Y.-X. Zhu, R.-H. Jiao, Y.-H. Lu and R.-X. Tan, *Bioresour. Technol.*, 2014, **159**, 112–117.
79. Y.-X. Zhu, L.-Y. Yao, R.-H. Jiao, Y.-H. Lu and R.-X. Tan, *Bioresour. Technol.*, 2014, **152**, 162–168.
80. H. Hulvova, P. Galuszka, J. Frebortova and I. Frebort, *Biotechnol. Adv.*, 2013, **31**, 79–89.
81. K. L. Ryan, C. T. Moore and D. G. Panaccione, *Toxins*, 2013, **5**, 445–455.
82. C. Wallwey and S. M. Li, *Methods Mol. Biol.*, 2012, **944**, 121–131.
83. E. C. Kornfeld, E. J. Fornefeld, G. B. Kline, M. J. Mann, D. E. Morrison, R. G. Jones and R. B. Woodward, *J. Am. Chem. Soc.*, 1956, **78**, 3087–3114.
84. E. C. Kornfeld, E. J. Fornefeld, G. B. Kline, M. J. Mann, R. G. Jones and R. B. Woodward, *J. Am. Chem. Soc.*, 1954, **76**, 5256–5257.
85. W. Oppolzer, E. Francotte and K. Bättig, *Helv. Chim. Acta*, 1981, **64**, 478–481.
86. V. W. Armstrong, S. Coulton and R. Ramage, *Tetrahedron Lett.*, 1976, **17**, 4311–4312.
87. J. Rebek Jr and D. F. Tai, *Tetrahedron Lett.*, 1983, **24**, 859–860.
88. T. Kurihara, T. Terada and R. Yoneda, *Chem. Pharm. Bull.*, 1986, **34**, 442–443.
89. J. B. Hendrickson and J. Wang, *Org. Lett.*, 2003, **6**, 3–5.
90. I. Ninomiya, C. Hashimoto, T. Kiguchi, T. Naito, D. H. R. Barton, X. Lusinchi and P. Milliet, *J. Chem. Soc. Perkin Trans. 1*, 1990, 707–713.
91. A. V. Kalinin, B. A. Chauder, S. Rakhit and V. Snieckus, *Org. Lett.*, 2003, **5**, 3519–3521.
92. S. Inuki, S. Oishi, N. Fujii and H. Ohno, *Org. Lett.*, 2008, **10**, 5239–5242.
93. S. Inuki, A. Iwata, S. Oishi, N. Fujii and H. Ohno, *J. Org. Chem.*, 2011, **76**, 2072–2083.
94. F. R. Petronijevic and P. Wipf, *J. Am. Chem. Soc.*, 2011, **133**, 7704–7707.
95. W. Wang, J. T. Lu, H. L. Zhang, Z. F. Shi, J. Wen and X. P. Cao, *J. Org. Chem.*, 2014, **79**, 122–127.
96. N. D. Jabre, T. Watanabe and M. Brewer, *Tetrahedron Lett.*, 2014, **56**, 197–199.
97. H. G. Floss, M. Tcheng-Lin, C. J. Chang, B. Naidoo, G. E. Blair, I. Abou-Chaar and J. M. Cassady, *J. Am. Chem. Soc.*, 1974, **96**, 1898–1909.
98. T. W. Stone, *Brain Res.*, 1974, **72**, 177–180.
99. M. G. Buzzi and M. A. Moskowitz, *Cephalalgia*, 1991, **11**, 165–168.

CHAPTER 15

Cyclic Peptides as Privileged Structures

PRABHAKAR CHERKUPALLY,[a,†] SUHAS RAMESH,[a,†] YAHYA E. JAD,[a,†] THAVENDRAN GOVENDER,[a] HENDRIK G. KRUGER,[a] BEATRIZ G. DE LA TORRE[a] AND FERNANDO ALBERICIO*[a,b,c,d,e]

[a] Catalysis and Peptide Research Unit, School of Health Sciences, University of KwaZulu-Natal, Durban 4001, South Africa; [b] Institute for Research in Biomedicine-Barcelona, Baldiri Reixac 10, 08028-Barcelona, Spain; [c] CIBER-BBN, Networking Centre on Bioengineering, Biomaterials and Nanomedicine, Barcelona Science Park, 08028-Barcelona, Spain; [d] Department of Organic Chemistry, University of Barcelona, 08028-Barcelona, Spain; [e] School of Chemistry and Physics, University of KwaZulu-Natal, Durban 4001, South Africa
*Email: fernando.albericio@irbbarcelona.org

15.1 Cyclic Peptides in Biology

Mother Nature provides bountiful molecules with a wide range of therapeutic properties. Once in the hands of medicinal chemists, these precursors can be tailor-made to obtain a given target of interest. A major challenge for chemical biology is how to exploit these compounds and study their interaction with biological systems. The last few decades have witnessed the introduction of many peptides with biological activities into

[†]The three authors have contributed equally to the present work.

RSC Drug Discovery Series No. 50
Privileged Scaffolds in Medicinal Chemistry: Design, Synthesis, Evaluation
Edited by Stefan Bräse

Published by the Royal Society of Chemistry, www.rsc.org

literature, and the structures, properties, and functions of these molecules have yielded a wealth of knowledge.[1,2] In addition to providing protection, peptides are crucial for bodily functions. Peptides that are simple in nature are consistently overlooked as potential drug candidates due to their susceptibility to enzyme degradation resulting in loss of activity and a very short half-life.[1,2] Furthermore, peptide drugs cannot be delivered orally as their amide bonds are unstable under physiological conditions and the gastrointestinal tract does not work synergistically for their absorption. It is precisely because of these shortcomings that peptides have received little attention in drug discovery.

Recent decades have seen renewed interest by pharma companies, as well as academics, in peptide chemistry. Research into peptides has now been embraced as these scaffolds enable easy access to relatively underexposed high molecular weight chemical matter. The growth of this field reflects the new status of peptides as "privileged scaffolds" in medicinal chemistry.

Peptides can be obtained from nature or they can be synthesized in the laboratory. The former approach is comparatively expensive and time-consuming. Synthetic peptides have the advantage that they exhibit high specificity and selectivity and also present low toxicity. However, these molecules are metabolically unstable and are unable to penetrate cell membranes.[3] In order to overcome these limitations, several chemical approaches have been developed, including cyclization,[4] substitution of L- by D-amino acids,[5] the replacement by unnatural amino acids,[6] formation of peptidomimetics by the isostere method, and peptoid synthesis.[7,8] Of these techniques, cyclization appears to be the method of choice, as it is a straightforward approach that attains peptides with metabolic stability and conformational constraints on the peptide backbone.[9] Peptide cyclization can be done in three ways, as depicted in Figure 15.1.

Cyclic peptides are a class of exciting and underexplored compounds that exhibit a wide spectrum of biological activity.[10] Compared to their linear counterparts, cyclic versions offer potential advantages as therapeutic candidates because they show increased enzymatic stability, receptor selectivity, and improved pharmacodynamic properties. Interest in these structures was

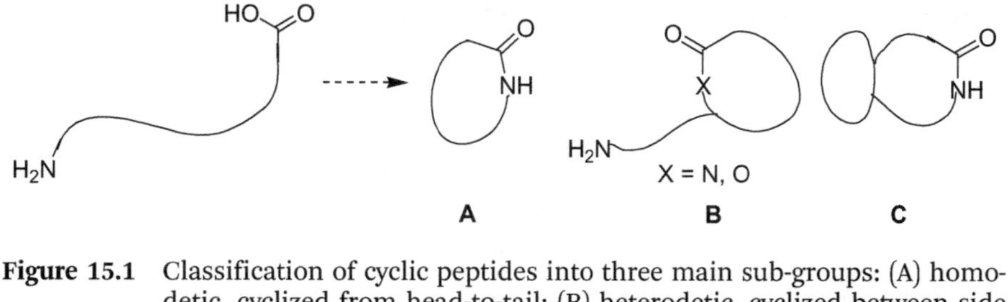

Figure 15.1 Classification of cyclic peptides into three main sub-groups: (A) homodetic, cyclized from head-to-tail; (B) heterodetic, cyclized between side chains or from a side chain to one of the termini; and (C) complex, comprised of a mixture of homodetic and heterodetic linkages.

first aroused early in 1940 and later grew considerably.[9] Though many biological activities of cyclic peptides have been reported, the mechanism of action and the molecular target of only a few have been addressed in depth. This research opens up challenging and interesting ways to study the molecular basis underlying the activity of these molecules, thus providing useful information for the development of drugs. Given the considerable interest in cyclic peptides, we have focused on discussing these molecules in this chapter. However, this is not a compilation of biologically active cyclic peptides; instead, we have turned our attention to three "privileged scaffolds", namely, diketopiperazines, benzodiazepines, and cyclotides, covering the key information that would enable the scientific community to expand their knowledge and develop new therapeutic agents.

15.2 Diketopiperazines

Diketopiperazines (DKPs) **1**, also known as piperazine-2,5-diones, are the smallest cyclic peptides, consisting of two amino acids (Figure 15.2). In 1888, Curtius and Goebel synthesized the first cyclic dipeptide, Cyclo(Gly-Gly).[11] Since its discovery, the DKP template has been identified in many bioactive molecules, including natural products, and those compounds that comprise combinational libraries. Some of the chemical characteristics of DKPs make them very interesting and attractive for medicinal chemistry purposes. In this regard, DKPs are small heterocyclic molecules, they are resistant to proteolysis, they show conformational rigidity, and they mimic peptide pharmacophoric groups and donor and acceptor groups for hydrogen bonding, thus favouring interactions with the biological system. Furthermore, DKPs are present in several natural products with biological properties, such as antitumour, antimicrobial, and antiviral activity, as well as in others with the capacity to modulate enzymes, receptors, and biochemical mediators.[12–20]

Examples of the multiple applications of DKPs in biology are numerous, and we will discuss some of them in this sub-section. For instance (Figure 15.3), cyclo(L-His-L-Phe) (**2**) shows antitumour activity and causes a significant slowing of heart rate and a decrease in coronary flow rate, while cyclo(L-His-L-Tyr) (**3**) significantly increases heart rate, in addition to showing antibacterial activity. However, both compounds cause an increase in ventricular pressure in isolated studies on the rat heart.[21]

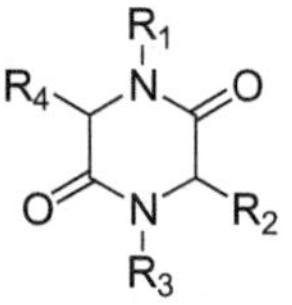

1

Figure 15.2 Basic structure of diketopiperazine.

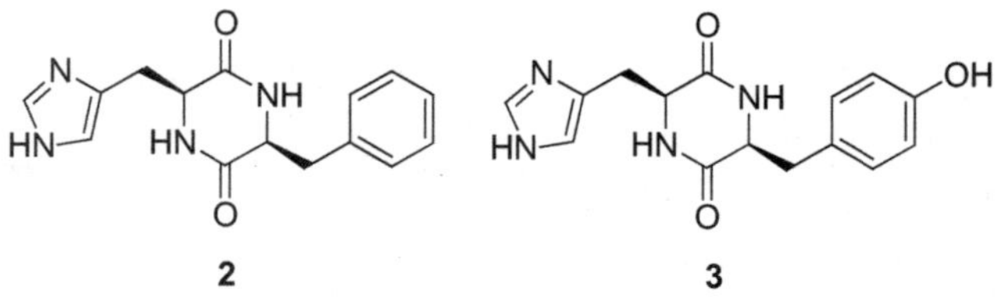

Figure 15.3 His-based DKPs.

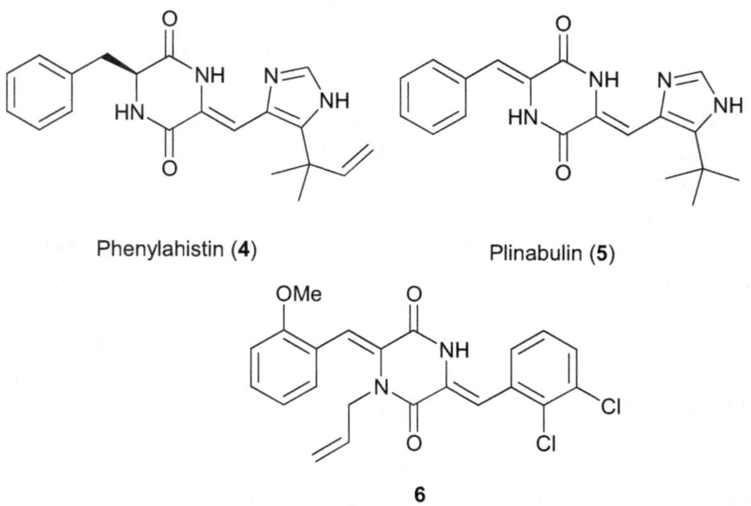

Figure 15.4 Structures of phenylahistin and plinabulin and its derivative.

Phenylahistin (**4**), a natural product isolated from *Aspergillus ustus,* showed antitumour activity.[22–24] Phenylahistin is a DKP derivative that consists of an L-phenylalanine and isoprenylated dehydrohistidine residue with a quaternary carbon at the 5-position of the imidazole ring.[24] Later, plinabulin (**5**) was derived from **4** and showed vascular disrupting activity. This compound is now under phase-II clinical trials as an anti-cancer drug.[25] Furthermore, Lieo *et al.*[26] reported a series of DKPs derived from **4** and **5** and tested them as anti-cancer agents. Among this series, 1-allyl-3-(2,3-dichlorobenzylidene)-6-(2-methoxybenzylidene)piperazine-2,5-dione (**6**) showed strong activity against all the cancer cell lines tested ($IC_{50} = 0.5–4.5\ \mu M$) (Figure 15.4).[26]

Another example that reveals the importance of DKPs as "privileged structures" in medicinal chemistry is brevianamide F [cyclo-(L-Trp-L-Pro), 7], a compound with antibacterial and antifungal activity[27] that is also used in the treatment of cardiovascular dysfunction.[28] Brevianamide F is recognized as a "privileged structure" and holds an indole ring on the Trp residue.[29] Moreover, tryprostatins A and B (**8** and **9,** respectively) were synthesized[30–32] by prenylation at the C2 position of the indole ring of 7 and exhibited

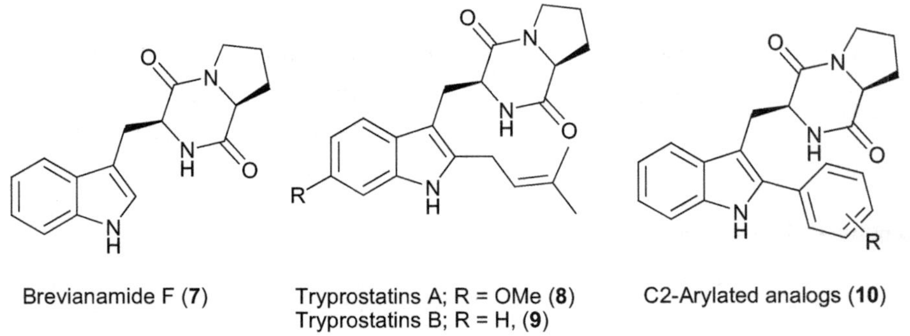

Figure 15.5 Structure of brevianamide F (7) and its derivatives.

cytotoxicity towards various cancer cell lines.[33] Furthermore, modification of 7 to prepare C2-arylated analogues (**10**) turned its activity from mildly antibiotic and antifungal into antitumoural (Figure 15.5).[34]

Numerous natural products with applications in medicinal chemistry contain hexahydropyrrolo[2,3-*b*]indole (HPI), in addition to a DKP core and indole ring as "privileged structures".[35] For instance, (+) leptosins D-F (**11–13**) isolated from the mycelium of a strain of *Leptosphaeria* sp. attached to the marine alga *Sargassum tortile* showed antitumour activity against cultured P388 cells. Furthermore, (+) leptosins A–C (**14–16**), which contain two DKP and HPI units in their structure, were also isolated from the same source and showed significant antitumour activities.[36] (+) Leptosins C (**16**) and F (**13**) showed inhibitory activities against topoisomerases I and II.[37] Plectosphaeroic acids A–C (**17–19**) have DKP and HPI units with an extra indole ring. Compounds (**17–19**) isolated from the fungus *Plectosphaerella cucumerina* showed inhibitory activities against indoleamine 2,3-dioxygenase (IDO).[38] WIN 64 821 (**20**) and WIN 64 745 (**21**), both isolated from a strain of *Aspergillus* sp., also contained a DKP dimer that acts as a neurokinin antagonist.[39] Antiviral agent asperdimin (**22**), isolated from *Aspergillus niger*, has a DKP dimer in its structure.[40] Another DKP dimer, verticillins A (**23**), which is obtained from *Verticillium* sp., exhibited antimicrobial activity against Gram-positive bacteria and antitumour activity in HeLa cell lines (Figure 15.6).[41–43]

DKPs are amongst the most common peptide derivatives found in nature.[13] Many of these molecules are endogenous to members of the animal and plant kingdoms, including marine sponges.[44] The only DKP shown to be endogenous to mammals is cyclo(L-His-L-Pro) (**24**) (Figure 15.7).[13] Cyclo(L-His-L-Pro) is endogenous to the blood, brain, and gastrointestinal tract of humans, and it exhibits a wide variety of effects on the central nervous, endocrine, electrophysiological, and cardiovascular systems.[45] Furthermore, it shows antifungal activity against Chitinases B ($IC_{50} = 1.1$ mM).[46]

A number of reviews have covered bioactive DKP compounds as natural products, some of which are shown in Table 15.1.[13]

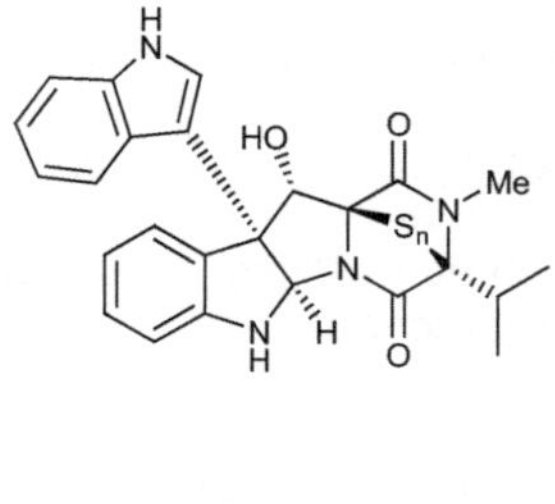

(+) leptosin D; n = 2 (**11**)
(+) leptosin E; n = 3 (**12**)
(+) leptosin F; n = 4 (**13**)

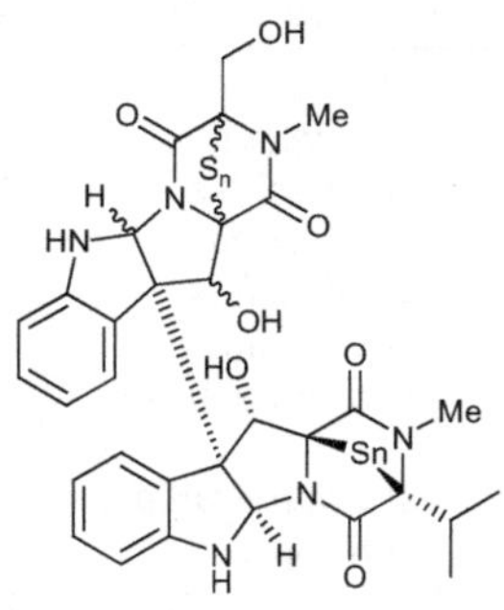

(+) leptosin A; n = 2 (**14**)
(+) leptosin B; n = 3 (**15**)
(+) leptosin C; n = 4 (**16**)

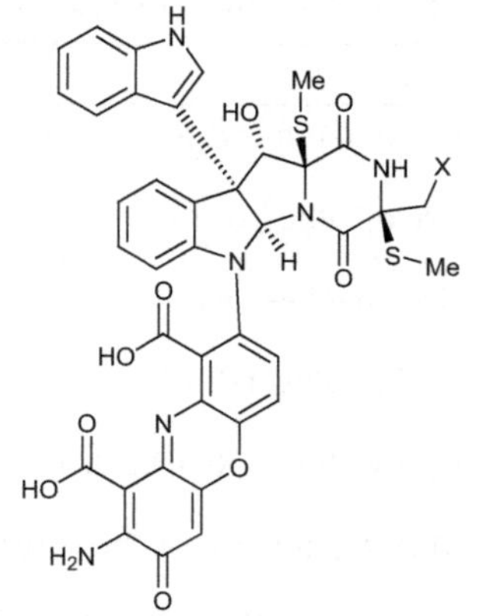

(+) plectosphaeroic acid A; X = OH (**17**)
(+) plectosphaeroic acid B; X = H (**18**)

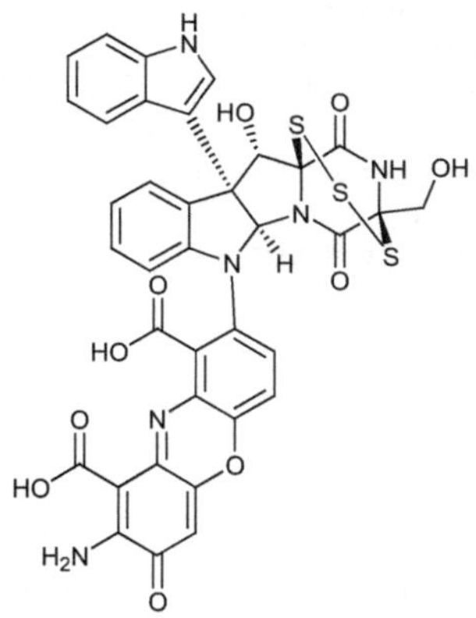

(+) plectosphaeroic acid C (**19**)

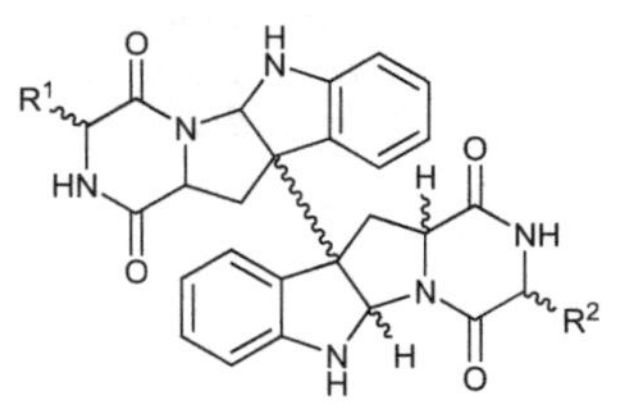

(+) WIN 64821; $R^1 = R^2 = Bn$ (**20**)
(+) WIN 64745; $R^1 = Bn$, $R^2 = iBu$ (**21**)
(+) asperdimin; $R^1 = iBu$, $R^2 = iPr$ (**22**)

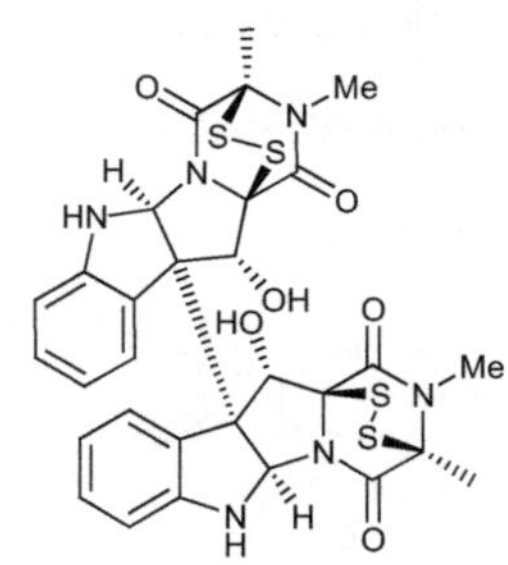

verticillin A (**23**)

Figure 15.6 Structure of some compounds that contain the HPI and DKP templates.

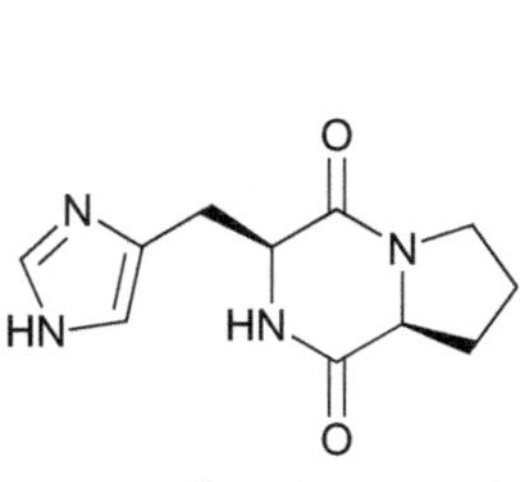

24

Figure 15.7 Cyclo-(L-His-L-Pro).

DKPs have also been used as a core scaffold to construct libraries in combinatorial chemistry. Some DKP-based bioactive compounds are summarized in Table 15.2.

Table 15.1 Bioactive natural DKP sequences/structures.

DKP	Isolated from	Activity	Ref.
Cyclo-(L-Pro-*trans*-4-OH-L-Pro)	*Suberites domuncula*	Antimicrobial	47
Cyclo(L-Phe–L-Pro)	*Lactobacillus plantarum*	Antifungal	48
Cyclo(L-Phe–*trans*-4-OH-L-Pro)	*Lactobacillus plantarum*	Antifungal	48
Cyclo(D-Phe–D-Pro)	Marine bacteria	Inhibition of *Vibrio anguillarum*	49, 50
Cyclo(L-Phe-4-R-OH-L-Pro)	*Pseudoalteromonas luteoviolacea*	Antibiotic	51
Maculosin [cyclo(-L-Pro-L-Tyr-)]	*Alternaria alternata*	Host-specific fungal phytotoxin	52, 53
cyclo(13,15-dichloro-L-Pro-L-Tyr)	*Leptoxyphium* sp.	Inhibitor of monocyte chemotactic protein-1 (CCL2)	54
Cyclo-(L-Pro-L-Met)	*Nocardiopsis* sp.	Anti-angiogenesis	55
Cyclo(L-Leu-L-Pro)	*Aspergillus parasiticus*	Antifungal	56, 57
Cyclo(L-Val-L-Pro)	*Pseudomonas rhizosphaerae*	Antibacterial and antilarval	58
Gliotoxin 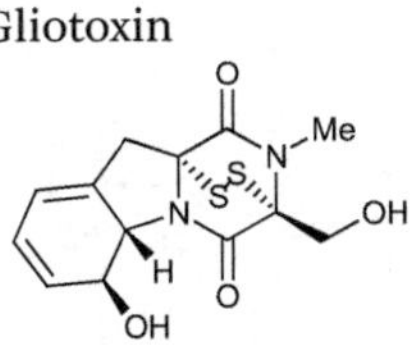 **25**	*Gliocladium virens*	Anti-TB agent	59, 60
Rostratins	*Exserohilum rostratum*	Antitumour	61–64
MPC1001 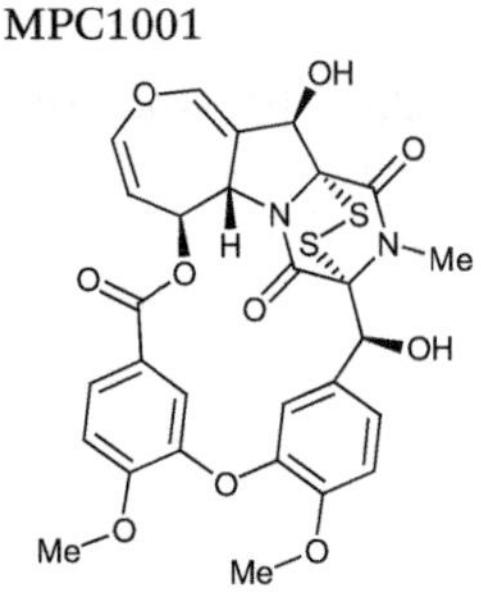 **26**	*Cladorrhinum* sp.	Antitumour	65, 66

15.3 Benzodiazepine

Over the last three decades, several structures and functionalities have been considered to be privileged. In the late 1980s, benzodiazepines (BZDs) were the first class of molecules to be acknowledged as "privileged structures" by Evans and co-workers.[83] Amongst the heterocyclic family, BZDs,

Table 15.2 DKP-based bioactive compounds and their activities.

DKP	Activity	Ref.
Tadalafil 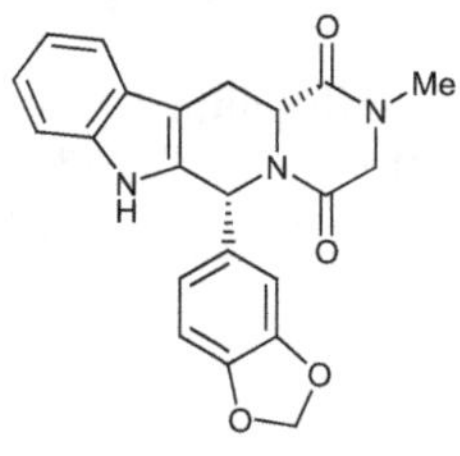 **27**	PDE5 inhibitor ($IC_{50} = 5$ nM)	67
GSK221149A (Retosiban) 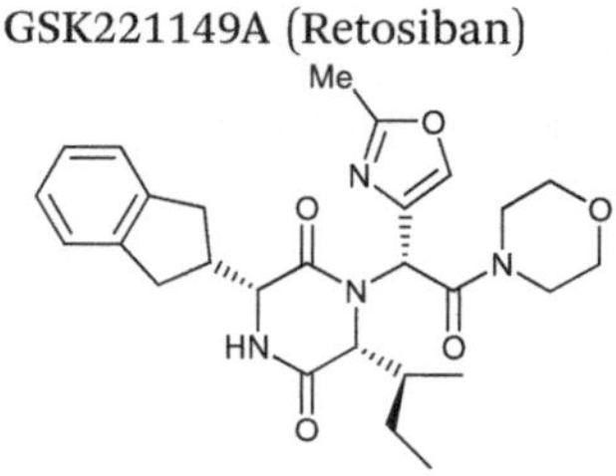 **28**	Oxytocin antagonist ($K_i = 0.65$ nM)	68
Epelsiban 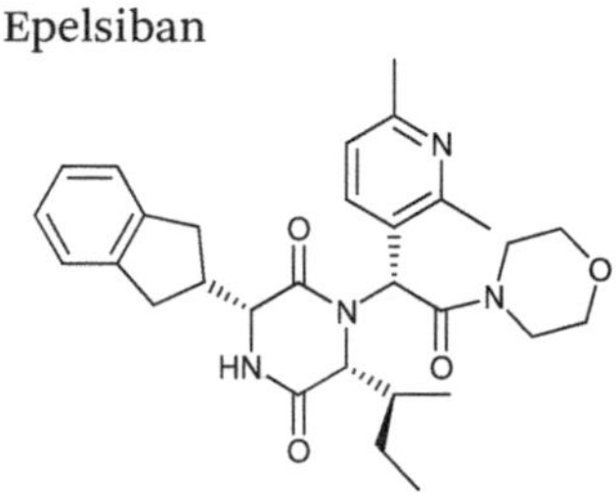 **29**	Oxytocin antagonist (p$K_i = 9.9$)	69
XR5967 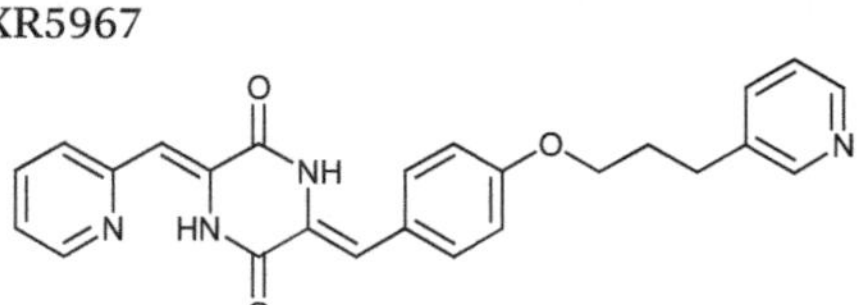 **30**	Anti-cancer ($IC_{50} = 800$ nM, human PAI-1)	70
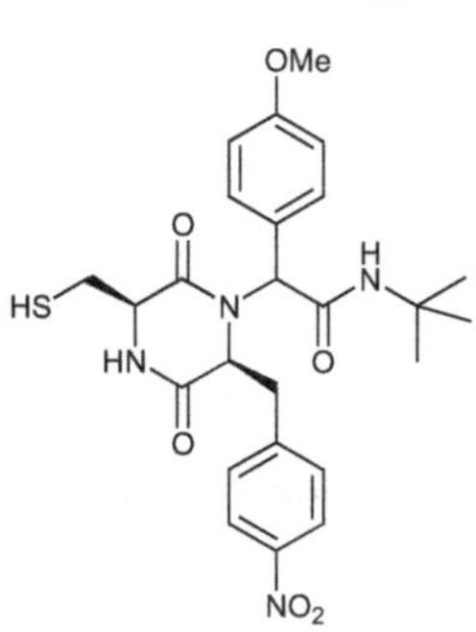 **31**	Matrix metalloproteinase (MMPs) inhibitor	71, 72

Table 15.2 (*Continued*)

DKP	Activity	Ref.
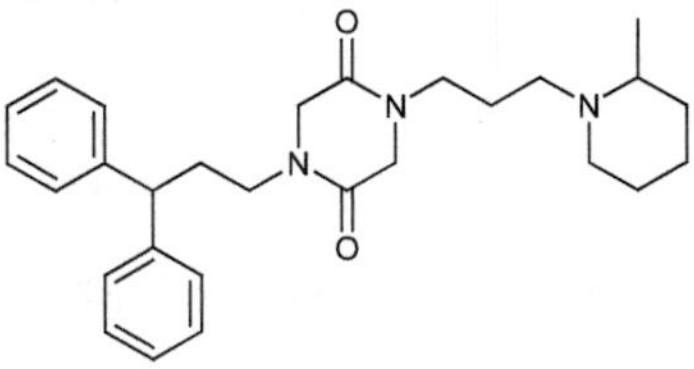 **32**	Viral haemorrhagic septicaemia virus (VHSV) inhibitor ($IC_{50} = 51$ μM)	73
Cairomycin B 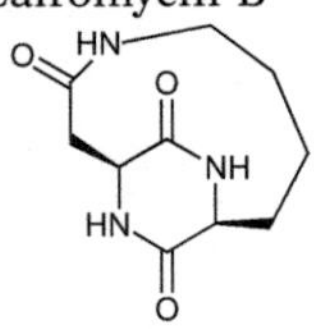 **33**	Antibacterial, mainly active against Gram-positive bacteria	74
PheDa4 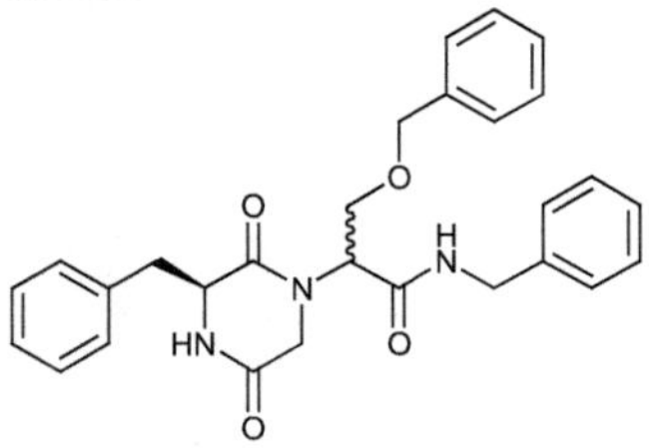 **34**	Antibacterial	75
NNZ 2591 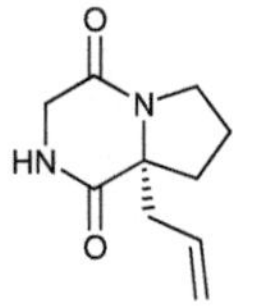 **35**	Neuroprotective agent	76
CSP-2503 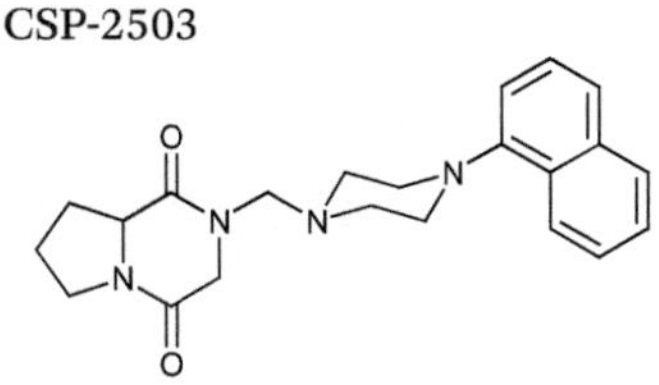 **36**	Anxiolytic agent	77, 78
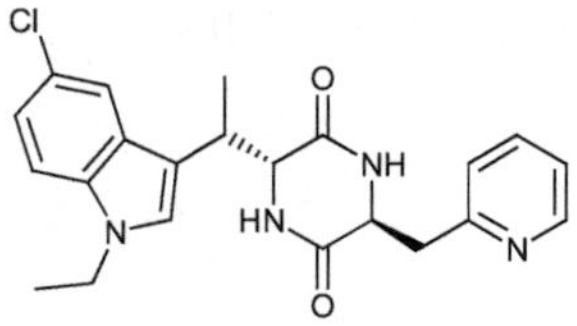 **37**	Inhibitor of platelet-activating factor (PAF) ($IC_{50} = 36$ nM)	79

Table 15.2 (*Continued*)

DKP	Activity	Ref.
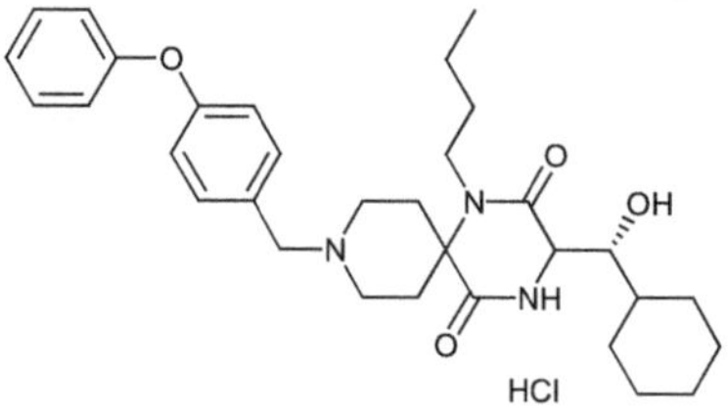 **38**	Anti-HIV ($IC_{50} = 0.6$ nM)	80, 81
AK602/ONO4128/GW873140 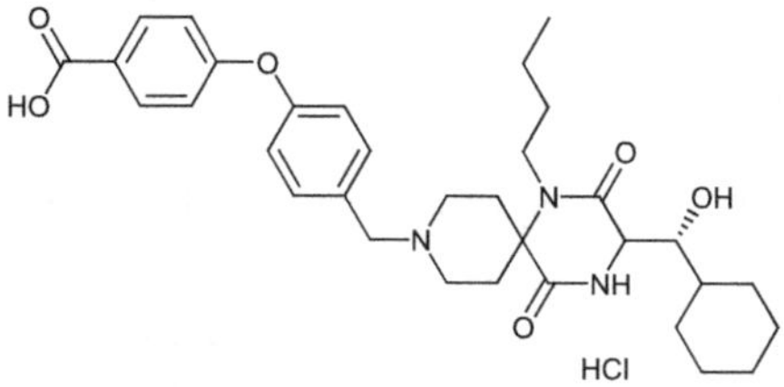 **39**	Anti-HIV	82

benzoaxazepines, and benzothiazepines, all with an amide/peptide template, have received much attention in the fields of medicine and pharmaceutical chemistry as a result of their broad spectrum of biological activities.[84,85] Several types of BZDs are known (Figure 15.8), such as 1,4-benzodiazepin-2-ones (**40**), 1,5-benzodiazepin-2-ones (**41**), 1,5-benzodiazepin-2,4-diones (**42**), 1,4-benzothiazepin-5-ones (**43**), pyrrolo[2,1-*c*][1,4]benzodiazepin-5-ones (**44**), pyrazolo[4,3-*e*][1,4]diazepine (**45**), and 5,11-dihydrobenzo[*e*]pyrido[3,2-*b*][1,4]diazepin-6-ones (**46**). Thousands of compounds belonging to these families have been produced by combinatorial synthesis, both in-solution and on solid-phase.[84]

15.3.1 1,4-Benzodiazepin-2-ones

BZDs are found in several types of central nervous system (CNS) agents and in ligands of both ion channel and G protein-coupled receptors (GPCRs). Derivatives of these compounds have the capacity to bind not only to BZD receptors of the CNS, but also to other receptors and enzymes.[86] These compounds exert their action on the CNS, acting selectively on γ-aminobutyric acid-A (GABA-A) receptors in the brain. They enhance response to the inhibitory neurotransmitter GABA by opening GABA-activated chloride channels and allowing chloride ions to enter the neuron, resulting in sedative, hypnotic (sleep-inducing), anxiolytic (anti-anxiety), anticonvulsant, and muscle relaxant properties.[87] These attributes have resulted in BZDs being explored for the treatment of anxiety, insomnia,[88] psychomotor agitation, epileptic seizure, and anticonvulsants.[89–91] Some heterocycles

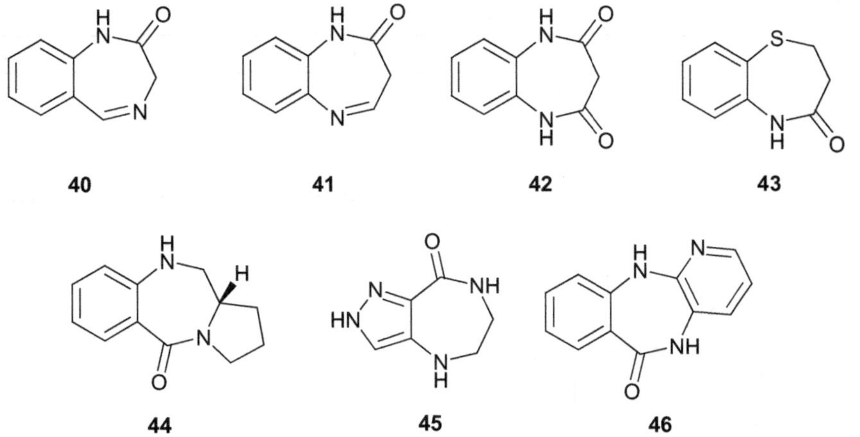

Figure 15.8 Fused privileged ring systems based on benzodiazepine and benzothiazepine scaffolds.

containing BZD moieties are reported to possess anti-inflammatory, antiviral, anti-HIV, antimicrobial, and antitumour activities. "Diazepam (Valium™, **47**)", invented by Dr Leo Sternbach of Hoffmann-La Roche (Nutley, New Jersey, US), was the second drug belonging to the BZD class to receive approval, which was granted in 1960. After this initial success, other pharmaceutical companies began to introduce BZD derivatives.[88,92] Some 1,4-benzodiazepine drugs, such as Diazepam (**47**), Lorazepam (**48**), Cinolazepam (**49**) and Clonazepam (**50**), are shown in Figure 15.9.

According to Evans' definition, "privileged structures" are compounds that can bind to various protein-receptor surfaces. BZDs are not only widely used in clinical practice as anxiolytics, they are also administered as cholecystokinin (CCK) A and B inhibitors. In 1986, Evans *et al.*[93] developed a new potent, selective, and orally effective non-peptidic antagonist for a peptide hormone CCK receptor for the treatment of gastrointestinal disorders (*e.g.* pancreatitis, dyspepsia, gastroparesis and gastric reflux) based on analogues of the natural product asperlicin (**51**), a compound that exhibited some activity. By making structural modifications to anxiolytic BZD drugs, such as diazepam (**47**), these scientists adapted its properties, either to increase the activity or binding strength of the original molecule **51**.[93]

In their work, the 5-phenyl-1,4-benzodiazepine ring and 3-hydroxyindoline subunits were inserted in the left and right half of parts of asperlicin (**51**), respectively (Figure 15.10). L-Tryptophan is the key amino acid for the carboxyl-terminal sequence of CCK-1. This research resulted in the development of devazepide (MK-329) (**52**) as the first specific non-peptidic BZD antagonist, which showed high selectivity for CCK-1 ($IC_{50} = 0.8$ nM).[93] Furthermore, they proposed that slight modifications of privileged structures could be a useful approach to develop receptor agonists or antagonists. An extensive number of non-peptide 1,4-benzodiazepine-containing ligands

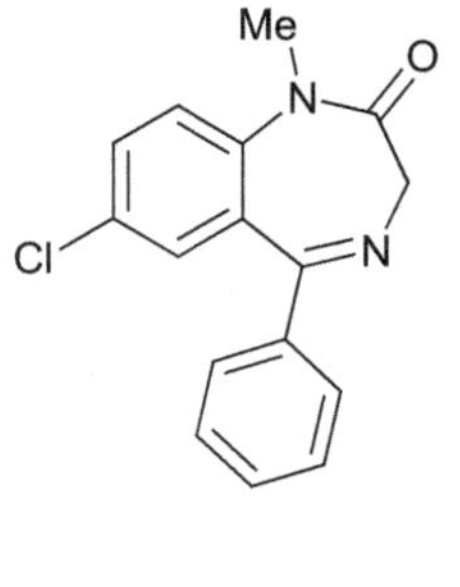

Diazepam (Valium) (**47**)
anxiolytic, anticonvulsant,
muscle relaxant

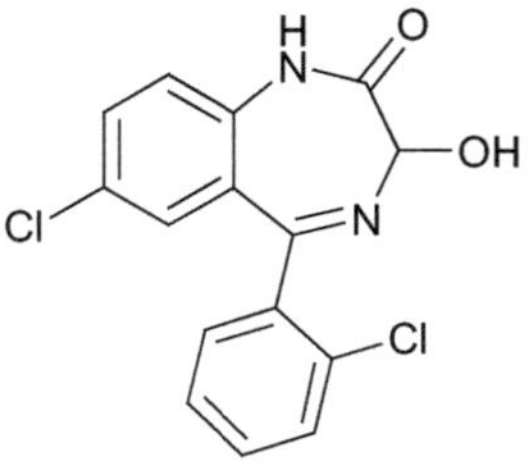

Lorazepam (**48**)
anxiolytic, anticonvulsant

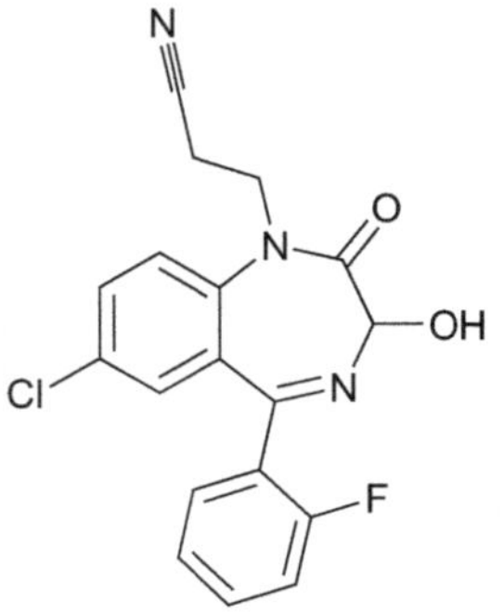

Cinolazepam (**49**)
anxiolytic, anticonvulsant,
sedative

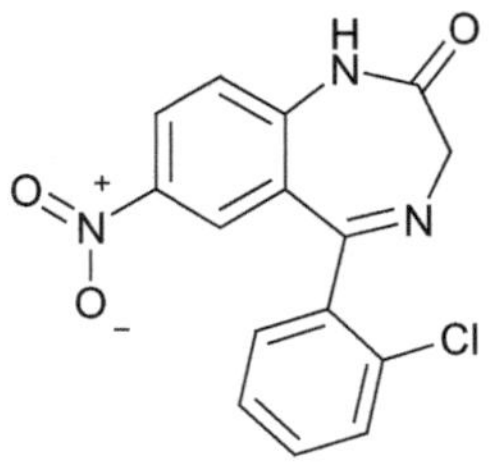

Clonazepam (**50**)
anxiolytic, anticonvulsant,
sedative

Figure 15.9 Structure of selected 1,4-benzodiazepine (BZD) drugs.

were developed for the gastrin/CCK (A and B) receptors.[94–106] Novel benzodiazepine (L-364,373) activated cardiac slow delayed rectifier K^+ currents, I_{ks}, is an important modulator of cardiac action potential repolarization,[107] while another BZD compound, (L-735,821) is a potent and selective blocker of cardiac I_{ks}.[108] Numerous other applications for ligands based on BZDs centroid to other GPCRs have been reported. These include k-selective opioid agonists like tifluadom for the treatment of visceral pain, antithrombotic platelet activation factor (PAF) antagonists,[109] analgesic and anti-inflammatory neurokinine (NK-1) receptor antagonists,[110] endothelin (ET) receptor antagonists,[111] class III anti-arrhythmic agents,[112] oxytocin antagonists,[31,113] bradykinin B_1/B_2 receptor agonists and antagonists,[114–117] vitronectin receptor antagonists,[118] fibrinogen receptor antagonists,[119] calcitonin gene-related peptide (CGRP) receptor antagonists,[120] and glycoprotein (GP) II_b/III_a receptor antagonists with antithrombotic profiles.[121–123] Also, multiple classes of enzyme inhibitors holding a BZD unit have been developed, including RAS-farnesyltransferase (R-FT) inhibitors for the treatment of cancer (*e.g.* BMS-214662120),[124,125] somatostatin receptor inhibitors,[126] and respiratory

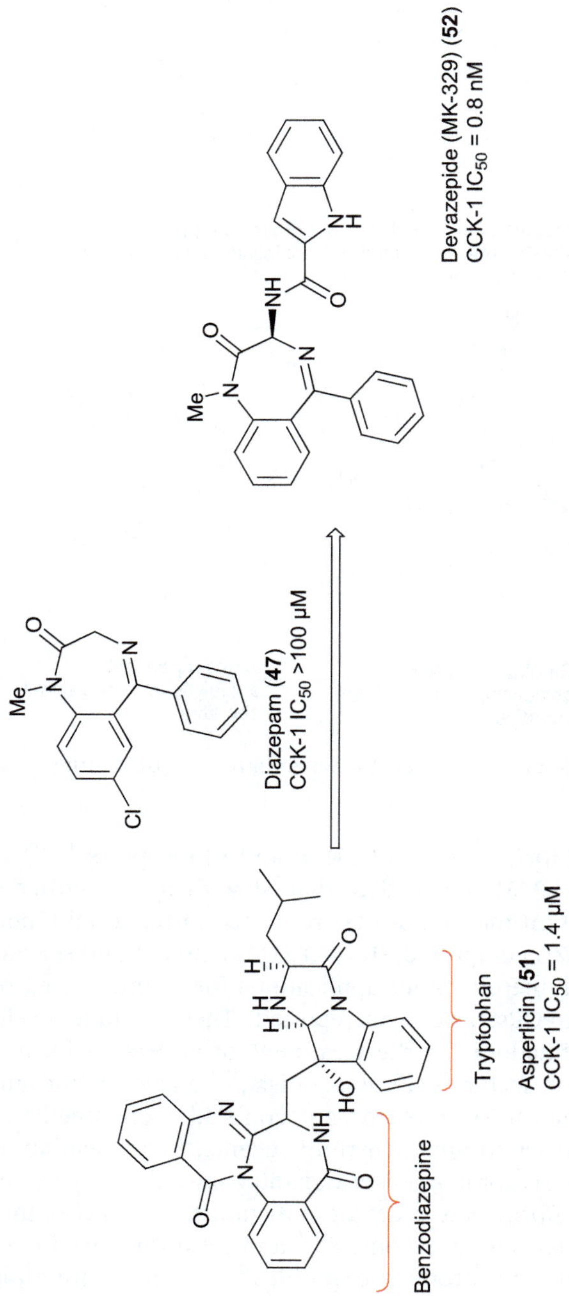

Figure 15.10 Natural product asperlicin guided for the development of CCK-1 antagonists.

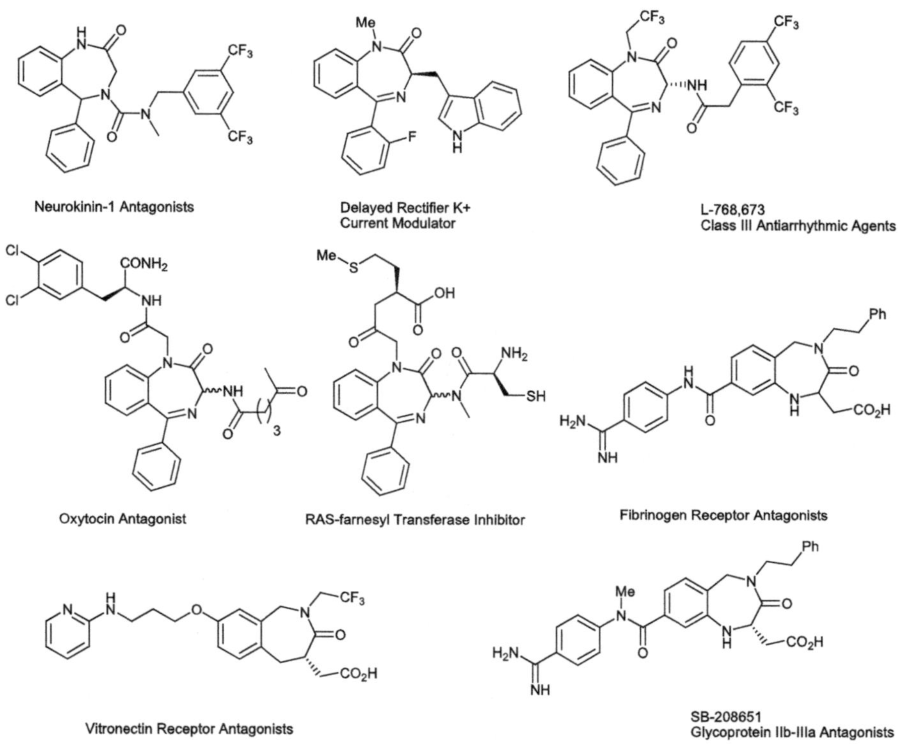

Figure 15.11 Some examples of benzodiazepine receptor agonists/antagonists and enzyme inhibitors.

syncytial virus (RSV) inhibitors.[127] A group of representative BZD receptor agonists/antagonists and enzyme inhibitors are shown in Figure 15.11.

Of note, many biologically active peptides and proteins exhibit β-turn motifs. Examination of why BZDs are privileged in this manner has led them to be identified as β-turn peptidomimetics.[128–130] In proteins, the presence of such structural motifs that are complementary to an array of primary and secondary structural elements offers a potential explanation for the promiscuous nature of the binding of many recurring scaffolds.

As peptidomimetics, BZDs are assumed to show intrinsically strong binding affinity to several proteins that interact similar regions of peptides or other proteins. For example, the peptide with sequence Ac-DEVD-H (Ac-Asp-Glu-Val-Asp-CHO, **53**) (Figure 15.12) is a potent and selective inhibitor of caspase-3. In this sequence, an aldehyde functional group reversibly forms a covalent bond with the thiol of a cysteine in the active-site.[131] While the Asp residue at P_1 is strictly required for activity, Asp at P_4 represents the most critical determinant of the inhibitor's specificity. The P_2 amide nitrogen is not used in a hydrogen bonding interaction with the

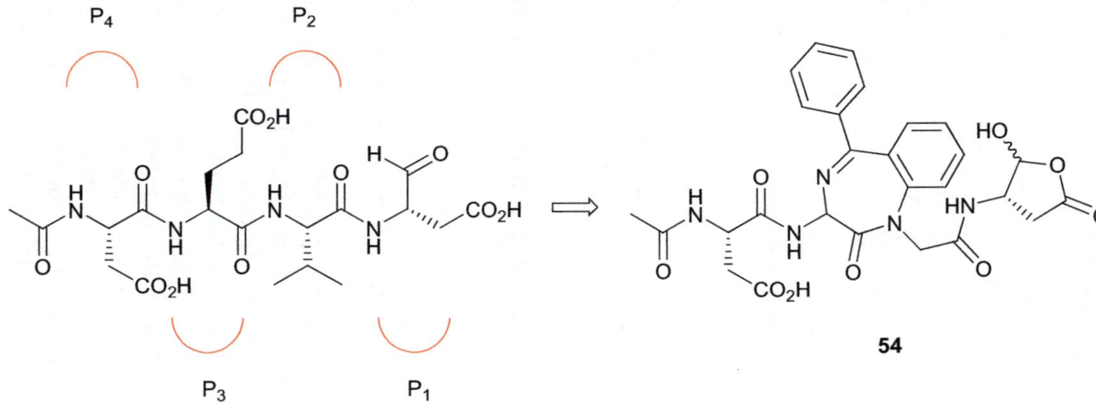

Ac-DEVD-H (**53**)

54

Figure 15.12 Benzodiazepine as a conformational constraint of the tetrapeptide Ac-DEVD-H.

enzyme, unlike the P_1 and P_3 amido hydrogens, which are best retained for high affinity binding. The BZD moiety as a conformational constraint for $P_3 - P_2$ dipeptide replacement led to the generation of a novel, potent, and specific inhibitor of caspase-3 (**54**). Although the inhibitory activity (K_i) of **54** was somewhat lower than the commonly used tetrapeptide, its selectivity for caspase-3 and capacity to inhibit apoptosis in living cells made it an attractive target.[131]

Non-classical peptidomimetic BZDs have also been reported as antifungal agents,[132] antimalarial agents for cysteine protease inhibitors,[133] vitamin D receptor (VDR) inhibitors,[134,135] and β-secretase (BACE-1) inhibitors.[136] In addition, BZDs also act as α-helix mimetics, *e.g.* HDM2 (BZD-containing compound) binds to an α-helix transactivation domain of p53, thereby inhibiting its tumour suppressive function.[137]

15.3.2 1,5-Benzodiazepin-2-ones and 1,5-Benzodiazepin-2,4-diones

Although significantly less research has been performed on 1,5-benzodiazepin-2-ones compared to that on 1,4-benzodiazepin-2-ones, molecules containing the former are also considered "privileged scaffolds". 1,5-Benzodiazepin-2-ones are associated with a wide range of biological activities, including the inhibition of interleukin-1β converting enzyme (ICE, caspase-1) and the delay of rectifier potassium current blocker (IK) (Figure 15.13).[138,139] Herpin *et al.*[138] used the Irori directed sorting system to synthesize a 10 000-member 1,5-benzodiazepin-2-one combinatorial library on solid-phase.[138] The non-peptidyl 1,5-benzodiazepin-2,4-diones and their derivatives act as peripheral CCK-A agonists (Figure 15.13).[140–142]

15.3.3 1,5-Benzothiazepin-2-ones

1,5-Benzothiazepin-2-one derivatives have received considerable attention for their potential use in the treatment of cardiovascular diseases. Some members of this class of compounds act as potent bradykinin agonists, growth hormone secretagogs, ligands for Src H_2 protein, spasmolytics, and squalene synthetase inhibitors.[143] Thiazesim (**55**) is a BZD derivative that shows antidepressant properties (Figure 15.14).[144] Diltiazem (**56**) is an important cardiovascular drug of this family that has been introduced for the treatment of a variety of cardiac conditions (Figure 15.14).[145] The BZD derivative JMV1116 (**57**) is an agonist of bradykinin and exhibits high affinity for human receptor ($K_i = 0.7$ nM), as described by Amblard *et al.* (Figure 15.14).[114,115] Moreover, a family of Src SH_2 inhibitors was designed, starting from a benzothioazepinone scaffold.[146]

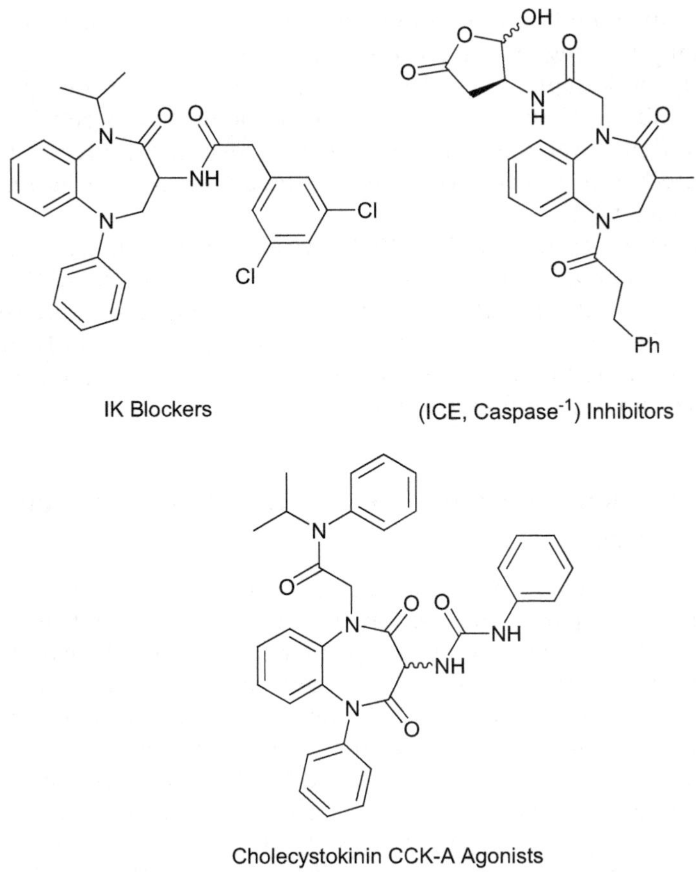

Figure 15.13 Biologically active 1,5-benzodiazepin-2-ones and 1,5-benzodiazepin-2,4-diones.

15.3.4 Pyrazolodiazepines

Research on a new class of BZD "privileged scaffold", pyrrolo[2,1-*c*][1,4]-benzodiazepines (PBDs), gained momentum as a result of the potential of these molecules as antitumour agents, gene regulators, and DNA probes. Examples of PBDs include abbeymycin (**58**) (Figure 15.15), anthramycin, tomaymycin, sibiromycin, neothramycin A and B, chicamycin, and DC-81.[147,148] When the native benzenoid ring was replaced with a pyrazole, pyridine, diazine, or pyrimidine ring to yield the novel corresponding pyrrolo[2,1-*c*][1,4] diazepine analogues (**59**), the resulting molecules showed cytotoxicity against L1210 leukaemia cell lines comparable to that reported for DC-81.[149,150] A library of tetrahydro-1,4-pyrazolodiazepin-8(2H)-one derivatives (**60**) was synthesized and assessed for activity against P2X_7R, BACE-1 and MC4R cell lines. The results indicated that the new class of

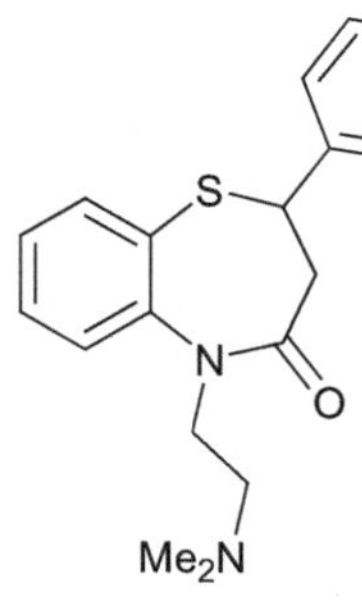

Thiazesim (**55**)

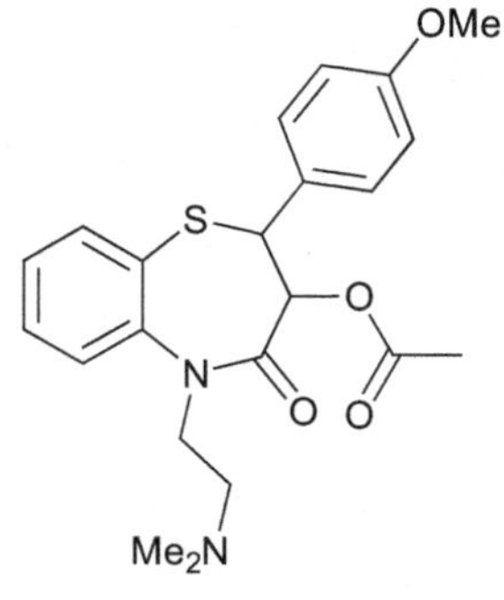

Diltiazem (**56**)

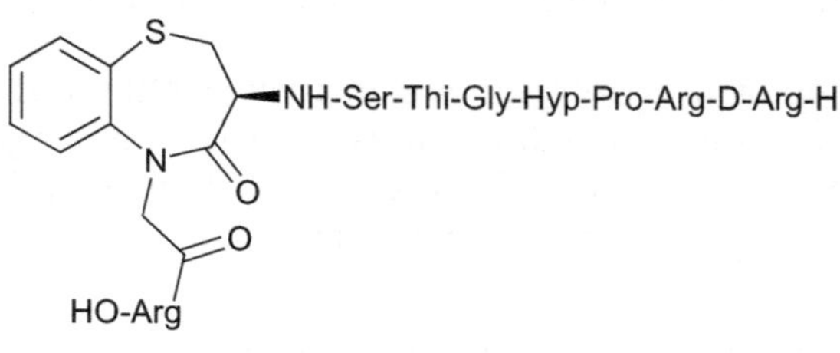

JMV1116 (**57**)

Figure 15.14 Some examples of 1,5-benzothiazepin-2-ones drugs.

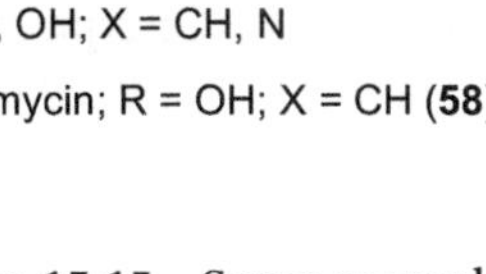

R= H, OH; X = CH, N

Abbymycin; R = OH; X = CH (**58**)

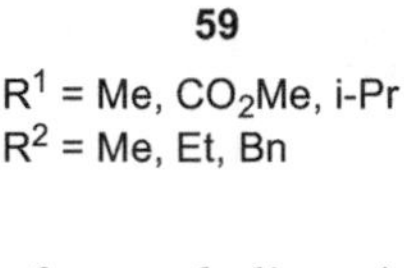

59

R^1 = Me, CO_2Me, i-Pr
R^2 = Me, Et, Bn

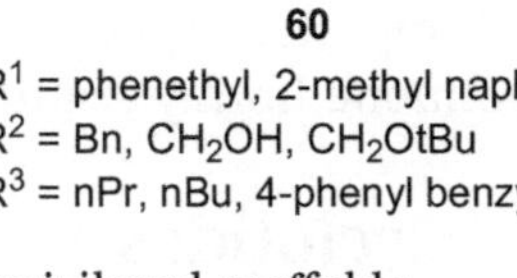

60

R^1 = phenethyl, 2-methyl naphthalene
R^2 = Bn, CH_2OH, CH_2OtBu
R^3 = nPr, nBu, 4-phenyl benzyl,

Figure 15.15 Some examples of pyrazolodiazepine privileged scaffolds.

pyrazolodiazepin-8-one derived moieties may find valuable applications in medicinal and pharmaceutical fields.[151]

15.3.5 5,11-Dihydro-benzo[*e*]pyrido[3,2-*b*][1,4]-diazepin-6-ones

5,11-Dihydro-benzo[*e*]pyrido[3,2-*b*][1,4]diazepin-6-ones show diverse therapeutic activities. Pirenzepine (gastrozepin, **61**), a prototypical M1-selective

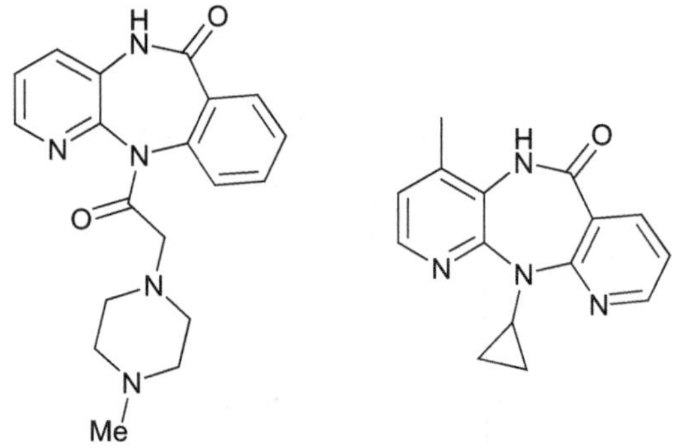

Pirenzepine (**61**) Nevirapine (**62**)

Figure 15.16 Structures of biologically active pirenzepine and nevirapine.

muscarinic receptor inhibitor, is used to treat peptic ulcers, as it reduces gastric acid secretion and reduces muscle spasm. Nevirapine (NVP, Viramune™, **62**) is a non-nucleoside reverse transcriptase inhibitor (NNRTI) used to treat HIV-1 infection and AIDS (Figure 15.16).[152–154]

15.3.6 Benzodiazepine-quinazolinones

Privileged 1,4-benzodiazepine-2,5-dione ring systems are the key intermediates for the synthesis of BZD-quinazolinone alkaloids (Figure 15.17). Sclerotigenin (**63**) was isolated from the sclerotia of *Penicillium sclerotigenum* and has shown promising anti-insectan activity. It is the simplest member of the BZD–quinazolinone natural alkaloid family. Other members of this family include circumdatins A–G (**64**), which are isolated from terrestrial fungus *Aspergillusochraceus,* and benzomalvins A–C (**65**), which is isolated from fungus *Penicillium* sp., also show biological activities of interest.[155] Derived from the fungus *Aspergillusalliaceus,* asperlicin (**51**) is a mycotoxin that acts as a selective antagonist for the CCK-A receptor.[156] Recently, Zhan *et al.*[157] reported a new synthetic protocol for sclerotigenin-type BZD–quinazolinone library scaffold.[157]

15.4 Cyclotides

Cyclotides (named from *Cycl*ic pep*tides*) are circular (head-to-tail) small proteins with 27–38 amino acids and are characterized by a unique Cys knot topology with six highly conserved Cys residues that are connected by means of three disulphide bonds, as exemplified in Figure 15.18. These S–S linkages (shown in yellow) join Cys residues (depicted as I–VI, Roman numerals), thus forming a ring and hence the knotted configuration that forms the six backbone segments (1–6 loops) between the consecutive residues. Almost every cyclotide has a β-hairpin bend in loop 5 that is aligned with its secondary structure.[158,159] It is worth mentioning that these are the only class of peptides that show fusion of the cyclic backbone and a Cys knot, a feature

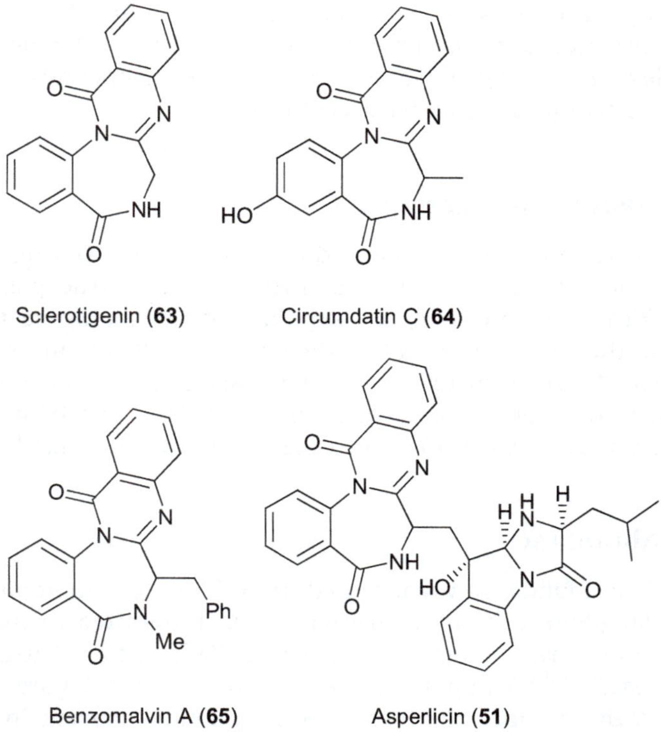

Figure 15.17 Some examples of benzodiazepine-quinazolinones.

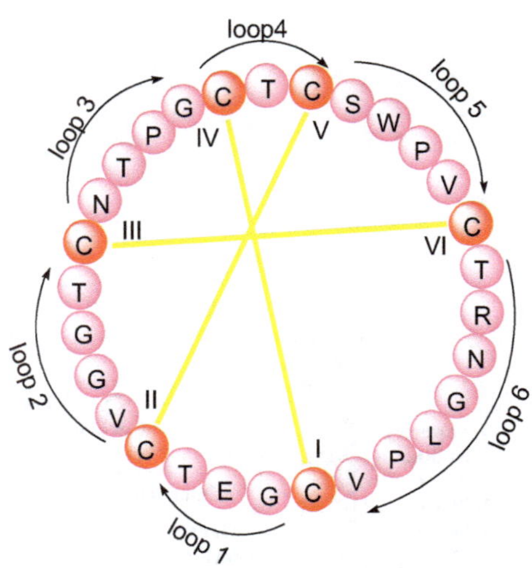

Figure 15.18 Structure of kalata B1; yellow lines indicate the linkage between Cys residues forming disulphide bonds; Cys (in red) residues are labelled serially I to VI; six backbone segments, termed as loops 1 to 6.

known as a "cyclic Cys knot" pattern. It is precisely this structural pattern that gives cyclotides extraordinary resistance to thermal, chemical, and enzymatic degradation. Such features make these compounds of primary interest for development as pharmaceuticals.

15.4.1 History and Structure

The first record of cyclotides appeared five decades ago. During the 1960s, these compounds were identified in medicinal and toxic plants of the Central African Republic and the Republic of Congo, where traditional healers use the plant *Oldenlandia affinis* to facilitate uterine contraction during labour.[160] The principal component responsible for this property was named kalata B1 after its source plant "Kalata-kalata" and B1 because this compound was chromatographically isolated from fraction B.[161]

15.4.2 Abundance

As far as their abundance is concerned, cyclotides are commonly found in the plant kingdom, and until now have been recognized in members of various families *viz.* Rubiaceae, Violaceae, Solanaceae, Fabaceae, and Cucurbitaceae.[162–164] A graphical representation shown in Figure 15.19 gives insight into the number of genera/species/cyclotides present in the plant kingdom.

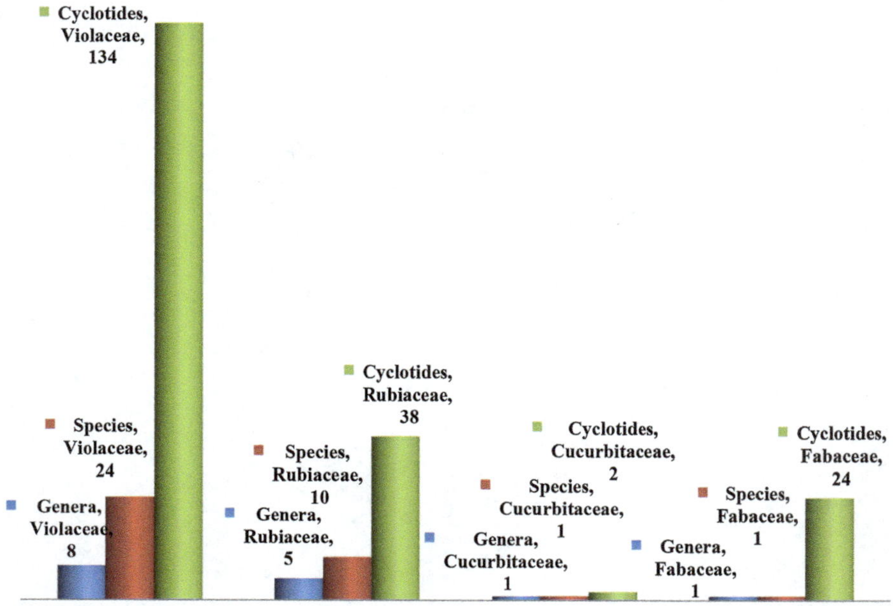

Figure 15.19 Graphical representation of abundance of cyclotides in different families.

15.4.3 Classification

Cyclotides are classified under the following three subfamilies: (a) Mobius, which has a *cis*-peptide bond before Pro in loop 5 which generates a coil in the conceptual ribbon of the peptide skeleton; (b) bracelet, which is characterized by the absence of this bond; and, (c) trypsin inhibitors, for which only two sequences have been discovered in this class to date. In general, cyclotides belonging to the bracelet subfamily show greater variation in loop size and amino acid sequences and more positively charged and more hydrophobic residues compared to the Mobius type.[165]

Achievement of the chemical synthesis[166–169] and folding mechanisms of "cyclic Cys knot" motifs opened up the route of solid-phase peptide synthesis of cyclotides, thus enabling their recognition as therapeutics.[167,170,171] The reader is encouraged to consult a number of excellent reviews on the discovery,[159,172,173] structures,[159,174,175] and applications[176–179] of these "privileged structures".

15.4.4 Cyclotides as Bioactive Candidates: Can Prospective Drugs be Foreseen?

In general, peptides are not considered therapeutic agents because they show poor stability and bioavailability. An exception is cyclotides, which are highly stable to proteolysis and also have superior sequence plasticity and a flexible backbone skeleton. Furthermore, cyclotides show several potential pharmacological activities, thus making them ideal candidates for drug development.[180,181] Since there are hundreds of cyclotides, mentioning all of them at this point is difficult due to space constraints. Hence, here we attempt to provide an overview of some of these molecules. A search using Scifinder was performed to gain insight into the number of articles in the literature reporting on the bioactivity of cyclotides (Figure 15.20). Furthermore, Table 15.3 provides "first-hand" information on cyclotides with regards to class, amino acid sequence, bioactivity, and reference. The rest of this section will be devoted to emphasizing the factors that affect the biological properties of these compounds and illustrating their possible applications as therapeutics.

15.4.5 Anti-HIV Activity

A report on the anti-HIV activity of cyclotides was produced as an initiative of the United States National Cancer Institute.[204] Initially, two compounds, presumed to be peptides, as revealed by NMR studies, were found to show anti-HIV activity. Surprisingly, these "peptides-to-be" were resistant to amino acid analysis/sequencing methods. These compounds were later confirmed as peptides and named circulin A and circulin B. They showed an EC_{50} of around 70 nM and also a cytotoxicity effect with an IC_{50} of about

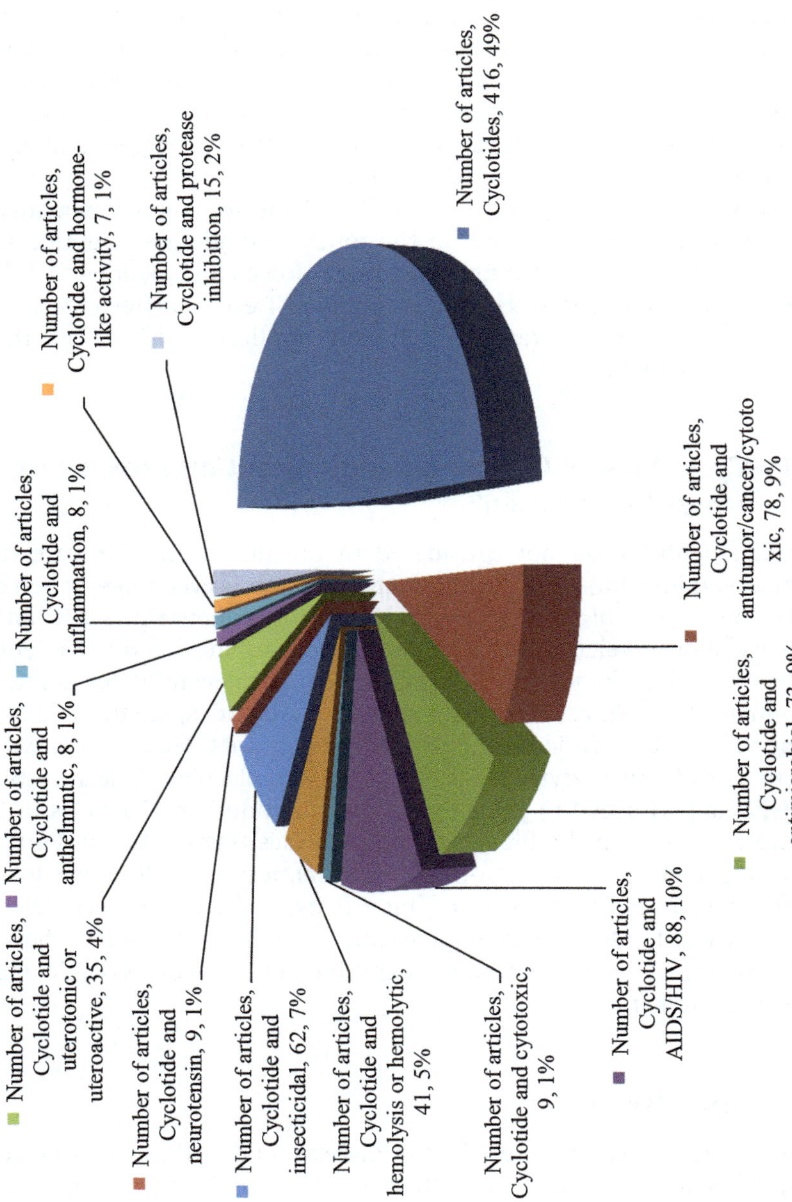

Figure 15.20 Schematic of the number of articles found using Scifinder with respect to the biological activities of cyclotides.

Table 15.3 Sequences and bioactivity of selected cyclotides along with references; a general schematic of a cyclotide is shown below, in which the blue line represents head to tail cyclization, yellow lines the disulphide bonds (three) between six Cys residues and "X" the amino acid residues.[a]

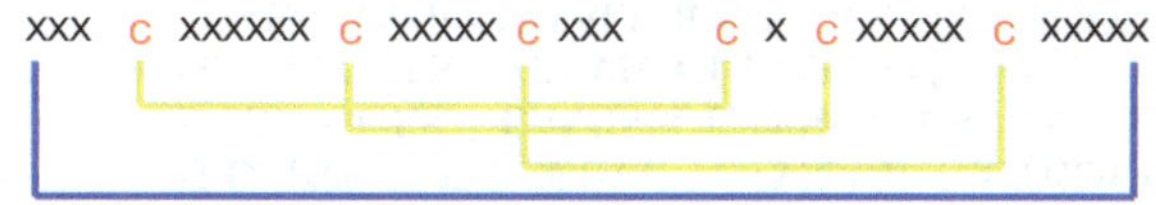

Name of the cyclotide	Class[b]	Species	Sequence[c]	Activity	Ref.
Circulin A	I	*C. parvifolia*	G. . .IP..CGES. . .CVWIP.CI.S.AAL.G.CSCKN. . .KVCYR..N	Antibacterial, Haemolytic, Anti-HIV	182, 183
Circulin B	I	*C. parvifolia*	GV..IP..CGES. . .CVFIP.CI.ST.LL.G.CSCKN. . .KVCYR..N	Antibacterial, Haemolytic, Anti-HIV	182, 183
Circulin C	I	*C. parvifolia*	G. . .IP..CGES. . .CVFIP.CI.TS.VA.G.CSCKS. . .KVCYR..N	Anti-HIV	183
Circulin D	I	*C. parvifolia*	K. . .IP..CGES. . .CVWIP.CV.TS.IF.N.CKCEN. . .KVCYH..D	Anti-HIV	183
Circulin E	I	*C. parvifolia*	K. . .IP..CGES. . .CVWIP.CL.TS.VF.N.CKCEN. . .KVCYH..D	Anti-HIV	183
Circulin F	I	*C. parvifolia*	A. . .IP..CGES. . .CVWIP.CI.S.AAI.G.CSCKN. . .KVCYR. . .	Anti-HIV	183
Cyclopsychotride A	I	*P. longipes*	S. . .IP..CGES. . .CVFIP.CTVT..ALLG.CSCKS. . .KVCYK..N	Antibacterial, Cytotoxic, Haemolytic, Neurotensin antagonist	168, 184
Cycloviolacin O1	I	*V. odorata*	..CAESCVYIP.CTVTALLGCSC. . .SNRVCY.NG.IP	Nematocidal, molluscicidal	158
Cycloviolacin O2	I	*V. odorata*	G. . .IP..CGES. . .CVWIP.CI.SSAI..G.CSCKS. . .KVCYR..N	Antibacterial, Cytotoxic, Haemolytic, Marine anti-fouling	185–188
Cycloviolacin O3	I	*V. odorata*	..CGESCVWIP.CISSA.IGCSC. . .KNKVCYRNG.IP	Anthelmintic	158
Cycloviolacin O4	I	*V. odorata*	..CGESCVWIP.CLTSA.IGCSC. . .KSKVCYRNG.IP	Host defence	158

Table 15.3 (*Continued*)

Name of the cyclotide	Class[b]	Species	Sequence[c]	Activity	Ref.
Cycloviolacin O5	I	*V. odorata*	..CGESCVWIP.CISSA.VGCSC...KNKVCYKNGT.P	Host defence	158
Cycloviolacin O6	I	*V. odorata*	..CGESCVWIP.CI.SAAVGCSC...KSKVCYKNGTLP	Host defence	158
Cycloviolacin O7	I	*V. odorata*	..CGESCVWIP.CTITALAGCKC...KSKVCY.NS.IP	Host defence	158
Cycloviolacin O8	I	*V. odorata*	..CGESCVWIP.CISS.VVGCSC...KSKVCYKNGTLP	Anthelmintic	158
Cycloviolacin O9	I	*V. odorata*	..CGESCVWIP.CLTSAV.GCSC...KSKVCYRNG.IP	Host defence	158
Cycloviolacin O10	I	*V. odorata*	..CGESCVYIP.CLTSAV.GCSC...KSKVCYRNG.IP	Host defence	158
Cycloviolacin O11	I	*V. odorata*	..CGESCVWIP.CI.SAVVGCSC...KSKVCYKNGTLP	Host defence	158
Cycloviolacin H1	I	*V. hederaceae*	..CGESCVYIP.CLTSA.IGCSC...KSKVCYRNG.IP	Host defence	158
Cycloviolacin Y1	I	*V. yedonesis*	GGT.I.FDCGET...CFLGT.CY.T.P...G.CSCGN..Y.GLCYGT.N	Haemolytic, Anti-HIV	189
Cycloviolacin Y2	I	*V. yedonesis*	GGT.I.FDCGES...CFLGT.CY.T.A...G.CSCGN..W.GLCYGT.N	Haemolytic, Anti-HIV	189
Cycloviolacin Y3	I	*V. yedonesis*	GGT.I.FDCGET...CFLGT.CY.T.A...G.CSCGN..W.GLCYGT.N	Haemolytic, Anti-HIV	189
Cycloviolacin Y4	I	*V. yedonesis*	G...VP..CGES...CVFIP.CITGVI...G.CSCSS...NVCY..LN	Haemolytic, Anti-HIV	189
Cycloviolacin Y5	I	*V. yedonesis*	G...IP..CAES...CVWIP.CT.TALV..G.CSCSD...KVCY...N	Haemolytic, Anti-HIV	189
Cycloviolin A	I	*L. cymosa*	GV..IP..CGES...CVFIP.CI.SAAI..G.CSCKN...KVCYR..N	Anti-HIV	190
Cycloviolin B	I	*L. cymosa*	GT.A...CGES...CYVLP.CF.T.V...G.CTCTS...SQ.CFK..N	Anti-HIV	190
Cycloviolin C	I	*L. cymosa*	G...IP..CGES...CVFIP.CL.TTVA..G.CSCKN...KVCYR..N	Anti-HIV	190
Cycloviolin D	I	*L. cymosa*	G...FP..CGES...CVFIP.CI.S.AAI.G.CSCKN...KVCYR..N	Anti-HIV	190
Kalata B5	I	*O. affinis*	..CGESCVYIP.CI.SGVIGCSC...TDKVCYLNGT.P	Nematocidal	158
Kalata B8	I	*O. affinis*	G.S.V.LNCGET...CLLGT.CY.TT...G.CTCNK..Y.RVCTK..D	Anti-HIV	191
Palicourein	I	*P. condensate*	G.D..PTFCGET...CRVIPVCTYS.AAL.G.CTCDDRS.DGLCKR..N	Anti-HIV	165
vhl-1	I	*V. hederacea*	S...I.S.CGES...CAMISFCF.TEVI..G.CSCKN...KVCY..LN	Anti-HIV	192
Vitri A	I	*V. tricolor*	G...IP..CGES...CVWIP.CI.TSAI..G.CSCKS...KVCYR..N	Cytotoxic	193
Hypa A	I	*H. parviflorus*	..CAESCVYIP.CTITALLGCSC...KNKVCY.NG.IP	Host defence	194
Cycloviolacin O14	II	*V. odorata*	G.SI.PA.CGES...CFKGK.CY.T.P...G.CSCSK..Y.PLCAK..N	Haemolytic	195
Cycloviolacin O15	II	*V. odorata*	GL.V.P..CGET...CFTGK.CY.T.P...G.CSCS...Y.PICKK..N	Haemolytic	195
Cycloviolacin O24	II	*V. odorata*	GL...PT.CGET...CFGGT.CN.T.P...G.CTCD..PW.PVCTH..N	Haemolytic	195

Kalata B1	II	*O. affinis, V. odorata*	GL...PV.CGET...CVGGT.CN.T.P...G.CTCS...W.PVCTR..D	Haemolytic, Insecticidal, Uterotonic, Anti-HIV	182, 196–198
Kalata B2	II	*O. affinis*	GL...PV.CGET...CFGGT.CN.T.P...G.CSCT...W.PICTR..D	Haemolytic, Insecticidal	182, 199
Kalata B3	II	*O. affinis*	.TCGETCFGGT.C...NTPGCTCD..PWPICTRDG.LP	Nematocidal	158
Kalata B6	II	*O. affinis*	.TCGETCFGGT.C...NTPGCSCS..SWPICTRNG.LP	Nematocidal	196
Kalata B7	II	*O. affinis*	.VCGETCTLGT.C...YTQGCTC...SWPICKRNG.LP	Nematocidal	196
Kalata S	II	*O. affinis*	.VCGETCVGGT.C...NTPGCSC...SWPVCTRNG.LP	Host defence	158
Varv A	II	*V. arvensis, V. odorata*	GL...PV.CGET...CVGGT.CN.T.P...G.CSCS...W.PVCTR..N	Cytotoxic, Haemolytic	200, 201
Varv B	II	*V. arvensis*	.VCGETCFGGT.C...NTPGCSCD..PWPMCSRNG.LP	Host defence	200
Varv C	II	*V. arvensis*	.ICGETCVGGT.C...NTPGCSC...SWPVCTRNGV.P	Host defence	200
Varv D	II	*V. arvensis*	.ICGETCVGGS.C...NTPGCSC...SWPVCTRNG.LP	Cytotoxic, Antitumour	200
Varv E	II	*V. arvensis, V. tricolor*	GL...PI.CGET...CVGGT.CN.T.P...G.CSCS...W.PVCTR..N	Cytotoxic	200
Varv F	II	*V. arvensis*	GV...PI.CGET...CTLGT.CY.T.A...G.CSCS...W.PVCTR..N	Cytotoxic	200
Varv G	II	*V. arvensis*	.VCGETCFGGT.C...NTPGCSCD..PWPVCSRNGV.P	Host defence	200
Varv H	II	*V. arvensis*	.VCGETCFGGT.C...NTPGCSCE..TWPVCSRNG.LP	Cytotoxic, Antitumour	200
Violapeptide 1	II	*V. arvensis*	.VCGETCVGGT.C...NTPGCSC...SRPVCTXNG.LP	Host defence	202
MCoTI-I	III	*M. cochinchinensis*	GG.V...CPKILQRCRRDSDC...P...GACICRG...NGYCGSGSD	Trypsin inhibitor	203
MCoTI-II	III	*M. cochinchinensis*	GG.V...CPKILKKCRRDSDC...P...GACICRG...NGYCGSGSD	Trypsin inhibitor	203

[a]Bold letters refer to Cys residues that mark the points of disulphide connectivity.
[b]Class: I = Bracelet; II = Mobius; III = Trypsin inhibitor.
[c]All peptides are cyclic and hence the choice of starting residue is arbitrary.

500 nM.[182] This characteristic dual behaviour is also observed in other cyclotides with anti-HIV activity. Prior to this discovery, the only peptides with anti-HIV activity belonging to the Mobius subfamily were kalata B1 and varv E.[189,205] Furthermore, it was observed that cyclotides of the subfamily Mobius are superior to those of bracelet in their capacity to inhibit HIV and as well as in their capacity as cytotoxic inducers.[205] It was demonstrated that variations in the amino acid sequences among the subfamilies did not affect anti-HIV activities, thus revealing that the overall peptide structure and not individual amino acids are essential for this activity. It is also important to note that linear versions of these peptides did not show anti-HIV activity, despite the fact that they are more flexible than the cyclic ones. On the basis of these results, it can be concluded that an intact "cyclic Cys knot" network is vital for HIV inhibition.

In another study, it was shown that there is a relationship between the hydrophobicity of certain loop regions and anti-HIV activity.[189] The presence of charged amino acids in the loops affected activities and is assumed to be caused by membrane binding interactions.[206]

15.4.6 Anti-cancer and Cytotoxic Activities

Cyclotides exhibit toxicity against various cell lines derived from different types of cancer types, such as myeloma, T-cell leukaemia, lung cancer, lymphoma, and adenocarcinoma.[207] This effect can be attributed mainly to the high specificity of these compounds to infected cells over normal counterparts. However, *in vivo* studies have not revealed promising results with cyclotides and hence these molecules warrant further evaluation.[186] In contrast, anti-angiogenic therapy has gained great attention in the field of cancer. In this regard, Gunasekara *et al.* used the concept of grafting biologically active peptide epitopes that carry polyarginine onto the kalata B1 scaffold.[181] The results revealed that the compounds acted as stabilized VEGF-A antagonists but did not merit clinical trials due to a lack of potency.

Another cyclotide scaffold, MCoTI-II, has been used in studies addressing the inhibition of tryptase and leucocyte elastase, both of which cause inflammatory disorders,[208] and the same motif has been used for the development of inhibitors for 3C protease of FMDV (foot and mouth disease virus).[209] Although micromolar values have been observed for this protease, this is apparently the first peptidic inhibitor known.

Tang and co-workers have shown that vila A and B cyclotides are the most cytotoxic agents against U251, MDA-MB-231, A549, DU145, and BEL-7402 cell lines.[210] Homology modelling studies revealed that hydrophobicity appears to be a key parameter in determining cytotoxicity. These results pave way for the cyclotides to be introduced into drug design. P. Lindholm *et al.* showed that varv A, varv F, and cycloviolacin O2 were highly cytotoxic, the latter being the strongest inhibitor, with an IC_{50} 0.1–0.3 μM.[207]

15.4.7 Antimicrobial Activity

The first report on the antimicrobial activity of synthetic cyclotides appeared in 1999.[168] It should be emphasized here that not all cyclotides are antimicrobials. In an initial attempt, kalata B1, circulin A and B and cyclopsychotride A showed promising activities against a panel of both Gram-positive and Gram-negative human pathogenic bacteria and fungi, with kalata B1 showing the highest potency (MIC = 0.26 μM). It was observed that these four peptides exhibited selective activity against bacteria and fungi. Although circulin A and circulin B have comparable sequences, the latter showed broader activities.[168] This difference in activities is relevant from a scientific perspective, since minor changes in the amino acid composition could result in selectivities for further development as pharmaceuticals. On the other hand, circulin A and kalata B1 showed differences in antimicrobial activities, thereby suggesting that hydrophobicity or charge distribution alone are not crucial for defining potency. Recently, cycloviolacin O2[187] and hedyotide B1[211] were shown to have anti-pathogenic effects on human bacteria. Further, no report on the effect of cyclotides on plant pathogens has been published. Another study addressed the effect of these peptides on soil bacteria, plants, and algae and determined their impact on the environment.[212]

Pranting *et al.* determined the efficacy of several cyclotides to inhibit bacteria and found that cycloviolacin O2 was highly potent in the series against *S. enterica serovar Typhimurium LT2*, *E. coli.*, *K. pneumoniae* and *P. aeruginosa.*[187] The need for charged species in the molecule was exhibited by the complete loss of activity when glutamic acid and lysine residues were masked in this peptide.

15.4.8 Anthelmintic Activity

The effects of cyclotide activity on a number of human parasites and livestock have been evaluated, and promising results have been obtained against the two most important gastrointestinal nematodes in sheep *viz.* *Haemonchus contortus* and *Trichostrongylus colubriformis.*[189] These peptides were also found to affect *Ancylostoma caninum*, a canine hookworm, and also the larvae of *Nector americanus.*[213] The property of anthelmintic activity could be an emerging field as cyclotide-producing plants can be considered natural medicines to fight against parasites that affect mainly third world countries.

15.4.9 Anti-insecticidal Activity

Significant social and economic importance has been given to the control of insect pests in crops. As a result of the growing human population, there is an urgent need to enhance the efficacy of food production, especially in crop plants. In this context, Gruber *et al.* took the advantage of cyclotides with insecticidal properties.[179] In an experiment conducted by this group, kalata B1 was inserted into an artificial bean-based diet, which was fed to neonates of *Helicoverpa punctigera*, a lepidopteran insect. It was observed that most of the food was left intact as the larvae failed to grow on the kalata B1 diets. In a control experiment, larvae reached the fifth instar stage of development in the 16-day period, while the larvae on the kalata B1 diet failed to grow. In the second trial, both kalata B1 and B2 were fed to a second *Helicoverpa* sp., *Helicoverpa armigera.* Larval growth was registered past the first instar stage of development but was still retarded by about 70%, and about 25% of the larvae failed to survive. These experiments showed that cyclotides affect the growth of insects and hence may be useful for crop treatment and protection.[199]

15.4.10 Application in Drug Design: A Ray of Hope!

Apart from being therapeutically active, cyclotides are striking molecules for drug development because of their remarkable stability. Kalata B1 is stable to chaotropic agents like 6M guanidine HCl and 8M urea, temperatures that approach boiling, acids, and also various proteases. In order to examine the structural significance of cyclotides for drug design and development, a SAR study on kalata B1 was undertaken by Colgrave *et al.*[214] The removal of one of the disulphide bonds not only resulted in reduced conformational rigidity but also made the molecule vulnerable for denaturation by chemical means. On the other hand, further loss of stability was observed when all of the S–S linkages were removed. Acyclic mutants of kalata B1 and native kalata B1 were treated with a number of enzymes and were less stable to proteolytic activity. These findings indicate that both the cyclic skeleton, as well as the "cyclic Cys knot" motif, contributes to the extra strong stability of cyclotides, thus opening up avenues for structural variation within the sequences and making them highly potential candidates in the field of drug design and associated fields.

15.4.11 Current Opinion and Future Outlook – is a New Scenario Emerging?

Cyclotides, because of their high stability, can be regarded as templates for the design and development of drug molecules. At present, exhaustive studies are underway to gauge their potential therapeutic applications. It is evident that solid-phase peptide synthesis is the strategy of choice that can generate a significant number of synthetic versions of these cyclic analogues. These molecules can then be subjected to a number of biological studies, including human clinical trials, in order to measure their bioavailability and toxicity. Although solid-phase synthesis is the best approach, other strategies for the biological production of cyclotides are progressing rapidly. In addition, proof-of-concept studies have revealed that pharmacologically approved bioactive peptide sequences can be implanted (grafted) in cyclotide templates, thus directing us towards a new horizon of agents to combat deadly diseases. However, a number of unmet challenges remain, such as lack of deep knowledge gained from pharmacokinetic studies on "drug-like" candidates and expansion of more potent leads. Furthermore, there is little information available on the oral bioavailability of these peptides. Nonetheless, it is apparent that this is a cutting-edge research field with the capacity to add new analogues to the list of molecules that are currently undergoing clinical trials.

Acknowledgements

The work carried out in the laboratories of the authors was partially supported by the National Research Foundation and the University of KwaZulu Natal (South Africa); CICYT (CTQ2012-30930), the *Generalitat de Catalunya* (2014 SGR 137), and the Institute for Research in Biomedicine Barcelona (IRB Barcelona) (Spain).

References

1. A. K. Sato, M. Viswanathan, R. B. Kent and C. R. Wood, *Curr. Opin. Biotechnol.*, 2006, **17**, 638–642.
2. P. Vlieghe, V. Lisowski, J. Martinez and M. Khrestchatisky, *Drug Discovery Today*, 2010, **15**, 40–56.
3. J. Hamman, G. Enslin and A. Kotzé, *BioDrugs*, 2005, **19**, 165–177.
4. H. Kessler, *Angew. Chem., Int. Ed. Engl.*, 1982, **21**, 512–523.
5. M. M. Shemyakin, Y. A. Ovchinnikov and V. T. Ivanov, *Angew. Chem., Int. Ed. Engl.*, 1969, **8**, 492–499.
6. A. Giannis and T. Kolter, *Angew. Chem., Int. Ed. Engl.*, 1993, **32**, 1244–1267.
7. R. J. Simon, R. S. Kania, R. N. Zuckermann, V. D. Huebner, D. A. Jewell, S. Banville, S. Ng, L. Wang, S. Rosenberg and C. K. Marlowe, *Proc. Natl. Acad. Sci. U. S. A.*, 1992, **89**, 9367–9371.
8. H.-K. Lee, M. D. Smith, B. J. Smith, J. Grussendorf, L. Xu, R. J. Gillies, H. S. White and G. Bulaj, *ACS Chem. Biol.*, 2009, **4**, 659–671.
9. C. J. White and A. K. Yudin, *Nat. Chem.*, 2011, **3**, 509–524.
10. F. Yokoyama, N. Suzuki, M. Haruki, N. Nishi, S. Oishi, N. Fujii, A. Utani, H. K. Kleinman and M. Nomizu, *Biochemistry*, 2004, **43**, 13590–13597.
11. T. Curtius and F. Goebel, *J. Prakt. Chem.*, 1888, **37**, 150–181.
12. S. Rajappa and M. V. Natekar, in *Advances in Heterocyclic Chemistry*, ed. R. K. Alan, Academic Press, 1993, vol. 57, pp. 187–289.
13. C. Prasad, *Peptides*, 1995, **16**, 151–164.
14. C. J. Dinsmore and D. C. Beshore, *Tetrahedron*, 2002, **58**, 3297–3312.
15. D. Horton, G. Bourne and M. Smythe, *J. Comput.-Aided Mol. Des.*, 2002, **16**, 415–431.

16. P. M. Fischer, *J. Pept. Sci.*, 2003, **9**, 9–35.
17. M. B. Martins and I. Carvalho, *Tetrahedron*, 2007, **63**, 9923–9932.
18. P. J. Milne and G. Kilian, in *Comprehensive Natural Products II*, ed. H.-W. Liu and L. Mander, Elsevier, Oxford, 2010, pp. 657–698.
19. A. D. Borthwick, *Chem. Rev.*, 2012, **112**, 3641–3716.
20. J. F. Gonzalez, I. Ortin, E. de la Cuesta and J. C. Menendez, *Chem. Soc. Rev.*, 2012, **41**, 6902–6915.
21. K. McCleland, P. J. Milne, F. R. Lucieto, C. Frost, S. C. Brauns, M. Van De Venter, J. Du Plessis and K. Dyason, *J. Pharm. Pharmacol.*, 2004, **56**, 1143–1153.
22. K. Kanoh, S. Kohno, T. Asari, T. Harada, J. Katada, M. Muramatsu, H. Kawashima, H. Sekiya and I. Uno, *Bioorg. Med. Chem. Lett.*, 1997, **7**, 2847–2852.
23. K. Kanoh, S. Kohno, J. Katada, J. Takahashi, I. Uno and Y. Hayashi, *Bioorg. Med. Chem.*, 1999, **7**, 1451–1457.
24. Y. Hayashi, S. Orikasa, K. Tanaka, K. Kanoh and Y. Kiso, *J. Org. Chem.*, 2000, **65**, 8402–8405.
25. A. V. Singh, M. Bandi, N. Raje, P. Richardson, M. A. Palladino, D. Chauhan and K. C. Anderson, *Blood*, 2011, **117**, 5692–5700.
26. S. Liao, X. Qin, D. Li, Z. Tu, J. Li, X. Zhou, J. Wang, B. Yang, X. Lin, J. Liu, X. Yang and Y. Liu, *Eur. J. Med. Chem.*, 2014, **83**, 236–244.
27. R. Ben Ameur Mehdi, K. A. Shaaban, I. K. Rebai, S. Smaoui, S. Bejar and L. Mellouli, *Nat. Prod. Res.*, 2009, **23**, 1095–1107.
28. H. Jamie, G. Kilian, K. Dyason and P. J. Milne, *J. Pharm. Pharmacol.*, 2002, **54**, 1659–1665.
29. F. Rodrigues de Sa Alves, E. J. Barreiro and C. A. M. Fraga, *Mini-Rev. Med. Chem.*, 2009, **9**, 782–793.
30. T. Gan and J. M. Cook, *Tetrahedron Lett.*, 1997, **38**, 1301–1304.
31. B. Evans, A. Pipe, L. Clark and M. Banks, *Bioorg. Med. Chem. Lett.*, 2001, **11**, 1297–1300.
32. S. Zhao, T. Gan, P. Yu and J. M. Cook, *Tetrahedron Lett.*, 1998, **39**, 7009–7012.
33. H. D. Jain, C. Zhang, S. Zhou, H. Zhou, J. Ma, X. Liu, X. Liao, A. M. Deveau, C. M. Dieckhaus, M. A. Johnson, K. S. Smith, T. L. Macdonald, H. Kakeya, H. Osada and J. M. Cook, *Bioorg. Med. Chem.*, 2008, **16**, 4626–4651.
34. S. Preciado, L. Mendive-Tapia, C. Torres-Garcia, R. Zamudio-Vazquez, V. Soto-Cerrato, R. Perez-Tomas, F. Albericio, E. Nicolas and R. Lavilla, *MedChemComm*, 2013, **4**, 1171–1174.
35. P. Ruiz-Sanchis, S. A. Savina, F. Albericio and M. Álvarez, *Chem. - Eur. J.*, 2011, **17**, 1388–1408.
36. C. Takahashi, A. Numata, Y. Ito, E. Matsumura, H. Araki, H. Iwaki and K. Kushida, *J. Chem. Soc., Perkin Trans. 1*, 1994, 1859–1864.
37. M. Yanagihara, N. Sasaki-Takahashi, T. Sugahara, S. Yamamoto, M. Shinomi, I. Yamashita, M. Hayashida, B. Yamanoha, A. Numata, T. Yamori and T. Andoh, *Cancer Sci.*, 2005, **96**, 816–824.

38. G. Carr, W. Tay, H. Bottriell, S. K. Andersen, A. G. Mauk and R. J. Andersen, *Org. Lett.*, 2009, **11**, 2996–2999.
39. C. J. Barrow, P. Cai, J. K. Snyder, D. M. Sedlock, H. H. Sun and R. Cooper, *J. Org. Chem.*, 1993, **58**, 6016–6021.
40. S. P. B. Ovenden, G. Sberna, R. M. Tait, H. G. Wildman, R. Patel, B. Li, K. Steffy, N. Nguyen and B. M. Meurer-Grimes, *J. Nat. Prod.*, 2004, **67**, 2093–2095.
41. H. Minato, M. Matsumoto and T. Katayama, *J. Chem. Soc., Perkin Trans. 1*, 1973, 1819–1825.
42. K. Katagiri, K. Sato, S. Hayakawa, T. Matsushi and H. Minato, *J. Antibiot.*, 1970, **23**, 420–422.
43. H. Minato, M. Matsumot and T. Katayama, *J. Chem. Soc. D*, 1971, 44–45.
44. X. Fu, M. L. G. Ferreira, F. J. Schmitz and M. Kelly-Borges, *J. Nat. Prod.*, 1998, **61**, 1226–1231.
45. Y. Takahara, F. Battaini, D. D. Ross, S. A. Akman, N. R. Bachur, K. R. Bailey, E. Lakatos and A. Peterkofsky, *J. Clin. Endocrinol. Metab.*, 1983, **56**, 312–319.
46. D. R. Houston, B. Synstad, V. G. H. Eijsink, M. J. R. Stark, I. M. Eggleston and D. M. F. van Aalten, *J. Med. Chem.*, 2004, **47**, 5713–5720.
47. M. Mitova, S. Popov and S. De Rosa, *J. Nat. Prod.*, 2004, **67**, 1178–1181.
48. K. Strom, J. Sjogren, A. Broberg and J. Schnurer, *Appl. Environ. Microbiol.*, 2002, **68**, 4322–4327.
49. F. Fdhila, V. Vázquez, J. L. Sánchez and R. Riguera, *J. Nat. Prod.*, 2003, **66**, 1299–1301.
50. L. Yang, L. Miao, O. Lee, X. Li, H. Xiong, K.-L. Pang, L. Vrijmoed and P.-Y. Qian, *Appl. Microbiol. Biotechnol.*, 2007, **74**, 1221–1231.
51. Z. Jiang, K. G. Boyd, A. Mearns-spragg, D. R. Adams, P. C. Wright and J. G. Burgess, *Nat. Prod. Lett.*, 2000, **14**, 435–440.
52. A. C. Stierle, J. H. Cardellina and G. A. Strobel, *Proc. Natl. Acad. Sci. U. S. A.*, 1988, **85**, 8008–8011.
53. C. C. Cain, D. H. Lee, R. H. Waldo, A. T. Henry, E. J. Casida, M. C. Wani, M. E. Wall, N. H. Oberlies and J. O. Falkinham, *Antimicrob. Agents Chemother.*, 2003, **47**, 2113–2117.
54. P. Klausmeyer, O. M. Z. Howard, S. M. Shipley and T. G. McCloud, *J. Nat. Prod.*, 2009, **72**, 1369–1372.
55. H. J. Shin, M. A. M. Mondol, T. K. Yu, H.-S. Lee, Y.-J. Lee, H. J. Jung, J. H. Kim and H. J. Kwon, *Phytochem. Lett.*, 2010, **3**, 194–197.
56. P. S. Yan, Y. Song, E. Sakuno, H. Nakajima, H. Nakagawa and K. Yabe, *Appl. Environ. Microbiol.*, 2004, **70**, 7466–7473.
57. K. H. Rhee, *J. Microbiol. Biotechnol.*, 2003, **13**, 984–988.
58. S.-H. Qi, Y. Xu, J. Gao, P.-Y. Qian and S. Zhang, *Ann. Microbiol.*, 2009, **59**, 229–233.
59. G. W. Kirby, G. V. Rao and D. J. Robins, *J. Chem. Soc., Perkin Trans. 1*, 1988, 301–304.

60. T. C. McMahon, S. Stanley, E. Kazyanskaya, D. Hung and J. L. Wood, *Tetrahedron Lett.*, 2011, **52**, 2262–2264.
61. R. X. Tan, P. R. Jensen, P. G. Williams and W. Fenical, *J. Nat. Prod.*, 2004, **67**, 1374–1382.
62. U. Gross, M. Nieger and S. Bräse, *Chem. - Eur J.*, 2010, **16**, 11624–11631.
63. B. M. Ruff, S. Zhong, M. Nieger, M. Sickert, C. Schneider and S. Bräse, *Eur. J. Org. Chem.*, 2011, 6558–6566.
64. K. C. Nicolaou, M. Lu, S. Totokotsopoulos, P. Heretsch, D. Giguere, Y. P. Sun, D. Sarlah, T. H. Nguyen, I. C. Wolf, D. F. Smee, C. W. Day, S. Bopp and E. A. Winzeler, *J. Am. Chem. Soc.*, 2012, **134**, 17320–17332.
65. N. Tsumagari, R. Nakai, H. Onodera, A. Hasegawa, E. S. Rahayu, K. Ando and Y. Yamashita, *J. Antibiot.*, 2004, **57**, 532–534.
66. H. Onodera, A. Hasegawa, N. Tsumagari, R. Nakai, T. Ogawa and Y. Kanda, *Org. Lett.*, 2004, **6**, 4101–4104.
67. A. Daugan, P. Grondin, C. Ruault, A.-C. Le Monnier de Gouville, H. Coste, J. M. Linget, J. Kirilovsky, F. Hyafil and R. Labaudinière, *J. Med. Chem.*, 2003, **46**, 4533–4542.
68. A. D. Borthwick, *J. Med. Chem.*, 2010, **53**, 6525–6538.
69. A. D. Borthwick, J. Liddle, D. E. Davies, A. M. Exall, C. Hamlett, D. M. Hickey, A. M. Mason, I. E. D. Smith, F. Nerozzi, S. Peace, D. Pollard, S. L. Sollis, M. J. Allen, P. M. Woollard, M. A. Pullen, T. D. Westfall and D. J. Stanislaus, *J. Med. Chem.*, 2012, **55**, 783–796.
70. T. D. Brooks, S. M. W. Wang, N. Brunner and P. A. Charlton, *Anti-Cancer Drugs*, 2004, **15**, 37–44.
71. A. K. Szardenings, D. Harris, S. Lam, L. Shi, D. Tien, Y. Wang, D. V. Patel, M. Navre and D. A. Campbell, *J. Med. Chem.*, 1998, **41**, 2194–2200.
72. A. K. Szardenings, V. Antonenko, D. A. Campbell, N. DeFrancisco, S. Ida, L. Shi, N. Sharkov, D. Tien, Y. Wang and M. Navre, *J. Med. Chem.*, 1999, **42**, 1348–1357.
73. V. Mas, A. Falco, I. Brocal, L. Perez, J. M. Coll and A. Estepa, *Antiviral Res.*, 2006, **72**, 107–115.
74. I. R. Shimi, N. Abedallah and S. Fathy, *Antimicrob. Agents Chemother.*, 1977, **11**, 373–375.
75. J. Campbell and H. E. Blackwell, *J. Comb. Chem.*, 2009, **11**, 1094–1099.
76. J. Guan, S. Mathai, P. Harris, J. Y. Wen, R. Zhang, M. Brimble and P. Gluckman, *Neuropharmacology*, 2007, **53**, 749–762.
77. M. Delgado, A. G. Caicoya, V. Greciano, B. Benhamú, M. L. López-Rodríguez, M. S. Fernández-Alfonso, M. A. Pozo, J. Manzanares and J. A. Fuentes, *Eur. J. Pharmacol.*, 2005, **511**, 9–19.
78. M. a. L. López-Rodríguez, M. J. Morcillo, E. Fernández, B. Benhamú, I. Tejada, D. Ayala, A. Viso, M. Olivella, L. Pardo, M. Delgado, J. Manzanares and J. A. Fuentes, *Bioorg. Med. Chem. Lett.*, 2003, **13**, 1429–1432.
79. N. Shimazaki, I. Shima, M. Okamoto, K. Yoshida, K. Hemmi and M. Hashimoto, *Lipids*, 1991, **26**, 1175–1178.

80. R. Nishizawa, T. Nishiyama, K. Hisaichi, N. Matsunaga, C. Minamoto, H. Habashita, Y. Takaoka, M. Toda, S. Shibayama, H. Tada, K. Sagawa, D. Fukushima, K. Maeda and H. Mitsuya, *Bioorg. Med. Chem. Lett.*, 2007, **17**, 727–731.
81. R. Nishizawa, T. Nishiyama, K. Hisaichi, C. Minamoto, N. Matsunaga, Y. Takaoka, H. Nakai, S. Jenkinson, W. M. Kazmierski, H. Tada, K. Sagawa, S. Shibayama, D. Fukushima, K. Maeda and H. Mitsuya, *Bioorg. Med. Chem. Lett.*, 2011, **21**, 1141–1145.
82. K. Maeda, H. Nakata, Y. Koh, T. Miyakawa, H. Ogata, Y. Takaoka, S. Shibayama, K. Sagawa, D. Fukushima, J. Moravek, Y. Koyanagi and H. Mitsuya, *J. Virol.*, 2004, **78**, 8654–8662.
83. B. E. Evans, K. E. Rittle, M. G. Bock, R. M. DiPardo, R. M. Freidinger, W. L. Whitter, G. F. Lundell, D. F. Veber and P. S. Anderson, *J. Med. Chem.*, 1988, **31**, 2235–2246.
84. D. A. Horton, G. T. Bourne and M. L. Smythe, *Chem. Rev.*, 2003, **103**, 893–930.
85. W. H. Miller, R. M. Keenan, R. N. Willette and M. W. Lark, *Drug Discovery Today*, 2000, **5**, 397–408.
86. R. Breinbauer, I. R. Vetter and H. Waldmann, *Angew. Chem., Int. Ed.*, 2002, **41**, 2878–2890.
87. C. Page, C. Michael, M. Sutter, M. Walker and B. B. Hoffman, *Integrated Pharmacology*, *C.V. Mosby*, 2nd edn, 2002.
88. K. T. Olkkola and J. Ahonen, in *Modern Anesthetics*, ed. J. Schüttler and H. Schwilden, Springer, Berlin Heidelberg, 2008, ch. 16, vol. 182, pp. 335–360.
89. D. F. Hanley Jr. and M. Pozo, *Int. J. Clin. Pract.*, 2000, **54**, 30–35.
90. J. R. T. Davidson, *J. Clin. Psychiatry*, 2001, **62**, 46–50.
91. W. V. McCall, *J. Clin. Psychiatry*, 2001, **62**, 27–32.
92. E. Shorter, *"Benzodiazepines". A Historical Dictionary of Psychiatry*, Oxford University Press, 2005, pp. 2041–2042.
93. B. E. Evans, M. G. Bock, K. E. Rittle, R. M. DiPardo, W. L. Whitter, D. F. Veber, P. S. Anderson and R. M. Freidinger, *Proc. Natl. Acad. Sci. U. S. A.*, 1986, **83**, 4918–4922.
94. M. G. Bock, R. M. DiPardo, B. E. Evans, K. E. Rittle, D. F. Veber, R. M. Freidinger, R. S. L. Chang and V. J. Lotti, *J. Med. Chem.*, 1988, **31**, 176–181.
95. M. G. Bock, R. M. DiPardo, B. E. Evans, K. E. Rittle, W. L. Whitter, D. F. Veber, R. M. Freidinger, R. S. L. Chang, T. B. Chen and V. J. Lotti, *J. Med. Chem.*, 1990, **33**, 450–455.
96. M. G. Bock, R. M. DiPardo, B. E. Evans, K. E. Rittle, W. L. Whitter, V. M. Garsky, K. F. Gilbert, J. L. Leighton and K. L. Carson, *J. Med. Chem.*, 1993, **36**, 4276–4292.
97. G. A. Showell, S. Bourrain, J. G. Neduvelil, S. R. Fletcher, R. Baker, A. P. Watt, A. E. Fletcher, S. B. Freedman and J. A. Kemp, *J. Med. Chem.*, 1994, **37**, 719–721.

98. M. S. Chambers, S. C. Hobbs, M. I. Graham, A. P. Watt, S. R. Fletcher, R. Baker, S. B. Freedman, S. Patel, A. J. Smith and V. G. Matassa, *Bioorg. Med. Chem. Lett.*, 1995, **5**, 2303–2308.
99. J. A. Lowe Iii, S. E. Drozda, S. McLean, D. K. Bryce, R. T. Crawford, S. Zorn, J. Morrone, T. A. Appleton and F. Lombardo, *Bioorg. Med. Chem. Lett.*, 1995, **5**, 1933–1936.
100. G. A. Showell, S. Bourrain, S. R. Fletcher, J. G. Neduvelil, A. E. Fletcher, S. B. Freedman, S. Patel, A. J. Smith, G. R. Marshall, M. I. Graham, B. Sohal and V. G. Matassa, *Bioorg. Med. Chem. Lett.*, 1995, **5**, 3023–3026.
101. M. B. Van Niel, S. B. Freedman, V. G. Matassa, S. Patel, R. R. Pengilley and A. J. Smith, *Bioorg. Med. Chem. Lett.*, 1995, **5**, 1421–1426.
102. J. L. Castro, R. G. Ball, H. B. Broughton, M. G. N. Russell, D. Rathbone, A. P. Watt, R. Baker, K. L. Chapman, A. E. Fletcher, S. Patel, A. J. Smith, G. R. Marshall, W. Ryecroft and V. G. Matassa, *J. Med. Chem.*, 1996, **39**, 842–849.
103. G. Semple, H. Ryder, D. A. Kendrick, M. Szelke, M. Ohta, M. Satoh, A. Nishida, S. Akuzawa and K. Miyata, *Bioorg. Med. Chem. Lett.*, 1996, **6**, 51–54.
104. J. L. Castro, H. B. Broughton, M. G. N. Russell, D. Rathbone, A. P. Watt, R. G. Ball, K. L. Chapman, S. Patel, A. J. Smith, G. R. Marshall and V. G. Matassa, *J. Med. Chem.*, 1997, **40**, 2491–2501.
105. G. Semple, H. Ryder, D. P. Rooker, A. R. Batt, D. A. Kendrick, M. Szelke, M. Ohta, M. Satoh, A. Nishida, S. Akuzawa and K. Miyata, *J. Med. Chem.*, 1997, **40**, 331–341.
106. E. M. Hadac, E. S. Dawson, J. W. Darrow, E. E. Sugg, T. P. Lybrand and L. J. Miller, *J. Med. Chem.*, 2006, **49**, 850–863.
107. J. J. Salata, N. K. Jurkiewicz, J. Wang, B. E. Evans, H. T. Orme and M. C. Sanguinetti, *Mol. Pharmacol.*, 1998, **54**, 220–230.
108. J. J. Salata, N. K. Jurkiewicz, M. C. Sanguinetti, P. K. S. Siegl, D. C. Claremon, D. C. Remy, J. M. Elliott and B. E. Libby, *Circulation*, 1996, I–529.
109. A. Walser, T. Flynn, C. Mason, H. Crowley, C. Maresca, B. Yaremko and M. O'Donnell, *J. Med. Chem.*, 1991, **34**, 1209–1221.
110. D. R. Armour, N. M. Aston, K. M. L. Morriss, M. S. Congreve, A. B. Hawcock, D. Marquess, J. E. Mordaunt, S. A. Richards and P. Ward, *Bioorg. Med. Chem. Lett.*, 1997, 7, 2037–2042.
111. M. H. Bolli, J. Marfurt, C. Grisostomi, C. Boss, C. Binkert, P. Hess, A. Treiber, E. Thorin, K. Morrison, S. Buchmann, D. Bur, H. Ramuz, M. Clozel, W. Fischli and T. Weller, *J. Med. Chem.*, 2004, **47**, 2776–2795.
112. H. G. Selnick, N. J. Liverton, J. J. Baldwin, J. W. Butcher, D. A. Claremon, J. M. Elliott, R. M. Freidinger, S. A. King, B. E. Libby, C. J. McIntyre, D. A. Pribush, D. C. Remy, G. R. Smith, A. J. Tebben, N. K. Jurkiewicz, J. J. Lynch, J. J. Salata, M. C. Sanguinetti, P. K. S. Siegl, D. E. Slaughter and K. Vyas, *J. Med. Chem.*, 1997, **40**, 3865–3868.

113. P. G. Wyatt, M. J. Allen, J. Chilcott, G. Hickin, N. D. Miller and P. M. Woollard, *Bioorg. Med. Chem. Lett.*, 2001, **11**, 1301–1305.
114. M. Amblard, I. Daffix, P. Bedos, G. Bergé, D. Pruneau, J.-L. Paquet, J.-M. Luccarini, P. Bélichard, P. Dodey and J. Martinez, *J. Med. Chem.*, 1999, **42**, 4185–4192.
115. M. Amblard, I. Daffix, G. Bergé, M. Calmès, P. Dodey, D. Pruneau, J.-L. Paquet, J.-M. Luccarini, P. Bélichard and J. Martinez, *J. Med. Chem.*, 1999, **42**, 4193–4201.
116. E. K. Dziadulewicz, M. C. Brown, A. R. Dunstan, W. Lee, N. B. Said and P. J. Garratt, *Bioorg. Med. Chem. Lett.*, 1999, **9**, 463–468.
117. M. R. Wood, J. J. Kim, W. Han, B. D. Dorsey, C. F. Homnick, R. M. DiPardo, S. D. Kuduk, T. MacNeil, K. L. Murphy, E. V. Lis, R. W. Ransom, G. L. Stump, J. J. Lynch, S. S. O'Malley, P. J. Miller, T.-B. Chen, C. M. Harrell, R. S. L. Chang, P. Sandhu, J. D. Ellis, P. J. Bondiskey, D. J. Pettibone, R. M. Freidinger and M. G. Bock, *J. Med. Chem.*, 2003, **46**, 1803–1806.
118. W. H. Miller, D. P. Alberts, P. K. Bhatnagar, W. E. Bondinell, J. F. Callahan, R. R. Calvo, R. D. Cousins, K. F. Erhard, D. A. Heerding, R. M. Keenan, C. Kwon, P. J. Manley, K. A. Newlander, S. T. Ross, J. M. Samanen, I. N. Uzinskas, J. W. Venslavsky, C. C. K. Yuan, R. C. Haltiwanger, M. Gowen, S.-M. Hwang, I. E. James, M. W. Lark, D. J. Rieman, G. B. Stroup, L. M. Azzarano, K. L. Salyers, B. R. Smith, K. W. Ward, K. O. Johanson and W. F. Huffman, *J. Med. Chem.*, 2000, **43**, 22–26.
119. R. M. Keenan, J. F. Callahan, J. M. Samanen, W. E. Bondinell, R. R. Calvo, L. Chen, C. DeBrosse, D. S. Eggleston, R. C. Haltiwanger, S. M. Hwang, D. R. Jakas, T. W. Ku, W. H. Miller, K. A. Newlander, A. Nichols, M. F. Parker, L. S. Southhall, I. Uzinskas, J. A. Vasko-Moser, J. W. Venslavsky, A. S. Wong and W. F. Huffman, *J. Med. Chem.*, 1999, **42**, 545–559.
120. T. M. Williams, C. A. Stump, D. N. Nguyen, A. G. Quigley, I. M. Bell, S. N. Gallicchio, C. B. Zartman, B.-L. Wan, K. D. Penna, P. Kunapuli, S. A. Kane, K. S. Koblan, S. D. Mosser, R. Z. Rutledge, C. Salvatore, J. F. Fay, J. P. Vacca and S. L. Graham, *Bioorg. Med. Chem. Lett.*, 2006, **16**, 2595–2598.
121. R. S. McDowell, B. K. Blackburn, T. R. Gadek, L. R. McGee, T. Rawson, M. E. Reynolds, K. D. Robarge, T. C. Somers and E. D. Thorsett, *J. Am. Chem. Soc.*, 1994, **116**, 5077–5083.
122. M. J. Fisher, B. Gunn, C. S. Harms, A. D. Kline, J. T. Mullaney, A. Nunes, R. M. Scarborough, A. E. Arfsten, M. A. Skelton, S. L. Um, B. G. Utterback and J. A. Jakubowski, *J. Med. Chem.*, 1997, **40**, 2085–2101.
123. R. M. Scarborough and D. D. Gretler, *J. Med. Chem.*, 2000, **43**, 3453–3473.
124. D. M. Leonard, *J. Med. Chem.*, 1997, **40**, 2971–2990.
125. J. T. Hunt, C. Z. Ding, R. Batorsky, M. Bednarz, R. Bhide, Y. Cho, S. Chong, S. Chao, J. Gullo-Brown, P. Guo, S. H. Kim, F. Y. F. Lee,

K. Leftheris, A. Miller, T. Mitt, M. Patel, B. A. Penhallow, C. Ricca, W. C. Rose, R. Schmidt, W. A. Slusarchyk, G. Vite and V. Manne, *J. Med. Chem.*, 2000, **43**, 3587–3595.
126. C. Papageorgiou and X. Borer, *Bioorg. Med. Chem. Lett.*, 1996, **6**, 267–272.
127. M. C. Carter, D. G. Alber, R. C. Baxter, S. K. Bithell, J. Budworth, A. Chubb, G. S. Cockerill, V. C. L. Dowdell, E. A. Henderson, S. J. Keegan, R. D. Kelsey, M. J. Lockyer, J. N. Stables, L. J. Wilson and K. L. Powell, *J. Med. Chem.*, 2006, **49**, 2311–2319.
128. W. C. Ripka, G. V. De Lucca, A. C. Bach Ii, R. S. Pottorf and J. M. Blaney, *Tetrahedron*, 1993, **49**, 3593–3608.
129. W. C. Ripka, G. V. De Lucca, A. C. Bach, Ii, R. S. Pottorf and J. M. Blaney, *Tetrahedron*, 1993, **49**, 3609–3628.
130. M. Hata and G. R. Marshall, *J. Comput.-Aided Mol. Des.*, 2006, **20**, 321–331.
131. N. Micale, R. Vairagoundar, A. G. Yakovlev and A. P. Kozikowski, *J. Med. Chem.*, 2004, **47**, 6455–6458.
132. K. Nakayama, H. C. Kawato, H. Inagaki and T. Ohta, *Org. Lett.*, 2001, **3**, 3447–3450.
133. N. Micale, A. P. Kozikowski, R. Ettari, S. Grasso, M. Zappalà, J.-J. Jeong, A. Kumar, M. Hanspal and A. H. Chishti, *J. Med. Chem.*, 2006, **49**, 3064–3067.
134. Y. Mita, K. Dodo, T. Noguchi-Yachide, H. Miyachi, M. Makishima, Y. Hashimoto and M. Ishikawa, *Bioorg. Med. Chem. Lett.*, 2010, **20**, 1712–1717.
135. Y. Mita, K. Dodo, T. Noguchi-Yachide, Y. Hashimoto and M. Ishikawa, *Bioorg. Med. Chem.*, 2013, **21**, 993–1005.
136. S. Butini, E. Gabellieri, M. Brindisi, A. Casagni, E. Guarino, P. B. Huleatt, N. Relitti, V. La Pietra, L. Marinelli, M. Giustiniano, E. Novellino, G. Campiani and S. Gemma, *Bioorg. Med. Chem. Lett.*, 2013, **23**, 85–89.
137. B. L. Grasberger, T. Lu, C. Schubert, D. J. Parks, T. E. Carver, H. K. Koblish, M. D. Cummings, L. V. LaFrance, K. L. Milkiewicz, R. R. Calvo, D. Maguire, J. Lattanze, C. F. Franks, S. Zhao, K. Ramachandren, G. R. Bylebyl, M. Zhang, C. L. Manthey, E. C. Petrella, M. W. Pantoliano, I. C. Deckman, J. C. Spurlino, A. C. Maroney, B. E. Tomczuk, C. J. Molloy and R. F. Bone, *J. Med. Chem.*, 2005, **48**, 909–912.
138. T. F. Herpin, K. G. Van Kirk, J. M. Salvino, S. T. Yu and R. F. Labaudinière, *J. Comb. Chem.*, 2000, **2**, 513–521.
139. D. J. Lauffer and M. D. Mullican, *Bioorg. Med. Chem. Lett.*, 2002, **12**, 1225–1227.
140. C. J. Aquino, D. R. Armour, J. M. Berman, L. S. Birkemo, R. A. E. Carr, D. K. Croom, M. Dezube, R. W. Dougherty, G. N. Ervin, M. K. Grizzle, J. E. Head, G. C. Hirst, M. K. James, M. F. Johnson, L. J. Miller, K. L. Queen, T. J. Rimele, D. N. Smith and E. E. Sugg, *J. Med. Chem.*, 1996, **39**, 562–569.

141. B. R. Henke, T. M. Willson, E. E. Sugg, D. K. Croom, R. W. Dougherty, K. L. Queen, L. S. Birkemo, G. N. Ervin, M. K. Grizzle, M. F. Johnson and M. K. James, *J. Med. Chem.*, 1996, **39**, 2655–2658.
142. T. M. Willson, B. R. Henke, T. M. Momtahen, P. L. Myers, E. E. Sugg, R. J. Unwalla, D. K. Croom, R. W. Dougherty, M. K. Grizzle, M. F. Johnson, K. L. Queen, T. J. Rimele, J. D. Yingling and M. K. James, *J. Med. Chem.*, 1996, **39**, 3030–3034.
143. A. G. Neo, C. F. Marcos, S. Marcaccini and R. Pepino, *Tetrahedron Lett.*, 2005, **46**, 7977–7979.
144. B. Blank, F. R. Pfeiffer, C. M. Greenberg and J. F. Kerwin, *J. Med. Chem.*, 1963, **6**, 554–560.
145. E. Grossman and F. H. Messerli, *Prog. Cardiovasc. Dis.*, 2004, **47**, 34–57.
146. D. Lesuisse, P. Deprez, E. Albert, T. T. Duc, B. Sortais, D. Gofflo, V. Jean-Baptiste, J.-P. Marquette, B. Schoot, E. Sarubbi, G. Lange, P. Broto and E. Mandine, *Bioorg. Med. Chem. Lett.*, 2001, **11**, 2127–2131.
147. D. E. Thurston and D. S. Bose, *Chem. Rev.*, 1994, **94**, 433–465.
148. D. R. Langley and D. E. Thurston, *J. Org. Chem.*, 1987, **52**, 91–97.
149. P. G. Baraldi, A. Leoni, B. Cacciari, S. Manfredini, D. Simoni, M. Bergomi, E. Menta and S. Spinelli, *J. Med. Chem.*, 1994, **37**, 4329–4337.
150. D. E. Thurston, D. S. Bose, P. W. Howard, T. C. Jenkins, A. Leoni, P. G. Baraldi, A. Guiotto, B. Cacciari, L. R. Kelland, M.-P. Foloppe and S. Rault, *J. Med. Chem.*, 1999, **42**, 1951–1964.
151. J.-Y. Lee and Y.-C. Kim, *ChemMedChem*, 2009, **4**, 733–737.
152. W. W. Engel, W. G. Eberlein, G. Mihm, R. Hammer and G. Trummlitz, *J. Med. Chem.*, 1989, **32**, 1718–1724.
153. K. D. Hargrave, J. R. Proudfoot, K. G. Grozinger, E. Cullen, S. R. Kapadia, U. R. Patel, V. U. Fuchs, S. C. Mauldin and J. Vitous, *J. Med. Chem.*, 1991, **34**, 2231–2241.
154. T. Watanabe, I. Kinoyama, A. Kakefuda, T. Okazaki, K. Takizawa, S. Hirano, H. Shibata and I. Yanagisawa, *Chem. Pharm. Bull.*, 1997, **45**, 996–1007.
155. M.-C. Tseng, C.-Y. Lai, Y.-W. Chu and Y.-H. Chu, *Chem. Commun.*, 2009, 445–447.
156. R. Chang, V. Lotti, R. Monaghan, J. Birnbaum, E. Stapley, M. Goetz, G. Albers-Schonberg, A. Patchett, J. Liesch, O. Hensens *et al.*, *Science*, 1985, **230**, 177–179.
157. W. Zhang, J. P. Williams, Y. Lu, T. Nagashima and Q. Chu, *Tetrahedron Lett.*, 2007, **48**, 563–565.
158. D. J. Craik, N. L. Daly, T. Bond and C. Waine, *J. Mol. Biol.*, 1999, **294**, 1327–1336.
159. D. J. Craik, N. L. Daly, J. Mulvenna, M. R. Plan and M. Trabi, *Curr. Protein Pept. Sci.*, 2004, **5**, 297–315.
160. F. Sandberg, *Chaier's de la Maboke*, Buffon, Paris, 1965, vol. III, p. 27.
161. L. Gran, *Lloydia*, 1973, **36**, 174–178.

162. J. P. Mulvenna, C. Wang and D. J. Craik, *Nucleic Acids Res.*, 2006, **34**, D192–D194.
163. U. Goeransson, R. Burman, S. Gunasekera, A. A. Stroemstedt and K. J. Rosengren, *J. Biol. Chem.*, 2012, **287**, 27001–27006.
164. A. G. Poth, L. Y. Chan and D. J. Craik, *Biopolymers*, 2013, **100**, 480–491.
165. H. R. Bokesch, L. K. Pannell, P. K. Cochran, R. C. Sowder II, T. C. McKee and M. R. Boyd, *J. Nat. Prod.*, 2001, **64**, 249–250.
166. J. P. Tam and Y.-A. Lu, *Protein Sci.*, 1998, **7**, 1583–1592.
167. N. L. Daly, S. Love, P. F. Alewood and D. J. Craik, *Biochemistry*, 1999, **38**, 10606–10614.
168. J. P. Tam, Y.-A. Lu, J.-L. Yang and K.-W. Chiu, *Proc. Natl. Acad. Sci. U. S. A.*, 1999, **96**, 8913–8918.
169. D. J. Craik, M. Cemazar and N. L. Daly, *Curr. Opin. Drug Discovery Dev.*, 2007, **10**, 176–184.
170. N. L. Daly, R. J. Clark and D. J. Craik, *J. Biol. Chem.*, 2003, **278**, 6314–6322.
171. U. Goransson and D. J. Craik, *J. Biol. Chem.*, 2003, **278**, 48188–48196.
172. U. Goransson, E. Svangard, P. Claeson and L. Bohlin, *Curr. Protein Pept. Sci.*, 2004, **5**, 317–329.
173. L. Gran, F. Sandberg and K. Sletten, *J. Ethnopharmacol.*, 2000, **70**, 197–203.
174. N. L. Daly, K. J. Rosengren and D. J. Craik, *Adv. Delivery Rev.*, 2009, **61**, 918–930.
175. P. B. Pelegrini, B. F. Quirino and O. L. Franco, *Peptides*, 2007, **28**, 1475–1481.
176. D. J. Craik, M. Cemazar and N. L. Daly, *Curr. Opin. Drug Discovery Dev.*, 2006, **9**, 251–260.
177. D. J. Craik, R. J. Clark and N. L. Daly, *Expert Opin. Invest. Drugs*, 2007, **16**, 595–604.
178. H. Sancheti and J. A. Camarero, *Adv. Delivery Rev.*, 2009, **61**, 908–917.
179. C. W. Gruber, M. Cemazar, M. A. Anderson and D. J. Craik, *Toxicon*, 2007, **49**, 561–575.
180. M. R. R. Plan, U. Goransson, R. J. Clark, N. L. Daly, M. L. Colgrave and D. J. Craik, *ChemBioChem*, 2007, **8**, 1001–1011.
181. S. Gunasekera, F. M. Foley, R. J. Clark, L. Sando, L. J. Fabri, D. J. Craik and N. L. Daly, *J. Med. Chem.*, 2008, **51**, 7697–7704.
182. K. R. Gustafson, R. C. Sowder II, L. E. Henderson, I. C. Parsons, Y. Kashman, J. H. Cardellina II, J. B. McMahon, R. W. Buckheit Jr., L. K. Pannell and M. R. Boyd, *J. Am. Chem. Soc.*, 1994, **116**, 9337–9338.
183. K. R. Gustafson, L. K. Walton, R. C. Sowder II, D. G. Johnson, L. K. Pannell, J. H. Cardellina and M. R. Boyd, *J. Nat. Prod.*, 2000, **63**, 176–178.
184. K. M. Witherup, M. J. Bogusky, P. S. Anderson, H. Ramjit, R. W. Ransom, T. Wood and M. Sardana, *J. Nat. Prod.*, 1994, **57**, 1619–1625.
185. U. Goransson, M. Sjogren, E. Svangard, P. Claeson and L. Bohlin, *J. Nat. Prod.*, 2004, **67**, 1287–1290.

186. R. Burman, E. Svedlund, J. Felth, S. Hassan, A. Herrmann, R. J. Clark, D. J. Craik, L. Bohlin, P. Claeson, U. Goeransson and J. Gullbo, *Biopolymers*, 2010, **94**, 626–634.
187. M. Pranting, C. Loov, R. Burman, U. Goransson and D. I. Andersson, *J. Antimicrob. Chemother.*, 2010, **65**, 1964–1971.
188. S. L. Gerlach, R. Rathinakumar, G. Chakravarty, U. Goransson, W. C. Wimley, S. P. Darwin and D. Mondal, *Biopolymers*, 2010, **94**, 617–625.
189. C. K. L. Wang, M. L. Colgrave, K. R. Gustafson, D. C. Ireland, U. Goransson and D. J. Craik, *J. Nat. Prod.*, 2008, **71**, 47–52.
190. Y. F. Hallock, R. C. Sowder II, L. K. Pannell, C. B. Hughes, D. G. Johnson, R. Gulakowski, J. H. Cardellina II and M. R. Boyd, *J. Org. Chem.*, 2000, **65**, 124–128.
191. N. L. Daly, R. J. Clark, M. R. Plan and D. J. Craik, *Biochem. J.*, 2006, **393**, 619–626.
192. B. Chen, M. L. Colgrave, N. L. Daly, K. J. Rosengren, K. R. Gustafson and D. J. Craik, *J. Biol. Chem.*, 2005, **280**, 22395–22405.
193. E. Svangard, U. Goransson, Z. Hocaoglu, J. Gullbo, R. Larsson, P. Claeson and L. Bohlin, *J. Nat. Prod.*, 2004, **67**, 144–147.
194. A. M. Broussalis, U. Göransson, J. D. Coussio, G. Ferraro, V. Martino and P. Claeson, *Phytochemistry*, 2001, **58**, 47–51.
195. D. C. Ireland, M. L. Colgrave and D. J. Craik, *Biochem. J.*, 2006, **400**, 1–12.
196. C. Jennings, J. West, C. Waine, D. Craik and M. Anderson, *Proc. Natl. Acad. Sci. U. S. A.*, 2001, **98**, 10614–10619.
197. L. Gran, *Acta Pharmacol. Toxicol.*, 1973, **33**, 400–408.
198. D. G. Barry, N. L. Daly, R. J. Clark, L. Sando and D. J. Craik, *Biochemistry*, 2003, **42**, 6688–6695.
199. C. V. Jennings, K. J. Rosengren, N. L. Daly, M. Plan, J. Stevens, M. J. Scanlon, C. Waine, D. G. Norman, M. A. Anderson and D. J. Craik, *Biochemistry*, 2005, **44**, 851–860.
200. U. Goransson, T. Luijendijk, S. Johansson, L. Bohlin and P. Claeson, *J. Nat. Prod.*, 1999, **62**, 283–286.
201. P. Claeson, U. Goeransson, S. Johansson, T. Luijendijk and L. Bohlin, *J. Nat. Prod.*, 1998, **61**, 77–81.
202. T. Schoepke, M. I. Hasan Agha, R. Kraft, A. Otto and K. Hiller, *Sci. Pharm.*, 1993, **61**, 145–153.
203. J.-F. Hernandez, J. Gagnon, L. Chiche, T. M. Nguyen, J.-P. Andrieu, A. Heitz, T. T. Hong, T. T. C. Pham and D. L. Nguyen, *Biochemistry*, 2000, **39**, 5722–5730.
204. K. R. Gustafson, T. C. McKee and H. R. Bokesch, *Curr. Protein Pept. Sci.*, 2004, **5**, 331–340.
205. N. L. Daly, K. R. Gustafson and D. J. Craik, *FEBS Lett.*, 2004, **574**, 69–72.
206. D. C. Ireland, C. K. L. Wang, J. A. Wilson, K. R. Gustafson and D. J. Craik, *Biopolymers*, 2008, **90**, 51–60.
207. P. Lindholm, U. Goransson, S. Johansson, P. Claeson, J. Gullbo, R. Larsson, L. Bohlin and A. Backlund, *Mol. Cancer Ther.*, 2002, **1**, 365–369.

208. P. Thongyoo, C. Bonomelli, R. J. Leatherbarrow and E. W. Tate, *J. Med. Chem.*, 2009, **52**, 6197–6200.
209. P. Thongyoo, N. Roque-Rosell, R. J. Leatherbarrow and E. W. Tate, *Org. Biomol. Chem.*, 2008, **6**, 1462–1470.
210. J. Tang, C. K. Wang, X. Pan, H. Yan, G. Zeng, W. Xu, W. He, N. L. Daly, D. J. Craik and N. Tan, *Helv. Chim. Acta*, 2010, **93**, 2287–2295.
211. G. K. T. Nguyen, S. Zhang, N. T. K. Nguyen, P. Q. T. Nguyen, M. S. Chiu, A. Hardjojo and J. P. Tam, *J. Biol. Chem.*, 2011, **286**, 24275–24287.
212. K. K. Brandt, R. G. Ovesen, U. Goransson, J. Nilesen, H. C. B. Hansen and N. Cedergree, *Environ. Toxicol. Chem.*, 2011, **30**, 119–1196.
213. M. L. Colgrave, A. C. Kotze, S. Kopp, J. S. McCarthy, G. T. Coleman and D. J. Craik, *Acta Trop.*, 2009, **109**, 163–166.
214. M. L. Colgrave and D. J. Craik, *Biochemistry*, 2004, **43**, 5965–5975.

CHAPTER 16

Spirocycles as Privileged Structural Motifs in Medicinal Chemistry

FELIX VOSS, STEFAN SCHUNK AND HENNING STEINHAGEN*

Grünenthal GmbH, 52099 Aachen, Germany
*Email: Henning.Steinhagen@grunenthal.com

16.1 Introduction

Spirocycles are important structural elements, which are frequently used as pharmacophores and scaffolds in modern drug discovery, offering structural complexity as well as structural rigidity. Intrinsic complexity and rigidity are favorable features in medicinal chemistry optimization programs against biological targets. The structural complexity offered by spirocycles is often advantageous to identify unchartered molecular space, enabling IP protection for the drug and drug synthesis. The rigidity can be beneficially used to position pharmacophores in an ideal spatial orientation maximizing H-Bond, π-stacking and hydrophobic interactions. Therefore, the spirocyclic moieties can be used to exchange or rescaffold flexible and highly entropic parts of bioactive molecules. This can be useful in order to achieve specific interaction of an exogenous ligand with the target as well as improve physiochemical properties like aqueous solubility.[1] Generally, spirocycles have been referred to as privileged structural motifs in drug discovery. Examples

RSC Drug Discovery Series No. 50
Privileged Scaffolds in Medicinal Chemistry: Design, Synthesis, Evaluation
Edited by Stefan Bräse

Published by the Royal Society of Chemistry, www.rsc.org

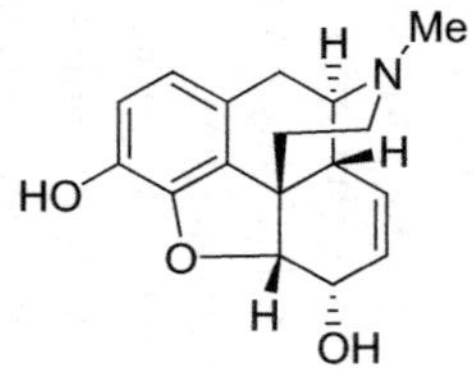

1: Morphine
Natural Product
Launched (analgesic)
Opioid Receptor Agonist

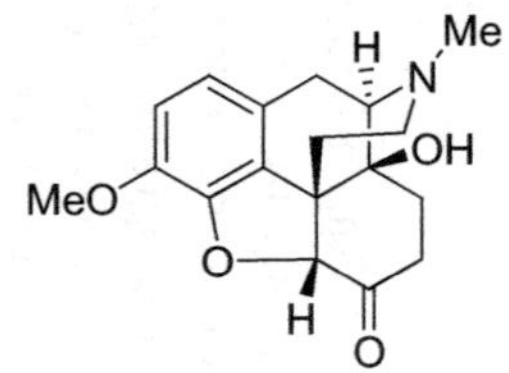

2: Oxycontin
Purdue Pharma LP
Launched (analgesic)
Opioid Receptor Agonist

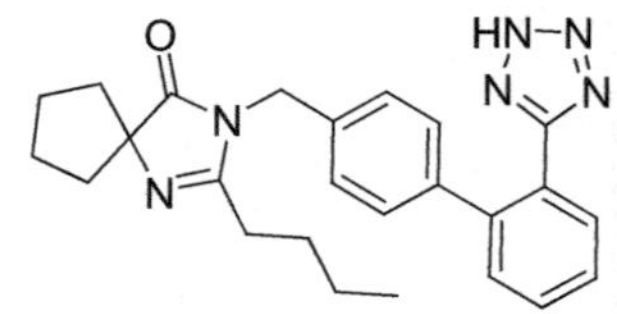

3: Irbersartan
Sanofi
Launched (hypertension)
Angiotensin AT1 Antagonists

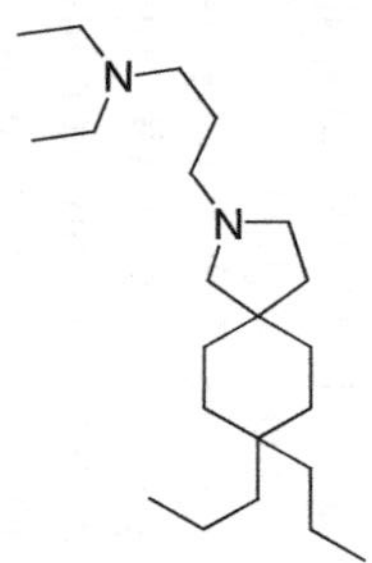

4: Atiprimod
Callisto/GSK
Orphan drug (cancer)
Angiogenesis Inhibitor

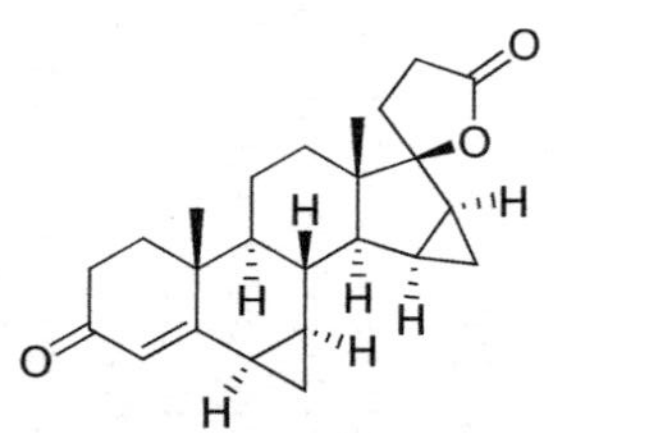

5: Drospirenone
Bayer Schering Pharma
Launched (contraceptive)
Corticoid Receptor Antagonists

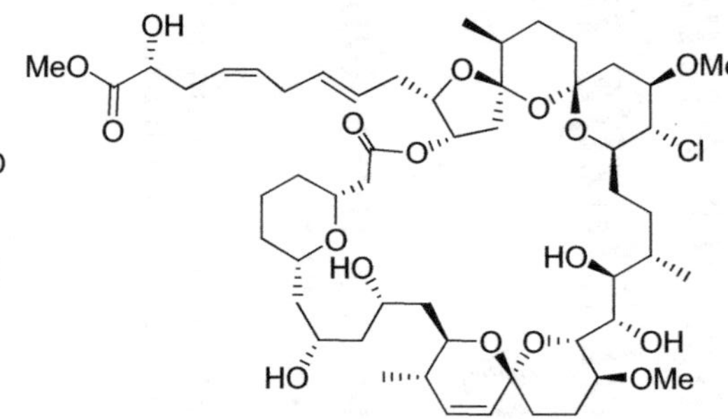

6: Spirastrellolide A methyl ester
Natural product
PP2A Inhibitor

Figure 16.1 Representative examples for bioactive spirocyclic compounds: **1** (morphine),[3] **2** (Oxycontin),[4] **3** (Irbesartan),[5] **4** (Atiprimod),[6] **5** (Drospirenone),[7] **6** (Spirastrellolide A methyl ester).[8]

are several spirocyclic drugs that have been launched for different target classes, *e.g.* GPCRs,[2] ion channels, nuclear hormone receptors. Representative examples (Figure 16.1) for successfully marketed drugs are the well-known opioid analgesics morphine (**1**)[3] Oxycontin (**2**)[4] and the Angiotensin-II-receptor antagonist Irbesartan (**3**),[5] which is used to control the blood pressure. Other examples are Atiprimod (**4**),[6] which is used as an orphan drug for the treatment of cancer and the contraceptive Drospirenone (**5**).[7] Spirastrellolide A methyl ester (**6**),[8] which shows a strong antimitotic activity against the human breast cancer line MCF-7 in the low nano-molar range, is a typical representative of a bioactive polyketide derived natural product containing multiple spiroketal motifs (Scheme 16.1).

In the following paragraphs, we will discuss selected examples of different spirocyclic motifs, namely, spiro-carbacycles (16.2), spiro-azacycles (16.3) and spiro-oxacycles (16.4) with regards to their biological activity and chemical synthesis, followed by a brief summary and an outlook to the field (16.5).

16.2 Spiro-carbacycles

Carbacyclic spirocycles appear to be the least prominent spirocyclic motifs in drug discovery. The reasons for this may be associated with intrinsic disadvantages on physiochemical and metabolic stability properties, resulting in a poor pharmacokinetic profile. Another reason could be both their often rather elaborate and complex synthesis as well as a lack of functional groups which are preferred for optimization of such motifs by medicinal chemists. On the other hand, spirocyclic carbacycles provide valuable frameworks which can serve as novel motifs and could help to further expand the chemical space in modern drug discovery. Figure 16.2 shows a structurally diverse set of biologically active small molecules containing various carbaspirocyclic frameworks.

Ingenol mebutate (7), which is derived from a plant extract, acts as a pan-activator of protein kinase C and effects several important cell functions. Aphidicolin (**8**) is a tetracyclic diterpene antibiotic isolated from the fungus *Cephalosporum aphidicola,* with antiviral and antimitotical properties. Aphidicolin acts as a reversible inhibitor of eukaryotic nuclear DNA replication. Another antibiotic drug based on a carbaspirocyclic framework is Platensimycin (**9**), a metabolite of *Streptomyces platensis*, which is an example of a unique structural class of natural antibiotics. This compound blocks enzymes involved in the condensation steps in fatty acid biosynthesis,[15] which Gram-positive bacteria need for biosynthesis. Allogibberic acid (**10**), derived from the plant growth hormone gibberellic acid is a tetracyclic anti-inflammatory agent in preclinical stage, acting through modulation of the transcription factor NF-κB. Sequosemperverin A (**11**), a plant derived natural diol displays antifungal activity. A synthetic example of a spiro tricyclic bioactive molecule is represented by **12** and acts as a dual ALX and FPRL2 agonist (Figure 16.2).

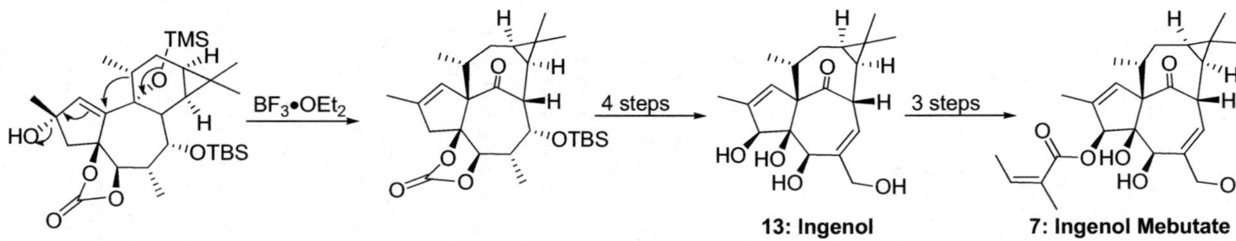

Scheme 16.1 Formation of the spirocyclic motif of ingenol (**13**) through a pivotal vinylogous pinacol rearrangement in the total synthesis of Baran *et al.*[17]

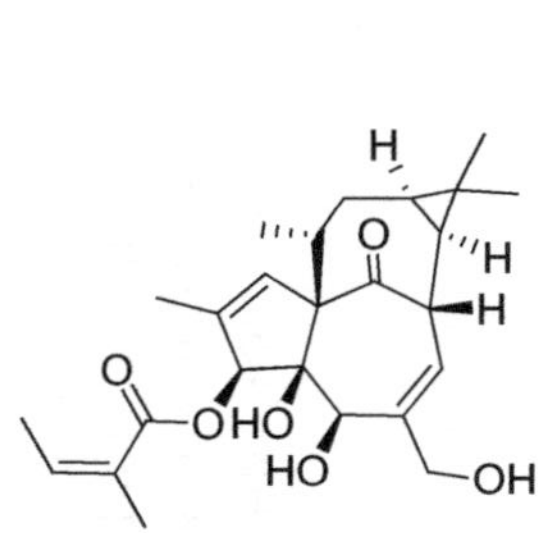

7: Ingenol Mebutate
Leo Pharma
Launched (actinic keratosis)
PKC Activator

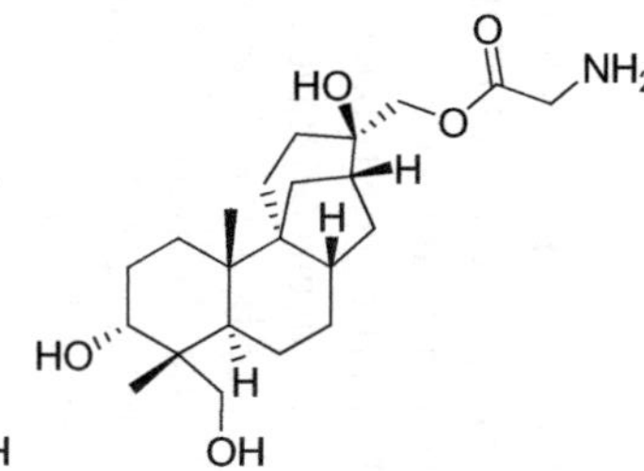

8: Aphidicolin Glycinate
Astra Zeneca
Phase 1
DNA Polymerase Inhibitor

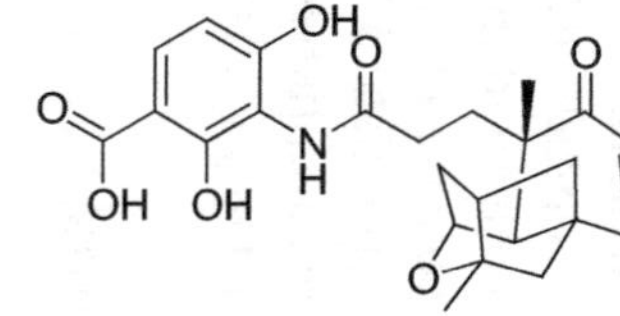

9: Platensimycin
Merck Sharp & Dohme
Preclinical (antiobiotic)
FabF Inhibitor

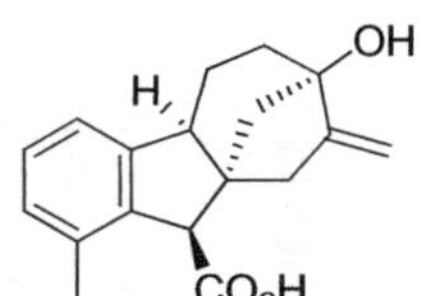

10: Allogibberic Acid
Harvard College
Preclinical
NFKB Modulator

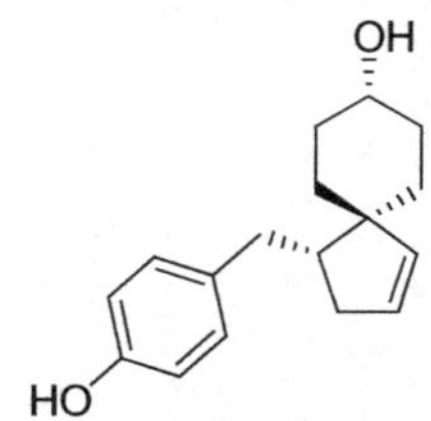

11: Sequosempervirin A
Natural product
Antifungal Activity

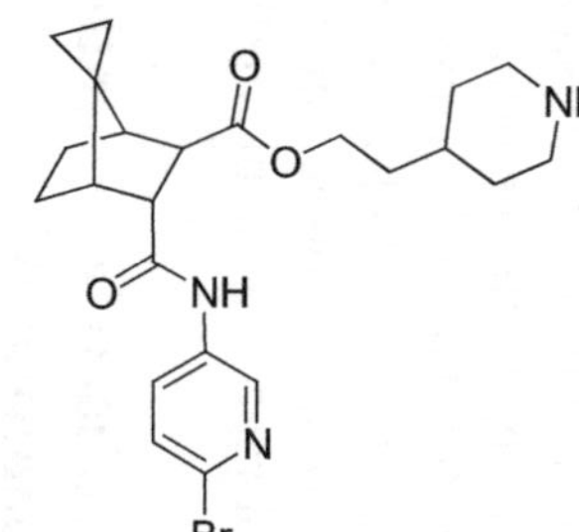

12: WO2012066488 A2
Actelion
Preclinical
ALX and FPRL2 receptor agonists

Figure 16.2 Representative examples for carbaspirocyclic bioactive compounds, **7** (Ingenol Mebutate),[9] **8** (Aphidicolin glycinate),[10] **9** (Platensimycin),[11] **10** gibberellic acid,[12] **11** Sequosemperivin,[13] **12** ALXR and FPRL2 agonists.[14]

16.2.1 Spiro-carbacycles – Synthetic Example 1: Ingenol (13)

The diterpene ester ingenol mebutate (7), an extract of the *Euphorbia peplus* plant, is a selective small-molecule pan-activator of the protein kinase C (PKC) isoenzym family which regulates proliferation, differentiation, apoptosis and other cellular processes. It contains a unique, highly condensed [5.7] carbacyclic framework, was approved by the FDA in 2012 as first-in-class drug for the treatment of actinic keratosis and commercialized under the brand name Picato® by Leo Pharma. The gel formulation of this drug is currently in Phase II clinical trials for the topical treatment of cutaneous squamous cell carcinoma *in situ* (SCCIS). Picato (7) is produced *via* a semi-synthetic route starting from ingenol (**13**), which is converted in 3 steps into ingenol mebutate (7).[16] Baran *et al.* accomplished a concise 14-step synthesis of ingenol (**13**), starting from (+)-3-carene which was published in 2013. The formation of the complex spirocyclic system was accomplished through a vinylogous pinacol rearrangement mediated by $BF_3 \cdot OEt_2$ (Scheme 16.1).[17]

16.2.2 Spiro-carbacycles – Synthetic Example 2: Platensimycin (9)

The antibiotic Platensimycin (**9**), a metabolite of *Streptomyces platensis*, was discovered during a screening program on FabF/H inhibitors at Merck.[11] The antibiotic mode of action is based on the inhibition of the elongation-condensing enzymes β-ketoacyl synthase I/II (FabF/B) in the type II bacterial fatty acid biosynthesis. Several total syntheses of Platensimycin (**9**) have been described in the literature to date. Nicolaou *et al.* described the first total synthesis of racemic Platensimycin (**9**) in 2006.[18] The formation of the spirocyclic moiety was accomplished through a ruthenium-catalysed enyne cycloisomerization following a methodology developed by Trost *et al.*[19] (Scheme 16.2).

16.3 Spiro-azacycles

Spiro-azacycles represent the largest class of diverse bioactive spirocyclic compounds reviewed in this chapter. An overview of selected examples is given in Figure 16.3 and Figure 16.4. This class of spirocycles are found in all phases of preclinical and clinical development, targeting a broad range of molecular target classes.

Tedisamil (**14**) and the spirolactam (**25**) are ion-channel modulators acting on potassium and sodium channels, whereas Aderbasib (**15**) and the synthetic amino imidazolinone (**28**) act as protease inhibitors on ADAM10/17 and BACE, respectively. Spiro-azacyles (**18–21**) represent a set of clinically tested enzyme inhibitors targeting the metabolic enzyme aldose reductase, which is mainly involved in sugar metabolism. The majority of all other examples target G-protein coupled receptors (Figure 16.3). Spiperone (**16**) represents the class of marketed Dopamine D2 receptor modulators used in

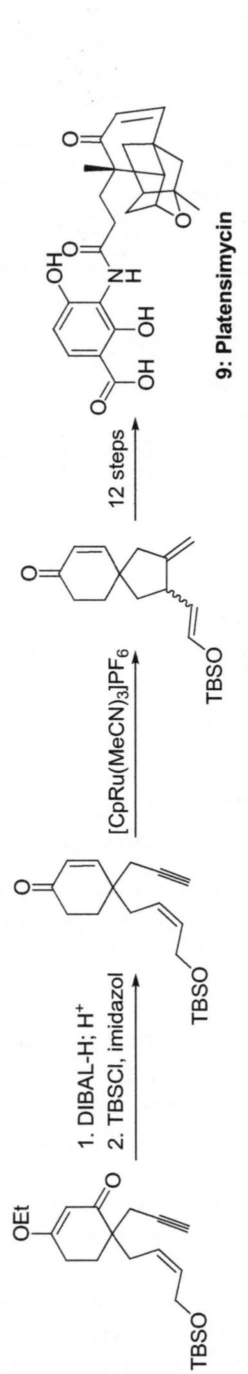

9: Platensimycin

Scheme 16.2 Assembly of the spirocyclic ring system through an enyne cycloisomerization in Nicolaou's total synthesis of Platensimycin (**9**).[18]

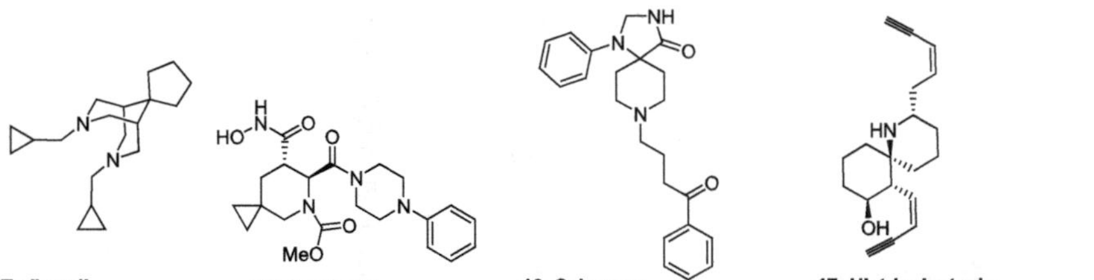

14: Tedisamil
AbbVie
Launched (atrial fibrillation)
Potassium Channel Inhibitor

15: Aderbasib
Incyte
Phase 2 (cancer)
ADAM10/17 Inhibitor

16: Spiperone
Janssen/Eisai
Launched (schizophrenia)
Dopamine D2 Antagonist

17: Histrionicotoxin
Natural products
Poison frog alkaloids
Acetylcholine Receptor Inhibitor

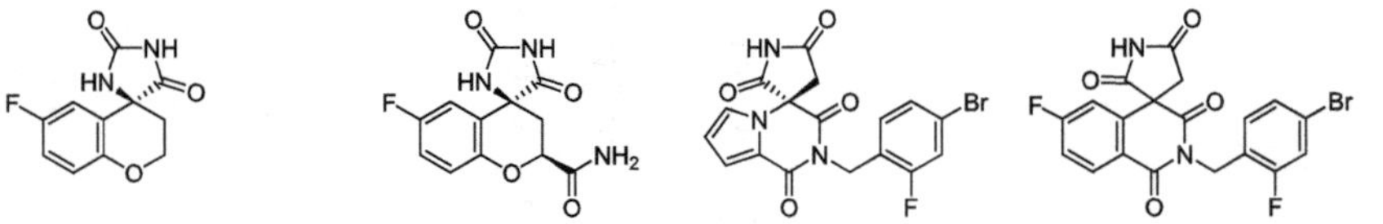

18: Sorbinil
Pfizer
Phase 3, discontinued
Aldose Reductase Inhibitor

19: Fidarestat
Sanwa
Phase 3, discontinued
Aldose Reductase Inhibitor

20: Ranirestat
Eisai/Dainippon Sumitomo
Phase 3 (diabetic neuropathy)
Aldose Reductase Inhibitor

21: Minalrestat
Pfizer
Phase 3 (diabetic neuropathy)
Aldose Reductase Inhibitor

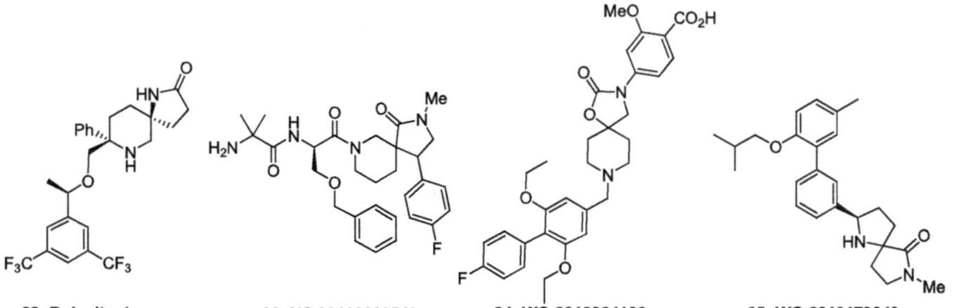

22: Rolapitant
Tesaro/Merck Sharp & Dohme
Phase 3 (nausea prophylaxis)
NK1 Antagonist

23: US 20120302540
Novartis
Preclinical
Ghrelin Receptor Agonists

24: WO 2012024183
Merck Sharp & Dohme
Preclinical
SST5 Antagonists

25: WO 2013179049
Convergence
Preclinical
Sodium Channel Blocker

Figure 16.3 Representative examples for aza-spirocyclic bioactive compounds: **14** (Tedisamil),[20] **15** (Aderbasib),[21] **16** (Spiperone),[22] **17** (Histrionicotoxins),[23] **18** (sorbinil),[24] **19** (Fidarestat),[25] **20** (Ranirestat),[26] **21** (Minalrestat),[27] **22** (Rolapitant),[28] **23**,[29] **24**,[30] **25**.[31]

CNS related disorders (*e.g.* schizophrenia) whereas the bis-enine containing spirocycle Histrionicotoxin (**17**) acts on the nicotinic acetylcholine receptor.

The Phase III drug candidate Rolapitant (**22**) and the preclinical compound **29** antagonize the Neurokinine NK1 receptor. Spirocycles **23** and **24**, which are both in the preclinical stage, modulate the activity of Ghrelin and Somatostatine SST5 receptors, respectively. Pentacycle **26** modulates the Histamine H3 receptor, whereas Satavaptan (**32**) antagonizes the Vasopressin V2 receptor and has been tested in Phase II clinical trials. ARN-509 (**33**) is an experimental drug targeting the nuclear Androgen receptor and is currently undergoing Phase III clinical trials in cancer indications. Strychnine (**30**) is a highly potent and toxic naturally occurring indole alkaloid derived *e.g.* from the seeds of the *Strychnos nux-vomica* tree. It has been used in very low doses as analeptic as well as rat poison and exhibits an LD_{50} in the mg kg^{-1} range in rat and human.[40] Horsfiline (**31**) is an oxindole alkaloid

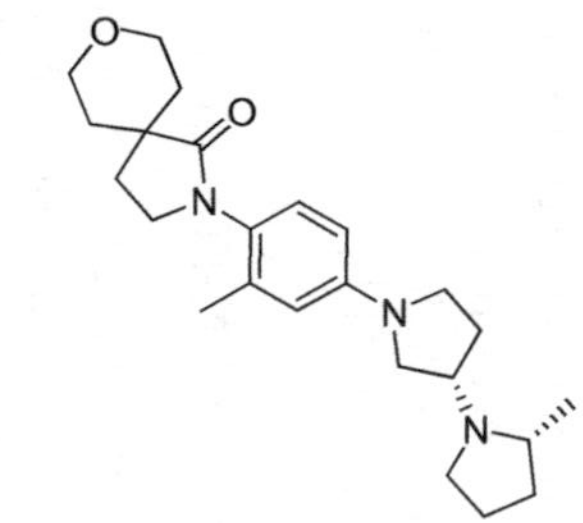

26: WO 2010065798
Sanofi
Preclinical
H3 Receptor Modulators

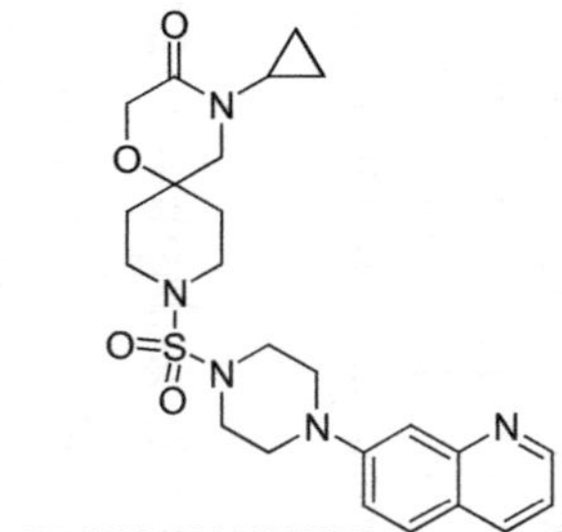

27: WO 2013177253
GlaxoSmithKline
Preclinical
Fatty Acid Synthase Inhibitors

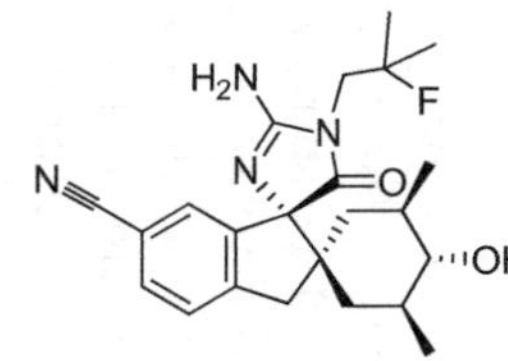

28: WO 2014035860
Boehringer Ingelheim / Vitae Phama
Preclinical
BACE Inhibitors

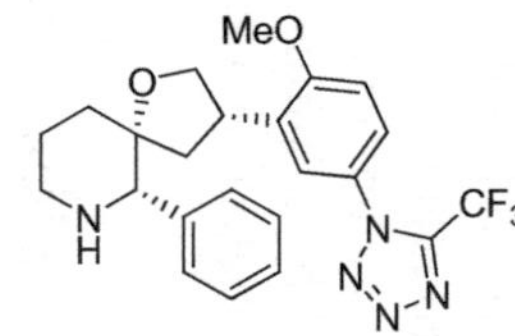

29: WO 9719084
Merck Sharp & Dohme
Preclinical
NK1 Antagonists

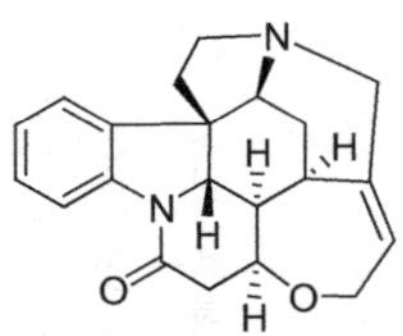

30: Strychnine
Natural product
Potent toxic alkaloid

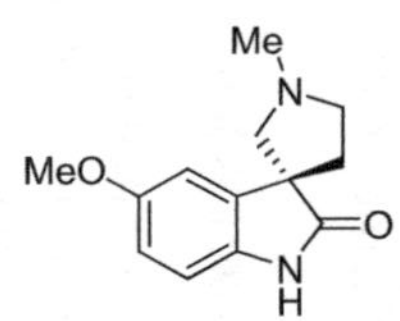

31: Horsfiline
Natural product
Analgesic

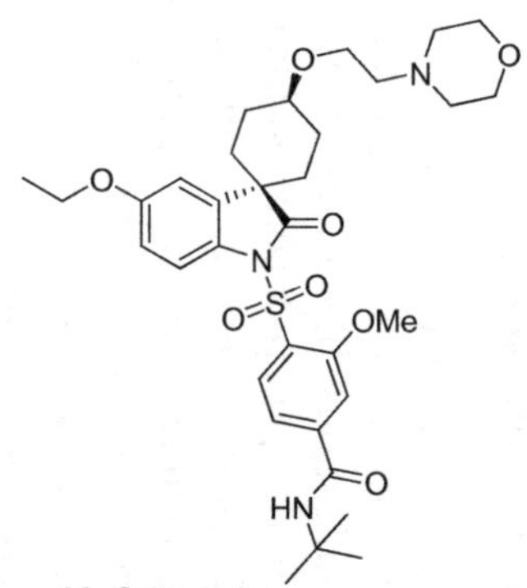

32: Satavaptan
Sanofi
Phase 2, discontinued (Hyponatremia)
Vasopressin 2 Receptor Antagonist

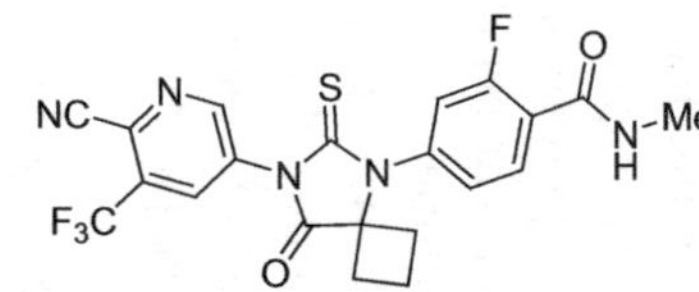

33: ARN-509
UC Oakland/Johnson & Johnson
Phase 3 (prostate cancer)
Androgen Receptor Degradation Enhancer

Figure 16.4 Representative examples for aza-spirocyclic bioactive compounds: **26**,[32] **27**,[33] **28**,[34] **29**,[35] **30** (Strychnin),[36] **31** (Horsfiline),[37] **32** (Satavaptan)[38] **33** (ARN-509).[39]

with analgetic properties that has been identified from the plant *Horsfieldia superba* and used in traditional herbal medicine (Figure 16.4).

16.3.1 Spiro-azacycles – Synthetic Example 1: Tedisamil (14)

The potassium channel blocker Tedisamil hydrochloride (**14**), which contains the rigid 3,7-diazaspiro[bicyclo[3.3.1]nonane skeleton,[41] was developed by AbbVie and was approved in 2008 in Europe for the intravenous treatment of atrial fibrillation. The synthesis of Tedisamil is described in Scheme 16.3. Starting from a cyclopentanone-derived *Knoevennagel* condensation product, the spirocycle is formed through a *Michael* addition/condensation cascade with 2-cyanoacetamide. The latter is cyclized mediated by sulphuric acid to give the desired 3,7-diazaspiro-bicyclo[3.3.1]nonane skeleton, which is then transformed in 3 steps into Tedisamil (**14**).[42]

16.3.2 Spiro-azacycles – Synthetic Example 2: Fidarestat (19) and Minalrestat (21)

The spiroimide derivatives **18–21** display a prominent pharmacophore within the class of the aldol reductase inhibitors, a target for the therapy of diabetic complications like diabetic neuropathy.[43] The spirocyclic imid functionality in this group of compounds plays a pivotal role as it serves as a carboxylic acid isoster. This results in improved oral absorption and tissue penetration as compared to the respective carboxylic acid containing drugs and therefore to an overall improved pharmacokinetic profile.[44] Spiro hydanthoins like Fidarestat (**19**) are accessible through a three component condensation reaction of the corresponding ketone with potassium cyanide and ammonium carbonate,[45] whereas spiro ureas like Minalrestat (**21**) are synthesized through the condensation of the corresponding 1,4-dicarbonyl compounds[46] (Scheme 16.4).

16.3.3 Spiro-azacycles – Synthetic Example 3: Rolapitant (22)

Rolapitant (**22**) is an antagonist of the G-protein coupled receptor tachykinin neurokinin 1 (NK1) and is currently in Phase III clinical trials for the prevention of chemotherapy-induced nausea and vomiting by Tesaro and OPKO. The spiro cyclic moiety of Rolipitant (**22**) is accessible through a conjugate 1,4-addition of the corresponding nitro piperidine to methyl acrylate, followed by a reductive lactamisation (Scheme 16.5).[47]

16.3.4 Spiro-azacycles – Synthetic Example 4: ARN-509 (33)

The spiro thioimid ARN-509 (**33**) is an androgen receptor degradation enhancer, currently in Phase III clinical trials at Johnson & Johnson for the treatment of progressive metastatic castration-resistant prostate cancer. The final spiro-forming step of the synthesis is a condensation of an isothiocynate with the corresponding α-amino nitrile (Scheme 16.6).[39]

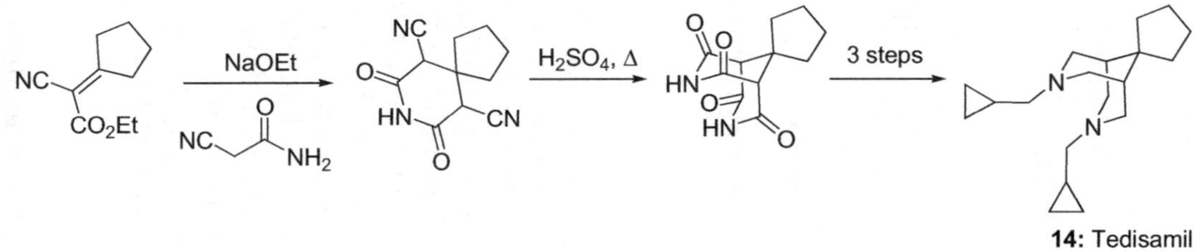

Scheme 16.3 Formation of the spirocycle and condensation to yield the 3,7-diazaspiro[bicyclo[3.3.1]nonane skeleton of Tedisamil (**14**).[42]

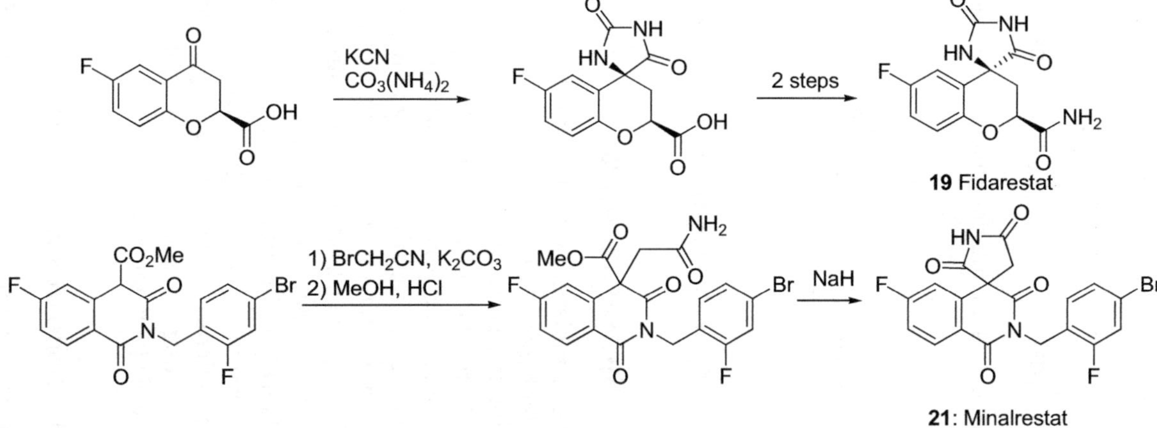

Scheme 16.4 Synthesis of aldose reductase inhibitors containing a sprio imid motif.

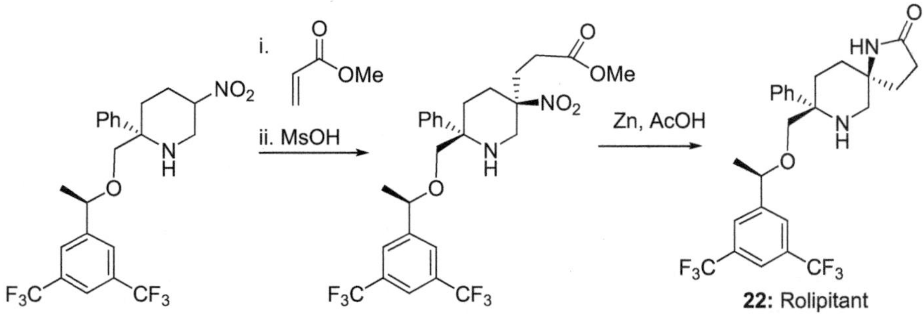

22: Rolipitant

Scheme 16.5 Spiro forming step in the synthesis of Rolapitant (**22**).[47]

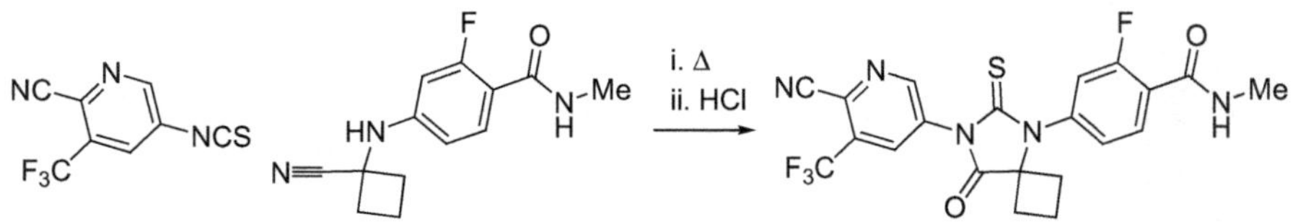

33: ARN-509

Scheme 16.6 Synthesis of the androgen receptor degradation enhancer ARN-509 (**33**).[39]

16.4 Spiro-oxacycles

Spirocyclic moieties, which contain one or more oxygen atoms, also show broad biological activity and are present in several examples (Figure 16.5), including marketed drugs like the anti-infective Fumagilin (**40**) and the anti-malarial drug Artemisinine (**41**). Fumagilin is a complex spiro epoxide based antibiotic isolated from *Aspergillus fumigatus*. It displays broad biological activity through irreversible binding to methionin-aminopeptidases (MetAPs).

In contrast, the interesting cyclic phosphate-based synthetic compound **36** described in WO2013174962 (preclinical stage) displays antiviral activity against the Hepatitis C virus (HCV). The polypropionate derived metabolite SNF-4435D (**37**) isolated from *Streptomyces spectabilis*, has been shown to exhibit potent immunosuppressive activity through inhibition of selective B-cell proliferation, whereas the spirocycles **38** and **39** display biological activity through release of neurotrophic factors and Acetyl-CoA inhibition, respectively (Scheme 16.5).

16.4.1 Spiro-oxacycles – Synthetic Example 1: Cebranopadol (34)

The nociceptin/orphanin FQ peptide (NOP) and opioid receptor agonist cebranopadol (**34**)[48,57] is an analgesic currently in clinical development for the treatment of severe chronic nociceptive and neuropathic pain. The spiro forming step of the cebranopadol synthesis is a trimethylsilyl trifluoromethanesulfonate mediated oxa-*Pictet-Spengler* reaction[58] of the respective cyclohexanone and fluoro indole derivatives, as described in Scheme 16.7.

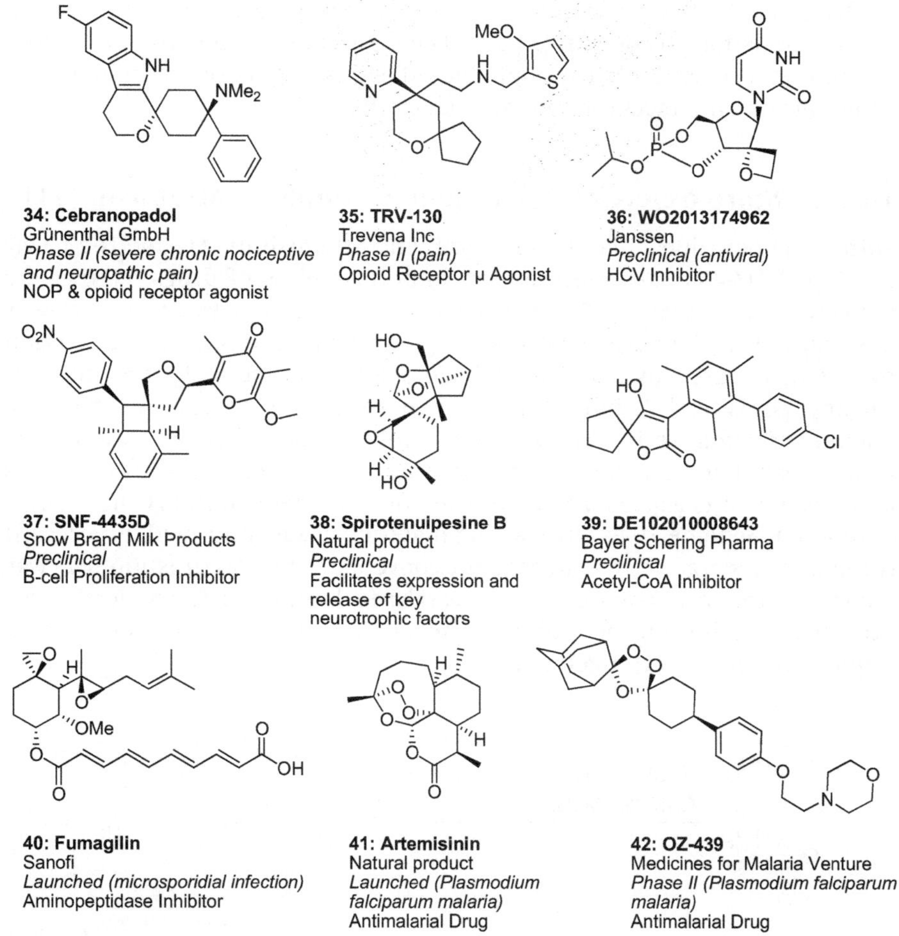

Figure 16.5 Representative examples for oxa-spirocyclic bioactive compounds: **34** (Cebranopadol),[48] **35** (TRV-130),[49] **36** (WO 2013174962),[50] **37** (SNF-4435D),[51] **38** (Spirotenuipesine B),[52] **39** (DE2010008643),[53] **40** (Fumagilin),[54] **41** (Artemisinin),[55] **42** (OZ-439).[56]

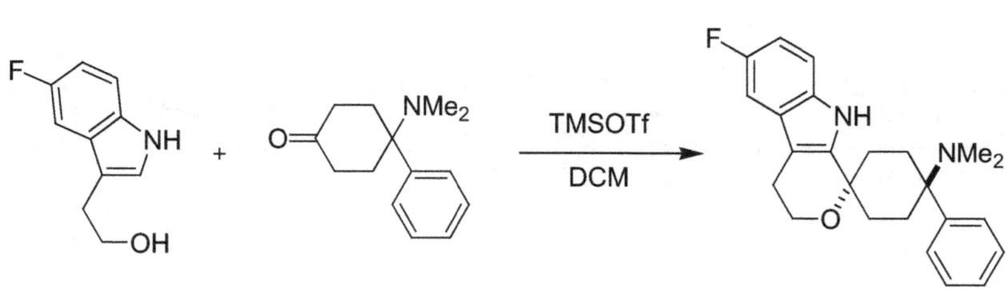

Scheme 16.7 Synthesis of cebranopadol (**34**) through an oxa-*Picet-Spengler* reaction.

TRV-130 (**35**) also represents an investigational drug currently in Phase II clinical trials for postoperative pain. The compound targets the mu opioid receptor in a specific way (through 'biased' signalling) to optimize analgesia while minimizing receptor-mediated side effects.[59]

16.4.2 Spiro-oxacycles – Synthetic Example 2: Artemisin (41)

Spirocyclic peroxides, like the semisynthetic Artemisinin (**41**) and the newly developed OZ-439 (**42**), represent an important class of drugs to treat the parasitic disease Malaria. The spiro peroxide function plays a pivotal role as the drugs exert their parsiticidal activity subsequent to reductive activation by haem, released as a result of haemoglobin digestion by the malaria-causing parasite.[60] A fully implemented industrial process for the production of Artemisinin (**41**) was only recently established by Sanofi,[61] replacing extraction as the only efficient access to Artemisinin (**41**).[62] This process, which is expected to deliver 60 tons of Artemisinin (**41**) in 2014, is outlined in Scheme 16.8. The starting material, artemisinic acid, is accessed through fermentation. Its subsequent conversion into Artemisinin (**41**) was realized through a homogeneously catalysed highly diastereoselective hydrogenation following by a photochemical singlet oxygenation and a subsequent complex rearrangement.

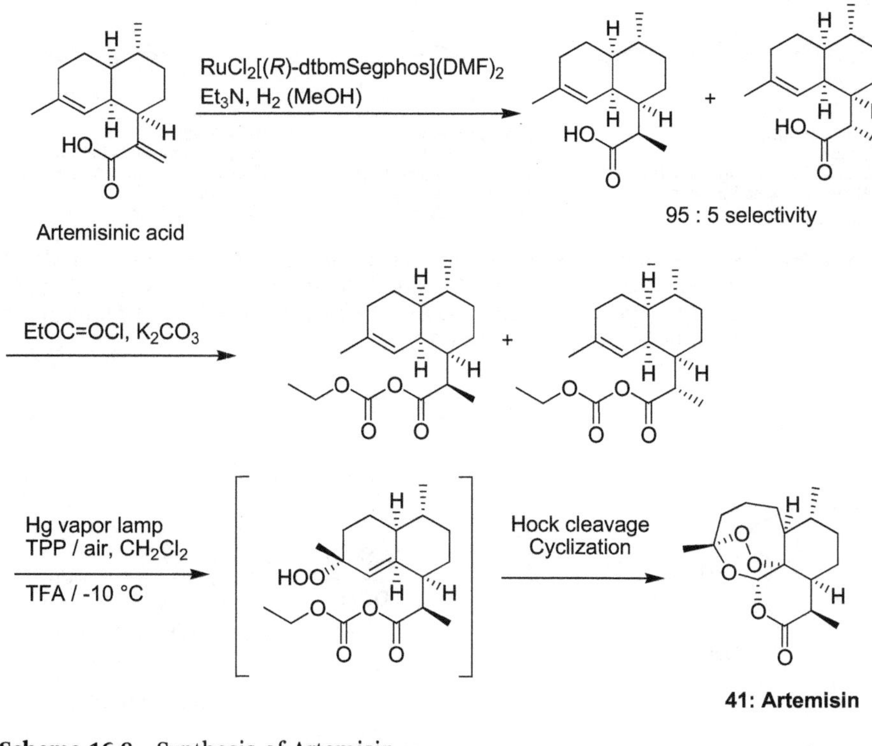

Scheme 16.8 Synthesis of Artemisin.

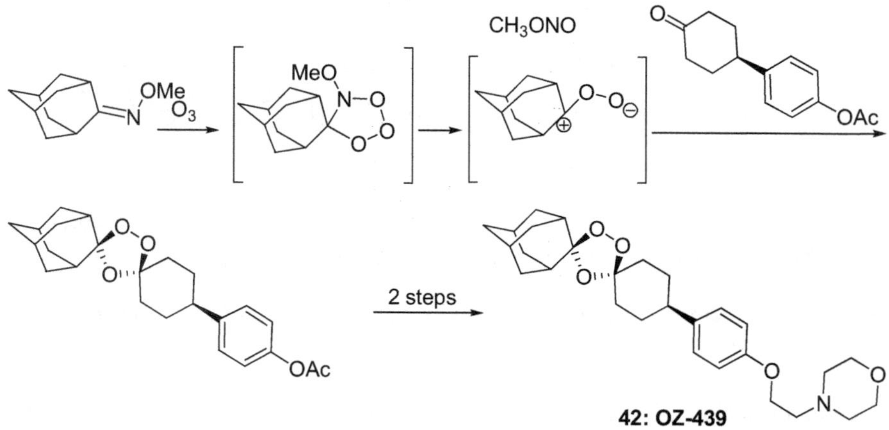

Scheme 16.9 Synthesis of OZ-439 utilizing mechanistically two consecutive 1,3-dipolar cycloadditions.

16.4.3 Spiro-oxacycles – Synthetic Example 3: OZ-439 (42)

The emergence of resistance to Artemisin (**41**) or its derivates Artemether and Artesunate[63] led to efforts to create new synthetic peroxide containing drugs like OZ-439 (**41**), which is currently undergoing Phase II clinical trials at Medicines for Malaria Venture. The synthesis of OZ-429 (**42**) is shown in Scheme 16.9. Starting from adamantan-2-one *O*-methyloxime, the spirocyclic moiety is formed in one step, which mechanistically consists of two consecutive 1,3-dipolar cycloadditions.[64]

16.5 Summary and Outlook

As has been demonstrated in this chapter, spirocylic based compounds span a broad range of unique structural diversity, resulting in biological activity on many target classes and subsequent potential medical applications. Many examples are currently being investigated in preclinical, as well as in clinical trials, while some have already reached the market. In this chapter, several synthetic examples from all three classes (Carba-, Oxa- and Aza-spirocycles) have been described, in which the spiro containing groups have to be constructed in a multistep and complex manner, partly representing a synthetic hurdle for broad variation. In contrast, spiro-containing building blocks for synthesis are becoming more and more attractive and have been used in medicinal chemistry for rapid optimization, circumventing some synthetic challenges. As an example, Carreira *et al.*, in collaboration with Hoffman-La Roche, developed a series of spirooxetanes and close analogs (*e.g.* oxetanones, sulfones) as building blocks for medicinal chemistry.[65] Spiro-oxetanes or related spirocyles can be *e.g.* utilised as isosters for the corresponding carbonyl compounds or can serve as surrogates for commonly employed heterocycles, such as morpholines or piperazines.[66]

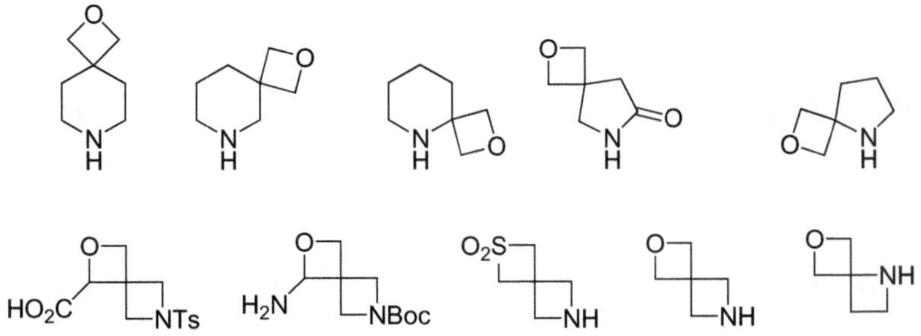

Figure 16.6 Examples for commercially available spirooxetane (and close analog) containing building blocks.

The introduction of such motifs has been demonstrated to improve some optimization parameters in a drug discovery program such as solubility, basicity, lipohilicity and metabolic stability. Due to an improved synthetic entry, many spirooxetane compounds have now become readily available, which further increases their attractiveness as building blocks in medicinal chemistry (Figure 16.6).

In summary, based on a both complex synthesis approaches, as well as the use of readily available spiro-containing building blocks, it will be possible in the future to create even more highly potent and selective synthetic agents against a broad range of biological targets to enable the development of novel break-through medicines in various indications for the overall benefit of patients.

References

1. P. Leeson and B. Springthorpe, *Nat. Rev. Drug Discovery*, 2007, **6**, 881–890; F. Lovering, J. Bikker and C. Humblet, *J. Med. Chem.*, 2009, **52**, 6752–6756.
2. T. Klabunde and G. Hessler, *ChemBioChem*, 2002, **3**, 919.
3. M. Gate and G. Tschudi, *J. Am. Chem. Soc.*, 1952, **74**, 1109; W. Sneader, *Drug Discovery: A History*, Wiley VCH, 2006, ch. 9.
4. R. B. Raffa, J. V. Pergolizzi, D. J. Segarnick and R. J. Tallarida, *Drugs Today*, 2010, **46**, 379.
5. C. A. Bernhart, J. C. Breliere, J. Clement, D. Nisato and P. Perreaut, WO 014679, 1991; C. A. Bernhart, P. M. Perreaut, B. P. Ferrari, Y. A. Muneaux, J. L. A. Assens, J. Clement, F. Haudricourt, C. F. Muneaux and J. E. Taillades, *J. Med. Chem.*, 1993, **36**, 3371.
6. M. Wang, L. Zhang, X. Han, J. Yang, J. Qian, S. Hong, F. Samaniego, J. Romaguera and Q. Yi, *Blood*, 2007, **109**, 5455.
7. P. Norman, J. Castaner and R. M. Castaner, *Drugs Future*, 2000, **25**, 1247.
8. D. E. Williams, M. Roberge, R. Van Soest and R. J. Andersen, *J. Am. Chem. Soc.*, 2003, **125**, 5296; D. E. Williams, M. Lapawa, X. Feng, T. Tarling,

M. Roberge and R. J. Andersen, *Org. Lett.*, 2004, **6**, 2607; D. E. Williams, R. A. Keyzers, K. Warabi, K. Desjardine, J. L. Riffell, M. Roberge and R. J. Andersen, *J. Org. Chem.*, 2007, **72**, 9842; K. Warabi, D. E. Williams, B. O. Patrick, M. Roberge and R. J. Andersen, *J. Am. Chem. Soc.*, 2007, **129**, 508.

9. G. Siller, R. Rosen, M. Freeman, P. Welburn, J. Katsamas and S. M. Ogbourne, *Australas. J. Dermatol.*, 2010, **51**, 99; M. Lebwohl, N. Swanson, L. L. Anderson, A. Melgaard, Z. Xu and B. Berman, *N. Engl. J. Med.*, 2012, **366**, 1010.
10. C. Sessa, M. Zucchetti, E. Davioli, R. Califano, F. Cavalli, S. Frustaci, L. Gumbrell, A. Sulkes, B. Winograd and M. D'Incalci, *J. Natl. Cancer Inst.*, 1991, **83**, 1160.
11. J. Wang, S. M. Soisson, K. Young, W. Shoop, S. Kodali, A. Galgoci, R. Painte, G. Parthasarathy, Y. S. Tang, R. Cummings, S. Ha, K. Dorso, M. Motyl, H. Jayasuriya, J. Ondeyka, K. Herath, C. Zhang, L. Hernandez, J. Allocco, Á. Basilio, J. R. Tormo, O. Genilloud, F. Vicente, F. Pelaez, L. Colwell, S. Ho Lee, B. Michael, T. Felcetto, C. Gill, L. L. Silver, J. D. Hermes, K. Bartizal, J. Barrett, D. Schmatz, J. W. Becker, D. Cully and S. B. Singh, *Nature*, 2006, **441**, 358.
12. A. N. Koehler, WO 062372, 2010.
13. Y.-M. Zhang, N.-H. Tan, Y.-B. Yang, Y. Lu, P. Cao and Y.-S. Wu, *Chem. Biodivers.*, 2005, **2**, 497.
14. D. Bur, O. Corminboeuf, S. Cren, C. Grisostomi, X. Leroy, D. Pozzi, and S. Richard-Bildstein, WO 066488, 2012, A2.
15. D. Häbich and F. Von Nussbaum, *ChemMedChem*, 2006, **1**, 951.
16. H. X. Dinga, C. A. Leverettb, R. E. Kyne Jr., K. K. C. Liuc, S. M. Sakya, A. C. Flick and C. J. O'Donnell, *Biorg. Med. Chem.*, 2014, **22**, 2005.
17. L. Jørgensen, S. J. McKerral, C. A. Kutruff, F. Ungeheuer, J. Felding and P. S. Baran, *Science*, 2013, **341**, 878.
18. K. C. Nicolaou, A. Li and D. J. Edmonds, *Angew. Chem.*, 2006, **118**, 7244.
19. B. M. Trost and F. D. Toste, *J. Am. Chem. Soc.*, 2004, **126**, 15592.
20. S. Hohnloser, P. Dorian, M. Straub, K. Beckmann and P. Kowey, *J. Am. Coll. Cardiol.*, 2004, **44**, 99.
21. J. Zhou, F. Zhang, B. Metcalf, M. Xu and W. Yao, WO 096139, 2004.
22. P. A. J. Janssen, US 3155670, 1964.
23. A. Sinclair and R. A. Stockman, *Nat. Prod. Rep.*, 2007, **24**, 298.
24. N. L. Dirlam, B. S. Moore and F. J. Urban, *J. Org. Chem.*, 1987, **52**, 3587.
25. M. Kurono, Y. Kondo, T. Yamaguchi, K. Miura, T. Usui, N. Terada, K. Asano, K. Mizuno, A. Matsubara, N. Kato, K. Sawai, R. Unno, H. Ozawa and M. Fukushima, US 5447946, 1995.
26. T. Negoro, M. Murata, S. Ueda, B. Fujitani and Y. Ono, US 5258382, 1993.
27. M. S. Malamas, EP 0519600, 1992.
28. I. Merglesberg, D. H. Scherer, M. E. Huttenloch, H. C. Tsui, S. Paliwal and N. Y. Shih, WO 118328, 2008.
29. A. V. Ambarkhane, G. Bhalay, M. Beckett, J. Dale, A. Hamadi, A. Mazzacani, J. McKenna and C. Thomson, WO 164473, 2012.

30. J. L. Duffy, J. Bao, D. L. Ondeyka, S. Tyagarian, P. Shao, F. Ye, R. Katipally, A. Zwicker, E. C. Sherer, M. A. Plotkin, R. Moningka, Z. Hussian, H. B. Wood, F. Ujjainwalla, A. Romero, P. Finke, Y. Zang and W. Liu, WO 024183, 2012.
31. G. M. P. Giblin, D. R. Witty, D. T. Macherson, S. J. Stanway, A. Wilson and L. Gleave, WO 179049, 2013.
32. Z. Gao, R. Hartung and D. Stefany, WO 065798, 2010.
33. N. D. Adams, T. J. Kiesow and K. Wiggal, WO 177253, 2013.
34. Y. Bukhtiyarov, S. Cacatian, L. W. Dillard, C. Dorner-Ciossek, K. Fuchs, U. Gross, N. Heine, L. Jia, D. S. Lala, A. Morales-Ramos, S. B. Singh, A. Sauer, S. Venkatraman, Z. Xu, J. Yuan, Y. Zhao and Y. Zheng, WO 035860, 2014.
35. R. Baker, T. Harrison, C. J. Swain and B. J. Wiliams, WO 019084, 1997.
36. R. B. Woodward, Michael P. Cava, W. D. Ollis, A. Hunger, H. U. Daeniker and K. Schenker, *J. Am. Chem. Soc.*, 1954, **76**, 4749.
37. A. Jossang, P. Jossang, H. A. Hadi, T. Sevenet and B. Bodo, *J. Org. Chem.*, 1991, **56**, 6527.
38. Di Malta, L. Foulon, G. Garcia, D. Nisato, R. Roux, C. Serradeil-Legal, G. Valette and J. Wagnon, EP 0636608, 1995.
39. M. E. Jung, C. L. Sawyers, S. Ouk, C. Tran, and J. Wongvipat, WO 126765, 2007.
40. http://toxnet.nlm.nih.gov.
41. R. Jeyaraman and S. Avila, *Chem. Rev.*, 1981, **81**, 149.
42. U. Schön, J. Antel, R. Brückner and J. Messinger, *J. Med. Chem.*, 1998, **41**, 318.
43. P. J. Oates, *Curr. Drug Targets*, 2008, **9**, 14.
44. R. Sarges, R. C. Schnur, J. L. Belletire and M. J. Peterson, *J. Med. Chem.*, 1988, **31**, 230; M. S. Malamas, T. C. Hohmann and J. Millen, *J. Med. Chem.*, 1994, **37**, 2043.
45. N. Mealy and J. Castaner, *Drugs Future*, 1996, **21**, 261.
46. M. S. Malamas and T. C. Hohman, *Drugs Future*, 1994, **19**, 442.
47. I. Mergelsberg, D. H. Scherer, M. E. Huttenloch, H. C. Tsui, S. Paliwal, and N. Y. Shih, US 20140024834, 2013.
48. K. Linz, T. Christoph, T. M. Tzschentke, T. Koch, K. Schiene, M. Gautrois, W. Schröder, B. Y. Kögel, H. Beier, W. Englberger, S. Schunk, J. De Vry, U. Jahnel and S. Frosch, *J. Pharmacol. Exp. Ther.*, 2014, **349**, 535.
49. S. M. DeWire, D. S. Yamashita, D. H. Rominger, G. Liu, C. L. Cowan, T. M. Graczyk, X.-T. Chen, P. M. Pitis, D. Gotchev, C. Yuan, M. Koblish, M. W. Lark and J. D. Violin, *Pharmacol. Exp. Ther.*, 2013, **344**, 708.
50. I. N. Houpis, T. H. Jonckers, P. J. Raboisson and A. Tahri, WO 174962, 2013.
51. K. A. Parker and Y.-H. Lim, *J. Am. Chem. Soc.*, 2004, **126**, 15968.
52. M. Dai, I. J. Krauss and S. J. Danishefsky, *J. Org. Chem.*, 2008, **73**, 9576.

53. N. Liu, K. Thede, P. Lienau, A. Scholz, M. Najjar, U. Boemer, K. Eis and R. Fischer, DE 102010008643, 2011.
54. N. Sin, L. Meng, M. Q. W. Wang, J. J. Wen, W. G. Bornmann and C. M. Crews, *Proc. Natl. Acad. Sci. U. S. A.*, 1997, **94**, 6099.
55. D. L. Klayman, *Science*, 1985, **228**, 1049.
56. S. A. Charman, S. Arbe-Barnes, I. C. Bathurst, R. Brun, M. Campbell, W. N. Charman, F. C. Chiu, J. Chollet, J. C. Craft, D. J. Creek, Y. Dong, H. Matile, M. Maurer, J. Morizzi, T. Nguyen, P. Papastogiannidis, C. Scheurer, D. M. Shackleford, K. Sriraghavan, L. Stingelin, Y. Tang, H. Urwyler, X. Wang, K. L. White, S. Wittlin, L. Zhou and J. L. Vennerstrom, *Proc. Natl. Acad. Sci. U. S. A.*, 2011, **108**, 4400.
57. S. Schunk, K. Linz, C. Hinze, S. Frormann, S. Oberbörsch, B. Sundermann, S. Zemolka, W. Englberger, T. Germann, A. Kless, T. Christoph, B. Y. Kögel, W. Schröder, S. Harlfinger, D. Saunders and H. Sonnenschein, *ACS Med. Chem. Lett.*, 2014, **5**, 857.
58. L. E. Larghi and T. S. Kaufman, *Synthesis*, 2006, **2**, 187.
59. X.-T. Chen, P. Pitis, G. Liu, C. Yuan, D. Gotchev, C. L. Cowan, D. H. Rominger, M. Koblish, S. M. DeWire, A. L. Crombie, J. D. Violin and D. Yamashita, *J. Med. Chem.*, 2013, **56**, 8019.
60. J. L. Vennerstrom, S. Arbe-Barnes, R. Brun, S. A. Charman, F. C. K. Chiu, J. Chollet, Y. Dong, A. Dorn, D. Hunziker, H. Matile, K. McIntosh, M. Padmanilayam, J. Santo Tomas, C. Scheurer, B. Scorneaux, Y. Tang, H. Urwyler, S. Wittlin and W. N. Charman, *Nature*, 2004, **430**, 900.
61. J. Turconi, F. Griolet, R. Guevel, G. Oddon, R. Villa, A. Geatti, M. Hvalla, K. Rossen, R. Gööller and A. Burgard, *Org. Process Res. Dev.*, 2014, **18**, 417; T. E. Wallaart, N. Pras, A. C. Beekmann and W. J. Quax, *Planta Med.*, 2000, **66**, 57; A. M. Dondorp, F. Nosten, P. Yi, D. Das, A. P. Phyo, J. Tarning, K. M. Lwin, F. Ariey, W. Hanpithakpong, S. J. Lee, P. Ringwald, K. Silamut, M. Imwong, K. Chotivanich, P. Lim, T. Herdman, S. S. An, S. Yeung, P. Singhasivanon, N. P. J. Day, N. Lindegardh, D. Socheat and N. J. White, *N. Engl. J. Med.*, 2009, **361**, 455; J. L. Vennerstrom, Y. Dong, S. A. Charman, S. Wittlin, J. Chollet, D. J. Creek, X. Wang, K. Sriraghavan, L. Zhou, H. Matile and W. N. Charman, WO 058859, 2009; G. Wuitschik, M. Rogers-Evans, K. Müller, H. Fischer, B. Wagner, F. Schuler, L. Polonchuk and E. M. Carreira, *Angew. Chem.*, 2006, **118**, 7900; J. A. Burkhard, G. Wuitschik, M. Rogers-Evans, K. Müller and E. Carreira, *Angew. Chem.*, 2010, **122**, 9236.
62. T. E. Wallaart, N. Pras, A. C. Beekmann and W. J. Quax, *Planta Med.*, 2000, **66**, 57.
63. A. M. Dondorp, F. Nosten, P. Yi, D. Das, A. P. Phyo, J. Tarning, K. M. Lwin, F. Ariey, W. Hanpithakpong, S. J. Lee, P. Ringwald, K. Silamut, M. Imwong, K. Chotivanich, P. Lim, T. Herdman, S. S. An, S. Yeung, P. Singhasivanon, N. P. J. Day, N. Lindegardh, D. Socheat and N. J. White, *N. Engl. J. Med.*, 2009, **361**, 455.

64. J. L. Vennerstrom, Y. Dong, S. A. Charman, S. Wittlin, J. Chollet, D. J. Creek, X. Wang, K. Sriraghavan, L. Zhou, H. Matile and W. N. Charman, WO 058859, 2009.
65. E. M. Carreira and T. C. Fessard, *Chem. Rev.*, 2014, **114**, 8257–8322.
66. J. A. Burkhard, C. Guérot, H. Knust and E. M. Carreira, *Org. Lett.*, 2012, **14**, 66.

Subject Index